Springer Series in Optical Sciences Volume 1

Springer Series in Optical Sciences

42 **Principles of Phase Conjugation**
By B. Ya. Zel'dovich, N.F. Pilipetsky, and V.V. Shkunov

43 **X-Ray Microscopy**
Editors: G. Schmahl and D. Rudolph

44 **Introduction to Laser Physics**
By K. Shimoda 2nd Edition

45 **Scanning Electron Microscopy**
Physics of Image Formation and Microanalysis
By L. Reimer

46 **Holography and Deformation Analysis**
By W. Schumann, J.-P. Zürcher, and D. Cuche

47 **Tunable Solid State Lasers**
Editors: P. Hammerling, A.B. Budgor, and A. Pinto

48 **Integrated Optics**
Editors: H.P. Nolting and R. Ulrich

49 **Laser Spectroscopy VII**
Editors: T.W. Hänsch and Y.R. Shen

50 **Laser-Induced Dynamic Gratings**
By H.J. Eichler, P. Günter, and D.W. Pohl

51 **Tunable Solid State Lasers for Remote Sensing**
Editors: R.L. Byer, E.K. Gustafson, and R. Trebino

52 **Tunable Solid-State Lasers II**
Editors: A.B. Budgor, L. Esterowitz, and L.G. DeShazer

53 **The CO_2 Laser** By W.J. Witteman

54 **Lasers, Spectroscopy and New Ideas**
A Tribute to Arthur L. Schawlow
Editors: W.M. Yen and M.D. Levenson

55 **Laser Spectroscopy VIII**
Editors: W. Persson and S. Svanberg

56 **X-Ray Microscopy II**
Editors: D. Sayre, M. Howells, J. Kirz, and H. Rarback

57 **Single-Mode Fibers** Fundamentals
By E.-G. Neumann

58 **Photoacoustic and Photothermal Phenomena**
Editors: P. Hess and J. Pelzl

Volumes 1–41 are listed on the back inside cover

Walter Koechner

Solid-State Laser Engineering

Second Completely Revised and Updated Edition

With 371 Figures

Springer-Verlag Berlin Heidelberg New York
London Paris Tokyo

Dr. Walter Koechner

Fibertek, Inc., 510-A Herndon Parkway,
Herndorn, VA 22070, USA

ISBN 3-540-18747-2 Springer-Verlag Berlin Heidelberg New York
ISBN 0-387-18747-2 Springer-Verlag New York Berlin Heidelberg

ISBN 3-540-90167-1 1. Auflage Springer-Verlag Berlin Heidelberg New York
ISBN 0-387-90167-1 1st edition Springer-Verlag New York Berlin Heidelberg

Printing: Druckhaus Beltz, 6944 Hemsbach/Bergstr.
Binding: J. Schäffer GmbH & Co. KG, 6718 Grünstadt.
2154/3150-543210 – Printed on acid-free paper

Preface to the Second Edition

This monograph, written from an industrial vantage point, provides a detailed discussion of solid-state lasers, their characteristics, design and construction, and practical problems. The title *Solid-State Laser Engineering* is chosen so as to convey the emphasis which is placed on engineering and practical considerations.

The author has tried to enhance the description of the engineering aspects of laser construction and operation by including numerical and technical data, tables, and curves.

The book is mainly intended for the practicing scientist or engineer who is interested in the design or use of solid-state lasers, but it is hoped that the comprehensive treatment of the subject makes the work useful also to students of laser physics who want to supplement their theoretical knowledge with the engineering aspects of lasers. Although not written in the form of a college textbook, the book might be used in an advanced college course on laser technology.

The aim was to present the subject as clearly as possible. Phenomenological descriptions using models were preferred to an abstract mathematical presentation, even though many simplifications had then to be accepted. Results are given in most cases without proof since the author tried to stress the application of the results rather than the derivation of the formulas. An extensive list of references is cited for each chapter to permit the interested reader to learn more about some particular subject.

This new edition has been updated and revised to include important new developments, concepts and technologies which have emerged during the last ten years.

Through a combination of new discoveries, successful implementation of well-known ideas, and new applications, research, development and engineering of solid-state lasers has greatly intensified during the last decade. Rather dramatic possibilities have appeared on the horizon which could bring about a revolutionary change in solid-state laser technology. We are just on the verge of seeing these technologies emerge into the commercial and military marketplace.

The most exciting prospect is that of solid-state laser materials pumped by laser-diode arrays and including high-efficiency converters. Efficient, compact, frequency agile solid-state lasers with average power levels of up to a few hundred watts and covering most of the visible, near- and mid-infrared regime could become a reality.

The major new developments in solid-state laser engineering which are described in this new edition are briefly summarized here.

Wavelength tunability: With the discovery of the alexandrite laser, followed by a number of other tunable lasers such as Ti : sapphire, great progress has been made, providing a tunable output from solid-state lasers for atmospheric and spectroscopic studies and for certain military applications.

Improved efficiency: Co-doped laser crystals, such as Cr : Nd : GSGG, which are highly efficient absorbers of flashlamp radiation have significantly improved overall system efficiency of flashlamp-pumped lasers. The introduction of Nd: phosphate glasses which have twice the gain of silicate glasses has helped to improve the efficiency of these lasers. The most dramatic impact on efficiency has come from laser-diode arrays employed as pump sources for solid-state lasers. Laser-diode pumping always looked attractive but was not practical due to such technological barriers as low efficiency, low power and short lifetime. The significant progress made in diode-laser technology, coupled with the emerging technology of linear and planar diode arrays, has removed these former technological barriers. Diode pumping is now being widely investigated and is considered by many researchers to be potentially practical even for high-power lasers.

Increased average power: Old concepts, such as the slab-laser design, have been revisited and engineering improvements have resulted in systems producing high average power with good beam quality. In particular, slab lasers pumped with laser-diode arrays look potentially very attractive.

Improved beam quality: The use of unstable resonators for solid-state lasers has increased the power extraction at low order modes. Also, optical phase conjugation employed to correct thermal distortions introduced by laser amplifiers has been shown to improve beam quality in high average power systems.

Frequency agility: Improved nonlinear materials such as KTP, BBO and several organic crystals as well as scaling of KDP to very large apertures have made harmonic processes more attractive as a means of extending the wavelength coverage of solid-state lasers. Also, stimulated Raman scattering is now being frequently used to shift the wavelength from a laser into a spectral region not covered by any other system.

Highly coherent systems: Excellent frequency stability and single axial mode operation of diode-pumped unidirectional ring lasers make it possible to design coherent Doppler systems in the near-infrared wavelength regime.

Laser Superstructures: The last 10 years have seen the emergence of Nd: glass laser systems employed for inertial-confinement fusion research with beam diameters of close to 1 m and unprecedented power and energy levels.

Although this edition has been expanded and updated, the organisation of the material has not changed compared to the first edition. The topics covered include the optical amplification process, properties of laser materials, laser oscillators and amplifiers, resonators, pump systems and heat removal, Q-switches and mode-lockers, nonlinear devices and optical damage.

Omitted from this second edition is the chapter on lasers designed for specific applications, since many publications are available which deal with the various uses of lasers in research, industry, medicine and the military.

This book would not have been possible without the many contributions to the field of laser engineering that have appeared in the open literature and which have been used here as the basic source material. I apologize to any of my colleagues whose work has not been acknowledged or adequately represented in this book.

My special thanks are due to the editor, Dr. H. Lotsch, for his support and assistance in preparing the new edition for printing.

Neither edition of this book could have been written without the encouragement, patience and support of my wife Renate.

Herndon, VA
January 1988

Walter Koechner

From the Preface to the First Edition

The first decade of solid-state laser technology has seen the development of an enormous number of lasing materials and a large variety of interesting design concepts. However, in recent years the technology has matured to a point where solid-state lasers have reached a plateau in their development.

To a major extent, the growth in importance of solid-state lasers for industrial and military applications and as a general research tool is due to the improvement in reliability and maintainability of these systems. The practical advances of these devices had several major consequences: A wealth of applications for solid-state lasers has emerged in materials processing, holography, rangefinding, target illumination and designation, satellite and lunar ranging, thermonuclear fusion, plasma experiments, and in general for scientific work requiring high power densities. Emphasis has shifted from research and innovation to cost reduction and system improvement. As a result, a standardization of the system designs has occurred. [...]

The author is indebted to Dr. M. Stitch and D. Smart who have carefully read the manuscript and suggested many corrections. A special note of gratitude for typing the manuscript on a tight schedule goes to Renate Koechner and Margaret Lochart. Thanks are due to the editor, Dr. H. Lotsch, for his assistance in preparing the work for printing.

The book is dedicated to my wife Renate, whose encouragement and understanding were a decisive factor in its timely completion. Without her encouragement and patience the final goal would not have been achieved.

Santa Monica, CA, 1976 *Walter Koechner*

Contents

1. Introduction 1
1.1 Optical Amplification 1
1.2 Interaction of Radiation with Matter 2
1.2.1 Blackbody Radiation 2
1.2.2 Boltzmann Statistics 3
1.2.3 Einstein Coefficients 4
1.2.4 Phase Coherence of Stimulated Emission 7
1.3 Absorption and Optical Gain 8
1.3.1 Atomic Lineshapes 8
1.3.2 Absorption by Stimulated Transitions 13
1.3.3 Population Inversion 15
1.4 Creation of a Population Inversion 17
1.4.1 The Three-Level System 17
1.4.2 The Four-Level System 19
1.4.3 The Metastable Level 20
1.5 Laser Rate Equations 22

2. Properties of Solid-State Laser Materials 28
2.1 Overview 28
2.1.1 Host Materials 29
2.1.2 Active Ions 34
2.2 Ruby 38
2.3 Nd : Lasers 47
2.3.1 Nd : YAG 48
2.3.2 Nd : Glass 54
2.3.3 Nd : Cr : GSGG 57
2.3.4 Nd : YLF 60
2.4 Er : Lasers 62
2.4.1 Er : YAG 63
2.4.2 Er : Glass 64
2.5 Tunable Lasers 66
2.5.1 Alexandrite Laser 71
2.5.2 Ti : Sapphire 76

3. Laser Oscillator 79
3.1 Operation at Threshold 80
3.2 Gain Saturation 84
3.3 Circulating Power 87
3.4 Output versus Input Calculations 88
3.4.1 Flashlamp-Pumped Lasers 89
3.4.2 Laser-Diode-Pumped Oscillators 94
3.5 Output Fluctuations 97
3.5.1 Relaxation Oscillations 98
3.5.2 Quantum Noise 102
3.6 Examples of Regenerative Oscillators 103
3.6.1 Ruby 103
3.6.2 Nd: Glass 111
3.6.3 Nd: YAG 114
3.6.4 Laser-Diode-Pumped Regenerative Oscillator 119
3.6.5 Alexandrite 124
3.7 Travelling-Wave Oscillator 126

4. Laser Amplifier 129
4.1 Pulse Amplification 131
4.1.1 Ruby Amplifiers 135
4.1.2 Nd: Glass Amplifiers 141
4.1.3 Nd: YAG Amplifiers 147
4.2 Steady-State Amplification 150
4.2.1 Ruby Amplifier 152
4.2.2 Nd: Glass Amplifier 154
4.3 Signal Distortion 155
4.3.1 Spatial Distortions 155
4.3.2 Temporal Distortion 160
4.4 Gain Limitation and Amplifier Stability 161
4.4.1 Spontaneous Decay, Superradiance, and Lateral Depumping 162
4.4.2 Prelasing and Parasitic Modes 165

5. Optical Resonator 168
5.1 Transverse Modes 168
5.1.1 Intensity Distribution of Transverse Modes 169
5.1.2 Characteristics of a Gaussian beam 172
5.1.3 Resonator Configurations 174
5.1.4 Stability of Laser Resonators 178
5.1.5 Diffraction Losses 180
5.1.6 Higher-Order Modes 182
5.1.7 Active Resonator 183
5.1.8 Resonator Sensitivity 185
5.1.9 Mode-Selecting Techniques 189
5.1.10 Examples of Advanced Stable Resonator Designs 196

5.2 Longitudinal Modes ... 203
5.2.1 Fabry-Perot Resonators ... 203
5.2.2 Spectral Characteristics of the Laser Output ... 211
5.2.3 Axial Mode Control ... 215
5.3 Temporal and Spectral Stability ... 226
5.3.1 Amplitude Fluctuations ... 226
5.3.2 Frequency Changes ... 229
5.4 Hardware Design ... 232
5.5 Unstable Resonators ... 236
5.5.1 Confocal Positive-Branch Unstable Resonator ... 240
5.5.2 Negative-Branch Unstable Resonators ... 242
5.6 Wavelength Selection ... 244

6. Optical Pump System ... 247
6.1 Pump Sources ... 247
6.1.1 Noble Gas Flashlamps ... 251
6.1.2 Continuous Arc Lamps ... 265
6.1.3 Tungsten-Filament Lamps ... 272
6.1.4 Laser Diodes ... 274
6.1.5 Nonelectric Pump Sources ... 289
6.2 Power Supplies ... 290
6.2.1 Operation of cw Pump Sources ... 290
6.2.2 Operation of Flashlamps ... 292
6.3 Pump Cavities ... 309
6.3.1 Pump Cavity Configurations ... 309
6.3.2 Energy Transfer Characteristics ... 321
6.3.3 Mechanical Design ... 335

7. Heat Removal ... 350
7.1 Thermal Effects in Laser Rods ... 350
7.1.1 cw Operation ... 351
7.1.2 Single-Shot Operation ... 369
7.1.3 Repetitively Pulsed Lasers ... 372
7.2 Cooling Techniques ... 382
7.2.1 Liquid Cooling ... 382
7.2.2 Air or Gas Cooling ... 385
7.2.3 Conductive Cooling ... 386
7.3 Noncylindrical Laser Elements ... 388

8. Q-Switches and External Switching Devices ... 402
8.1 Pulse-Reflection Mode Q-Switches ... 402
8.1.1 Q-Switch Theory ... 403
8.1.2 Mechanical Devices ... 412
8.1.3 Electrooptical Q-Switches ... 414
8.1.4 Acoustooptic Q-Switches ... 431
8.1.5 Dye Q-Switch ... 437

8.2 Pulse-Transmission Mode Q-Switches 442
8.3 Optical Gates External to Resonator 446

9. Mode Locking 451
9.1 Passive Mode Locking 455
9.1.1 Design and Performance Characteristics of Passively Mode-Locked Solid-State Lasers 460
9.2 Active Mode Locking 466
9.2.1 Design of Actively Mode-Locked Laser Systems 469
9.3 Active-Passive Mode Locking 476

10. Nonlinear Devices 477
10.1 Harmonic Generation 479
10.1.1 Basic Equations of Second-Harmonic Generation 479
10.1.2 Parameters Affecting the Doubling Efficiency 487
10.1.3 Properties of Nonlinear Crystals 492
10.1.4 Intracavity Frequency Doubling 504
10.1.5 Third-Harmonic Generation 511
10.1.6 Examples of Harmonic Generation 514
10.2 Parametric Oscillators 518
10.3 Raman Laser 526
10.4 Optical Phase Conjugation 535

11. Damage of Optical Elements 540
11.1 Surface Damage 541
11.2 Inclusion Damage 542
11.3 Self-focusing 543
11.4 Damage Threshold of Optical Materials 549
11.4.1 Scaling Laws 550
11.4.2 Laser Materials 551
11.4.3 Damage in Optical Glass 552
11.4.4 Damage Levels for Nonlinear Materials 553
11.4.5 Laser Induced Damage in Dielectric Thin Films 554
11.5 System Design Considerations 556

Appendix A 559

Appendix B 564

References 567

Subject Index 601

1. Introduction

In this introductory chapter we shall outline the basic ideas underlying the operation of solid-state lasers. Assuming familiarity with laser physics [1.1] we shall sketch some of the principles governing the interaction of radiation with matter.

1.1 Optical Amplification

In this chapter we will outline the basic ideas underlying laser action. To understand the operation of a laser we have to know some of the principles governing the interaction of radiation with matter.

Atomic systems such as atoms, ions, and molecules can exist only in discrete energy states. A change from one energy state to another, called a transition, is associated with either the emission or the absorption of a photon. The wavelength of the absorbed or emitted radiation is given by Bohr's frequency relation

$$E_2 - E_1 = h\nu_{21} \quad , \tag{1.1}$$

where E_2 and E_1 are two discrete energy levels, ν_{21} is the frequency, and h is Planck's constant. An electromagnetic wave whose freqency ν_{21} corresponds to an energy gap of such an atomic system can interact with it. To the approximation required in this context, a solid-state material can be considered an ensemble of very many identical atomic systems. At thermal equilibrium, the lower energy states in the material are more heavily populated than the higher energy states. A wave interacting with the substance will raise the atoms or molecules from lower to higher energy levels and thereby experience absorption.

The operation of a laser requires that the energy equilibrium of a laser material be changed such that energy is stored in the atoms, ions, or molecules of this material. This is achieved by an external pump source which transfers electrons from a lower energy level to a higher one. The pump radiation thereby causes a "population inversion." An electromagnetic wave of appropriate frequency, incident on the "inverted" laser material, will be amplified because the incident photons cause the atoms in the higher level to drop to a lower level and thereby emit additional photons. As a result, energy is extracted from the atomic system and supplied to the radiation field. The release of the stored energy by interaction with an electromagnetic wave is based on stimulated or induced emission.

Stated very briefly, when a material is excited in such a way as to provide more atoms (or molecules) in a higher energy level than in some lower level, the material will be capable of amplifying radiation at the frequency corresponding to the energy level difference. The acronym "laser" derives its name from this process: "**L**ight **A**mplification by **S**timulated **E**mission of **R**adiation."

A quantum mechanical treatment of the interaction between radiation and matter demonstrates that the stimulated emission is, in fact, completely indistinguishable from the stimulating radiation field. This means that the stimulated radiation has the same directional properties, same polarization, same phase, and same spectral characteristics as the stimulating emission. These facts are responsible for the extremely high degree of coherence which characterizes the emission from lasers. The fundamental nature of the induced or stimulated emission process was already described by A. Einstein and M. Planck.

In solid-state lasers, the energy levels and the associated transition frequencies result from the different quantum energy levels or allowed quantum states of the electrons orbiting about the nuclei of atoms. In addition to the electronic transitions, multiatom molecules in gases exhibit energy levels that arise from the vibrational and rotational motions of the molecule as a whole.

1.2 Interaction of Radiation with Matter

Many of the properties of a laser may be readily discussed in terms of the absorption and emission processes which take place when an atomic system interacts with a radiation field. In the first decade of this century Planck described the spectral distribution of thermal radiation, and in the second decade Einstein, by combining Planck's law and Boltzmann statistics, formulated the concept of stimulated emission. Einstein's discovery of stimulated emission provided essentially all of the theory necessary to describe the physical principle of the laser.

1.2.1 Blackbody Radiation

When electromagnetic radiation in an isothermal enclosure, or cavity, is in thermal equilibrium at temperature T, the distribution of radiation density $\varrho(\nu)\,d\nu$, contained in a bandwidth $d\nu$, is given by Planck's law

$$\varrho(\nu)\,d\nu = \frac{8\pi\nu^2\,d\nu}{c^3}\frac{h\nu}{e^{h\nu/kT}-1} \quad , \tag{1.2}$$

where $\varrho(\nu)$ is the radiation density per unit frequency [Js/cm^3], k is Boltzmann's constant, and c is the velocity of light. The spectral distribution of thermal radiation vanishes at $\nu = 0$ and $\nu \rightarrow \infty$, and has a peak which depends on the temperature.

The factor

$$\frac{8\pi\nu^2}{c^3} = p_n \tag{1.3}$$

in (1.2) gives the density of radiation modes per unit volume and unit frequency interval. The factor p_n can also be interpreted as the number of degrees of freedom associated with a radiation field, per unit volume, per unit frequency interval. The expression for the mode density p_n [modes s/cm^3] plays an important role in connecting the spontaneous and the induced transition probabilities.

For a uniform, isotropic radiation field, the following relationship is valid

$$W = \frac{\varrho(\nu)c}{4} \quad , \tag{1.4}$$

where W is the blackbody radiation [W/cm^2] which will be emitted from an opening in the cavity of the blackbody. Many solids radiate like a blackbody. Therefore, the radiation emitted from the surface of a solid can be calculated from (1.4).

According to the Stefan-Boltzmann equation, the total black body radiation is

$$W = \sigma T^4 \quad , \tag{1.5}$$

where $\sigma = 5.68 \times 10^{-12}\,\mathrm{W/cm^2\,K^4}$. The emitted radiation W has a maximum which is obtained from Wien's displacement law

$$\frac{\lambda_{\max}}{\mu\mathrm{m}} = \frac{2893}{T/\mathrm{K}} \quad . \tag{1.6}$$

For example, a blackbody at a temperature of 5200 K has its radiation peak at 5564 Å, which is about the center of the visible spectrum.

A good introduction to the fundamentals of radiation and its interaction with matter can be found in [1.2].

1.2.2 Boltzmann Statistics

According to a basic principle of statistical mechanics, when a large collection of similar atoms is in thermal equilibrium at temperature T, the relative populations of any two energy levels E_1 and E_2, such as the ones shown in Fig. 1.1, must be related by the Boltzmann ratio

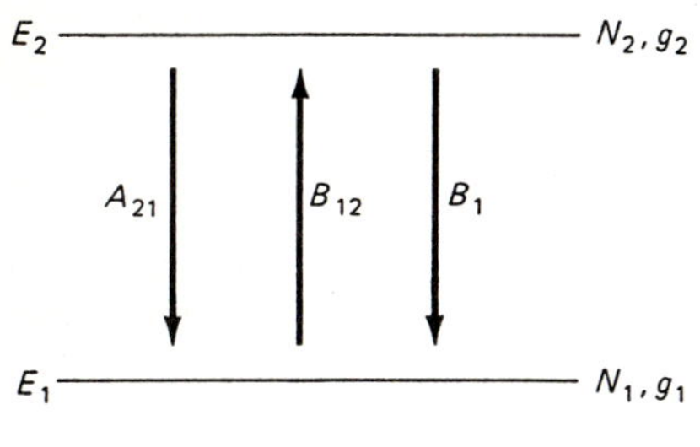

Fig. 1.1. Two energy levels with population N_1, N_2 and degeneracies g_1, g_2, respectively

$$\frac{N_2}{N_1} = \exp\left(\frac{-(E_2 - E_1)}{kT}\right) \quad , \tag{1.7}$$

where N_1 and N_2 are the number of atoms in the energy levels E_1 and E_2, respectively. For energy gaps large enough that $E_2 - E_1 = h\nu_{21} \gg kT$, the ratio is close to zero, and there will be very few atoms in the upper energy level at thermal equilibrium. The energy kT at room temperature ($T \approx 300\,\text{K}$) corresponds to an energy gap $h\nu$ with $\nu \approx 6 \times 10^{12}\,\text{Hz}$, which is equivalent in wavelength to $\lambda \approx 50\,\mu\text{m}$. Therefore, for any energy gap whose transition frequency ν_{21} lies in the near-infrared or visible regions, the Boltzmann exponent will be $\gg 1$ at normal temperatures. The number of atoms in any upper level will then be very small compared to the lower levels. For example, in ruby the ground level E_1 and the upper laser level E_2 are separated by an energy gap corresponding to a wavelength of $\lambda \approx 0.69\,\mu\text{m}$. Let us put numbers into (1.7). Since $h = 6.6 \times 10^{-34}\,\text{Ws}^2$, then $E_2 - E_1 = h\nu = 2.86 \times 10^{-19}\,\text{Ws}$. With $k = 1.38 \times 10^{-23}\,\text{Ws}\,\text{K}$ and $T = 300\,\text{K}$, it follows that $N_2/N_1 \approx 10^{-32}$. Therefore at thermal equilibrium virtually all the atoms will be in the ground level.

Equation (1.7) is valid for atomic systems having only non-degenerate levels. If there are g_i different states of the atom corresponding to the energy E_i, then g_i is defined as the degeneracy of the ith energy level.

We recall that atomic systems, such as atoms, ions, molecules, can exist only in certain sationary states, each of which corresponds to a definite value of energy and thus specifies an energy level. When two ore more states have the same energy, the respective level is called degenerate, and the number of states with the same energy is the multiplicity of the level. All states of the same energy level will be equally populated, therefore the number or atoms in levels 1 and 2 is $N_1 = g_1 N_1^{'}$ and $N_2 = g_2 N_2^{'}$, where $N_1^{'}$ and $N_2^{'}$ refer to the population of any of the states in levels 1 and 2, respectively. It follows then from (1.7) that the populations of the energy levels 1 and 2 are related by the formula

$$\frac{N_2}{N_1} = \frac{g_2}{g_1}\frac{N_2^{'}}{N_1^{'}} = \frac{g_2}{g_1}\exp\left(\frac{-(E_2 - E_1)}{kT}\right) \quad . \tag{1.8}$$

At absolute zero temperature, Boltzmann's statistics predicts that all atoms will be in the ground state. Thermal equilibrium at any temperature requires that a state with a lower energy be more densely populated than a state with a higher energy. Therefore N_2/N_1 is always less than unity for $E_2 > E_1$ and $T > 0$. This will turn out to mean that optical amplification is not possible in thermal equilibrium.

1.2.3 Einstein Coefficients

We can most conveniently introduce the concept of Einstein's A and B coefficients by loosely following Einstein's original derivation. To simplify the discussion, let us consider an idealized material with just two nondegenerate energy levels, 1 and 2, having populations of N_1 and N_2, respectively. The total number of atoms in these two levels is assumed to be constant

$$N_1 + N_2 = N_{\text{tot}} \quad . \tag{1.9}$$

Radiative transfer between the two energy levels which differ by $E_2 - E_1 = h\nu_{21}$ is allowed. The atom can transfer from state E_2 to the ground state E_1 by emitting energy; conversely, transition from state E_1 to E_2 is possible by absorbing energy. The energy removed or added to the atom appears as quanta of $h\nu_{21}$. We can identify three types of interaction between electromagnetic radiation and a simple two-level atomic system:

Absorption. If a quasimonochromatic electromagnetic wave of frequency ν_{21} passes through an atomic system with energy gap $h\nu_{21}$, then the population of the lower level will be depleted at a rate proportional both to the radiation density $\varrho(\nu)$ and to the population N_1 of that level

$$\frac{\partial N_1}{\partial t} = -B_{12}\varrho(\nu)N_1 \quad , \tag{1.10}$$

where B_{12} is a constant of proportionality with dimensions $\mathrm{cm^3/s^2\,J}$.

The product $B_{12}\varrho(\nu)$ can be interpreted as the probability per unit frequency that transitions are induced by the effect of the field.

Spontaneous Emission. After an atom has been raised to the upper level by absorption, the population of the upper level 2 decays spontaneously to the lower level at a rate proportional to the upper level population.

$$\frac{\partial N_2}{\partial t} = -A_{21}N_2 \quad , \tag{1.11}$$

where A_{21} is a constant of proportionality with the dimensions s^{-1}. The quantity A_{21}, being a characteristic of the pair of energy levels in question, is called the spontaneous transition probability because this coefficient gives the probability that an atom in level 2 will spontaneously change to a lower level 1 within a unit of time.

Spontaneous emission is a statistical function of space and time. With a large number of spontaneously emitting atoms there is no phase relationship between the individual emission processes; the quanta emitted are incoherent. Spontaneous emission is characterized by the lifetime of the electron in the excited state, after which it will spontaneously return to the lower state and radiate away the energy. this can occur without the presence of an electromagnetic field.

Equation (1.11) has a solution

$$N_2(t) = N_2(0)\exp\left(\frac{-t}{\tau_{21}}\right) \quad , \tag{1.12}$$

where τ_{21} is the lifetime for spontaneous radiation of level 2. This radiation lifetime is equal to the reciprocal of the Einstein's coefficient,

$$\tau_{21} = A_{21}^{-1} \quad . \tag{1.13}$$

In general, the reciprocal of the transition probability of a process is called its lifetime.

Stimulated Emission. Emission takes place not only spontaneously but also under stimulation by electromagnetic radiation of appropriate frequency. In this case, the atom gives up a quantum to the radiation field by "induced emission" according to

$$\frac{\partial N_2}{\partial t} = -B_{21}\varrho(\nu_{21})N_2 \quad , \tag{1.14}$$

where B_{21} again is a constant of proportionality.

Radiation emitted from an atomic system in the presence of external radiation consists of two parts. The part whose intensity is proportional to A_{21} is the spontaneous radiation; its phase is independent of that of the external radiation. The part whose intensity is proportional to $\varrho(\nu)B_{21}$ is the stimulated radiation; its phase is the same as that of the stimulating external radiation.

The probability of induced transition is proportional to the energy density of external radiation in contrast to spontaneous emission. In the case of induced transition there is a firm phase relationship between the stimulating field and the atom. The quantum which is emitted to the field by the induced emission is coherent with it.

But we shall see later, the useful parameter for laser action is the B_{21} coefficient; the A_{21} coefficient represents a loss term and introduces into the system photons that are not phase-related to the incident photon flux of electric field. Thus the spontaneous process represents a noise source in a laser.

If we combine absorption, spontaneous, and stimulated emission, as expressed by (1.10, 11, and 14), we can write for the change of the upper and lower level populations in our two-level model

$$\frac{\partial N_1}{\partial t} = -\frac{\partial N_2}{\partial t} = B_{21}\varrho(\nu)N_2 - B_{12}\varrho(\nu)N_1 + A_{21}N_2 \quad . \tag{1.15}$$

The relation

$$\frac{\partial N_1}{\partial t} = -\frac{\partial N_2}{\partial t} \tag{1.16}$$

follows from (1.9).

In thermal equilibrium, the number of transitions per unit time from E_1 to E_2 must be equal to the number of transitions from E_2 to E_1. Certainly, in thermal equilibrium

$$\frac{\partial N_1}{\partial t} = \frac{\partial N_2}{\partial t} = 0 \quad . \tag{1.17}$$

Therfore we can write

$$\begin{array}{llll} N_2 A_{21} & + \quad N_2 \dot{\varrho}(\nu) B_{21} & = \quad N_1 \varrho(\nu) B_{12} & . \\ \text{Spontaneous} & \text{Stimulated} & \text{Absorption} & \\ \text{emission} & \text{emission} & & \end{array} \tag{1.18}$$

Using the Boltzmann equation (1.8) for the ratio N_2/N_1, we then write the above expression as

$$\varrho(\nu_{21}) = \frac{(A_{21}/B_{21})}{(g_1/g_2)(B_{12}/B_{21}) \exp(h\nu_{21}/kT) - 1} \quad . \tag{1.19}$$

Comparing this expression with the black body radiation law (1.2), we see that

$$\frac{A_{21}}{B_{21}} = \frac{8\pi\nu^2 h\nu}{c^3} \quad \text{and} \quad B_{21} = \frac{g_1 B_{12}}{g_2} \quad . \tag{1.20}$$

The relations between the A's and B's are known as Einstein's relations. The factor $8\pi\nu^2/c^3$ in (1.20) is the mode density p_n given by (1.3).

In solids the speed of light is $c = c_0/n$, where n is the index of refraction and c_0 is the speed of light in vacuum.

For a simple system with no degeneracy, that is, one in which $g_1 = g_2$, we see that $B_{21} = B_{12}$. Thus, the Einstein coefficients for stimulated emission and absorption are equal. If the two levels have unequal degeneracy, the probablity for stimulated absorption is no longer the same as that for stimulated emission.

1.2.4 Phase Coherence of Stimulated Emission

The stimulated emission provides a phase-coherent amplification mechanism for an applied signal. The signal extracts from the atoms a response that is directly proportional to, and phase-coherent with, the electric field of the stimulating signal. Thus the amplification process is phase-preserving. The stimulated emission is, in fact, completely indistinguishable from the stimulating radiation field. This means that the stimulated emission has the same directional properties, same polarization, same phase, and same spectral characteristics as the stimulating emission. these facts are responsible for the extremely high degree of coherence which characterizes the emission from lasers. The proof of this fact is beyond the scope of this elementary introduction, and requires a quantum mechanical treatment of the interaction between radiation and matter. However, the concept of induced transition, or the interaction between a signal and an atomic system, can be demonstrated, qualitatively, with the aid of the classical electron-oscillator model.

Electromagnetic radiation interacts with matter through the electric charges in the substance. Consider an electron which is elastically bound to a nucleus. One can think of electrons and ions held together by spring-type bonds which are capable of vibrating around equilibrium positions. An applied electric field will cause a relative displacement between electron and nucleus from their equilibrium position. They will execute an oscillatory motion about their equilibrium position. Therefore, the model exhibits an oscillatory or resonant behavior and a response to an applied field. Since the nucleus is so much heavier than the electron, we assume that only the electron moves. The most important model for understanding the interaction of light and matter is that of the harmonic oscillator. We take as our model a single electron, assumed to be bound to its equilibrium position by a linear restoring force. We may visualize the electron as a point of mass suspended by springs. Classical electromagnetic theory asserts that any oscillating electric charge will act as a miniature antenna or dipole and will continuously radiate away electromagnetic energy to its surroundings.

A detailed description of the electric dipole transition and the classical electron-oscillator model can be found in [1.3].

1.3 Absorption and Optical Gain

In this section we will develop the quantitative relations that govern absorption and amplification processes in substances. This requires that we increase the realism of our mathematical model by introducing the concept of atomic lineshapes. Therefore, the important features and the physical processes which lead to different atomic lineshapes will be considered first.

1.3.1 Atomic Lineshapes

In deriving Einstein's coefficients we have assumed a monochromatic wave with frequency ν_{21} acting on a two-level system with an infinitely sharp energy gap $h\nu_{21}$. We will now consider the interaction between an atomic system having a finite transition linewidth $\Delta\nu$ and a signal with a bandwidth $d\nu$.

Before we can obtain an expression for the transition rate for this case, it is necessary to introduce the concept of the atomic lineshape function $g(\nu,\nu_0)$. The distribution $g(\nu,\nu_0)$, centered at ν_0, is the equilibrium shape of the linewidth-broadened transitions. Suppose that N_2 is the total number of ions in the upper energy level considered previously. The spectral distribution of ions per unit frequency is then

$$N(\nu) = g(\nu,\nu_0)N_2 \quad . \tag{1.21}$$

If we integrate both sides over all frequencies we have to obtain N_2 as a result:

$$\int_0^\infty N(\nu)d\nu = N_2 \int_0^\infty g(\nu,\nu_0)d\nu = N_2 \quad . \tag{1.22}$$

Therefore the lineshape function must be normalized to unity:

$$\int_0^\infty g(\nu,\nu_0)d\nu = 1 \quad . \tag{1.23}$$

If we know the function $g(\nu, \nu_0)$, we can calculate the number of atoms $N(\nu)d\nu$ in level 1 which are capable of absorbing in the frequency range ν to $\nu + d\nu$, or the number of atoms in level 2 which are capable of emitting in the same range. From (1.21) we have

$$N(\nu)\,d\nu = g(\nu, \nu_0)\,d\nu\,N_2 \quad . \tag{1.24}$$

From the foregoing it follows that $g(\nu, \nu_0)$ can be defined as the probability of emission or absorption per unit frequency. Therefore $g(\nu)\,d\nu$ is the probability that a given transition will result in an emission (or absorption) of a photon with energy between $h\nu$ and $h(\nu + d\nu)$. The probability that a transition will occur between $\nu = 0$ and $\nu = \infty$ has to be 1.

It is clear from the definition of $g(\nu, \nu_0)$ that we can, for example, rewrite (1.11) in the form

$$-\frac{\partial N_2}{\partial t} = A_{21} N_2 g(\nu, \nu_0)\,d\nu \quad , \tag{1.25}$$

where N_2 is the total number of atoms in level 2, and $\partial N_2/\partial t$ is the number of photons spontaneously emitted per second between ν and $\nu + d\nu$.

The linewidth and lineshape of an atomic transition depends on the cause of line broadening. Optical frequency transitions in gases can be broadened by lifetime, collision, or Doppler broadening, whereas transitions in solids can be broadened by lifetime, dipolar, thermal broadening, or by random inhomogeneities. All these linewidth-broadening mechanisms lead to two distinctly different atomic lineshapes, the homogeneously and the inhomogeneously broadened line [1.4].

The Homogeneously Broadened Line

The essential feature of a homogeneously broadened atomic transition is that every atom has the same atomic lineshape and frequency response, so that a signal applied to the transition has exactly the same effect on all atoms in the collection. This means that within the linewidth of the energy level each atom has the same probability function for a transition.

Differences between homogeneously and inhomogeneously broadened transitions show up in the saturation behavior of these transitions. This has a major effect on the laser operation. The important point about a homogeneous lineshape is that the transition will saturate uniformly under the influence of a sufficiently strong signal applied anywhere within the atomic linewidth.

Mechanisms which result in a homogeneously broadened line are lifetime broadening, collision broadening, dipolar broadening, and thermal broadening.

Lifetime Broadening. This type of broadening is caused by the decay mechanisms of the atomic system. Spontaneous emission or fluorescence has a radiative lifetime. Broadening of the atomic transition due to this process is related to the fluorescence lifetime τ by $\Delta\omega_a \tau = 1$, where ω_a is the bandwidth.

Actually, physical situations in which the lineshape and linewidth are determinded by the spontaneous emission process itself are vanishingly rare. Since the natural or intrinsic linewidth of an atomic line is extremely small, it is the linewidth that would be observed from atoms at rest without interaction with one another.

Collision Broadening. Collision of radiating particles (atoms or molecules) with one another and the consequent interruption of the radiative process in a random manner leads to broadening. As an atomic collision interrupts either the emission or the absorption of radiation, the long wave train which otherwise would be present becomes truncated. The atom restarts its motion after the collision with a completely random initial phase. After the collision the process is restarted without memory of the phase of the radiation prior to the collision. The result of frequent collisions is the presence of many truncated radiative or absorptive processes.

Since the spectrum of a wave train is inversely proportional to the length of the train, the linewidth of the radiation in the presence of collision is greater than that of an individual uninterrupted process.

Collision broadening is observed in gas lasers operated at higher pressures, hence the name pressure broadening. At higher pressures collisions between gas atoms limit their radiative lifetime. Collision broadening, therefore, is quite similar to lifetime broadening, in that the collisions interrupt the initial state of the atoms.

Dipolar Broadening. Dipolar broadening arises from interactions between the magnetic or electric dipolar fields of neighboring atoms. This interaction leads to results very similar to collision broadening, including a linewidth that increases with increasing density of atoms. Since dipolar broadening represents a kind of coupling between atoms, so that excitation applied to one atom is distributed or shared with other atoms, dipolar broadening is a homogeneous broadening mechanism.

Thermal Broadening. Thermal broadening is brought about by the effect of the thermal lattice vibrations on the atomic transition. The thermal vibrations of the lattice surrounding the active ions modulate the resonance frequency of each atom at a very high frequency. This frequency modulation represents a coupling mechanism between the atoms, therefore a homogeneous linewidth is obtained. Thermal broadening is the mechanism responsible for the linewidth of the ruby laser and Nd : YAG laser.

The lineshape of homogeneous broadening mechanisms lead to a Lorentzian lineshape for atomic response. For the normalized Lorentz distribution, the equation

$$g(\nu) = \left(\frac{\Delta\nu}{2\pi}\right)\left[(\nu-\nu_0)^2 + \left(\frac{\Delta\nu}{2}\right)^2\right]^{-1} \tag{1.26}$$

is valid. Here, ν_0 is the center frequency, and $\Delta\nu$ is the width between the half-power points of the curve. The factor $\Delta\nu/2\pi$ assures normalization of the area under the curve according to (1.23). The peak value for the Lorentz curve is

$$g(\nu_0) = \frac{2}{\pi\Delta\nu} \quad . \tag{1.27}$$

The Inhomogeneous Broadened Line

Mechanisms which cause inhomogeneous broadening tend to displace the center frequencies of individual atoms, thereby broadeing the overall response of a collection without broadening the response of individual atoms. Different atoms have slightly different resonance frequencies on the same transition, for example, owing to Doppler shifts. As a result, the overall response of the collection is broadened. An applied signal at a given frequency within the overall linewidth interacts strongly only with those atoms whose shifted resonance frequencies lie close to the signal frequency. The applied signal does not have the same effect on all the atoms in an inhomogeneously broadened collection.

Since in an inhomogeneously broadened line interaction occurs only with those atoms whose resonance frequencies lie close to the applied signal frequency, a strong signal will eventually deplete the upper laser level in a very narrow frequency interval. The signal will eventually "burn a hole" in the atomic absorption curve. Examples of inhomogeneous frequency-shifting mechanisms include Doppler broadening and broadening due to crystal inhomogeneities.

Doppler Broadening. The apparent resonance frequencies of atoms undergoing random motions in a gas are shifted randomly so that the overall frequency response of the collection of atoms is broadened. A particular atom moving with a velocity component ν relative to an observer in the z direction will radiate at a frequency measured by the observer as $\nu_0(1+v/c)$. When these velocities are averaged, the resulting lineshape is Gaussian. Doppler broadening is one form of inhomogeneous broadening, since each atom emits a different frequency rather than one atom having a probability distribution for emitting any frequency within the linewidth. In the actual physical situation, the Doppler line is best visualized as a packet of homogeneous lines of width $\Delta\nu_n$, which superimpose to give the observed Doppler shape. The He-Ne laser has a Doppler-broadened linewidth. Most visible and near-infrared gas laser transitions are inhomogeneously broadened by Doppler effects.

Line Broadening Due to Crystal Inhomogeneities. Solid-state lasers may be inhomogeneously broadened by crystalline defects. This happens only at low temperatures where the lattice vibrations are small. Random variations of dislocations, lattice strains, etc., may cause small shifts in the exact energy level spacings and transition frequencies from ion to ion. Like Doppler broadening, these variations do not broaden the response on an individual atom, but they do cause the exact resonance frequencies of different atoms to be slightly different. Thus random crystal imperfection can be a source of inhomogeneous broadening in a solid-state laser crystal.

A good example of an imhomogeneously broadened line occurs in the fluorescence of neodymium-doped glass. As a result of the so-called glassy state, there are variations, from rare earth site to rare earth site, in the relative atomic positions occupied by the surrounding lattice ions. This gives rise to a random distribution of static crystalline fields acting on the rare-earth ions. Since the line shifts corresponding to such crystal-field variations are larger, generally speaking, than the width contributed by other factors associated with the transition, an inhomogeneous line results.

The inhomogeneous-broadened linewidth can be represented by a Gaussian frequency distribution. For the normalized distribution, the equation

$$g(\nu) = \frac{2}{\Delta\nu}\left(\frac{\ln 2}{\pi}\right)^{1/2} \exp\left[-\left(\frac{\nu-\nu_0}{\Delta\nu/2}\right)^2 \ln 2\right] \tag{1.28}$$

is valid. Where ν_0 is the frequency at the center of the line, and $\Delta\nu$ is the linewidth at which the amplitude falls to one-half. The peak value of the normalized Gaussian curve is

$$g(\nu_0) = \frac{2}{\Delta\nu}\left(\frac{\ln 2}{\pi}\right)^{1/2} \quad . \tag{1.29}$$

In Fig. 1.2 the normalized Gaussian and Lorentz lines are plotted for a common linewidth.

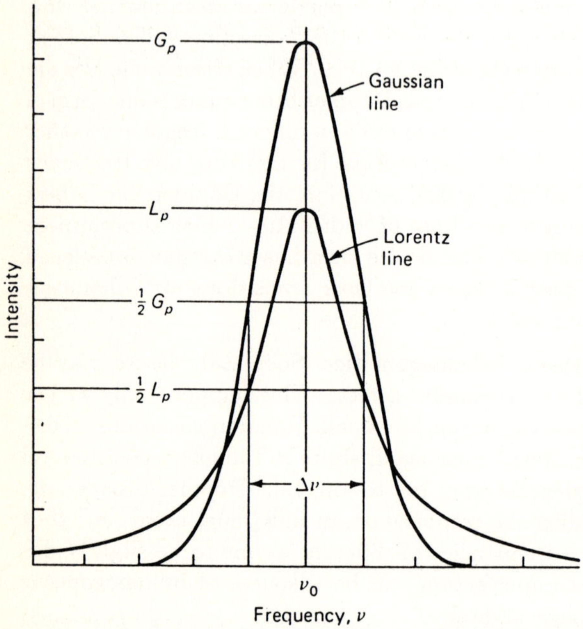

Fig. 1.2. Gaussian and Lorentz lines of common linewidth (G_p and L_p are the peak intensities)

1.3.2 Absorption by Stimulated Transitions

We assume a quasicollimated beam of energy density $\varrho(\nu)$ incident on a thin absorbing sample of thickness dx; as before, we consider the case of an optical system that operates between only two energy levels as illustrated schematically in Fig. 1.1. The populations of the two levels are N_1 and N_2, respectively. Level 1 is the ground level and level 2 is the excited level. We consider absorption of radiation in the material and emission from the stimulated processes but neglect the spontaneous emission. From (1.15 and 1.20) we obtain

$$-\frac{\partial N_1}{\partial t} = \varrho(\nu) B_{21}\left(\frac{g_2}{g_1} N_1 - N_2\right). \tag{1.30}$$

As we recall, this relation was obtained by considering infinitely sharp energy levels separated by $h\nu_{21}$ and a monochromatic wave of frequency ν_{21}.

We will now consider the interaction between two linewidth-broadened energy levels with an energy separation centered at ν_0, and a half-width of $\Delta\nu$ characterized by $g(\nu, \nu_0)$ and a signal with center frequency ν_s and bandwidth $d\nu$. The situation is shown schematically in Fig. 1.3. The spectral width of the signal is narrow, as compared to the linewidth-broadened transition. If N_1 and N_2 are the total number of atoms in level 1 and level 2, then the number of atoms capable of interacting with a radiation of frequency ν_s and bandwidth $d\nu$ are

$$\left(\frac{g_2}{g_1} N_1 - N_2\right) g(\nu_s, \nu_0) d\nu \quad . \tag{1.31}$$

The net change of atoms in energy level 1 can be expressed in terms of energy density $\varrho(\nu)d\nu$ by multiplying both sides of (1.30) with photon energy $h\nu$ and dividing by the volume V. We will further express the populations N_1 and N_2 as population densities n_1 and n_2.

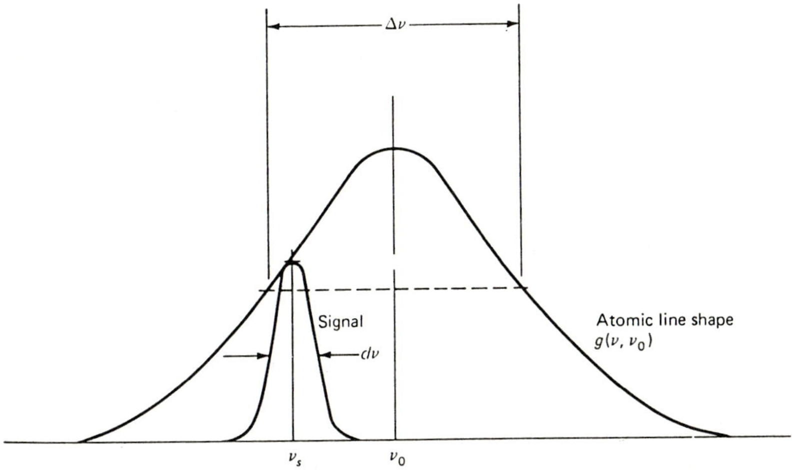

Fig. 1.3. Linewidth-broadened atomic transition line centered at ν_0 and narrow band signal centered at ν_s

Equation (1.30) now becomes

$$-\frac{\partial}{\partial t}[\varrho(\nu_s)d\nu] = \varrho(\nu_s)d\nu\, B_{21}h\nu g(\nu_s,\nu_0)\left(\frac{g_2}{g_1}n_1 - n_2\right) \quad . \tag{1.32}$$

This equation gives the net rate of absorbed energy in the frequency interval $d\nu$ centered around ν_s. In an actual laser system the wavelength of the emitted radiation, corresponding to the signal bandwidth $d\nu$ in our model, is very narrow as compared to the natural linewidth of the material. Ruby, for example, has a fluorescent linewidth of 5 Å, whereas the linewidth of the laser output is typically 0.1 to 0.01 Å. The operation of a laser, therefore, can be fairly accurately characterized as the interaction of linewidth-broadened energy levels with a monochromatic wave. The photon density of a monochromatic radiation of frequency ν_0 can then be represented by a delta function $\delta(\nu - \nu_0)$. After integrating (1.32) in the interval $d\nu$, we obtain, for a monochromatic signal of frequency ν_s and a linewidth-broadened transition,

$$-\frac{\partial\varrho(\nu_s)}{\partial t} = \varrho(\nu_s)B_{21}h\nu_s g(\nu_s,\nu_0)\left(\frac{g_2}{g_1}n_1 - n_2\right) \quad . \tag{1.33}$$

The signal will travel through the material of thickness dx in the time $dt = dx/c = (n/c_0)dx$. Then, as the wave advances from x to $x + dx$, the decrease of energy in the beam is

$$-\frac{\partial\varrho(\nu_s)}{\partial x} = h\nu_s\varrho(\nu_s)g(\nu_s,\nu_0)B_{21}\left(\frac{g_2}{g_1}n_1 - n_2\right)\frac{1}{c} \quad . \tag{1.34}$$

Integration of (1.34) gives

$$\frac{\varrho(\nu_s)}{\varrho_0(\nu_s)} = \exp\left[-h\nu_s g(\nu_s,\nu_0)B_{21}\left(\frac{g_2}{g_1}n_1 - n_2\right)\frac{x}{c}\right] \quad . \tag{1.35}$$

If we introduce an absorption coefficient $\alpha(\nu_s)$,

$$\alpha(\nu_s) = \left(\frac{g_2}{g_1}n_1 - n_2\right)\sigma_{21}(\nu_s) \quad , \qquad \text{where} \tag{1.36}$$

$$\sigma_{21}(\nu_s) = \frac{h\nu_s g(\nu_s,\nu_0)B_{21}}{c} \quad . \tag{1.37}$$

Then we can write (1.35) as

$$\varrho(\nu_s) = \varrho_0(\nu_s)\exp[-\alpha(\nu_s)x] \quad . \tag{1.38}$$

Equation (1.38) is the well-known exponential absorption equation for thermal equilibrium condition $n_1g_2/g_1 > n_2$. The energy of the radiation decreases exponentially with the depth of penetration into the substance. The maximum possible absorption occurs when all atoms exist in the ground state n_1. For equal population of the energy states $n_1 = (g_1/g_2)n_2$, the absorption is eliminated and the material is transparent. The parameter σ_{21} is the cross section for the radiative transition $2 \rightarrow 1$. The cross section for stimulated emission σ_{21} is related to the absorption cross section σ_{12} by the ratio of the level degeneracies,

$$\sigma_{21}/\sigma_{12} = g_1/g_2 \quad . \tag{1.39}$$

The cross section is a very useful parameter to which we will refer in the following chapters. If we replace B_{21} by the Einstein relation (1.20), we obtain σ_{21} in a form which we will find most useful:

$$\sigma_{21}(\nu_s) = \frac{A_{21}\lambda_0^2}{8\pi n^2} g(\nu_s, \nu_0) \quad . \tag{1.40}$$

As we will see later, the gain for the radiation building up in a laser resonator will be highest at the center of the atomic transitions. Therefore, in lasers we are mostly dealing with stimulated transitions which occur at the center of the linewidth.

If we assume $\nu \approx \nu_s \approx \nu_0$, we obtain, for the spectral stimulated emission cross section at the center of the atomic transition for a Lorentzian lineshape,

$$\sigma_{21} = \frac{A_{21}\lambda_0^2}{4\pi^2 n^2 \Delta\nu} \quad , \tag{1.41}$$

and for a Gaussian lineshape,

$$\sigma_{21} = \frac{A_{21}\lambda_0^2}{4\pi n^2 \Delta\nu}\left(\frac{\ln 2}{\pi}\right)^{1/2} \quad . \tag{1.42}$$

Here we have introduced into (1.40) the peak values of the lineshape function, as given in (1.27 and 1.29) for the Lorentzian and Gaussian curves respectively. For example, in the case of the R_1 line of ruby, where $\lambda_0 = 6.94 \times 10^{-5}$ cm, $n = 1.76, A_{21} = 2.5 \times 10^2\,\mathrm{s}^{-1}$, and $\Delta\nu = 11.2\,\mathrm{cm}^{-1} = 3.4 \times 10^{11}$ Hz at 300 K, one finds, according to (1.41), $\sigma_{21} = 2.8 \times 10^{-20}\,\mathrm{cm}^2$. The experimental value σ_{21} at the center of the R_1 line equals $2.5 \times 10^{-20}\,\mathrm{cm}^2$.

1.3.3 Population Inversion

According to the Boltzmann distribution (1.7), in a collection of atoms at thermal equilibrium there are always fewer atoms in a higher-lying level E_2 than in a lower level E_1. Therefore the population difference $N_1 - N_2$ is always positive, which means that the absorption coefficient $\alpha(\nu_s)$ in (1.36) is positive and the incident radiation is absorbed (Fig. 1.4).

Suppose that it were possible to achieve a temporary situation such that there are more atoms in an upper energy level than in a lower energy level. The normally positive population difference on that transition then becomes negative, and the normal stimulated absorption as seen from an applied signal on that transition is correspondingly changed to stimulated emission, or amplification of the applied signal. That is, the applied signal gains energy as it interacts with the atoms and hence is amplified. The energy for this signal amplification is supplied by the atoms involved in the interaction process. This situation is characterized by a negative absorption coefficient $\alpha(\nu_s)$ according to (1.36). From (1.34) it follows that $\partial\varrho(\nu)/\partial x > 0$.

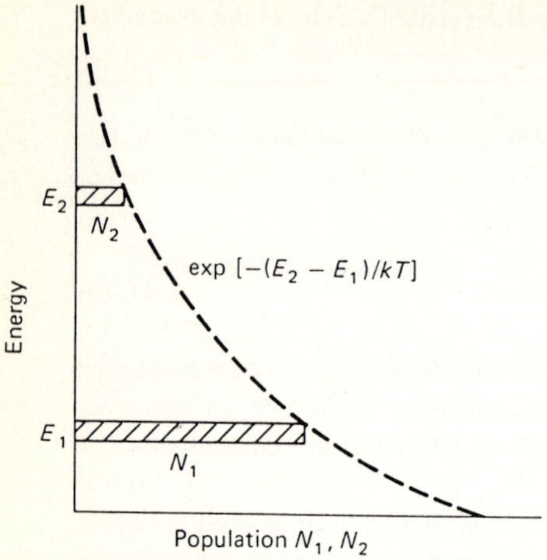

Fig. 1.4. Relative populations in two energy levels as given by the Boltzmann relation for thermal equilibrium

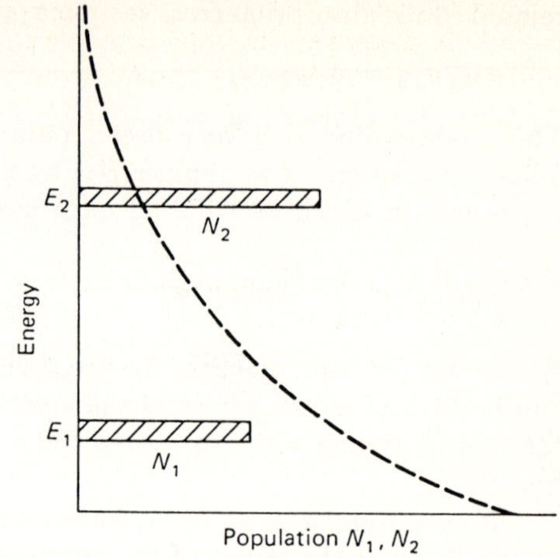

Fig. 1.5. Inverted population difference required for optical amplification

The essential condition for amplification is thus that somehow we must have, at a given instant, more atoms in an upper energy level than in a lower energy level; i.e., for amplification,

$$N_2 > N_1 \quad \text{if} \quad E_2 > E_1 \quad , \tag{1.43}$$

as illustrated in Fig. 1.5. The resulting negative sign of the population difference $(N_2 - g_2 N_1/g_1)$ on that transition is called a population inversion. Population inversion is clearly an abnormal situation; it is never observed at thermal equilibrium. The point at which the population of both states is equal is called the "inversion threshold."

Stimulated absorption and emission processes always occur side by side independently of the population distribution among the levels. So long as the population of the higher energy level is smaller than that of the lower energy level, the number of absorption transitions is larger than that of the emission transitions, so that there is an overall attenuation of the radiation. When the numbers of atoms in both states are equal, the number of emissions becomes equal to the number of absorptions; the material is then transparent to the incident radiation. As soon as the population of the higher level becomes larger than that of the lower level, emission processes predominate and the radiation is enhanced collectively during passage through the material. In order to produce an inversion, we must have a source of energy to populate a specified energy level; we call this energy the pump energy.

In Sect. 1.4 we will discuss the type of energy level structure an atomic system must possess in order to make it possible to generate an inversion. Techniques by which the atoms of a solid-state laser can be raised or pumped into upper energy levels are discussed in Sect. 6.1. Depending on the atomic system involved, an inverted population condition may be obtainable only on a transient basis, yielding intermittent or pulsed laser action; or it may be possible to maintain the population inversion on a steady-state basis, yielding continuous-wave (cw) laser action.

The total amount of energy which is supplied by the atoms to the light wave is

$$E = \Delta N h\nu \quad , \tag{1.44}$$

where ΔN is the total number of atoms which are caused to drop from the upper to the lower energy level during the time the signal is applied. If laser action is to be maintained, the pumping process must continually replenish the supply of upper-state atoms. The size of the inverted population difference is reduced not only by the amplification process but also by spontaneous emission which always tends to return the energy level populations to their thermal equilibrium values.

1.4 Creation of a Population Inversion

We are concerned in this section with how the necessary population inversion for laser action is obtained in solid-state lasers. We can gain considerable understanding on how laser devices are pumped and how their population densities are inverted by studying some simplified but fairly realistic models.

The discussion up to this point has been based on a hypothetical $2 \leftrightarrow 1$ transition and has not been concerned with how the levels 2 and 1 fit into the energy level scheme of the atom. This detached point of view must be abandoned when one tries to understand how laser action takes place in a solid-state medium. As already noted, the operation of the laser depends on a material with narrow energy levels between which electrons can make transitions. Usually these levels are due to impurity atoms in a host crystal. The pumping and laser processes in real laser systems typically involve a very large number of energy levels, with complex excitation processes and cascaded relaxation processes among all these levels. Operation of an actual laser material is properly described only by a many-level energy diagram. The main features can be understood, however, through the familiar three-level or four-level idealizations of Figs. 1.6 and 1.7. More detailed energy level diagrams of some of the most important solid-state laser materials are presented in Chap. 2.

1.4.1 The Three-Level System

Figure 1.6 shows a diagram which can be used to explain the operation of an optically pumped three-level laser, such as ruby. Initially, all atoms of the laser material are in the lowest level 1. Excitation is supplied to the solid by radiation of frequencies which produce absorption into the broad band 3. Thus, the pump light raises atoms from the ground state to the pump band, level 3. In general, the "pumping" band, level 3, is actually made up of a

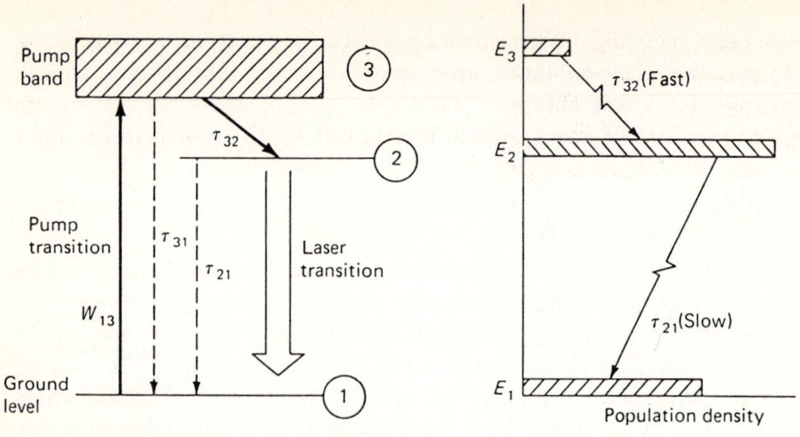

Fig. 1.6. Simplified energy level diagram of a three-level laser

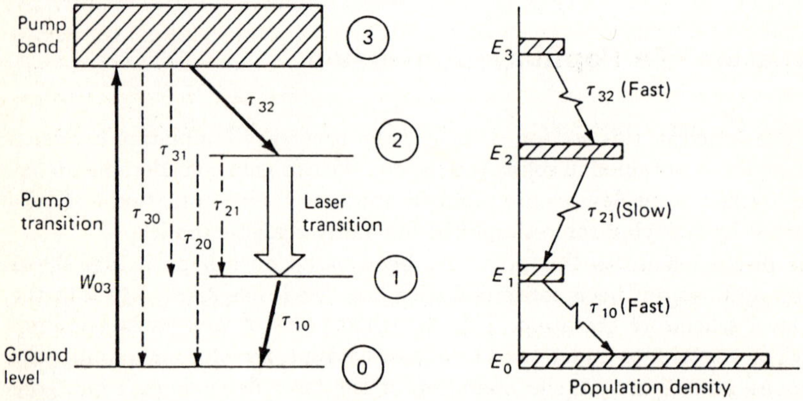

Fig. 1.7. Simplified energy level diagram of a four-level laser

number of bands, so that the optical pumping can be accomplished over a broad spectral range. In practice, xenon, krypton, mercury, and tungsten lamps are used for optically pumping solid-state lasers. Most of the excited atoms are transferred by fast radiationless transitions into the intermediate sharp level 2. In this process the energy lost by the electron is transferred to the lattice. Finally, the electron returns to the ground level by the emission of a photon. It is this last transition that is responsible for the laser action. If pumping intensity is below laser threshold, atoms in level 2 predominantly return to the ground state by spontaneous emission. Ordinary fluorescence acts as a drain on the population of level 2. After the pump radiation is extinguished, level 2 is emptied by fluorescence at a rate that varies from material to material. In ruby, at room temperature, the lifetime of level 2 is 3 ms. When the pump intensity is above laser threshold, the decay from the fluorescent level consists of stimulated as well as spontaneous radiation; the stimulated radiation produces the laser output beam. Since the terminal level of the laser transition is the highly populated ground state, a very high population must be reached in the E_2 level before the $2 \rightarrow 1$ transition is inverted.

It is necessary, in general, that the rate of radiationless transfer from the uppermost level to the level at which the laser action begins be fast compared with the other spontaneous transition rates in a three-level laser. Therefore, the lifetime of the E_2 state should be large in comparison with the relaxation time of the $3 \rightarrow 2$ transition, i.e.,

$$\tau_{21} \gg \tau_{32} \quad . \tag{1.45}$$

The number of atoms N_3 in level E_3 is then negligible compared with the number of atoms in the other two states, i.e., $N_3 \ll N_1, N_2$. Therefore,

$$N_1 + N_2 \approx N_{\text{tot}} \quad . \tag{1.46}$$

A vital aspect of the three-level system is that the atoms are in effect pumped directly from level 1 into the metastable level 2 with only a momentary pause as they pass through level 3. With these conditions, we can calculate as if only two levels were present. In order that an equal population is achieved between the E_2 and E_1 levels, one-half of all atoms must be excited to the E_2 level:

$$N_2 = N_1 = \frac{N_{\text{tot}}}{2} \quad . \tag{1.47}$$

In order to maintain a specified amplification, the population of the second level must be larger than that of the first level. In most cases which are of practical importance, however, the necessary inversion $(N_2 - N_1)$ is small compared with the total number of all atoms. The pump power necessary for maintaining this inversion is also small compared with the inversion power necessary for equal population of the level.

The disadvantage of a three-level system is that more than half of the atoms in the ground state must be raised to the metastable level E_2. There are thus many atoms present to contribute to the spontaneous emission. Moreover, each of the atoms which participate in the pump cycle transfer energy into the lattice from the $E_3 \rightarrow E_2$ transition. This transition is normally radiationless, the energy being carried into the lattice by phonons.

1.4.2 The Four-Level System

The four-level laser system, which is characteristic of the rare earth ions in glass or crystalline host materials, is illustrated in Fig. 1.7. Note that a characteristic of the three-level laser material is that the laser transition takes place between the excited laser level 2 and the final ground state 1, the lowest energy level of the system. This leads to low efficiency. The four-level system avoids this disadvantage. The pump transition extends again from the ground state (now level E_0) to a wide absorption band E_3. As in the case of the three-level system, the atoms so excited will proceed rapidly to the sharply defined level E_2. The laser transition, however, proceeds now to a fourth, terminal level E_1, which is situated above the ground state E_0. From here the atom undergoes a rapid nonradiative transition to the ground level. In a true four-level system, the terminal laser level E_1 will be empty. To qualifiy as a four-level system a material must possess a relaxation time between the terminal laser level and the ground level which is fast compared to the fluorescent lifetime, i.e., $\tau_{10} \ll \tau_{21}$. In addition the terminal laser level must be far above the ground state so that its thermal population is small. The equilibrium population of the terminal laser level 1 is determined by the relation

$$\frac{N_1}{N_0} = \exp\left(\frac{-\Delta E}{kT}\right) \quad , \tag{1.48}$$

where ΔE is the energy separation between level 1 and the ground state, and T is the operating temperature of the laser material. If $\Delta E \gg kT$, then $N_1/N_0 \ll 1$, and the intermediate level will always be relatively empty. In some laser materials the energy gap between the lower laser level and the ground state is relatively small and, therefore, they must be cooled to function as four-level lasers. In a four-level system an inversion of the $2 \to 1$ transition can occur even with vanishingly small pump power, and the high pump rate, necessary to maintain equilibrium population in the aforementioned three-level system, is no longer needed. In the most favorable case, the relaxation times of the $3 \to 2$ and $1 \to 0$ transitions in the four-level system are short compared with the spontaneous emission lifetime of the laser transition τ_{21}. Here we can also carry out the calculations as if only the E_1 and E_2 states were populated.

By far the majority of lasers materials operate, because of the more favorable population ratios, as four-level systems. The only laser of practical importance which operates as a three-level system is ruby. By a combination of favorable circumstances, it is possible in this unique case to overcome the disadvantages of the three-level scheme.

1.4.3 The Metastable Level

After this brief introduction to the energy level structure of solid-state lasers we can ask the question, "what energy level scheme must a solid possess to make it a useful laser?" As we have seen in the previous discussion, the existence of a metastable level is of paramount importance for laser action to occur. The relatively long lifetime of the metastable level provides a mechanism by which inverted population can be achieved. Most transitions of atoms show rapid nonradiative decay, because the coupling of the internal atomic oscillations to the surrounding lattice is strong. Nonradiative decay processes can occur readily, and characteristically have short lifetimes and broad linewidths. A few transitions of selected atoms in solids turn out to be decoupled from the lattice vibration. These transitions have a radiative decay which leads to relatively long lifetimes.

In typical laser systems with energy levels, such as illustrated by Fig. 1.6 and 7, the $3 \rightarrow 2$ transition frequencies, as well as the $1 \rightarrow 0$ transition frequencies, all fall within the frequency range of the vibration spectrum of the host crystal lattice. Therfore, all these transitions can relax extremely rapidly by direct nonradiative decay, i.e., by emitting a phonon to the lattice vibrations, with $\tau_{32}, \tau_{10} \approx 10^{-8}$ to 10^{-11} s. However, the larger $3 \rightarrow 0$, $3 \rightarrow 1$, $2 \rightarrow 0$, and $2 \rightarrow 1$ energy gaps in these atoms often correspons to transition frequencies that are higher than the highest possible vibration frequency of the crystal lattice. Such transitions cannot relax via simple single-phonon spontaneous emission, since the lattice simply cannot accept phonons at those high frequencies. These transitions must then relax either by radiative (photon) emission or by multiple-phonon processes. Since both these processes are relatively weak compared to direct single-phonon relaxation, the high-frequency transitions will have much slower relaxation rates ($\tau_{21} \approx 10^{-5}$ to 10^{-3} s in many cases). Therfore the various levels lumped into level 3 will all relax mostly into level 2 while level 2 itself is metastable and long-lived because there are no other levels located close below it into which it can decay directly.

The existence of metastable levels follows from quantum mechanical considerations that will not be discussed here. However, for completeness we will at least explain the term "forbidden transition". As we have seen in Sect. 1.2.4, the mechanism by which energy exchange takes place between an atom and the electromagnetic fields is the dipole radiation. As a consequence of quantum-mechanical considerations and the ensuing selection rules, transfer between certain states cannot occur due to forbidden transitions. The term "forbidden" means that a transition among the states concerned does not take place as a result of the interaction of the electric dipole moment of the atom with the radiation field. As a result of the selection rules, an atom may get into an excited state from which it will have difficulty returning to the ground state. A state from which all dipole transitions to lower energy states are forbidden is metastable; an atom entering such a state will generally remain in that state much longer than it would in an ordinary excited state from which escape is comparatively easy.

In the absence of a metastable level, the atoms which become excited by pump radiation and are transferred to a higher energy level will return either directly to the ground state by spontaneous radiation or by cascading down on intermediate levels, or they may release energy by phonon interaction with the lattice. In order for the population to increase at the metastable laser level, several other conditions have to be met. Let us consider the more general case of a four-level system illustrated in Fig. 1.7. (Note that a three-level system can be thought of as a special case of a four-level scheme where level 1 and level 0 coincide). Pumping takes place between two levels and laser action takes place between two other levels. Energy from the pump band is transferred to the upper laser level by fast radiative transitions. Energy is removed from the lower laser level again by fast radiationless transitions.

For electrons in the pump band at level 3 to transfer to level 2 rather than return directly to the ground state, it is required that $\tau_{30} \gg \tau_{32}$. For population to build up, relaxation out of the lower level 1 has to be fast, $\tau_{21} \gg \tau_{10}$. Thus, as a first conclusion, we may say that if the right relaxation time ratio exists between any two levels (such as 3 and 2) in an energy level system, a population inversion should be possible. If so, then obtaining a large enough inversion for successful laser operation becomes primarily a matter of the right pumping method. The optical pumping method is generally applicable only by the availability of systems which combine a narrow laser emission line with a broad absorption transition, so that a broad-band intense light source can be used as the pump source.

Having achieved population inversion in a material by correct combination of relaxation times and the existence of broad pump bands, the linewidth of the laser transition becomes very important. In the following chapter we will see that the optical gain for a given population inversion is inversely proportional to linewidth. Therfore, the metastable level should have a sufficiently narrow linewidth.

1.5 Laser Rate Equations

The dynamic behavior of a laser can be described with reasonable precision by a set of coupled rate equations [1.5]. In their simplest forms, a pair of simultaneous differential equations describe the population inversion and the radiation density within a spatially uniform laser medium. We will describe the system in terms of the energy-level diagrams shown in Figs. 1.6 and 1.7. As we have seen in the preceding discussions, two energy levels are of prime importance in laser action: the excited upper laser level E_2 and the lower laser level E_1. Thus for many analyses of laser action an approximation of the three- and four-level systems by a two-level representation is very useful.

The rate-equation approach used in this section involves a number of simplifying assumptions; in using a single set of rate equations we are ignoring longitudinal and radial variations of the radiation within the laser medium. In spite of these limitations, the simple rate-equation approach remains a useful tool and, properly used, provides a great deal of insight into the behavior of real solid-state laser deviced. We will derive from the rate equations the threshold condition for laser actions, and obtain a first-order approximation of the relaxation oscillations in a solid-state laser. Furthermore, in Chap. 4 we will use the rate equations to calculate the gain in a laser amplifier.

In general, the rate equations are useful in predicting the gross features of the laser output, such as average and peak power, Q-switched pulse-envelope shape, threshold condition, etc. On the other hand, many details of the nature of the laser emission are inacessible from the point of view of a simple rate equation. These include detailed descriptions of the spectral, temporal, and spatial distributions of the laser emission. Fortunately, these details can often be accounted for independently.

In applying the rate equations to the various aspects of laser operation, we will find it more convenient to express the probability for stimulated emission $\varrho(\nu)B_{21}$ by the photon density ϕ and the stimulated emission cross section σ.

With (1.37) we can express the Einstein coefficient for stimulated emission B_{21} in terms of the stimulated emission cross section $\sigma(\nu)$,

$$B_{21} = \frac{c}{h\nu g(\nu)} \sigma_{21}(\nu) \quad , \tag{1.49}$$

where $c = c_0/n$ is the speed of light in the medium. The energy density per unit frequency $\varrho(\nu)$ is expressed in terms of the lineshape factor $g(\nu)$, the energy $h\nu$, and the photon density ϕ [photons/cm^2] by

$$\varrho(\nu) = h\nu g(\nu)\phi \quad . \tag{1.50}$$

From (1.49 and 50) we obtain

$$B_{21}\varrho(\nu) = c\sigma_{21}(\nu)\phi \quad . \tag{1.51}$$

Three Level System

In order to approximate the three-level system with a two-level scheme, we assume that the transition from the pump band to the upper laser level is so fast that $N_3 \approx 0$. Therefore pumping does not affect the other processes at all except to allow a mechanism of populating the upper level and thereby obtaining population inversion ($N_2 > N_1$).

Looking at Fig. 1.6, this assumption requires that the relaxation time ratio τ_{32}/τ_{21} be very small. In solid-state lasers $\tau_{32}/\tau_{21} = 0$ is a good approximation. Spontaneous losses from the pump band to the ground state can be expressed by a pumping efficiency factor η_0. This parameter, defined as

$$\eta_0 = \left(1 + \frac{\tau_{32}}{\tau_{31}}\right)^{-1} \leq 1 \quad , \tag{1.52}$$

specifies what fraction of the total atoms excited to level 3 drop from there to level 2, thus becoming potentially useful for laser action. A small η_0 obviously requires a correspondingly larger pump power.

The changes in the electron population densities in a three-level system, based on the assumption that essentially all of the laser ions are in either level 1 or level 2, are

$$\frac{\partial n_1}{\partial t} = \left(n_2 - \frac{g_2}{g_1} n_1\right) c\phi\sigma + \frac{n_2}{\tau_{21}} - W_\mathrm{p} n_1 \tag{1.53}$$

and

$$\frac{\partial n_2}{\partial t} = -\frac{\partial n_1}{\partial t} \quad , \tag{1.54}$$

since

$$n_{\mathrm{tot}} = n_1 + n_2 \quad , \tag{1.55}$$

where W_{p} is the pumping rate [s^{-1}].

The terms of the right-hand side of (1.53) express the net stimulated emission, the spontaneous emission, and the optical pumping.

The time variation of the population in both levels due to absorption, spontaneous, and stimulated emission is obtained from (1.15). Note that the populations N_1 and N_2 are now expressed in terms of population densities n_1 and n_2. To take into account the effect of pumping, we have added the term $W_{\mathrm{p}}n_1$, which can be thought of as the rate of supply of atoms to the metastable level 2. More precisely, $W_{\mathrm{p}}n_1$ is the number of atoms transferred from the ground level to the upper laser level per unit time per unit volume. The pump rate W_{p} is related to the pump parameter W_{13} in Fig. 1.6 by

$$W_{\mathrm{p}} = \eta_0 W_{13} \quad . \tag{1.56}$$

The negative sign in front of $W_{\mathrm{p}}n_1$ in (1.53) indicates that the pump mechanism removes atoms from the ground level 1 and increases the population of level 2.

If we now define the inversion population density by

$$n = n_2 - \frac{g_2 n_1}{g_1} \tag{1.57}$$

we can combine (1.53, 54, and 57) to obtain

$$\frac{\partial n}{\partial t} - \gamma n \phi \sigma c - \frac{n + n_{\mathrm{tot}}(\gamma - 1)}{\tau_f} + W_{\mathrm{p}}(n_{\mathrm{tot}} - n), \tag{1.58}$$

where

$$\gamma = 1 + \frac{g_2}{g_1} \quad \text{and} \quad \tau_f = \tau_{21} \quad . \tag{1.59}$$

In obtaining (1.58) we have used the relations

$$n_1 = \frac{n_{\mathrm{tot}} - n}{1 + g_2/g_1} \quad \text{and} \quad n_2 = \frac{n + (g_2/g_1)n_{\mathrm{tot}}}{1 + g_2/g_1} \quad . \tag{1.60}$$

Another equation, usually regarded together with (1.58), describes the rate of change of the photon density within the laser resonator,

$$\frac{\partial \phi}{\partial t} = c\phi\sigma n - \frac{\phi}{\tau_c} + S, \tag{1.61}$$

where τ_c is the decay time for photons in the optical resonator and S is the rate at which spontaneous emission is added to the laser emission.

If we consider for the moment only the first term on the right, which is the increase of the photon density by stimulated emission, then (1.61) is identical to (1.33). However, for the time variation of the photon density in the laser resonator we must also take into account the decrease of radiation due to losses in the system and the increase of radiation due to a small amount of spontaneous emission which is added to the laser emission. Although very small, this term must be included because it provides the source of radiation which initiates laser emission.

An important consideration for initiation of laser oscillation is the total number p of resonant modes possible in the laser resonator volume V_R, since in general only a few of these modes are initiated into oscillations. This number is given by the familiar expression (1.3),

$$p = 8\pi\nu^2 \frac{\Delta\nu V_R}{c^3} \quad , \tag{1.62}$$

where ν is the laser optical frequency, and $\Delta\nu$ is the bandwidth of spontaneous emission. Let p_L be the number of modes of the laser output. Then S can be expressed as the rate at which spontaneous emission contributes to stimulated emission, namely,

$$S = \frac{p_L n_2}{p \tau_{21}} \quad . \tag{1.63}$$

The reader is referred to Chap. 3 for a more detailed description of the factor τ_c which appears in (1.61). For now we only need to know that τ_c represents all the losses in an optical resonator of a laser oscillator. Since τ_c has the dimension of time, the losses are expressed in terms of a relaxation time. The decay of the photon population in the cavity results from transmission and absorption at the end mirrors, "spillover" diffraction loss due to the finite apertures of the mirrors, scattering and absorptive losses in the laser material itself, etc. In the absence of the amplifying mechanism, (1.61) becomes

$$\frac{\partial\phi}{\partial t} = -\frac{\phi}{\tau_c} \quad , \tag{1.64}$$

the solution of which is $\phi(\tau) = \phi_0 \exp(-t/\tau_c)$.

The importance of (1.61) should be emphasized by noting that the right-hand side of this equation describes the net gain per transit of an electromagnetic wave passing through a laser material

Four-Level System

We will assume again that the transition from the pump band into the upper laser level occurs very rapidly. Therefore the population of the pump band is negligible, i.e., $n_3 \approx 0$. With this assumption the rate of change of the two laser levels in a four-level system is

$$\frac{dn_2}{dt} = W_p n_0 - \left(n_2 - \frac{g_2}{g_1} n_1\right) \sigma\phi c - \frac{n_2}{\tau_{21} + \tau_{20}} \quad , \tag{1.65}$$

$$\frac{dn_1}{dt} = \left(n_2 - \frac{g_2}{g_1} n_1\right)\sigma\phi c + \frac{n_2}{\tau_{21}} - \frac{n_1}{\tau_{10}} \quad , \tag{1.66}$$

$$n_{\text{tot}} = n_0 + n_1 + n_2 \quad . \tag{1.67}$$

From (1.65) follows that the upper laser level population in a four-level system increases due to pumping and decreases due to stimulated emission and spontaneous emissions into level 1 and level 0. The lower level population increases due to stimulated and spontaneous emission and decreases by a radiationless relaxation process into the ground level. This process is characterized by the time constant τ_{10}. In an ideal four-level system the terminal level empties infinitely fast to the ground level. If we let $\tau_{10} \approx 0$, then it follows from (1.66) that $n_1 = 0$. In this case the entire population is divided between the ground level 0 and the upper level of the laser transition. The system appears to be pumping from a large source that is independent of the lower laser level. With $\tau_{10} = 0$ and $n_1 = 0$, we obtain the following rate equation for the ideal four-level system

$$n = n_2 \quad \text{and} \tag{1.68}$$

$$n_{\text{tot}} = n_0 + n_2 \quad . \tag{1.69}$$

Therefore, instead of (1.58), we have

$$\frac{\partial n}{\partial t} = -n\sigma\phi c - \frac{n}{\tau_{\text{f}}} + W_{\text{p}}(n_{\text{tot}} - n). \tag{1.70}$$

The fluorescence decay time τ_{f} of the upper laser level is given by

$$\frac{1}{\tau_{\text{f}}} = \frac{1}{\tau_{21}} + \frac{1}{\tau_{20}} \quad , \tag{1.71}$$

where $\tau_{21} = A_{21}^{-1}$ is the effective radiative lifetime associated with the laser line. In the equation for the rate of change of the upper laser level we have again taken into account the fact that not all atoms pumped to level 3 will end up at the upper laser level. It is

$$W_{\text{p}} = \eta_0 W_{03} \quad , \tag{1.72}$$

where η_0 depends on the branching ratios which are the relative relaxation rates for the atoms along the various possible downward paths,

$$\eta_0 = \left(1 + \frac{\tau_{32}}{\tau_{31}} + \frac{\tau_{32}}{\tau_{30}}\right)^{-1} \leq 1. \tag{1.73}$$

The equation which describes the rate of change of the photon density within the laser resonator is the same as in the case of the three-level system.

Summary

The rate equation applicable to three-and four-level systems can be expressed by a single pair of equations, namely, (1.58 and 61), where $\gamma = 1 + g_2/g_1$ for a three-level system and $\gamma = 1$ for a four-level system. The parameters τ_f and W_p are defined by (1.56, 59, 72, and 73) for the different systems. The factor S in (1.61), which represents the initial noise level of ϕ due to spontaneous emission at the laser frequency, is small and needs to be considered only for initial starting of the laser action. It will be dropped from this point on.

A more detailed analysis of the laser rate equations can be found in [1.3,6].

2. Properties of Solid-State Laser Materials

Materials for laser operation must possess sharp fluorescent lines, strong absorption bands, and a reasonably high quantum efficiency for the fluorescent transition of interest. These characteristics are generally shown by solids (crystals or glass) which incorporate in small amounts elements in which optical transitions can occur between states of inner, incomplete electron shells. Thus the transition metals, the rare earth (lanthanide) series, and the actinide series are of interest in this connection. The sharp fluorescence lines in the spectra of crystals doped with these elements result from the fact that the electrons involved in transitions in the optical regime are shielded by the outer shells from the surrounding crystal lattice. The corresponding transitions are similar to those of the free ions. In addition to a sharp fluorescence emission line, a laser material suitable for optical pumping should possess broad-band pump transitions since, as a rule, only broad-band light sources, i.e., incandescent lamps, cw arc lamps, or flashlamps are available as pump sources for optically pumped solid-state lasers.

The three principal elements leading to gain in a laser are:

- *The host material* with its macroscopic mechanical, thermal and optical properties, and its unique microscopic lattice properties.
- *The activator/sensitizer ions* with their distinctive charge states and free-ion electronic configerations.
- *The optical pump source* with its particular geometry, spectral irradiance, and duration.

These elements are interactive and must be selectetd self-consistently to achieve a given system performance.

In this chapter we consider the properties of various host materials and activator/sensitizer combinations. Pump sources for solid-state lasers are treated in Chap. 6.

2.1 Overview

The conditions for laser action at optical frequencies were first described by *Schawlow* and *Townes* [2.1] in 1958. The first demonstration of laser action by *Maiman* [2.2] was achieved in 1960 using ruby ($Cr^{3+}: Al_2O_3$), a crystaline solid system. The next step in the development of solid-state lasers was the operation of trivalent uranium in CaF_2 and divalent samarium in CaF_2 by *Sorokin* and *Stevenson* [2.3]. In 1961 *Snitzer* [2.4] demonstrated laser action in neodymium-doped glass. The first continuously operating crystal laser was reported in 1961 by *Johnson* and *Nassau* [2.5] using $Nd^{3+}:CaWO_4$. Since then laser action has been achieved from trivalent rare earths (Nd^{3+}, Er^{3+}, Ho^{3+}, Ce^{3+}, Tm^{3+}, Pr^{3+}, Gd^{3+}, Eu^{3+}, Yb^{3+}), divalent rare earths (Sm^{2+}, Dy^{2+}, Tm^{2+}), transition metals (Cr^{3+}, Ni^{2+}, Co^{2+}, Ti^{3+}, V^{2+}), and the actinide ion U^{3+} embedded in various host materials. Optically pumped laser action has been demonstrated in hundreds of ion-host crystal combinations covering a spectral range from the visible to the mid-infrared.

The exceptionally favorable characteristics of the trivalent neodymium ion for laser action were recognized at a relatively early stage of the search for solid-state laser materials. Thus, Nd^{3+} was known to exhibit a satisfactorily long fluorescence lifetime and narrow fluorescence linewidths, and to possess a terminal state for the laser transition sufficiently high above the ground state so that cw operation at room temperature was readily feasible. Therefore, this ion was incorporated as a dopant in a variety of host materials, i.e., glass, $CaWO_4$, $CaMoO_4$, CaF_2, LaF_3, etc., in an effort to make use of its great potential. However, most of these early hosts displayed undesirable shortcomings, either from the standpoint of their intrinsic physical properties or because of the way in which they interacted with the Nd^3 ions. Finally, yttrium aluminum garnet ("YAG") was explored by *Geusic* et al. [2.6] as a host for Nd^3 and its superiority to other host materials was quickly demonstrated. Nd:YAG lasers displayed the lowest thresholds for cw operation at room temperature of any known host-dopant combination.

2.1.1 Host Materials

Solid-state host materials may be broadly grouped into crystalline solids and glasses. The host must have good optical, mechanical and thermal properties to withstand the severe operating conditions of practical lasers. Desirable properties include hardness, chemical inertness, absence of internal strain and refractive index variation, resistance to radiation-induced color centers, and ease of fabrication.

Several interactions between the host crystal and the additive ion restrict the number of useful material combinations. These include size disparity, valence, and spectroscopic properties. Ideally the size and valence of the additive ion should match that of the host ion it replaces.

In selecting a crystal suitable for a laser ion host one must consider the following key criteria:

i) The crystal must possess favorable optical properties. Variations in the index of refraction lead to inhomogeneous propagation of light through the crystal with consequent poor beam quality.
ii) The crystal must possess a set of mechanical and thermal properties that will permit repetitively pulsed operation without suffering excessive stress under the operational thermal load.

iii) The crystal must have lattice sites that can accept the dopant ions and that have local crystal fields of symmetry and strength needed to induce the desired spectroscopic properties. In general, ions placed in a crystal host must achieve high radiative lifetime with cross sections near $10^{-20}\,\mathrm{cm}^2$.

iv) It must be possible to scale the growth of the impurity-doped crystal, while maintaining high optical quality and high yield. It appears that the greatest prospect for successful growth scaling is for crystals that melt congruently at temperaturs below 1300°C. This relatively low melting temperature permits the use of a wide variety of crucible materials and growth techniques.

Glasses

Glasses form an important class of host materials for some of the rare earths, particularly Nd^{3+}. The outstanding practical advantage compared to crystalline materials is the tremendous size capability for high-energy applications. Rods up to 1 m in length and over 10 cm in diameter and disks up to 90 cm in diameter and several cm thick are currently available. The optical quality can be excellent, and beam angles approaching the diffraction limit can be achieved. Glass, of course, is easily fabricated and takes a good optical finish. Laser ions placed in glass generally show a larger fluorescent linewidth than in crystals as a result of the lack of a unique and well-defined crystalline surrounding for the individual active atom. Therefore, the laser thresholds for glass lasers have been found to run higher than their crystalline counterparts. Also, glass has a much lower thermal conductivity than most crystalline hosts. The latter factor leads to thermally induced birefringence and optical distortion in glass laser rods when they are operated at high average powers. Ions which have been made to lase in glass include Nd^{3+}, which will be discussed in detail in Sect. 2.3.2, Yb^{3+}, Er^{3+}, Tm^{3+}, and Ho^{3+}. Glass doped with erbium is of special importance, because its radiation of 1.55 μm does not penetrate the lens of the human eye, and therefore cannot destroy the retina. Because of the three-level behavior of erbium and the small absorption of pump light by Er^{3+}, multiple doping with neodymium and ytterbium is necessary to obtain satisfactory system efficiency. In this technique, called sensitization, the amplifying ion either absorbes radiation at wavelengths other than those of the laser ion and then radiates within the pump band of the laser ion or transfers its excitation energy directly to the laser ion. Because of its three-level operation, the Nd^{3+}-Yb^{3+}-Er^{3+} : glass is at least an order of magnitude less efficient than the Nd : glass lasers [2.7]. For specific details on laser glasses, the reader is referred to [2.8].

A large number of crystalline host materials have been investigated since the discovery of the ruby laser. Crystalline laser hosts generally offer as advantages over glasses their higher thermal conductivity, narrower fluorescence linewidths, and, in some cases, greater hardness. However, the optical quality and doping homogeneity of crystalline hosts are often poorer, and the absorption lines are generally narrower. For an overview of crystalline lasers, see [2.9].

Oxides

Sapphire. The first laser material to be discovered (ruby laser) employed sapphire as a host. The Al_2O_3 (sapphire) host is hard, with high thermal conductivity, and transition metals can readily be incorporated substitutionally for the Al. The Al site is too small for rare earths, and it is not possible to incorporate appreciablc concentrations of these impurities into sapphire. Besides ruby which is still used today, Ti-doped sapphire has gained significance as a tunable-laser material. The properties of ruby and Ti-sapphire will be discussed in Sect. 2.2 and 2.5.2.

Garnets. Some of the most useful laser hosts are the synthetic garnets: yttrium aluminum, $Y_3Al_5O_{12}$ (YAG), gadolinium gallium garnet , $Gd_3Ga_5O_{12}$ (GGG) [2.10], and gadolinium scandium aluminum garnet $Gd_3Sc_2Al_3O_{12}$ (GSGG) [2.6, 11]. These garnets have many properties that are desirable in a laser host material. They are stable, hard, optically isotropic, and have good thermal conductivities, which permits laser operation at high average power levels.

In particular, yttrium aluminum garnet doped with neodymium (Nd : YAG) has achieved a position of dominance among solid-state laser materials. YAG is a very hard, isotropic crystal, which can be grown and fabricated in a manner that yields rods of high optical quality. At the present time, it is the best commercially available crystalline laser host for Nd^{3+}, offering low threshold and high gain. The properties of Nd : YAG are discussed in more detail in Sect. 2.3.1. Besides Nd^{3+}, the host crystal YAG has been doped with Tm^{3+}, Er^{3+}, Ho^{3+}, and Yb^{3+}. Laser action in these materials are reported in [2.12, 13].

Aluminate. In 1970 a crystal host derived from the same Y_2O_3-Al_2O_3 system as YAG became commercially available [2.15]. The crystal yttrium ortho aluminate ($YAlO_3$), termed YAlO or YAP, is the only solid-state material other than Nd : YAG to exhibit the high conductivity and hardness combined with low threshold necessary to achieve high average power operation in the cw-pumped mode at room temperature. Measurements of the physical and spectroscopic properties of $YAlO_3$ reveal that they are generally comparable to those of YAG. There are, however, several significant differences which enhance its potential value as a laser host. These include a faster growth rate and near-ideal distribution coefficients for rare earth dopants. Also, in contrast to YAG, which is cubic and isotropic, YAP is orthorhombic and anisotropic. The anisotropy of the spectral properties of $YAlO_3$ enables one to select crystallographic orientations of the laser rod which optimize particular performance characteristics. Thus rod orientations can be chosen for high gain and low thresholds or, alternatively, for low gain and high energy storage required for Q-switching operation.

The fluorescence in Nd : $YAlO_3$ occurs predominantly in three strong lines at 1.0645, 1.0725, and 1.0795 μm. For light propagating along the b axis of the crystal, the gain is maximum at 1.0795 μm with the electric field E parallel to the c direction, and is comparable to that of the 1.064-μm line in Nd : $YAlO_3$, the gain at 1.0645 μm is larger than at 1.0729 and 1.0795 μm, but it is only about one-half that in Nd : YAG. Laser action in YAP has been achieved with ions including Nd^{3+}, Er^{3+}, Ho^{3+}, and Tm^{3+} [2.15–18].

Despite several major potential advantages of YAP over YAG, such as a polarized output, the capability of accepting a higher concentration of Nd, the possibility of varying gain by changing the rod axis, and lower costs because of faster crystal growth, the material disappeared from the market only two years after its introduction. YAP's disadvantages were a lower efficiency than expected and erratic performance. It was found, for example, that during the growth process Fe^{3+} impurities enter the single crystal [2.16]. YAP crystals grown or annealed in H_2 contain both Fe^{3+} and OH^-, which increase the absorption loss at 1.06 μm and cause fluorescence quenching. Recently, however, Heraeus, FR Germany, has been able to grow high-optical-quality YAP crystals, and there is renewed interest in this material [2.14].

Oxysulfide. The application of rare earth oxysulfides as laser host materials has also been explored. The entire oxysulfide series, from lanthanum oxysulfide through lutetium oxysulfide and yttrium oxysulfide, possess the same (uniaxial) crystal structure. Thus, solid solutions of any concentration of rare earth activator in any other rare earth oxysulfide host are possible. Host materials which are transparent from 0.35 to 7 μm include lanthanum, gadolinium, yttrium, and lutetium oxysulfide. Laser action was observed at 1.075 μm for Nd and La_2O_2S. The lasing transition cross section of Nd : LOS was measured to be about one-third of that of Nd : YAG [2.19]. At present the optical quality of LOS crystals is far below the quality of YAG crystals.

Phosphates and Silicates

Laser oscillations have been produced in crystals of Nd^{3+}-doped calcium fluorophosphate or $Ca_5(PO_4)_3F$. The host crystal has the mineral name fluorapatite, from which the name FAP was coined [2.20]. This material is unique in that the Nd^{3+} fluorescence spectrum is predominantly concentrated in a single narrow and intense line, whereas the absorption spectrum is relatively broad and intense. Fluorapatites have low oscillation thresholds and high slope efficiencies, but they are soft, are susceptible to the formation of color centers, and their low thermal conductivity leads to strong thermal distortions. Because of these latter properties, FAP did not become a popular laser material. In 1972 another apatite became available: silicate oxyapatite or CaLaSOAP. In contrast to FAP, CaLaSOAP is considerably harder; however, its thermal conductivity is nearly equal to FAP's and only one-ninth the thermal conductivity of YAG.

The energy storage in Nd : SOAP is about five times that of YAG. In addition to SOAP's high energy storage, the material has the potential that large crystals can be grown inexpensively. Crystals 15 cm long with diameters as large as 1.2 cm have been grown at growth rates of 2 to 3 mm/h. It was reported that the optical quality of SOAP rods 7.5 by 0.6 cm was comparable to the quality of YAG rods with undetectable scattering and less than 0.5 fringe. The main disadvantages of SOAP are a low laser damage threshold and low thermal conductivity. Physical, spectroscopic, and laser characteristics of neodymium-doped silicate oxyapatite are reported in [2.21–23].

Laser action has been achieved in the neodymium pentaphosphates $YNdP_5O_{14}$, $LaNdP_5O_{14}$, and $ScNdP_5O_{14}$ [2.24–28]. An outstanding property of these materials is the high gain which can be achieved as a result of the favorable position of the $^4I_{15/2}$ manifold relative to the upper laser states. As a result of the high gain, cw oscillation at room temperature has been obtained in $ScNdP_5O_{14}$ with a pump power of only 4 mW using an argon laser as a pump source [2.28].

Tungstates, Molybdates, Vanadates, and Beryllates

$CaWO_4$ was the most popular material for Nd before YAG became commercially available. The rare earth substitutes for Ca, but only in the trivalent oxidation state, and hence charge compensation is needed. For optimum laser performance, substitution of Na^+ for Ca^{2+} was found to be best [2.29]. The material is very prone to fracture, even when well annealed, and thus considerable care is required when the boules are being fabricated into laser rods. The thermal conductivity is three to four times greater than that of the glasses. The absorption spectra of Nd^{3+} in this material consists of a large number of rather fine lines.

Sodium rare earth molybdates and tungstates have served as host materials for active ions. In these materials, which are similar to $CaWO_4$ and $CaMoO_4$, one-half of the calcium atoms are randomly replaced with sodium and the other half with rare earth. Laser action has been observed from Nd^{3+} in $NaLa(MoO_4)_2$ and $NaGd(WO_4)_2$ and from $NaNd(WO_4)_2$. Interest in $Nd:NaLa(MoO_4)_2$ stems from the fact that this material has a low gain, intermediate to Nd : glass and Nd : YAG and thus is capable of higher efficiencies than the former material and greater energy storage than the latter material [2.30]. The thermal conductivity of $Nd:NaLa(MoO_4)$ is three times that of Nd : glass. However, the lower thermal conductivity and the higher thermal expansion coefficient of $Nd:NaLa(MoO_4)_2$ as compared to Nd : YAG results in a $Nd:NaLa(MoO_4)_2$ laser rod having much greater thermal stress than a Nd : YAG rod when both are subjected to the same average input power during lasing action.

Nd^{3+}-doped yttrium orthovanadate (YVO_4) has shown relatively low threshold at pulsed operation [2.31]. However, early studies of this crystal were hampered by severe crystal growth problems, and as a result YVO_4 was discarded as a host. Although larger and better samples became available eventually [2.10a], the poor thermal properties precluded any further interest. Very recently the material was pumped by a laser diode and a 12 % overall conversion efficiency was achieved [2.32a]. This is the highest wall-plug efficiency reported to date for any solid-state laser. For $Nd:YVO_4$ the laser threshold is lower and the stimulated emission is 2.7 times higher than that of Nd : YAG [2.32b].

A candidate of the beryllates is Nd^{3+}-doped lanthanum beryllate ($Nd:La_2Be_2O_5$) which is commercially available as a laser material [2.33]. Since La^{3+} is the largest of the rare-earth ions, $La_2Be_2O_5$ (BEL) has large distribution coefficients for other trivalent rare-earth ions. For this reason boules of BEL with high concentrations of rare-earth ions are much more readily grown than are YAG and other hosts based on the yttrium ion. Since the thermal conductivity and the cross section are considerably lower than those of YAG [2.32b] the material has not found acceptance in the market place.

An interesting property was reported by *Chin* et al. [2.32c] regarding this material. Nd : BEL is optically bi-axial, and has positive and negative thermal coefficients for the refractive index. Therefore, an optical path can be selected which minimizes thermal lensing, thus leading to athermal behavior.

Fluorides

The divalent fluorides are relatively soft, isotropic crystals. Rare earth-doped CaF_2 crystals have been studied extensively [2.3, 34], since this material was the host of many early solid-state lasers. The doping of trivalent rare earth into fluoride hosts requires charge compensation, which represents a major drawback. In recent years yttrium lithium fluoride ($YLiF_4$), a uniaxial crystal, has received attention as a host for Ho^{3+}, Er^{3+} [2.35] and Nd^{3+} [2.36]. $YLiF_4$ is transparent to 1500 Å, therefore, high-current-density xenon flashlamps which emit strongly in the blue and near-ultraviolet can be used as pump sources without damage to the material. The linewidth of Er : YLF is only $10\,cm^{-1}$, indicating fairly high gain. In order to obtain efficient operation from a Ho : YLF laser, the material must be sensitized with Er-Tm. Nd : YLF offers a reduction in thermal lensing and birefringence combined with improved energy storage relative to Nd : YAG. The pumping efficiency and thermomechanical properties of Nd : YLF, however, are not as good as those of Nd : YAG. Considerable development of Nd : YLF has taken place in the areas of crystal growth, spectroscopy, material characterization and laser physics. The Nd : YLF laser will be discussed in Sect. 2.3.4.

Ceramics

Laser action has also been achieved by doping optical ceramic with Nd [2.37, 38]. Advantages of this type of host material over glass or crystals include low cost and higher thermal conductivity, and better thermal shock resistance compared with glass. However, glass ceramics are plagued with high scattering losses.

2.1.2 Active Ions

Before proceeding to a discussion of the active laser ions, we will review briefly the nomenclature of atomic energy levels.

Different energy states of electrons are expressed by different quantum numbers. The electrons in an atom are characterized by a principal quantum number n, an orbital angular momentum l, the orientation of the angular momentum vector m, and a spin quantum number s. A tradition from the early days of line-series allocation has established the following method of designating individual electronic orbits: a number followed by a letter symbolizes the principal quantum number n and the angular number l, respectively. The letters s, p, d, f stand for $l = 0, 1, 2, 3$, respectively. For example a $3d$ electron is in an orbit with $n = 3$ and $l = 2$.

To designate an atomic energy term one uses by convention capital letters with a system of subscripts and superscripts. The symbol characterizing the term is of the form ${}^{2S+1}L_J$, where the resultant orbital quantum numbers $L = 0, 1, 2, 3, 4$ are expressed by the capital letters S, P, D, F, G, H. A superscript to the left of the letter indicates the value $(2S + 1)$, i.e., the multiplicity of the term due to possible orientation of the resultant spin S. Thus a one-electron system ($S = \frac{1}{2}$) has a multiplicity 2. L and S can combine to various resultants J, indicated by a subscript to the right of the letter. Thus the symbol ${}^2P_{3/2}$ shows an energy level with an orbital quantum number $L = 1$, a resultant spin of $S = \frac{1}{2}$, and a total angular momentum of $J = \frac{3}{2}$. The complete term description must include the configuration of the excited electron, which precedes the letter symbol. Thus the ground state of Li has the symbol $2s^2S_{1/2}$.

When an atom contains many electrons, the electrons that form a closed shell may be disregarded and the energy differences associated with transitions in the atom may be calculated by considering only the electrons outside the closed shell.

In describing the state of a multielectron atom, the orbital angular momenta and the spin angular momenta are added separately. The sum of the orbital angular momenta are designated by the letter L, and the total spin is characterized by S. The total angular momentum J of the atom may then be obtained by vector addition of L and S. The collection of energy states with common values of J, L, and S is called a term.

In the following section, a qualitative description is given of some of the prominent features of the most important rare earth, actinide, and transition metal ions.

Rare Earth Ions

The rare earth ions are natural candidates to serve as active ions in solid-state laser materials because they exhibit a wealth of sharp fluorescent transitions representing almost every region of the visible and near-infrared portions of the electromagnetic spectrum. It is a characteristic of these lines that they may be very sharp, even in the presence of the strong local fields of crystals, as a result of the shielding effect of the outer electrons. The outermost electrons of these ions form a complete rare gas shell, which is the xenon shell with two $5s$ and six $5p$ electrons. This shell is optically inactive. Next inside the xenon shell is the $4f$ shell, which is filled successively in passing from one element to the next. Trivalent cerium, Ce^{3+}, has one $4f$ electron, and trivalent ytterbium, Yb^{3+}, has 13. As long as the $4f$ shell is not completely filled with 14 electrons, a number of $4f$ levels are unoccupied, and electrons already present in the $4f$ shell can be raised by light absorption into these empty levels. The sharp lines observed in rare earth absorption and emission spectra are ascribed to these transitions, and the sharpness of the lines is explained by the fact that the electrons making the transition lie inside the xenon shell and thus interact only weakly with outside ions. Table 2.1 shows the population of the outermost electron shells of the rare earths.

Table 2.1. Electronic structure of elements 59 to 71. (For comparison, Xe has been added to the list)

Number	Element	Outermost electron shell					
54	Xenon, Xe	$4d^{10}$	–	$5s^2$	$5p^6$	–	–
59	Praseodymium, Pr	$4d^{10}$	$4f^3$	$5s^2$	$5p^6$	–	$6s^2$
60	Neodymium, Nd	$4d^{10}$	$4f^4$	$5s^2$	$5p^6$	–	$6s^2$
61	Promethium, Pm	$4d^{10}$	$4f^5$	$5s^2$	$5p^6$	–	$6s^2$
62	Samarium, Sm	$4d^{10}$	$4f^6$	$5s^2$	$5p^6$	–	$6s^2$
63	Europium, Eu	$4d^{10}$	$4f^7$	$5s^2$	$5p^6$	–	$6s^2$
64	Gadolinium, Gd	$4d^{10}$	$4f^8$	$5s^2$	$5p^6$	$5d$	$6s^2$
65	Terbium, Tb	$4d^{10}$	$4f^9$	$5s^2$	$5p^6$	–	$6s^2$
66	Dysprosium, Dy	$4d^{10}$	$4f^{10}$	$5s^2$	$5p^6$	–	$6s^2$
67	Holmium, Ho	$4d^{10}$	$4f^{11}$	$5s^2$	$5p^6$	–	$6s^2$
68	Erbium, Er	$4d^{10}$	$4f^{12}$	$5s^2$	$5p^6$	–	$6s^2$
69	Thulium, Tm	$4d^{10}$	$4f^{13}$	$5s^2$	$5p^6$	–	$6s^2$
70	Ytterbium, Yb	$4d^{10}$	$4f^{14}$	$5s^2$	$5p^6$	–	$6s^2$
71	Lutetium, Lu	$4d^{10}$	$4f^{14}$	$5s^2$	$5p^6$	$5d$	$6s^2$

Rare earth ions usually exist in solids in either the trivalent or the divalent state. A divalent rare earth ion is formed when the atom gives up its outermost $6s$ electrons. When a trivalent ion is formed the atom also loses its $5d$ electron if it has one; otherwise, one of the $4f$ electrons is lost.

Neodymium. Nd^{3+} was the first of the trivalent rare earth ions to be used in a laser, and it remains by far the most important element in this group. Stimulated emission has been obtained with this ion incorporated in at least 40 different host materials, and a higher power level has been obtained from Nd lasers than from any other four-level material. The principal host materials are YAG and glass. In these hosts stimulated emission is obtained at a number of frequencies within three different groups of transitions centered at 0.9, 1.06, and 1.35 μm. Radiation at these wavelengths results from ${}^4F_{3/2} \rightarrow {}^4I_{9/2}, {}^4I_{11/2}, {}^4I_{13/2}$ transitions, respectively.

The nomenclature of the energy levels may be illustrated by a discussion of the Nd^{3+} ion. This ion has three electrons in the $4f$ subshell. In the ground state their orbits are so aligned that the orbital angular momentum adds up to $3+2+1=6$ atomic units. The total angular momentum $L=6$ is expressed by the letter I. The spins of the three electrons are aligned parallel to each other, providing an additional $\frac{3}{2}$ units of angular momentum, which, when added antiparallel to the orbital angular momentum, gives a total angular momentum of $6-\frac{3}{2}=\frac{9}{2}$ units. According to the quantum rules for the addition of angular momenta, the vector sum of an orbital angular momentum of 6 and a spin angular momentum of $\frac{3}{2}$ may result in the following four values of the total angular momentum: $\frac{9}{2}, \frac{11}{2}, \frac{13}{2}$, and $\frac{15}{2}$. The levels corresponding to these values are ${}^4I_{9/2}$, ${}^4I_{11/2}$, ${}^4I_{13/2}$, and ${}^4I_{15/2}$. The first of these, which has the lowest energy, is the ground state; the others are among the first few excited levels of Nd^{3+}. These levels are distinguished by the orientation of the spins with respect to the resultant orbital angular momentum. Other excited levels are obtained when another combination of the orbital angular momenta is chosen.

Erbium. Numerous studies of the absorption and fluorescence properties of erbium in various host materials have been conducted to determine its potential as an active laser ion. Laser oscillation was observed most frequently in the wavelength region 1.53 to 1.66 μm arising from transitions between the $^4I_{13/2}$ state and the $^4I_{15/2}$ ground state Er^{3+}. Stimulated emission in the vicinity of 1.6 μm is of interest, because the eye is less subject to retinal damage by laser radiation at these wavelengths due to the greatly reduced transmissivity of the ocular media.

Host materials of erbium have included YAG [2.12], YLF [2.35], $YALO_3$ [2.39], LaF_3 [2.40], $CaWO_4$ [2.41], CaF_2 [2.42], and various glasses [2.43, 44]. The terminal level in Er^{3+} is between 525 cm^{-1} for Er : YAG and 50 cm^{-1} for Er : glass. At room temperature all levels of the terminal $^4I_{15/2}$ manifold are populated to some degree, thus this transition forms a three-level laser scheme with a correspondingly high threshold. Laser action is generally achieved either by lowering the temperature to depopulate the higher-lying levels of the $^4I_{15/2}$ manifold, or by codoping the materials with trivalent ytterbium to improve the optical pumping efficiency via $Yb^{3+} \rightarrow Er^{3+}$ energy transfer.

Since YAG possesses the largest ground-state splitting of all host materials doped with Er and has other properties which make it the best available host for many applications, emphasis is placed mainly on the optimization of sensitized Er : YAG. Of particular interest is YAG, highly doped with Er which produces an output around 2.9 μm.

The erbium glass lasers which have been developed have used phosphate and silicate glass base materials and have been doped with neodymium and ytterbium in addition to erbium. The neodymium and ytterbium act is sensitizing agents by absorbing pump light in regions where the erbium is relatively transparent. Er : YAG and Er : glass lasers are discussed in greater detail in Sect. 2.4.

Holmium. Laser action in Ho^{3+} has been reported in at least 20 different host materials [2.45–47]. Because the terminal level is only about 250 cm^{-1} above ground level, the lower laser level has a relatively high thermal population at room temperature. While Ho : YAG and Ho : YLF have proven to be efficient lasers, operation has been limited in most cases to cryogenic temperatures, which will depopulate the laser level. At these temperatures efficient operation has been achieved for both continuous and pulsed operation. Slope efficiencies in excess of 6 % have been demonstrated [2.45]. Previous efforts in flashlamp pumped 2 μm lasers have concentrated on Er : Tm : Ho doped YAG and YLF. At 77 K, Er is an effective sensitizer because it provides absorption bands at the flashlamp output spectrum that very efficiently transfer energy through thulium to the Ho $^5I_7$. At room temperature the threshold is very high due to increased Er fluorescence, non-radiative decay processes in Er and a thermally populated terminal laser level.

Very recently it was discovered that Cr-sensitized Tm : Ho : YAG offers several advantages over the Er sensitized materials. Using Cr instead of Er as the primary sensitizer, efficient room-temperature flashlamp pumped laser operation of the 2.1 laser Ho $^5I_7$–$^5I_8$ transition was achieved [2.47d–f]. Laser diode pumping of a Tm:Ho:YAG laser via an absorption line in Tm^{3+} also produced laser operation at room temperature [2.47f].

Thulium, Praseodymium, Gadolinium, Europium, Ytterbium, Cerium. Laser action in all these triply ionized rare earths has been reported; however, only marginal performance was obtained in hosts containing Tm^{3+}, Pr^{3+}, Gd^{3+}, Eu^{3+}, Yb^{3+} and Ce^{3+}. On the other hand, some of these rare earth ions are important as sensitizers in Er and Ho lasers.

Samarium, Dysprosium, Thulium. The divalent rare earths Sm^{2+}, Dy^{2+}, and Tm^{2+} have an additional electron in the $4f$ shell, which lowers the energy of the $5d$ configuration. Consequently, the allowed $4f$-$5d$ absorption bands fall in the visible region of the spectrum. These bands are particularly suitable for pumping the laser systems. Tm^{2+}, Dy^{2+}, and Sm^{2+} have been operated as lasers, all in a CaF_2 host. For laser operation, these crystals must be refrigerated to at least 77 K. Pulsed laser action at 2.35 μm in Cr^{3+}-sensitized Tm^{3+}:$YAlO_3$ has also been reported [2.48]. In this transition the terminal state is unpopulated at room temperature, assuring four-level operation.

Actinide Ions

The actinides are similar to the rare earths in having the $5f$ electrons partially shielded by $6s$ and $6p$ electrons. Most of the actinide elements are radioactive, and only uranium in CaF_2 has been successfully used in a laser [2.3]. The host was doped with 0.05 % uranium. Laser action occurred at 2.6 μm between a metastable level and a terminal level some 515 cm^{-1} above the ground state.

Transition Metals

Important members of the transition metal group include the ruby (Cr^{3+} : Al_2O_3), alexandrite (Cr^{3+} : $BeAL_2O_4$) and Ti : sapphire (Ti^{3+} : Al_2O_3) laser, which are discussed in separate sections. Laser action has been observed in most other transition metals and particularly in Ni^{2+} and Co^{2+}. Considerable effort has gone into investigations of the Co:MgF_2 laser. Earlier work has been extended in recent years by *Moulton* et al. who have presented results for laser pumped crystals under a variety of operating conditions [2.52]. The need for cryogenic cooling is a major drawback for most applications.

Summary

Excellent compilations of useful materials, parameters and references on laser host and impurity ions can be found in [2.8, 9, 55a, 86, 87].

2.2 Ruby

The ruby laser, although a three-level system, still remains in use today for certain applications. From an application point of view, ruby is attractive because its output lies in the visible range, in contrast to most rare earth four-level lasers, whose outputs are in the near-infrared region. Photodetectors and photographic emulsions are much more sensitive at the ruby wavelength than in the infrared. Spectroscopically, ruby possesses an unusually favorable combination of a relatively narrow linewidth, a long fluorescent lifetime, a high quantum efficiency, and broad and well-located pump absorption bands which make unusually efficient use of the pump radiation emitted by available flashlamps.

Physical Properties

Ruby chemically consists of sapphire (Al_2O_3) in which a small percentage of the Al^{3+} has been replaced by Cr^{3+}. This is done by adding small amounts of Cr_2O_3 to the melt of highly purified Al_2O_3. The pure single host crystal is uniaxial and possesses a rhombohedral or hexagonal unit cell, as shown in Fig. 2.1. The crystal has an axis of symmetry, the so-called c axis, which forms the major diagonal of the unit cell. Since the crystal is uniaxial, it has two indices of refraction, the ordinary ray having the E vector perpendicular to the c (optic) axis, and the extraordinary ray having the E vector parallel to the c axis.

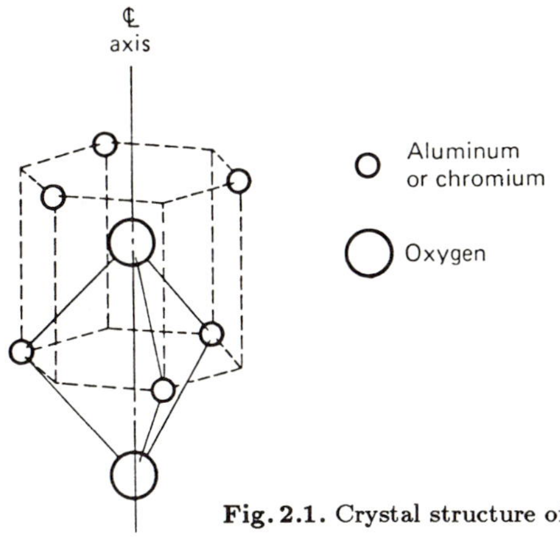

Fig. 2.1. Crystal structure of sapphire

As a laser host crystal, sapphire has many desirable physical and chemical properties. The crystal is a refractory material, hard and durable. It has good thermal conductivity, is chemically stable, and is capable of being grown to very high quality. Ruby is grown by the Czochralski method. In this procedure the solid crystal is slowly pulled from a liquid melt by initiation of growth on high-quality seed material. Iridium crucibles and rf heating are used to contain the melt and control the melt temperature, respectively. The crystal boules can be grown in the 0°, 60°, or 90° configuration, where the term refers to the angle between the growth axis and the crystallographic c axis. For laser grade ruby the 60° type is commonly used.

As has already been stated, the active material in ruby is the Cr^{3+} ion. This ion has three d electrons in its unfilled shell; the ground state of the free

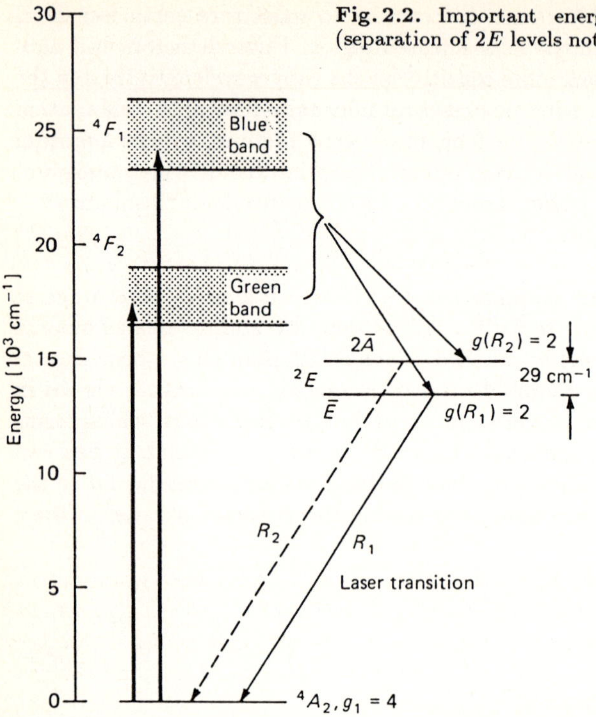

Fig. 2.2. Important energy levels of Cr^{3+} in ruby (separation of $2E$ levels not to scale)

ion is described by the spectroscopic symbol ^4A. The amount of doping is nominally 0.05 weight percent Cr_2O_3. However, in some applications it is desirable to lower the Cr^{3+} concentration to approximately 0.035 weight percent to obtain maximum beam quality in ruby oscillators.

Laser Properties of Ruby

A simplified energy level diagram of ruby is given in Fig. 2.2. In ruby lasers, population inversion with respect to the so-called ^2E level is obtained by optically pumping Cr^{3+} ions from the $^4A_2$ ground state to the broad pump bands $^4F_2$ and $^4F_1$. The lifetime at the pump bands, which are each about 1000 Å wide, located in the green (18,000 cm^{-1}) and in the violet (25,000 cm^{-1}), is extremely short, with the ions returning to a metastable state ^2E. This metastable level is split into two sublevels with a separation of $\Delta E = 29\,\mathrm{cm}^{-1}$. The upper one is the $2\overline{A}$ and the lower one the $\overline{E}$ sublevel. The two transitions ($\overline{E} \rightarrow {}^4A_2$ and $2\overline{A} \rightarrow {}^4A_2$) are referred to as the R_1 and R_2 lines. Each is approximately 5 Å wide at room temperature, and the lines lie at the end of the visible, at 6943 and 6929 Å.

At thermal equilibrium the difference in population between the $\overline{E}$ and $2\overline{A}$ level is

$$\frac{N(2\overline{A})}{N(\overline{E})} = \exp\left(\frac{\Delta E}{kT}\right) = K \quad . \tag{2.1}$$

At room temperature the Boltzmann factor is $K = 0.87$. The fluorescence in ruby consists of the R_1 and R_2 lines. However, laser action takes place only at the R_1 line, i.e., between the $\overline{E}$ and $^4A_2$ level. The R_1 line attains laser threshold before the R_2 line because of the higher inversion. Once laser action commences in the R_1 line, the $\overline{E}$ level becomes depleted and population transfer from the nearby $2\overline{A}$ level proceeds at such a fast rate that the threshold level is never reached for the R_2 line.

The relaxation time between the $2\overline{A}$ and $\overline{E}$ levels is very short, on the order of a nanosecond or less. For laser pulses which are long compared to this time constant the population ratio of the two states is kept unchanged, but since $\overline{E}$ decays much faster, almost the entire initial population of the two states decays through R_1 emission. If we compare the energy level diagram of ruby with our simplified scheme of a three-level system in Fig. 1.6, the levels $^4F_1$ and $^4F_2$ jointly constitute level 3, whereas the ^2E and $^4A_2$ states represent level 2 and level 1, respectively. We can write

$$N_2 = N_2(R_1) + N_2(R_2) \tag{2.2}$$

for the metastable level and N_1 for the ground level. Threshold and gain in ruby depends only on the population of level $N_2(R_1)$. However, in relating gain and threshold to the population of the ground level N_1 or to the total number of Cr^{3+} ions N_{tot}, one has to take the population of $N_2(R_2)$ into account. In ruby all levels are degenerate, that is,

$$g(N_1) = 4 \quad , \quad g(R_1) = g(R_2) = 2 \quad . \tag{2.3}$$

Because of the higher degeneracy of the ground state, amplification occurs when the R_1 level is at least one-half as densely populated as the ground state.

$$N_2(R_1) = \tfrac{1}{2} N_1 \quad . \tag{2.4}$$

The two upper levels are related by

$$N_2(R_2) = K N_2(R_1) \quad . \tag{2.5}$$

Since

$$N_2(R_1) + N_2(R_2) + N_1 = N_{tot} \quad , \tag{2.6}$$

we have the threshold (300 K)

$$\begin{aligned} N_2(R_1) &= \frac{N_{tot}}{3+K} = 0.26 N_{tot} \quad , \\ N_2(R_2) &= \frac{K N_{tot}}{3+K} = 0.22 N_{tot} \quad , \\ N_1 &= \frac{2 N_{tot}}{3+K} = 0.52 N_{tot} \quad . \end{aligned} \tag{2.7}$$

Thus we must have just under one-half of the atoms in the two upper levels in order to reach threshold. At complete inversion we have

$$N_2(R_1) = \frac{N_{\rm tot}}{K+1} = 0.53 N_{\rm tot} \quad ,$$

$$N_2(R_2) = \frac{K N_{\rm tot}}{K+1} = 0.47 N_{\rm tot} \quad , \tag{2.8}$$

$$N_1 = 0 \quad .$$

For normal laser operation the population densities of the various levels are between those given for threshold and total inversion. The amount of energy per unit volume which can be extracted from the inverted ruby depends on the population of levels $N_2(R_1)$ and $N_2(R_2)$, provided that the pulse duration is long enough that these two levels remain in thermal equilibrium. With a Cr^{3+} concentration of $n_{\rm tot} = 1.58 \times 10^{19}\,\rm cm^{-3}$ and a photon energy of $h\nu = 2.86 \times 10^{-19}$ Ws, we obtain for the maximum upper-state energy density, when all atoms exist in the excited energy states $N_2(R_1)$ and $N_2(R_2)$, a value of $E = n_{\rm tot} h\nu = 4.52\,\rm J/cm^3$. The maximum energy which can be extracted, assuming complete inversion, is, according to (2.7, 8),

$$E_{\rm ex\,max} = \frac{2h\nu n_{\rm tot}}{3+K} = 2.35\,\rm J/cm^3 \quad .$$

For pulses which are short compared to the relaxation time between levels $N_2(R_1)$ and $N_2(R_2)$, only energy stored in the level $N_2(R_1)$ can be extracted. Again, assuming an initial complete inversion of the material, we can extract a maximum energy per unit volume of

$$E'_{\rm ex\,max} = \frac{2h\nu n_{\rm tot}}{3(1+K)} = 1.6\,\rm J/cm^3 \quad .$$

The upper-state energy density at the inversion level $E_{\rm uth} = h\nu n_{\rm tot}(1+K)/(3+K)$ has a numerical value of $E_{\rm uth} = 2.18\,\rm J/cm^3$. The small-signal gain coefficient in ruby is

$$g_0 = \sigma_{21}\left[n_2(R_1) - \frac{g(R_1)n_1}{g(N_1)}\right] \quad , \tag{2.9}$$

where σ_{21} is the stimulated emission cross section of the R_1 line and $n_2(R_1)$ is the population density of the E level. With (2.2, 5 and 6) we can write

$$g_0 = \sigma_{21}\left(\frac{3+K}{2(1+K)} n_2 - \frac{n_{\rm tot}}{2}\right) \quad . \tag{2.10}$$

It is customary to express the gain coefficient in terms of the absorption coefficient. Since $\sigma_{21} = \sigma_{12} g(N_1)/g(R_1)$, the gain coefficient can be expressed as

$$g_0 = \alpha_0 \left(\frac{(3+K)n_2}{(1+K)n_{\text{tot}}} - 1 \right) \quad , \tag{2.11}$$

where $\alpha_0 = \sigma_{12} n_{\text{tot}}$ is the absorption coefficient of ruby. With the approximation $K \approx 1$ the expressions above can be simplified to

$$g_0 = \alpha_0 \left(\frac{2n_2}{n_{\text{tot}}} - 1 \right) = \sigma_{21} \left(n_2 - \frac{n_{\text{tot}}}{2} \right) = \sigma_{12}(2n_2 - n_{\text{tot}}) \quad . \tag{2.12}$$

With all the chromium ions in the ground state ($n_2 = 0$), the gain of the unexcited ruby crystal is $g_0 = -\alpha_0$. The maximum gain achieved at total inversion ($n_2 = n_{\text{tot}}$) is $g_0 = 2\alpha_0/(1+K) \approx \alpha_0$. Numerical values for α_0 and σ_{12} are obtained from the absorption data for ruby. Figure 2.3 shows the absorption coefficient and absorption cross section for the R lines of ruby as a function of wavelength [2.55b]. As one can see from these curves, the absorption coefficient of the R lines (and, therefore, gain in the presence of inversion) for light having its E vector normal to the c axis is greater than that for light with E parallel to the c axis. This accounts for the polarization of the output from lasers employing ruby rods with c axis orientations away from the rod axis. The absorption cross section at the peak of the R_1 line is seven times higher for a beam polarized perpendicular to the c axis as for a beam polarized parallel to the c axis. From Fig. 2.3 follows an absorption cross section of $\sigma_{12} = 1.22 \times 10^{-20}\,\text{cm}^2$ and an absorption coefficient of $\alpha_0 = \sigma_{12} n_{\text{tot}} =$

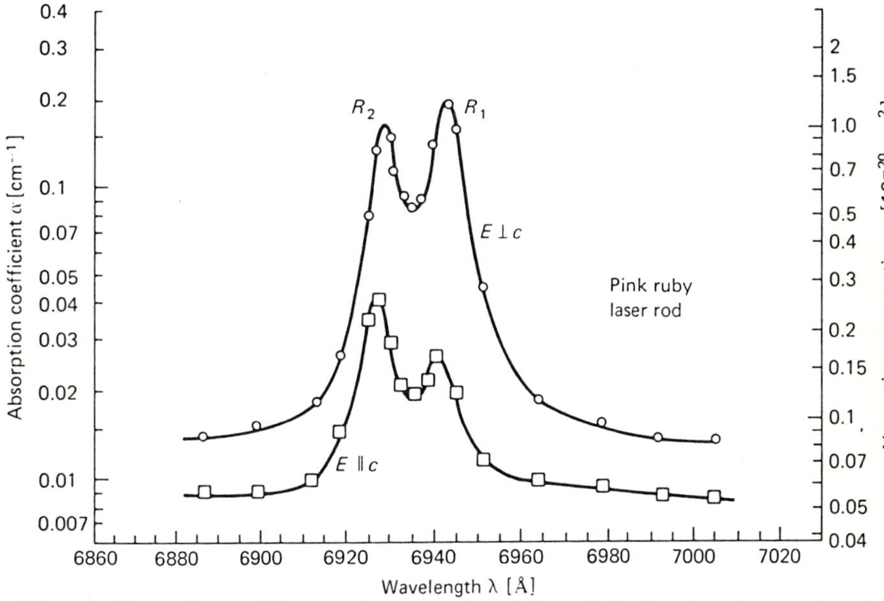

Fig. 2.3. Absorption coefficient and cross section of R_1 and R_2 lines in ruby as a function of wavelength, incident beam polarized parallel ($E \parallel c$) and orthogonal ($E \perp c$) to c axis of crystal. Cr^{3+} concentration is $1.58 \times 10^{19}\,\text{cm}^{-3}$ [2.55b]

$0.2\,\mathrm{cm}^{-1}$ for the peak of the R_1 line ($E \perp c$) at a doping concentration of $n_{\mathrm{tot}} = 1.58 \times 10^{19}\,\mathrm{cm}^{-3}$. Other published values are $\sigma_{12} = 1.7 \times 10^{-20}\,\mathrm{cm}^2$ [2.56] and $\alpha_0 \approx 0.28\,\mathrm{cm}^{-1}$ [2.57]. Using Cronemeyer's data we obtain for the stimulated emission cross section in ruby a value of $\sigma_{21} = 2.5 \times 10^{-20}\,\mathrm{cm}^2$. The maximum gain coefficient for complete inversion in a 0.05 % Cr-doped rod is then $g_{0(\mathrm{max})} = 0.215\,\mathrm{cm}^{-1}$. The latter two values are in agreement with the data published by *Kriukov* [2.58]. The various parameters for maximum gain and absorption in ruby are related according to

$$g_{0(\mathrm{max})} = \frac{\sigma_{21} n_{\mathrm{tot}}}{K+1} \approx \frac{2\sigma_{12} n_{\mathrm{tot}}}{K+1} \approx \frac{2\alpha_0}{K+1} \quad . \tag{2.13}$$

Despite the 20 % lower stimulated emission cross section and the lower population density, laser action at the R_2 line can be achieved in ruby either by employing dielectrically coated resonator mirrors which have a sufficiently higher transmission at R_1 than at the R_2 line [2.59], or by inserting a polarizer and a retardation plate inside the resonator [2.60].

The absorbed pump energy required to obtain threshold is

$$E_{\mathrm{ab}} \approx \frac{n_{\mathrm{tot}} h\nu_{\mathrm{p}}}{2\eta_1} \quad , \tag{2.14}$$

where $\nu_{\mathrm{p}} = 6.25 \times 10^{14}\,\mathrm{s}^{-1}$ is the average frequency of the two main pump bands and η_1 is the quantum efficiency. Since the average quantum efficiency in ruby for the two main pump bands is $\eta_1 \approx 0.70$, one obtains $E_{\mathrm{ab}} = 4.55\,\mathrm{J/cm}^3$.

The absorption spectrum of ruby is given in Fig. 2.4 [2.57]. The best-quality ruby is grown with the crystal c axis at 60° to the boule axis. From Fig. 2.4 follows that in this case the absorption spectrum is different for a light

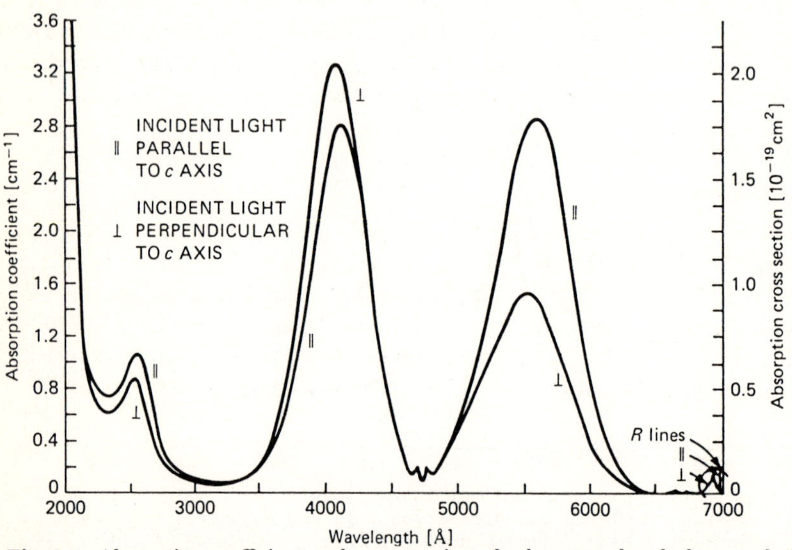

Fig. 2.4. Absorption coefficient and cross section of ruby pump bands for unpolarized light as a function of wavelength. Cr^{3+} concentration is $1.88 \times 10^{19}\,\mathrm{cm}^{-3}$ [2.57]

Table 2.2. Optical and laser properties of ruby at room temperature

Property	Values and units
Cr_2O_3 doping	0.05 wt. %
Cr^{3+} concentration	1.58×10^{19} ions cm^{-3}
Output wavelengths, 25 C	R_1, 14403 cm^{-1}, 6943 Å R_2, 14432 cm^{-1}, 6929 Å
Fluorescent lifetime	3.0 ms at 300 K
Spectral linewidth	11 cm^{-1}, 5.3 Å
Photon energy ($h\nu$)	2.86×10^{-19} Ws
Quantum efficiency	0.7
Separation of R_1 and R_2 lines	29 cm^{-1}, 870 GHz, 14 Å
Absorption coefficient and cross section of laser line (R_1 level, $E \perp c$)	$\alpha_{R_1} = 0.2\,cm^{-1}$; $\sigma_{R_1} = 1.22 \times 10^{-20}\,cm^2$
Stimulated emission cross section	$\sigma_{21} = 2.5 \times 10^{-20}\,cm^2$
Scatter losses	$\alpha_{sc} \approx 0.001\,cm^{-1}$
Major pump bands	
Blue (4040 Å)	$\alpha_{\parallel} = 2.8\,cm^{-1}$; $\alpha_{\perp} = 3.2\,cm^{-1}$
Green (5540 Å)	$\alpha_{\parallel} = 2.8\,cm^{-1}$; $\alpha_{\perp} = 1.4\,cm^{-1}$
Refractive index at 6943 Å	1.763 ordinary ray $E \perp c$ 1.755 extraordinary ray $E \parallel c$
Brewster angle	60° 37′ at 6943 Å
Maximum extractable energy	2.35 J/cm^3 (complete inversion)
Maximum upper-state energy density	4.52 J/cm^3 (complete inversion)
Upper-state energy at threshold	2.18 J/cm^3

beam incident parallel or normal to the c axis of the crystal. When such a ruby is pumped in a diffuse reflecting pump cavity, the loci of constant inversion are elliptical as a result of the differences in absorption of the pump light in directions parallel and perpendicular to the c axis. This effect, which causes the beam pattern in the near and far field to be elliptical, can be observed experimentally. Important ruby laser parameters obtained from (2.55, 57, 61, 62 are listed in Table 2.2.

Thermal Properties

Variation of temperature causes a change in wavelength and linewidth of the fluorescent lines [2.63, 64]. At 300 K the laser line is centered at $\approx$ 6943 Å and has a linewidth of $\approx 11\,cm^{-1}$, while at 77 K the line has moved to $\approx$ 6934 Å and has a linewidth of only $\approx 0.15\,cm^{-1}$. Near room temperature the peak of the R_1 line shifts $\approx$ 0.05 Å/C. In the temperature region from −80 to +20 C the wavelength of the ruby laser is given approximately by the formula

$$\lambda(T) = 6943.25 + 0.068(T - 20) \quad ,$$

where T is the temperature in degrees centigrade. The reduction of the linewidth at lower temperatures causes a large increase of the gain coefficient in accordance with (1.41). The temperature dependence of pertinent materials param-

Table 2.3. Temperature dependence of ruby parameters

Temperature dependence of wavelength	0.065 Å/C, 25–80 C 0.045 Å/C, −180–0 C
Temperature dependence of refractive index (dn/dT)	$+12.6 \times 10^{-6}\,C^{-1}$
Quantum efficiency	1.0 at 77 K 0.7 at 300 K 0.1 at 500 K
Thermal population of $2A$ and E levels	$N(2A)/N(E) = 0.87$ at 300 K $N(2A)/N(E) = 0.58$ at 77 K
Fluorescent lifetime	3.0 ms at 300 K 4.3 ms at 77 K
Linewidth	5.3 Å at 300 K 0.07 Å at 77 K
Thermal conductivity	$0.42\,W\,cm^{-1}\,K^{-1}$ at 300 K $10.0\,W\,cm^{-1}\,K^{-1}$ at 77 K
Thermal expansion coefficient	$5.8 \times 10^{-6}\,C^{-1}$
Specific heat	$0.18\,g\,cal\,K^{-1}$
Diffusivity	$0.13\,cm^2\,s^{-1}$

eters for ruby is listed in Table 2.3. Note the strong temperature dependence of the thermal conductivity [2.65].

Commercially Available Ruby Laser Rods

Ruby can be grown in relatively large boules with high optical quality. The boules are cut into smaller cylindrical sections. Core-free cylindrical rods can be obtained up to 30 cm in length and 2.5 cm in diameter. Typical commercially available rods with flat ends are fabricated to the following specifications: ends flat to $\lambda/10$, ends parallel to ± 4 arc seconds, perpendicularity to rod axis to ± 5 minutes, rod axis parallel to growth axis (0°, 60°, 90°) to $\pm 5°$. Common rod geometries may be plane parallel, wedged surfaces, Brewster angle, or prismatic. The best way to inspect a laser rod is by means of an interferometer, such as a Twyman-Green. Optical inhomogeneities, strain, distortion, etc., will show up as fringes. Scattering centers or regions of high defects in a laser rod are revealed by illuminating the crystal from the side with a He-Ne gas laser.

Commercially available ruby laser rods are offered in three optical grades: standard, standard internal quality, and select. The grade is determined by the number of fringes across the rod aperture if a double-pass Twyman-Green interferometric analysis is performed. Standard rods have less than 2 fringes, standard internal quality rods less than 1 fringe and select rods less than 0.5 fringe per 2.5 cm of rod length. In addition to a Twyman-Green, a passive beam divergence measurement of the laser rods is usually performed. In this case, a diffraction-limited and collimated He-Ne gas laser beam is passed through the crystal, projecting a far field pattern which is analyzed to ascertain the light flux distribution from the rod. The specification of select-grade ruby is that 80% of the output energy be with 0.1 milliradians per 2.5 cm of rod length.

The Ruby *c* Plane

In ruby lasers containing optical elements which are dependent on the polarization of the beam, the *c* plane of the ruby crystal must be properly aligned. The proper alignment of a ruby laser rod in a system can be done as follows: View the ruby in a direction perpendicular to the cylinder axis, and rotate the rod about this axis. When the transmitted light is deepest red, you are viewing parallel to the *c* plane. The *E* vector of the laser radiation will be perpendicular to this plane. The light and dark zones, which one sees by looking at a ruby crystal from the side and rotating it, are easily explained with the aid of Fig. 2.4. From this figure follows that incident light polarized parallel to the *c* axis is absorbed stronger in the 5000 to 6000 Å region than light polarized perpendicular to the *c* axis. At the laser wavelength, however, light polarized perpendicular to the *c* axis is absorbed more strongly (Fig. 2.3).

To achieve fine adjustment of the installed and prealigned laser rod, one can place a polarizer in front of and behind the rod. The axes of the polarizer must be parallel and normal to the desired plane of polarization of the laser radiation. The rod can be viewed between these crossed polarizers by placing a diffuse reflector (such as a white piece of paper) behind the polarizer in back and illuminating the paper so that it may be seen from the front end of the ruby. As the ruby is rotated at small angles (i.e., $\pm 20°$) about its original position, variations in the amount of transmitted light will occur. If the ruby is adjusted for maximum absorption, the ruby *c* plane will then be oriented parallel to one of the polarizer axes.

Ruby properties have been summarized in [2.66], and useful parameters listed in [2.67].

2.3 Nd : Lasers

From the large number of Nd : doped materials the four lasers discussed in more detail in this chapter have gained prominence. Nd : YAG, because of its high gain and good thermal and mechanical properties, is by far the most important solid-state laser for scientific, medical, industrial and military applications. Nd : glass is important for laser fusion drivers because it can be produced in large sizes. Very recently, Nd : Cr : GSGG has received considerable attention because of the good spectral match between the flashlamp emission and the absorption of the Cr ions. An efficient energy transfer between the Cr and Nd ions results in a highly efficient Nd : laser. Nd : YLF is a good candidate for certain specialized application, because the output is polarized, and the crystal exhibits lower thermal birefringence. Nd : YLF has a higher energy storage capability (due to its lower gain coefficient) compared to Nd : YAG and its output wavelength matches that of phosphate Nd : glass, therefore modelocked and Q-switched Nd : YLF lasers have become the standard oscillators for large glass lasers employed in fusion research.

2.3.1 Nd : YAG

The Nd : YAG laser is by far the most commonly used type of solid-state laser. Neodymium-doped yttrium aluminum garnet (Nd : YAG) possesses a combination of properties uniquely favorable for laser operation. The YAG host is hard, of good optical quality, and has a high thermal conductivity. Furthermore, the cubic structure of YAG favors a narrow fluorescent linewidth, which results in high gain and low threshold for laser operation. In Nd : YAG, trivalent neodymium substitutes for trivalent yttrium, so charge compensation is not required.

Five years after *Geusic* et al. [2.6] reported the first successful lasing of Nd : YAG, rapid strides had been made in improving both the quality of the material and pumping technique so that available cw power outputs rose from the fractional watt level initially obtained to several hundred watts from a single laser rod [2.68]. On the other hand, in single-crystal Nd : YAG fiber lasers, threshold was achieved with absorbed pump powers as small as 1 mW [2.69]. Today, more than twenty years after its first operation, the Nd : YAG laser has emerged as the most versatile solid-state system in existence.

Physical Properties

In addition to the very favorable spectral and lasing characteristics displayed by Nd : YAG, the host lattice is noteworthy for its unusually attractive blend of physical, chemical, and mechanical properties. The YAG structure is stable from the lowest temperatures up to the melting point, and no transformations have been reported in the solid phase. The strength and hardness of YAG are lower than ruby but still high enough so that normal fabrication procedures do not produce any serious breakage problems.

Pure $Y_3Al_5O_{12}$ is a colorless, optically isotropic crystal which possesses a cubic structure characteristic of garnets. In Nd : YAG about 1 % of Y^{3+} is substituted by Nd^{3+}. The radii of the two rare earth ions differ by about 3 %. Therefore, with the addition of large amounts of neodymium, strained crystals are obtained – indicating that either the solubility limit of neodymium is exceeded or that the lattice of YAG is seriously distorted by the inclusion of neodymium. Some of the important physical properties of YAG are listed in Table 2.4, together with optical and laser parameters [2.70–76].

Laser Properties

The Nd : YAG laser is a four-level system as depicted by a simplified energy level diagram in Fig. 2.5. The laser transition, having a wavelength of 10641 Å, originates from the R_2 component of the $^4F_{3/2}$ level and terminates at the Y_3 component of the $^4I_{11/12}$ level. At room temperature only 40 % of the $^4F_{3/2}$ population is at level R_2; the remaining 60 % are at the lower sublevel R_1 according to Boltzmann's law. Lasing takes place only by R_2 ions whereby the R_2 level population is replenished from R_1 by thermal transitions. The ground level of Nd : YAG is the $^4I_{9/2}$ level. There are a number of relatively broad en-

Table 2.4. Physical and optical properties of Nd:YAG

Chemical formula	$Nd:Y_3Al_5O_{12}$
Weight % Nd	0.725
Atomic % Nd	1.0
Nd atoms/cm^3	1.38×10^{20}
Melting point	1970 C
Knoop hardness	1215
Density	4.56 g/cm^3
Rupture stress	$1.3–2.6 \times 10^3$ kg/cm^3
Modulus of elasticity	3×10^3 kg/cm^2
Thermal expansion coefficient	
[100] orientation	$8.2 \times 10^{-6}\,C^{-1}$, 0 –250 C
[110] orientation	$7.7 \times 10^{-6}\,C^{-1}$, 10 –250 C
[111] orientation	$7.8 \times 10^{-6}\,C^{-1}$, 0 –250 C
Linewidth	4.5 Å
Stimulated emission cross section	$\sigma_{21} = 6.5 \times 10^{-19}$ cm^2
Relaxation time ($^4I_{11/2} \rightarrow {}^4I_{9/2}$)	30 ns
Radiative lifetime ($^4F_{3/2} \rightarrow {}^4I_{11/2}$)	550 μs
Spontaneous fluorescence lifetime	230 μs
Photon energy at 1.06 μm	$h\nu = 1.86 \times 10^{-19}$ J
Index of refraction	1.82 (at 1.0 μm)
Scatter losses	$\alpha_{sc} \approx 0.002\,cm^{-1}$

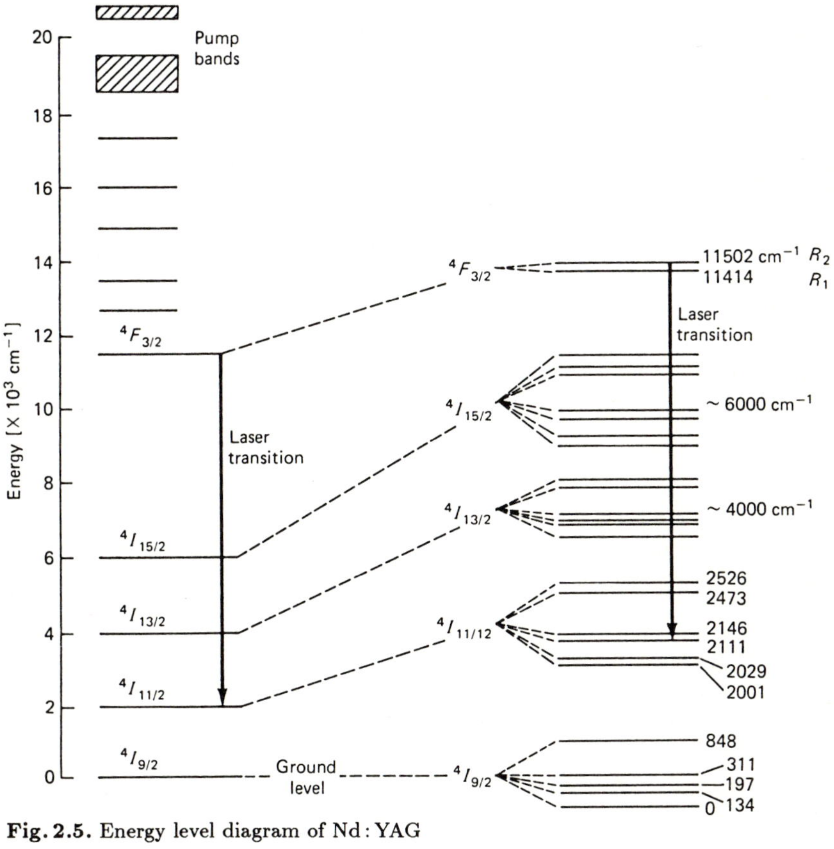

Fig. 2.5. Energy level diagram of Nd:YAG

ergy levels, which together may be viewed as comprising pump level 3. Of the main pump bands shown, the 0.81 and 0.75 μm bands are the strongest. The terminal laser level is 2111 cm^{-1} above the ground state and thus the population is a factor of $\exp(\Delta E/kT) \approx \exp(-10)$ of the ground-state density. Since the terminal level is not populated thermally, the threshold condition is easy to obtain.

The upper laser level, $^4F_{3/2}$, has a fluorescent efficiency greater than 99.5 % [2.70] and a radiative lifetime of 230 μs. The branching ratio of emission from $^4F_{3/2}$ is as follows [2.72]: $^4F_{3/2} \rightarrow {}^4I_{9/2} = 0.25$, $^4F_{3/2} \rightarrow {}^4I_{11/2} = 0.60$, $^4F_{3/2} \rightarrow {}^4I_{13/12} = 0.14$, and $^4F_{3/2} \rightarrow {}^4I_{15/2} < 0.01$. This means that almost all the ions transferred from the ground level to the pump bands end up at the upper laser level, and 60 % of the ions at the upper laser level cause fluorescent output at the $^4I_{11/2}$ manifold.

At room temperature the main 1.06-μm line in Nd : YAG is homogeneously broadened by thermally activated lattice vibrations. The published values of the stimulated emission cross section for the extensively measured 1.064–μm line have ranged from 2.7 to 8.8 $\times$ 10^{-19} cm^2 [2.19, 72–74]. *Ranon* et al. [2.73] reported that the inaccurate determination of the Nd concentration is responsible for these conflicting results. Using the concentration value obtained from wet chemistry, the researchers obtained a value of 6.5 $\times$ 10^{-19} cm^2 for the cross section of Nd : YAG.

Figure 2.6 shows the fluorescence spectrum [2.75] of Nd^{3+} in YAG near the region of the laser output with the corresponding energy levels for the various transitions. The absorption of Nd : YAG in the range 0.3 to 0.9 μm is given in Fig. 2.7. Thermal properties of Nd : YAG [2.77] are summarized in Table 2.5.

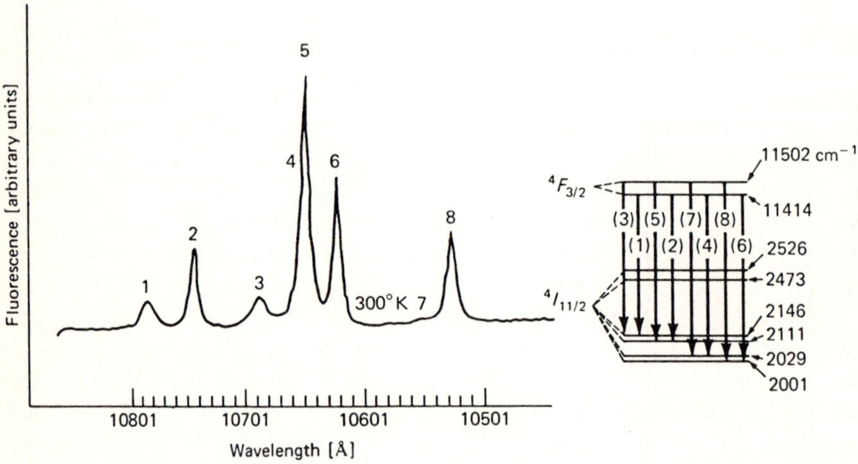

Fig. 2.6. Fluorescence spectrum of Nd^{3+} in YAG at 300 K in the region of 1.06 μm [2.75]

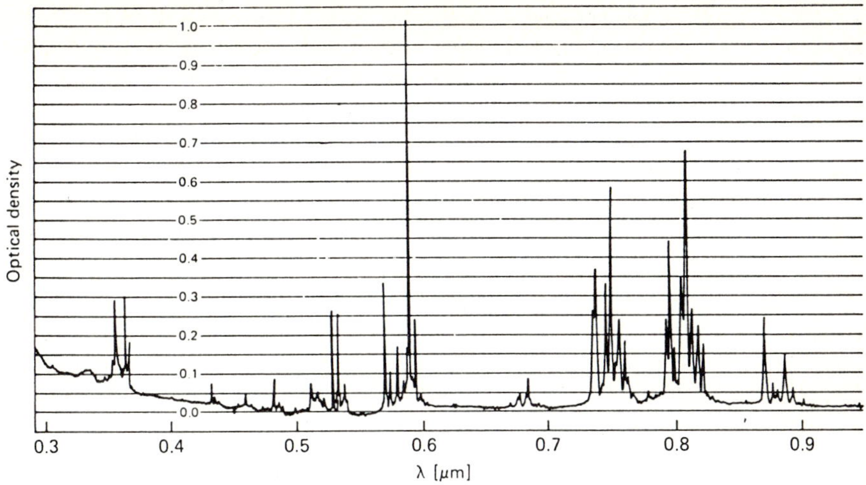

Fig. 2.7. Absorption spectrum of Nd : YAG at 300 K

Table 2.5. Thermal properties of Nd : YAG

Property	Units	300 K	200 K	100 K
Thermal conductivity	$W\ cm^{-1}\ K^{-1}$	0.13	0.21	0.58
Specific heat	$W\ s\ g^{-1}\ K^{-1}$	0.59	0.43	0.13
Thermal diffusivity	$cm^2\ s^{-1}$	0.046	0.10	0.92
Thermal expansion	K^{-1}	7.5	5.8	4.25
$\partial n/\partial T$	K^{-1}	7.3×10^{-6}	–	–

Nd : YAG Laser Rods

Commercially available laser crystals are grown exclusively by the Czochralski method. Growth rates, dopants, annealing procedures, and final size generally determine the manufacturing rate of each crystal. The boule axis or growth direction is customarily in the [111] direction. The high manufacturing costs of Nd : YAG are mainly caused by the very slow growth rate of Nd : YAG, which is of the order of 0.5 mm/h. Typical boules of 10 to 15 cm in length require a growth run of several weeks.

Boules grown from Nd : YAG typically contain very few optically observable scattering centers, and show negligible absorption in the lasing wavelength. However, all Nd : YAG crystals grown by Czochralski techniques show a bright core running along the length of the crystal when positioned between crossed polarizers. Strain flares are also visible, radiating from the core toward the surfaces of the crystal. Electron microprobe studies have revealed that in the core region the Nd concentration can run as much as twice as high as in the surrounding areas. The cores originate from the presence of facets on the growth interface which have a different distribution coefficient for neodymium than the surrounding growth surface. These compositional differences cause corresponding differences in thermal expansion coefficients which, in turn, give rise to the observed stress patterns during the cool down of the crystals from the growth temperature. Annealing does not seem to eliminate the cores and, thus far, no way has been found of avoiding the formation of facets on the growth interface. However, by choosing the growth direction carefully and by maintaining as steep an interface angle as is practical, the stresses can be confined to a very localized region and high-optical quality rods can be fabricated from the surrounding material. This means, of course, that in order to provide rods of a given diameter, the crystal must be grown with a diameter that is somewhat more than twice as large. The boules are processed by quartering into sections. At the present time, rods can be fabricated with maximum diameters of about 10 mm and lengths of up to 150 mm. The optical quality of such rods is normally quite good and comparable to the best quality of Czochralski ruby or optical glass. For example, 6 mm by 100 mm rods cut from the outer sections of 20 mm by 150 mm boules typically may show only 1 to 2 fringes in a Twyman-Green interferometer.

Neodymium concentration by atom percent in YAG has been limited to 1.0 to 1.5 %. Higher doping levels tend to shorten the fluorescent lifetime, broaden the linewidth, and cause strain in the crystal, resulting in poor optical quality. The Nd doping level in YAG is sometimes expressed in different concentration units: A concentration of 1.0 % Nd atoms in the lattice is equivalent to 0.727 % Nd or 0.848 % Nd_2O_3 by weight, respectively. The concentration of Nd^{3+} sites in these cases is $1.386 \times 10^{20}\,cm^{-3}$.

In specifying Nd : YAG rods, the emphasis is on size, dimensional tolerance, doping level, and passive optical tests of rod quality. Cylindrical rods with flat ends are typically finished to the following specifications: end flat to $\lambda/10$, ends parallel to ± 4 arc seconds, perpendicularity to rod axis to ± 5 minutes, rod axis parallel to within $\pm 5°$ to [111] direction. Dimensional tolerances typically are ± 0.5 mm on length and ± 0.025 mm on diameter. Most suppliers furnish the laser crystals with a photograph showing the fringe pattern of the crystal as examined by a Twyman-Green interferometer. A double-pass Twyman-Green interferometer quickly reveals strained areas, small defects, or processing errors.

In a particular application the performance of a Nd : YAG laser can be somewhat improved by the choice of the optimum Nd concentration. As a general guideline, it can be said that a high doping concentration (approximately 1.2 %) is desirable for Q-switch operation because this will lead to high energy storage. For cw operation, a low doping concentration (0.6 to 0.8 %) is usually chosen to obtain good beam quality.

It is worth noting that in contrast to a liquid or a glass, a crystal host is not amenable to uniform dopant concentration. This problem arises as a result of the crystal-growth mechanism. In the substitution of the larger Nd^{3+} for a Y^{3+} in $Y_3Al_5O_{l12}$, the neodymium is preferentially retained in the melt. The increase in concentration of Nd from the seed to the terminus of a 20-cm long boule is about 20 to 25 %. For a laser rod 3 to 8 cm long, this end-to-end variation may be 0.05 to 0.10 % of Nd_2O_3 by weight.

Different Laser Transitions in Nd:YAG

Under normal operating conditions the Nd:YAG laser oscillates at room temperature on the strongest ${}^4F_{3/2} \rightarrow {}^4I_{11/2}$ transition at 1.0641 μm. It is possible, however, to obtain oscillation at other wavelengths by inserting etalons or dispersive prisms in the resonator, by utilizing a specially designed resonant reflector as an output mirror [2.78], or by employing highly selective dielectrically coated mirrors [2.79]. These elements suppress laser oscillation at the undesirable wavelength and provide optimum conditions at the wavelength desired. With this technique over 20 transitions have been made to laser in Nd:YAG [2.80]. The relative output of those transitions at which room-temperature cw operation has been achieved is listed in Table 2.6 [2.80d]. The relative performance is compared to 1.064 μm emission for which 71.5 W output was obtained.

Table 2.6. Main room-temperature transitions in Nd:YAG

Wavelength ([μm], air)	Transition	Relative Performance
1.05205	$R_2 \rightarrow Y_1$	46
1.06152	$R_1 \rightarrow Y_1$	92
1.06414	$R_2 \rightarrow Y_3$	100
1.0646	$R_1 \rightarrow Y_2$	~50
1.0738	$R_1 \rightarrow Y_3$	65
1.0780	$R_1 \rightarrow Y_4$	34
1.1054	$R_2 \rightarrow Y_5$	9
1.1121	$R_2 \rightarrow Y_6$	49
1.1159	$R_1 \rightarrow Y_5$	46
1.12267	$R_1 \rightarrow Y_6$	40
1.3188	$R_2 \rightarrow X_1$	34
1.3200	$R_2 \rightarrow X_2$	9
1.3338	$R_1 \rightarrow X_1$	13
1.3350	$R_1 \rightarrow X_2$	15
1.3382	$R_2 \rightarrow X_3$	24
1.3410	$R_2 \rightarrow X_4$	9
1.3564	$R_1 \rightarrow X_4$	14
1.4140	$R_2 \rightarrow X_6$	1
1.4440	$R_1 \rightarrow X_7$	0.2

The 1.064- and 1.061-μm transitions, ${}^4F_{3/2} \rightarrow {}^4I_{11/2}$, provide the lowest threshold laser lines in Nd:YAG. At room temperature the 1.064-μm line $R_2 \rightarrow Y_3$ is dominant, while at low temperatures the 1.061-μm line $R_1 \rightarrow Y_1$ has the lower threshold [2.81]. If the laser crystal is cooled, additional laser transitions are obtained, most notably the 1.839-μm line and the 0.946-μm line [2.82, 83].

Recently the design and operation of a high-power Nd:YAG laser operating at 1.3 μm was reported [2.84]. The laser achieved in the long-pulse mode an output in excess of 8 J per pulse at 5 Hz, and up to 165 W of laser output at 50 Hz.

2.3.2 Nd : Glass

There are a number of characteristics which distinguish glass from other solid-state laser host materials. Its properties are isotropic. It can be doped at very high concentrations with excellent uniformity, and it can be made in large pieces of diffraction-limited optical quality. In addition, glass lasers have been made, in a variety of shapes and sizes, from fibers a few micrometers wide supporting only a single dielectric waveguide mode, to rods 2 m long and 7.5 cm in diameter and disks up to 90 cm in diameter and 5 cm thick.

There is a wide variety of Nd-doped laser glasses depending on the compositions of the glass network former and the network-modifying ions. Among various laser glasses only silicates and phosphates are commercially available at present with sufficient optical, mechanical and chemical properties. The Nd : phosphates are generally characterized by a large stimulated emission cross section ($\sigma = 3.7$–$4.5 \times 10^{-20}\,\mathrm{cm}^2$) and a relatively small nonlinear index ($n_2 = 0.91$–1.15×10^{-13} ESU). Typically, the cross section of a phosphate laser glass is 50 % higher than a comparable silicate. The Nd : phosphate glasses have been adopted in large laser systems employed for fusion research [2.85–87].

Since the advent of lasers, thousands of glasses have been formulated to investigate the effects of changes in glass network and network-modifier ions on the spectroscopic and lasing parameters of neodymium. The host glass has an important influence on the ability of the lasing ion to absorb light from the optical pumping source, to store this energy, and to release it to amplify the laser beam. Energy storage by the lasing ion is governed by its absorption

Table 2.7. Physical and optical properties of Nd-doped glasses

Glass Type Spectroscopic Properties	$Q-246$ Silicate (Kigre)	$Q-88$ Phosphate (Kigre)	$LHG-5$ Phosphate (Hoya)	$LHG-8$ Phosphate (Hoya)	$LG-670$ Silicate (Schott)	$LG-760$ Phosphate (Schott)
Peak Wavelength [nm]	1062	1054	1054	1054	1061	1054
Cross Section [$\times 10^{20}$ cm]	2.9	4.0	4.1	4.2	2.7	4.3
Fluorescent Lifetime [μs]	340	330	290	315	330	330
Linewidth FWHM [nm]	27.7	21.9	18.6	20.1	27.8	19.5
Density [gm/cc]	2.55	2.71	2.68	2.83	2.54	2.60
Index of Refraction [Nd]	1.568	1.545	1.539	1.528	1.561	1.503
Nonlinear Index n_2 [10^{-13} esu]	1.4	1.1	1.28	1.13	1.41	1.04
dn/dt (20°–40° C [10^{-6}/° C]	2.9	−0.5	8.6	−5.3	2.9	−6.8
Thermal Coefficient of Optical Path (20°–40° C) [10^{-6}/° C]	+8.0	+2.7	+4.6	+0.6	8.0	–
Transformation Point [° C]	518	367	455	485	468	–
Thermal Expansion coeff. (20°–40°) [10^{-7}/° C]	90	104	86	127	92.6	138
Thermal Conductivity [w/m]	1.30	0.84	1.19	–	1.35	0.67
Specific Heat [J/g · K]	0.93	0.81	0.71	0.75	0.92	0.57
Knoop Hardness	600	418	497	321	497	–
Young's Modulus [kg/mm²]	8570	7123	6910	5109	6249	–
Poisson's Ratio	0.24	0.24	0.237	0.258	0.24	0.27

properties, excited-state lifetimes, and quantum efficiency. For rare-earth laser glasses, the energy-storage capability varies only slightly with the host glass. The rate of energy extraction, on the other hand, is governed by a product of the optical intensity of the extracting beam and the stimulated-emission cross section σ of the lasing ions. Both of these factors are strongly influenced by the characteristics of the host glass. Hence, by appropriate choice of the host glass, one can produce lasers with wide varying performance.

The most common commercial optical glasses are oxide glasses, principally silicates and phosphates, i.e., SiO_2 and P_2O_5 based. Table 2.7 summarizes some important physical and optical properties of commercially available silicate and phosphate glasses. The 1053-nm gain cross sections of available phosphates range from 3.0×10^{-20} to $4.2 \times 10^{-20}\,\mathrm{cm}^2$, and are generally larger than the 1064-nm cross sections of silicate glasses. Silicate and phosphate glasses have fluorescent decay times of around $300\,\mu s$ at doping levels of 2×10^{20} Nd atoms/cm^3. Nonradiative processes account for 50–60 % of the excited decay.

In high-power lasers, with intense radiant fluxes such as employed in fusion research, nonlinear contributions to the refractive index and two-photon absorption become critical. It has been shown that those two nonlinear effects are minimized in fluoride glasses such as fluorphosphate and fluorberyllate.

Laser Properties

There are two important differences between glass and crystal lasers. First, the thermal conductivity of glass is considerably lower than that of most crystal hosts. Second, the emission lines of ions in glasses are inherently broader than in crystals. A wider line increases the laser threshold value of amplification. Nevertheless, this broadening has an advantage. A broader line offers the possibility of obtaining and amplifying shorter light pulses, and, in addition, it permits the storage of larger amounts of energy in the amplifying medium for the same linear amplification coefficient. Thus, glass and crystalline lasers complement each other. For continuous or very high repetition-rate operation, crystalline lasers provide higher gain and greater thermal conductivity. Glasses are more suitable for high-energy pulsed operation because of their large size, flexibility in their physical parameters, and the broadened fluorescent line.

Unlike many crystals, the concentration of the active ions can be very high in glass. The practical limit is determined by the fact that the fluorescence lifetime and, therefore the efficiency of stimulated emission, decreases with higher concentrations. In silicate glass, this decrease becomes noticeable at a concentration of 5 % Nd_2O_3.

Figure 2.8 shows a simplified energy level diagram of Nd : glass. The Nd^{3+} ion in glass represents a four-level system. The upper laser level indicated in Fig. 2.8 is the lower-lying component of the $^4F_{3/2}$ pair with a several-hundred microsecond spontaneous emission lifetime. The terminal laser level is the lower-lying level of the pair in the $^4I_{11/2}$ multiplet. The $^4I_{11/2}$ group empties spontaneously by a radiationless phonon transition to the $^4I_{9/2}$ ground state about

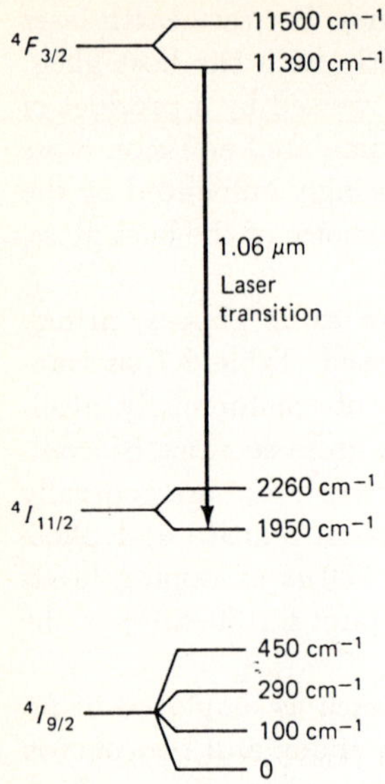

Fig. 2.8. Partial energy level diagram of Nd^{3+} in glass

1950 cm^{-1} below. Published values for the $^4I_{11/2}$ level lifetime vary from 10 to 100 ns [2.91–95]. This lifetime is difficult to measure because the transition $^4I_{11/2} \rightarrow {}^4I_{9/2}$, at a wavelength of 5 μm, is absorbed by the glass host. The degeneracy of either member of the $^4F_{3/2}$ level is 1. Measurements performed by several researchers [2.93, 94] seem to indicate that the degeneracy of the terminal laser level is likely to be 1 or 2.

Due to the large separation of the terminal laser level from the ground state, even at elevated temperatures there is no significant terminal-state population and, therefore, no degradation of laser performance. In addition, the fluorescent linewidth of the neodymium ion in glass is quite insensitive to temperature variation; only a 10 % reduction is observed in going from room temperature to liquid nitrogen temperature. As a result of these two characteristics, it is possible to operate a neodymium-doped glass laser with little change in performance over a temperature range of −100° to +100° C.

Figure 2.9 shows the pump bands of Nd : glass. In comparing Fig. 2.9 with Fig. 2.7, one notes that the locations of the absorption peaks in Nd : YAG and Nd : glass are about the same; however, in Nd : glass the peaks are much wider and have less fine a structure compared to Nd : YAG.

Commercially Available Laser Rods. Glass is a mixture of oxides. Its main constituents are nonmetal oxides, such as SiO_2, B_2O_3, and P_2O_5. Different metal oxides alter the structure in various ways and make it possible to obtain a large variety of properties. The components are mixed before melting, with the laser activators also added to the batch. The mixture is heated in a heat-resistant crucible. The principal laser glass manufacturers use either platinium

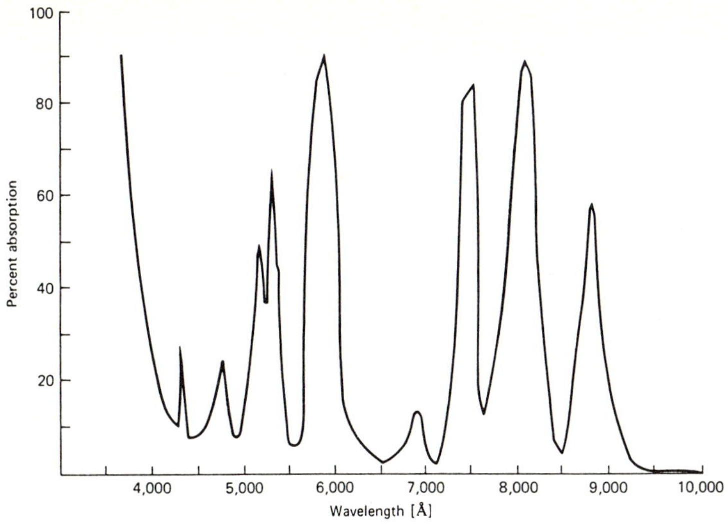

Fig. 2.9. Absorption versus wavelength of Nd : glass. (Material: ED-2; thickness: 6.3 mm)

or ceramic crucibles, clay pots, or ceramic continuous tanks to contain the melt. When the melt has reached a high viscosity it is cast into a mold. Finally, the glass in the mold is placed into an annealing furnace where it is very slowly cooled down.

A brief survey of laser glass fabrication techniques can be found in [2.96]. Glass laser rods are fabricated in a large variety of sizes. Typical rod sizes are between 10 and 50 cm in length, with diameters from 1 to 3 cm. However, rods up to 1 m in length and 10 cm in diameter are commercially available. Standard rod end configurations are the same as those mentioned for ruby and Nd : YAG rods.

Data Sources on Laser Glasses. Useful reviews on glass lasers can be found in [2.8, 9, 97–101]. Pertinent materials parameters are listed in the data sheets of the major Nd : glass manufacturers [2.90]. A detailed characterization of hundreds of glasses can be found in the Laser Glass Handbook prepared by Lawrence Livermore Laboratory [2.9a].

2.3.3 Nd : Cr : GSGG

Soon after the invention of the Nd : YAG laser attempts were made to increase the efficiency of transferring radiation from the pump source to the laser crystal by utilizing a second dopant called a "sensitizer". A particularly attractive sensitizer is Cr^{3+} because the broad absorption bands of chromium can efficiently absorb light throughout the whole visible region of the spectrum. The concept of improving efficiency by co-doping a Nd laser crystal with Cr^{3+} ions is based on transferring excitation, absorbed by the broad Cr^{3+} absorption bands, over to the Nd^{3+} ions. No improvement was achieved with the host crystal YAG because all Cr^{3+} excitation was deposited in the $_2E$ level and the spin-forbidden nature of the $^2E \rightarrow {}^4A_2$ transition resulted in an inefficient transfer process.

However, researchers at the Lebedev Inst. and the University of Hamburg [2.102–106] discovered that nearly 100% transfer efficiency could be achieved in the co-doped garnet crystal GSGG. Unlike YAG, a large percentage of the Cr^{3+} excitation in GSGG appears in the $^4T_2$ state, nonradiative transfer to Nd^{3+} ions can occur via the $^4T_2 \rightarrow {}^4A_2$ transition, which is spin-allowed and has a good spectral overlap with the Nd^{3+} levels. Experiments performed in several laboratories showed, with flashlamp-pumped operation, nearly a factor-of-three improvement in slope efficiency for the doubly doped garnet compared to a Nd:YAG crystal. It was found that the pumping efficiency improvement is not a strong function of the Cr concentration for the range $1\text{–}2 \times 10^{20}/\text{cm}^3$ [2.107].

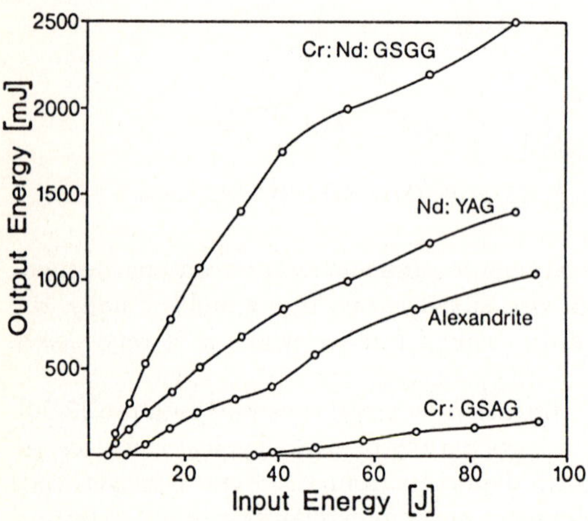

Fig. 2.10. Output energy versus input energy for Cr:Nd:GSGG, Nd:YAG, alexandrite, and Cr:GSAG [2.108]

Figure 2.10 illustrates the high efficiency which can be achieved with this material by comparing it to Nd:YAG and to tunable lasers such as alexandrite and Cr:GSGG [2.108]. The experiments were carried out in a silver plated single-ellipse pump cavity using a $5 \times 75\,\text{mm}^2$ flashlamp. The laser rods were $6 \times 75\,\text{mm}^2$ in an identical laser-resonator configuration but with output coupling optimized for the particular laser. Another comparison between Nd:YAG and co-doped GSGG is shown in Fig. 2.11 [2.109]. The efficiency factors indicated on the curves are close to state-of-the art performance for both materials.

The higher pump efficiency of Nd:Cr:GSGG does not automatically translate into better system performance because Nd:Cr:GSGG does exhibit much stronger thermal focusing and stress birefringence, as compared to Nd:YAG. The absorption efficiency and the heat-deposition rate for the Cr:Nd:GSGG rod is almost three times those of Nd:YAG, a consequence of the broad red and blue absorption bands of the Cr^{3+} sensitizer [2.110]. As a consequence, thermal focusing power as a function of lamp input power has been reported to be several times larger in Cr:Nd:GSGG than in Nd:YAG [2.110, 111]. Therefore, if beam brightness is the criteria, rather than output energy, some of the advantage of GSGG is offset, particularly at high average powers. Measurements at

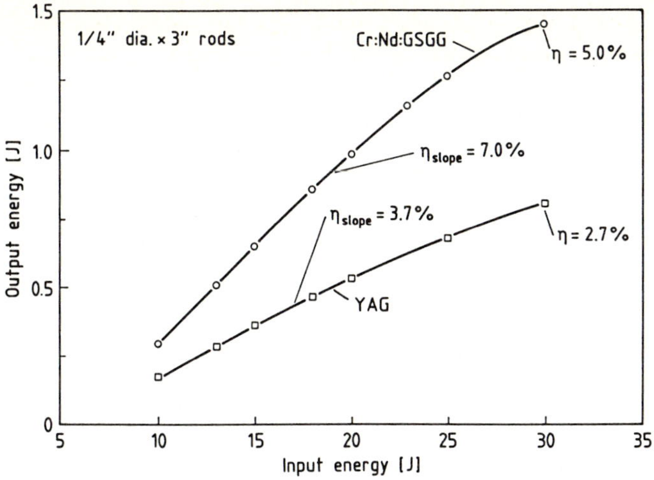

Fig. 2.11. Efficiency comparison of Nd:YAG and Cr:Nd:GSGG [2.109]

the author's laboratory have shown that under single-shot and low-repetition-rate operation, a factor 2 higher beam brightness is achieved in a GSGG system as compared to YAG. However, as the repetition rate is increased, this difference became smaller and beam brightness for a given input was about equal in both systems at 25 Hz for a Q-switched system at 100 mJ output.

The explanation for the high efficiency of co-doped GSGG follows from Fig. 2.12 which shows the absorption spectrum of this material. Two high-intensity Cr^{3+} absorption bands having maxima at 450 and 640 nm covering an appreciable part of the visible region, and narrow Nd lines can be observed. In contrast to Nd:Cr:YAG the transfer-time in the GdScGa-Garnet $(Cr \rightarrow Nd) = 10\,\mu s$ is about 3 orders of magnitude faster. This is due to a favor-

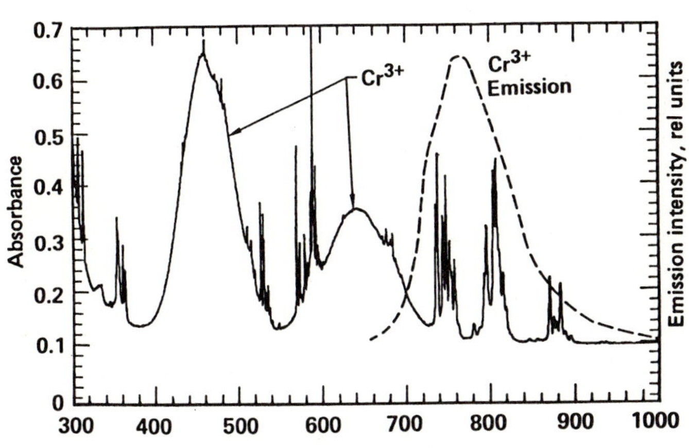

Fig. 2.12. Absorption spectra of Cr:Nd:GSGG [2.109]

Table 2.8. Material properties of GSGG

Property	Cr : Nd : GSGG
Thermal conductivity, κ	6.0 W/m°C
Heat capacity, C_P	0.40 J/g°C
Thermal expansion, α	7.4×10^{-6}/°C
Young's modulus, E	212 MPa
Poisson's ratio, ν	0.28
Index of refraction, 1060 nm	1.942
Change of index with temp. at 1152 nm, $\partial n/\partial T$	10.5×10^{-6}/°C
Density	6.44
Elasto-optic coefficients	
P_{11}	−0.012
P_{12}	0.019
P_{44}	−0.0665

able red shift of the $^4T_2$ energy band of Cr^{3+} and the high radiative transition probability of the $^4T_2 \rightarrow {}^4A_2$ transition. Both the enhancement of energetic 3*d*-4*f* overlap and the increased 3*d*-3*d* oscillator strength result in the observed strong multipole coupling of the sensitizer Cr^{3+} and acceptor Nd^{3+}.

Material parameters of Cr : Nd : GSGG are listed in Table 2.8. A comparison with Nd : YAG reveals a lower thermal conductivity and a lower heat capacity for GSGG but, in general, the materials parameters are fairly close. The gain of Cr : Nd : GSGG is slightly less than that of Nd : YAG, however, since it has a higher saturation fluence, its stored energy before saturation is reached is actually 1.7 times that of YAG [2.107]. Current efforts are aimed at improving the optical quality of the material and at increasing the size of the boules suitable for fabrication of slabs and large rods [2.112, 113].

2.3.4 Nd : YLF

During the last 10 years, the crystal quality of the scheelite-structured host lithium yttrium fluoride has been dramatically improved and the material has gained a firm foothold for certain applications.

The laser transition of Nd : $LiYF_4$ (Nd : YLF) at 1053 nm matches well to the peak gain of Nd doped phosphate and fluorophosphate glasses. As a result, it is currently being used in master oscillators for amplifier chains using these glasses. Nd : YLF is also employed in medium energy *Q*-switched oscillator and oscillator-amplifiers because it requires fewer stages as compared to Nd : YAG lasers for the same output energy.

Figure 2.13 shows a simplified energy level diagram of Nd : YLF. Depending on the polarization, two lines each are obtained around 1.05 and 1.3 μm. For example, with an intracavity polarizer one can select either the 1047 nm (extraordinary) or 1053 (ordinary) transition. The same can be done for the two 1.3 μm transitions, however, in addition lasing at the 1.05 μm lines has to be suppressed. All lines originate on the same Stark split $^4F_{3/2}$ upper level.

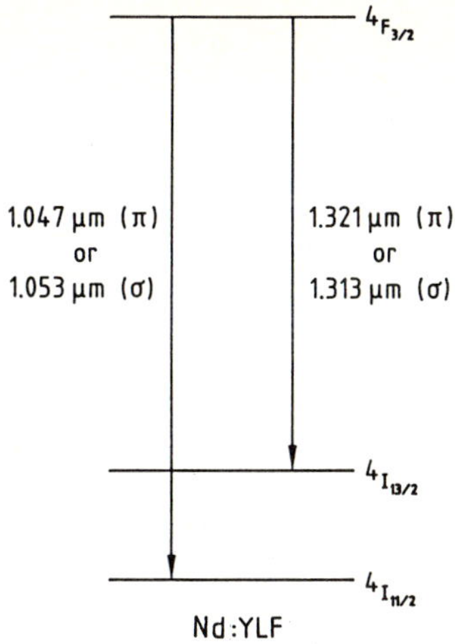

Fig. 2.13. Simplified energy level diagram of Nd : YLF

Table 2.9. Properties of Nd doped lithium yttrium fluoride (YLF)

Lasing wavelength [nm]	1053
Index of Refraction $\lambda = 1.06\ \mu m$	$n_0 = 1.4481$ $n_e = 1.4704$
Fluorescent lifetime	520 μs
Density [gm/cm^3]	3.99 (undoped)
Hardness [Mohs]	4–5
Elastic modulus [N/m^2]	7.5×10^{10}
Strength [N/m^2]	3.3×10^7
Poisson's ratio	0.33
Thermal conductivity [W/cm-K]	0.06
Thermal Expansion coefficient [° C^{-1}]	a axis: 13×10^{-6} c axis: 8×10^{-6}
Melting point [° C]	825

Material properties of Nd : YLF are listed in Table 2.9. The relatively large thermal conductivity allows efficient heat extraction, and its natural birefringence overwhelms thermally induced birefringence eliminating the thermal depolarization problems of optically isotropic hosts like YAG. The cross-section for YLF is a factor of 2 lower than YAG. For certain lasers requiring moderate Q-switch energies, the lower gain offers advantages in system architecture compared to the higher gain material Nd : YAG.

The energy storage in Q-switched operation of Nd oscillators and amplifiers is constrained by the onset of parasitic oscillations. To first order, the energy storage limit of two materials is inversely proportional to the ratio of the stimulated emission cross-section. Therefore, higher storage densities are obtained in the lower cross-section material Nd : YLF as compared to Nd : YAG. Figure 2.14 shows the performance of a Q-switched YLF oscillator, producing an output of 400 mJ [2.114]. This output level would require the addition of an amplifier

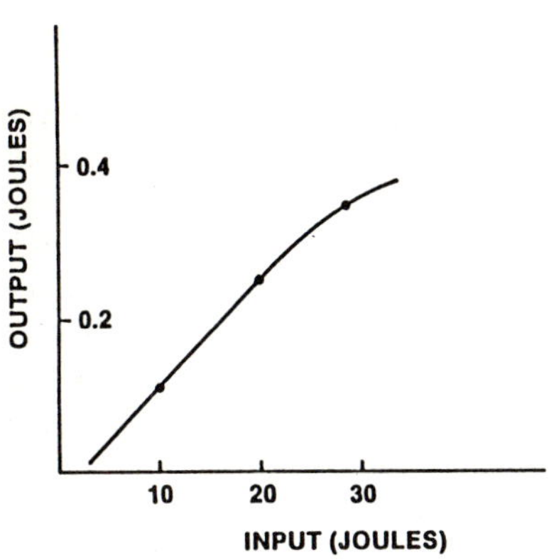

Fig. 2.14. Performance of a Q-switched YLF oscillator (Laser rod 0.25 × 2.5 inch, output coupler 30 % reflectivity, wavelength 1.053 μm) [2.114]

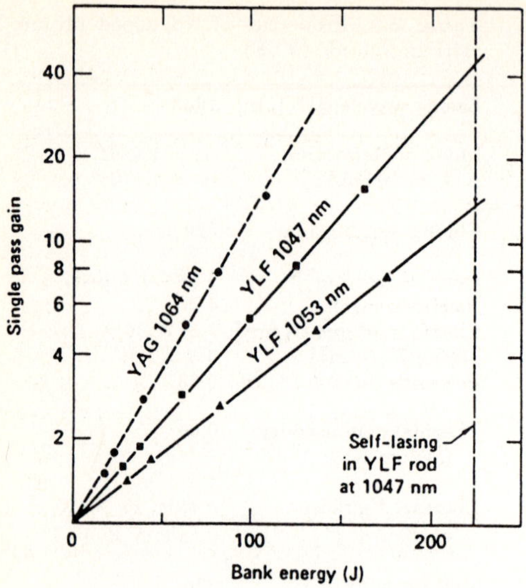

Fig. 2.15. Single pass gain versus pump energy for YLF and YAG [2.117]

in the case of a Nd : YAG laser. Single oscillators with up to 0.5 J Q-switched output, and oscillator-amplifier configurations with several Joules output usually result in less complicated structures if designed with Nd : YLF rods as compared to Nd : YAG [2.115, 116]. The weaker thermal lensing in Nd : YLF, coupled with its natural birefringence, provide additional advantages in certain applications compared to Nd : YAG [2.117, 118].

Figure 2.15 shows measured single-pass gain versus pump energy for the two transitions [2.117]. The measurements of the Nd : YLF amplifier were performed on a 10 mm diameter × 76 mm YLF rod with 1 % Nd doping. The crystal was pumped in a silver-coated double elliptical pumping head. The YAG data were obtained by replacing the Nd : YLF with a 10 mm diameter Nd : YAG rod in the same amplifier head.

2.4 Er : Lasers

Erbium-laser performance is not very impressive in terms of efficiency or energy output. For these reasons, the lasers have not gained acceptance for industrial or military applications. However, recently erbium has attracted renewed attention because of two particular wavelengths of interest. A crystal, such as YAG, highly doped with erbium produces an output around 2.9 μm, and Er-doped phosphate glass generates an output at 1.54 μm. Both of these wavelengths are absorbed by water, which leads to interesting medical applications in the case of the 2.9 μm lasers, and to eye safe military rangefinders and optical fiber-fault location systems in the case of the shorter wavelength.

2.4.1 Er:YAG

In 1975 Soviet researchers [2.119–121] discovered laser operation at room temperature in highly doped Er:YAG, i.e. erbium concentrations around 50%. Laser emission occurs at 2940 nm. In this country, interest in this material has been shown only very recently as a laser for medical applications and as a direct source of infrared light. The spectroscopy of Er^{3+} in YAG indicated that the upper laser level of the 2940 nm transition is pumped by light at wavelengths less than 600 nm. Therefore, the pump efficiency is not very high in this material. Also, due to a very long, lower-level lifetime the Er:YAG laser cannot be Q-switched. The 2940 nm lasing transition is between the $^4I_{11/2}$ and $^4I_{13/2}$ states of the Er^{3+} ion. The lower level has a much longer lifetime than the upper level (2 ms as compared to 0.1 ms) and so this transition can be blocked by the accumulation of population in the $^4I_{13/2}$ state. Despite these above mentioned drawbacks, the laser is of particular interest due to its wavelength which coincides with a water absorption line. Optical absorption by the water in tissue is extremely large ($> 3000\,cm^{-1}$), making potential medical applications such as plastic surgery very attractive.

Figure 2.16 shows the output vs. input energy of an oscillator containing a 75 mm long by 6.25 mm diameter rod. The pump pulse duration was 170 μs FWHM. At a 2 Hz repetition rate, an output of 820 mW was generated at an efficiency of 0.1% [2.122]. The pump cavity was a double ellipse, silver coated and containing two flashlamps with 7 mm bore diameters.

Recent improvements with co-doping the crystals have increased the efficiency to 1% at output energies of up to 1 J for a 200 μs flashlamp pulse [2.123]. TEM_{00} mode performance yielded 200 mJ per pulse. The results were obtained

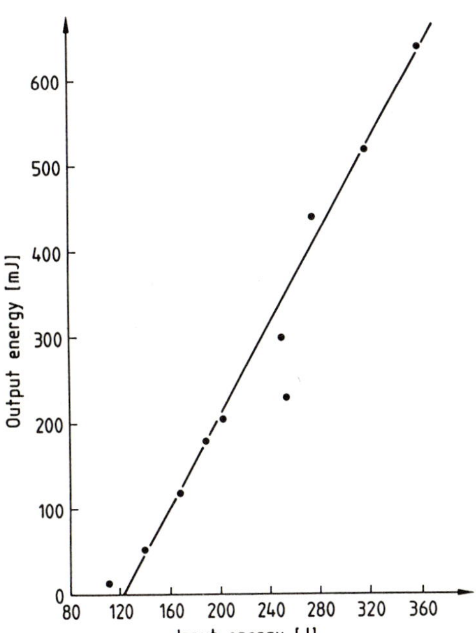

Fig. 2.16. Output versus input energy for 50% Er:YAG pumped in a double elliptical cavity by two xenon filled flashlamps [2.122]

with a 3″ × 1/4″ laser rod, containing 50 % erbium. Both output and overall efficiency were reduced for rods with lower erbium concentration.

Besides Er : YAG, other erbium lasers such as Er : $YALO_3$, Er : YLF and Er : Cr : YSGG are the subject of current investigations [2.124, 125]. These lasers provide additional laser lines between 2.71 to 2.92 μm, or at 1.7 μm or they have the potential for improved performance. Also laser emission from Er : YLF pumped by a laser diode array for both pulsed and CW modes has been reported recently [2.126].

2.4.2 Er : Glass

Erbium has been made to lase at 1.54 μm in both silicate and phosphate glasses [2.127, 128]. Due to the three-level behavior of erbium, and the weak direct pumping of erbium, co-doping with other rare earth ions is necessary to obtain satisfactory system efficiency. In fact, the erbium must be sensitized with ytterbium if the glass is to lase at all at room temperature. The Yb acts as sensitizing agent by absorbing pump light in regions where the erbium ion is relatively transparent. The principal absorption band is the Yb^{3+} band at 1 μm.

The flash-lamp current density required to match this band is about 1000–2000 A/cm^2. Such current densities require pulse times of the order of several milliseconds which are, however, compatible with the long fluorescent transition lifetimes of the dopant ions. Xenon flashlamps have the spectral output which best matches this pumping requirement of the erbium laser. Besides Yb a small amount of Nd is usually added to further reduce threshold for laser operation. The energy level scheme relevant to the Er^{3+} is between the $^4I_{13/2}$ multiplet in the region of 6500 cm^{-1} (1.54 μm) above the ground state, and the ground state $^4I_{15/2}$ levels (Fig. 2.17).

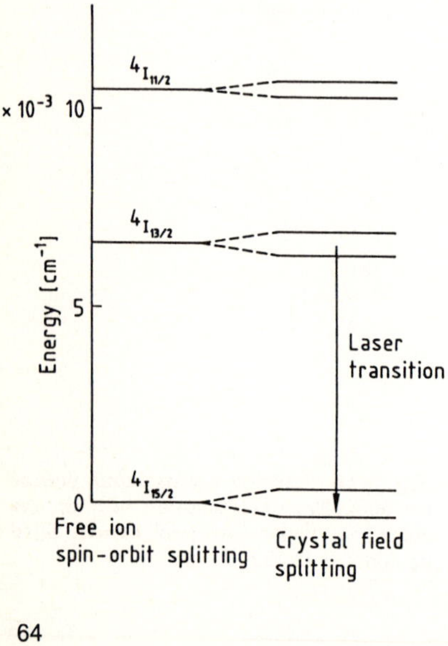

Fig. 2.17. Er^{3+} energy level scheme

The "optimum" erbium concentration is usually found to fall in the range of 0.2 to 0.5 wt.% Er_2O_3 (or about 2 to 5 $\times 10^{19}$ Er^{3+} ions per cubic centimeter). This "optimum" is a balance between several factors. The Er^{3+} concentration must be sufficiently low to produce a reasonable threshold for lasing. The Er^{3+} ion is a three-level laser and about 60% of the ions must be excited before threshold is reached. However, the Er^{3+} concentration must also be sufficiently high to produce a high energy-transfer efficiency from the Yb^{3+} to the Er^{3+}. Erbium-doped glass lasers emit at a wavelength of about 1.54 μm. This wavelength is desirable for some applications since it is strongly absorbed by liquid water and therefore has a high threshold for damage to the retina of the eye.

The spectral region between 1.35 and 2.2 μm is an eye-safe region. For wavelengths shorter than 1.35 μm, the transmissivity of the eye rapidly rises to a value six orders of magnitude higher than it is in the eye-safe region, with a corresponding increase in the danger to the retina. For wavelengths longer than 2.2 μm, the eye becomes sensitive to corneal burns. Thus, the erbium glass laser, which emits at a wavelength of 1.54 μm, is of interest from this point of view. It has an energy threshold for corneal damage at least 10^8 higher and a corresponding safe eye-exposure distance of at least a factor of 10^3 greater than the neodymium (1.06 μm) and ruby (0.694 μm) lasers.

Table 2.10. Material properties of an erbium doped phosphate glass

Emission wavelength	1.54 μm
Fluorescent lifetime	8 ms
Index of refraction at 1.54 μm	1.531
dn/dT	$63 \times 10^{-7}/°C$
Thermal expansion (α)	$124 \times 10^{-7}/°C$
Thermo-optic coefficient (W)	$-3 \times 10^{-7}/°C$

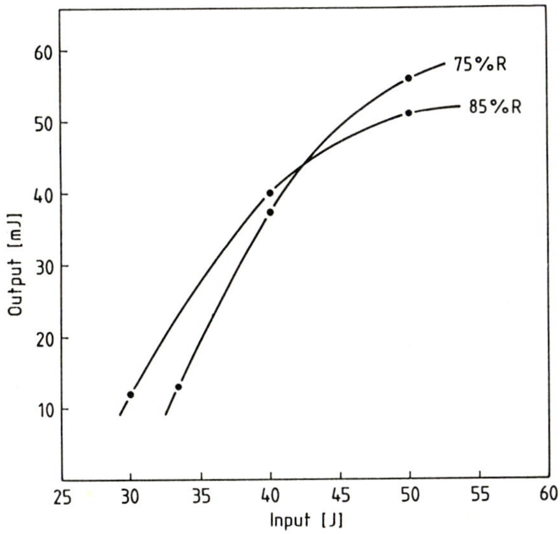

Fig. 2.18. Q-switched performance from an erbium phosphate glass oscillator [KIGRE, INC]

Material properties of Er : Glass are listed in Table 2.10. Q-switched performance from an erbium phosphate laser glass is illustrated in Fig. 2.18. The laser contains a 3×50 mm glass-rod, and a rotating prism as Q-switch. Flashlamp pulse length is 3 ms. This laser is typical for eye-safe range finders.

Employing an athermal phosphate glass, a high repetition-rate Er : Glass laser has been built and operated at 30 Hz [2.129] for fiber-fault location in communications fiber networks. Also an Er-doped low-loss single-mode fiber was recently made to lase at 1.55 μm for possible applications in fiber sensors [2.130].

2.5 Tunable Lasers

In most lasers, all of the energy released via stimulated emission by the excited medium is in the form of photons. Tunability of the emission in solid-state lasers is achieved when the stimulated emission of photons is intimately coupled to the emission of vibrational quanta (phonons) in a crystal lattice. In these "vibronic" lasers, the total energy of the lasing transition is fixed, but can be partitioned between photons and phonons in a continuous fashion. The result is broad wavelength tunability of the laser output. In other words, the existence of tunable solid-state lasers is due to the subtle interplay between the Coulomb field of the lasing ion, the crystal field of the host lattice, and electron-phonon coupling permitting broad-band absorption and emission. Therefore, the gain in vibronic lasers depends on transitions between coupled vibrational and electronic states; that is, a phonon is either emitted or absorbed with each electronic transition.

To put the current research activity in this area in perspective, we note that the history of tunable solid-state lasers can be traced back to work performed over 25 years ago at Bell Laboratories. In 1963, Bell-Labs researchers reported the first vibronic laser, a nickel-doped magnesium fluoride ($Ni:MgF_2$) device [2.131]. The same group later built a series of vibronic lasers using nickel, cobalt or vanadium as the dopant and MnF_2, MgO, MgF_2, ZnF_2, or $KMgF_3$ as the host crystal [2.132–134]. Tunable in the 1.12 to 2.17 μm range, these flashlamp-pumped lasers had a serious drawback: they operated only when cooled to cryogenic temperatures. Despite these early results, this field lay dormant for more than a decade, probably because of the need for cryogenic cooling of the laser material, the emergence of Nd : YAG and the concentration on dye lasers and, more recently, on color-center lasers [2.135]. The first room-temperature vibronic laser, reported by Bell Labs in 1974, was a flashlamp-pumped $Ho:BaY_2F_8$ device emitting at 2.17 μm [2.136].

In 1977, the crystal alexandrite (chromium doped $BeAl_2O_4$) was made to lase on a vibronic transition at the Allied Corporation [2.137–139]. The crystal field strength at the chromium site in alexandrite is sufficiently low to permit thermal activation of broad $^4T_2$ fluorescence. The result is a low-gain, four-level laser tunable in the range of 0.7 to 0.8 μm. After an ambitious development program at Allied, alexandrite has become available on a commercial basis.

Shortly after the discovery of alexandrite it became apparent that tunable Cr^{3+}-doped lasers can be constructed in hosts other than alexandrite. The emerald laser, Cr^{3+} in $Be_3Al_2(SiO_3)_6$, appears to have a higher-gain cross-section than alexandrite and if large, high-optical-quality crystals can be grown, could be superior in performance for some applications [2.140–144].

During the mid-1970s research started on a variety of divalent transition-metal doped crystals at MIT Lincoln Laboratory. The Co : MgF_2 laser, optically pumped by a 1.3 μm neodymium laser showed continuous tuning from 1.63 to 2.08 μm. Other research included the Ni : MgF_2 and Ni : MgO [2.145–148] lasers. This earlier work resulted also in the discovery of the Ti : sapphire laser by *Moulton* in 1982 [2.149]. In the sapphire host, Ti^{3+} exhibits no appreciable excited-state absorption, and therefore can exhibit a tuning range that, in principle, would span virtually its entire fluorescence emission wavelength range.

In the early 1980s, work carried out at the University of Hamburg and at the Lebedev Institute in Moscow demonstrated Cr^{3+} laser operation in several garnet structured crystals. The most promising crystal being flashlamp pumped Cr^{3+}-doped $Gd_3Sc_2Ga_3O_{12}$ [2.150–153]. The garnets chosen have ligand fields at the chromium site low enough to permit the vibronic, tunable mode of laser operation exhibited by alexandrite.

Interest in tunable solid-state lasers has grown strongly in recent years and is best illustrated by the fact that there is now an annual meeting on Tunable Solid State Lasers. The proceedings of these meetings provide a wealth of information on material and device research in this area (see, e.g., [2.154]).

The tunable ranges of representative vibronic lasers operating at room temperature are shown in Fig. 2.19. As one can see from this figure, most tunable lasers contain Cr^{3+} as the active ion.

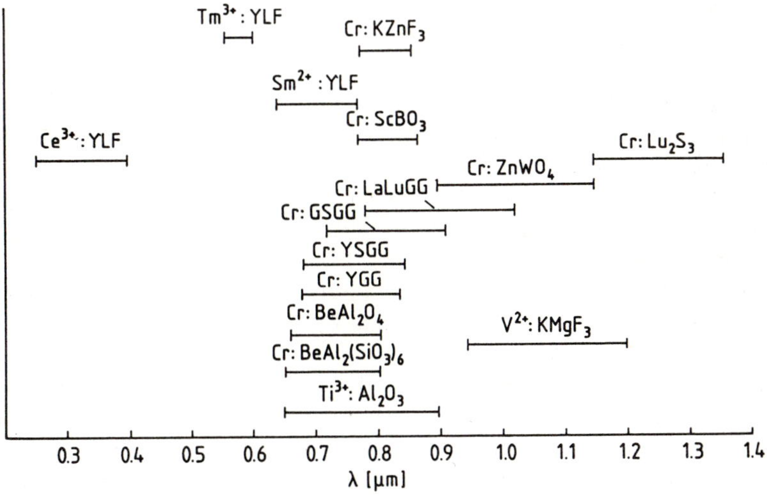

Fig. 2.19. Tunable ranges of representative vibronic lasers

Chromium Laser

Chromium has enjoyed considerable success as a tunable-laser ion due to the special nature of its trivalent d^3 electronic configuration. Particular attributes of Cr over other transition metals are its chemical stability, broad pump bands, large energy-level splitting, and reduced excited state absorption (ESA). These advantages have led to at least a dozen crystals being demonstrated as good laser hosts for Cr. Chromium has a laser tuning range of about 100 nm, with the center wavelength shifted by the particular crystal host.

Alexandrite. Cr in chrysoberyl ($BeAl_2O_4$) occupies the 700–800 nm range and was the first tunable ion solid-state laser considered for practical use. Details of this commercially available laser material are given in Sect. 2.5.1.

Cr : GdScGa-Garnet. Cr^{3+}-doped Ga garnets present a new class of transition metal ion laser which can operate pulsed as well as CW at room temperature [2.150–153]. Due to a low-crystal field at the Cr site, the $^4A_2$-$^4T_2$ splitting is nearly equal to the $^4A_2$-$^4T_2$ separation. Thus, in contrast to ruby or alexandrite, the fluorescence is totally dominated by the broadband four-level $^4T_2 \rightarrow {}^4A_2$ transition.

Garnet crystals are described by the formula $C_3A_2D_3O_{12}$, where one chooses large ions A = Ga, Sc, Lu for the octahedral site, D = Ga for the tetrahedral site, and C = Y, Gd, La, Lu for the dodecahedral site. Broadband four-level fluorescence has been obtained in the 700–950 nm spectral range from Cr-doped YGG, YScGG (YSGG), GGG, GdScGG(GSGG), and LaLuGG(LLGG) with lifetimes ranging from 240 to 70 μs, respectively. The crystals are grown by the standard Czochralski technique at $\approx$ 1750° C.

One compound, Cr^{3+}-doped GdScGa-garnet, has received particular attention. In Cr^{3+}-doped GdScGa-garnet, the R-line fluorescence is totally absent and due to a large 2400 cm^{-1} Stoke shift of the $^4T_2$ level, the whole fluorescence is channelled into the broad band (4 level) $^4T_2 \rightarrow {}^4A_2$ transition. CW and pulsed laser action has been obtained from this material.

Figure 2.20 shows the spectrum which extends from 660 up to 950 nm with the peak intensity at about 760 nm [2.151]. The lifetime is $t = 120\,\mu$s, which is also the radiative lifetime (quantum efficiency: 1). Figure 2.21 shows the transmission spectrum, typical for Cr^{3+}. The valleys at 640 and 450 nm correspond to the $^4A_2 \rightarrow {}^4T_2$ and $^4A_2 \rightarrow {}^4A_1$ transitions, respectively.

Pulsed laser action was achieved at 790 nm by pumping the Cr^{3+} : GSGG rod with a dye laser at 590 nm. CW laser action was achieved by pumping crystals longitudinally with a Kr laser beam. Slope efficiencies of up to 11 % were obtained. From the observed threshold pump powers, an effective emission cross-section of 0.9×10^{-20} cm^2 was calculated [2.150].

Emerald Laser. Ever since alexandrite was reported as a room-temperature, tunable laser material, researchers have been looking for other materials with similar properties. A promising candidate is emerald, a chromium-doped beryllium aluminum silicate Cr^{3+} in $Be_3Al_2Si_6O_{18}$. Emerald exhibits a wide fluorescence spectra from 700 to 850 nm [2.140–144]. The observed tuning range

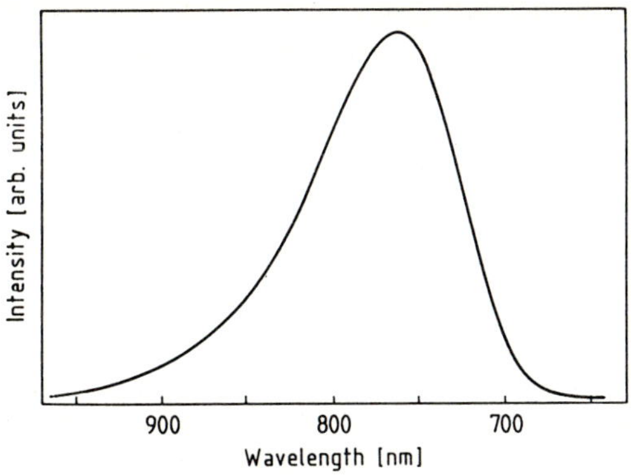

Fig. 2.20. Fluorescence of a Cr^{3+} : GSGG laser crystal [2.151]

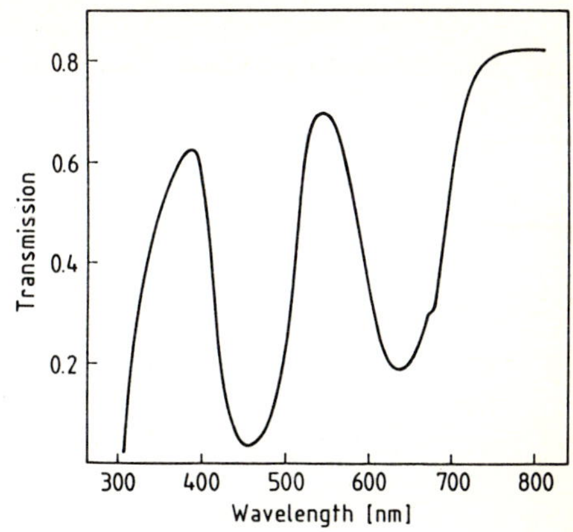

Fig. 2.21. Transmission spectrum of a 7 mm thick Cr^{3+} : GSGG crystal [2.151]

of the emerald laser is 729 to 809 nm which does not cover the entire fluorescence range of emerald because of excited state absorption of the laser photons [2.141].

The broad absorption spectra of emerald are from the visible region up to 700 nm, which makes it suitable for pumping by readily available flashlamps. Emerald is similar to ruby and alexandrite in electronic structure. Contrary to the emission from ruby and alexandrite the R-line is smaller and is only indicated by a small peak in the broad emission spectra. This is caused by the weaker crystal field which alters the energy states of Cr^{3+} ions in emerald. As a result, the two ^4T levels ($^4T_2$ and $^4T_1$) lie at slightly lower energies, but the position of the ^2E band is essentially unaltered. The energy difference at room temperature between $^4T_2$ (responsible for broadband emission) and ^2E (R-line radiation) is $\approx 400\,\mathrm{cm}^{-1}$. This is in contrast to the energy differences for ruby and alexandrite of $2300\,\mathrm{cm}^{-1}$ and $800\,\mathrm{cm}^{-1}$, respectively. This energy difference allows the repopulation of the $^4T_2$ level due to thermalization from the long-lived ^2E level. Therefore, emerald has a relatively larger inversion of population of vibronic modes at room and higher temperatures.

The emission cross-section at room temperature was measured to be $2.0 \times$ - $10^{-20}\,\mathrm{cm}^2$ at the peak of the vibronic gain curve. For comparison, alexandrite at room temperature has an emission cross-section of $0.7 \times 10^{-20}\,\mathrm{cm}^2$.

Poor optical quality is a major drawback of the emerald laser rods fabricated so far. The crystal cannot be grown by the conventional Czochralski technique. When light travels parallel to the crystal growth planes, beam breakup occurs due to the inhomogeneity of the index of refraction. To minimize this effect, the beam has been directed normal to these planes. Although, from a performance point of view, this laser has the potential of being superior to alexandrite, it is not clear whether the growth problems will be solved.

$Cr:KZnF_3$. This fluorine perovskite has become the second tunable solid-state laser material (after alexandrite) to be offered commercially [2.155a, 154a].

$Cr:KZnF_3$ operates from 785 to 865 nm, with good slope efficiency when laser pumped between 650 to 700 nm. Up to 14 % efficiency has been reported, while this efficiency is substantially lower than the 30 to 50 % range found for emerald and alexandrite, its wavelength coverage does extend further into the infrared.

Scandium Borate. Chromium in scandium borate ($ScBO_3$) shows potential for tunability in the 780–890 nm range. This recently discovered material already exhibits a laser slope efficiency of 29 % [2.154b].

It is likely that host crystals will be found suitable for tunable operation with the Cr^{3+} ion; there are, for example, an almost unlimited number of garnet-structured crystals with the appropriate ligand fields. Many of these have been grown in large sizes for application as bubble-memory substrates, and doping with chromium should be relatively straightforward. One is reminded of the early years of solid-state laser research when many crystals were evaluated as hosts for the Nd^{3+} ion, with YAG finally emerging as desirable for most applications. Unlike the Nd^{3+} ion, which will always exhibit laser operation in a narrow region around 1.06 μm independent of the host crystal, the operating wavelength range and other characteristics of Cr^{3+} ion lasers are sensitive to the particular host. It is not clear whether one particular crystal will emerge as superior to others in all respects after all the current research efforts in this area have ended.

Vanadium Laser

It has been shown that $V:MgF_2$ does act as a tunable vibronic laser in the near-IR. The laser, however, has very low gain towards the upper end of its tuning range due to excited state absorption. Consistent with experimental observation, the 1.12 μm peak (8930 cm^{-1}) of the $V:MgF_2$ fluorescence lies in the high energy wing of the $^4T_2 \rightarrow {}^4T_1$ excited state absorption transition. $V:MgF_2$ has been mentioned as a candidate for a fusion driver because of its high absorption in the 0.5–0.6 and 0.8–0.9 μm bands, where xenon lamp outputs are highest. Further, magnesium fluoride has high thermal conductivity, low nonlinear index, and high-energy extraction properties. However, excited-state absorption has reduced the net gain to an unacceptably low value. Research is continuing into other combinations of V and hosts in the hope that a crystal with the right properties can be found. It is expected that $V:KMgF_3$ shows less excited state absorption, therefore attention has turned to this material recently.

Titanium Laser

The Ti : sapphire laser has an exceptionally wide tuning range and a large gain cross-section, i.e., about 50 % of Nd : YAG. Also, large, high quality crystals are commercially available. These attributes have led to the use of this material for specialized laser applications. Further details will be discussed in Sect. 2.5.2.

Rare-Earth Tunable Lasers

Rare earth ions doped in appropriate host crystals also exhibit vibronic lasing; Ce, Eu, and Sm have lased from $5f$-$4f$ transitions, with emission in the blue.

The main difference between transition metal and rare earth ions is that the former is crystal field sensitive and the latter is not. As distinct from transition metal ions, the broad-band transition for rare earth ions are quantum mechanically allowed and therefore have short lifetimes and high cross-sections. The Ce^{3+} laser, using a $5d$-$4f$ transition, has operated in the host crystals, $LiYF_4$ and LaF_3 at wavelengths around 325 and 286 nm, respectively, when pumped by an excimer laser [2.156, 157]. Lamp-pumped operation of the Ce^{3+} laser may be possible; such a system would be alternative to the excimer laser as a uv source, with the added advantage of broad tunability.

2.5.1 Alexandrite Laser

Alexandrite ($BeAl_2O_4 : Cr^{3+}$) is the best-characterized commercially-developed vibronic laser. Alexandrite is the common name for chromium-doped chrysoberyl – one of the olivine family of crystals – with four units of $BeAl_2O_4$ forming an orthorhombic structure. The crystal is grown in large boules by the Czochralski method much like ruby and YAG. Laser rods up to 1 cm in diameter and 10 cm long with a nominal 2-fringe total optical distortion are commercially available. The chromium concentration of alexandrite is expressed in terms of the percentage of aluminum ions in the crystal which have been replaced by chromium ions. The Cr^{3+} dopant concentration, occupying the Al^{3+} sites, can be as high as 0.4 atomic percent and still yield crystals of good optical quality. A concentration of 0.1 atomic percent represents 3.51×10^{19} chromium ions per cubic centimeter.

Alexandrite is optically and mechanically similar to ruby, and possesses many of the physical and chemical properties of a good laser host. Hardness, strength, chemical stability and high thermal conductivity (two-thirds that of ruby and twice that of YAG) enables alexandrite rods to be pumped at high average powers without thermal fracture. Alexandrite has a thermal fracture limit which is 60 % that of ruby and five times higher than YAG. Surface damage tests using focussed 750 nm radiation indicate that alexandrite is at least as damage resistant as ruby. Table 2.11 lists the chrysoberyl material properties [2.158].

Due to its orthorhombic structure, alexandrite is biaxial with the principal axes of the index ellipsoid along the crystallographic axes. Light emitted from the laser is polarized with the E vector parallel to the b axis. The gain in the $E \parallel b$ polarization is 10 times that in the alternate polarizations. Alexandrite lases at room temperature with flashlamp pumping throughout the range 701 to 818 nm. The alexandrite absorption bands are very similar to those of ruby, and span the region from about 380 to 630 nm with peaks occurring at 410 and 590 nm. Figure 2.22 shows the absorption bands of alexandrite. The laser gain cross-section increases from $7 \times 10^{-21}\,cm^2$ at 300 K to $2 \times 10^{-20}\,cm^2$ at 475 K which results in improved laser performance at elevated temperature. The 262 μs, room-temperature fluorescence lifetime permits effective energy storage and Q-switched operation.

Table 2.11. Material parameters of alexandrite

Laser wavelength [nm]	700–818
Stimulated emission cross-section [cm^2]	1.0–5.0 $\times$ 10^{-20}
Spontaneous lifetime [μs]	260 ($T = 298$ K)
Doping density [at %]	0.05–0.3
Fluorescent linewidth [Å]	1000
Inversion for 1% gain per cm [cm^{-3}]	2–10 $\times$ 10^{17}
Stored energy for 1% gain per cm [J/cm^3]	0.05–0.26
Gain coefficient for 1 J/cm^3 stored energy [cm^{-1}]	0.038–0.19
Index of refraction: (750 nm)	$E \parallel a$ 1.7367 $E \parallel b$ 1.7421 $E \parallel c$ 1.7346
Thermal Expansion:	$\parallel a$ 5.9 $\times$ 10^{-6}/K $\parallel b$ 6.1 $\parallel c$ 6.7
Thermal conductivity:	0.23 W/cm-K
Melting point:	1870° C
Hardness:	2000kg/mm^2

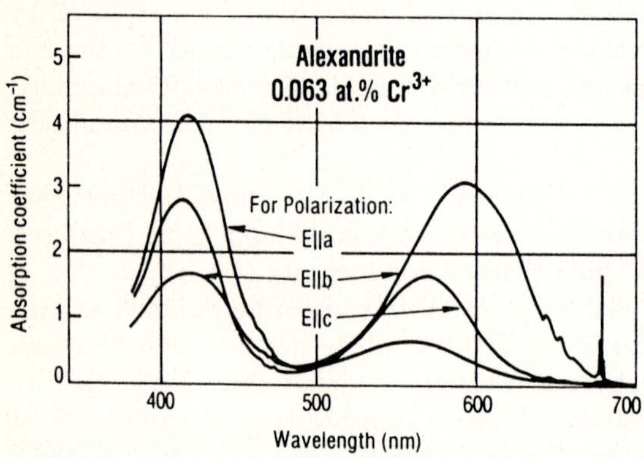

Fig. 2.22. Absorption spectrum of alexandrite [Allied Corp. Data Sheet]

Laser action has also been demonstrated on the R line at 680.4 nm. This three-level mode is analogous to the lasing in ruby except that the stimulated emission cross-section in alexandrite (3×10^{-19} cm^2) is ten times larger than for ruby.

Kinetics. As noted above, alexandrite can operate both as a four-level vibronic laser and as a three-level system analogous to ruby. As a three-level laser, it has a high threshold, fixed output wavelength (680.4 nm at room temperature) and relatively low efficiency. Obviously, the primary interest of alexandrite lies in its vibronic nature.

The basic physics of the 4-level alexandrite laser can be discussed with reference to the energy level diagram (Fig. 2.23). The $^4A_2$ level is the ground state, and the $^4T_2$ and $^4T_1$ are the two absorption state energy continua. Vibronic las-

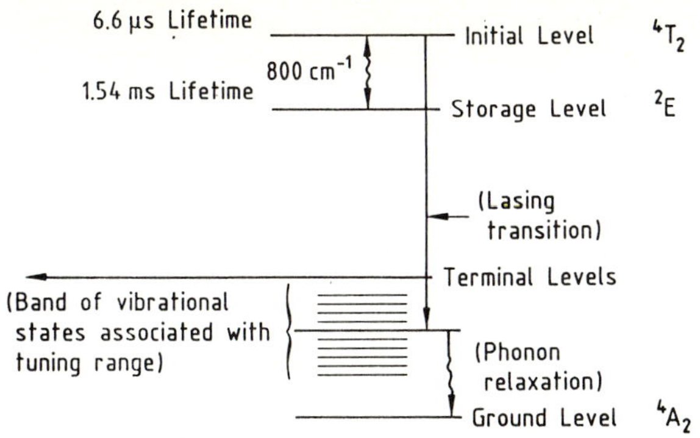

Fig. 2.23. Energy level diagram for chromium ions in alexandrite. The variable partition in de-excitation energy between photons and phonons leads to wavelength tunability

ing is due to emission from the $^4T_2$ state to excited vibronic states within $^4A_2$. Subsequent phonon emission returns the system to equilibrium. Since alexandrite is an intermediate crystal field material ($E \sim 800\,\mathrm{cm}^{-1}$) there is coupling between the ^2E state and the $^4T_2$. The lifetimes of each of these states is 1.5 ms and 6.6 μs, respectively. The two phosphorescent R lines emitted from ^2E occur in the vicinity of 680 nm, as for ruby. The terminal laser level is a set of vibrational states well above the ground state. The initial laser level is a level $800\,\mathrm{cm}^{-1}$ above a long lived storage level and in thermal equilibrium with it.

Due to the vibronic nature of the alexandrite laser, the emission of a photon is accompanied by the emission of phonons. These phonons contribute to thermalization of the ground-state vibrational levels. The laser wavelength depends on which vibrationally excited terminal level acts as the transition terminus; any energy not released by the laser photon will then be carried off by a vibrational phonon, leaving the chromium ion at its ground state.

Of fundamental importance in the kinetics of alexandrite is the cross-sectional probability σ_{2a} that the excited chromium ions will themselves absorb laser photons circulating in the cavity. This probability must be small compared to the effective emission cross section, otherwise the excited ions will absorb a significant amount of the emitted laser power before it escapes into the resonator. The excited Cr^{3+} ions which absorb the laser emission decay immediately back to their original excited state distribution, so that there is no net loss of excited state population due to excited state absorption. The intracavity flux, of course, suffers a loss due to this absorption contributing to additional heating of the laser rod. In alexandrite the excited-ion absorption band has a deep broad minimum just where the laser emission gain is maximum. At the band center, σ_{2a} is less than 10 % of σ. If σ_{2a} were greater than σ then lasing could not occur at all. In fact, the latter is responsible for the long-wavelength tuning limit in alexandrite [2.159a].

Temperature Effects in Alexandrite. As the temperature increases, the gain of alexandrite increases, the gain peak shifts to longer wavelength and the fluorescence lifetime decreases. The four-level model can be used to predict the temperature dependence of the laser performance. The ^2E state acts as a storage level for $^4T_2$. Thus, as the temperature of alexandrite increases, the vibronic continua in $^4T_2$ are successively populated from ^2E in accordance with the Boltzmann distribution, and the stimulated emission cross-section increases.

However, raising the temperature also tends to populate the terminal levels – especially those which lie closest to the ground level and which therefore correspond to the highest-energy (shortest-wavelength) photons. Since laser performance is highest with a maximally populated initial level and a minimally populated terminal level, it can be seen that increasing the temperature has two conflicting effects. The result is that performance is positively affected by temperature increases only for wavelengths above 730 nm.

Another adverse effect of higher temperature on laser performance is the reduction of the fluorescent decay time. The fluorescent lifetime is 260 μs at room temperature and 130 μs at 100° C. The total radiative quantum efficiency is nearly constant and equal to unity in the regime of interest. As the temperature is increased, the initial laser level ($^4T_2$) has an increased share of the excited population; since this level has a much higher decay rate than the storage level ^2E the overall fluorescence lifetime of the upper level (the combined storage and initial levels) is reduced. Therefore, at some higher temperature the storage time becomes much shorter than the flashlamp pulse duration and much energy is lost in fluorescence. This situation limits the advantage derived from the increasing population of the initial level.

There is also a shift of the peak of the gain curve to longer wavelength as the temperature increases. This is due to an increased optical phonon population with temperature and therefore a relatively lower population inversion at shorter wavelengths; and also due to a shift of the R lines with temperature with consequent enhancement of the long wavelength vibronics.

From the foregoing discussion, one can draw the following conclusions: The laser gain is the product of the population inversion between the initial and final levels and the cross section. The gain for a given excited state population will increase with temperature, especially for the longer wavelengths where the thermal population of the terminal level is negligible. It seems therefore that operation at elevated temperatures should be advantageous. For flashlamp-pumped, high-energy, Q-switched laser operation, however, a long storage time in the upper level is desirable. With increased thermal population in the initial level, the fluorescent lifetime decreases. A compromise between gain and storage time therefore has to be made. For the shorter wavelengths the thermal population of the terminal level is much higher and one expects less improvement with increasing temperature.

Guch [2.159b] evaluated the performance of a flashlamp pumped alexandrite laser at temperatures from ambient to 310° C. As illustrated in Fig. 2.24a, the variation in laser output over the 34°–310° C range is dramatic. The threshold performance indicates that laser gain rises significantly, as temperature increases. Figure 2.24b illustrates laser output energy as a function of temperature for a fixed input to the flashlamp. The ability of alexandrite lasers to

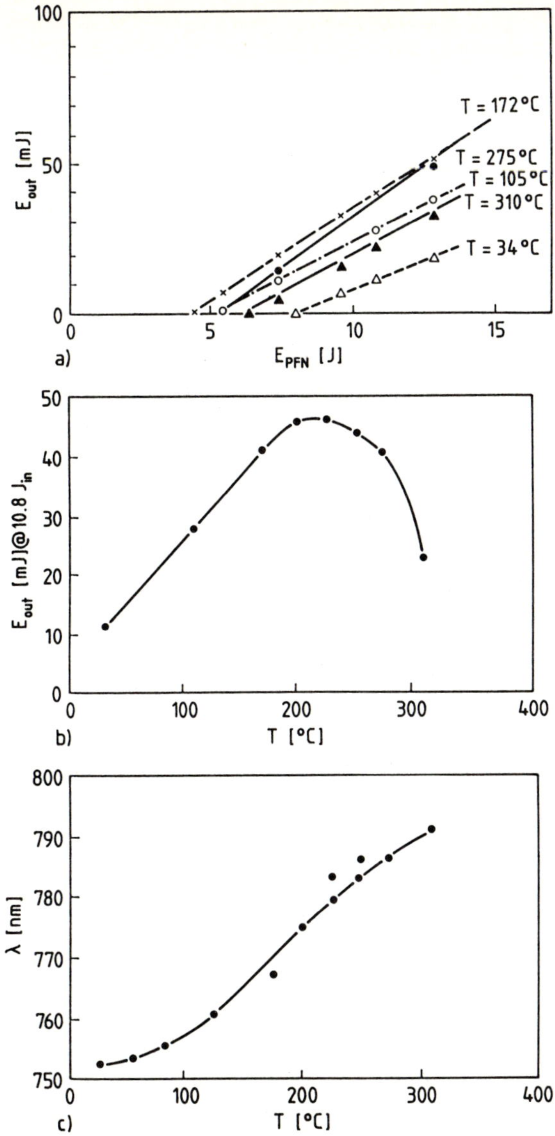

Fig. 2.24a–c. Temperature dependence of alexandrite (**a**) Alexandrite laser energy as a function of pump energy at temperatures from 34 to 310° C. (**b**) Alexandrite laser output energy as a function of temperature for fixed 10.8-J flashlamp output. (**c**) Alexandrite laser wavelengths as a function of temperature. [2.159b]

sustain high gain and efficiency at temperatures above those normally encountered in solid-state lasers is particularly striking. The spectral dependence of the output wavelength is shown in Fig. 2.24c. The wavelength increases almost monotonically with temperature. The behavior of the alexandrite laser depicted in Fig. 2.24 is qualitatively in good agreement with the predictions given above, which are based on the energy-level diagram.

Alexandrite Laser Performance. The development of the alexandrite laser has reached maturity after nearly 10 years of efforts. Its current high average-power performance is 100 W if operated at 100 Hz. Overall efficiency is close to 0.5 %. Tunability over the range of approximately 700 to 818 nm has been demonstrated with tuning accomplished in a manner similar to dye lasers: a combination of etalons and birefringent filters. With these standard spectral control devices 0.5 cm^{-1} linewidths and tunability over 150 nm has been achieved. Alexandrite has been lased in pulsed and cw modes; it has been Q-switched and mode locked.

A rod 10 cm long and 0.63 cm in diameter, when lased in a stable resonator, yields over 5 J long-pulsed, and as much as 2 J with pulse duration less than 30 ns when Q-switched [2.159b]. The reason for such high output energies is the fact that alexandrite is a low-gain medium, $g = 0.04\,\mathrm{cm}^{-1}$–$0.1\,\mathrm{cm}^{-1}$ at room temperature [2.160, 161]. Highly stable frequency locking (without loss of bandwidth control) was achieved with an injection power ten orders of magnitude smaller than the oscillator output [2.162].

Commercially available alexandrite lasers feature continuous, automatic tuning and a minimum 100 mJ of Q-switched output with 0.2 nm bandwidth over the wavelength range from 730–780 nm.

Because of alexandrite's physical strength and thermal properties, cw operation is possible at room temperature. The bulk of the experimentation to date has been with cw xenon arc lamps. CW operation with arc lamp pumping has proved difficult to achieve, yet output powers of up to 40 W with good transverse-mode quality are now being generated. CW lasers were also acousto-optically Q-switched at rates greater than 10 kHz, with peak powers as high as 300 W and pulsewidth of 1 μs [2.163, 164].

2.5.2 Ti : Sapphire

The Ti : Al_2O_3 laser is one of the more promising tunable solid-state systems, combining a broad tuning range (800 nm peak, 300 nm bandwidth) with a relatively large gain cross-section ($\sim$ 50 % of Nd : YAG's value at its peak). Also the energy level structure of the Ti^{3+} ion is unique among transition-metal laser ions in that there are no d-state energy levels above the upper laser level. The simple energy-level structure ($3d^1$ configuration) eliminates the possibility of excited-state absorption of the laser radiation, an effect which has limited the tuning range and reduced the efficiency of other transition-metal-doped lasers [2.165].

One of the greatest advantages is the material properties of the sapphire host itself, namely very high thermal conductivity, exceptional chemical inertness and mechanical rigidity. Titanium sapphire is presently available from commercial vendors in sizes of 3.5 cm diameter by 15 cm long and, due to the well-developed growth technology for sapphire, of good optical quality. The absorption and fluorescence spectra for Ti : Al_2O_3 are shown in Fig. 2.25. The broad, widely separated absorption and fluorescence bands are caused by the

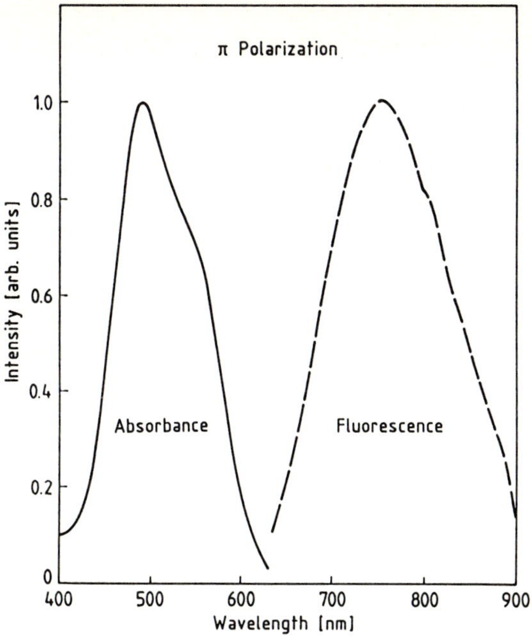

Fig. 2.25. Absorption and fluorescence spectra of the Ti^{3+} ion in Al_2O_3 (sapphire) [2.165]

strong coupling between the ion and host lattice, and are the key to broadly tunable laser operation. The laser parameters of Ti : Al_2O_3 are listed in Table 2.12.

Development of the Ti : Al_2O_3 laser has been confined mainly to laser pumping because of the material's short (3.2 μs) spontaneous-emission lifetime. *Lacovara* et al. [2.166] recently reported successful flashlamp pumping of Ti : Al_2O_3 with efficiency enhanced by using a dye surrounding the laser rod

Table 2.12. Laser parameters of Ti : Al_2O_3

Index of refraction	$n = 1.76$
Fluorescent lifetime	$\tau = 3.2\,\mu s$
Fluorescent linewidth (FWHM)	$\Delta\lambda \sim 1220$ Å
Peak emission wavelength	$\lambda_p \sim 7350$ Å
Peak stimulated emission cross section	
parallel to c axis	$\sigma_p \parallel \sim 4.1 \times 10^{-19}\,cm^2$
perpendicular to c axis	$\sigma_p \perp \sim 2.0 \times 10^{-19}\,cm^2$
Stimulated emission cross section	
at 0.795 μm (‖ c axis)	$\sigma \parallel = 2.8 \times 10^{-19}\,cm^2$
Absorption at 0.795 μm	
parallel to c axis	$\alpha_{\parallel} = 0.15\,cm^{-1}$
perpendicular to c axis	$\alpha_{\perp} = 0.10\,cm^{-1}$
Quantum efficiency of converting a 0.53 μm pump photon into an inverted site	$\eta_Q \approx 1$
Saturation fluence at 0.795 μm	$E_{sat} = 0.9\,J/cm^2$

to convert near-uv light from the flashlamp into blue-green fluorescence. This spectral region is within the titanium absorption band. Laser action [2.167] in Ti : sapphire has been reported from 660 to 1,178 nm, although the laser is usually operated between 700 and 900 nm for maximum efficiency. Output powers of up to 1.6 W have been achieved. State-of-the-art performance results in 40 % overall efficiency if pumped with a frequency doubled Nd : YAG laser, and pulse energies of 100 mJ in 4 ns at a repetition rate from 1 to 10 Hz [2.154b].

In a more general sense, the Ti : Al_2O_3 laser may be the first of a series of broadly-tunable solid-state lasers based on the Ti^{3+} ion. Since the energy levels of the ion depend strongly on the particular host, crystals other than Al_2O_3 will likely exhibit different tuning ranges.

3. Laser Oscillator

In Chap. 1 we studied the processes which lead to optical amplification in substances. The regenerative laser oscillator is essentially a combination of two basic components: an optical amplifier, and an optical resonator. The optical resonator, comprised of two opposing plane-parallel or curved mirrors at right angles to the axis of the active material, performs the function of a highly selective feedback element by coupling back in phase a portion of the signal emerging from the amplifying medium.

Figure 3.1 shows the basic elements of a laser oscillator. The pump lamp inverts the electron population in the laser material, leading to energy storage in the upper laser level. If this energy is released to the optical beam by stimulated emission, amplification takes place. Having been triggered by some spontaneous radiation emitted along the axis of the laser, the system starts to oscillate if the feedback is sufficiently large to compensate for the internal losses of the system. The amount of feedback is determined by the reflectivity of the mirrors. Lowering the reflectivity of the mirrors is equivalent to decreasing the feedback factor. The mirror at the output end of the laser must be partially transparent for a fraction of the radiation to "leak out" or emerge from the oscillator.

An optical structure composed of two plane-parallel mirrors is called a Fabry-Pérot resonator. In Chap. 5 we will discuss the temporal and spatial mode structures which can exist in such a resonator. For the purpose of this discussion it is sufficient to know that the role of the resonator is to maintain an electromagnetic field configuration whose losses are replenished by the amplifying medium through induced emission. Thus, the resonator defines the

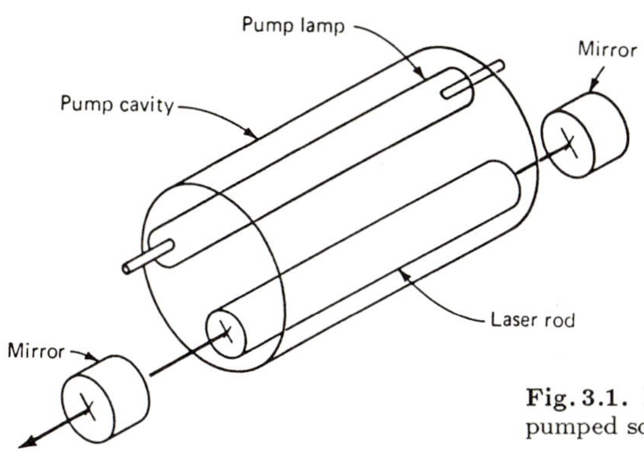

Fig. 3.1. Major components of an optically pumped solid-state laser oscillator

spectral, directional, and spatial characteristics of the laser radiation, and the amplifying medium serves as the energy source.

In this chapter we will develop an analytical model of a laser oscillator that is based mainly on laser systems parameters.

3.1. Operation at Threshold

We will calculate the threshold condition of a laser oscillator composed of two mirrors having the reflectivities R_1 and R_2, and an active material of length l. We assume a gain per unit length of g in the inverted laser material. In each passage through the material the intensity gains by a factor of $\exp(gl)$. At each reflection a fraction $1 - R_1$ or $1 - R_2$ of the energy is lost. Starting at one point, the radiation will suffer two reflections before it can pass the same point in the original direction. The threshold condition is established by requiring that the photon density – after the radiation has traversed the laser material, been reflected by mirror with R_1, and returned through the material to be reflected by mirror with R_2 – be equal to the initial photon density. Then on every complete two-way passage of the light through the laser the loss will just equal the gain. We can express the threshold condition by

$$R_1 R_2 \exp(2gl) = 1 \quad . \tag{3.1}$$

The regenerative amplifier becomes unstable when the amplification per transit exceeds the losses. In this case oscillations will build up, starting from a small disturbance. Clearly, if the loop gain

$$G = R_1 R_2 \exp(2gl) \tag{3.2}$$

is larger than 1, radiation of the proper frequency will build up rapidly until it becomes so large that the stimulated transitions will deplete the upper level and reduce the value of g. The condition of steady state is reached if the gain per pass exactly balances the internal and external losses. This process, called gain saturation, will be discussed in Sect. 3.2. In an oscillator a number of loss mechanisms are instrumental in attenuating the beam; the most important ones are reflection, scattering, and absorption losses in the mirrors, the amplifying medium, and all other elements in the resonator, and diffraction losses.

We will find it convenient to lump all of the non-output losses into a single parameter, the absorption coefficient per unit length α. The condition for oscillation is then

$$R_1 R_2 \exp(g - \alpha)2l = 1 \quad . \tag{3.3}$$

In Sect. 1.4 we characterized all the loss mechanisms by a single parameter τ_c which is equal to the decay time constant of the radiation in a passive resonator. Resonators are characterized by a quality factor Q, which is defined as the ratio of energy stored in the resonator to power dissipated from the resonator per unit angular frequency ω_0. The resonator Q defined in this way is equal to

$$Q = 2\pi\left[1 - \exp\left(\frac{-T_0}{\tau_c}\right)\right]^{-1} \approx \frac{2\pi\tau_c}{T_0} = 2\pi\nu_0\tau_c \quad , \tag{3.4}$$

where $\omega_0 = 2\pi\nu_0 = 2\pi/T_0$.

The loss mechanism, besides limiting the lifetime of the oscillation, causes a broadening of the resonance frequency. The width $\Delta\nu$ of the resonance curve at which the intensity has fallen off to half the maximum value is

$$\Delta\nu = (2\pi\tau_c)^{-1} \quad . \tag{3.5}$$

If we introduce this expression into (3.4) we obtain for the Q value

$$Q = \frac{\nu_0}{\Delta\nu} \quad . \tag{3.6}$$

The decay time constant of the radition τ_c can also be defined as the average lifetime of the photons in the resonator. A photon in the cavity will have some average lifetime in the cavity before being scattered or emitted or lost in other ways to the optical system. If we relate τ_c to the fractional power loss ε per round trip, we obtain

$$\varepsilon = \frac{t_R}{\tau_c} \quad , \tag{3.7}$$

where $t_R = 2l'/c$ is the round-trip time of a photon in a resonator having an optical length l'. Rearranging (3.3) yields

$$2gl = (\ln R_1R_2)^{-1} + 2\alpha l \quad . \tag{3.8}$$

The expression on the right is the total fractional power loss per round trip. Since $2gl = \varepsilon = t_R/\tau_c$, we obtain

$$\tau_c = \frac{2l'}{c}[\ln(R_1R_2)^{-1} + 2\alpha l]^{-1} \quad . \tag{3.9}$$

Miscellaneous losses, such as absorption and scattering at the mirrors and diffraction losses of the resonator, can be thought of as leakage from the rear mirror. Hence the reduced reflectivity R_2 of the rear mirror $R_2 = 1 - L_M$ takes into account the miscellaneous losses. In practice, L_M does not exceed a few percent. With the approximation

$$\ln(1 - L_M) \approx -L_M \quad , \tag{3.10}$$

one can combine the optical losses in the cavity with the losses in the crystal:

$$L = 2\alpha l + L_M \quad . \tag{3.11}$$

With this approximation (3.9) reduces to

$$\tau_c = \frac{2l'}{c(L - \ln R_1)} \quad . \tag{3.12}$$

In a typical pulsed laser, the transmission of the output mirror is around 50 % and the combined losses are around 10 %. If we assume a typical resonator length of 50 cm, we obtain a cavity photon lifetime of $\tau_c = 5.5$ ns. In a continuous Nd:YAG laser, transmission of the output mirror is typically 90 %, therefore $\tau_c \approx 17$ ns, all other parameters being the same.

With the aid of (3.10, 11) we can express the threshold condition (3.3) in the following form:

$$2gl = L - \ln R_1 \approx T + L \quad . \tag{3.13}$$

The approximation $-\ln R_1 \approx T$ is valid only for values of R_1 close to one.

We turn now to the rate equation (1.61), which gives the photon density in the amplifying medium. It is clear from this equation that for onset of laser emission the rate of change of the photon density must be equal to or greater than zero. Thus at laser threshold for sustained oscillation the condition

$$\frac{\partial \phi}{\partial t} \geq 0 \tag{3.14}$$

must be fulfilled, which enables us to obtain from (1.61) the required inversion density at threshold,

$$n \geq \frac{1}{c\sigma\tau_c} \quad . \tag{3.15}$$

In deriving this expression we have ignored the factor S, which denotes the small contribution from spontaneous emission to the induced emission. The reader should note that by introducing $g(\nu_s) = -\alpha(\nu_s) = n\sigma_{21}(\nu_s)$ according to (1.36) and expressing τ_c by (3.9), this equation is identical to the threshold condition (3.3).

We may write the threshold condition in terms of the fundamental laser parameters. Upon substitution of $\sigma_{21}(\nu_s)$ from (1.40), we obtain

$$n_2 - \frac{g_2 n_1}{g_1} > \frac{\tau_{21} 8\pi\nu^2}{\tau_c c^3 g(\nu_s, \nu_0)} \quad . \tag{3.16}$$

The lineshape factor $g(\nu_s, \nu_0)$ and therefore the stimulated emission cross section σ are largest at the center of the atomic line. Thus from (3.16) we can see qualitatively how the linewidth of the laser output is related to the linewidth of the atomic system. Self-sustained oscillation which develops from noise will occur in the neighborhood of the resonant frequency, because only at a narrow spectral range at the peak will the amplification be large enough to offset losses. Consequently, the output of the laser will be sharply peaked, and its linewidth will be much narrower than the atomic linewidth.

It is also obvious from this equation that an increase of the inversion by stronger pumping will increase the laser linewidth because threshold condition can now be met for values of $g(\nu_s, \nu_0)$ farther away from the center. As we will see in Chap. 5 the linewidth of an actual laser system is related to the linewidth of the active material, the level of pump power, and the properties of the optical resonator. The threshold condition at the center of the atomic line is obtained by introducing the peak values of the amplification curve into (3.16). If $g(\nu_s, \nu_0)$ has a Lorentzian shape with full width at half-maximum of $\Delta\nu$ centered about ν_s, then $g(\nu_0) = 2/\pi\Delta\nu$ and

$$n_2 - \frac{g_2 n_1}{g_1} > \frac{\tau_{21} 4\pi^2 \Delta\nu \nu_0^2}{\tau_c c^3} \quad . \tag{3.17}$$

For a Gaussian lineshape, $g(\nu_0) = 2(\pi \ln 2)^{1/2}/\pi\Delta\nu$, and the start-oscillation condition is still given by (3.17) with $\Delta\nu$ replaced by $\Delta\nu/(\pi \ln 2)^{1/2}$. Again, we have assumed that the laser threshold will be reached first by a resonator mode whose resonant frequency lies closest to the center of the atomic line.

From (3.17) we can infer those factors favoring high gain and low threshold for a laser oscillator. In order to achieve a low threshold inversion, the atomic linewidth $\Delta\nu$ of the laser material should be narrow. Furthermore, the incidental losses in the laser cavity and crystal should be minimized to increase the photon lifetime τ_c. It is to be noted that the critical inversion density for threshold depends only on a single resonator parameter, namely τ_c. A high reflectivity of the output mirror will increase τ_c and therefore decrease the laser threshold. However, this will also decrease the useful radiation coupled out from the laser. We will address the question of optimum output coupling in Sect. 3.4.

We will now calculate the pumping rate W_p which is required to maintain the oscillator at threshold. For operation at or near threshold the photon density ϕ is very small and can be ignored. Setting $\phi = 0$ in the rate equation (1.58) and assuming a steady-state condition of the inversion, $\partial n/\partial t = 0$, as is the case in a conventional operation of the laser oscillator, we obtain for a three-level system

$$\frac{n}{n_{tot}} = \frac{W_p \tau_{21} - g_2/g_1}{W_p \tau_{21} + 1} \quad , \tag{3.18}$$

and for a four-level system

$$\frac{n}{n_{tot}} = \frac{W_p \tau_f}{W_p \tau_f + 1} \approx W_p \tau_f \quad . \tag{3.19}$$

Other factors being equal, four-level laser systems have lower pump-power thresholds than three-level systems. In a four-level system where $n = n_2$, an inversion is achieved for any finite pumping rate W_p. In a three-level system we have the requirement that the pumping rate W_p exceeds a minimum or threshold value given by

$$W_{\mathrm{p(th)}} = \frac{g_2}{\tau_{21} g_1} \tag{3.20}$$

before any inversion at all can be obtained. Whereas for a four-level material the spontaneous lifetime has no effect on obtaining threshold inversion, in a three-level material the pump rate required to reach threshold is inversely proportional to τ_{21}. Thus, for three-level oscillators only materials with long fluorescence lifetimes are of interest.

The reader is reminded again that (3.18, 19) are valid only for a negligible photon flux ϕ. This situation occurs at operation near threshold; it will later be characterized as the regime of small-signal amplification. We will now calculate the minimum pump power which has to be absorbed in the pump bands of the crystal to maintain the threshold inversion. This will be accomplished by first calculating the fluorescence power at threshold. While just above threshold almost all the pump power supplied to the active material goes into spontaneous emission, the fluorescence power of the laser transition in a four-level system is

$$P_{\mathrm{f}} = \frac{h\nu n_{\mathrm{th}}}{\tau_{\mathrm{f}}} \quad . \tag{3.21}$$

where $n_2 = n_{\mathrm{th}}$ is the inversion at threshold.

In a three-level system at threshold, $n_2 \approx n_1 \approx n_{\mathrm{tot}}/2$ and

$$P_{\mathrm{f}} \approx \frac{h\nu n_{\mathrm{tot}}}{2\tau_{21}} \quad . \tag{3.22}$$

In order that the critical inversion may be continuously maintained, the loss by fluorescence from the upper laser level must be supplied by the pump energy. As a result, we obtain for the absorbed pump power P_{ab} needed to compensate for population loss of the laser level by spontaneous emission

$$P_{\mathrm{ab}} = \frac{\nu_{\mathrm{p}} P_{\mathrm{f}}}{\nu_0 \eta_0} = \frac{P_{\mathrm{f}}}{\eta_1} \quad . \tag{3.23}$$

The factor ν_{p}/ν_0 represents the ratio of the pump-band energy $h\nu_{\mathrm{p}}$ and the laser energy $h\nu_0$, and η_0 is the efficiency factor of pumping, as defined in (1.56 and 73). The difference between the pump power and the fluorescence power represents the thermal power which is released to the lattice of the crystal.

3.2 Gain Saturation

In the previous section we considered the conditions for laser threshold. Threshold was characterized by a steady-state population inversion, i.e., $\partial n/\partial t = 0$, in the rate equations. In doing this we neglected the effect of stimulated emission by setting $\phi = 0$. This is a perfectly legitimate assumption at threshold, where the induced transitions are small compared with the number of spontaneous processes.

As the threshold is exceeded, however, stimulated emission and photon density in the resonator build up. Far above threshold we have to consider a large photon density in the resonator. From (1.58) we can see that $\partial n/\partial t$ decreases for increasing photon density. Steady state is reached when the population inversion stabilizes at a point where the upward transitions supplied by the pump source equal the downward transitions caused by stimulated and spontaneous emission. With $\partial n/\partial t = 0$ one obtains for the steady-state inversion population in the presence of a strong photon density ϕ

$$n = n_{\text{tot}}\left(W_{\text{p}} - \frac{(\gamma - 1)}{\tau_{\text{f}}}\right)\left(\gamma c\sigma\phi + W_{\text{p}} + \frac{1}{\tau_{\text{f}}}\right)^{-1} \quad . \tag{3.24}$$

The photon density ϕ is given by the sum of two beams travelling in opposite directions through the laser material.

We will now express (3.24) in terms of operating parameters. From Sect. 1.2 we recall that the gain coefficient $g = -\alpha$ is defined by the product of stimulated emission and inversion population. Furthermore, we will define a gain coefficient which the system would have at a certain pump level in the absence of stimulated emission. Setting $\phi = 0$ in (3.24), we obtain the small-signal gain coefficient

$$g_0 = \sigma_{21} n_{\text{tot}}[W_{\text{p}}\tau_{\text{f}} - (\gamma - 1)](W_{\text{p}}\tau_{\text{f}} + 1)^{-1} \tag{3.25}$$

which an active material has when pumped at a level above threshold and when lasing action is inhibited by blocking the optical beam or by removing one or both of the resonator mirrors. If feedback is restored, the photon density in the resonator will increase exponentially at the onset with g_0. As soon as the photon density becomes appreciable, the gain of the system is reduced according to

$$g = g_0\left(1 + \frac{\gamma c\sigma_{21}\phi}{W_{\text{p}} + (1/\tau_{\text{f}})}\right)^{-1} \tag{3.26}$$

where g is the saturated gain coefficient. Equation (3.26) was obtained by introducing (3.25) into (3.24) and using $g = \sigma_{21} n$.

We can express ϕ by the power density I in the system. With $I = c\phi h\nu$ we obtain

$$g = \frac{g_0}{1 + I/I_{\text{s}}} \,, \qquad \text{where} \tag{3.27}$$

$$I_{\text{s}} = \left(W_{\text{p}} + \frac{1}{\tau_{\text{f}}}\right)\frac{h\nu}{\gamma\sigma_{21}} \quad . \tag{3.28}$$

The saturation density I_{s} defines a power density in the active material at which the small-signal gain coefficient g_0 is reduced by one-half [3.1].

In a four-level system $W_{\text{p}} \ll 1/\tau_{\text{f}}$ and $\gamma = 1$, so (3.28) reduces to

$$I_{\text{s}} = \frac{h\nu}{\sigma_{21}\tau_{\text{f}}} \quad . \tag{3.29}$$

For a three-level system the saturation density is

$$I_s = \frac{h\nu[W_p + (1/\tau_{21})]}{\sigma_{21}[1 + (g_2/g_1)]} \quad . \tag{3.30}$$

As we can see from (3.25), the small-signal gain depends only on the material parameters and the amount of pumping power delivered to the active material. The large-signal or saturated gain depends in addition on the power density in the resonator.

In a four-level system, a very interesting relationship between the stimulated emission lifetime τ_{st} and the saturation power density I_s can be obtained. Since the total number of downward transitions per second depends on the radiation density, the lifetime of decay for the excitation of the upper level will show a similar dependence. We may write for the total number of downward transitions per second

$$\frac{\partial n_2}{\partial t} = n_2\left(\frac{1}{\tau_f} + \frac{1}{\tau_{st}}\right) \quad , \tag{3.31}$$

where τ_f is the fluorescence decay time of the upper laser level, and τ_{st} is the stimulated emission lifetime.

From (1.10, 51) it follows

$$\tau_{st} = \frac{h\nu}{\sigma_{21} I} \quad , \tag{3.32}$$

where I is the power density in the active material. Thus, as the excitation power is increased beyond the oscillation threshold, the portion of power going into stimulated emission increases. Comparing (3.29) and (3.32), we obtain

$$\tau_{st} = \tau_f\left(\frac{I_s}{I}\right) \quad . \tag{3.33}$$

For a power density in the laser material which equals the saturation power density, the stimulated emission lifetime equals the fluorescence decay time of the upper laser level.

Gain saturation as a function of steady-state radiation intensity must be analyzed for lasers with homogeneous and inhomogeneous line broadening. Equation (3.27) is valid only for the former case, in which the gain decreases proportionately over the entire transition line. As we have seen in Chap. 2, a ruby laser has a homogeneously broadened bandwidth, whereas in Nd: glass the interaction of the active ion with the electrostatic field of the host leads to an inhomogeneous line. However, in solid-state materials such as Nd: glass, the cross-relaxation rate is very fast. The latter is associated with any process characteristic of the laser medium that affects the transfer of excitation within the atomic spectral line so as to prevent or minimize the departure of this line from the equilibrium distribution. It has been shown by *Cabezas* and *Treat* [3.2] that in the case of a very fast cross-relaxation within the inhomogeneous line, the saturated gain is in agreement with that of a homogeneously broadened bandwidth.

3.3 Circulating Power

For a single-pass laser amplifier we can write the equation for energy density at each point (x) in the material:

$$\frac{dI(x)}{dx} = \frac{g_0 I(x)}{1 + I(x)/I_s} - \alpha I(x). \tag{3.34}$$

If we assume unsaturated operation, that is, $I(x) \ll I_s$, the single-pass gain is

$$G = \frac{I(l)}{I(0)} = \exp(g_0 - \alpha)l \tag{3.35}$$

for a laser amplifier of length l. Due to the loss coefficient α, the gain equation (3.34) becomes transcendental and hence only solvable numerically. This is a case which we will study in Chap. 4.

Now consider the situation where we have two waves propagating through the amplifier in opposite directions, as shown in Fig. 3.2. The saturation factor is a function of the total power density in the medium, hence the gain equation for the two separate beams can be written as

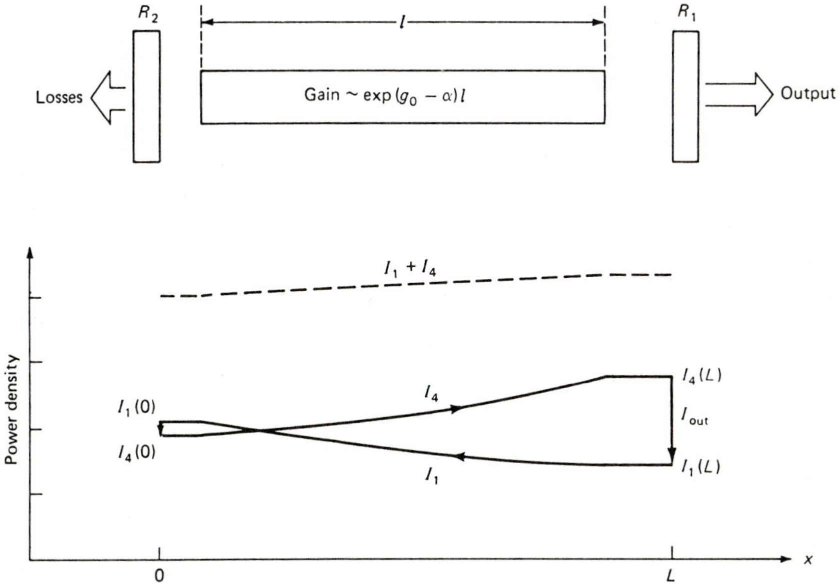

Fig. 3.2. Circulating power in a laser oscillator

$$\frac{dI_{1,4}(x)}{dx} = \frac{\mp g_0 I_{1,4}(x)}{1 + [I_1(x) + I_4(x)]/I_s} \pm \alpha I_{1,4}(x) \quad . \tag{3.36}$$

The two equations for $I_1(x)$ and $I_4(x)$ are coupled because the beams "share" the same atoms in the stimulated emission process. By multiplying the first equation by $I_1(x)$ and the second by $I_4(x)$ and adding, one obtains

$$\frac{d}{dx}[I_1(x)I_4(x)] = 0 \quad . \tag{3.37}$$

Thus, rather than the average power density of $I_4(x)$ and $I_1(x)$, it is the geometric mean power density

$$I = [I_1(x)I_4(x)]^{1/2} \tag{3.38}$$

which is constant.

Assume that mirrors are placed at the ends of the laser rod with R_2 at $x = 0$ and R_1 at $x = l$, where R denotes the power reflectivity of the mirrors. Then

$$\frac{I_1(l)}{I_4(l)} = R_1 \quad \text{and} \quad I_4(l) = \frac{I}{(R_1)^{1/2}} \quad . \tag{3.39}$$

With $I_4(l)$ and $I_1(l)$ being the power densities of the beam incident and reflected at the front mirror, the laser output is given by

$$P_{\text{out}} = A[I_4(l) - I_1(l)] \quad . \tag{3.40}$$

Combining (3.39 and 40) yields

$$P_{\text{out}} = AI(1 - R_1)R_1^{-1/2} \quad . \tag{3.41}$$

This equation relates the output power obtained from the oscillator to the geometric mean power density I inside the optical resonator. For the values of R_1 close to one, one can simply write $P_{\text{out}} \approx AIT$.

3.4 Output versus Input Calculations

In this section we will develop a model for the laser oscillator. It is important to emphasize that the point-by-point agreement between experiment and computations is of relatively little importance. Since some of the parameters in the computations are not known accurately, it is necessary to fit theoretical points to the experimental data. The importance of comparing observations with a theoretical model, and the main utility of the model, lie in the fact that once agreement is obtained using a given set of parameters, the theory can be fairly reliably used to predict the performance of the system even if one or several operating parameters are changed.

The oscillations which build up from noise will increase rapidly. As a result of the increasing radiation flux, the gain coefficient decreases and finally stabilizes at the value required by (3.27). Substitution of (3.8) into (3.27) gives the steady-state power density in the resonator:

$$I = I_s\left(\frac{2lg_0}{2\alpha l - \ln R_1 R_2} - 1\right) \approx I_s\left(\frac{2lg_0}{L - \ln R_1} - 1\right) \quad . \tag{3.42}$$

In this equation I_s is a materials parameter, and R_1 and R_2 are system parameters. These quantities are usually known, whereas the small-signal gain coefficient g_0 and the losses α are normally not. We will now discuss methods to determine the small-signal gain coefficient g_0 and the losses in a laser oscillator.

3.4.1 Flashlamp-Pumped Lasers

The Four-level System

The population inversion in a four-level system as a function of pump rate is given by (3.19). Making the assumption that $W_p\tau_p \ll 1$ and multiplying both sides of this equation by the stimulated emission cross section yields

$$g_0 = \sigma_{21} n_{tot} W_p \tau_f \quad . \tag{3.43}$$

Now we relate the total pump power absorbed in the crystal to the electrical input of the pump source:

$$P_{ab} = \frac{P_f}{\eta_1} = \eta_2\eta_3\eta_4 P_{in} \quad . \tag{3.44}$$

The efficiency factors are defined as follows: η_2 is the ratio of lamp radiation within absorption bands of the laser material to electrical input power, that is, η_2 is the fraction of the electrical input power which results in potentially useful radiation; η_3 is the efficiency obtained in transferring the useful radiation from the pump source to the laser rod; and η_4 is the fraction of useful pump light which is actually absorbed by the laser material. This last parameter depends on the doping level and the diameter of the rod and on the reflection losses of the pump light on the rod surface. All parameters on the right-hand side of (3.44) are system parameters associated with the design of the laser. Combining (3.19, 21, 43, and 44), we can express the small-signal gain coefficient in terms of lamp input power P_{th}:

$$g_0 = \frac{\sigma_{21}\tau_f\eta_1\eta_2\eta_3\eta_4 P_{th}}{h\nu_0 V} \quad . \tag{3.45}$$

After introducing the small-signal, single-pass gain of the laser rod $\ln G_0 = g_0 l$, l being the length of the rod, (3.45) can be written as

$$\ln G_0 = g_0 l = K P_{th} \quad , \quad \text{where} \tag{3.46}$$

$$K = \frac{\eta_1 \eta_2 \eta_3 \eta_4}{I_s A} \tag{3.47}$$

can be considered a pumping coefficient, with A being the rod's cross-sectional area.

Substituting K from (3.46) into (3.13) yields

$$P_{th} \approx \frac{L - \ln R_1}{2K} \tag{3.48}$$

for operation at threshold.

Equation (3.48) can be written as

$$-\ln R_1 = 2KP_{th} - L \quad . \tag{3.49}$$

Using two different output mirrors and measuring the threshold input power for each mirror, K and L can be determined from

$$K = \frac{\ln(R_1'/R_1'')}{2(P_{th}'' - P_{th}')} \quad \text{and} \quad L = \frac{P_{th}'' \ln(R_1') - P_{th}'(\ln R_1'')}{P_{th}'' - P_{th}'} \quad . \tag{3.50}$$

If we plot a curve of P_{th} versus $-\ln R_1$, then from (3.49) it follows that the slope of the curve is given by

$$\frac{d(-\ln R_1)}{dP_{th}} = 2K \quad . \tag{3.51}$$

Extrapolation of the curve to $P_{th} = 0$ gives the combined optical loss of the resonator,

$$\ln R_1 = L \quad . \tag{3.52}$$

With the value of K known, one can plot the small-signal, single-pass rod gain as a function of lamp input power. Furthermore, with the knowledge of α and losses L, one can calculate the power density in the cavity as a function of input power for any value of mirror reflectivity R_1.

Now, we will consider operation above threshold. Introducing (3.46, 27) into (3.13) yields

$$\frac{2KP_{in}}{1 + I/I_s} = L - \ln R_1 \quad . \tag{3.53}$$

The lamp input power P_{th} for threshold operation has been replaced by P_{in}, the input power above threshold. The circulating power density I can be expressed in terms of output power P_{out} according to (3.41). We obtain after some algebra the following expression for the laser output power

$$P_{out} = \sigma_s(P_{in} - P_{th}) \quad , \tag{3.54}$$

where P_{th} is the lamp input power required to achieve threshold

$$P_{\mathrm{th}} = \frac{L - \ln R_1}{2K} \quad . \tag{3.55}$$

The slope efficiency σ_{s} of the laser output versus lamp input curve is

$$\sigma_{\mathrm{s}} = K I_{\mathrm{s}} A \eta_5 \quad , \quad \text{where} \tag{3.56}$$

$$\eta_5 = \frac{2(1 - R_1)}{(R_1)^{1/2}(L - \ln R_1)} \tag{3.57}$$

is an output coupling term.

As expected, high optical losses L, caused by reflection, scattering, or absorption, increase the threshold input power and decrease the slope efficiency. The coefficient K is proportional to the slope efficiency and inversely proportional to threshold. A highly efficient system indicated by a large A will have a large slope efficiency and a low threshold. If one introduces the full expression for K into (3.56), a very simple relationship is obtained for the slope efficiency:

$$\sigma_{\mathrm{s}} = \eta_1 \eta_2 \eta_3 \eta_4 \eta_5 \quad . \tag{3.58}$$

The slope efficiency is simply the product of all the individual efficiencies of the system. The output coupling factor η_5 can be optimized by proper selection of the output mirror reflectivity.

Equation (3.54) can be readily differentiated with respect to R_1 and set equal to zero in order to determine the output coupling which maximizes the laser output power

$$R_{1(\mathrm{opt})} \approx 1 - \frac{(2KP_{\mathrm{in}}L)^{1/2} - L}{1 + L} \quad . \tag{3.59}$$

This expression is useful if one wants to optimize the performance of the system for various input powers. As we can see from this expression, the reflectivity of the output mirror must be decreased for increasing input powers. In pulsed solid-state lasers the optimum reflectivity of the output mirror is typically 30–50 % reflection. In cw-pumped systems the optimum output mirror reflectivity is usually between 80 and 98 %. The reason for the large difference in the output mirror reflectivity for the two modes of operation lies in the fact that pulsed systems are operated at much higher input powers P_{in}. Pulsed systems therefore have a correspondingly higher gain. For example, flashlamp input powers for pulsed systems range typically from 100 kW to 10 MW. The smaller value would be realized in a typical military-type high-repetition-rate Nd : YAG system pumped by a flashlamp pulse of 20 J energy and 200 μs duration. A large Nd : glass oscillator would typically be operated at 5000 J input at a duration of the flashlamp pulse of approximately 500 μs. In contrast to these high flashlamp peak powers, continuous lasers such as Nd : YAG lasers are normally driven at input powers between 1 and 12 kW.

The Three-level System

The population inversion in a three-level system as a function of the pump rate is given by (3.18). If we multiply both sides of this equation by the stimulated emission cross section σ_{21} we obtain

$$g_0 = \frac{\alpha_0}{g_2/g_1} \frac{W_p\tau_{21} - (g_2/g_1)}{W_p\tau_{21} + 1} \quad , \tag{3.60}$$

where $\alpha_0 = \sigma_{21}(g_2/g_1)n_{tot}$ is the absorption coefficient of the material when all atoms exist in the ground state. In the absence of pumping, (3.60) simply becomes $g_0 = -\alpha_0$.

We assume now that the pump rate W_p is a linear function of lamp input P_{in}:

$$W_p\tau_{21} = KP_{in} \quad . \tag{3.61}$$

Introducing this expression into (3.60) and assuming $g_2 = g_1$ to simplify our analysis, we obtain

$$g_0 = \alpha_0(KP_{in} - 1)(KP_{in} + 1)^{-1} \quad . \tag{3.62}$$

Combining (3.61 and 60) with the threshold condition (3.13) and expressing the losses in the same way as we did in the four-level case, we obtain

$$-\ln R_1 = \frac{2l\alpha_0}{g_2/g_1}\left(\frac{KP_{th} - (g_2/g_2)}{KP_{th} + 1}\right) - L \quad . \tag{3.63}$$

P_{th} is the lamp input energy required to achieve threshold operation of the oscillator. By measuring the threshold energy as a function of the mirror output reflectivity for at least three points, one can determine the unknowns K, L, and α_0 in (3.63). However, the value of α_0 is normally known for the three-level material; in this case only two measurements must be made to determine the losses in the optical resonator and the pumping coefficient.

It should be pointed out that this technique for computing K and L provides only a very rough estimate of these quantities. The nonlinearity of (3.63) and the fact that in a three-level system threshold changes very little with output mirror reflectivity makes an accurate determination of K and L difficult.

The equations derived for the circulating power inside the resonator described in the previous section are applicable to the three- and four-level systems. The saturation density I_s is obtained by introducing (3.60) into (3.30):

$$I_s = \frac{h\nu}{\sigma_{21}\tau_{21}} \frac{1}{[1 - (g_0 g_2/\alpha_0 g_1)]} \quad . \tag{3.64}$$

By combining (3.42, 62, and 64), expressions for the laser output versus input are obtained which are identical to (3.54 and 58). However, the output coupling efficiency is now

$$\eta_5 = \frac{(1 - L_{\mathrm{M}})(1 - R_1)}{(L - \ln R_1)(R_1)^{1/2}} \quad , \tag{3.65}$$

where $L_{\mathrm{M}} = (L - \ln R_1)/2l\alpha_0$ and the threshold input power is

$$P_{\mathrm{th}} = \frac{(1 + L_{\mathrm{M}})}{K(1 - L_{\mathrm{M}})} \quad . \tag{3.66}$$

The factor K and, therefore, the gain in the system as well as the losses L, can be calculated from the threshold input power and slope efficiency:

$$L = A + \ln R_1 \quad \text{and} \quad K = \frac{2l\alpha_0 + A}{P_{\mathrm{th}}(2l\alpha_0 - A)} \quad , \tag{3.67}$$

where

$$A = \left(\frac{\sigma_{\mathrm{s}} P_{\mathrm{th}} (R_1)^{1/2}}{P_{\mathrm{f}}(1 - R_1)} - \frac{1}{2l\alpha_0} \right)^{-1} \quad .$$

In these expressions σ_{s}, P_{th}, R_1, and l are measurable parameters, whereas P_{f} and α_0 must be calculated from the basic materials parameters. For example, for ruby $\alpha_0 = 0.2\,\mathrm{cm}^{-1}$ and $P_{\mathrm{f}} = 727\,\mathrm{W/cm}^3$.

We derived (3.67) by using the relationship

$$K = \eta_1\eta_2\eta_3\eta_4 P_{\mathrm{f}}^{-1} = \sigma_{\mathrm{s}}(P_{\mathrm{f}}\eta_5)^{-1} \tag{3.68}$$

which is obtained from (3.44, 58), and noting that at population inversion the input power P'_{in} is related to K by $P'_{\mathrm{in}} = K^{-1}$ according to (3.62). Combining (3.66 and 68) yields the expression (3.67).

If we compare these expressions with the equations derived for a four-level laser, some of the basic differences between these systems can be illustrated. For example, in a four-level system which has no absorption losses ($L = 0$) and no coupling losses ($R_1 = 1$), the threshold is $P_{\mathrm{th}} = 0$. In the three-level system threshold becomes, according to (3.66, 68), $P_{\mathrm{th}} = P'_{\mathrm{in}} = P_{\mathrm{f}}/\eta_1\eta_2\eta_3\eta_4$, which is the input power required to achieve inversion. Comparing (3.55 and 66) it is also quite apparent that changes in reflectivity R_1 affect the threshold in a three-level system much less than in a four-level system. For example, if we use output mirror reflectivities of $R_1 = 0.93$, 0.53, 0.29, and 0.18 in a laser oscillator and assume a loss of $L = 0.20$, then according to (3.66) in a ruby oscillator with a 20-cm-long crystal ($2l\alpha_0 = 8$) the relative change in threshold is 1 : 1.1 : 1.3 : 1.5, whereas in a four-level system changes in threshold are 1 : 2.7 : 4.8 : 6.3.

The calculations carried out so far are valid only for a three-level system operated at steady state, i.e., a system which is operated either cw or pulsed with a pump pulse long compared to the fluorescence time τ_{21}. In practice, a three-level system such as ruby is normally pumped with a 1-ms-long pump pulse, which is short compared to the 3-ms spontaneous emission time. For this case we will replace the fluorescence power at inversion P_{f} by the energy E_{ui}

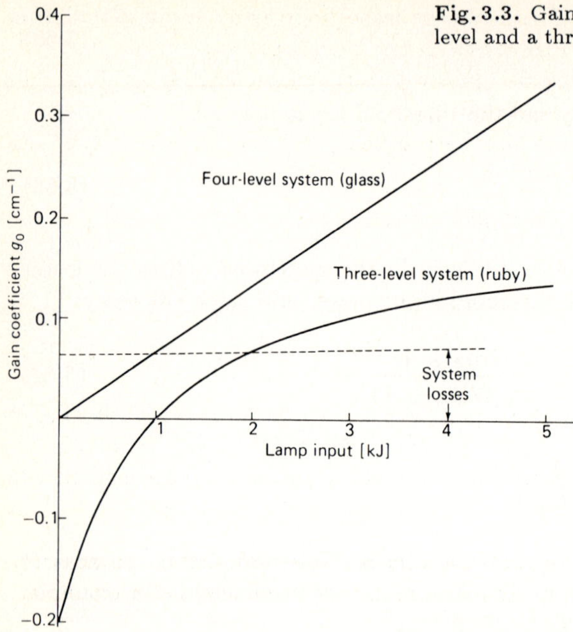

Fig. 3.3. Gain versus lamp input for a four-level and a three-level system

which is stored in the upper level at inversion. Furthermore, the lamp input power P_{in} will be substituted by the input energy E_{in}. Equation (3.62) is plotted in Fig. 3.3 together with the expression $g_0 = (K/l)E_{in}$ which is valid for a four-level system. The numerical parameters were chosen to be $K = 10^{-3}\,J^{-1}$, $l = 15\,cm$, $\alpha = 0.2\,cm^{-1}$.

If we want to apply the expressions derived in this chapter to the ruby system, we have to recall that the inversion population density $n_2 - (g_2/g_1)n_1$, which appears in (1.53, 57, 16, and 17), must be replaced by $(n_2 - n_1)/2$ according to (2.9). Therefore, the multiplicities in the case of ruby are $g_2/g_1 = 1$, and a factor of 2 enters into most equations. In particular, in (3.28, 30, and 64), the right-hand side has to be multiplied by 2, and in (3.18 and 25) the right side has to be divided by 2. In (3.24 and 26) the term $W_p + (1/\tau_f)$ has to be multiplied by 2.

In performing these kinds of model comparisons it has to be pointed out that in a conventional-mode ruby oscillator the output consists of a series of random spikes rather than a smooth pulse. Therefore, the parameters calculated above are averaged values obtained by integrating over the whole pulse length.

3.4.2 Laser-Diode-Pumped Oscillators

In this section we will describe the principal factors of a laser diode pumped solid-state laser that contribute to the output of the system.

The energy transfer from electrical input to the laser diodes to laser output from the solid-state medium can conveniently be expressed as a three-step process:

- Conversion of electrical input to the laser-diode array to pump radiation. This will be expressed as the laser-diode efficiency η_D
- Transfer of pump radiation to energy in the upper laser level of the gain medium (upper state efficiency η_u)
- Conversion of the stored upper state energy into useful laser output (expressed as output efficiency η_{out})

With the definitions given above, we can write

$$E_{out} = \eta_D \eta_u \eta_{out} E_{EL} \tag{3.69}$$

where E_{out} is the laser output and E_{EL} is the electrical input into the diode arrays.

Each of the efficiency factors given above include several contributing elements and steps which are involved in the pump and lasing process. The parameters which determine these efficiencies will be discussed below and specific examples will be given.

Laser-diode Efficiency. From the stand point of the laser designer the parameter η_D is a given, and depends on the fabrication process of the laser-diode arrays.

Upper State Efficiency. The rate of change of the upper laser level, assuming no stimulated emission, is given by

$$\frac{dN_u}{dt} = N_l W_p - \frac{N_u}{\tau_{sp}} \quad , \tag{3.70}$$

where N_u and N_l are the population densities of the upper and ground state, W_p is the pumping rate, and τ_{sp} is the spontaneous decay rate of the upper level.

In steady state the upper-level population density is given by

$$N_u = N_l W_p \tau_{sp} \quad . \tag{3.71}$$

The stored energy density in the upper laser level is

$$E_s = h\nu_L N_u \tag{3.72}$$

wher $h\nu_L$ is the energy per photon of the solid-state laser output.

The upper-state population density and pumping rate can be related to the total absorbed pump power P_{abs}. It is

$$P_{abs} = N_l W_p h\nu_D V \tag{3.73}$$

where $h\nu_D$ is the energy per photon of the laser-diode pump, and V is the volume of the solid-laser medium.

Combining (3.70, 71) and substituting $N_l W_p$ from (3.72) yields for the stored energy density

$$E_s = \left(\frac{h\nu_c}{h\nu_D}\right)\left(\frac{\tau_{sp}}{V}\right)P_{abs} \quad . \tag{3.74}$$

The pump power absorbed in the solid-state laser is related to the laser-diode output power P_D as follows:

$$P_{abs} = P_D[1 - \exp(-\alpha_D l](1 - r) \quad , \tag{3.75}$$

where α_D is the absorption coefficient at the diode wavelength in the solid medium, l is the path length of the pump radiation in the medium, and r summarizes the reflection losses occurring between the pump source and the solid medium.

If we introduce (3.75) into (3.74) and note that

$$P_D = E_D/t_D$$

where E_D and t_D are the laser-diode energy per pulse and pulse length, respectively, we obtain

$$E_{st} = \left(\frac{h\nu_L}{h\nu_D}\right)\left(\frac{\tau_{sp}}{t_D}\right)[1 - \exp(-\alpha_D l)](1 - r)E_D \quad . \tag{3.76}$$

Instead of the energy density, we introduced E_{st} which is the total energy stored in the upper level

$$E_{st} = E_s V \quad . \tag{3.77}$$

We will examine briefly the terms of (3.76). The first term is the Stokes efficiency which accounts for the photon-energy ratio of the laser and pump emissions

$$\eta_\nu = \nu_L/\nu_D \quad . \tag{3.78}$$

The second term expresses the fraction of pump power remaining prior to extraction by the Q-switched pulse

$$\eta_p = \tau_{sp}/t_D \quad . \tag{3.79}$$

For a pump pulse longer than the fluorescence lifetime, gain is reduced because of the depletion of the upper level by spontaneous emission. The last two terms in (3.76), determine the transfer efficiency from the pump array into the active medium.

$$\eta_T = [1 - \exp(-\alpha_D l)](1 - r) \quad . \tag{3.80}$$

A high efficiency is achieved by making the active medium thick enough to absorb the laser-diode radiation, even in the presence of temperature variations in the array. Reflection losses and spillover losses at the edges of the active medium expressed by the parameter r have to be minimized for efficient transfer of energy. Equation (3.76) can now be expressed as follows

$$E_{\mathrm{st}} = \eta_{\nu}\eta_{\mathrm{P}}\eta_{\mathrm{T}}\eta_{\mathrm{Q}}E_{\mathrm{D}} \quad \text{or} \tag{3.81}$$

$$E_{\mathrm{st}} = \eta_{\mathrm{u}}E_{\mathrm{D}} \quad . \tag{3.82}$$

We added to the above equation the quantum efficiency η_{Q}, which expresses the fraction of pump photons reaching the upper laser level.

Output Efficiency. Conversion of the stored upper-state energy into useful laser output depends on the following factors:

The spatial overlap of the resonator modes with the upper-state inversion density is usually expressed by the beam fill factor η_{B}. Losses due to amplified spontaneous emission (ASE) will reduce the available stored energy. ASE losses can be taken into account by a factor η_{F}.

In Q-switched operation, the amount of stored energy extracted by the pulse depends on the inversion above threshold, as will be discussed in Chap. 8. We will express the fraction of energy extracted by η_{ex}. Optical losses L in the resonator due to scattering, absorption or reflection further diminish the laser output. We have

$$\eta_{\mathrm{R}} = \frac{1}{1 + L/(1 - R_1)} \quad , \tag{3.83}$$

where R_1 is the reflectivity of the output mirror. The efficiency factors discussed above can be grouped together

$$\eta_{\mathrm{out}} = \eta_{\mathrm{B}}\eta_{\mathrm{F}}\eta_{\mathrm{ex}}\eta_{\mathrm{R}} \quad ; \tag{3.84}$$

therefore,

$$E_{\mathrm{out}} = \eta_{\mathrm{out}}E_{\mathrm{st}} \quad . \tag{3.85}$$

After having discussed the various steps involved in the pump chain of a laser-diode-pumped solid-state laser we will give specific examples of the performance characteristics of such systems in Sect. 3.6.

3.5 Output Fluctuations

So far in this chapter we have considered only the steady-state behavior of the laser oscillator. Let us now consider some aspects of transient or dynamic behavior. Generally speaking, one can classify the output fluctuations which occur in pulsed and cw-pumped solid-state lasers in the following categories:

Relaxation Oscillation. This is by far the most predominant mechanism causing fluctuations in the output of a solid-state laser. Instead of being a smooth pulse, the output of a pumped laser is comprised of characteristic spikes. In cw-pumped solid-state lasers the relaxation oscillations, rather than causing spiking of the output, manifest themselves as damped, sinusoidal oscillations with a well-defined decay time. Relaxation oscillations will be discussed in more detail in the next section.

Mode Beating. This type of amplitude modulation, readily observed on the envelope of a Q-switched pulse or on continuous lasers, is caused by the interaction of transverse or longitudinal modes. The output is modulated with the beat frequency of adjacent modes. Partial mode locking is another source of amplitude modulation. These phenomena will be discussed in more detail in Chap. 5. At this point we will only note that operating the laser in a single transverse and longitudinal resonator mode will eliminate this type of modulation.

Parasitic Noise. Under this term we can include all external disturbances such as ripple and random fluctuations in the output of the pump source, mechanical vibrations introduced to the laser, variation in the cooling conditions of the laser rod, turbulence in the air inside the resonator, etc., which lead to an amplitude modulation of a cw laser. Parasitic noise typically causes low-frequency fluctuations in the range from 0 to 100 kHz.

Quantum Noise. The ultimate limit of the output fluctuations in a laser is given by the quantum nature of the emitted light (Sect. 3.5.2).

3.5.1 Relaxation Oscillations

The most notable transient effect in solid-state laser oscillators is the phenomenon of relaxation oscillations.

In flashlamp-pumped ruby and many other solid-state lasers the output is a highly irregular function of time. The output consists of individual bursts with random amplitude, duration, and separation (see, for example, Fig. 3.9). These lasers typically exhibit what is termed "spiking" in their output. We will explain the phenomena of the spike formation with the aid of Fig. 3.4. When the laser pump source is first turned on there are a negligible number of photons in the cavity at the appropriate frequency. The pump radiation causes a linear buildup of excited atoms and the population is inverted.

Although under steady-state oscillation conditions N_2 can never exceed $N_{2,\mathrm{th}}$, under transient conditions the pump can raise N_2 above the threshold level, because no laser oscillation has yet been built up and no radiation yet exists in the cavity to pull N_2 back down by means of stimulated emission.

The laser oscillation does not begin to build up, in fact, until after N_2 passes $N_{2,\mathrm{th}}$, so that the net round-trip gain in the laser exceeds unity. Then, however, because N_2 is considerably in excess of $N_{2,\mathrm{th}}$, the oscillation level will actually build up very rapidly to a value of the photon flux ϕ substantially in excess of the steady-state value for the particular pumping level.

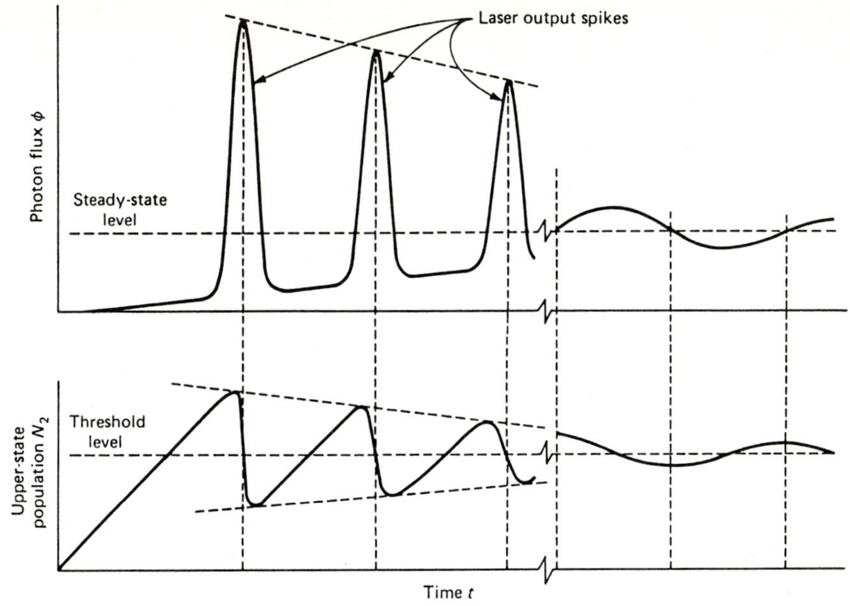

Fig. 3.4. Spiking behavior of a laser oscillator

But, when $\phi(t)$ becomes very large, the rate of depletion of the upper-level atoms due to stimulated emission becomes correspondingly large, in fact considerably larger than the pumping rate W_p. As a result, the upper-level population $N_2(t)$ passes through a maximum and begins to decrease rapidly, driven downward by the large radiation density. The population $N_2(t)$ is driven back below the threshold level $N_{2,th}$; the net gain in the laser cavity becomes less than unity, and so the existing oscillation in the laser cavity begins to die out.

To complete the cycle of this relaxation process, once the radiation level has decreased below the proper steady-state level, the stimulated emission rate again becomes small. At this point the pumping process can begin to build the population level N_2 back up toward and through the threshold value again. This causes the generation of another burst of laser action, and the system can again go through a repeat performance of the same or a very similar cycle.

Turning now to the rate equation, we can interpret these curves as follows: At the beginning of the pump pulse we can assume that the induced emission is negligible because of the low photon density. During this time we may neglect the term containing ϕ in (1.53) and write

$$\frac{dn}{dt} = W_p n_{tot} \quad . \tag{3.86}$$

The population inversion therefore increases linearly with time before the development of a large spiking pulse. As the photon density builds up, the stimulated emission terms become important and for the short duration of one pulse the effect of the pumping can be neglected. Therefore, during the actual spiking pulse the rate equations can be written by neglecting both the pumping rate for the excess population and the cavity loss rate in (1.58, 61):

$$\frac{dn}{dt} = -\gamma c\sigma n\phi \quad , \quad \frac{d\phi}{dt} = +c\sigma n\phi \quad . \tag{3.87}$$

The photon density thus grows with time and the population inversion decreases with time. The photon density reaches a peak when the decreasing inversion reaches the threshold value n_{th}. The inversion reaches a minimum for which we have $\gamma nc\sigma\phi \approx W_{\text{p}}n_{\text{tot}}$, at which point the pump is able to maintain the then-existing small population inversion. The cycle repeats itself, forming another spike. The inversion fluctuates in a zigzag fashion around the threshold value n_{th}. As time passes, the peaks become smaller and the curve becomes damped sinusoidal.

The computer solutions of the laser rate equations predict a train of regular and damped spikes at the output of the laser [3.3–5]. Most lasers, however, show completely irregular, undamped spikes. This discrepancy between theory and experiment is due to the fact that the spiking behavior dies out very slowly in many real solid-state lasers and therefore persists over the complete pump cycle. Furthermore, mechanical and thermal shocks and disturbances present in many real lasers act to continually reexcite the spiking behavior and keep it from damping out. Hence many lasers, especially the ruby laser, spike continuously without ever damping down to the steady state. Depending on the system parameters such as mode structure, resonator design, pump level, etc., the spiking may be highly irregular in appearance or it may be regular. Conditions for regular spiking in pulsed ruby lasers are summarized in [3.5–6].

Spike Suppression in Solid-State Lasers

For some laser applications it is desirable to have a smooth output pulse with very little modulation. Nonspiking operation can be obtained by introducing active or passive feedback into the resonator or by use of nonresonant laser oscillators. Active feedback has been achieved in ruby lasers [3.7–11] by sampling a portion of the output beam with a photodiode and using the detector signal to control the voltage applied to an electro-optic shutter such as a Pockels or Kerr cell located inside the resonator. In order to obtain satisfactory spike and fluctuation suppression, the delay time between the detector and the shutter must be much smaller than the spike width. Since the individual spikes have a duration of approximately 0.1–1 μs, this requires a bandwidth of at least 20–50 MHz for the feedback system. Spike-free operation for a Nd : glass laser system has been achieved by mutual coupling of two oscillators [3.12]. The mechanism of stabilization of the emission is provided by coupling of the resonator Q factors. Since the Q of an active resonator depends on the gain, one can recognize in this mechanism a certain analogy with stabilization of laser emission by negative feedback.

Passive negative feedback using a Rayleigh-active material has been used successfully to suppress ruby laser spikes [3.13]. This was achieved by placing a cell containing benzene inside the laser resonator. Stimulated Rayleigh scattering, a nonlinear effect that slightly shifts the frequency of some incident photons, occurs in the benzene due to a reorientation of the liquid's molecules by the light's intense electric field. The result is a small intensity-dependent loss that is used to eliminate spiking.

In a nonresonant laser system [3.14, 15], the laser rod is operated without cavity mirrors. The emitted spontaneous emission has a smooth envelope. The operation of such a system depends critically on elimination of the off-axis spontaneous emission through the use of relatively long and small diameter rods. Since the output energy in a given beam angle will be very small due to the poor directionality of the spontaneous emission, these types of oscillators are usually followed by amplifiers.

In cw-pumped lasers, such as Nd:YAG, the relaxation oscillations are much weaker and usually consist of damped sinusoidal oscillations around the steady-state value. These oscillations may be treated as perturbations of the steady-state population inversion and photon density given in the rate equations (1.58, 61). Compared to the fluorescence time τ_f, the relaxation oscillations have a much shorter period, therefore the term including τ_f in the rate equations can be ignored (that is, $\tau_f \to \infty$).

We now introduce a small perturbation Δn into the steady-state value of the population inversion n; similarly, a perturbation $\Delta\phi$ is introduced into the steady state of the photon density ϕ. Thus we may write

$$\tilde{n} = n + \Delta n \quad \text{and} \quad \tilde{\phi} = \phi + \Delta\phi \quad . \tag{3.88}$$

We now proceed to eliminate the population inversion n from (1.61). This is done by first differentiating the equation and then substituting $\partial n/\partial t$ from (1.58). The differential equation is then linearized by introducing n and ϕ from (3.71). Neglecting products of $(\Delta n \Delta\phi)$, we finally obtain

$$\frac{d^2(\Delta\phi)}{dt^2} + c\sigma\phi\frac{d(\Delta\phi)}{dt} + (\sigma c)^2\phi n(\Delta\phi) = 0 \quad . \tag{3.89}$$

The solution of this equation gives the time variation of the photon density

$$\Delta\phi \approx \exp\left(-\frac{\sigma c\phi}{2}\right)t \sin\left[\sigma c(\phi n)^{1/2}t\right] \quad . \tag{3.90}$$

The frequency $\omega_s = \sigma c(\phi n)^{1/2}$ and the decay time constant $\tau_R = 2/\sigma c\phi$ of this oscillation can be expressed in terms of laser parameters by noting that $I = c\phi h\nu$ and $n = 1/c\sigma\tau_c$. Therefore

$$\omega_s = \sqrt{\frac{\sigma I}{\tau_c h\nu}} \quad \text{and} \quad \tau_R = \frac{2h\nu}{I\sigma} \quad . \tag{3.91}$$

These expressions can be further simplified for the case of a four-level system by introducing the saturation power density I_s, leading to

$$\omega_s = \sqrt{\frac{I}{I_s \tau_f \tau_c}} \quad \text{and} \quad \tau_R = 2\tau_f \left(\frac{I_s}{I}\right) \quad . \tag{3.92}$$

Note that the greater the circulating power density I and therefore the output power from the laser, the higher the oscillation frequency. The decay time τ_R will decrease for higher output power.

From these equations it follows that the damping time is proportional to the spontaneous lifetime. This is the reason that relaxation oscillations are observed mainly in solid-state lasers where the upper-state lifetime is relatively long. Investigations of the output fluctuations of cw-pumped Nd:YAG lasers can be found in [3.16–23] (see also Sect. 3.6). Recently, relaxation oscillations in an actively modelocked Nd:YLF laser have been eliminated by electronic feedback stabilization [3.24].

3.5.2 Quantum Noise

The fundamental limit of fluctuations in the laser output is set by the spontaneous emission. The spontaneous emission noise power is approximately given by the fluorescence power at threshold,

$$P_f = \frac{N_{2,\text{th}} h\nu}{\tau_{21}} \quad . \tag{3.93}$$

This noise power corresponds to an effective noise temperature T_n,

$$T_n = T \frac{h\nu}{kT \ln 2} \quad . \tag{3.94}$$

Taking as an example a Nd:YAG laser operating at 1.06 μm and 290 K, the ratio $h\nu/kT \approx 50$ so that $T_n \sim 15{,}000$ K. The noise temperature is related to the noise figure F by

$$F = \frac{1 + T_n}{290\,\text{K}} \quad . \tag{3.95}$$

The quantum nature of radiation causes random fluctuations in the output of optical detectors which can be described by the shot noise formula. The basic photoelectric process in a square-law detector is characterized by a linear proportionality between the number of incident photons and the number of electrons produced. In terms of optical power P and primary photocurrent I_s, the proportionality becomes

$$I_s = \frac{\eta e P}{h\nu} \quad , \tag{3.96}$$

where e is the magnitude of the electronic charge and η is the quantum efficiency of the detector. In a shot noise-limited detector the ratio of input signal to input noise power is proportional to the square root of the average number of received photons per second, or

$$I_n = \sqrt{2eI_s\Delta\nu} \quad , \tag{3.97}$$

where $\Delta\nu$ is the postdetection bandwidth and I_n is the noise current. Combining (3.79 and 80), one obtains

$$I_n = \sqrt{\frac{2e^2\Delta\nu\eta P}{h\nu}} \quad \text{and} \quad \frac{I_s}{I_n} = \sqrt{\frac{\eta P}{2\Delta\nu h\nu}} \tag{3.98}$$

for a quantum-noise-limited detection system.

3.6 Examples of Regenerative Oscillators

In this section we will discuss the performance characteristics of typical ruby, Nd : glass, and Nd : YAG oscillators.

3.6.1 Ruby

Figure 3.5 shows a photograph of the head of a typical commercially available ruby laser. The primary components of the laser head are the laser rod, a helical flashlamp, and a reflector. The laser rod is surrounded by a close-coupled helical flashlamp. A cylindrical reflector around the flashlamp aids in efficiently directing the flashlamp pump light into the laser rod. The entire laser housing, including the region occupied by the flashlamp, is filled with cooling water. Readily available laser heads contain ruby rods ranging in size from 7.5 cm by 1 cm to 20 cm by 2 cm. Depending on the size of the flashlamp, input energies are between 3 and 20 kJ. In addition to the laser head, shown in Fig. 3.5, a complete laser system contains the following modules: an optical rail which supports the laser head, resonator mirrors, and other optical components; a power supply; a flashlamp trigger unit; an energy-storage capacitor bank; and a water cooler.

We will now examine the performance of an oscillator composed of a ruby rod (0.05 % doping) which is 10.4 cm long and 0.95 cm in diameter. The optical resonator consists of two flat mirrors with a separation of 71 cm. The performance of the system around threshold as a function of output coupling is shown in Fig. 3.6. As one can see, a mirror reflectivity around 50 % is about optimum for this system. A reflectivity considerably higher than this value results in an output versus input curve which has a very low slope efficiency. On the other hand, a reflectivity which is appreciably lower than 50 % results in a very high threshold. The curvature in the curves near the threshold is caused by nonuniform pumping. More pump light is absorbed in the areas immediately below the surfaces than in the center of the rod. Therefore the rod starts to lase with a ring-like structure. At a somewhat higher energy input the full cross section of the rod will start to lase. The curve for $R = 0.92$ is straight even at threshold, which indicates that at this high reflectivity, pumping uniformity is insignifi-

Fig. 3.5. Cutaway of a ruby laser head

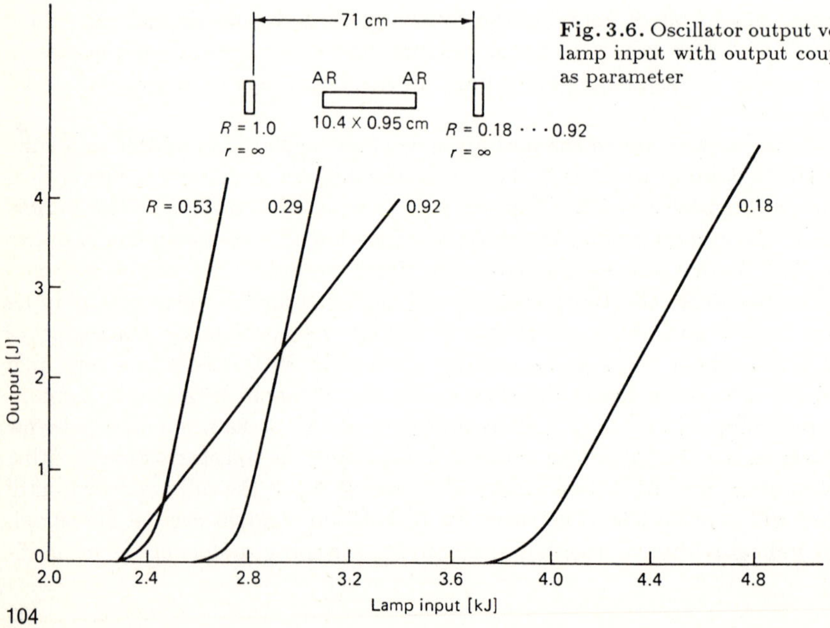

Fig. 3.6. Oscillator output versus lamp input with output coupling as parameter

cant and the whole rod starts to lase almost uniformly. From these curves we can calculate the parameters K and L according to (3.67). Taking, for example, the curve with the 92 % mirror, we have $R_1 = 0.92$. The slope efficiency of this curve is $\sigma_s = 3.6 \times 10^{-3}$ and the rod has a length of 10 cm and a volume of $V = 7.8\,\mathrm{cm}^3$. In Chap. 2 we calculated the upper-level energy density at inversion $E_{ui} = 2.18\,\mathrm{J/cm}^3$, therefore the total stored energy is $E_{ui} = 17\,\mathrm{J}$. We need also the absorption coefficient of ruby, which is $\alpha_0 = 0.2\,\mathrm{cm}^{-1}$. Introducing these values into (3.67) yields $L = 10\,\%$ and $K = 4.7 \times 10^{-4}\,\mathrm{J}^{-1}$. Therefore the gain in this rod as a function of input energy can be expressed as

$$g_0 = 0.2\frac{0.47E_{in}[\mathrm{kJ}] - 1}{0.47E_{in}[\mathrm{kJ}] + 1}\ [\mathrm{cm}^{-1}] \quad .$$

From the value for K it follows that inversion in this rod occurs at the flashlamp input energy of $E = 1/K = 2.1\,\mathrm{kJ}$.

The performance of the same oscillator operated at higher input energies and with the optimum output coupler is shown in Fig. 3.7 for a large number of rods. The curves show energy output versus flashlamp input for different rod sizes, doping levels, and pump pulse length. As the curves illustrate, rods doped with 0.05 % Cr^{3+} yield higher outputs than the lower-doped rods. Furthermore,

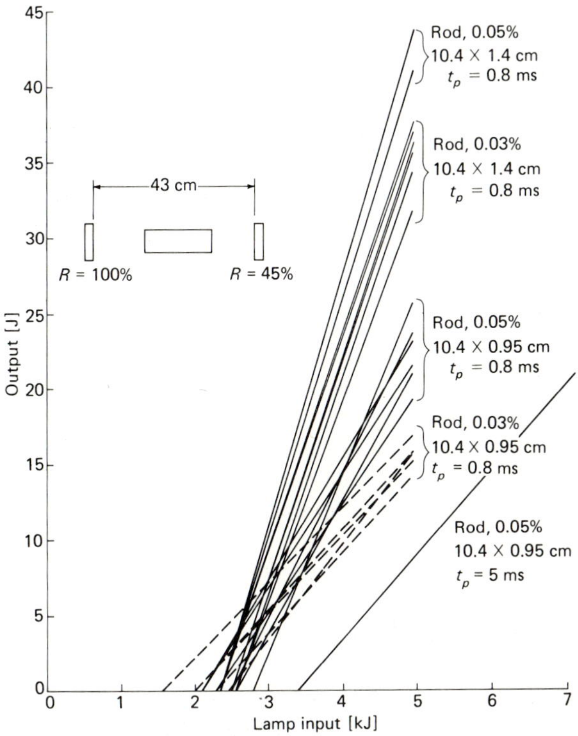

Fig. 3.7. Performance of a ruby oscillator employing different rods

the large-diameter rods provide more output energy for the same input than the smaller rods. This is explained by the fact that rods with a high doping concentration and/or larger diameter absorb more pump light in a diffusely reflecting cavity. As can be seen from Fig. 3.7, the rods with a diameter of 1.4 cm have about the same threshold as the rods with 0.95 cm diameter, although the rod volume differs by a factor of 2.2.

The curve to the far right in Fig. 3.7 was obtained by stretching the pump pulse to 5 ms by adding capacitors and inductors to the pulse-forming network. The higher value of the storage capacitor resulted in a lower bank voltage and a lower lamp peak current for the same input energy than for the shorter lamp pulses. Consequently, the lamp brightness was lower than in the other cases, which explains the relatively poor performance obtained in the long pulse mode. Furthermore, the pump pulse was longer than the spontaneous emission time, which means that in this mode of operation higher fluorescence losses occur.

The performance of an average 10.4-cm by 1.4-cm rod doped with 0.03 % Cr^{3+} can be approximated by

$$E_{\text{out}} = 0.013(E_{\text{in}}[\text{kJ}] - 2.5)[\text{kJ}] \quad .$$

For the 10.4-cm by 0.95-cm rods doped with 0.05 % Cr^{3+}, a typical output versus input curve can be described by

$$E_{\text{out}} = 0.0075(E_{\text{in}}[\text{kJ}] - 2.2)[\text{J}] \quad .$$

If we introduce the values of slope efficiency $\sigma_s = 7.5 \times 10^{-3}$ and threshold energy $E_{\text{th}} = 2.2 \times 10^3$ J together with the other pertinent rod parameters in (3.67), we obtain $L = 0.27$ and $K = 7.3 \times 10^{-4}\,\text{J}^{-1}$. The most important factors which contribute to losses L are the scattering losses, the ruby absorption losses caused by the rod holders which shadow the ends of the rods from the pump radiation, and possible losses in the coatings. The gain as a function of input energy is

$$g_0 = 0.2\frac{0.73E_{\text{in}}[\text{kJ}] - 1}{0.73E_{\text{in}}[\text{kJ}] + 1}[\text{cm}^{-1}] \quad .$$

At maximum input energy of 5 kJ, this rod possesses a small-signal gain coefficient of $g_0 = 0.11\,\text{cm}^{-1}$ and a single-pass gain of $G_0 = \exp(g_0 l) = 3.3$.

In Fig. 3.8 the beam angle at which half of the output energy is contained is plotted versus the total output energy for different ruby rods. The difference in beam divergence between rods of "standard quality" and "superior internal quality" is quite obvious.

The typical time dependence of the oscillator output pulses is shown in Fig. 3.9.

Figure 3.10 shows performance data of 15-cm-long ruby rods of different doping and diameter. In Fig. 3.10a output versus lamp input energy is plotted for two rods of different size and doping level. As can be seen from Fig. 3.10b, the lower-doped rod produces more output energy in beam angles smaller than 4 mrad.

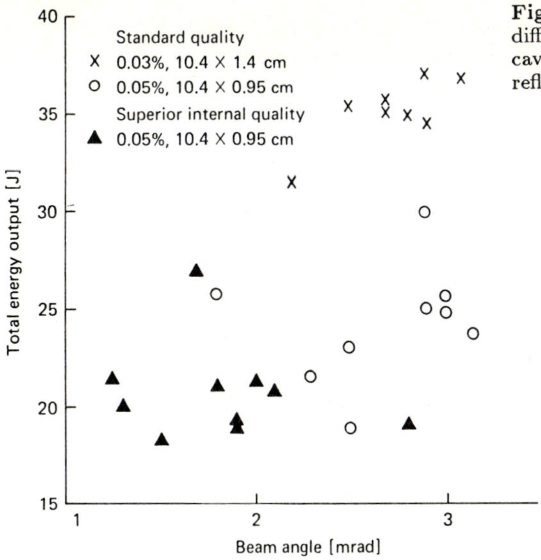

Fig. 3.8. Beam angle obtained from different ruby rods. Lamp input: 5 kJ; cavity length: 43 cm; front mirror: 45 % reflective; pulse width: 800 μs

Energy output versus lamp input curves for ruby oscillators employing even larger rods are plotted in Fig. 3.11. The various curves indicate typical variations of output energy due to different rod qualities. Performance data of different ruby systems normalized to the rod volume are summarized in Table 3.1. As one can see, the input energy per cm^3 of rod required to achieve threshold varies from 92 to 315 J/cm^3, depending on rod size and doping level. The variation for the slope and overall efficiencies is 0.5–1.5 % and 0.3–1.0 %, respectively. Generally speaking, the rods with the smallest diameter and the lowest doping level – being optically thin – require the highest lamp input for threshold and yield the lowest efficiencies. From the data published in this table and from our model derived in Sect. 3.4 we can estimate the various factors which contribute to the overall efficiency of the ruby laser oscillator.

The slope efficiency listed in Table 3.1 is the product of the efficiency factors η_1 and η_5 described in Sect. 3.4. For a typical ruby crystal pumped by a xenon lamp in a diffuse reflecting cavity, the following numerical values can be assumed: $\eta_1 = 0.50$; $\eta_2 = 0.25$; $\eta_3 = 0.30$–0.40; $\eta_4 = 0.30$–0.60; $\eta_5 = 0.70$–0.60. Therefore $\sigma_s = 0.006$–0.015. In a diffusely reflecting cavity a large-diameter rod is able to capture a larger fraction of the pump light than a smaller rod. Furthermore, a large-diameter rod provides a longer absorption path for the pump light and can therefore absorb a larger fraction of the pump radiation. In ruby we assume an average absorption coefficient in the pump bands of $\alpha = 1.5\,\mathrm{cm}^{-1}$ for a 0.05 %-doped rod and $\alpha = 1.0\,\mathrm{cm}^{-1}$ for a 0.03 % Cr^{3+} concentration. The absorbed pump radiation as a function of rod diameter is $P_{ab}/P_{in} = 1 - \exp(-\alpha D)$. A rod doped with 0.03 % which is 1 cm in diameter will absorb 63 % of the incident pump radiation; for a 2-cm-diameter rod doped at 0.05 %, this fraction increases to 95 %. Therefore, η_4 will be larger for large-diameter and/or higher-doped crystals. The factor η_5 depends mainly

(a)

(b)

(c)

(d)

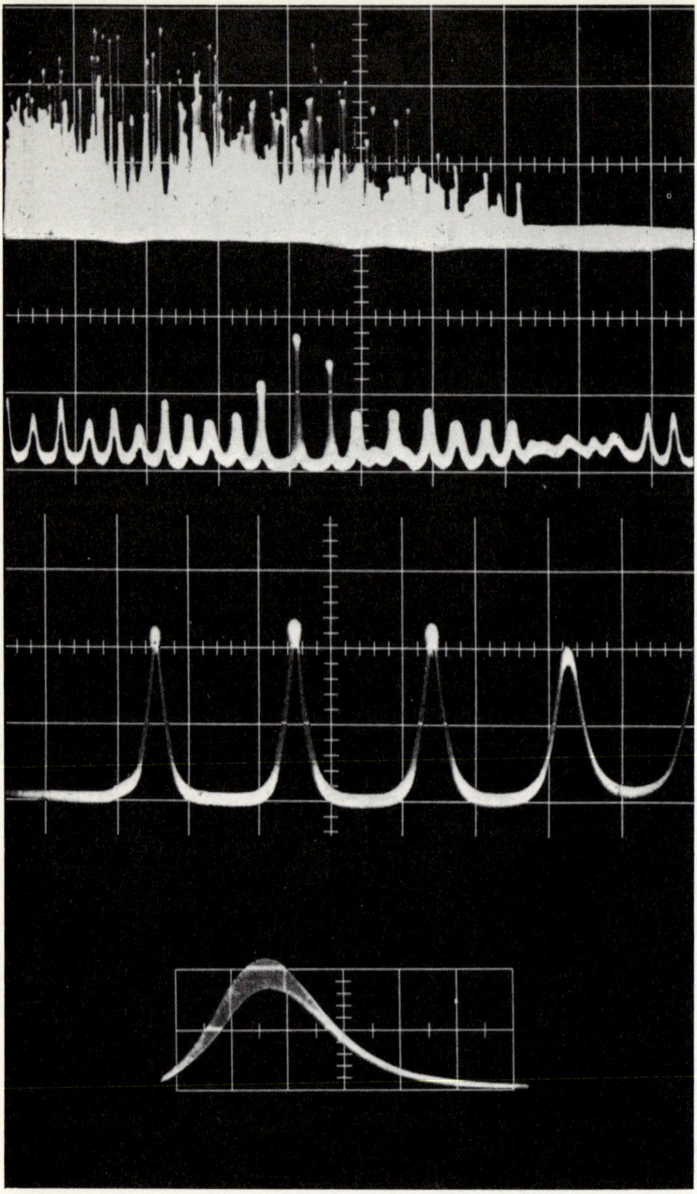

Fig. 3.9a–d. Temporal structure of the ruby laser output. Oscillator is comprised of a 10-cm by 1-cm ruby rod pumped at 4.75 kJ and two flat mirrors separated by 60 cm. Output is 25 J. The upper trace **(a)** shows the complete laser output. The lower traces (b–d) represent portions of the output pulse taken at an expanded scale. The time scale is **(a)** 200 μs, **(b)** 5 μs, **(c)** 1 μs, and **(d)** 100 ns/div. The energy of the individual spike shown in **(d)** is 7.5 mJ

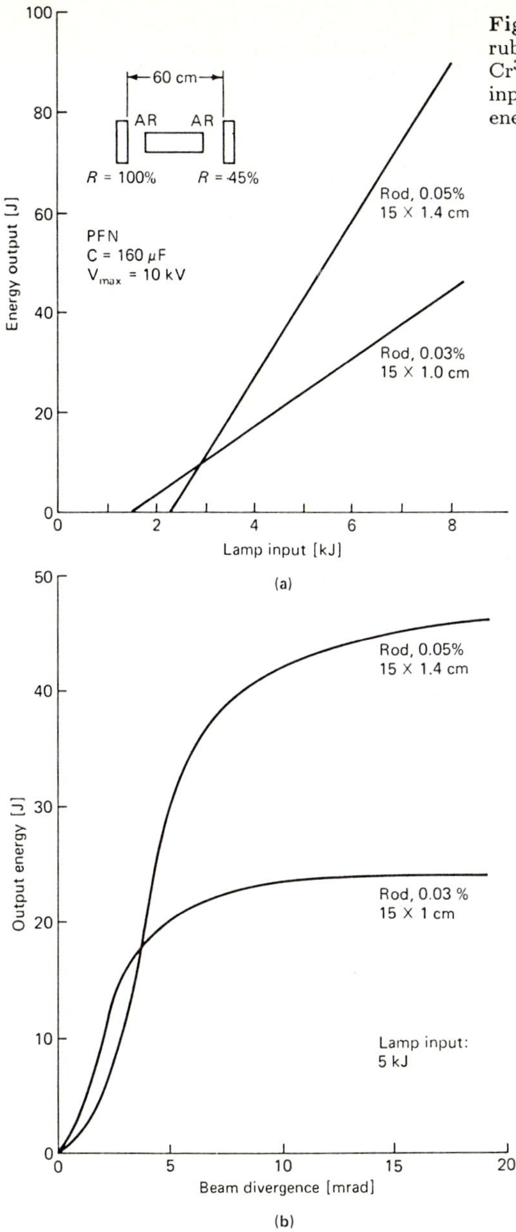

Fig. 3.10. Performance of a conventional ruby oscillator employing rods of different Cr^{3+} concentrations. **(a)** Output versus input. **(b)** Beam divergence versus output energy

on the optical losses in the system. For example, a combined loss of $L = 0.20$ and an output reflector of $R_1 = 0.45$ for the case of a 10-cm-long ruby crystal results in a coupling efficiency of $\eta_5 = 0.61$. The longer rods will have higher scattering and absorption losses, which will decrease η_5. The efficiency factors can be related to the energy levels per cubic centimeter of ruby crystal. For a ruby crystal at population inversion we can write

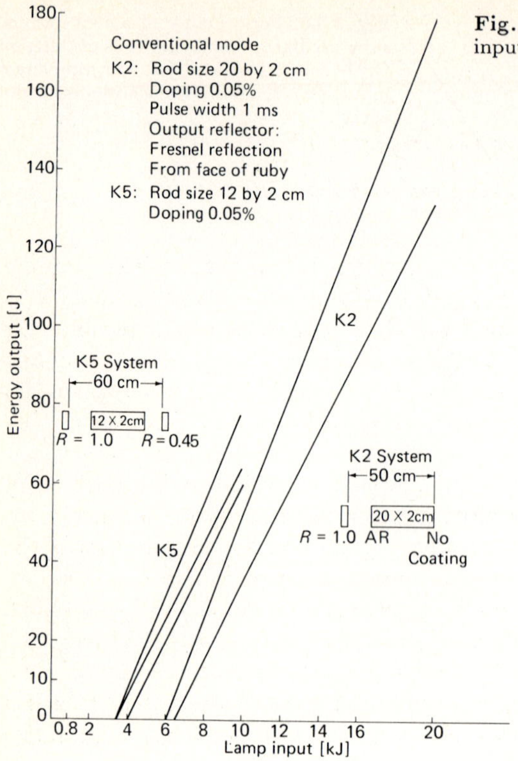

Fig. 3.11. Output energy as a function of input energy for large ruby oscillators

Table 3.1. Conventional-mode ruby oscillators. Ruby rods pumped by helical flashlamps

	Threshold		At maximum input			
Rod size [cm] and doping level	Lamp input/ Rod volume [J/cm^3]	Slope efficiency [%]	Lamp input/ Rod volume [J/cm^3]	Energy output/ Rod volume [J/cm^3]	Overall efficiency	Rod volume [cm^3]
10.4 × 0.95 0.03 %	263	0.5	658	1.98	0.3	7.6
10.4 × 0.95 0.05 %	315	0.8	658	2.99	0.45	7.6
10.4 × 1.4 0.03 %	143	1.3	300	2.03	0.7	16.7
10.4 × 1.4 0.05 %	143	1.5	300	2.45	0.8	16.7
15 × 0.95 0.03 %	148	0.7	740	4.1	0.55	10.8
15 × 1.4 0.05 %	95	1.5	330	3.5	1.0	24.2
12 × 2 0.05 %	92	1.1	256	1.8	0.7	39
20 × 2 0.05 %	94	1.1	307	2.5	0.8	65

Stored energy in upper level: $E_{u,th} = 2.18\,J/cm^3$

Minimum absorbed energy: $E_{ab} = E_{u,th}/\eta_1 = 4.4\,J/cm^3$

Lamp input at inversion: $E_{in} = E_{ab}/\eta_2\eta_3\eta_4 = 70–190\,J/cm^3$

Lamp input at threshold: $E'_{in} = E_{in}/\eta_5 = 100–320\,J/cm^3$

We will conclude this subsection on ruby oscillators by briefly summarizing the performance of a cw-pumped oscillator. Since ruby operates as a three-level system, about one-half of the Cr^{3+} ions must be excited into the upper laser level before inversion is obtained. At population inversion the fluorescence power is, according to Sect. 3.4, $P_f = 726\,W/cm^3$. In order to sustain population inversion, at least $P_{ab} = P_f/\eta_1 = 1.4\,W/cm^3$ of pump power has to be absorbed by the crystal. Assuming the same efficiency factors η_2, η_3, η_4 as we had in the pulsed case, the electrical input power required into the lamp is about $70\,kW/cm^3$ of active material. Since it is not possible to concentrate the power of an incoherent light source into a column smaller than that of the source itself, the necessary pump radiation must be produced in a light source having an extremely high power density. Continuous operation can be expected, therefore, only in very small volumes of ruby. All ruby lasers which have been operated continuously used mercury arc lamps as pump sources [3.25, 26]. In a capillary mercury arc lamp which has typically a 1-mm bore, a wall-stabilized arc contracts toward the center of the tube at an internal pressure of 200 atm and forms an extremely brilliant filamentary light source of 0.3–0.6 mm effective diameter.

A cw-pumped ruby laser, which used a rod 2 mm in diameter and 50 mm in length, generated an output of 1.3 W at an input of 2.9 kW [3.25]. Only a small part of the crystal's cross section was excited by the filament arc, and lasing action occurred only in the small volume of $6 \times 10^{-3}\,cm^3$. Using this value, the lamp input power per unit volume of active material required to obtain threshold is approximately $230\,kW/cm^3$. The individual efficiency factors of the system were estimated to be $\eta_1 = 0.25$, $\eta_2 = 0.04$, $\eta_3 = 0.80$, $\eta_4 = 0.50$, and $\eta_5 = 0.30$. The main reason for the poor overall efficiency was the low absorption of useful pump light by the small lasing volume.

3.6.2 Nd : Glass

We shall now examine the performance of a Nd : glass oscillator. The laser rod, 15 cm long and 1 cm in diameter, is pumped by two linear flashlamps in a highly polished double elliptical cylinder. The flashlamps, matching the laser rod in size, generate a 600 μs-long pulse with a total energy of up to 1 kJ. The oscillator starts to lase at a threshold input energy of 300 J and produces an output of 10 J at an input of 1 kJ. In order to determine the gain and the inherent losses of the oscillator, threshold was measured as a function of mirror reflectivity, as shown in Fig. 3.12. The measurement reveals a total loss in the

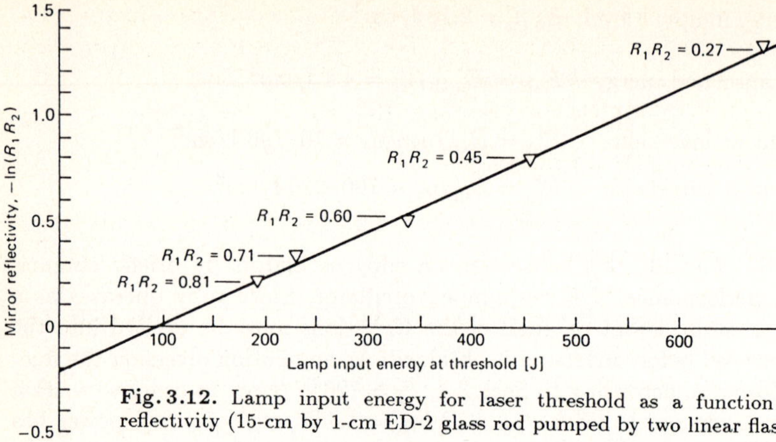

Fig. 3.12. Lamp input energy for laser threshold as a function of mirror reflectivity (15-cm by 1-cm ED-2 glass rod pumped by two linear flashlamps)

resonator of $L = 0.21$ and a value of $1.1 \times 10^{-3}\,\mathrm{J}^{-1}$ for the parameter K. With K measured, we can express the small-signal coefficient as a function of input energy according to $g_0 = KE_{\mathrm{in}}$. For the 15-cm-long laser rod we obtain $g_0 = 7 \times 10^{-5} E_{\mathrm{in}}$ [J]. For example, for a lamp input of 1 kJ, the gain coefficient is $g_0 = 0.07\,\mathrm{cm}^{-1}$ and the single-pass gain is $G_0 = 2.9$.

Typical performance data obtained from water-cooled laser heads containing helical flashlamps and ED-2 glass rods are displayed in Fig. 3.13. The vari-

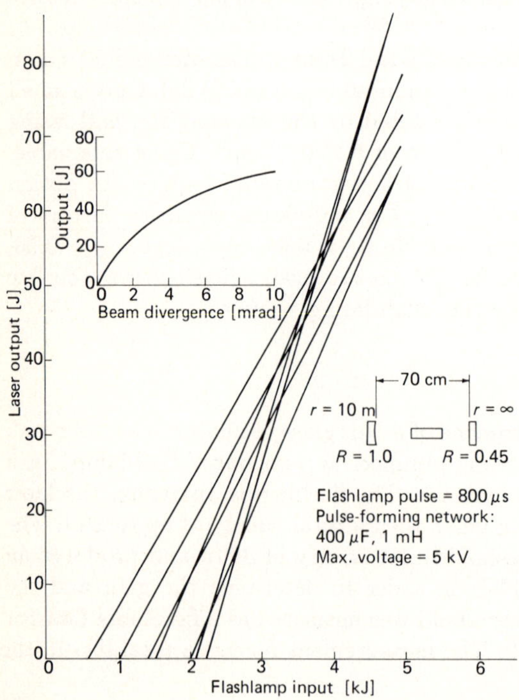

Fig. 3.13. Performance of 15-cm by 1.2-cm Nd : glass rods pumped by a helical flashlamp (pump length 10 cm). Insert: Average beam divergence versus energy output

ous curves obtained by testing eight different glass rods illustrate the spread in performance depending on the glass rod quality. The average rod performances can be characterized by a threshold of 1.5 KJ, a slope efficiency of 2%, and a total efficiency at maximum input of 1.4%, that is,

$$E_{\text{out}} = 0.02(E_{\text{in}}[\text{kJ}] - 1.5)[\text{J}] \quad .$$

Given the distribution of performance data shown in Fig. 3.13, the laser manufacturer's minimum guaranteed performance is 55 J, allowing a margin for component variations and system degradation. The insert shows a measurement of beam divergence at maximum lamp input energy. For most rods, half of the output energy is emitted in a beam with a 3–4-mrad beam divergence. From larger Nd : glass rods pumped at higher input energies, correspondingly larger outputs are obtainable, as is illustrated in Fig. 3.14. For example, from an ED-2 laser rod 25 cm long and 2 cm in diameter pumped at 20,000 J, an output of 200 J can be achieved.

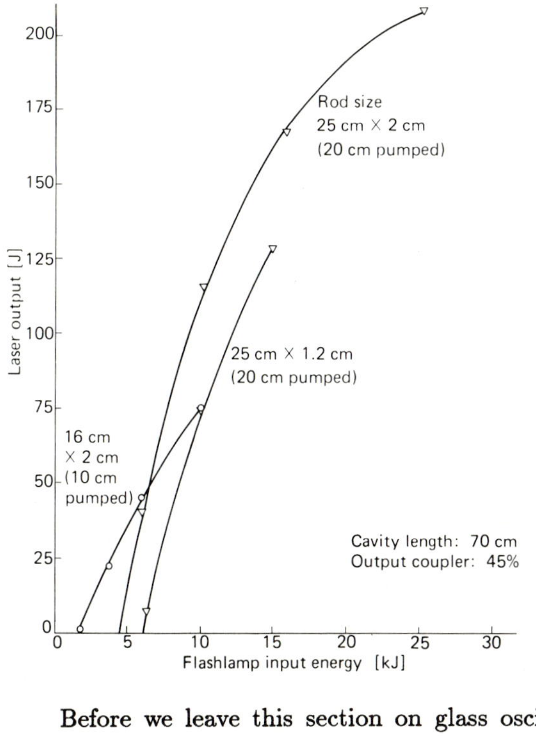

Fig. 3.14. Performance of medium-sized Nd : glass conventional-mode oscillators

Before we leave this section on glass oscillators we will consider a few examples of very large systems. Figure 3.15 shows the performance of a system composed of a 94-cm-long and 1-cm-diameter ED-2 glass rod, a single linear flashlamp (EG&G FX-77-35), and silver foil wrapped closely around the rod and the lamp [3.27].

The highest output energy from a single-element laser oscillator was obtained from a 95-cm-long and 3.8-cm-diameter glass rod pumped by four 1-m

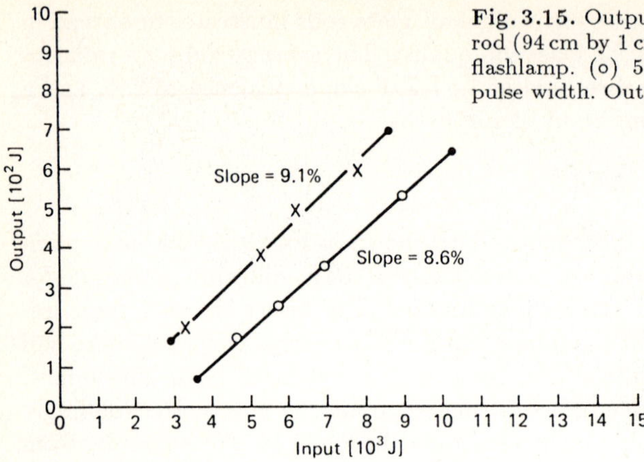

Fig. 3.15. Output versus input of a Nd:glass rod (94 cm by 1 cm) pumped by a single linear flashlamp. (o) 5 ms pulse width; (x) 2.2 ms pulse width. Output reflection 28%

linear flashlamps in a close-wrap configuration. The output from the system was 5 kJ at an input of 180 kJ [3.28].

3.6.3 Nd:YAG

To illustrate the application of the equations which we derived in this chapter, let us consider a cw-pumped Nd:YAG laser. The system contains a laser rod 75 mm in length and 6.2 mm in diameter. The laser rod is pumped by two krypton arc lamps capable of a maximum electrical input power of 12 kW. The laser rod and the pump lamps are contained in a highly polished, gold-plated, double-elliptical cylinder. The major and minor axes of a single ellipse are 38 and 33 mm, respectively. Cooling of the rod and lamps is accomplished by circulating water in flowtubes which surround the crystal and lamps. The interior of the reflector is also water-cooled. The optical resonator is composed of two dielectrically coated mirrors which have a separation of 40 cm. The rear mirror has a concave curvature of 2 m, whereas the front mirror is flat. The krypton arc lamps, having a cold-fill pressure of 2 atm, operate at a maximum current level of 55 A. Phase-controlled SCRs in the power supply permit operation of the lamps at any desired input level.

Figure 3.16 shows a photograph of a typical, commercially available, cw pumped Nd:YAG laser. The laser head contains a single arc lamp. The mechanical shutter is used to stop laser oscillations for short periods of time without having to turn off the arc lamp. The purpose of the mode selector and Q-switch will be discussed in Chaps. 5 and 8. Bellows are employed between each optical element in the laser head in order to seal out dust and dirt particles from the optical surfaces. The laser head cover is sealed with a gasket in order to further reduce environmental contamination.

In Fig. 3.17 the output power of the two lamp systems mentioned above is plotted against lamp input power for different Nd:YAG crystals and front-mirror reflectivities. The highest output from a 7.5-cm-long crystal was achieved with a front mirror of 85% reflectivity. The output versus input curve for this

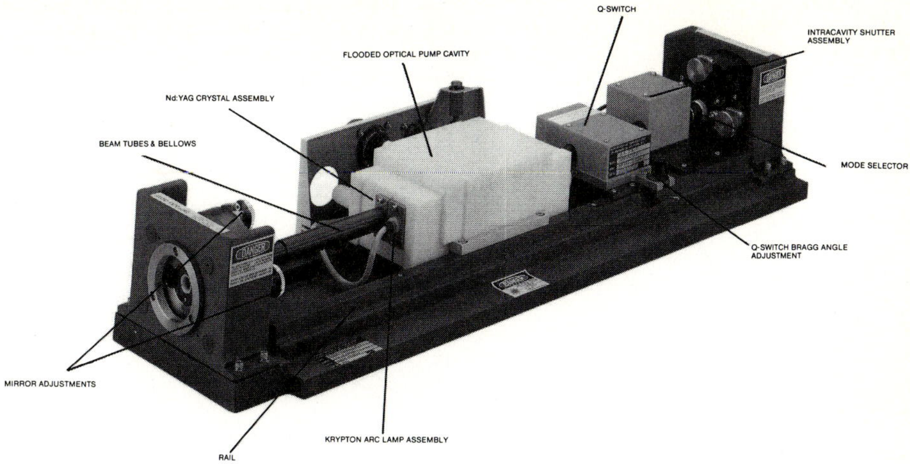

Fig. 3.16. Photograph of a commercial cw-pumped Nd : YAG laser (Quantronix, Series 100, High power laser systems)

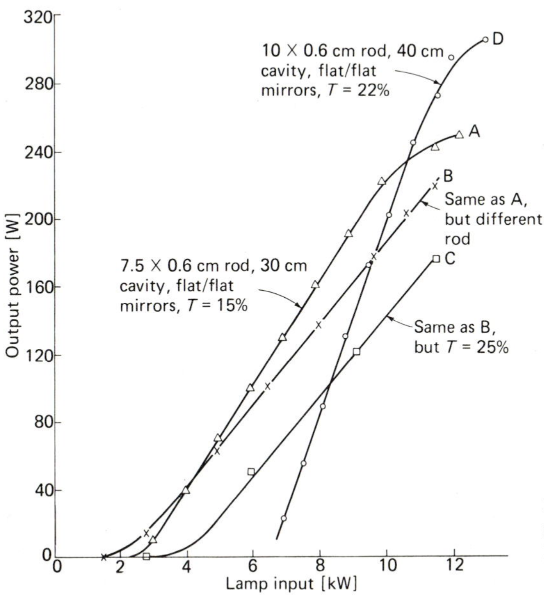

Fig. 3.17. Continuous output versus lamp input of a powerful Nd : YAG laser

mirror shows a slope efficiency of $\sigma_s = 0.026$ and an extrapolated threshold of $P_{th} = 2.8\,kW$. The nonlinear portion of the curve close to threshold is due to the focusing action of the mirror-finished elliptical cylinder. At first, only the center of the rod lases. In Fig. 3.18 the lamp input power required to achieve laser threshold was measured for different mirror reflectivities. If one plots $\ln(1/R_1)$ rather than R_1, one obtains a linear function according to (3.49). From this measurement follows a value of the pumping coefficient of $K = 72 \times 10^{-6}\,W^{-1}$ and a combined loss of $L = 0.075$. With these two

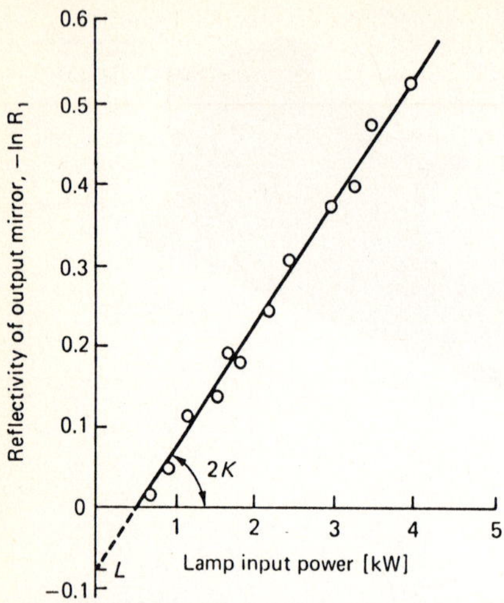

Fig. 3.18. Threshold power input as a function of mirror reflectivity

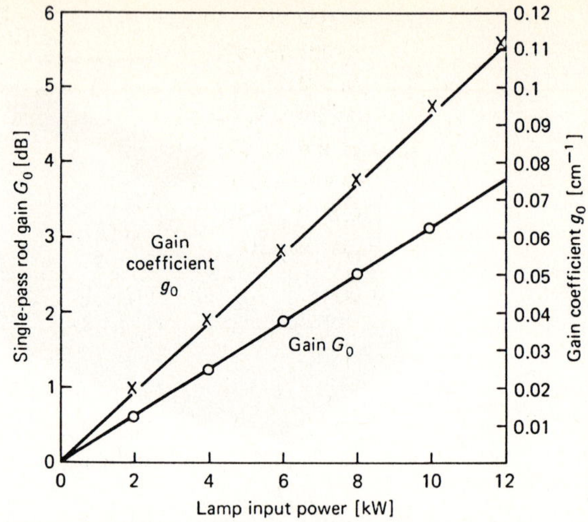

Fig. 3.19. Small-signal, single-pass rod gain and gain coefficient as a function of lamp input

values known, it is possible to plot the small-signal, single-pass rod gain as a function of lamp input power, as shown in Fig. 3.19. Assuming a ground-state population density of $6.0 \times 10^{19}\,\mathrm{cm}^{-3}$ and a stimulated emission cross section of $8.8 \times 10^{-19}\,\mathrm{cm}^2$, it follows from Fig. 3.19 that at maximum lamp input only 0.2% of the total neodymium ion concentration is inverted. At threshold, which occurs at around 2 kW input, the small-signal gain coefficient is about $0.01\,\mathrm{cm}^{-1}$, which corresponds to only a 0.02% inversion of the total ground-state population.

Using (3.21) we can calculate the total fluorescence output of the laser at threshold. With $V = 2.3\,\mathrm{cm}^3$, $\tau_\mathrm{f} = 230\,\mu\mathrm{s}$, $h\nu = 1.86 \times 10^{-19}\,\mathrm{Ws}$, and $n_\mathrm{th} = 1.1 \times 10^{16}\,\mathrm{cm}^{-3}$, one obtains $P_\mathrm{f} = 20\,\mathrm{W}$.

The saturation parameter I_s can be calculated if one introduces the measured values of σ_s, K, A, R_1, and L into (3.56). With $\sigma_\mathrm{s} = 0.026$, $K = 72 \times 10^{-6}\,\mathrm{W}^{-1}$, $A = 0.31\,\mathrm{cm}^2$, $L = 0.075$, and $R_1 = 0.85$, one obtains $I_\mathrm{s} = 810\,\mathrm{W/cm}^2$. The theoretical value for Nd:YAG which follows from (3.29) is $I_\mathrm{s} = 920\,\mathrm{W/cm}^2$.

The numerical values for K, I_s, A, and L were introduced into (3.54–57), and the output power as a function of the front mirror reflectivity was calculated for different lamp input powers. The result of the computerized calculations is shown in Fig. 3.20. As can be seen, relatively broad maxima of the optimum mirror reflectivities are obtained. The mirrors which give the highest output power for the different input powers are located along the dashed curve. This curve can also be obtained from (3.59). Also indicated in Fig. 3.20 are the experimentally determined reflectivities which gave the highest output. Figure 3.21 shows the intracavity power density as a function of laser output power for a fixed lamp input power of 12 kW. The parameter is the reflectivity of

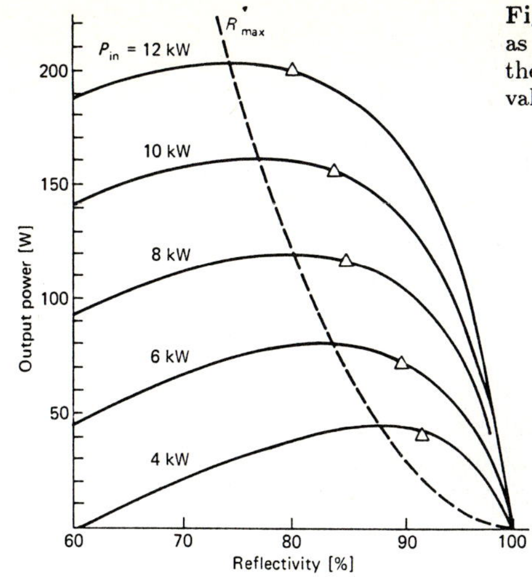

Fig. 3.20. Optimum output-mirror reflectivity as a function of lamp input power. Line shows theoretical values; triangles are experimental values

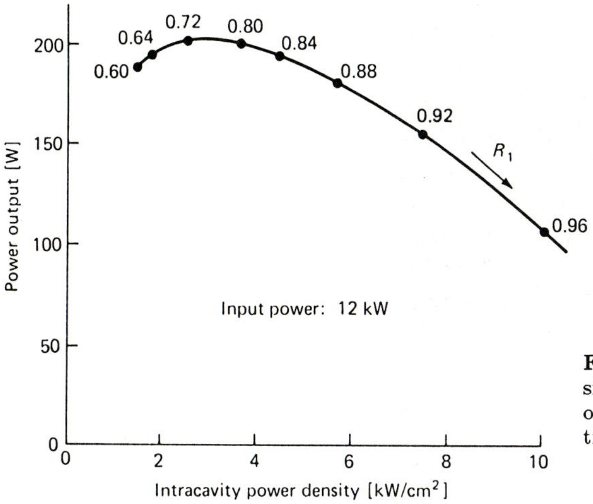

Fig. 3.21. Intracavity power density as a function of laser power output. Parameter is the reflectivity of the output coupler

the front mirror. This curve is obtained from Fig. 3.20 (uppermost curve) and using (3.41). As we can see from this figure, the circulating power density in the cavity increases for the higher reflectivities despite the reduction of output power.

Like most solid-state lasers, Nd:YAG exhibits relaxation oscillation. Figure 3.22 exhibits oscilloscope traces of the relaxation oscillations of a small cw-pumped Nd:YAG laser. This laser has a 3-by-63-mm crystal pumped by two 1000-W halogen-cycle tungsten filament lamps in a double-elliptical reflector configuration. The maximum output which can be achieved with this system is about 10 W at an input of 2 kW. The optical resonator is formed by two 5-m concave mirrors which are 35 cm apart.

In practice, relaxation oscillations in solid-state lasers have amplitudes and intervals between oscillations which appear to be almost random. Most

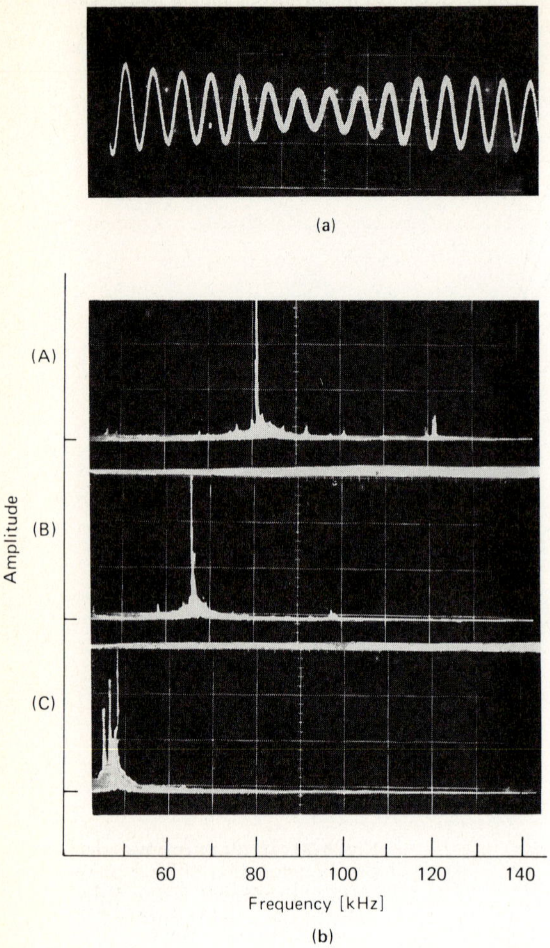

Fig. 3.22a,b. Relaxation oscillation of a cw-pumped Nd:YAG laser. **(a)** Oscilloscope trace showing the temporal behavior of a relaxation oscillation: time scale: 20 μs/div. **(b)** Frequency spectrum of relaxation oscillations at different output power levels: *(A)* 1.3 W, *(B)* 1.0 W, *(C)* 0.25 W

of the irregularities are caused by multimode operation, and mechanical instabilities of the resonator, and pumping nonuniformities. Under very carefully controlled conditions and single-mode operation, regular periodic relaxation oscillations can be observed. Figure 3.22a shows an oscilloscope trace of a relaxation oscillation. The oscillation is a damped sine wave with only a small content of harmonics. Figure 3.22b displays the spectrum of the relaxation oscillations, as obtained by a spectrum analyzer. From (3.74), it follows that the resonant frequency is proportional to $(P_{\text{out}})^{1/2}$ of the laser. Figure 3.22b illustrates this dependence. With (3.74, 75) we can calculate the center frequency f_{s} and the time constant τ_{R} of the relaxation oscillations. With the laser operated at 1 W output, a mirror transmission of $T = 0.05$, and a beam diameter of 0.12 cm, one obtains $I = 2.7 \times 10^3\,\text{W/cm}^2$; $l = 35\,\text{cm}$ is the length of the cavity, and $L = 0.03$ are the combined cavity losses. With these values, and $I_{\text{s}} = 920\,\text{W/cm}^2$ and $\tau_{\text{f}} = 230\,\mu\text{s}$, it follows that $\tau_{\text{R}} = 150\,\mu\text{s}$ and $f_{\text{s}} = 90\,\text{kHz}$.

In cw-pumped systems one can reduce the amplitude fluctuations caused by relaxation oscillations by orders of magnitude by proper design procedures, which we will discuss in Chap. 5.

The largest Nd:YAG crystals which are readily available are about 10 cm long and 1 cm in diameter, although rods of up to 15 cm have been fabricated. The output from a 10-cm-long crystal pumped by a helical flashlamp is shown in Fig. 3.23. The overall efficiency of the system at maximum input is between 0.8 and 1% depending on the particular crystal. The output energy as a function of beam angle is plotted in the insert of Fig. 3.23. As can be seen, for most crystals half of the total output energy is contained in a beam angle between 3 and 4 mrad, and 90% of the energy is contained in a 6- to 8-mrad angle.

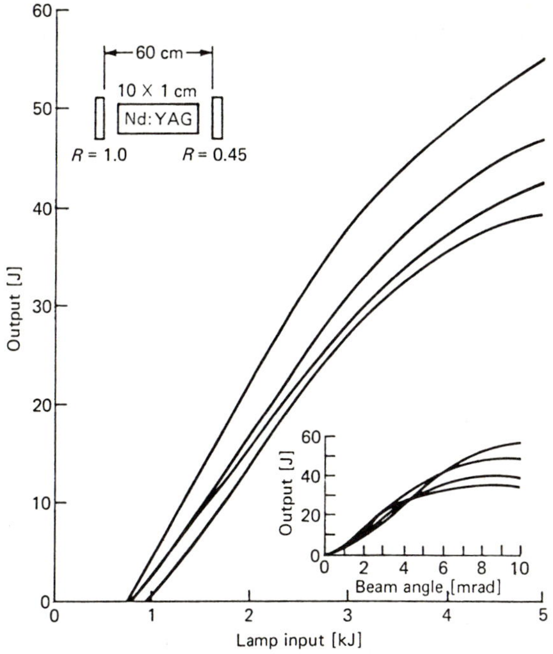

Fig. 3.23. Performance of a pumped pulsed Nd:YAG laser oscillator. Pulse width is 1 ms. Major illustration shows output versus input. Insert shows beam divergence versus output

3.6.4 Laser-Diode-Pumped Regenerative Oscillator

As will be discussed in Chap. 6, the main attraction of laser diodes over arc lamps in pumping a solid-state laser is the efficient excitation of ground-state ions very closely to the upper-state laser level. Besides the excellent spectral match, one other advantage of the diode laser over flashlamps is that the diode's output can be collimated and focused. In this section we will concentrate on one particular type of oscillator which takes advantage of the spectral as well as the spatial properties of laser diodes.

In this so-called end-pumped configuration, the radiation from a laser-diode array is focused to a small spot on the end of a laser rod. With a suitable choice of focusing optics, the diode pump radiation can be varied from 50–

100 μm in diameter to coincide with the diameter of the TEM_{00} resonator mode. The pump radiation penetrates deeply into the rod. The end pumping configuration thus allows the maximum use of the energy from the laser diodes.

Using this longitudinal pumping scheme, the fraction of the active laser volume excited by the diode laser can be matched quite well to the TEM_{00} lasing volume. A solid-state laser pumped in this manner runs naturally in the fundamental spatial mode without intracavity apertures. This end-pumping concept was the subject of considerable interest during the mid 1970s for use as transmitters for optical-fiber communications [3.29, 30]. Only recently, however, have laser diodes of sufficient output power been available to fully exploit this highly efficient regime of operation.

Endpumping of a Nd:YAG crystal with a powerful GaAlAs laser-diode array, operating at about 200 mW output at 810 μm, by *Sipes* [3.31] produced a solid-state laser with an unprecedented 7% overall efficiency. The work by *Sipes* caused a great deal of excitement and a number of systems have been developed commercially utilizing the end-pumped and mode matching concept. As an illustration of the performance of an endpumped oscillator we will use the data from the original work.

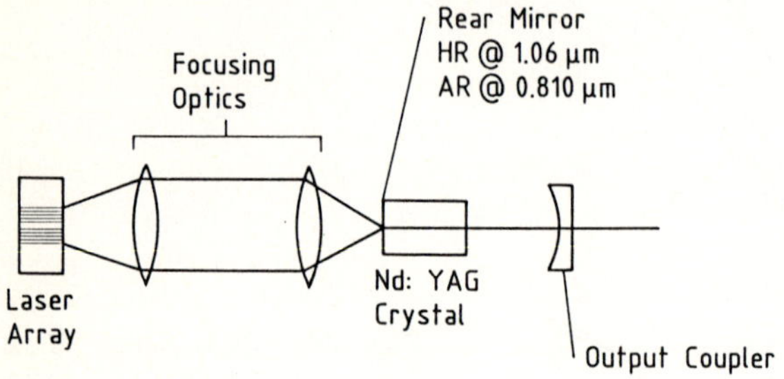

Fig. 3.24. Schematic diagram of a laser-diode array end-pumped Nd:YAG laser [3.31]

The experimental setup is illustrated in Fig. 2.24. It utilizes a single Spectra Diode Labs Model SDL-2410-A GaAlAs laser-diode array operating at approximately 200 mW cw output at 810 μm with approximately 20% electrical to optical efficiency. The dual lobed output is then collimated and focused into a 1 cm long $\times$ 0.5 cm diameter 1% Nd:YAG sample. The resonator configuration is plano-concave, with the pumped end of the Nd:YAG rod being coated for high reflection at 1.06 μm, and with an output coupler having a 5 cm radius of curvature and a reflectivity at 1.06 μm of 95%.

This configuration has a number of advantages compared to side-pumped laser rods: First, the absorption length can be made as long as necessary to absorb practically all the pump light. Secondly, the pump light can be focused to provide the intensities needed for efficient lasing. Finally, the beams can be adjusted to overlap for optimum mode matching.

The efficiency with which pump photons are converted to lasing photons can be calculated as a function of input pump power, cavity loss and pump and mode radii from the steady-state rate equations described earlier.

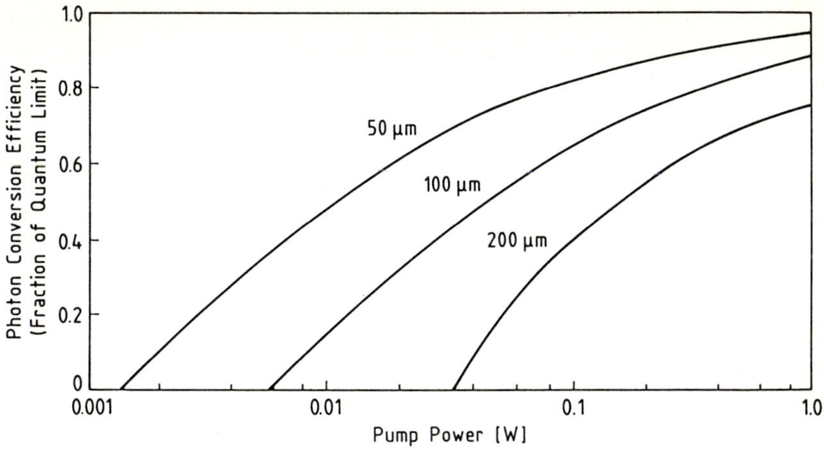

Fig. 3.25. Photon conversion efficiency as a function of pump power for various focused pump spot sizes [3.31]

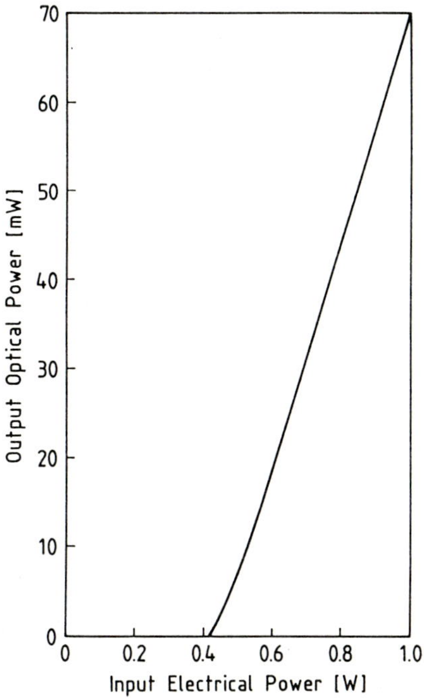

Fig. 3.26. Nd : YAG laser output power versus laser-diode pump power [3.31]

These equations have been solved numerically [3.32] to find the absorbed photon to laser output photon conversion efficiency as a function of pump power for various focused pump spot sizes. Figure 3.25 illustrates that in the case where the modes are matched, the photon conversion efficiency approaches the quantum efficiency (the ratio of pump wavelength to lasing wavelength) for an input power of one Watt and a pump spot size of around 50 μm (a single pass loss of 1 % and $\sigma = 7.6 \times 10^{19}\,\mathrm{cm}^2$ were assumed).

Figure 3.26 shows input electrical power versus 1.06 μm output power for the configuration illustrated in Fig. 3.24. We see that for approximately 1 W of electrical input power, 70 mW of Nd : YAG output is measured. This corresponds to a measured laser diode efficiency of approximately 20 % and an optical conversion efficiency of approximately 35 %.

As mentioned earlier, in order to achieve this kind of performance, two key elements in the design of the laser-diode array-pumped solid-state laser are longitudinal pumping of the active medium, and careful design of the laser resonator to provide good overlap between the pumped region and the TEM_{00} laser-mode volume. The outputs of several laser diode arrays can be combined in the longitudinal pumping geometry by using polarization beam combining techniques and pumping from both ends, thus multiplying the output power.

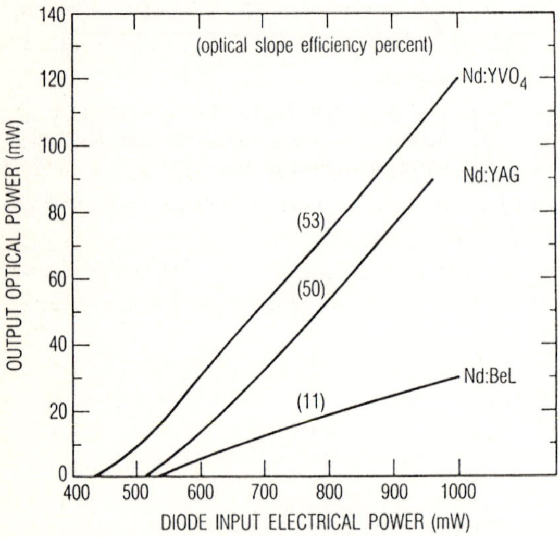

Fig. 3.27. Laser output as a function of pump laser diode electrical input. Optical slope efficiency percents are in parentheses. (YAG and YVO_4 at 97 % and BeL at 99 % output couplers) [3.33]

Laser-diode end-pumped cw operation has been extended to a number of laser materials in the meantime. Figure 3.27 compares the output from Nd : BEL, Nd : YAG and Nd : YVO_4 [3.33]. The spectral differences of the materials studied have a considerable impact on laser threshold and slope efficiency. The comparative power curves of Fig. 3.27 were compiled based on the optimized value of output coupler and diode-pump wavelength for each crystal. Nd : YVO_4 clearly exhibits the lowest threshold despite its larger optical losses and short fluorescence lifetime (95 μs). The 12 % overall efficiency of this laser is the highest number reported for a solid-state laser to date. As follows from Fig. 3.27 an output of 120 mW for 1 W electrical input was obtained.

In recent years laser action by means of laser-diode end-pumping has been demonstrated in a number of solid-state laser materials and in a number of configurations such as Q-switching, frequency doubling and mode-locking. For example, laser action was reported in Nd : Glass at 1.05 μm [3.34], Nd : YLF at 1.05 μm [3.35], HoErTm : YAG at 2.1 μm [3.36], in Nd : YAG at 946 nm [3.37], in Ho : YLF at 2.06 μm [3.38], in Tm : YLF at 2.31 μm [3.39], and in Q-switched Nd : YAG [3.40], and Nd : YLF [3.41], as well as in intracavity frequency doubled Nd : YAG [3.40], and mode-locked Nd : Glass [3.42].

In the diode-pumped lasers discussed so far, the active element is a thin slab or rod a few millimeters long. An extreme case of diode-array end-pumped laser is the fiber laser.

The development of low-loss rare-earth-doped fibers has led to the construction of a number of single-mode fiber lasers [3.43–54]. Due to the small volume of the active core, low threshold and efficient operation has been achieved.

Single-mode fiber lasers possess a number of advantages over their bulk counterparts. As a consequence of the high pump intensity within the small ($> 8\,\mu$m) core, very low threshold ($100\,\mu$W) and large gains can be achieved. Moreover, the small fiber diameter minimizes the thermal effects which plague bulk-glass lasers, and high levels of pump power can be absorbed. Silica, the laser medium, has good power-handling properties and, in addition, broadens the rare-earth transitions, thus enabling tunable lasers [3.45, 53] and broadband amplifiers to be constructed.

As a result of the above attributes, laser action in fibers can be observed using a number of less commonplace rare-earth dopants and transitions, even when the transitions involved are weak. Moreover, continuous laser operation is possible in three-level laser systems such as erbium which have previously only operated in a pulsed mode [3.52, 54].

A typical fiber-laser configuration is shown in Fig. 3.28. For Nd^{3+} doped fibers, a lasing threshold as low as $100\,\mu$W can be obtained using a semiconduc-

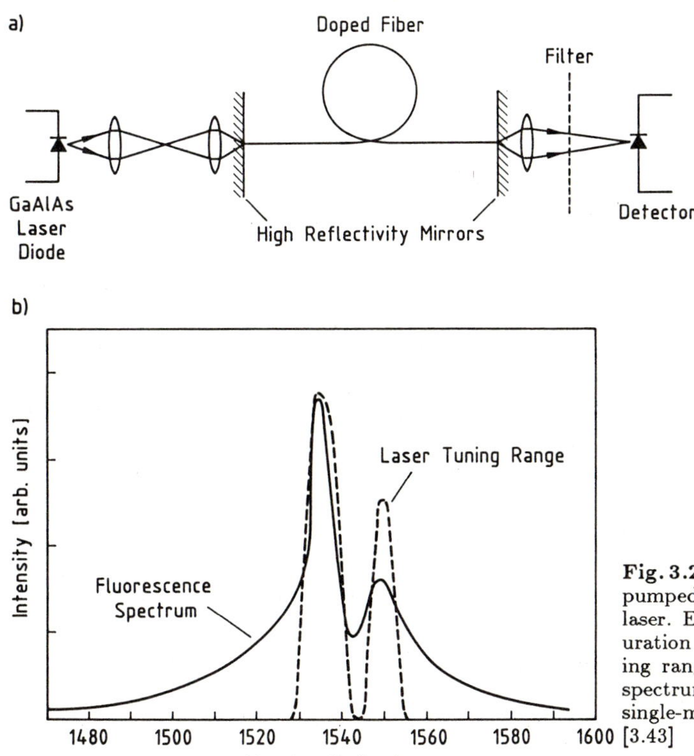

Fig. 3.28a,b. Laser diode pumped single-mode fiber laser. Experimental configuration (**a**) and laser tuning range and fluorescence spectrum of Er^{3+} doped single-mode fiber laser (*b*) [3.43]

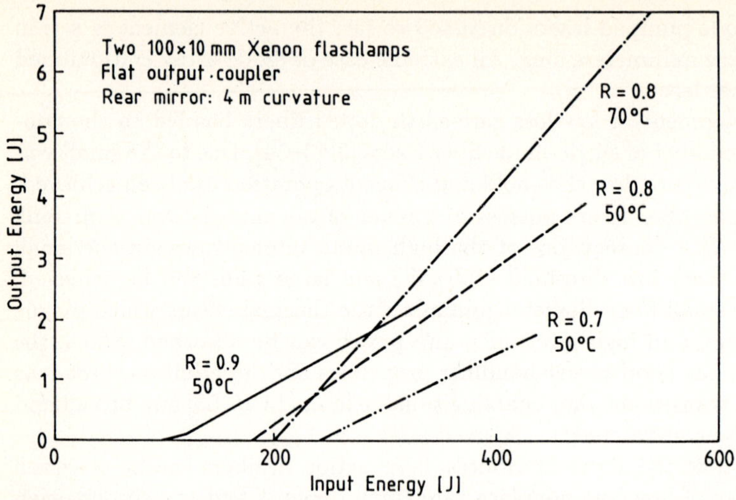

Fig. 3.29. Output energy versus input energy for alexandrite laser at elevated temperatures, [3.63]

tor laser end-pump [3.43]. In an optimized cavity an output exceeding 1 mW at a wavelength of 1088 nm has been observed [3.47] with a slope efficiency of 30 %. Tuning of the output wavelength can be accomplished by substituting a grating for one of the mirrors.

Q-switching of fiber lasers [3.45, 52] using an acousto-optic modulator or rotating chopper is also possible and peak powers of several watts have been observed in pulses ranging from 50 ns and 1 μs.

3.6.5 Alexandrite

Figure 3.29 shows the output performance of a 9.5 mm diameter by 76 mm long alexandrite rod contained in a double-elliptical pump cavity and pumped by two flashlamps. The resonator length was 42 cm. The figure shows the dependence of output pulse energy on input energy with various rod coolant temperatures and output mirror reflectivities. With a mirror reflectivity of 80 % and a rod coolant temperature of 70° C, the threshold energy was 200 J and the slope efficiency was 2 %. The system was operated at a repetition rate of 15 Hz. The temperature dependence of the emission and excited state absorption cross-section of alexandrite is illustrated in Fig. 3.30. As part of an effort to compare theoretical predictions with experimental results, computer predictions and measured output vs. input data are plotted in Fig. 3.31 for a single elliptical pump cavity system. The experimental conditions are listed in Fig. 3.31, the predictions were modelled based on the equations given in Sect. 3.4.

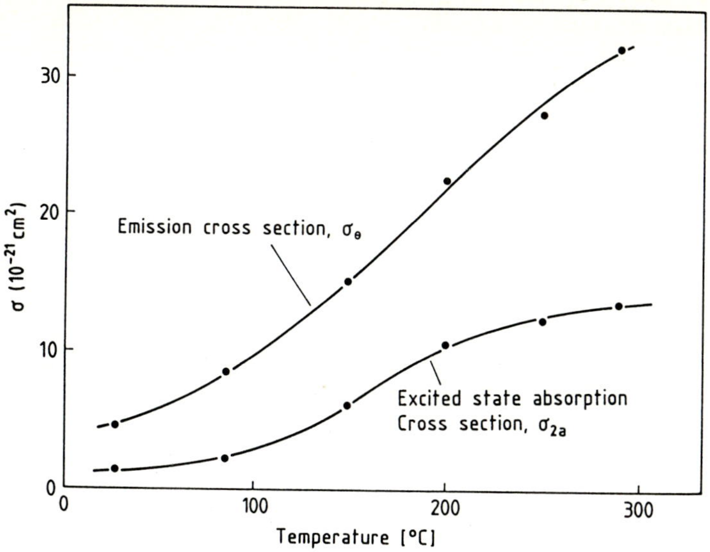

Fig. 3.30. Temperature dependence of emission and excited-state absorption cross section at 790 nm [3.63]

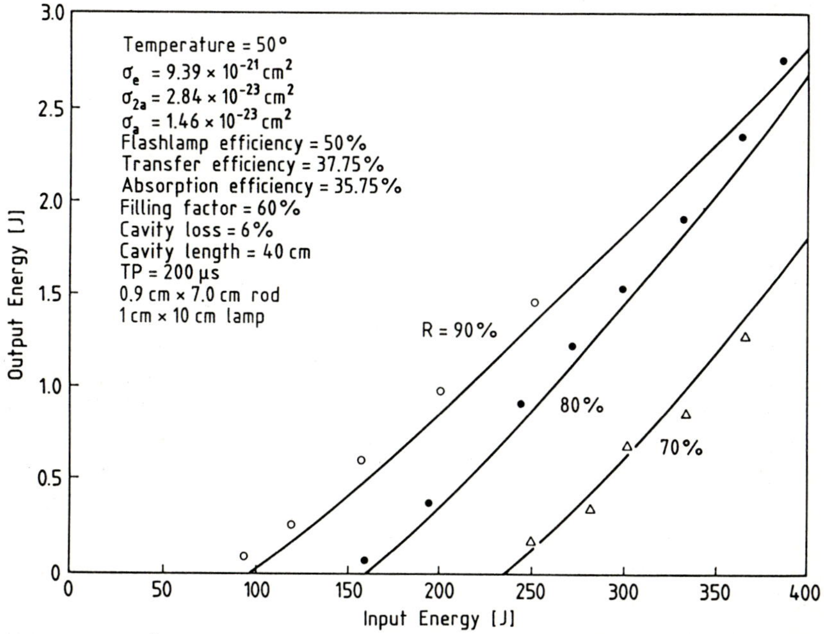

Fig. 3.31. Energy output vs input energy for an alexandrite laser as function of output coupling [3.63]

3.7 Travelling-Wave Oscillator

The oscillators which we have discussed so far are characterized by standing waves in the resonator. In an oscillator consisting of a ring-like cavity utilizing 3 or 4 mirrors and a nonreciprocal optical gate, a travelling wave can be generated [3.55, 56]. The optical gate provides a high loss for one of the two countercirculating travelling waves. The wave with the high loss is suppressed, and a unidirectional output from the laser is obtained. A typical laser cavity configuration of a travelling-wave oscillator is shown in Fig. 3.32 [3.57]. The system consists of a four-mirror rectangular resonator, a Brewster-ended laser rod, a $\lambda/2$ plate, and a Faraday rotator. Three mirrors are coated for maximum reflectivity at the laser wavelength and the fourth mirror is partially transparent. The unidirectional optical gate is formed by a half-wave plate and a Faraday rotator, consisting of a glass rod located within a solenoid-generated axial magnetic field.

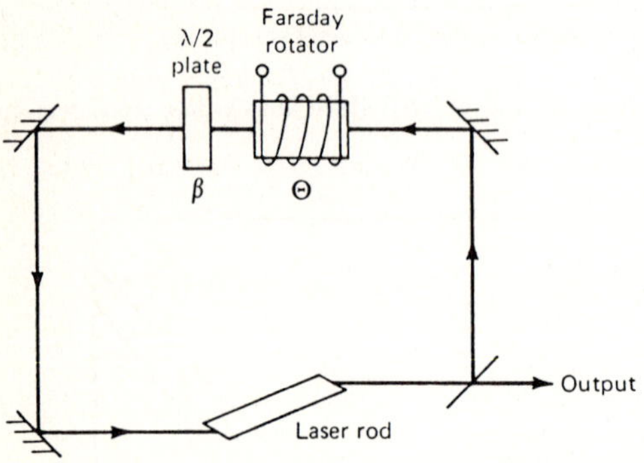

Fig. 3.32. Optical schematic of a travelling-wave oscillator [3.57]

The Faraday element, which should possess a high Verdet constant, rotates the plane of polarization (defined by the Brewster-ended rod) of the two circulating beams by a small angle $\pm\Theta$, the sign being dependent on the propagation direction and polarity of the magnetic field. The half-wave plate is orientated with one of its axes at an angle $\beta/2$ with respect to the polarization of the beams. The magnetic field is adjusted such that for the wave travelling in the clockwise direction, the half-wave plate rotates the plane of polarization by an angle β and the Faraday cell by an angle $+\Theta$ so that the total rotation is $\beta+\Theta$. The differential loss $\Delta\alpha$ between the counterrotating waves is then

$$\Delta\alpha = \sin^2(\beta+\Theta) - \sin^2(\beta-\Theta) \quad . \tag{3.99}$$

For maximum power output, the half-wave plate and the intensity of the magnetic field on the Faraday rotator are adjusted so that $\beta = \Theta$. Then the entire loss, proportional to $4\Theta^2$, is experienced only in the clockwise direction, leaving the wave in the counterclockwise direction unattenuated. As a result the laser will oscillate in a unidirectional wave.

Instead of the rectangular four-mirror cavity, a three-mirror cavity can be employed equally well. Also, the Brewster-ended rod can be replaced with a polarizer and a flat-ended rod. In some designs, rather than using a Faraday rotator, the travelling mode is produced by means of an asymmetric resonator incorporating a return mirror [3.58]. Travelling-wave oscillators have generated interest mainly as a way to eliminate "spatial hole burning" (Sect. 5.2) caused by the standing-wave distribution of the intensity in a conventional oscillator. Since travelling-wave oscillators of the type depicted in Fig. 3.32 are more complicated to construct and require more optical components compared to standing-wave oscillators, these systems have not found any real applications in the past. However, the interest in travelling-wave lasers has dramatically increased with the emergence of laser diodes as practical pumps for Nd : YAG lasers. The compact designs made possible with end-pumped diode-laser geometries have resulted in monolithic ring lasers where the functions of the elements shown in Fig. 3.32 are performed by a single Nd : YAG crystal. Unidirectional ring lasers contain three essential elements: a polarizer, a half-wave plate, and a Faraday rotator. The polarizer, half-wave plate equivalent, and the Faraday rotator are all embodied in the nonplanar ring Nd : YAG laser first proposed by *Kane* et al. [3.59], and illustrated in Fig. 3.33. With a magnetic field H present in the direction shown, the YAG crystal itself acts as the Faraday rotator, the out-of-plane total internal reflection bounces (labelled A and C) act as the half-wave plate, and the output coupler (mirror D) acts as a partial polarizer.

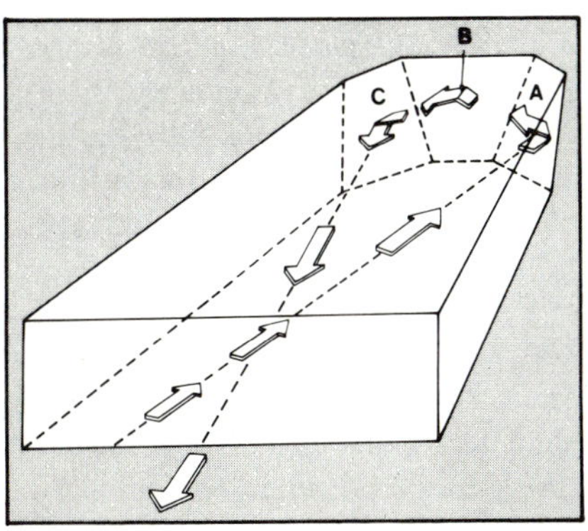

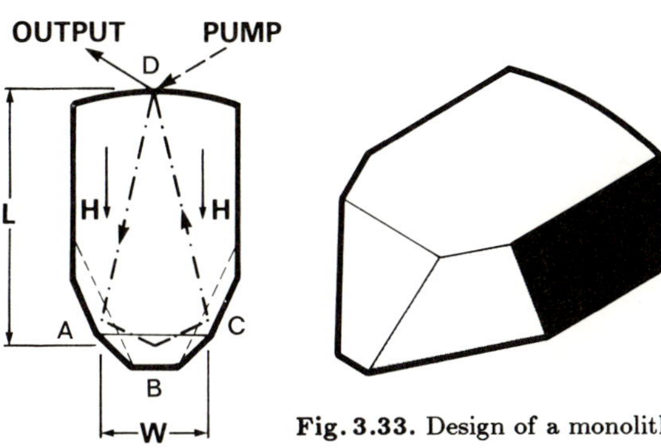

Fig. 3.33. Design of a monolithic ring Nd : YAG laser [3.59]

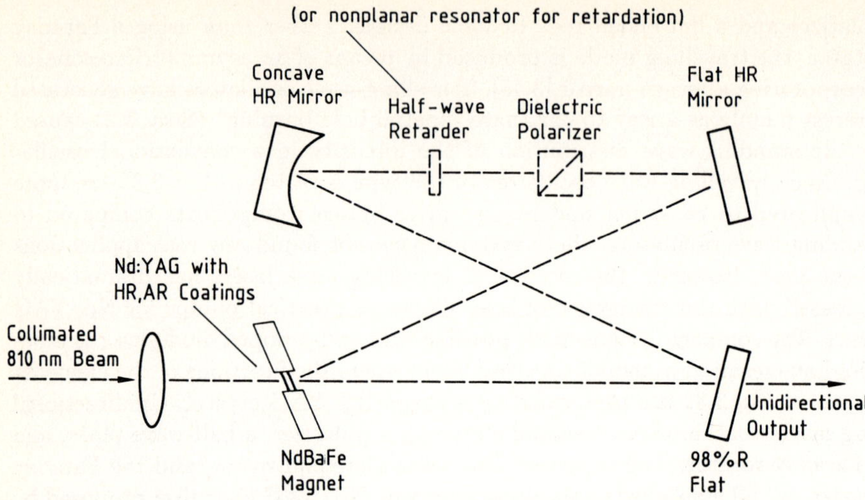

Fig. 3.34. Design of a discrete element diode pumped Nd : YAG ring laser [3.62]

The basic idea is to provide the equivalent of a discrete element design of a half-wave plate with a fast axis rotation angle which is half of the Faraday rotation angle. Since the Faraday rotation is small, the equivalent wave-plate rotation angle is also made small. Modified versions of the original monolithic ring laser have also been reported [3.60] as well as operation at 1.32 μm [3.61].

Laser-diode pumped Nd : YAG ring lasers, comprising discrete elements, can be made very small and compact. These devices find applications where it is necessary to insert intracavity elements in order to provide a tunable, single-frequency output. An example of such a laser pumped by a commercial 200 mW-output diode array is shown in Fig. 3.34 [3.62]. The laser crystal serves as one of the HR mirrors and is placed at the beam waist produced by the spherical power of the concave mirror. A magnetic field provided by a ring shaped permanent magnet is established along the axis of the laser crystal, producing a Faraday rotation of the cavity-mode polarization.

4. Laser Amplifier

In this chapter we will discuss the gain in energy for a light beam making a single pass through an optically active material. The use of lasers as pulse amplifiers is of great interest in the design of high-energy, high-brightness light sources. The generation of high-energy pulses is based on the combination of a master oscillator and multistage power amplifier. For the purpose of illustrating the amplifier concept and principles we assume a straightforward system, as shown in Fig. 4.1. In this scheme an amplifier is driven by an oscillator which generates an initial light pulse of moderate power and energy. In the power amplifier with a large volume of active material the pulse power can grow, in extreme cases, up to 100 times.

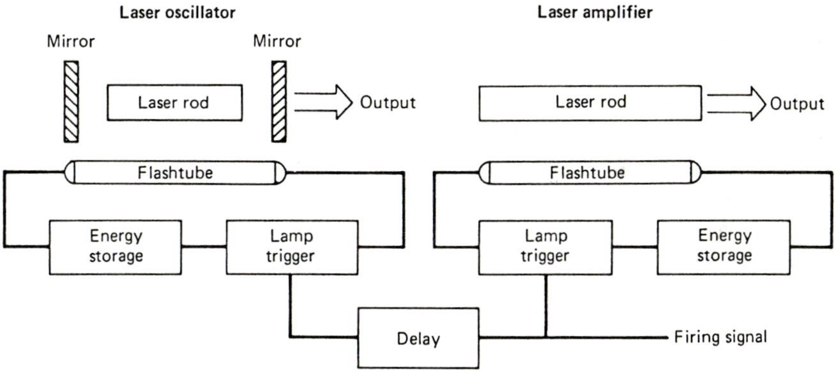

Fig. 4.1. Schematic diagram of a laser oscillator-amplifier configuration

In an oscillator-amplifier system, pulse width, beam divergence, and spectral width are primarily determined by the oscillator, whereas pulse energy and power are determined by the amplifier. Operating an oscillator at relatively low energy levels reduces beam divergence and spectral width. Therefore, from an oscillator-amplifier combination one can obtain either a higher energy than is achievable from an oscillator alone or the same energy in a beam which has a smaller beam divergence and narrower linewidth. Generally speaking, the purpose of adding an amplifier to a laser oscillator is to increase the brightness $B[W\,\mathrm{cm}^{-2}\mathrm{sr}^{-1}]$ of the output beam

$$B = \frac{P}{A\Omega} \ , \tag{4.1}$$

where P is the power of the output beam emitted from the area A, and Ω is the solid-angle divergence of the beam. Multiple-stage amplifier systems can be built if higher amplifications are required. In extreme cases Nd : glass laser systems have been constructed in which a 1-mJ output from an oscillator is amplified to a 1-kJ beam.

In the design of laser amplifiers the following aspects must be considered:

- Gain and energy extraction.
- Wavefront and pulse-shape distortions introduced by the amplifier.
- Energy and power densities at the optical elements of the amplifier system.
- Feedback in the amplifier which may lead to superradiance or prelasing.

Of primary interest in the design of amplifiers is the gain which can be achieved and the energy which can be extracted from the amplifier. The rod length in an amplifier is determined primarily by the desired gain, while the rod diameter, set by damage threshold considerations, is dependent on the output energy. We shall see in the following sections that the gain of an amplifier pumped at a certain inversion level depends on the intensity and duration of the input pulse.

To a first approximation we can assume the growth of energy in the wave to be exponential, for the amount of stimulated emission is proportional to the exciting photon flux. It will be seen, however, that exponential amplification will occur only at low photon-flux levels. A deviation from the exponential gain regime in an amplifier occurs when an optical pulse traveling in the inverted medium becomes strong enough to change the population of the laser levels appreciably. The optical amplifier will exhibit saturation effects as a result of depletion of the inversion density by the driving signal. Taking an extreme case, we can see that if a high-intensity light pulse is incident on a laser rod, the stimulated emission can completely deplete the stored energy as it progresses. Then the gain can be expected to be linear with the length of the rod rather than exponential.

Let us assume that a low-level signal is incident on a long amplifier which is uniformly inverted. At the beginning the signal increases exponentially with distance, then after a transition region the signal amplitude grows linearly with length, and finally the signal no longer increases at all. This happens when the gain per unit length just balances the absorption losses per unit length in the rod.

So far we have discussed the effect of the input signal level on the gain of the amplifier. We will also have to consider the pulse length and its influence on the amplification mechanism. In an amplifier where the input pulse is considerably shorter than the fluorescence lifetime of the medium, such as a Q-switch pulse or a mode-locked pulse, energy is extracted from the amplifier which was stored in the amplifying medium prior to the arrival of the pulse. If the driving signal is long compared to the fluorescent time, such as the output from a conventional-mode oscillator or a CW laser, a steady-state gain best characterizes the amplification mechanism. Hence the effect on the inversion density of spontaneous emission and pumping cannot be neglected.

The two regimes of light amplification – short pulse and steady state or long pulse – will be discussed in Sects. 4.1 and 4.2. The phenomenon of gain saturation is common to both time domains. We will see, however, that in the case of pulses which are short compared to the fluorescence lifetime of the material the amplification depends on the energy density, whereas in the case of long pulses or CW mode of operation the gain depends on power density.

Any amplification process is associated with some kind of distortion. In Sect. 4.3 we will summarize wavefront and pulse-shape distortions associated with the amplification of optical pulses. In multiple-stage amplifier systems the stability of the system is of prime concern to the laser designer. In Sect. 4.4 the conditions for stable operation of amplifiers will be discussed.

4.1 Pulse Amplification

The events during the amplifier action are assumed to be fast compared with the pumping rate W_p and the spontaneous emission time τ_f. Therfore $t_p \ll \tau_f, W_p^{-1}$, t_p being the width of the pulse which passes through the laser rod.

Thus the amplification process is based on the energy stored in the upper laser level prior to the arrival of the input signal. As the input pulse passes through the rod, the atoms are stimulated to release the stored energy. The amplification process can be described by the rate equations (1.58, 61). If we ignore the effect of fluorescence and pumping during the pulse duration, we obtain for the population inversion

$$\frac{\partial n}{\partial t} = -\gamma n c \sigma \phi. \tag{4.2}$$

The growth of a radiation pulse traversing a medium with an inverted population is described by the nonlinear, time-dependent photon-transport equation, which accounts for the effect of the radiation on the active medium and vice versa,

$$\frac{\partial \phi}{\partial t} = c n \sigma \phi - \frac{\partial \phi}{\partial x} \quad . \tag{4.3}$$

The rate at which the photon density changes in a small volume of material is equal to the net difference between the generation of photons by the stimulated emission process and the flux of photons which flows out from that region. The latter process is described by the second term on the right of (4.3). This term which characterizes a traveling-wave process is absent in (1.61).

Consider the one-dimensional case of a beam of monochromatic radiation incident on the front surface of an amplifier rod of lenth L. The point at which the beam enters the laser rod is designated the reference point, $x = 0$. The two differential equations (4.2, 3) must be solved for the inverted electron population n and the photon flux ϕ. *Frantz* and *Nodvik* [4.1] and others [4.2, 3] solved these nonlinear equations for various types of input pulse shapes.

If we take for the input to the amplifier a square pulse of duration t_p and initial photon density ϕ_0, the solution for the photon density is

$$\frac{\phi(x,t)}{\phi_0} = \left\{1 - [1 - \exp(-\sigma n x)] \exp\left[-\gamma\sigma\phi_0\left(t - \frac{x}{c}\right)\right]\right\}^{-1} , \tag{4.4}$$

where n is the inverted population density, assumed to be uniform throughout the laser material at $t = 0$. The energy gain for a light beam passing through a laser amplifier of length $x = l$ is given by

$$G = \frac{1}{\phi_0 t_p} \int_{-\infty}^{+\infty} \phi(l,t)dt \quad . \tag{4.5}$$

After introducing (4.4) into (4.5) and integrating, we obtain

$$G = \frac{1}{\gamma\sigma\phi_0 t_p} \ln\{1 + [\exp(\gamma\sigma\phi_0\tau_0) - 1]\, e^{n\sigma l}\} \quad . \tag{4.6}$$

We shall cast this equation in a different form such that it contains directly measurable laser parameters. The input energy per unit area can be expressed as

$$E_{in} = \phi_0 t_p h\nu \quad . \tag{4.7}$$

A saturation energy density can be defined by

$$E_s = \frac{h\nu}{\gamma\sigma} \quad . \tag{4.8}$$

The physical meaning of the parameter E_s can be seen by multiplying the right-hand side of (4.8) by n. The saturation energy E_s becomes the ratio of the energy which can be extracted from the amplifying medium divided by the small-signal gain coefficient

$$E_{ex} = g_0 E_s = \frac{h\nu n}{\gamma} = \frac{E_{st}}{\gamma} \quad . \tag{4.9}$$

From (4.9) follows that when an input pulse has an energy density $E_{in} = E_s$, the small-signal gain coefficient times the input energy density just equals the maximum energy per volume which can be extracted from the amplifier. As indicated in (4.9), in a four-level system $\gamma = 1$ and all the stored energy $E_{st} = h\nu n$ can theoretically be extracted by a signal. In a three-level system such as ruby $\gamma = 1 + g_2/g_1$, and only a fraction of the stored energy will be released because as the upper laser level is depleted, the lower-level density is building up.

Introducing (4.7, 8) into (4.6), one obtains

$$G = \frac{E_s}{E_{in}} \ln\left\{1 + \left[\exp\left(\frac{E_{in}}{E_s}\right) - 1\right] G_0\right\} \quad . \tag{4.10}$$

This expression represents a unique relationship between the gain G, the input pulse energy density E_{in}, the saturation parameter E_s, and the small-signal, single-pass gain $G_0 = \exp(g_0 l)$.

Equation (4.10), which is valid for rectangular input pulses, encompasses the regime from small-signal gain to complete saturation of the amplifier. The equation can be simplified for these extreme cases. Consider a low-input signal E_{in} such that $E_{in}/E_s \ll 1$, and furthermore $G_0 E_{in}/E_s \ll 1$; then (4.10) can be approximated to

$$G \approx G_0 \equiv \exp(g_0 l) \quad . \tag{4.11}$$

In this case, the "low-level gain" is exponential with rod length, and no saturation effects occur. This, of course, holds only for rod lengths up to a value where the output energy density $G_0 E_{in}$ is small compared to E_s.

For high-level energy densities such that $E_{in}/E_s \gg 1$, (4.10) becomes

$$G \simeq 1 + \left(\frac{E_s}{E_{in}}\right) g_0 l \quad . \tag{4.12}$$

Thus, the energy gain is linear with the length of the rod, implying that every excited state contributes its stimulated emission to the beam. Such a condition obviously represents the most efficient conversion of stored energy to beam energy, and for this reason amplifier designs which operate in saturation are used wherever practical, with the major limitation being laser rod damage thresholds.

In laser amplifier technology the small-signal gain coefficient $g_0 = n\sigma_{21}$ is often expressed as

$$g_0 = \beta E_{st} \quad , \tag{4.13}$$

where E_{st} is the previously discussed stored energy per unit volume and

$$\beta = \frac{\sigma_{21}}{h\nu} \tag{4.14}$$

is a parameter relating the gain to the stored energy. Equation (4.10) can be rearranged to take the form

$$E_{out} = E_s \ln\left\{1 + \left[\exp\left(\frac{E_{in}}{E_s}\right) - 1\right] \exp(\beta E_{st} l)\right\} \quad . \tag{4.15}$$

Equations (4.10, 15) permit one to calculate the gain of an amplifier as a function of the input energy density, provided that the small-signal gain or the energy stored in the amplifier is known. These parameters, which depend on the input energy, the volume of the active material, the efficiency of the pump structure, etc., normally must be estimated. Later in this section we will list typical performance data for ruby, Nd: glass, and Nd: YAG amplifiers which can be used as guidelines.

One of the significant points of (4.10) for the design of laser amplifiers lies in the fact that if one data point of an existing amplifier is known, the performance of the amplifier under different operating conditions can be calculated. Also, the effect on the performance of changes in design, such as amplifier length or diameter or the use of multiple stages in a system, can be studied with the aid of this equation. Before we discuss practical examples, it should be noted that in deriving (4.10) two assumptions were made:

1. It was assumed that the pulse shape of the incident pulse was rectangular. However, it should be mentioned that despite this assumption, the above analysis holds, to a good approximation, for a symmetrical triangular-shaped pulse. This approximation becomes less valid in cases where amplified pulse shapes differ significantly from incident pulse shapes as a result of the higher gain experienced by the leading edge of the pulse. In such cases, more accurate gain equations, given in [4.4, 5], should be used.
2. We have assumed a lossless amplifier. In real solid-state laser amplifiers there inevitably exist linear losses of radiation as a result of absorption and scattering caused by defects and impurities in the active medium. A linear loss limits the energy growth in the saturation regime. There the energy tends to grow linearly as a result of the amplification and to decrease exponentially as a result of loss.

Avizonis and *Grotbeck* [4.6] have derived an expression which describes the gain process in an amplifier without approximation. They obtained

$$\frac{dE(x)}{dx} = E_s g_0 \left[1 - \exp\left(\frac{-E(x)}{E_s}\right)\right] - \alpha E(x) \quad , \tag{4.16}$$

where $E(x)$ is the pulse energy at point x, α is the loss coefficient per unit length, and x is the amplifier length coordinate.

Equation (4.16) can be solved analytically only if a zero loss is assumed ($\alpha = 0$). In this case the result is identical to (4.10). From (4.16) follows that, in the presence of losses, gain in the saturating regime occurs only for pulse energies below a limiting value E_{max}. For $\alpha \ll g_0$ this limiting value is

$$E_{max} = \frac{g_0 E_s}{\alpha} \quad . \tag{4.17}$$

Modern laser materials such as ruby, Nd: glass, and Nd: YAG have sufficiently small losses that they can be neglected in most cases in the design of amplifiers. Loss coefficients for these materials are typically less than 0.001 cm^{-1}. For values of E_s between 5–10 J/cm^2 and g_0 between 0.1 and 0.5 cm^{-1}, the theoretical value of the limiting energy $E_{\max}$ is in the range of several hundreds of joules per square centimeter. In practice, the realizable pulse output energy is limited to significantly lower values because of self-damage in the active medium. Therefore, only in optically very poor host materials, where scattering or absorption losses are high (such as Verneuill-grown rubies or early laser glasses containing platinum inclusions), need one take α into account. Additional references treating laser amplifiers including losses can be found in [4.7, 8].

The amplification efficiency (i.e., the ratio of the energy extracted by the signal to the energy stored in the active medium) in laser amplifiers can be increased by passing the signal twice through the active medium. The double-pass technique yields a considerable increase in the small-signal gain. This advantage must be weighted against the added complexity of a multiple-pass amplifier brought about by the addition of beam-deflecting optics. Double-pass amplifiers are described in [4.9, 10].

In order to illustrate the usefulness of (4.15), we will calculate the gain for several amplifiers and compare the calculations with measured data.

4.1.1 Ruby Amplifiers

As mentioned in Sect. 2.2, laser action in ruby occurs on the R_1 emission line. However, since energy is stored in both the $\overline{E}$ and $2\overline{A}$ levels, a transfer of energy from the $2\overline{A}$ to the $\overline{E}$ level can take place during the amplification process. Whether energy transfer between the two levels actually occurs depends on whether the relaxation time between the $\overline{E}$ and $2\overline{A}$ levels is short or long compared with the length of the amplified pulse. The relaxation time between the two excited levels in ruby is on the order of 1 ns or less.

Amplification of Q-Switched Pulses

First we will consider the amplification mechanisms for pulses which are longer than 1 ns. In this case, the two upper levels (i.e., E and $2A$ levels) remain in thermal equilibrium and energy can be extracted from both levels. The stored energy in the combined upper levels is

$$E_{\mathrm{st}} = h\nu n = h\nu(2n_2 - n_{\mathrm{tot}}) \quad . \tag{4.18}$$

The maximum energy which one may hope to extract from ruby is

$$E_{\mathrm{ex}} = h\nu\left(n_2 - \frac{n_{\mathrm{tot}}}{2}\right) = E_{\mathrm{st}}(\overline{E}) \quad , \tag{4.19}$$

where n_2 is the upper-state population prior to the arrival of the pulse. The ratio of stored energy to extractable energy defined by (4.9) becomes

$$\gamma = \frac{E_{\mathrm{st}}(\overline{E})}{E_{\mathrm{ex}}} = 1 \quad , \tag{4.20}$$

and from (2.12 and 4.13, 18), we obtain for ruby

$$g_0 = \beta E_{st}(\overline{E}) = \beta' E_{st} \quad \text{with} \quad \beta' = \beta/2 \quad . \tag{4.21}$$

The saturation energy, defined as the ratio of extractable energy to the small-signal gain coefficient, is

$$E_s = \frac{h\nu}{\sigma_{21}} \quad . \tag{4.22}$$

If we introduce the parameters listed in Table 2.2 into (4.21, 22), we obtain $E = 11.0\,\mathrm{J/cm^2}$ and $\beta' = 0.044\,\mathrm{cm^2/J}$. The expressions which relate the upper-state population density to the gain coefficient, stored energy, extractable energy, and total upper-state energy density are presented in graphical form in Fig. 4.2.

Having summarized the pertinent laser parameters for ruby, we turn now to (4.15). Introducing the materials parameter E_s and β' into this equation and assuming values for E_{st} and E_{in}, one can plot the output energy density of a ruby amplifier as a function of amplifier length. The curves in Fig. 4.3 were obtained by assuming a value of $E_{in} = 0.1\,\mathrm{J/cm^2}$ and three different values of the stored energy $E_{st} = 3.0$, 4.0, and $4.5\,\mathrm{J/cm^3}$ for total inversion.

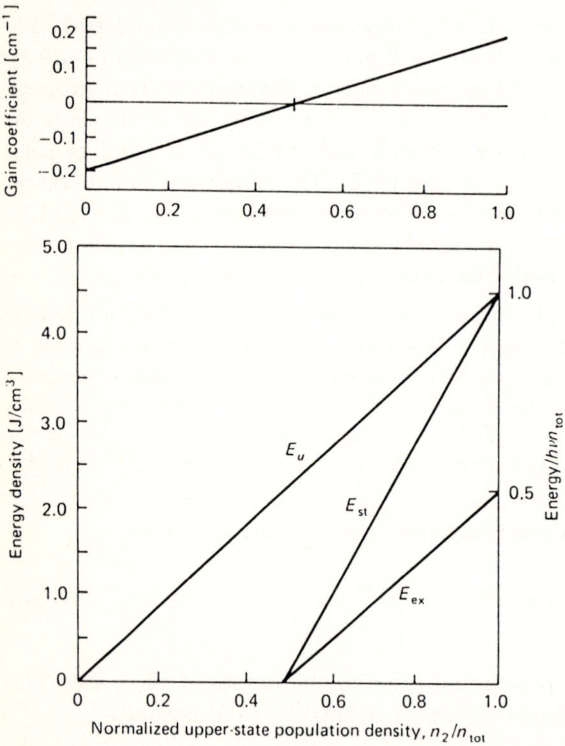

Fig. 4.2. Gain coefficient and upper-state energy in optically pumped ruby. E_u is the upper-state energy density, E_{st} is the stored energy, E_{ex} is the maximum extractable energy density, n_{tot} is the total Cr^{3+} concentration in ruby, and n_2 is the upper-state population density

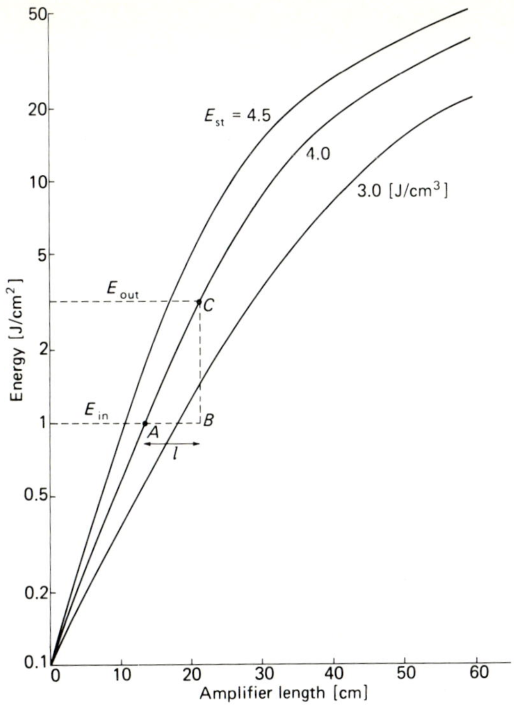

Fig. 4.3. Ruby amplifier gain as a function of amplifier length and stored energy density

The practical significance of these curves lie in the fact that they can be used for any input energy and amplifier length. For example, assume that there is a total of $1\,\mathrm{J/cm^2}$ being delivered by the oscillator to a 10-cm-long amplifier with a stored energy of $E_{st} = 4\,\mathrm{J/cm^3}$. A line drawn through the ordinate at $1\,\mathrm{J/cm^2}$ intersects the curve at A, adding to that the amplifier length, we obtain point B. Point C, which has the same abscissa as point B, intersects the curve at $4\,\mathrm{J/cm^2}$, which is the output from the amplifier.

The only parameter that is difficult to determine is the rod energy storage E_{st} per unit volume. This parameter is dependent on the flashlamp input energy and pulse width, the design and efficiency of the pumping geometry, and the laser rod geometry. From the rate equations (1.53) we can obtain a relationship between the pump pulse intensity and the stored energy. Since $\phi = 0$ prior to the arrival of the signal, (1.53) has the solution

$$\frac{n_1}{n_{tot}} = \left\{1 + W_p\tau_f \exp\left[-\left(W_p + \frac{1}{\tau_f}\right)t_p\right]\right\}(1 + W_p\tau_f)^{-1} \quad . \tag{4.23}$$

The pump rate W_p can be expressed by the lamp input energy per cubic centimeter of rod volume E_p, the pulse duration t_p, and an adjustable parameter f which is essentially the pump efficiency [4.11–13].

$$W_p = \frac{fE_p}{t_p} \quad . \tag{4.24}$$

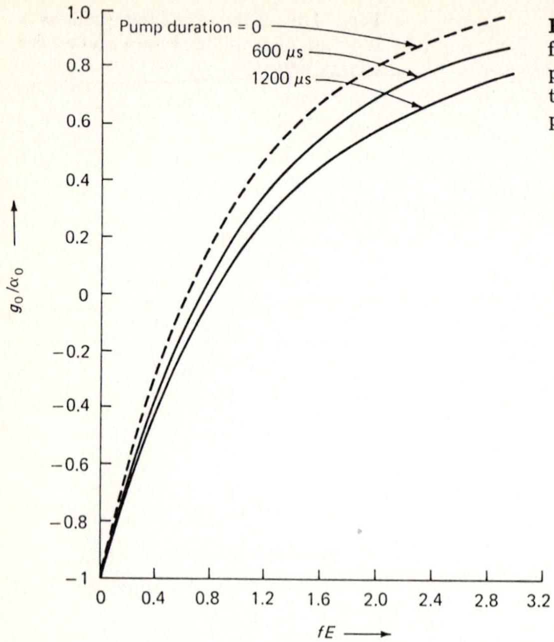

Fig. 4.4. Ruby gain coefficient as a function of pump energy and pump pulse length. The parameter f is the pump efficiency factor; E is the parameter E_p [4.12]

From (4.23 and 24) we obtain and expression of the population inversion as a function of flashlamp input energy per cubic centimeter of rod volume:

$$n = n_{tot}\left[1 - 2\left(\frac{1 + (fE_p\tau_f/t_p)\exp(-fE_p - t_p/\tau_f)}{1 + fE_p\tau_f/t_p}\right)\right] \quad . \tag{4.25}$$

The expression $g_0 = \sigma_{12}n$ normalized to the absorption coefficient α_0 is plotted in Fig. 4.4 as a function of fE_p with t_p as a parameter. As illustrated, the gain which can be achieved for a given lamp input energy depends on the ratio of pulse width to fluorescence lifetime. For a longer pump pulse, gain is reduced because of the depletion of the upper level by spontaneous emission. For an infinitely short pump pulse, $t_p = 0$ and the gain coefficient becomes

$$g_0 = \alpha_0[1 - 2\exp(-fE_p)] \quad . \tag{4.26}$$

The parameter f depends only on the efficiency of the pump geometry and the spectral output of the lamp. In Fig. 4.5 the stored energy E_{st} as a function of lamp input energy per cubic centimeter of rod volume is plotted for a practical pulse length of $t_p = 1$ ms. The parameter is the pumping coefficient f.

Figures 4.3 and 5 completely describe the performance of a ruby amplifier in the Q-switch regime. Starting with the total lamp input power and the rod volume, the parameter E_p can be calculated. For a given E_p, there follows from Fig. 4.5 a value for the upper-state stored energy E_{st}. With this value known, from Fig. 4.3 the output of the amplifier for any input energy density can be obtained. The only assumption which has to be made is the numerical value

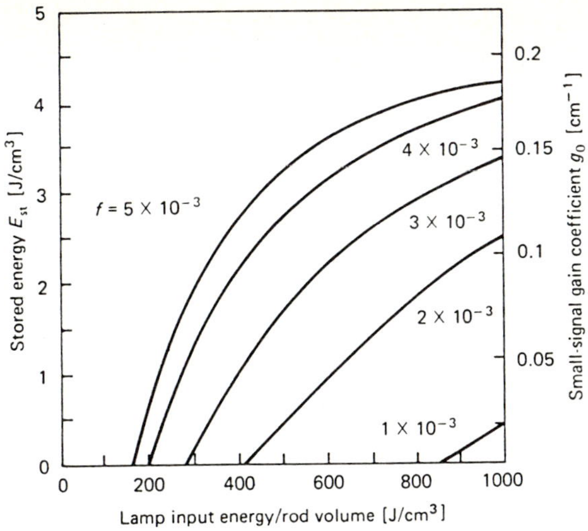

Fig. 4.5. Stored energy density in ruby as a function of lamp input energy per cubic centimeter of rod volume. Parameter is the pump efficiency factor f [cm^3/J]

of the pump coefficient f. A good estimate for most helical lamp geometries is $f = (4-8) \times 10^{-3}$ [cm^3/J] and $f = (6-8) \times 10^{-3}$[cm^3/J] for focusing pump cavities.

Once an amplifier has been built and one data point $(E_{\text{in}}, E_{\text{out}}, E_{\text{p}}, t_{\text{p}})$ has been measured the amplifier can, in most cases, be completely described for a variety of different operating conditions using the equations and graphs presented in this section. For example, if E_{out} is measured for a single input power E_{in}, then we can calculate the small-signal single-pass gain according to (4.10)

$$G_0 = \frac{\exp(E_{\text{out}}/E_{\text{s}}) - 1}{\exp(E_{\text{in}}/E_{\text{s}}) - 1} \quad , \tag{4.27}$$

and with G_0 known, the small-signal gain coefficient g_0 and the stored energy E_{st}. If the flashlamp input energy and pulse duration was recorded during the gain measurement, the pump parameter f can be determinded from (4.26). We will now compare the theoretical results with measurements performed on actual ruby amplifiers. In Fig. 4.6, the small-signal gain coefficient g_0 is plotted as a function of flashlamp input per cubic centimeter of rod volume for a number of different ruby laser amplifiers. The systems, varying in ruby rod size, are all pumped by helical flashlamps. The measured values can be approximated by (4.25) if one assumes a pump pulse of 0.8 ms and a pump parameter of $f = 8 \times 10^{-3}$[cm^3/J]. The flashlamp pulse in these systems is typically 0.8–1.0 ms wide at the 50 % power points. The signal input is normally delayed by about the same time with respect to the triggering of the flashlamp. In accordance with the calculated value, the gain curve bends over for high lamp input energies as a result of depletion of the ground level. Measurements on about 20 helical flashlamp pumped ruby laser amplifiers have shown that energy storages of $E_{\text{st}} = 3.5 - 3.8$ J/cm^3 are typical for a flashlamp input of 350 J/cm^3 of active material. The small-signal gain coefficient under these

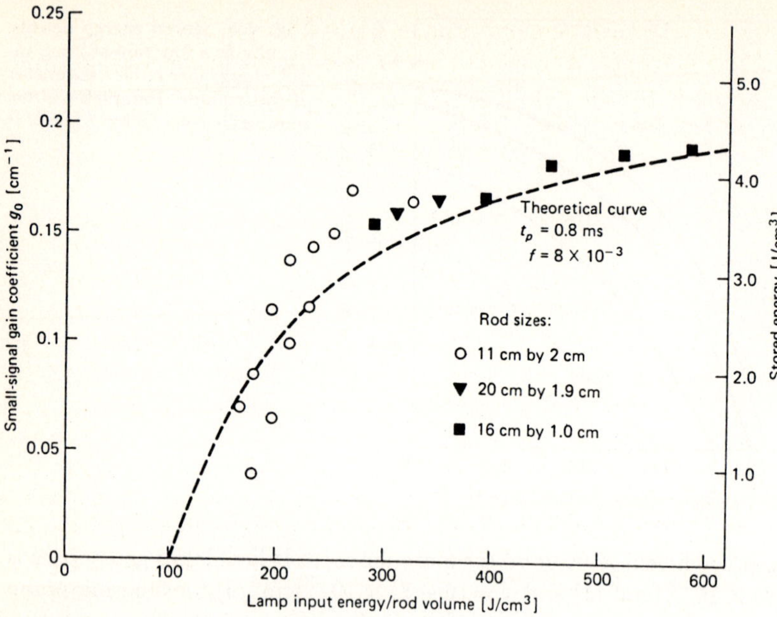

Fig. 4.6. Small-signal gain coefficient and stored energy versus lamp input energy per cubic centimeter of rod for several commercial ruby laser heads

conditions is, therefore, approximately $g_0 = 0.15 - 0.17\,\mathrm{cm}^{-1}$ for rods doped at 0.05% Cr^{3+}. Since the small-signal gain coefficient is proportional to the doping concentration, see (2.13), variations in the peformance of different ruby rods can often be traced to differences in doping concentration.

Amplification of Mode-Locked Pulses

For input pulses which are so short ($t_p \ll 1$ ns) that no thermalization between the $\overline{E}$ and $2\overline{A}$ level can take place during the amplification process, the energy which can be extracted from the ruby crystal depends on the population of the $\overline{E}$ level only. In Sect. 2.2, for the maximum energy which can be extracted from ruby in a subnanosecond pulse if prior to the arrival of the pulse complete inversion exists, we obtained

$$E'_{\mathrm{ex,max}} \sim \frac{n_{\mathrm{tot}} h\nu}{3} \quad . \tag{4.28}$$

Since the small-signal coefficient is the same as for the longer pulse, we obtain after introducing (4.28) and (2.13) into (4.9),

$$E'_{\mathrm{s}} = \frac{2h\nu}{3\sigma_{21}} \quad . \tag{4.29}$$

The numerical value for the saturation density is $E_s = 7.7\,\mathrm{J/cm^2}$. Gain saturation in connection with the amplification of mode-locked pulses is only of academic interest, because energy densities attainable in practice are considerably below E_s as a result of the limitation in peak power imposed by material damage. The performance of ruby amplifiers employed to amplify picosecond pulses have been described in [4.14].

4.1.2 Nd:Glass Amplifiers

An enormous data basis exists regarding the design of Nd:Glass amplifiers since these systems have become the lasers of choice for laser fusion research. Motivated by requirements to drive inertial confinement fusion targets at ever higher powers and energies, very large Nd:glass laser systems have been designed, built, and operated at a number of laboratories throughout the world over the past fifteen years.

During the 1960's and into the early 1970's, large glass laser systems consisted of pulsed oscillators followed by rod amplifiers. The introduction of face-pumped disk amplifiers relieved the inherent aperture constraint of rod amplifiers. Initially the systems employed silicate glass as the host material for neodymium. The lower nonlinear refractive index and higher gain coefficient of phosphate glasses has provided a powerful incentive to build systems based upon these materials.

The architectural design of these master oscillator-pulsed amplifier systems is determined by the nonlinear aspects of propagation, namely self-focusing and gain saturation. Most large Nd:glass fusion laser systems constructed to date have utilized the master-oscillator power-amplifier chain (MOPA) as the basic building block. For example, the component layout of a laser MOPA chain is shown schematically in Fig. 4.7. A MOPA chain consists of a master oscillator, which generates a well-controlloed low-energy pulse for amplification, and a series of power amplifiers to increase the beam energy. The clear apertures of

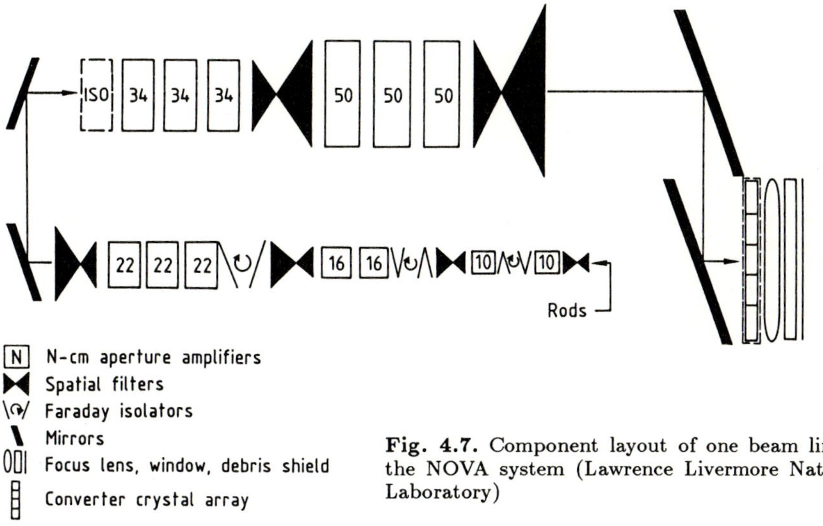

Fig. 4.7. Component layout of one beam line of the NOVA system (Lawrence Livermore National Laboratory)

the power amplifiers increase stepwise down the chain to avoid optical damage as the beam energy grows. To minimize the effects of small-scale self-focusing and to maximize the output fluency, optical-relay spatial filters are placed between the amplifiers along the chain. Laser oscillation in the chain is prevented by the appropriate placement along the chain of Faraday rotators and polarizer plates. To construct laser systems that provide more energy than can be obtained from a single MOPA chain, or to provide for multibeam irradiation geometries, the ouptut from a small MOPA chain is split into the desired number of beams, and each of these beams is used to drive a full MOPA chain. The system shown in Fig. 4.7 is an example of such a design. Shown is one 10 KJ beam line of the NOVA glass laser system built by Lawrence Livermore Laboratory. The complete system has 10 such identical beam lines producing a total output energy in excess of 100 KJ. An alternative system concept that uses components of MOPA chains consists of division of the initial beam into two beams, followed by amplification, splitting, amplification, etc., until the desired number of beams and total energy are obtained.

It is outside the scope of this book to address the complex issues surrounding the design, development and operation of glass lasers employed in fusion target irradiation facilities. The reader interested in the design, modeling, and materials aspects of these lasers is referred to the extensive literature which covers the whole aspect of lasers for inertial confinement fusion. The monograph by *Brown* [4.15], the special issue on lasers for fusion of the IEEE Journal of Quantum Electronics [4.16], the annual reports issued by LLNL [4.17] as well as several overview papers on fusion glass lasers [4.18, 19] are recommended as an introduction to this area of laser technology.

In this section we will proceed to provide several basic guidelines which are important in the design of Nd : glass oscillator-amplifier systems.

In order to calculate gain and energy extraction from a glass amplifier for a given input energy we have to know the saturation density E_s and the small signal gain G_0 according to (4.10).

The saturation fluency E_s of a laser glass depends inversely on the gain cross-section τ and can be written as

$$E = (h\nu/\sigma)k \quad .$$

The 1053 nm gain cross-section of phosphate glasses range from 3.0 to 4.2 × 10^{-20} cm^2 and k is a parameter which is dependent on the output fluency and the duration of the amplified laser pulse. Saturation fluency can depend on the pulse duration if it is less than the lifetime of the lower laser level (less than 1 ns for most glasses). The dependence of E_s on output fluency has been attributed to a hole-burning mechanism [4.20]. It is believed that the saturating pulse couples more strongly to one fraction of the inverted ions in the glass.

Figure 4.8 shows representative values of saturation fluence versus output fluences for several phosphate glasses [4.20, 21].

Fig. 4.8 a–d. Output versus input fluence, and saturation versus output fluences for a number of laser glasses. (**a**) OIH-9, a silicate glass made by Owens Illinois Inc., and LG-56, a silicate glass made by Schott Optical Glass Inc. (**b**) ED-2, a silicate glass made by Owens Illinois Inc. (**c**) Q-88, a phosphate glass made by Kigre Inc. (**d**) LHG-8, a phosphate glass made by Hoya Corp. [4.20a, 21]

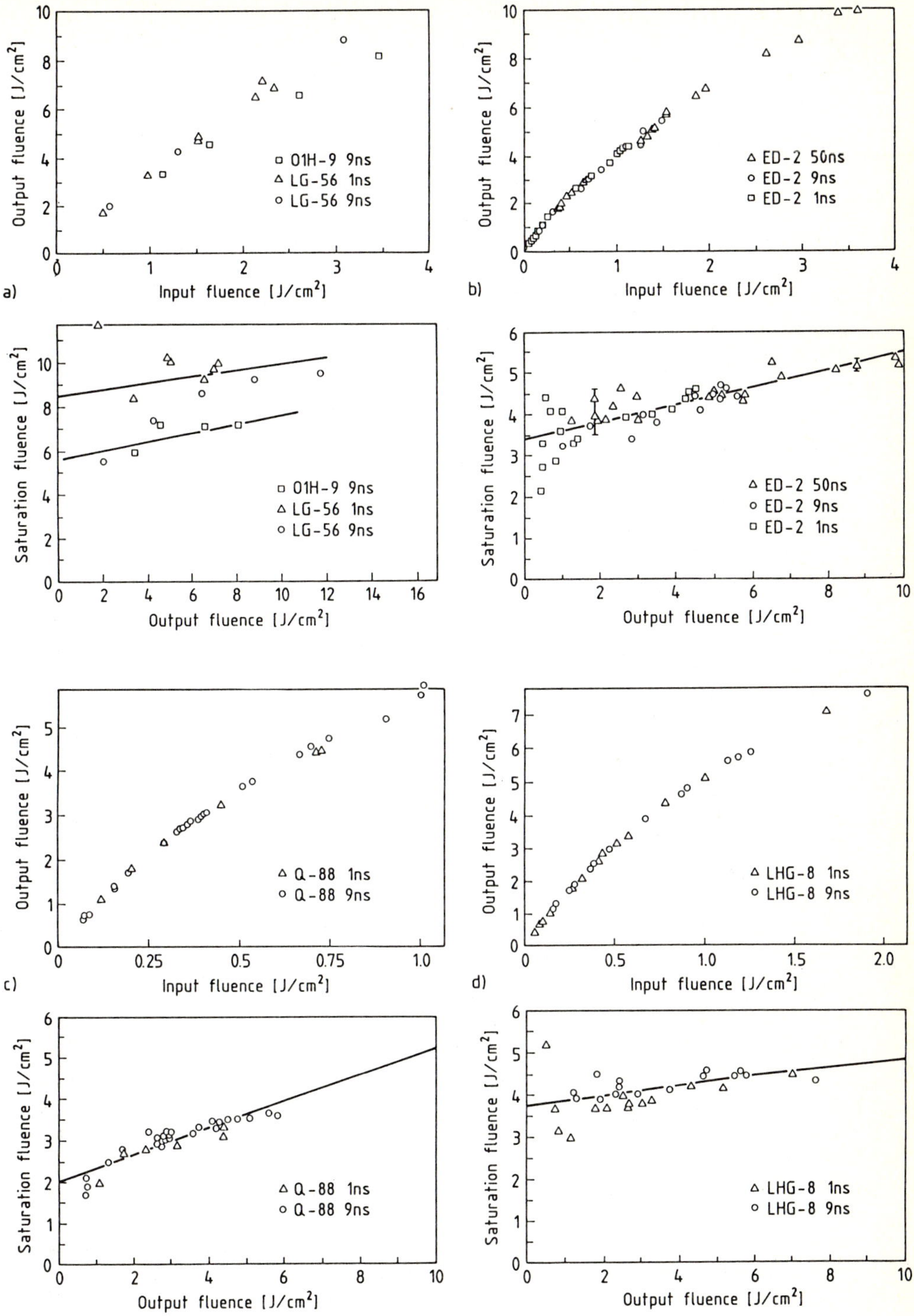

Fig.4.8 a–d. Caption see opposite page

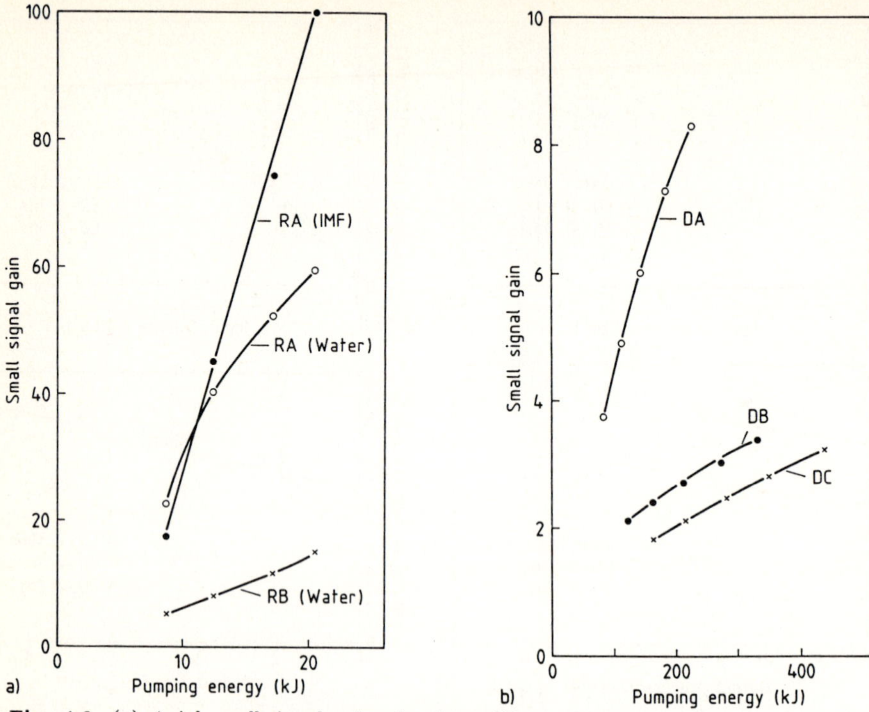

Fig. 4.9. (**a**) Axial small-signal gain of rod amplifiers with 2.5 cm (RA) and 5 cm clear aperture (Hoya laser glass LHG-7). The small rod was cooled either by water or by an index matching fluid (IMF) [4.22]. (**b**) Small-signal gain of several disk amplifiers (Hoya, LHG-7) with clear apertures of 10 cm (DA), 15 cm (DB) and 20 cm (DC) [4.22]

Pulsewidth ranges from 1 to 20 ns and the effect on saturation fluency is barely noticeable, however the increase of E_s with output fluency is quite pronounced.

The next step in the design of a glass amplifier is the calculation of the small signal gain. According to (4.13) we need to determine the stored energy density in the rod, which depends on the flashlamp operating parameters and energy output as well as on the pump cavity design and transfer efficiency.

Figure 4.9a shows the axial small-signal gains of rod amplifiers with diameters of 25 mm (RA) and 50 mm (RB), respectively. The 380 mm long glass rods are pumped by six 15 mm I.D. xenon flashlamps. the glass material is Hoya LHG-7 with Nd_2O_3 concentration of either 1.2 (RA) or 0.6 wt%(RB), respectively. The rod is in contact with the cooling liquid which flows in a Pyrex jacket that surrounds the laser rod. The gain of the amplifer RA depends on the cooling liquid which suppresses the parasitic oscillation. The axial gain at the pumping energy of 22.5 kJ is 59 with water as the coolant and 100 with the index-matching fluid (aqueous solution of $ZnCl_2$ and $SmCl_3$.) The axial gain of the amplifier RB is 15.5 at the pumping energy of 22.5 kJ. Gain coefficient and stored energies are given in Table 4.1.

Table 4.1. System and performance parameters of rod (RA, RB) and disk (DA, DB, DC) Nd : phosphate glass amplifiers [4.22]

		Clear aperture [cm]	Laser glass dimension [cm]	Nd_2O_3 concent. [wt•%]	Flashlamp arc length [cm]	Pump energy [KJ]	Small signal gain	Gain coefficient [cm^{-1}]	Stored energy [J/cm^3]	Efficiency[a] [%]
ROD	RA	2.5	38 × 2.5	1.2	30 (6 lamps)	22.5	59	0.136	0.66	0.5
	RB	5.0	38 × 5.0	0.6	30 (6 lamps)	22.5	15.5	0.091	0.44	1.2
DISK	DA	10	21.4 × 11.4 × 2.4 (6 disks)	2.9	127 (16 lamps)	220	8.5	0.124	0.60	0.75
	DB	15	32.5 × 18.4 × 3 (4 disks)	2.0	127 (24 lamps)	330	3.4	0.085	0.41	0.70
	DC	20	40 × 21.4 × 3.2 (3 disks)	1.9	127 (32 lamps)	440	3.2	0.101	0.49	0.72

[a] Stored energy/pump energy

For comparison, Table 4.1 and Fig. 4.4 show also data on these three types of disk amplifiers having clear apertures ranging from 10 to 20 cm. The numbers given in Table 4.1 for the stored energy density, and pumping efficiency E_s/E_p (energy stored in laser glass divided by the pumping energy can be used as guidelines for the design of glass amplifiers).

For a critically damped lamp pulse of about 300 μs, one can expect typically a 0.5 to 1.5 % conversion of pump input energy per unit rod volume to energy stored per unit rod volume. The numerical value for the stored energy normally ranges from 0.2 to 0.8 J/cm^3.

the spread in the ratio of E_{st}/E_p and in the numerical value of E_{st} is due mainly to the following:

1) The conversion efficiency E_{st}/E_p increases for large-diameter amplifiers because they are optically thicker. However, because of the large rod volume, the pump energy density and the stored energy per volume decrease for large amplifiers, i.e., large amplifiers have low gain but high storage.

2) As the amplifying medium becomes large, fluorescence depumping effects such as amplified spontaneous emission and parasitic oscillations tend to decrease the pumping efficiency.

The effect of fluorescence losses is demonstrated rather dramatically in Fig. 4.9a. The use of an index-matching fluid eliminates the onset of radial parasitic modes thereby allowing a much higher energy storage compared to a rod with a glass/water interface. Fluorescence losses are also the reason for the lower efficiency of the large 20 cm disk compared to the small 10 cm disk amplifier in Table 4.1.

3) Flashlamp pulse-width and -shape is critical in glass amplifiers because the lamp pulse is comparable to, or larger than, the fluorescence time of the gain storage medium. As a result, considerable depletion of the upper level can take place as a result of fluorescence.

The fluorescence time in Nd : glasses is between 200 and 300 μs. It is very difficult to obtain flashlamp pulses which are shorter than 300 μs without sacrificing lamp life to an intolerable value.

4) The main absorption bands for glass are between 0.6 and 0.9 μm. If on increases the electrical input to a flashlamp, then the higher current density will cause a blue shift of the radiation. As a result, the pump source becomes less efficient and one observes a gain foldover in the amplifier if one plots gain as a function of lamp input [4.7, 23, 24].

5) Of course, in addition to the factors mentioned above, the design of the pump chamber, flashlamp configuration, reflectivity of the walls etc. will have a major influence on peak energy storage and gain uniformity.

Equation (4.10) can be graphically represented, as shown in Fig. 4.10. The energy output is plotted as a function of small signal gain G_0, with the saturation density as parameter. Depending on the glass type and desired output energy the saturation fluence can be obtained from Fig. 4.8. The small-signal gain G_0 follows from the stored energy density and rod length.

To obtain the gain or the output energy for a given input, the graph is used in the following manner. As shown in the insert, the input energy density

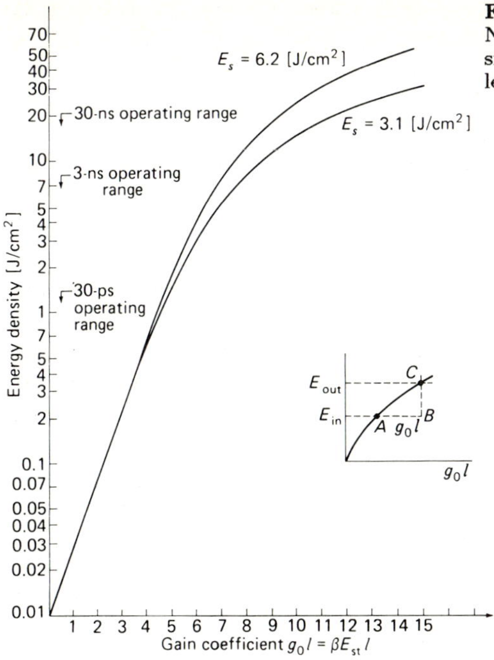

Fig. 4.10. Energy output from a Nd:glass amplifier as a function of small-signal gain coefficient g_0 and amplifier length

determines the point A on the curve. The value for the gain factor $g_0 l$ for the amplifier must then be added to obtain point B. The abscissa of B intersects the curve at C. Also indicated on this graph are the maximum permissible output energy densities in Nd:glass amplifiers for several typical operating ranges. These energy densities are determined by the damage threshold of Nd:glass (Chap. 11). Although these energy levels are not strongly defined, most commercial Nd:glass laser manufacturers will not recommend operation of systems beyond these limits. The maximum power densities for the indicated pulse widths of 30 ps, 3 ns, and 30 ns are 50 GW/cm^2, 3 GW/cm^2, and 0.7 GW/cm^2, respectively.

The two curves of Fig. 4.10 are for two different values of E_s, corresponding to a typical value of a phosphate glass such as Q-88 at an output fluence of around 3 J/cm^2, whereas the high value of E_s is more typical for a silicate glass such as 01H9 at the same output fluence.

4.1.3 Nd:YAG Amplifiers

Amplified spontaneous emission and self-oscillation due to the high gain properties of Nd:YAG effectively limit the energy storage density and therefore the useful energy which can be extracted from a given rod [4.25–27]. The small amount of energy which can be extracted from a Nd:YAG rod and the limits on rod size imposed by crystal-growth constraints have restricted the use of

Nd : YAG amplifiers to a few specialized applications. The small-signal gain coefficient g_0 and stored energy are related according to $g_0 = \beta E_{st}$. With the materials parameters listed in Table 2.4, we obtain for Nd : YAG, β=4.73 cm^2/J, a value 30 times higher than that for Nd : glass and 105 times higher than that for ruby. If we want to extract 500 mJ from a Nd : YAG rod 6.3 mm in diameter and 7.5 cm long, the minimum stored energy density has to be $E_{st} = 0.21\,\mathrm{J/cm^3}$. The small-signal single-pass gain in the rod will be $G_0 = \exp(\beta E_{st} l) = 1720$. A Nd : glass rod and a ruby rod of the same dimensions would have a gain of 1.3 and 1.1, respectively.

As a result of the high gain in Nd : YAG, only small inversion levels can be achieved. Once the gain reaches a certain level, amplification of spontaneous emission effectively depletes the upper level. Furthermore, small reflections from the end of the rod or other components in the optical path can lead to oscillations. These loss mechanisms, which will be discussed in more detail in Sect. 4.4, lead to a leveling off of the output energy versus pump input energy curve in a Nd : YAG amplifier. Figure 4.11 shows plots of energy extracted from a Nd : YAG amplifier versus lamp input energy [4.25]. As can be seen from these curves, the maximum energy which can be extracted from the different rods reaches a saturation level. The data show that the saturation limit of the energy output density from a Nd : YAG amplifier tends to be fairly independent of rod length for rods longer than 50 mm. This contrasts with the strong dependence of energy storage density on rod diameter. In fact, rods with lengths longer than 50 mm tend to decrease the saturation limit, as shown in Fig. 4.11. A thin, long rod will provide a long path and therefore high gain for the spontaneous emission to build up, whereas in a relatively short rod of large diameter more total energy can be stored for the same total gain. Measurements have shown that for rods 5, 6.3, and 9 mm in diameter the maximum extractable energy is 0.3, 0.5, and 0.9 J, respectively. Increasing the temper-

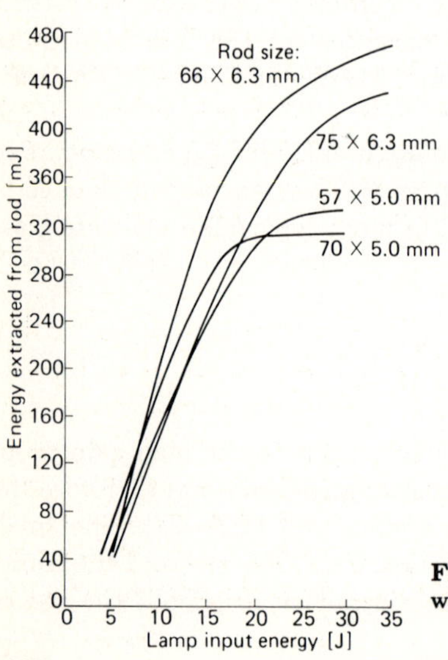

Fig. 4.11. Energy extraction from Nd : YAG amplifiers with different rod sizes [4.25]

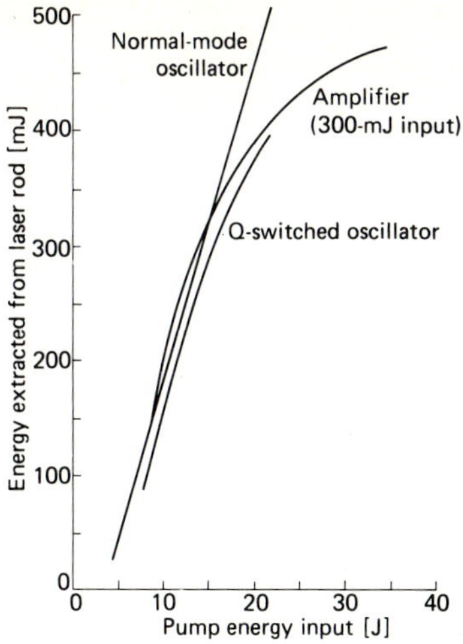

Fig. 4.12. Energy extraction from Nd:YAG operated as a normal mode and Q-switched oscillator, and amplifier of Q-switched pulses. [4.25]

ature of a Nd:YAG rod will reduce its gain, and therefore more energy can be stored before depletion occurs, because of amplified spontaneous emission. For example, the extracted energy from an amplifier was increased from 770 to 926 mJ/cm^2 by raising the temperature from 26 to 96°C.

The relative performances of a laser rod when used as a normal-mode oscillator, as a Q-switched oscillator, and as a single-pass amplifier are displayed in Fig. 4.12 [4.25]. A 0.63- by 6.6-cm Nd:YAG laser rod in a silver-plated, single-ellipse, single-lamp pump cavity was used in all modes of operation. Normal-mode performance was achieved with two plane-parallel dielectric-coated mirrors optimized for performance at the 500-mJ output level. The energy input waveform had a duration of approximately 100 μs at the half-power points, and was obtained with a single-mesh discharge circuit. The Q-switched performance was obtained with a rotating prism switch in which the prism speed and mirror reflectivity had been optimized for maximum efficiency in the output range of 100 to 400 mJ. The single-pass amplifier performance represents the energy extracted from the rod with a 300-mJ input from an oscillator. These data show that all modes of operation are approximately equivalent until the 320-mJ output level is reached. Above this level the modes of operation which require energy storage become much less efficient. The Q-switched oscillator operates at a slightly lower efficieny than does the same rod in the amplifier mode.

If the Nd:YAG crystal is pumped with an input of 14 J, a total of 300 mJ can be extracted from the amplifier. Since the signal input is 300 mJ, the amplifier has a saturated gain of $G_s = 2$ and a total output of 600 mJ. In order to extract 300 mJ from this rod, at least 0.15 J/cm^3 must be stored in the upper level. This corresponds to a small-signal gain of $G_0 = 108$. Using (4.12) we can compare the measured saturated gain with the theoretical gain. For an amplifier operated in the saturation regime $E_{in} \ll E_s$, the theoretical gain is $G = 1 + (E_s/E_{in})g_0 l$. where the saturation energy for Nd:YAG is

$E_s = 1/\beta = 0.2\,\mathrm{J/cm^2}$. With $E_{in} = 1\,\mathrm{J/cm^2}$, $g_0 = 0.71\,\mathrm{cm^{-1}}$, and $l = 6.6\,\mathrm{cm}$ for this amplifier, we obtain $G = 2$, in agreement with the measurement.

As was mentioned in the beginning of this chapter, the fraction of stored energy which can be extracted from a signal depends on the energy density of the incoming signal. In this particular case the amplifier is completely saturated, since $E_{in} = 5E_s$. According to our previous discussion, this should result in a high extraction efficiency. The expression for the saturated gain can be written $G_s = 1 + (E_s/E_{in})\beta E_{st} l$. As we have seen, if we substitute the stored energy E_{st} in this equation with the extracted energy $E_{ex} = 300\,\mathrm{mJ}$, we obtain exactly the measured gain of the amplifier. This suggests that the extraction efficiency of this amplifier, $\eta_a = E_{ex}/E_{st}$, is close to 100%. Figure 4.13 shows the measured extraction efficiency of this amplifier as a function of signal input.

The transition region in which the amplifier output as a function of signal input changes from an exponential (small-signal) relationship to a linear (saturated) relationship occurs for input energy densities of approximately $0.2\,\mathrm{J/cm^2}$, in accordance with the theoretical value. The pulse width for the data presented in Fig. 4.13 ranged from 15 to 22 ns. The curve in this figure shows that for efficient energy extraction a Nd:YAG amplifier should be operated with an input signal of around $1\,\mathrm{J/cm^2}$.

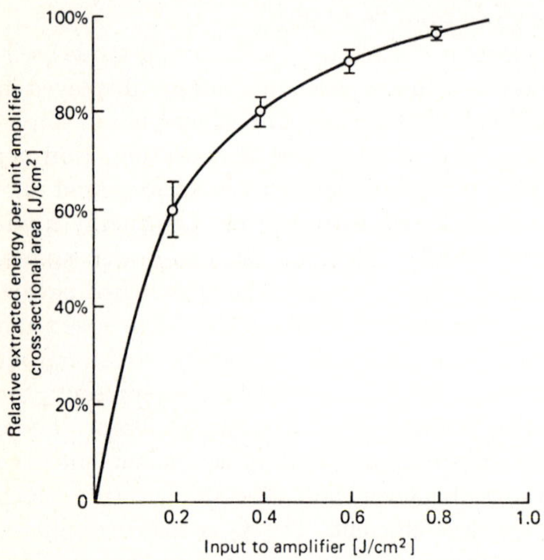

Fig. 4.13. Amplifier energy extraction as a function of oscillator energy density [4.25]

4.2 Steady-State Amplification

If the pulse duration is long compared to the fluorescent lifetime τ_f, the population inversion and amplification coefficient follow the intensity in a quasistationary way. This situation is realized, for example, with the amplification of conventional-mode laser pulses. The nonlinear gain characteristics for laser amplifiers as a function of steady-state radiation have been analyzed for media with homogeneous and inhomogeneous line broadening [4.28–30]. In the former case, the gain decreases proportionately over the entire transition line. In the case of an inhomogeneous line, spectral hole burning can take place if the driving signal is a narrow-band signal. However, in solid-state lasers, materials with an inhomogeneous line such as Nd : glass have a cross-relaxation which is fast compared to the long pulses discussed in this section. For example, the cross-relaxation rate is of the order of nanoseconds in Nd : glass. The following analysis is applicable to systems with homogeneous line broadening, such as in ruby, and with inhomogeneous line broadening, such as in Nd : glass, provided in the latter case that the driving signal is broad band and therefore taps the entire inversion density, or in the case of a narrow-band signal it is assumed that cross-relaxation is infinitely fast.

In Sect. 3.2 we derived an expression for the saturated gain in an oscillator. In the case of an amplifier, the gain depends on the coordinate z measured in the direction of the axis of the amplifier.

$$g(z)=\frac{g_0}{1+I(z)/I_s} \quad , \tag{4.30}$$

where g_0 is the small-signal gain, $g(z)$ is the gain at position z in the amplifier obtained at a signal level $I(z)$, and I_s is the saturation density defined as the signal power, which reduces the small-signal gain by a factor of one-half.

Assuming that the inversion n does not change much over a photon transit time in the amplifier, the power density $I(z)$ builds up along z in the following manner

$$\frac{dI}{dz}=\sigma n(z)I(z)=g(z)I(z) \quad . \tag{4.31}$$

We have ignored losses in this expression. Substituting g from (4.30) into (4.31) and then integrating over the power density I results, after some algebra, in [4.29, 30]

$$\frac{I_{in}}{I_s}=\frac{\ln(G_0/G)}{G-1} \quad , \tag{4.32}$$

where $G_0=\exp(g_0 l)$ is the small-signal, single-pass gain of the amplifier of length l and $G=I_{out}/I_{in}$ is the effective gain. Equation (4.32), which is plotted in Fig. 4.14, permits one to calculate the power gain of solid-state lasers in the steady-state regime. For cases where $G \gg 1$, we can simplify (4.32) and write

$$\frac{I_{out}}{I_s}=\ln\left(\frac{G_0}{G}\right) \quad . \tag{4.33}$$

The gain drops to one-half for $I_{out}=0.69 I_s$ and to $1/e$ for $I_{out}=I_s$.

For small-signal levels $G I_{in} \ll I_s$ we obtain $G=G_0$, or

$$I_{out}=I_{in}\exp(g_0 l) \quad . \tag{4.34}$$

Accordingly, the small-signal gain is identical for short-pulse and steady-state

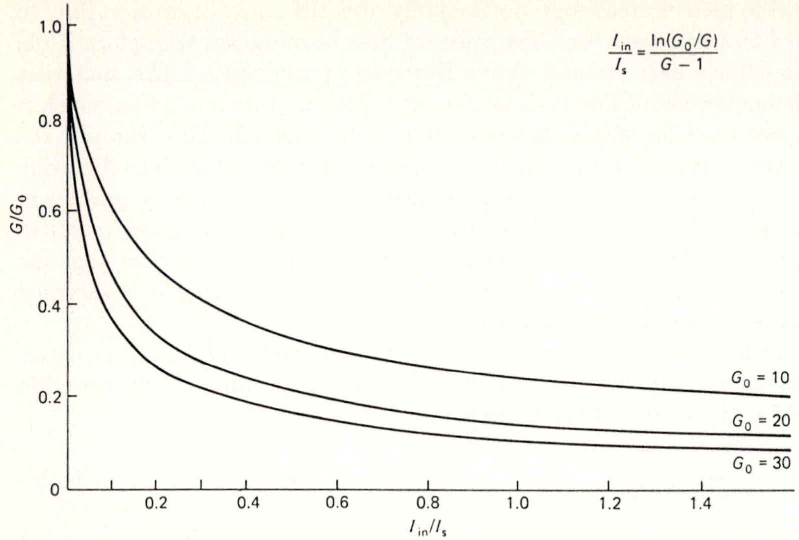

Fig. 4.14. Steady-state gain versus input power density in laser amplifiers. Parameters are the small-signal gain G_0 and the saturation power density I_s

amplification. At large signal levels $I_{in} \gg I_s$, one obtains

$$I_{out} = I_{in} + I_s g_0 l \quad . \tag{4.35}$$

In this energy regime the output increases linearly with length. In a lossless amplifier, the gain decreases with increasing beam intensity to a small but finite value. In the presence of scattering or absorption loss, however, the gain approaches one at a beam intensity given by [4.5].

$$\frac{I_{out}}{I_s} \approx \frac{g_0}{\alpha} \quad , \tag{4.36}$$

where α is the loss coefficient in the amplifying medium. For higher power densities the overall gain no longer increases with amplifier length.

4.2.1 Ruby Amplifier

The calculation of the steady-state gain in a ruby amplifier is somewhat complicated because the saturation density I_s depends not only on fundamental material parameters but also on the pump rate W_p. In principle, this is true for a four-level system as well. As we have seen in Chap. 3, in a four-level system $W_p \tau_f \ll 1$, and the effect of the pump rate on gain saturation can be ignored. In a three-level system $W_p \tau \geq 1$, and the pump rate must be considered in the term for saturation. In the pulse regime we defined a saturation energy E_s, in the case of steady-state amplification a power density I_s will be used to describe saturation. According to (3.64), we can write (Fig. 4.15)

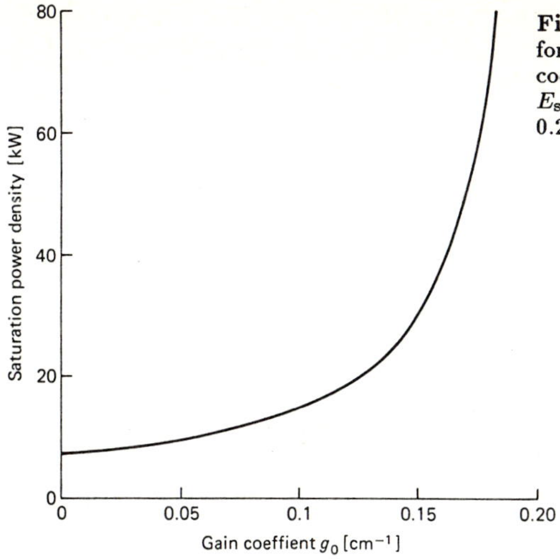

Fig. 4.15. Saturation power density I_s for ruby as a function of small-signal gain coefficient g_0, calculated from (4.37) with $E_s = 11.0\,\mathrm{J/cm^2}, \tau_f = 3\,\mathrm{ms}$, and $\alpha_o = 0.2\,\mathrm{cm^{-1}}$

$$I_s = \frac{2E_s}{\tau_f(1 - g_0/\alpha_0)} \quad . \tag{4.37}$$

From Figs. 4.14 and 15, the gain as a function of signal input power can be obtained for ruby lasers, provided that the small-signal gain G_0 or the gain coefficient g_0 is known.

In order to illustrate the application of these calculations, we will consider the input versus output characteristics of a 20- by 2-cm ruby rod pumped by a helical flashlamp. The small-signal gain for this ruby laser amplifier in the pulsed mode was measured to be 29 for an input energy of 20 kJ. In the long-pulse mode the small-signal gain was $G_0 = 20$ for the same flashlamp input. Figure 4.16 shows the measured amplifier output as a function of input. The

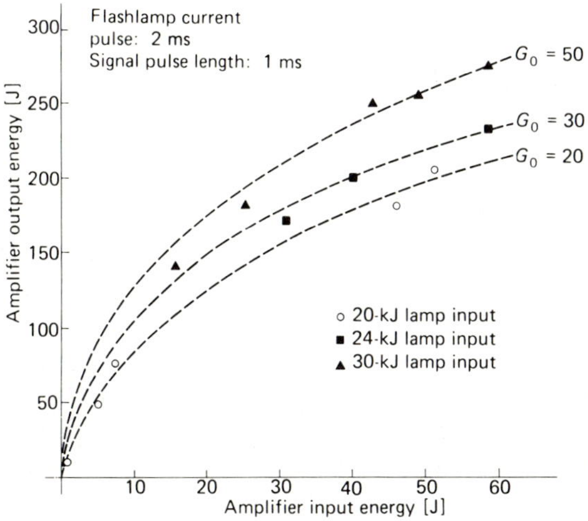

Fig. 4.16. Output versus input energy for a steady-state ruby amplifier. Signal pulse length was 1 ms and flashlamp current pulse had a duration of 2 ms

parameter is the flashlamp pump energy. For an input of 24 kJ, the amplifier can be described by (4.32) if one uses $G_0 = 30$ and $I_s = 50\,\mathrm{kW/cm^2}$, which follows from Fig. 4.15 for a gain coefficient of $0.17\,\mathrm{cm^{-1}}$. A best fit through the measured data for $G_0 = 20$ and $G_0 = 50$ is obtained if one uses also $I_s = 50\,\mathrm{kW/cm^2}$ for these gain curves. It turns out that the saturation parameter might have to be fitted to experimental data. As mentioned before, the equations derived in this subsection are based on the condition that the signal duration is long compared to the fluorescence time. In the typical pulse-width regime of about 1 ms, this condition is not met.

4.2.2 Nd : Glass Amplifier

With the materials parameter of Nd:glass $h\nu = 1.89 \times 10^{-19}\,\mathrm{Ws}$, $\gamma = 1$, $\sigma_{21} = 3.5 \times 10^{-20}\,\mathrm{cm^2}$, and $\tau_f = 300\,\mu\mathrm{s}$, we obtain $I_s = 20\,\mathrm{kW/cm^2}$ from (3.29). Introducing this parameter into (4.32), we can calculate the normalized gain as a function of the normalized input power for the amplification of long pulses. The gain as a function of input power level can also be obtained from Fig. 4.14. The small-signal gain must be known for a particular amplifier. If no experimental data are available, the small-signal gain can be estimated using the graphs and formulas derived in Sect. 4.1.

The performance of a Nd : glass amplifier [4.31] employed in the amplification of long pulses is shown in Fig. 4.17. The unsaturated gain for the amplifier

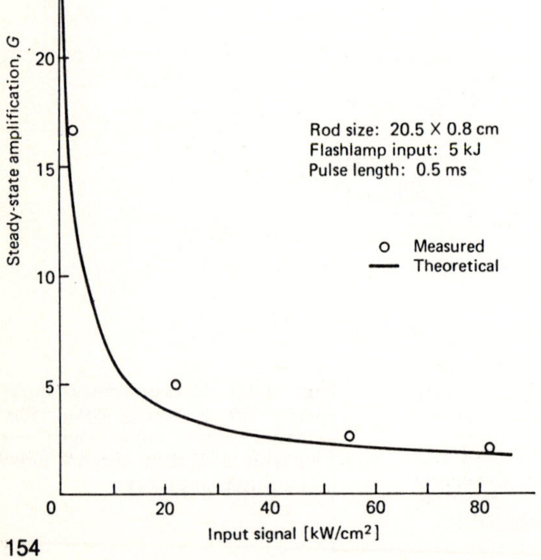

Fig. 4.17. Steady-state amplification in a 20.5-cm-long and 0.8-cm-diameter Nd : glass rod. The signal input had a duration of $300\,\mu\mathrm{s}$; $I_s = 20\,\mathrm{kW}$; $G_0 = 63$. [4.31]

was measured to be 63. At a maximum driving power density of $82.5\,\mathrm{kW/cm^2}$, the gain decreased to 2.2. As seen from Fig. 4.17, there is a reasonably good agreement between theory and experiment.

4.3 Signal Distortion

As an optical signal propagates through a laser amplifier, distortions will arise as the result of a number of physical processes. We can distinguish between spatial and temporal distortions.

4.3.1 Spatial Distortions

The development of high-power laser oscillator-amplifier systems has led to considerable interest in the quality of the output beam attainable from these devices. Beam distortions produced during amplification may lead either to an increase in divergence or to localized high energy densities which can cause laser rod damage. We will consider the main phenomena producing a spatical distortion, which is analogous to a wavefront or intensity distortion.

Nonuniform Pumping. Due to the exponential absorption of pump light, the center of the rod is pumped less than the edges. Figure 4.18 shows the stored energy profile of several Nd : glass amplifier stages of the Omega system built at the University of Rochester [4.32]. The curves were obtained by measuring the small-signal gain with a narrow probe beam across the rod on five amplifier stages of increasing diameter. In order to provide a uniform gain profile as much as possible, the Nd concentration is decreased for increasing rod diameters.

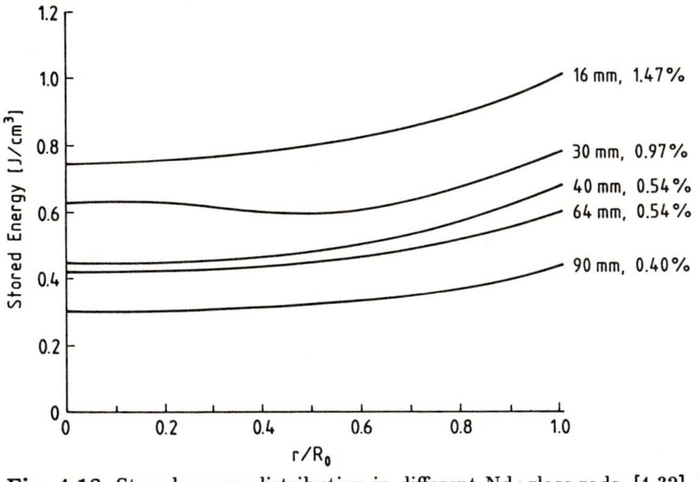

Fig. 4.18. Stored energy distribution in different Nd : glass rods. [4.32]. Rod diameters and doping levels (wt.%) are indicated on the curves

Nonuniformities in the Active Material. Even laser rods of excellent optical quality contain a small amount of inherent stress, index of refraction variations, gradients in the active ion concentration, contaminants, inclusions, etc. These nonuniformities will significantly modify the energy distribution of an incoming signal. An initially smoot energy distribution usually takes on the form illustrated in Fig. 4.20 after passing through a laser amplifier.

Gain Saturation. A beam propagating through an amplifier suffers distortion because of the saturation-induced change in the distribution of gain. The weaker portions of the signal are amplified relatively more than the stronger portions because they saturate the medium to a lesser degree. The spreading of a Gaussian input beam due to homogeneous gain saturation obtained by integrating (4.30) numerically step by step in the z direction of the amplifier is shown in Fig. 4.19 [4.33]. The figure shows the change in the radial intensity distribution for a TEM_{00} beam (dashed line) entering a homogeneously broadened laser amplifier with a radially uniform small-signal gain of 40 dB. Each curve corresponds to a specific value of the normalized signal parameter I_{in}/I_s. The output distribution has been normalized to its value on the amplifier axis to make the distortion more evident. Figure 4.19 shows that even for input signals 10^{-4} below the saturation flux (i.e., a signal that begins to saturate only near the amplifier output), there is about a 10 % expansion of the e^{-2} radius. When the input has increased to a value that begins to saturate the amplifier at its input ($I_{in}/I_s = 1$), the e^{-2} radius has just about doubled. Calculations presented in [4.33] reveal that for a further increase in input signal the beam expansion falls below the $I_{in}/I_s = 1$ curve. This is to be expected, since for very large input signals the amplifier is so highly saturated that it adds very little to the signal passing through and thus cannot distort it. Hence, it would seem that the maximum distortion occurs when the input signal is about equal to the saturation flux. For either extreme, $I_{in} \ll I_s$ and $I_{in} \gg I_s$, the output resembles the input in shape.

Diffraction Effects. An amplifier rod represents a limiting aperture for an incoming beam. Diffraction effects at the edges of the amplifier rod will give rise to Fresnel rings, which can strongly disturb the beam uniformity. One method

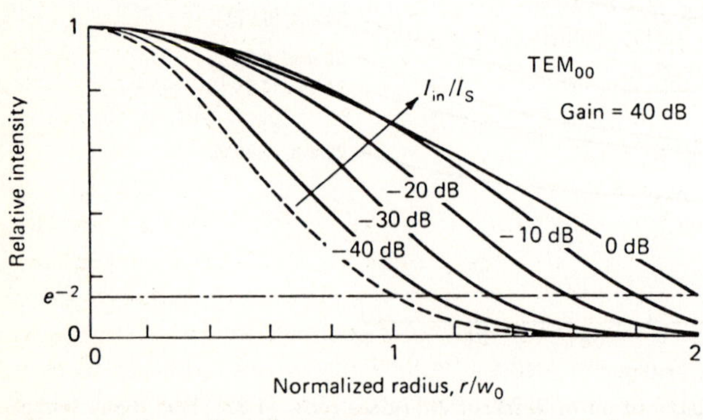

Fig. 4.19. Spreading of a TEM_{00} input beam due to homogenous gain saturation. [4.33]

to minimize the effect of diffraction is to utilize a Gaussian-shaped beam. If the beam size is chosen properly with respect to the limiting aperture, the energy density on the edges will be low enough to produce a negligible spatial modulation across the beam. *Campbell* and *DeShazer* [4.34] calculated the near-field beam pattern for Gaussian beams truncated by circular apertures for various combinations of aperture radius a and beam radius w. (For $w = a$ the amplitude of the beam is reduced to $1/e$ at the edge of the aperture, and 86.5 % of the energy is transmitted through the aperture). It was shown that truncation of a Gaussian beam by a circular aperture introduced structure in the near-field beam profile. However, if the ratio of aperture radius to beam radius is 2 or greater ($a \geq 2w$), then the structure is negligible and the transverse-beam profile remains very nearly Gaussian. For $a = 2w$ the power at the edges of the aperture is -33 dB as compared to the maximum intensity.

Trenholme [4.35] has demonstrated empirically that if the beam is apertured so as to cut off portions of intensities below K, where K is a specified fraction of the beam's maximum intensity, then the peak-to-peak magnitude of the diffraction ripple will be on the order of $0.40\,\mathrm{K}^{1/2}$.

Methods to reduce diffraction effects, in addition to tailoring the Gaussian beam profile to the limiting aperture, consist of using spatial filtering. A spatial filter consists of two positive lenses and a pinhole at the common focal plane. Spatial filtering in a amplifier chain is obtained if we match an aperture diameter to the first dark ring of the far-field pattern produced by the preceeding stage. No modulation will apear in the regime of the Airy disc. In a more general sense, low-pass spatial filters in the amplifier chain filter the beam noise resulting from inperfection of the optics and diffraction. By choosing the pinhole diameter one can control which spatial frequencies are passed. The use of spatial filters in high peak power systems leads to the requirement that they be operated in vacuum, because the high intensities reached at the focal spot lead to optical breakdown in air.

Figure 4.20 illustrates spatial modulation due to diffraction and ripples induced by dust and scattering particles of a high power beam. (Shiva beam line operated at 700 J in 600 ps [4.36]). Figure 4.20b shows the relative intensity of one Shiva beam line before and after installation of a spatial filter at the output [4.36]. As one can see from these densitometer scans, the intensity of the high frequency spike was reduced by a factor 4 by the low-pass filter. Besides the removal of unwanted ripples in the beam, spatial filters can at the same time be utilized for imaging or beam relaying and for magnification to match different diameter amplifiers [4.37].

Thermal Distortions. The nonuniform pumping which leads to a higher gain coefficient on the edges of a laser rod also causes a nonuniform temperature profile across the rod. The absorbed pump power raises the temperature of the surface of the rod above the temperature of the rod center. As a result, a negative thermal lens is created, which distorts the wavefront of the beam. Associated with the thermal gradients are stresses in the active material which lead to birefringence. Therefore the thermal effect leads to strong optical aber-

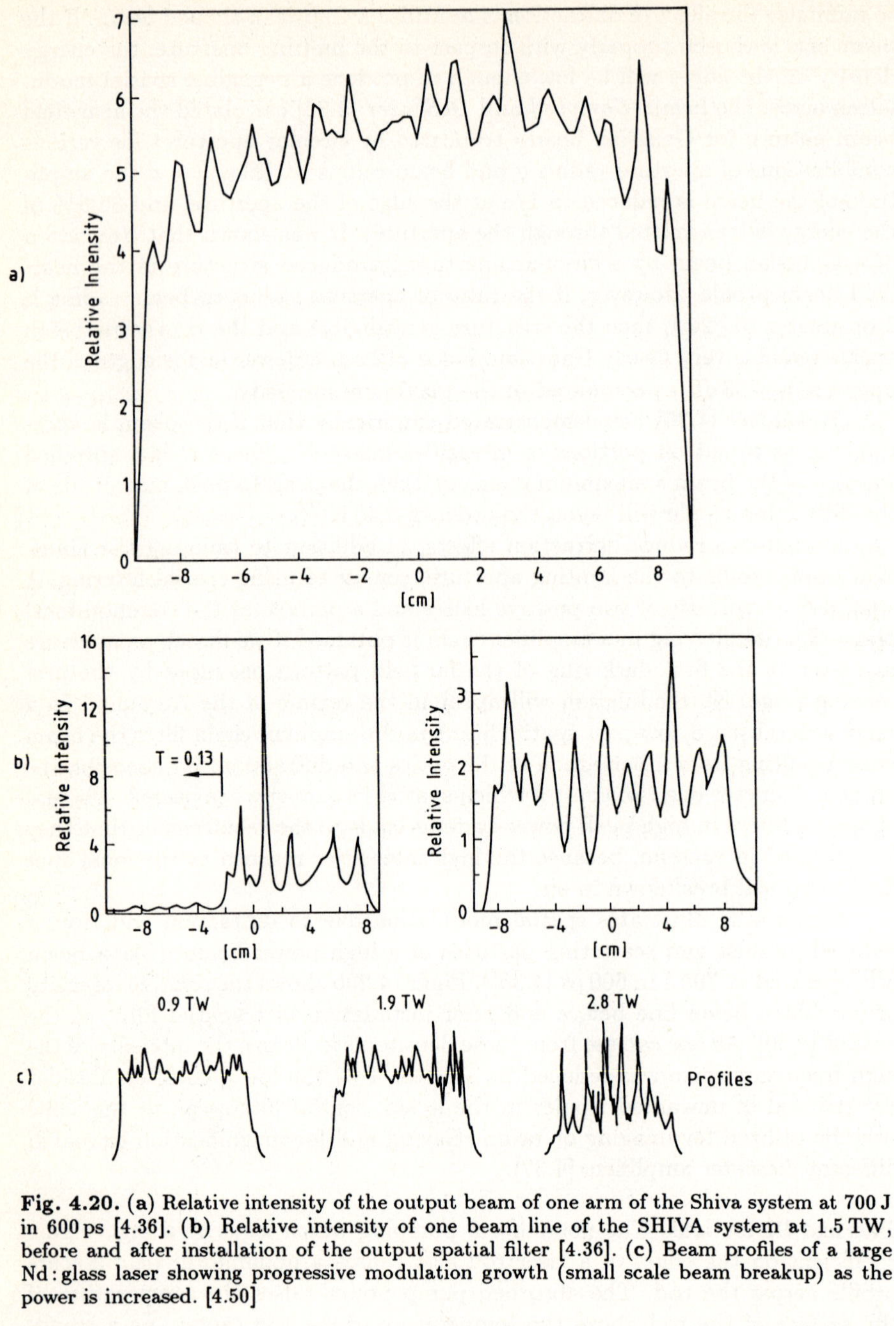

Fig. 4.20. (**a**) Relative intensity of the output beam of one arm of the Shiva system at 700 J in 600 ps [4.36]. (**b**) Relative intensity of one beam line of the SHIVA system at 1.5 TW, before and after installation of the output spatial filter [4.36]. (**c**) Beam profiles of a large Nd: glass laser showing progressive modulation growth (small scale beam breakup) as the power is increased. [4.50]

rations which can be only partially compensated for by insertion of correcting elements. Thermal effects in laser rods are discussed in detail in Chap. 7.

Index Nonlinearity. When a powerful laser pulse propagates through an isotropic material, it induces changes in the refractive index described to first order by the equation

$$n = n_0 + n_2|E|^2 \quad , \tag{4.38}$$

where n_0 is the linear and n_2 is the nonlinear index of refraction, and $|E|^2$ is the time average square of the electric field. From (4.38) it follows that a laser beam propagating in a transparent medium induces an increase in the index of refraction by an amount proportional to the laser intensity. This effect which causes a self-induced aberration of the beam is known as whole-beam phase distortion [4.38, 39]. The index nonlinearity gives also rise to so-called small-scale ripple growth, in which case the amplitude of small perturbations grows exponentially with propagation distance [4.40–43].

The principal limit on the performance of a high-power glass laser such as employed in laser fusion research is the nonlinear refractive index of transparent dielectrics [4.44–49]. In the presence of an intense light wave, with a power density of gigawatts per centimer squared, or more, the refractive index of a transparent dielectric is increased. This intensity-induced index change is small, on the order of 5 ppm at $10^{10}\,\mathrm{W/cm^2}$, but is has profound effects.

The first problem caused by the nonlinear index is the accumulation of self-induced phase front distortions on the spatial envelope of the pulse (whole-beam phase distortion) which alter the focusing properties of the system output. In order to maintain the optical quality of the beam the integrated nonlinear index along the optical path given by the parameter

$$B = \frac{2\pi}{\lambda_0}\int \frac{\Delta n}{n}\,dl \tag{4.39}$$

must be kept below a critical value. In (4.39), where n is the refractive index, Δn is the nonlinear index change, and λ_0 is the wavelength of the light in vacuum, the integral is taken along any ray passing through the laser. The value of B at any position across the beam gives the radians of phase delay which that portion has undergone when compared to a low-intensity beam. For fusion applications, values of B at the beam center above about 9 (1.5 waves of delay) are unacceptable.

The second problem caused by the nonlinear index is small-scale ripple growth. In this process, a small region of higher intensity on a beam raises the index locally, which tends to focus light toward it and raise the intensity even more. It has been shown that ripples grow from their original amplitude by a factor of

$$G = \mathrm{e}^{B} \tag{4.40}$$

which involves the same B integral as the whole-beam distortions. Experimentally, it is found that fusion-quality beams can be produced only when B is less than 4 or 5, because of the small-scale growth.

Figure 4.20c illustrates the progressive growth of beam modulation as the power is increased in a high power fusion glass laser [4.50].

High-power solid-state laser systems must not only be carefully designed and built to minimize beam irregularities caused by dirt particles, material imperfections, diffraction from apertures, and other "noise" sources; some provision must also be made for removing the remaining small-scale structure, before it becomes large enough to cause severe degradation of the laser's focusing properties. Apodizing apertures and vacuum spatial filters inserted in the beam have resulted in improved performance [4.49–51].

4.3.2 Temporal Distortions

Pulse-shape Distortion. For a square pulse traversing an amplifier, the leading pulse edge sees a larger inverted population than does the trailing edge. This occurs simply because the leading edge stimulates the release of some of the stored energy and decreases the population inversion, thereby causing the trailing edge to see a different inverted population. Thus less energy is added to the final portions of a pulse than to the leading regions. The change of an arbitrary input pulse shape after propagation is described in [4.1, 2]. If we consider for simplicity a square pulse, we can write for the power of the pulse

$$P(x,t) = P_{\text{in}}\left\{1 - [1 - \exp(-g_0 x)] \exp\left[-\left(\frac{t - x/c}{t_{\text{p}}}\right)\right] \frac{E_{\text{in}}}{E_{\text{s}}}\right\}^{-1} , \quad (4.41)$$

where g_0 is the small-signal gain coefficient, E_{in} and E_{s} are the input and saturation energy densities, respectively, and t_{p} is the length of the pulse. Note that the power is dependent on time and location x within the amplifying medium and also on the input energy $E_{\text{in}} = P_{\text{in}} t_{\text{p}}$.

At the input of the amplifying medium ($x = 0$) we have $P(0,t) = P_{\text{in}}$. For the square-pulse case considered here, however, we note that for any value of input the gain at the leading edge of the pulse, that is, for the position $x = ct$, is exponential with distance through the amplifier. The trailing edge of the pulse is characterized by $t_{\text{p}} = t - x/c$. In Fig. 4.21 the change of a rectangular input pulse is shown. The pulse undergoes a sharpening of the leading edge and a reduction of its duration. Pulse-shape distortions for input pulses which have profiles other than rectangular are treated in [4.4, 5]. The pulse-shape evolution depends appreciably on the rise time and shape of the leading edge of the input pulse. In general, one observes a forward shift of the peak as the pulse transverses the amplifying medium.

Frequency Modulation. The intensity-induced index changes are not only responsible for the catastrophic self-focusing problems in high-power glass laser amplifiers but also cause a frequency shift of the pulse in the amplifying medium. In mode-locked pulses, this frequency modulation causes a broadening of the

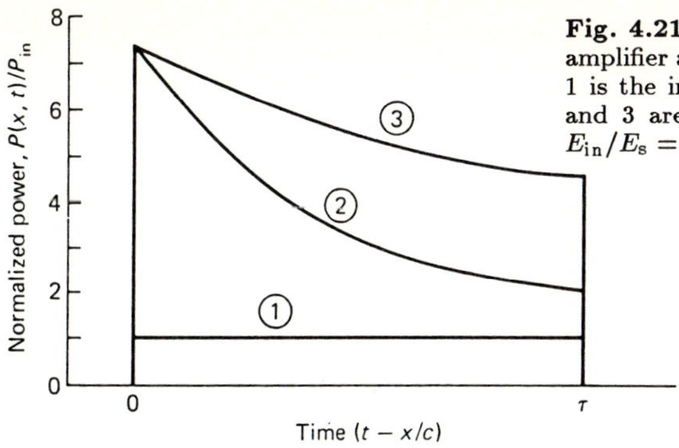

Fig. 4.21. Change of the pulse shape in an amplifier as a result of gain saturation. Curve 1 is the initially rectangular pulse, curves 2 and 3 are for $g_0 = 0.1\,\mathrm{cm}^{-1}$, $L = 20\,\mathrm{cm}$, $E_{in}/E_s = 0.5$, and $E_{in}/E_s = 0.1$, respectively

pulse width [4.41, 42, 52–57]. The frequency shift $\Delta\nu$ impressed on a wave going through a medium of length l, whose refractive index changes at the rate $\delta n/\delta t$, is given by

$$\Delta\nu = -\frac{l}{\lambda}\frac{\delta n}{\delta t} \quad , \tag{4.42}$$

where $\delta n = n_2|E|^2$, λ is the wavelength in vacuum, and δt is the rise time of the pulse. An experimental demonstration of this Doppler shift is given in [4.52]. The δn term is approximated for Nd : glass by

$$\delta n \approx (10^{-15}\,\mathrm{cm}^2/\mathrm{W}) \times I \quad ,$$

where I is the intensity in watts per square centimeter. Thus a mode-locked pulse with a 5-ps rise time and a peak power of $3\,\mathrm{GW/cm}^2$ will experience a frequency shift of $\Delta\nu/\nu = 10^{-3}$ or 10 Å in a 50-cm-long Nd : glass amplifier. *Duguay et al.* [4.56] measured a mean spectral broadening of $5\,\mathrm{cm}^{-1}$ per round trip in a mode-locked glass amplifier.

4.4 Gain Limitation and Amplifier Stability

One of the main considerations in the design of an amplifier or amplifier chain is its stability under the expected operating conditions. In a laser operating as an amplifier or as a Q-switched oscillator the active material acts as an energy storage device, and to a large extent its utility is determined by the amount of population inversion which can be achieved. As the inversion is increased in the active material, a number of different mechanisms begin to depopulate the upper level. We can divide these depumping processes into two categories.

4.4.1 Spontaneous Decay, Superradiance, and Lateral Depumping

The principal factor limiting the degree of population inversion is the depletion of metastable laser atoms as a result of both spontaneous and amplified spontaneous emission. The latter process is described in the literature either as superradiance [4.12, 58], superfluorescence [4.59, 60], superluminescence [4.61, 62], or photon avalanche [4.63, 64]. As soon as the pump rate reaches a value which leads to high inversion, amplification of spontaneous emission sets in. The amplification manifests itself in a strong increase of spontaneous emission. In the absence of internal reflections, one can assume that the only spontaneous photons which are effective in generating superradiance are those which propagate at small angles with respect to the axis and which transverse the entire length of the rod. From each end of the rod an intense beam of light is emitted having approximately a solid angle.

$$\Omega = \frac{n^2 A}{L^2} \quad , \tag{4.43}$$

where n is the refractive index, and L and A are the rod length and cross-sectional area, respectively. As a result of refraction at the end faces, the geometrical aperture angle of the rod A/L^2 is increased by n^2. The favorable geometry for the formation of superradiance is a slender rod. This geometry provides long passes which, combined with a high gain, can cause a stimulated buildup from individual incoherent spontaneous decays. *Tonks* [4.63] has developed a criterion for the maximum useful gain relative to rod length and rod cross section. He concluded that internal reflections from the cylindrical rod surfaces can drastically increase the effective aspect ratio, thus reducing the maximum gain which can be obtained from a particular rod.

Near the ends of high-gain rods, amplified spontaneous emission can give rise to photon fluxes sufficiently intense to cause appreciable or even complete depumping. Calculations on the photon flux created by superradiance in three- and four-level systems can be found in [4.12, 63]. Experimental results for ruby habe been reported in [4.59]. Figure 4.22 shows the superradiance obtained from an uncoated Nd : glass rod, 1.25 cm in diameter and 25 cm long, pumped by a helical flashlamp. At a flashlamp input of 6 kJ, an output of 60 mJ in each direction is obtained. As can be seen from Fig. 4.22, the calculated full beam angle $\Theta = Dn/L$, which is $\Theta \approx 4.5°$ for the parameters mentioned above, contains 75 % of the total energy.

Superradiance has been used in a few specialized applications as the source of output energy from a laser. Because of the high gain obtainable in long Nd : glass laser rods, several nonresonant oscillators (no laser cavity mirrors) have been built. In these systems the spontaneous emission was amplified in one or two passes to generate a very powerful beam. In one such system [4.64], a photon avalanche was created by optically pumping six laser rods to a high inversion and then suddenly connecting them in series. The cumulative 155-dB gain wass sufficient to generate a 1-GW, 70-ns photon avalanche pulse at the

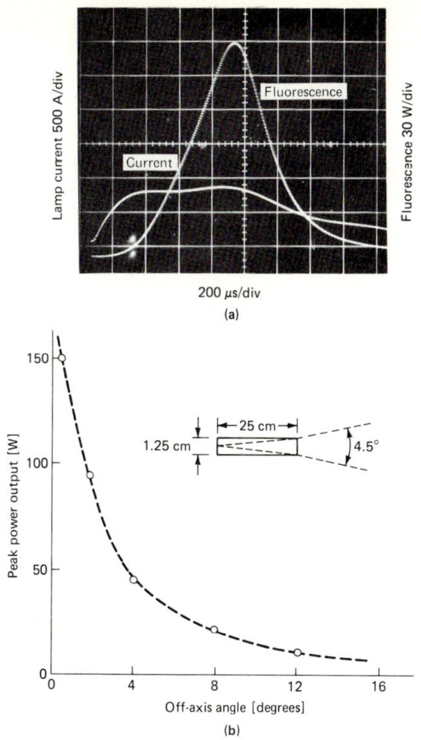

Fig. 4.22. (a) Flashlamp current pulse and superradiant output from a 25-cm Nd : glass rod pumped at 5 kJ. (b) Directionality of superradiance

output. Because of the absence of any resonant structure, the output from a superradiant laser is spectrally and temporally smooth, i.e., longitudinal and transverse-mode structures as well as spiking of the output do not occur in these systems. In contrast to spontaneous emission, superradiance has some temporal coherence since each spontaneously emitted light avalanche is itself coherent.

Depopulation caused by incident pump intensities at the wavelength corresponding to that of the laser transition can lead to a reduction of stored energy. Gain saturation resulting from amplification of pump radiation in the active medium has been observed mainly in Nd : glass and Nd : YAG amplifiers [4.26, 60]. In addition to the depumping process caused by the lamp radiation, fluorescence being emitted from the laser rod escaping out into the laser pump cavity can be reflected back into the laser rod, and thereby stimulating further off-axis emission. Thus the presence of pump-cavity walls which have a high effective reflectivity at the fluorescence wavelength causes a transverse depumping action which depletes the energy available for on-axis stimulated emission. Elimination of these effects may require the use of optical filters in the pump cavity, cladding of the laser rod with a material which absorbs at the laser wavelength, or the addition of chemicals to the cooling fluid which serve the same purpose.

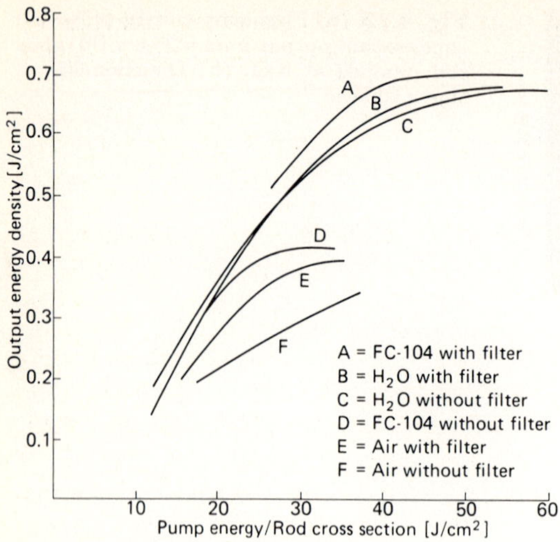

Fig. 4.23. Energy extraction density of YAG : Nd as a function of coolant properties. [4.25]

The influence of coolant properties upon the performance of a Nd : YAG laser amplifier is shown in Fig. 4.23. Two properties of the cooling fluid are of interest. These are the refractive index and the 1.06-μm absorption. The data is shown for a cored 2-cm-diameter by 10-cm-long Nd : YAG rod operated with water, FC-104, and air as cooling media [4.25]. FC-104 is a fluorcarbon flashlamp coolant made by the 3M Company. All fluids were used with and without a 1.06-μm absorbing filter (samarium-doped glass) in the pump cavity which was located between the rod and lamp. The addition of a 1.06-μm absorbing filter when water was used as a coolant made little difference in amplifier performance. Water possesses an appreciable intrinsic absorption at 1.06 μm. The performance obtained with FC-104, which is highly transparent at 1.06 μm, was greatly affected by the use of 1.06-μm absorbing filters. The performance obtained with air cooling was also affected by the use of a 1.06-μm absorbing filter, but the change was not nearly as marked as the change noted with FC-104. The primary difference between air and FC-104 in this experiment is their respective refractive indices. FC-104 has a low refractive index of approximately 1.28, but this is still significantly better than that of air. With air cooling the reflection at the rod surface is so high that it exceeds the return from the pump reflector, thereby limiting the effectiveness of the samarium filter. The dependence of the saturation level on absorption in the pump cavity at 1.06 μm and on the refractive index of the material surrounding the rod is attributed to the presence of amplified spontaneous emission from the rod as well as the presence of significant reflections of 1.06 μm radiation form both the rod surfaces and pump reflectors. This energy storage limiting mechanism, although most pronounced in Nd : YAG rods because of the high gain obtained in this material, is not unique to Nd : YAG lasers.

Gain limitations in Nd : Glass rods resulting from amplified pump radiation and reflection of spontaneous emission from the pump cavity walls back into the rod have led to the development of samarium-clad Nd : glass rods. Cladding of the rod also suppresses internal reflection from the rod walls when the refractive indices of the rod and cladding are matched.

4.4.2 Prelasing and Parasitic Modes

Prelasing and parasitic modes are consequences of the interaction of an inverted material with highly reflecting surroundings and are susceptible to design variation.

Prelasing. Normal laser action, occurring during the pump phase in an amplifier, results from the residual feedback of the various dielectric interfaces in the optical path. The condition for stable operation can be derived by considering the gain in an amplifier rod of length L pumped to an inversion level characterized by a gain coefficient g per unit length. Let the end faces of the rod exhibit reflectivities of r_1 and r_2. Then spontaneous emission emanating from any given location in the rod, traveling normal to the rod end faces, will exhibit a loop gain after one round trip in the rod of approximately $r_1 r_2 \exp(2gL)$. If this loop gain is greater than unity, oscillations will build up until the usable rod-stored energy is depleted. The requirement for an amplifier is thus

$$r_1 r_2 \exp(2gL) < 1 \quad , \tag{4.44}$$

which sets an upper limit on rod length.

Parasitic Modes. Because of the dielectric discontinuity at the surface of the laser rod, a fraction of the spontaneous-decay radiation is reflected back into the active region. At sufficiently high inversion, the lateral gain in the rod will exceed the wall losses and there will be net gain inside the active material. This will preclude higher inversion. Further, in rods with polished walls there will exist internal modes completely contained by total internal reflection. These are particularly troublesome.

One type of internal mode, the whisper mode, propagates circumferentially and has no longitudinal component. This mode can penetrate into the rod as far as $r = R/n$, where n is the index of refraction [4.12, 60]. Another type of internal mode, the light-pipe mode, propagates down the rod in a zigzag fashion. The onset of light-pipe modes has been noted by *Sooy* et al. [4.12] in pumped rubies which had no external mirrors. In these experiments the ruby was placed in a helical lamp and the spontaneous decay was observed. Below the threshold for the internal modes the spontaneous emission increased as the ruby was pumped and, thereafter, dropped off with its characteristic decay time. After threshold was reached, the spontaneous emission increased quite suddenly.

The gain limitations due to internal radial modes are considerably improved in immersed rods particularly in case of an index matching fluid, or rods with rough surfaces, grooves or anti-reflection coatings or chemical treatments. [4.13]. Longitudinal modes can be suppressed if the faces of the rod are either wedged or anti-reflection coated. As the size of the disk becomes large, parasitic processes also begin to become significant. Several researchers have discussed parasitic processes in laser disks [4.65–67]. The threshold of the parasitic oscillation depends on the edge-coating material, which reduces the feedback by absorbing as much fluorescence as possible.

The threshold for parasitic oscillation sets an upper limit on the energy storage attainable in Nd : glass laser discs. It has been shown [4.65] that the lowest-loss path lies in a plane across the diameter of the disc. The path includes one reflection at each face of the disc at the smallest angle from the normal face which still allows total internal reflection. The only losses on such a path come at the edge reflections. At the oscillation threshold this edge loss must just cancel the path gain. For very thin discs the length for this path is approximately the disc refractive index n times the diameter D (assuming that the disc is in air). At the oscillation threshold we have

$$R\mathrm{e}^{nDg} = 1 \quad , \tag{4.45}$$

where g is the gain coefficient in the laser material and R is the edge reflectivity. For typical design parameters, such as $D = 15\,\mathrm{cm}$, $g = 0.2\,\mathrm{cm}^{-1}$, and $n = 1.56$, the reflectivity must be less than 1 % to avoid oscillation. Cladding the edge of the disks with an index-matched absorbing glass results in a dramatic reduction in parasitic losses. Besides these bulk parasitic modes which include those rays that total internally reflect between the faces of the disk there are surface modes. These occur primarily across the disk face because of the larger gain coefficient at the surface due to pump light absorption [4.65, 68]. Without the onset of the parasitic oscillation the stored energy is limited by the increased loss rate due to the fluorescence amplification.

Amplified spontaneous emission (ASE) strongly influences the performance of large aperture Nd : glass disk amplifiers. Even with perfectly absorbing edge claddings on the disks, depumping caused by single pass ASE increases rapidly with aperture, reducing peak gain and storage efficiency. ASE becomes the principal depumping process in large, edge-clad disks when the product of the gain coefficient and disk major axis gD becomes large (greater than 3) [4.66, 69].

Techniques Used to Prevent or Reduce Feedback in Amplifiers. In order to minimize prelasing, rod ends are cut at an oblique angle to the rod axis, often at the Brewster angle, or they are coated with antireflection layers. Reflections for off-axis spontaneous emission from the cylindrical rod surfaces are minimized by use of index-matching fluids around the laser rod or by rough grinding of the cylindrical walls. If more than one stage of amplification is required, the separate stages are in part decoupled by one or more of the following measures. The distance between amplifier stages may be made as large as possible. Since spontaneous emission and superradiance occur in relatively large beam angles, a large distance between stages very effectively reduces feedback between stages. If the small-signal gain of an amplifier chain is very high (30 dB or more), optical isolators are placed between amplifier stages. As decoupling elements one may employ electrooptical shutters, isolators based on the Faraday effect, or dye cells which bleach under powerful laser light. The design and application for these devices is described in detail in Chap. 8.

5. Optical Resonator

The light emitted by most lasers contains several discrete optical frequencies, separated from each other by frequency differences which can be associated with different modes of the optical resonator. It is common practice to distinguish two types of resonator modes: "Longitudinal" modes differ from one another only in their oscillation frequency; "transverse" modes differ from one another not only in their oscillation frequency, but also in their field distribution in a plane perpendicular to the direction of propagation. Corresponding to a given transverse mode are a number of longitudinal modes which have the same field distribution as the given transverse mode but which differ in frequency.

To describe the electromagnetic field variations inside optical resonators, the symbols TEM_{mnq} or TEM_{plq} are used. The small capital letters stand for "transverse electromagnetic waves" and the first two indices identify a particular transverse mode, whereas q describes a longitudinal mode. Because resonators that are used for typical lasers are long compared to the laser wavelength, they will, in general, have a large number of longitudinal modes. Therefore, the index q which specifies the number of modes along the axis of the cavity will be very high. The indices for the transverse modes, which specify the field variations in the plane normal to the axis, are very much lower and sometimes may be only the first few integers.

The spectral characteristics of a laser, such as linewidth and coherence length, are primarily determined by the longitudinal modes; whereas beam divergence, beam diameter, and energy distribution are governed by the transverse modes. In general, lasers are multimode oscillators unless specific efforts are made to limit the number of oscillating modes. The reason for this lies in the fact that a very large number of longitudinal resonator modes fall within the bandwidth exhibited by the laser transition and a large number of transverse resonator modes can occupy the cross section of the active material.

5.1 Transverse Modes

The theory of modes in optical resonators has been treated in [5.1–3]; comprehensive reviews of the subject can also be found in [5.4, 5].

5.1.1 Intensity Distribution of Transverse Modes

In an optical resonator electromagnetic fields can exist whose distribution of amplitudes and phases reproduce themselves upon repeated reflections between the mirrors. These particular field configurations comprise the transverse electromagnetic modes of a passive resonator.

Transverse modes are defined by the designation TEM_{mn} for Cartesian coordinates. The integers m and n represent the number of nodes of zeros of intensity transverse to the beam axis in the vertical and horizontal directions. In cylindrical coordinates the modes are labeled TEM_{pl} and are characterized by the number of radial nodes p and angular nodes l. The higher the values of m, n, p, and l, the higher the mode order. The lowest-order mode is the TEM_{00} mode, which has a Gaussian-like intensity profile with its maximum on the beam axis. For modes with subscripts of 1 or more, intensity maxima occur that are off-axis in a symmetrical pattern. To determine the location and amplitudes of the peaks and nodes of the oscillation modes, it is necessary to employ higher-order equations which either involve Hermite (H) or Laguerre (L) polynomials. The Hermite polynomials are used when working with rectangular coordinates, while Laguerre polynomials are more convenient when working with cylindrical coordinates.

In cylindrical coordinates, the radial intensity distribution of allowable circularly symmetric TEM_{pl} modes is given by the expression

$$I_{pl}(r,\phi,z) = I_0 \varrho^l [\mathrm{L}_p^l \varrho]^2 (\cos^2 l\phi) \exp(-\varrho) \tag{5.1}$$

with $\varrho = 2r^2(z)/w^2(z)$, where z is the propagation direction of the beam, and r, ϕ are the polar coordinates in a plane transverse to the beam direction. The radial intensity distributions are normalized to the spot size of a Gaussian profile; that is, $w(z)$ is the spot size of the Gaussian beam, defined as the radius at which the intensity of the TEM_{00} mode is $1/e^2$ of its peak value on the axis. L_p is the generalized Laguerre polynomial of order p and index l.

The intensity distribution given in (5.1) is the product of a radial part and an angular part. For modes with $l = 0$ (i.e., TEM_{p0}), the angular dependence drops out and the mode pattern contains p dark concentric rings, each ring corresponding to a zero of$L_p^0(\varrho)$. The radial intensity distribution decays due to the factor $\exp(-\varrho)$. The center of a pl mode will be bright if $l = 0$, but dark otherwise because of the factor ϱ^l. These modes, besides having p zeros in the radial direction, also have $2l$ nodes in azimuth.

The only change in a (pl) mode distribution comes through the dependence of the spot size $w(z)$ on the axial position z. However, the modes preserve the general shape of their electric field distributions for all values of z. As w increases with z, the transverse dimensions increase so that the sizes of the mode patterns stay in constant ratio to each other.

From (5.1) it is possible to determine any beam mode profile. Figure 5.1a depicts various cylindrical transverse intensity patterns as they would appear in the output beam of a laser. Note that the area occupied by a mode increases

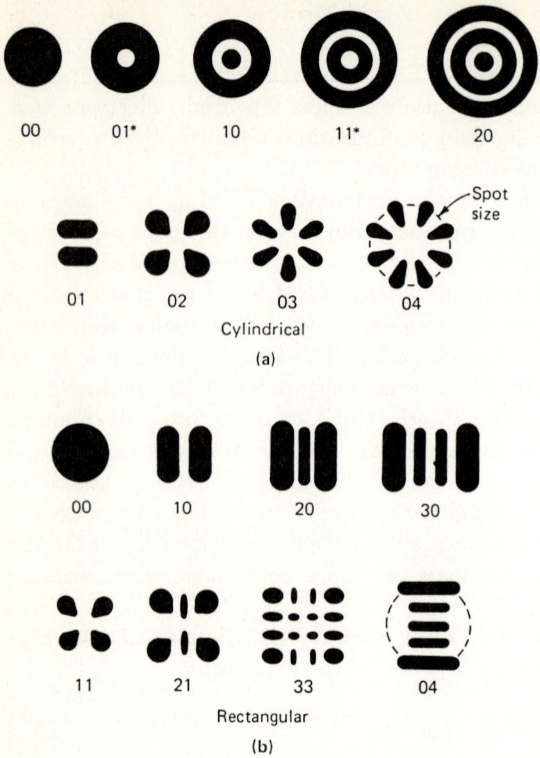

Fig. 5.1. Examples of (**a**) cylindrical and (**b**) rectangular transverse mode patterns. For cylindrical modes, the first subscript indicates the number of dark rings, whereas the second subscript indicates the number of dark bars across the pattern. For rectangular patterns, the two subscripts give the number of dark bars in the x and y directions [5.6]

with the mode number. A mode designation accompanied by an asterisk indicates a mode which is a linear superposition of two like modes, one rotated 90° about the axis relative to the other. For example, the TEM mode designated 01^* is made up of two TEM_{01} modes.

The intensity distribution of the modes shown in Fig. 5.1a can be calculated if we introduce the appropriate Laguerre polynomials into (5.1), i.e.,

$$\mathrm{L}_0^l(\varrho) = 1; \quad \mathrm{L}_1^0(\varrho) = 1 - \varrho; \quad \mathrm{L}_2^0(\varrho) = 1 - 2\varrho + \tfrac{1}{2}\varrho^2 \quad .$$

A plot of the intensity distributions of the lowest-order mode and the next two higher-order transverse modes, i.e., TEM_{00}, TEM_{01^*} and TEM_{10}, is illustrated in Fig. 5.2.

In rectangular coordinates the intensity distributions of a (m, n) mode is given by

$$I_{mn}(x,y,z) = I_0 \left[\mathrm{H}_m\left(\frac{x(2)^{1/2}}{w(z)}\right) \exp\left(\frac{-x^2}{w^2(z)}\right)\right]^2 \times \left[\mathrm{H}_n\left(\frac{y(2)^{1/2}}{w(z)}\right) \exp\left(\frac{-y^2}{w^2(z)}\right)\right]^2 \quad . \tag{5.2}$$

As before, $w(z)$ is the spot size at which the transverse intensity decreases

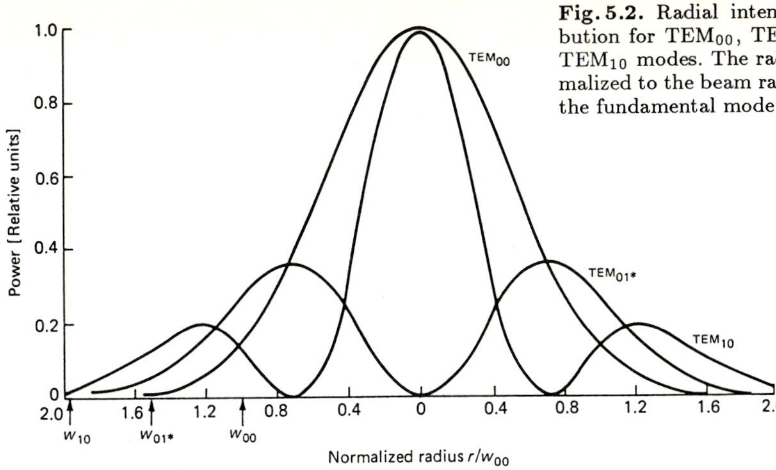

Fig. 5.2. Radial intensity distribution for TEM_{00}, TEM_{01^*}, and TEM_{10} modes. The radii are normalized to the beam radius w_{00} of the fundamental mode [5.7]

to $1/e^2$ of the peak intensity of the lowest-order mode. The function $\mathrm{H}_m(s)$ is the mth-order Hermite polynomial, for example, $\mathrm{H}_0(s) = 1, \mathrm{H}_1(s) = 2s$, and $\mathrm{H}_2(s) = 4s^2 - 2$. At a given axial position z, the intensity distribution consists of the product of a function of x alone and a function of y alone. The intensity patterns of rectangular transverse modes are sketched in Fig. 5.1b. The m, n values of a single spatial mode can be determined by counting the number of dark bars crossing the pattern in the x and y directions. Note that the fundamental mode ($m = n = 0$) in this geometry is identical with the fundamental mode in cylindrical geometry.

The transverse modes shown in Fig. 5.1 can exist as linearly polarized beams, as illustrated by Fig. 5.3. By combining two orthogonally polarized modes of the same order, it is possible to synthesize other polarization configurations; this is shown in Fig. 5.4 for the TEM_{01} mode.

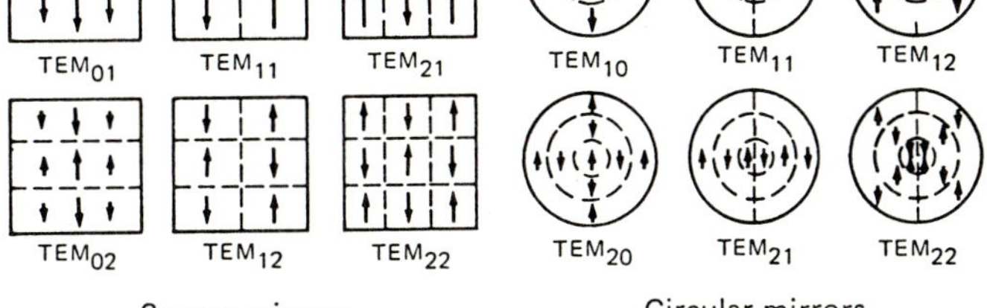

Fig. 5.3. Linearly polarized resonator mode configurations for square and circular mirrors [5.1]

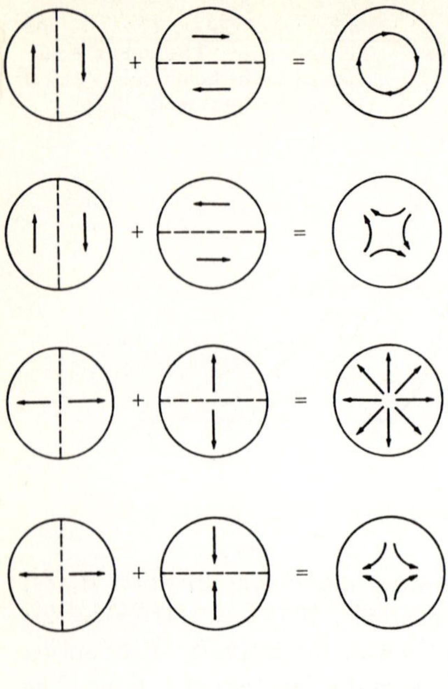

Fig. 5.4. Synthesis of different polarization configurations from the linearly polarized TEM_{01} mode

5.1.2 Characteristics of a Gaussian Beam

A light beam emitted from a laser with a Gaussian intensity profile is called the "fundamental mode" or TEM_{00} mode. Because of its importance it is discussed here in greater detail. The decrease of the field amplitude with distance r from the axis in a Gaussian beam is described by

$$E(r) = E_0 \exp\left(\frac{-r^2}{w^2}\right) \quad . \tag{5.3}$$

Thus, the distribution of power density is

$$I(r) = I_0 \exp\left(\frac{-2r^2}{w^2}\right) \quad . \tag{5.4}$$

The quantity w is the radial distance at which the field amplitude drops to $1/e$ of its value on the axis and the power density is decreased to $1/e^2$ of its axial value. The parameter w is often called the beam radius or "spot size," and $2w$, the beam diameter. The fraction of the total power of a Gaussian beam which is contained in a radial aperture of $r = w, r = 1.5w$, and $r = 2w$ is 86.5%, 98.9%, and 99.9%. If a Gaussian beam is passed through a radial aperture of $3w$, then only 10^{-6}% of the beam power is lost due to the obstruction. For our subsequent discussion, an "infinite aperture" will mean a radial aperture in excess of three spot sizes.

If we consider now a propagating Gaussian beam, we note that although the intensity distribution is Gaussian in every beam cross section, the width of the intensity profile changes along the axis. The Gaussian beam contracts to a minimum diameter $2w_0$ at the beam waist where the phase front is planar. If one measures z from this waist, the expansion laws for the beam assume a simple form. The spot size a distance z from the beam waist expands as a hyperbola, which has the form

$$w(z) = w_0\left[1+\left(\frac{\lambda z}{\pi w_0^2}\right)^2\right]^{1/2} . \tag{5.5}$$

Its asymptote is inclined at an angle $\Theta/2$ with the axis, as shown in Fig. 5.5, and defines the far-field divergence angle of the emerging beam. The full divergence angle for the fundamental mode is given by

$$\Theta = \lim_{z\to\infty}\frac{2w(z)}{z} = \frac{2\lambda}{\pi w_0} = 1.27\frac{\lambda}{(2w_0)} . \tag{5.6}$$

From these considerations it follows that at large distances, the spot size increases linearly with z, and the beam diverges at a constant cone angle Θ. A most interesting point here is that the smaller the spot size w_0 at the beam waist, the greater the divergence.

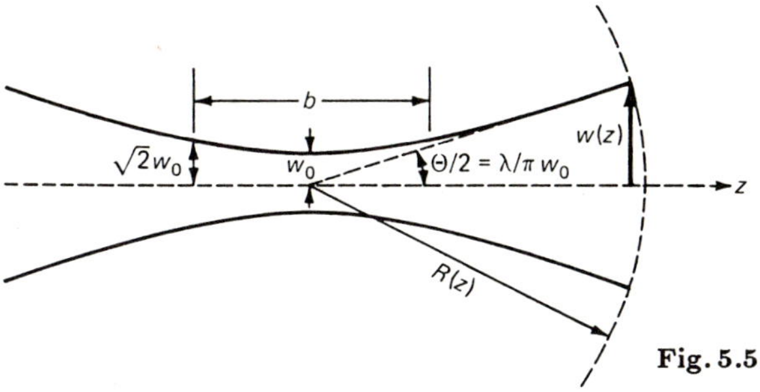

Fig. 5.5. Contour of a Gaussian beam

At sufficiently large distances from the beam waist, the wave has a spherical wavefront appearing to emanate from a point on the beam axis at the waist. If $R(z)$ is the radius of curvature of the wavefront that intersects the axis at z, then

$$R(z) = z\left[1+\left(\frac{\pi w_0^2}{\lambda z}\right)^2\right] . \tag{5.7}$$

It is important to note that in a Gaussian beam the wavefront has the same phase across its entire surface.

Sometimes the properties of a TEM_{00} mode beam are described by specifying a confocal parameter

$$b = \frac{2\pi w_0^2}{\lambda} , \tag{5.8}$$

where b is the distance between the points at each side of the beam waist for which $w(z) = (2)^{1/2} w_0$ (Fig. 5.5).

Before we leave the subject of Gaussian beams, we will briefly point out the difference in the definition of beam divergence for Gaussian beams and plane waves. A laser operating at the TEM_{00} mode will have a beam divergence according to (5.6). For a plane wavefront incident upon a circular aperture of diameter D, the full cone angle of the central (Airy) disc, defined at the first minimum of the Fraunhofer diffraction pattern, is given by

$$\Theta_p = \frac{2.44\lambda}{D} \quad . \tag{5.9}$$

The energy contained within this angle is about 84% of the total energy transmitted by the aperture.

Equations (5.6, 9) are often confused, because various conventions have been adopted by different authors with the equation

$$\Theta_\mathrm{R} = \frac{1.22\lambda}{D} \quad , \tag{5.10}$$

which represents the *half*- cone angle of the Fraunhofer diffraction pattern and also happens to be the "Rayleigh criterion" for the angular resolution of an optical instrument.

In laboratory work, a beam size is often obtained by measuring the diameter of the illuminated spot with a scale. This is not the spot size $2w_0$ as defined by (5.5). There is no obvious visual cue to the magnitude of the spot size in the appearance of the illuminated spot. Thus, "spot size" and "size of the illuminated spot" are totally different concepts. The former is a property of the laser cavity; the latter is a subjective estimate. To measure the spot size, the illuminated spot should be scanned with a photodetector behind a small pinhole. The resulting curve of intensity versus position of the pinhole will yield a Gaussian curve from which the spot size w_0 must be extracted by mathematical methods.

5.1.3 Resonator Configurations

The most commonly used laser resonators are composed of two spherical or flat mirrors facing each other. We will first consider the generation of the lowest-order mode by such a resonant structure. Once the parameters of the TEM_{00} mode are known, all higher-order modes simply scale from it in a known manner. Diffraction effects due to the finite size of the mirrors will be neglected in this section.

The Gaussian beam indicated in Fig. 5.6 has a wavefront curvature of R_1 at a distance t_1 from the beam waist. If we put a mirror at t_1 whose radius of curvature equals that of the wavefront, then the mode shape has not been altered. To proceed further, we can go along the z axis to another point t_2 where the TEM_{00} has a radius of curvature R_2, and place there a mirror whose radius of curvature R_2 equals that of the spherical wavefront at t_2. Again the mode shape remains unaltered.

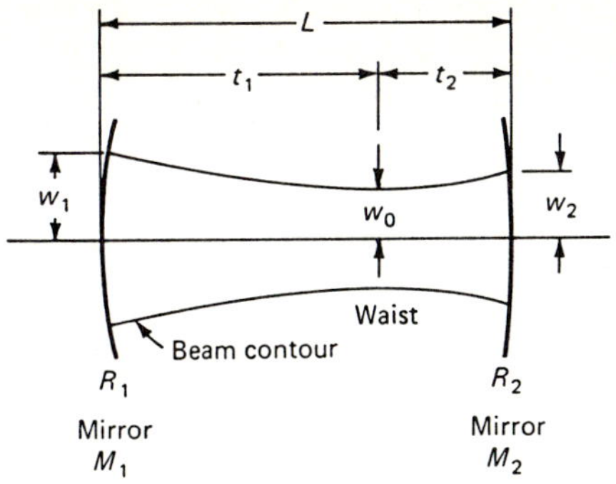

Fig. 5.6. Mode parameters of interest for a resonator with mirrors of unequal curvature

Therefore, to make a resonator, we simply insert two reflectors which match two of the spherical surfaces defined by (5.7). Alternatively, given two mirrors separated by a distance L, if the position of the plane $z = 0$ and the value of the parameter w_0 can be adjusted so that the mirror curvatures coincide with the wavefront surfaces, we will have found the resonator mode.

We will now list formulas, derived by *Kogelnik* and *Li* [5.4], which relate the mode parameters w_1, w_2, w_0, t_1, and t_2 to the resonator parameters R_1, R_2, and L. As illustrated in Fig. 5.6, w_1 and w_2 are the spot radii at mirrors M_1 and M_2, respectively; t_1 and t_2 are the distances of the beam waist described by w_0 from mirrors M_1 and M_2, respectively; and R_1 and R_2 are the curvatures of mirrors M_1 and M_2 which are separated a distance L. Labeling conventions are that concave curvatures are positive.

The beam radii at the mirrors are given by

$$w_1^4 = \left(\frac{\lambda R_1}{\pi}\right)^2 \frac{R_2 - L}{R_1 - L}\left(\frac{L}{R_1 + R_2 - L}\right) ,$$

$$w_2^4 = \left(\frac{\lambda R_2}{\pi}\right)^2 \frac{R_1 - L}{R_2 - L}\left(\frac{L}{R_1 + R_2 - L}\right) . \tag{5.11}$$

The radius of the beam waist, which is formed either inside or outside the resonator, is given by

$$w_0^4 = \left(\frac{\lambda}{\pi}\right)^2 \frac{L(R_1 - L)(R_2 - L)(R_1 + R_2 - L)}{(R_1 + R_2 - 2L)^2} . \tag{5.12}$$

The distances t_1 and t_2 between the waist and the mirrors, measured positive (Fig. 5.6), are

$$t_1 = \frac{L(R_2 - L)}{R_1 + R_2 - 2L} , \quad t_2 = \frac{L(R_1 - L)}{R_1 + R_2 - 2L} . \tag{5.13}$$

These equations treat the most general case of a resonator. There are many optical resonator configurations for which (5.11–13) are greatly simplified. Figure 5.7 shows some of the most commonly used geometries.

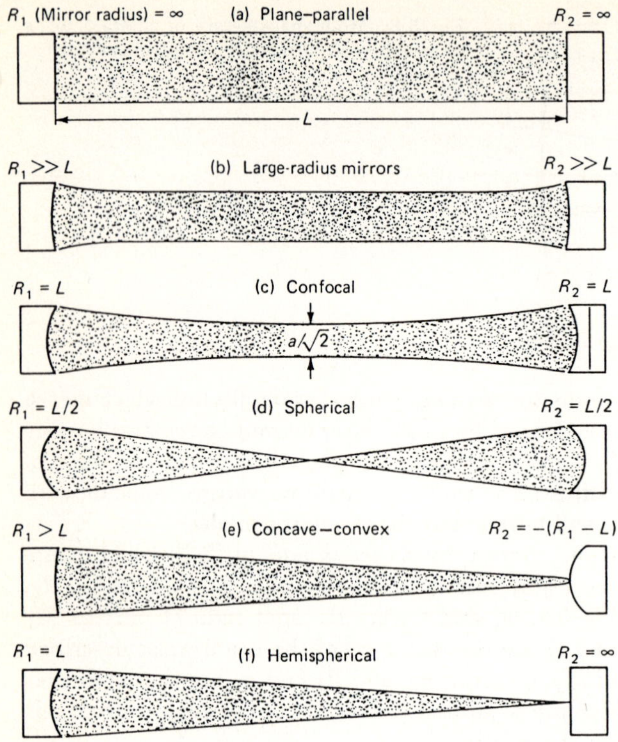

Fig. 5.7. Resonator configurations giving uniphase wavefronts (intracavity radiation pattern is shaded)

Mirrors of Equal Curvature

With $R_1 = R_2 = R$ we obtain from (5.11)

$$w_{1,2}^2 = \frac{\lambda R}{\pi}\left(\frac{L}{2R-L}\right)^{1/2} \quad . \tag{5.14}$$

The beam waist which occurs at the center of the resonator $t_1 = t_2 = R/2$ is

$$w_0^2 = \frac{\lambda}{2\pi}[L(2R-L)]^{1/2} \quad . \tag{5.15}$$

If we further assume that the mirror radii are large compared to the resonator length $R \gg L$, the above formula simplifies to

$$w_{1,2}^2 = w_0^2 = \left(\frac{\lambda}{\pi}\right)\left(\frac{RL}{2}\right)^{1/2} \quad . \tag{5.16}$$

As follows from (5.16), in a resonator comprised of large-radius mirrors, the beam diameter changes very little as a function of distance.

A special case of a symmetrical configuration is the spherical resonator which consists of two mirrors separated by twice their radius, that is, $R = L/2$. The corresponding beam consists of a mode whose dimensions are fairly large at each mirror and which focus down to a diffraction-limited point at the center of the resonator.

Another very important special case of a resonator with mirrors of equal curvature is the confocal resonator. For this resonator the mirror separation equals the curvature of the identical mirrors, that is, $R = L$. From (5.14, 15) we obtain the simplified relation

$$w_{1,2} = \left(\frac{\lambda R}{\pi}\right)^{1/2} \quad \text{and} \quad w_0 = \frac{w_{1,2}}{(2)^{1/2}} \quad . \tag{5.17}$$

The confocal configuration gives the smallest possible mode dimension for a resonator of given length.

Plano-Concave Resonator

For a resonator with one flat mirror ($R_1 = \infty$) and one curved mirror we obtain

$$w_1^2 = w_0^2 = \left(\frac{\lambda}{\pi}\right)[L(R_2 - L)]^{1/2} \quad \text{and} \quad w_2^2 = \left(\frac{\lambda}{\pi}\right) R_2 \left(\frac{L}{R_2 - L}\right)^{1/2} \quad . \tag{5.18}$$

The beam waist w_0 occurs at the flat mirror (that is $t_1 = 0$ and $t_2 = L$). A special case of this resonator configuration is the hemispherical resonator. The hemispherical resonator consists of one spherical mirror and one flat mirror placed approximately at the center of curvature of the sphere. The resultant mode has a relatively large diameter at the spherical mirror and focuses to a diffraction-limited point at the plane mirror. The wavefront outputs from the two ends of the laser behave approximately as if they came from a difraction-limited point at the surface of the flat mirror. Equation (5.18) illustrates the large disparity in dimensions of the mode at the two mirrors if R_2 approaches the value of L.

It is also clear that, in practice, the actual separation of the mirrors can never be quite as large as MR_2, since this would result in extremely high diffraction losses at the spherical mirror, together with a "zero" mode dimension at the flat mirror, which is physically impossible. In practice, one makes the mirror separation L slightly less than R_2 so that a value of w_1 is obtained that gives reasonably small diffraction losses.

Concave-Convex Resonator

The pertinent beam parameters for concave-convex resonators can be calculated if we introduce a negative radius ($-R_2$) for the convex mirror into (5.11–13).

Plane-Parallel Resonator

The plane-paralles resonator, which can be considered a special case of a large-radius mirror configuration if $R_1 = R_2 = \infty$, has been thoroughly analyzed by

Fox and *Li* [5.1] who showed that the output of the plane-parallel resonator does not consist of a plane-parallel wavefront. Instead, owing to fixed and rather large diffraction losses around the edges, there is a phase lag of approximately 30° near the edges which gives the wavefront a slight curvature.

In calculating the mode size in a resonator, it must be noted that in most cases the resonator is formed by spherical mirrors with the reflecting surfaces deposited on plane-concave mirror blanks of index n. The front mirror of such a resonator acts as a negative lens and will change the characteristics of the emerging beam. The beam appears to have a different waist diameter and location. Taking the negative lens effect into account, a beam emerging from a resonator of equal mirror radii R can be described as having a beam waist of

$$w'_0 = w_0\left(1 + \frac{L(n^2-1)}{2R}\right)^{-1/2} , \tag{5.19}$$

with the waist located at

$$t_2 = \left(\frac{L}{2}\right)\frac{nR}{R + L(n^2-1)/2} . \tag{5.20}$$

The parameter w_0 refers to the beam waist calculated previously for the resonator of equal mirror radii. As we can see from (5.19), the beam leaving the cavity appears to have a smaller beam waist, which indicates larger divergence, and the location of the waist appears to be moved toward the front mirror. The negative lens effect of the output mirror disappears if the outer surface of the output mirror has a radius of curvature given by

$$R'_1 = \frac{R_1(n-1)}{n} . \tag{5.21}$$

5.1.4 Stability of Laser Resonators

For certain combinations of R_1, R_2, and L the equations summarized in the previous sub-section give nonphysical solutions (i.e., imaginary spot sizes). This is the region where low-loss modes do not exist in the resonator.

Light rays that bounce back and forth between the spherical mirrors of a laser resonator experience a periodic focusing action. The effect on the rays is the same as in a periodic sequence of lenses [5.8, 9]. Rays passing through a stable sequence of lenses are periodically refocused. For unstable systems the rays become more and more dispersed the further they pass through the sequence. In an optical resonator operated in the stable region, the waves propagate between reflectors without spreading appreciably. This fact can be expressed by a stability criterion [5.10]

$$0 < \left(1 - \frac{L}{R_1}\right)\left(1 - \frac{L}{R_2}\right) < 1 . \tag{5.22}$$

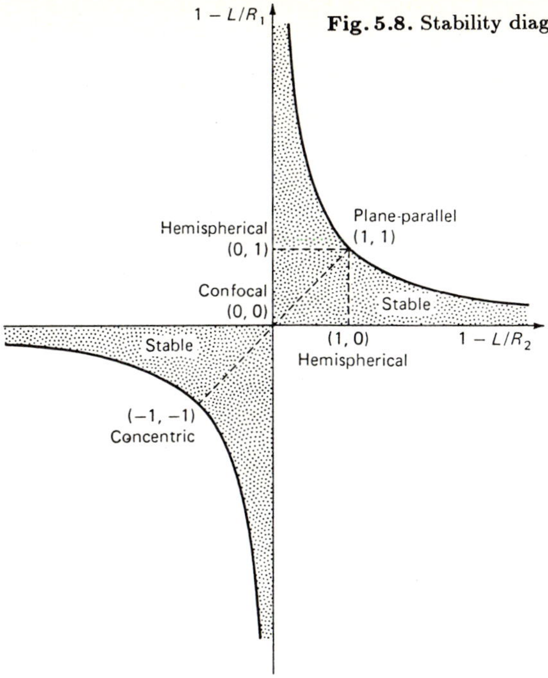

Fig. 5.8. Stability diagram for the passive laser resonator

To show graphically which type of resonator is stable and which is unstable, it is useful to plot a stability diagram on which each particular resonator geometry is represented by a point. This is shown in Fig. 5.8, where the parameters

$$g_1 = 1 - \frac{L}{R_1} \quad , \quad g_2 = 1 - \frac{L}{R_2} \tag{5.23}$$

are drawn as the coordinate axes.

All cavity configurations are unstable unless they correspond to points lying in the area enclosed by a branch of the hyperbola $g_1 g_2 = 1$ and the coordinate axes. The origin of the diagram represents the confocal system. The low-loss regions of Fig. 5.8 represent the following three sets of possible mirror geometries:

1. Both mirrors are concave, and the center of curvature of each mirror lies beyond the other mirror.
2. Both mirrors are concave, and the center of curvature of each mirror lies between the other mirror and the center of curvature of the other mirror.
3. One mirror is convex and one is concave, and the center of curvature of the concave mirror lies between the convex mirror and the center of curvature of the convex mirror. It turns out that the mirror configuration which satisfy the stability criterion are those for which the wavefront and mirror curvatures are matched.

5.1.5 Diffraction Losses

In any real laser resonator some part of the laser beam will be lost either by spillover at the mirrors or by limiting apertures, such as the lateral boundaries of the active material. These losses will depend on the diameter of the laser beam in the plane of the aperture and the aperture radius. If we take a finite aperture of radius a within the resonator into account, the diffraction losses depend on four parameters, R_1, R_2, L, and a, which describe the resonator; and on three parameters λ, m, and n, characterizing the particular optical beam present in the resonator. Fortunately, the losses depend only on certain combinations of these parameters. These combinations are the so-called Fresnel number,

$$N = \frac{a^2}{\lambda L} \quad , \tag{5.24}$$

and the quantities g_1 and g_2 which were defined in (5.23). The parameter N can be thought of as the ratio of the acceptance angle (a/L) of one mirror as viewed from the center of the opposing mirror to the diffraction angle (λ/a) of the beam. Therefore, when N is small, especially if $N<1$, the loss factor will be high because only a portion of the beam will be intercepted by the mirrors. When N is large, the losses will be low for the stable resonator configurations and large for the unstable resonators. If two resonators have the same values of N, g_1, and g_2, then they have the same diffraction loss, the same resonant frequency, and the same mode patterns. The fractional energy loss per transit due to diffraction effects for the lowest-order mode of a resonator with identical mirrors ($g_1 = g_2 = g$) is given in Fig. 5.9. For the plane-parallel resonator with

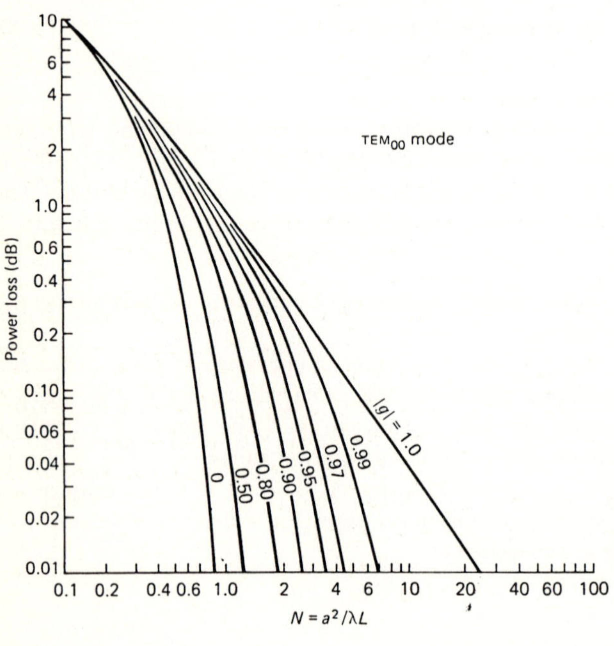

Fig. 5.9. Diffraction losses per transit for the TEM_{00} mode of various symmetrical and stable resonators [5.10]

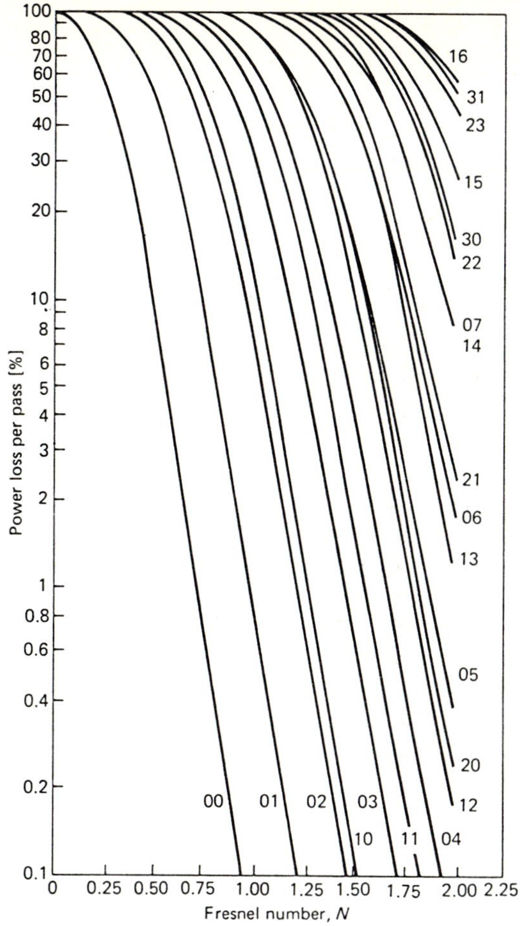

Fig. 5.10. Diffraction loss for various resonator modes as a function of the Fresnel number [5.12]

circular aperture, an analytical expression of the diffraction losses for large Fresnel numbers ($F \geq 1$) has been derived in [5.11], namely

$$\alpha = 8\kappa_{pl}^2 \frac{\delta(M+\delta)}{[(M+\delta)^2+\delta^2]^2} \quad , \tag{5.25}$$

where $\delta = 0.824$, $M = (8\pi N)^{1/2}$, and k_{pl} is the $(p+l)$th zero of the Bessel function of order l. The diffraction losses in a confocal resonator for a number of low-order modes are plotted in Fig. 5.10 [5.12]. The mode designations are in the cylindrical coordinate system. Note that all modes have very large losses for small N and that the losses drop with increasing N. However, the losses for the higher-order modes drop less rapidly than the losses for the lower-order modes.

5.1.6 Higher-Order Modes

The mode radius and the divergence angle for modes of cylindrical symmetry are usually defined in terms of the $1/e^2$ intensity points of the outermost lobe. As indicated in Fig. 5.2, the radius of each mode increases with increasing mode number. The ratio of the radii of different modes with respect to that of the lowest-order mode is designated c_{pl}. From (5.1) follows that the ratios of different mode diameters remain constant at any plane inside or outside of the cavity, in the near or far field. Therefore, the mode radius and the beam divergence of a higher-order mode can be related to the fundamental mode parameters as follows [5.7 13]:

$$W_{pl}(z) = C_{pl} w_0; \qquad \Theta_{pl} = C_{pl} \Theta \quad , \tag{5.26}$$

where Θ is obtained from (5.6) and w_0 can be calculated for a particular resonator from the formulas summarized in Sect. 5.1.3. Numerical values for C_{pl}, as obtained from (5.1) for several low-order modes are $C_{00} \equiv 1$; $C_{01^*} = 1.5$; $C_{10} = 1.9$ $C_{11^*} = 2.15$; $C_{20} = 2.42$; $C_{21^*} = 2.63$.

For modes of rectangular symmetry, defining the mode radius and the divergence angle in terms of the $1/e^2$ intensity points of the outermost lobe is not adequate, particularly for those rectangular modes whose pattern is far from square. A possible description of rectangular mode sizes is in terms of a radius which contains something like 90% of the beam power. This is a highly functional definition from an applications standpoint. The relative size of some rectangular modes have been calculated by *Kruger* [5.6, 14]. Figure 5.11 shows the fraction of the total beam power for a particular rectangular mode (TEM_{mn}) within a circular cross section of the beam. From Fig. 5.11 it is also possible to determine the relative beam divergence for any mode. If we define the mode size at the 90% power points, the divergence angle for a particular

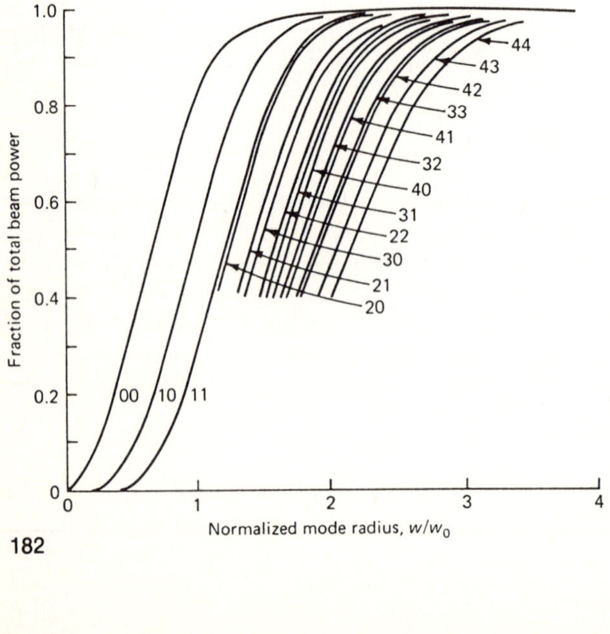

Fig. 5.11. Fraction of total power for a particular mode TEM_{mn} within a circular section of radius R. The number on each curve corresponds to the individual modes [5.6]

mode is the TEM_{00} divergence angle times the normalized radius at the 90 % level. For example, since the cross section containing 90 % of the power for the TEM_{44} mode has a radius $r = 2.74$, the divergence angle for TEM_{44} is 2.74 times the TEM_{00} divergence angle.

5.1.7 Active Resonator

So far we have discussed the modes in a passive resonator consisting of a pair of mirrors. Introducing an active element into the resonator, such as a laser crystal, in addition to altering the optical length of the cavity, will perturb the mode configuration, since the active material possesses a saturable, nonuniform gain and exhibits thermal lensing and birefringence. In high-gain, giant-pulse lasers, gain saturation at the center of a TEM_{00} mode can lead to a flattening of the intensity profile [5.15, 16]. Also, pump nonuniformities leading to a nonuniform gain distribution across the beam will lead to non-Gaussian output intensity profiles. Theoretical and experimental investigations have shown that in solid-state lasers the governing mechanisms which distort the mode structure in the resonator are the thermal effects of the laser rod. As will be discussed in more detail in Chap. 7, optical pumping leads to a radial temperature gradient in the laser rod. As a result, in CW and high average power systems, the rod is acting like a positive thick lens of an effective focal length f, which is inversely proportional to the pump power.

The theory necessary to analyze resonators that contain optical elements other than the end mirrors has been developed by *Kogelnik* [5.17]. We will apply this theory to the case of a resonator containing an internal thin lens. To a first approximation, this lens can be thought of as representing the thermal lensing introduced by the laser rod. The more complex case of a distributed thick lens which more adequately describes thermal lensing has been treated in [5.17–20].

Beam properties of resonators containing internal optical elements are described in terms of an equivalent resonator composed of only two mirrors. The pertinent parameters of a resonator equivalent to one with an internal thin lens are

$$g_1 = 1 - \frac{L_2}{f} - \frac{L_0}{R_1} \quad ; \quad g_2 = 1 - \frac{L_1}{f} - \frac{L_0}{R_2} \quad , \tag{5.27}$$

where $L_0 = L_1 + L_2 - (L_1 L_2 / f)$ and f is the focal lenth of the internal lens; L_1 and L_2 are the spacings between mirrors M_1, M_2 and the lens, as shown in Fig. 5.12a.

In any resonator, the TEM_{00} mode spot size at one mirror can be expressed as a function of the resonator parameters

$$w_1^2 = \frac{\lambda L}{\pi} \left(\frac{g_2}{g_1(1 - g_1 g_2)} \right)^{1/2} \quad . \tag{5.28}$$

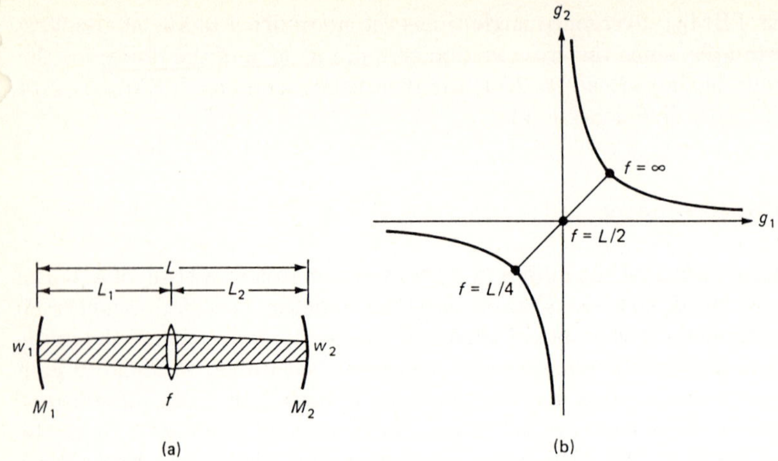

Fig. 5.12. (a) Geometry and (b) stability diagram of a resonator containing a thin positive lens

The ratio of the spot sizes at the two mirrors is

$$\frac{w_1^2}{w_2^2} = \frac{g_2}{g_1} \quad . \tag{5.29}$$

The stability condition (5.22) remains unchanged.

As an example we will consider a resonator with flat mirrors ($R_1 = R_2 = \infty$) and a thin lens in the center ($L_1 = L_2 = L/2$). From (5.27) we obtain

$$g = g_1 = g_2 = 1 - \frac{L}{2f} \quad , w_1^2 = w_2^2 = \left(\frac{\lambda L}{\pi}\right)(1 - g^2)^{-1/2} \quad . \tag{5.30}$$

For $f = \infty$ the resonator configuration is plane-parallel; for $f = L/2$ we obtain the equivalent of a confocal resonator; and for $f = L/4$ the resonator corresponds to a spherical configuration.

The mode size in the resonator will grow to infinity as the mirror separation approaches four times the focal length of the laser rod. Figure 5.12b shows the location of a plane-parallel resonator with an internal lens of variable focal length in the stability diagram. A very detailed analysis of a resonator consisting of a flat and a curved mirror containing an internal lens has been presented in [5.19].

The discussion of resonators has so far been restricted to devices having axial symmetry with respect to the beam axis. Resonators that contain inclined surfaces, such as Brewster-ended rods, polarizers, prisms, etc., lack axial symmetry. The effect of these asymmetric devices is to produce astigmatic beams, i.e., beams that have different spot sizes, wavefront curvatures, and beam waist positions in two orthogonal directions. Axially asymmetric laser cavities have been analyzed in [5.21, 22].

5.1.8 Resonator Sensitivity

There are two contexts in which the term "stability" is used. First, laser resonators are said to be optically stable or unstable depending on the value of g_1, g_2. Second, the mode size and position are sensitive to mechanical and optical perturbations of the optical elements. We will refer to the stability of the mode against these perturbations as the resonator sensitivity.

When designing resonators for an optimum mode size, it will be of the utmost importance to consider the resonator sensitivity to these mechanical and optical perturbations. Usually one is interested in the sensitivity of the resonator to two common types of perturbations: first, a time-varying thermal lensing effect caused by the laser rod, and second, misalignments of the resonator mirrors. The former perturbation leads mainly to a change in mode size and beam divergence, whereas the latter perturbation leads to a lateral displacement and angular tilt of the output beam which causes an increase of the diffraction losses and, therefore, a reduction of output power. First-order effects on the modes as a function of cavity perturbation are usually obtained by evaluation of a sensitivity matrix [5.19, 20].

Considering first the resonator's sensitivity to lensing effects, we note from (5.29 30) that a resonator is insensitive to axial perturbations if the spot size w_1 is insensitive to changes of g_1 and g_2. A calculation of the relative sensitivities of various resonators to small changes in mirror radii of curvature has been carried out by *Chesler* and *Maydan* [5.23] and *Steffen* et al. [5.18]. A perturbation within the resonator producing changes in mirror radii is equivalent to the introduction of a lens of some focal length f. In order for the resonator to have a low sensitivity to axial perturbations, i.e., small spot size changes for large changes of g_1 and g_2, it is necessary that $dw_1/df = 0$. This condition is met for resonator geometries which satisfy the following equation [5.18]

$$2g_2g_1 - 1 + \left(\frac{g_1}{g_2}\right)\left(\frac{L_1}{L_2}\right)^2 + \frac{2g_1L_1}{L_2} = 0 \quad . \tag{5.31}$$

One particular resonator satisfying (5.31) is determined by

$$g_1g_2 = \tfrac{1}{2} \quad ; \quad L_1 = 0 \quad . \tag{5.32}$$

In this case the internal lens is either absent or located at the surface of mirror R_1. Figure 5.13 presents the resonator stability diagram discussed in Sect. 5.1.4 with curves of constant TEM_{00} mode spot sizes [5.18]. The curves, which are obtained from (5.28), reveal that the spot size is fairly insensitive to variations of g_1 and g_2 for resonator configurations which can be represented by points

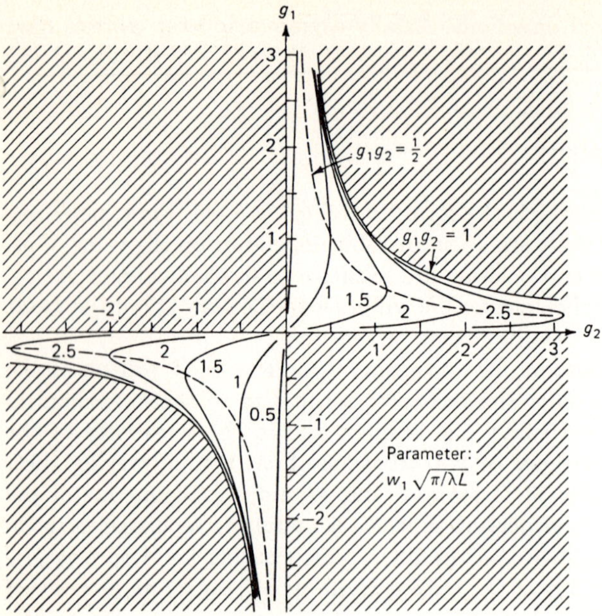

Fig. 5.13. Resonator stability diagram with curves for constant values of the normalized TEM_{00} mode spot size $w_1(\pi/\lambda L)^{1/2}$ at one mirror [5.18]

on the hyperbola $g_1g_2 = 0.5$. Note that large spot sizes w_1 are obtained for resonators with large g_2 values. From (5.23) follows that in order for $g_2>1$, the radius of curvature of mirror R_2 has to become negative, which indicates a convex mirror according to our labeling convention.

The resonator sensitivity to mirror misalignment is related to the fact that the mode axis must be normal to each of the two mirrors. This can be satisfied only if the ray is incident along a line which passes through the center of curvature of the mirrors. Considering a typical resonator of mirror radii R_1 and R_2 and separation L, as shown in Fig. 5.14, a rotation of mirror M_1 through an angle Θ rotates the line joining the centers of curvatures of the two mirrors through an angle ϕ and causes a linear displacement x and y. Small-angle approximations are valid, and we have

$$x = \frac{R_1(R_2 - L)\Theta}{R_1 + R_2 - L} \quad , \quad y = \frac{R_1R_2\Theta}{R_1 + R_2 - L} \quad . \tag{5.33}$$

For a resonator with large radius mirrors of equal radii $R_1 = R_2 = R$ we obtain, from (5.33),

$$x = y = \frac{R\Theta}{2} \quad . \tag{5.34}$$

Note that, if one of the mirrors is slightly tilted, the entire mode is displaced parallel to the resonator axis.

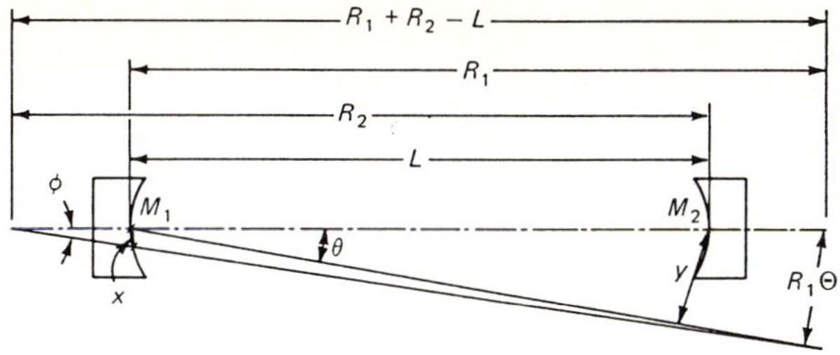

Fig. 5.14. Mirror alignment parameters

For a confocal resonator ($R_1 = R_2 = L$) we have

$$x = 0 \quad \text{and} \quad y = L\Theta \quad . \tag{5.35}$$

In this case the mirror being tilted represents the pivot point for the mode axis.

If the flat mirror of a hemispherical resonator ($R_1 \approx \infty, R_2 \approx L$) is tilted, we obtain

$$x = (R_2 - L)\Theta \approx 0, \qquad y = L\Theta \quad , \tag{5.36}$$

which is similar to that of a confocal resonator. On the other hand, if the spherical mirror is tilted ($R_1 \approx L, R_2 \approx \infty$), then

$$x \approx y \approx L\Theta \quad . \tag{5.37}$$

Comparing the sensitivity to angular tilt of the various resonator configurations, we note that, for example, a large-radius mirror resonator with $R = 10L$ is five times more sensitive to tilt than the confocal and hemispherical resonators.

In accordance with this theory, measurements performed on various lasers have shown that the alignment tolerances of a resonator with relatively short-radius mirrors is less stringent than those imposed with long-radius mirrors. Furthermore, a plano-concave resonator is more sensitive to misalignment than a resonator with two curved mirrors. Also, alignment tolerances become progressively less stringent for higher-order modes. Figure 5.15 gives experimental

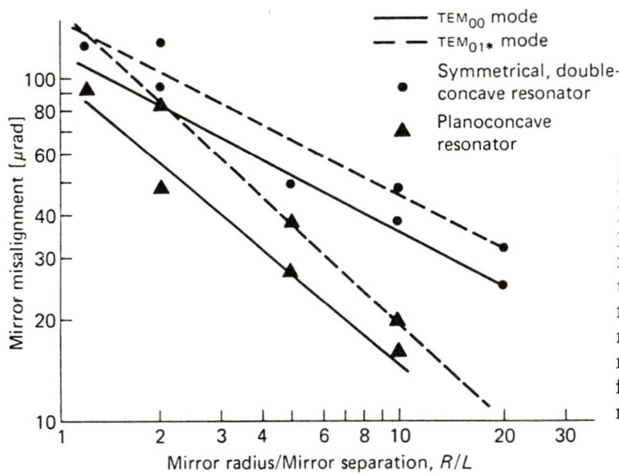

Fig. 5.15. Mirror misalignment in an argon laser which will produce a 10% drop in output power versus the normalized mirror radius. The curves show the characteristics for a symmetrical resonator with two curved mirrors and for a plano-concave resonator. Sensitivity is plotted for the TEM_{00} and TEM_{01^*} mode [5.7]

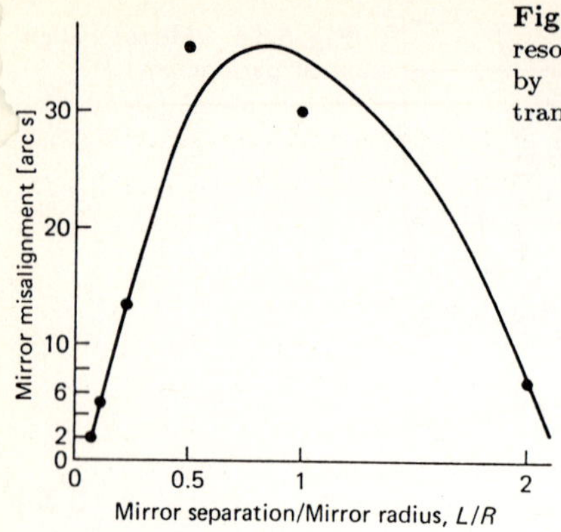

Fig. 5.16. Alignment tolerance of various types of resonators. Curve shows a reduction of output power by 50% of a He-Ne laser operated in a single transverse mode [5.26]

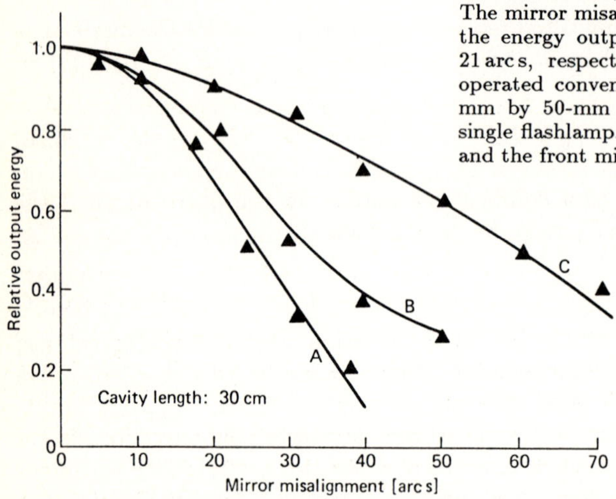

Fig. 5.17. Mirror misalignment sensitivity versus relative output power of a Nd:YAG laser. (**A**: Two flat mirrors, **B**: One flat, one 10-m-radius mirror, **C**: Two 10-m-radius mirrors). The mirror misalignment sensitivity to reduce the energy output by 10% was 12, 15, and 21 arc s, respectively. The laser, which was operated conventional mode, comprised a 5-mm by 50-mm Nd:YAG rod pumped by a single flashlamp. The cavity length was 30 cm and the front mirror was 70% reflective [5.27]

results obtained with an argon laser [5.25]. Clearly, from these curves it follows that the higher-order mode is less sensitive to mirror misalignment and, furthermore, that the confocal resonator has the highest tolerance in terms if misalignment. Figure 5.16 shows the alignment tolerance of various types of resonators for a HeNe laser operating single mode [5.26]. Again, the confocal resonator is far more forgiving for mirror misalignment than the other types. As is apparent from this figure, the alignment tolerances for a concentric-type resonator ($L/R = 2$) and a resonator having large-radius mirrors is about the same.

Figure 5.17 shows the drop in outpt power of a pulsed Nd:YAG laser as a function of mirror misalignment for different mirror combinations [5.27]. As is to be expected, the resonator containing two curved mirrors is least sensitive to misalignment.

5.1.9 Mode-Selecting Techniques

Many applications of solid-state lasers, such as micromachining, nonlinear optical experiments, holography, and rangefinding, quite often require operation of the laser at the TEM_{00} mode since this mode produces the smallest beam divergence, the highest power density, and, hence, the highest brightness. Furthermore, the radial intensity profile is uniform and uniphase. The latter property, i.e., the spatial coherence of the TEM_{00} mode, is particularly important for holographic applications. Focusing a fundamental-mode beam by an optical system will produce a diffraction-limited spot of maximum power per unit area. Generally speaking, in many applications it is a high brightness (power/unit area/solid angle) rather than large total emitted power that is desired from the laser.

Transverse mode selection generally restricts the area of the laser cross section over which oscillation occurs, thus decreasing the total output power. However, mode selection reduces the beam divergence so that the overall effect of mode selection is an increase in the brightness of the laser. For example, the beam diameter and beam divergence for a TEM_{pl} mode increases with the factor C_{pl} introduced in Sect. 5.1.6, which means that for the same output power the brightness decreases by a factor of $(C_{pl})^{-4}$ for the higher-order modes.

Most practical lasers tend to oscillate not only in higher-order transverse modes, but in many such modes at once. Because of the fact that higher-order transverse modes have a larger spatial extent than the fundamental mode, a given size aperture will preferentially discriminate against higher-order modes in a laser resonator. As a result, the question of whether or not a laser will operate only in the lowest-order mode depends on the size of this mode and the diameter of the smallest aperture in the resonator. If the aperture is much smaller than the TEM_{00} mode size, large diffraction losses will occur which will prevent the laser from oscillating. If the aperture is much larger than the TEM_{00} mode size, then higher-order modes will have sufficiently small diffraction losses to be able to oscillate.

The diffraction losses caused by a given aperture and the transverse mode selectivity achievable with an aperture of radius a is illustrated in Fig. 5.18

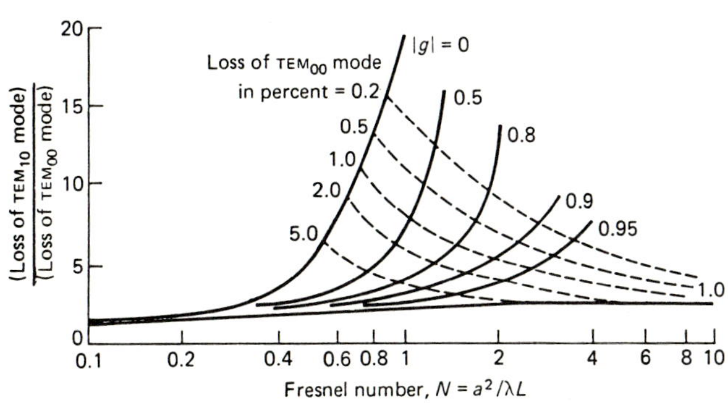

Fig. 5.18. Ratio of the losses per transit of the two lowest-order modes for the symmetric geommetry. The dotted curves are contours of constant loss for the TEM_{00} mode [5.10]

[5.10]. In this figure the ratio of the loss of the TEM_{10} mode to the loss of the TEM_{00} mode is plotted as a function of the Fresnel number for a symmetrical resonator. Note that the mode selectivity is strongly dependent on the resonator geometry, and is greatest for a confocal resonator and smallest for the plane-parallel resonator. From Fig. 5.18 it follows that the resonators of lasers operating in the TEM_{00} mode will have Fresnel numbers on the order of approximately 0.5 to 2.0. For Fresnel numbers much smaller than these, the diffraction losses will become prohibitively high, and for much larger values of N mode discrimination will be insufficient.

These predictions are in agreement with the experimental observations. For example, typical ruby and Nd:YAG lasers have cavity lengths of 50 to 100 cm and TEM_{00} operation typically requires the insertion of an aperture in the cavity with a diameter between 1 and 2 mm. Without an aperture, a 50-cm-long resonator with a 0.62-cm-diameter Nd:YAG rod as the limiting aperture will have a Fresnel number of 19. In ruby lasers, where oscillator rods of 15-mm diameter are not uncommon, the Fresnel number would be 160 for the same resonator length.

We will now discuss typical resonator configurations which are used in the generation of fundamental mode output from solid-state lasers. Because the TEM_{00} mode has the smallest beam diameter of all the resonator modes, a number of techniques have been developed to increase the TEM_{00} mode volume in the active material, which is normally considerably larger in diameter than the mode size. We will find that a resonator designed for TEM_{00} mode operation will represent a compromise between the conflicting goals of large mode radius, insensitivity to perturbation, good mode discrimination, and compact resonator length.

Large-Radius Mirror Configuration. The most common technique to produce TEM_{00} mode output is the use of a nearly plane-parallel resonator with an internal aperture for mode selection. A typical example of this type of resonator is sketched in Fig. 5.19 [5.28]. The resonator has half the diffraction losses of a resonator which is 2.5 m long and has two 10-m-radius mirrors. The relevant parameters for the latter resonator operated at the ruby wavelength are $N = 0.58$ and $g = 0.75$. With N and g known, the diffraction losses for the symmetrical resonator can be obtained from [5.10]. We find for the resonator illustrated in Fig. 5.19 a single-pass diffraction loss of 20% for the TEM_{00} mode and 50% for the TEM_{10} mode.

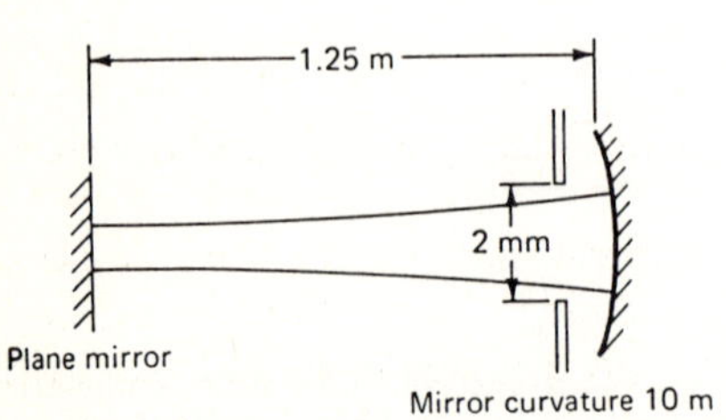

Fig. 5.19. Resonator employed to produce TEM_{00} mode from a ruby laser

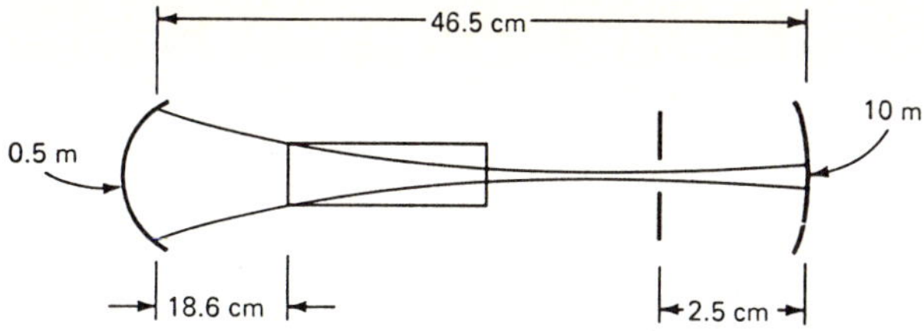

Fig. 5.20. Focusing resonator geometry

Resonators Providing Internal Beam Focusing. Spherical and hemispherical resonators and systems containing internal lenses have in common that they support large mode size differences in the resonator due to their focusing action. For example, in a hemispherical cavity the spot size in the limit can theoretically become zero at the flat mirror and grow to infinity for $L = R$. Location of the laser rod close to the curved mirror permits utilization of a large active volume. An example of this type of resonator is indicated in Fig. 5.20 [5.29]. Mode selection in this resonator, which was employed in a CW Nd:YAG laser, is achieved by axially moving the laser rod until it becomes the limiting aperture for TEM_{00} operation.

A simple resonator scheme, which can be used in CW experiments, is to operate the laser with two flat mirrors which are symmetrically moved farther apart until the TEM_{00} mode power is optimized. The thermal lensing of the rod makes this resonator equivalent to a symmetrical system with strongly curved mirrors. As was discussed in Sect. 5.1.8, theoretically the mode size in the crystal will grow to infinity as the mirror separation approaches four times the focal length of the laser rod.

Resonators with strong internal focusing action suffer from several disadvantages which make them unattractive candidates for field use. In particular, since it is necessary to operate quite close to the edge of the optically stable region, the configurations are extremely sensitive to mechanical and optical perturbations.

Convex-Concave Resonators. *Chesler* and *Maydan* [5.23] have described in some detail the use of convex-concave resonators for efficient and stable generation of TEM_{00} power in solid-state lasers. With these resonators, one can make $g_2/g_1>1$ with the resultant increased size of w_1, and independently set $g_1g_2 = 0.5$ so that the axial sensitivity to thermal focusing is minimized. Operation at the positive branch ($g_2>1$) rather than the negative branch ($g_2<1$) is usually preferred because the cone angle of the mode is smaller and allows a higher degree of overlap with the cylindrically shaped laser rod.

Optical resonators with one convex and one concave mirror offer significant advantages for TEM_{00} mode operation of solid-state lasers. High efficiency and low sensitivity to perturbations and compactness have been obtained in comparison to other resonator types.

As an example we will calculate the parameters for a concave-convex resonator which has been used for a high-repetition-rate Nd:YAG laser [5.18] (Fig. 5.21a). The laser rod, with a diameter of 5 mm, was measured to have a

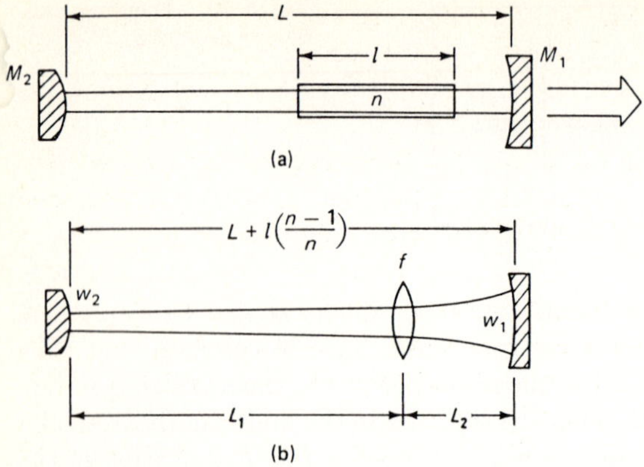

Fig. 5.21. Convex-concave resonator containing (**a**) a laser rod and (**b**) an equivalent passive resonator

focal length of 6 m for a particular input. The optical length of the cavity was restricted to 0.8 m. The optimum value of w_1 has been found empirically to be equal to one-half the laser rod radius. With the rod as close as possible to the output mirror, the following design parameters are obtained for the equivalent resonator (Fig. 5.21b).

$$w_1 = 1.25\,\mathrm{mm}, \quad f = 6\,\mathrm{m}, \quad L_1 = 0.1\,\mathrm{m}, \quad L_2 = 0.7\,\mathrm{m} \quad .$$

Introducing the stability criterion $g_1 g_2 = 0.5$ into (5.29) yields an expression for the mode size as a function of g_1

$$w_1 = \left(\frac{\lambda L}{\pi g_1}\right)^{1/2} \quad . \tag{5.38}$$

Introducing the values for L and w_1 into this equation, one obtains $g_1 = 0.16$, and from (5.32), $g_2 = 3.12$. From (5.27) it follows that $R_1 = 1.1\,\mathrm{m}$, $R_2 = -0.36\,\mathrm{m}$; and from (5.29) we obtain $w_2 = 0.28\,\mathrm{mm}$. The mirror M_1 and the rod surface can be combined by grinding a curvature $R_1' = nR_1$ onto the end of the rod.

Thermally Compensated Resonators. As we will discuss in more detail in Chap. 7, a cylindrical laser rod with uniform internal heating and surface cooling assumes a radial parabolic temperature gradient. This gradient and the associated thermal stresses lead to a radial variation of the index of refraction which transforms the laser rod into a positive lens with large aberrations and causes birefringence. The optical power of the crystal varies directly with pump power.

In the presence of thermal lensing, the TEM_{00} mode volume is reduced by the focusing action of the rod [5.30] and by thermal aberrations of the Gaussian wavefront [5.31]. By measuring the focal length of the laser rod as a function of lamp power, which is best done by passing a collimated HeNe-laser beam through the pumped rod, a first-order compensation of the thermal distortions can be achieved. Enhanced TEM_{00} mode output has been obtained by grinding a concave surface on each end of the rod surface [5.30 ,32], which negates the positive lensing effect of the rod for one particular lamp input. Uniform pumping and the reduction of nonusable pump radiation will minimize wavefront distortions for the TEM_{00} mode.

The primary effect of thermally induced birefringence is to reduce the amount of power which can be extracted from the laser in a linearly polarized beam. The latter property is desired in lasers which are Q-switched or modulated with electrooptical devices or frequency doubled. The use of Nd: $YAlO_3$ or Nd :YLF with its natural birefringence minimizes the problem of polarization conversion and allows a large fraction of the multimode power of a given crystal to be extracted in a polarized TEM_{00} mode. If the host is not naturally birefringent, as is the case of YAG or glass, control of the polarization problems is much more difficult and requires elaborate multicomponent compensation schemes [5.33]. Birefringence compensation techniques usually include an in-cavity polarization rotation between two laser rods. This causes each part of the beam to pass through nearly identical regions of the two rods, with the result that the birefringence induced by one rod is cancelled by the other. This allows a Brewster plate to discriminate against one linear polarization without simultaneously attenuating the other via the birefringent coupling mechanism. Figure 5.22 shows a technique which was used to achieve high TEM_{00} mode output in a linearly polarized beam [5.33]. The resonator configuration employed consisted of a 50-cm convex mirror, two flat-ended Nd: YAG rods separated by a 90° crystal quartz rotator, a Brewster plate, and a 10-m output mirror.

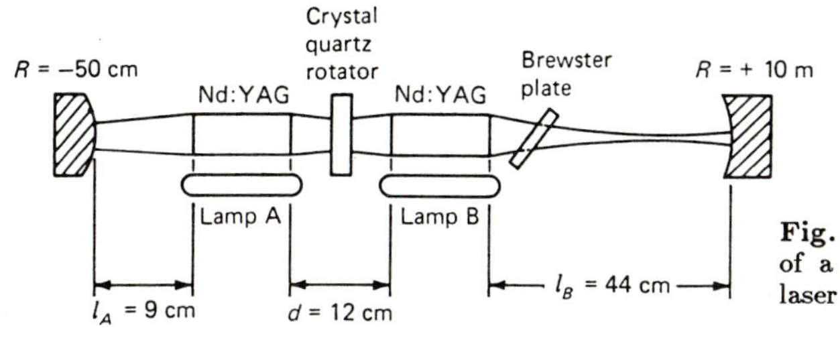

Fig. 5.22. Schematic diagram of a birefringence-compensated laser

Angular-Sensitive Resonators. Experiments have shown that reduction of the angular divergence of the beam is a means of transverse-mode control. Mode selectivity can be obtained by the sharp angular variations of the internal reflectivity near the critical angle of prisms or Lummer-Gehrke plates [5.34].

Performance of Diffraction-Limited Systems. Before leaving the subject of mode selection, it should be pointed out that besides the mode selecting techniques which were discussed here, it is important to have a perfectly aligned system with clean and damage-free optical surfaces. Slight misalignments, tilts, or imperfections (dust particles) of laser reflectors can cause changes in the mode character of the output, favoring higher-order modes. For example, a particle or a damage spot located at the center of the beam can prevent oscil-

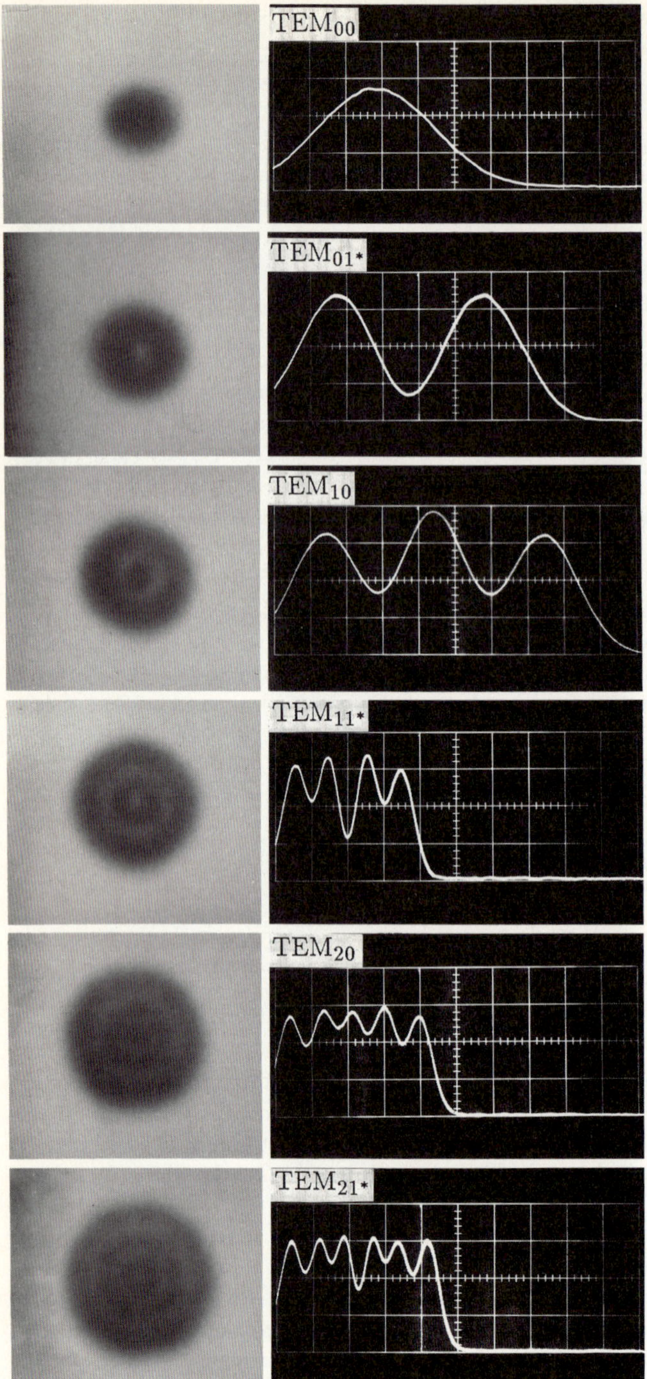

Fig. 5.23. Mode patterns and scans of the far-field beam intensity distribution of a CW-pumped Nd : YAG laser

lation in the TEM_{00} mode and cause oscillation at the TEM_{01^*} mode (for a theoretical discussion of this effect see, for example, [5.35]).

To illustrate the effect of mode selection on the performance of a laser system, Fig. 5.23 presents mode patterns and radial intensity distributions from a CW-pumped Nd: YAG laser. Shown are the first six modes from the laser, which were obtained by successively increasing the mode-selecting aperture. The mode patterns were taken in the far field (20 m away from the laser) by photographing the light striking a ground-glass surface through an infrared image intensifier. The radial intensity distribution of the transverse mode patterns was observed by sweeping the laser beam across a small-aperture detector using a rotating plane mirror and displaying the detector signal on an oscilloscope. Knowing the distance from the rotating mirror to the detector and the rate of rotation of the mirror, the absolute diameter of the beam can be accurately determined. The laser uses a 3-mm by 63-mm Nd: YAG rod which is pumped by two tungsten-filament lamps. The resonator comprises two mirrors with 4-m curvatures separated by 50 cm. With an input of 2 kW, a flat-ended rod was measured to have a focal length of 2 m. In this particular system the rod has a concave surface on each end with a radius of (−5) m. With $n = 1.78$ for Nd: YAG, it follows from $f = R/2(n-1)$ that the focal length $f = -3.2$ m for the unpumped rod. The combination of thermal lensing and negative curvature results in an effective focal length of 5.3 m at 2 kW input, at which the beam-divergence data were taken. The system delivered a maximun of 8.5 W of multimode power. Mode selection was accomplished by insertion of different-size apertured into the cavity. The various aperture sizes and the corresponding output powers, beam divergences, and mode structures are summarized in Table 5.1. From this table follows in accordance with theory that the brightness for higher-order modes decreases despite an increase in total output power.

Employing thermal compensation schemes, such as concave-convex resonators, CW-pumped Nd: YAG lasers using a 4-mm by 75-mm Nd: YAG rod, pumped by two krypton arc lamps in a double-elliptical cavity, will produce between 20 and 25 W TEM_{00} mode power at an input power of 5.5 kW. In a high-repetition-rate Nd: YAG system operating at 20 pps, a Q-switched energy of 60 mJ was achieved in the fundamental mode [5.19]. The resonator, which was 42 cm long, consisted of a rear convex mirror and a flat front mirror which

Table 5.1. Performance of a CW-pumped Nd: YAG laser at different transverse modes

	TEM_{00}	TEM_{01^*}	TEM_{10}	TEM_{11^*}	TEM_{20}	TEM_{21^*}
Aperture size [mm]	1.4	1.6	1.8	2.0	2.2	2.3
Laser output power [W]	1.5	2.4	3.5	4.5	5.5	6.0
Beam divergence [mrad] ($1/e^2$ points)	1.9	2.3	2.8	3.1	3.4	3.6
Brightness [MW/cm^2 sr]	28	18	20	18	15	15

was actually one end of the Nd:YAG rod. The flat-flat laser rod was 5 mm in diameter and 50 mm long; a 30 % reflective coating was used on the output end. The rod was pumped at 42 J input by a xenon flashlamp in a silver-plated single-elliptical cavity. The same system produced a non-Q-switched output of 500 mJ in a 100 μs-long-pulse.

Ruby lasers employing large-radius mirrors and mode-selecting apertures have achieved an output of 200 mJ in a Q-switched pulse and 500 mJ in a 1-ms-long pulse. This performance was achieved in a 75-cm-long cavity with mirrors of 10-m curvature and a 1.5-mm aperture. The rod, 10 cm by 1 1cm, was pumped by a helical flashlamp at an input of 5 kJ. A diffraction-limited Nd: glass laser was reported to generate a pulse of 300 mJ of energy in 25 ns [5.36].

5.1.10 Examples of Advanced Stable Resonator Designs

A large amount of research has been devoted in the recent past to the design of new optical resonator configurations, which could optimize the efficiency of energy extraction from solid state lasers. In fact, operation with stable cavities in the TEM_{00} mode, while producing a beam with a smooth and well-controlled spatial profile, in general results in a poor filling of the active volume and hence in a large waste of the stored energy. Recent developments can be divided into the following optical designs: stable telescopic resonators and concave-convex resonators. Both schemes were extensively studied and experimentally tested and found also commercial exploitations, exhibiting somewhat competing characteristics.

In stable telescopic resonators a magnifying telescope is added to a conventionals stable cavity to expand the mode cross section in the arm of the cavity where the field interacts with the active medium. In the concave-convex resonator, the same effect is achieved by the particular choice of mirror curvature and resonator length. In both cases, the beam quality remains good, but the mode volume is still limited, for the TEM_{00} mode. Furthermore, at the highest intensities, damage problems arise for the optical elements in the resonator section where the beam gets its smaller dimension. For solid-state lasers with low gain, or for systems where a Gaussian profile in the near field is required, the stable resonator is the only choice. Almost all laser applications require a small beam divergence, either to obtain a small spot size at a large range, or a high power density at the focal plane of a lens. Therefore, the challenge in designing a stable resonator is to maximize low-order mode power extraction. More specifically we can establish the following design criteria:

- The diameter of the TEM_{00} mode should be limited by the active material.
- The resonator should be dynamically stable, i.e., insensitive to pump-induced fluctuations of the rod's focal length.
- The resonator modes should be fairly insensitive to mechanical misalignments.

Laser operating in the fundamental mode usually require the insertion of an aperture in the resonator to prevent oscillations of higher-order modes. In this case, the efficiency of the laser is generally lower, compared with multimode operation, due to the small volume of active material involved in the laser action. Large-diameter TEM_{00} modes can be obtained using special resonator configurations, but, if proper design criteria are not applied, the resonator becomes quite sensitive to small perturbations in the mirror curvatures and in the alignment. Also, in solid-state lasers, thermal focusing of the rod greatly modifies the modes and the pump-induced fluctuations of the focal length may strongly perturb the laser output, even preventing any practical or reliable use of the laser. For efficient exploitation of the rod of a solid-state laser operating in the fundamental mode, two conflicting problems have to be solved. The mode volume in the rod has to be maximized, but the resonator should remain as insensitive as possible to focal length and alignment perturbations. Early solutions proposed compensation of the thermal lens by a convex mirror or by negative lenses ground at the ends of the rod, that exactly eliminate the focusing effect of the rod. With these methods high power in a singlemode beam can be obtained; the compensation, however, is effective only for one particular value of the focal length. Large fundamental mode volume and good stability against thermal lens fluctuations have been achieved by a particular choice of mirror curvatures or by insertion of a telescope in the resonator.

In the following subsections we will discuss these two approaches which lead to the design of convex-concave and telescopic resonators.

The Concave-Convex Resonator

The design procedure for resonators known as dynamic stable, in which the fluctuation of the mode volume in the rod is kept under control by an appropriate choice of mirror curvatures, has been described by (5.31,32,38). This earlier work on dynamic stable concave-convex resonators [5.18,23,37] has been expanded considerably in recent years [5.38–40]. *Magni* [5.39] has shown that, for a given resonator, as the focal length of the laser rod varies, there are always two zones of stability. These two zones have the same width in terms of the dynamic optical power of the rod. The fundamental mode volume in the laser rod is inversely proportional to this width. As far as tolerance to mirror alignment is concerned, the two zones have different misalignment sensitivities.

In the following sub-section, we will summarize some of the results presented in [5.39].

A number of new variables are introduced to describe the resonator parameters

$$u_1 = L_1\left(1 - \frac{L_1}{R_1}\right) \quad , \quad u_2 = L_2\left(1 - \frac{L_2}{R_2}\right) \quad , \quad x = \frac{1}{f} - \frac{1}{L_1} - \frac{1}{L_2} . \tag{5.39a}$$

Using the above equations g_1 and g_2 can be expressed as

$$g_1 = -\frac{L_2}{L_1}(1 + xu_1) \quad , \quad g_2 = -\frac{L_1}{L_2}(1 + xu_2) \quad . \tag{5.39b}$$

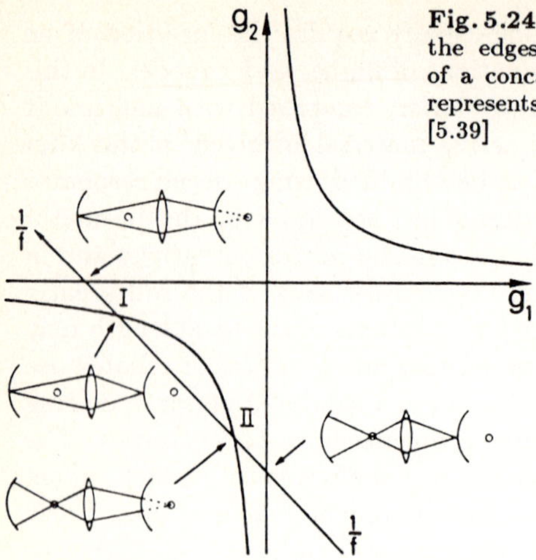

Fig. 5.24. Stability diagram and mode profiles at the edges of the stability zones (marked I and II) of a concave-convex resonator. The straight line $1/f$ represents a resonator with a variable internal lens [5.39]

By eliminating x in (5.39b) a linear relationship between g_1 and g_2 is derived which describes a given resonator configuration for changes in focal length.

$$g_2 = \left(\frac{L_1}{L_2}\right)^2 \frac{u_2}{u_1} g_1 + \frac{L_1}{L_2}\left(\frac{u_2}{u_1} - 1\right) \quad . \tag{5.39c}$$

The intersections of this line with the axes and with the hyperbola $g_1 g_2 = 1$ defines two distinct stability zones indicated, as I and II in Fig. 5.24. The two zones have the same width in terms of x, as illustrated in Fig. 5.25. The parameter w_{30} is the spot size at the principal planes of the laser rod and is the same in both stability zones. The value of w_{30} is given by

$$w_{30}^2 = \frac{2\lambda}{\pi}|u_1| \quad , \quad |u_1|>|u_2| \quad . \tag{5.40a}$$

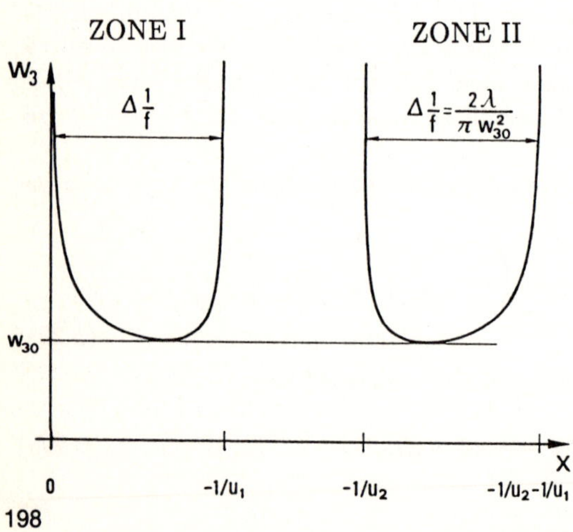

Fig. 5.25. Spot size w_3 inside the laser rod as a function of the optical power $1/f$ of the rod [5.39]

The relationship between the width of the stability zones and the minimum spot sizes on the lens is given by

$$w_{30}^2 = \frac{2\lambda}{\pi} \frac{1}{|\Delta \frac{1}{f}|} \quad . \tag{5.40b}$$

The volume of active material involved in the laser action is inversely proportional to the width of the stability zone and hence to the range of input power to the lamp for which the resonator remains stable. The centers of the two stability zones in Fig. 5.25 are at

$$x_I = -\frac{1}{2u_1} \text{ and } x_{II} = -\frac{1}{u_2} - \frac{1}{2u_1} \quad . \tag{5.40c}$$

At the edges of the stable zones, defined by $g_1 g_2 = 0$ and $g_1 g_2 = 1$, the spot size w_3 goes to infinity.

Magni [5.39] treated in his analysis not only the stability of the resonator to small fluctuations of the focal length of the laser material, but also considered the mechanical stability of the resonator in terms of a misalignment sensitivity.

For example, the misalignment sensitivity $S_{1/2}$ in the middle of zone I can be expressed as

$$S_{1/2} = \frac{1}{w_{30}} \frac{2L_1 L_2}{(4L_1^2 + L_2^2)^{1/2}} \quad . \tag{5.41}$$

To a very rough approximation, it may be assumed that $1/S_{1/2}$ is the tilt angle for about a 10% increase in resonator losses [5.41,42]. It was found, that the tolerance to mirror tilt is higher in zone I as compared to zone II.

The design of a resonator described in [5.39] proceeds by specifying a focal length $f_{1/2}$ of the laser rod, a spot size w_{30} inside the laser material and a total resonator length L. With these parameters given, the mirror curvatures R_1 and R_2, the separation of the mirrors from the principal planes of the rod L_1 and L_2, and the sensitivity parameter $S_{1/2}$ can be calculated. Please note that $L = L_1 + L_2$, the physical length is different, since it also includes the distance between the principal planes of the rod. In Fig. 5.26 the values of L_1, R_1, R_2, and $S_{1/2}$ for optimized resonators are plotted vs $f_{1/2}$ for a few values of w_{30}. From this figure the values of the curvature radii and the position of the rod can be readily obtained for a given focal length and for a given spot size of the mode in the rod.

Figure 5.27 illustrates a resonator design based on the results presented in Fig. 5.26. We assume a mode radius of $w_{30} = 3$ mm in the active medium. This is the optimum mode radius for a 6 mm diameter rod. The length of the resonator is given as 150 cm.

For a focal length of $f = 17$ cm ($1/f = 6\,\text{m}^{-1}$) we obtain $R_1 = -14$ cm, $R_2 = 55$ cm and $L_1 = 130$ cm. The resonator has a misalignment sensitivity of about $S_{1/2} = 80$, which would indicate that for a tilt angle of about 12 mrad, the losses will increase by 10%. In Fig. 5.27 the distance from the end of the rod to the principal plane is $h = l/2n$, where l is the length of the laser rod.

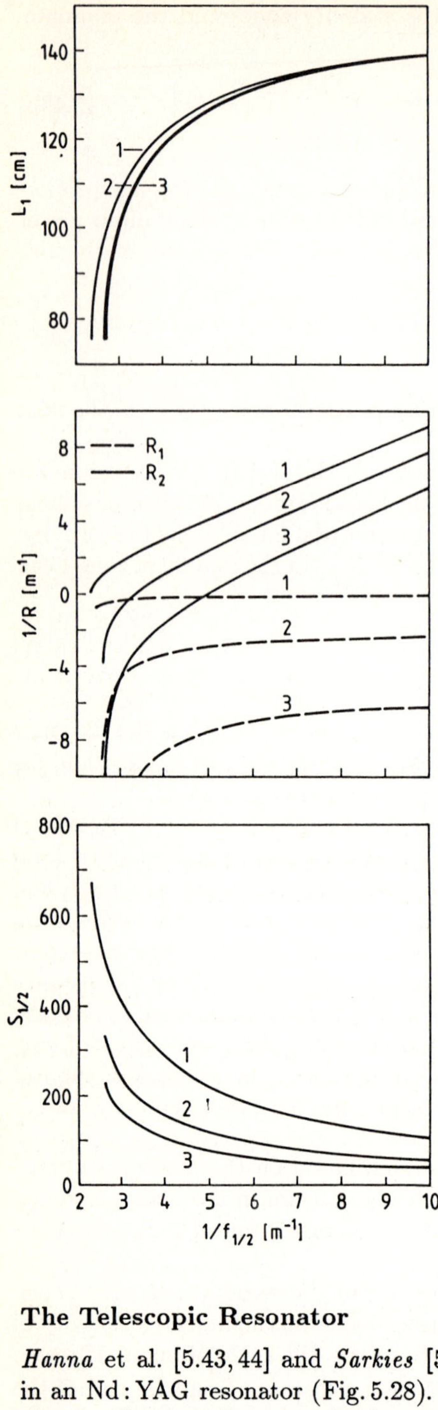

Fig. 5.26. Distance L_1, mirror curvatures $1/R_1$ and $1/R_2$, and misalignment sensitivity $S_{1/2}$ vs the optical power $1/f_{1/2}$ for a specific concave-convex resonator. Resonator length fixed at $L = 150\,\mathrm{cm}$, wavelength $\lambda = 1.064\,\mu\mathrm{m}$. The parameter on the curves is the spot size $w_{30} = 1, 2, 3\,\mathrm{mm}$ [5.39]

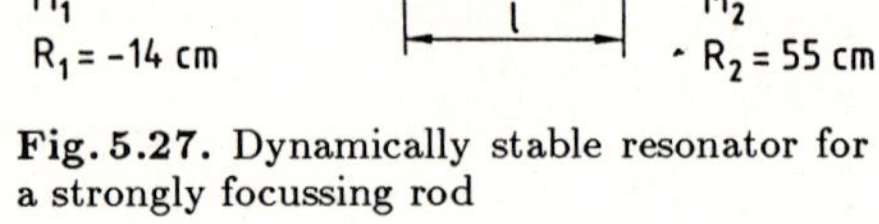

Fig. 5.27. Dynamically stable resonator for a strongly focussing rod

The Telescopic Resonator

Hanna et al. [5.43, 44] and *Sarkies* [5.45], reported on the use of a telescope in an Nd:YAG resonator (Fig. 5.28). An attractive feature of the telescope is that it allows easily controllable adjustment to compensate thermal lensing

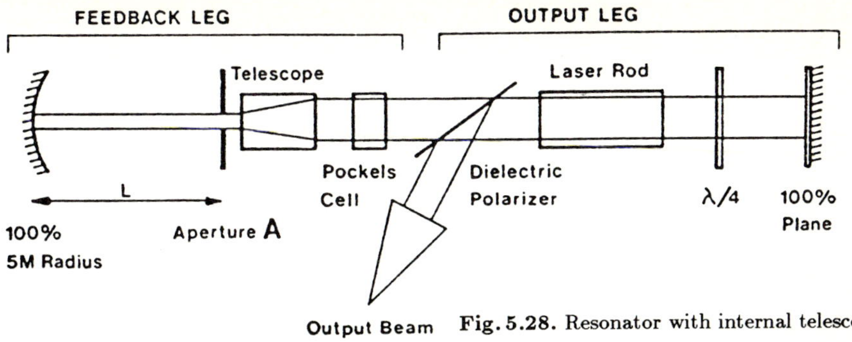

Fig. 5.28. Resonator with internal telescope

under varying pumping conditions. In addition, the telescopic resonator avoids the very small spot on the convex mirror of the convex-concave mirror design. This is particularly important at the high power levels typical for Q-switched Nd:YAG lasers.

By introducing a suitably adjusted telescope into a Q-switched Nd:YAG laser resonator, the investigators mentioned above, have been able to obtain reliable operation with a large-volume TEM_{00} mode. The basic principle behind the resonator design is that of choosing a telescope adjustment which compensates the thermal lensing in the laser rod (thus permitting a large spot size) and at the same time ensuring that the spot size is insensitive to fluctuations in focal length of the thermal lens.

The telescope performs two distinctly separate functions. Firstly, it reduces the size of the beam to increase the diffraction per unit length. Since the beam size on the input side is always the same as the rod diameter, the diffraction is constant and dependent only on the telescope magnification. The aperture at the telescope output is set to be D/M, D being the rod diameter. Secondly, the telescope is an element of variable focal length. It can therefore be adjusted to place the resonator anywhere on the stability diagram. Because the ratio of the diffraction losses of the higher-order modes to the lower-order ones increases as the telescope output beam decreases, the telescope can be adjusted to ensure that modes above a certain order do not reach threshold. Thus the mode selection process is controlled by two telescope parameters, the magnification M and the focal length f. Clearly sufficient mode selection can be achieved by either parameter alone, but, on the one hand, too high a magnification may result in a very high power density in the feedback beam, which could exceed the damage threshold of the components. On the other hand, too much bias introduced by the telescope could result in a laser threshold that is very high. Thus the correct balance must be established to ensure optimum operation.

The telescope adjustment is chosen to minimize the effect of focal-length variations in the laser rod and at the same time ensures the optimum mode-selection properties of a confocal resonator. *Hanna* et al. [5.43] performed a detailed analysis of the telescopic resonator. The analysis provided simple design equations relating the spot size, resonator length, telescope magnification, and defocusing and diffraction losses. A short summary of the key design parameters is given below.

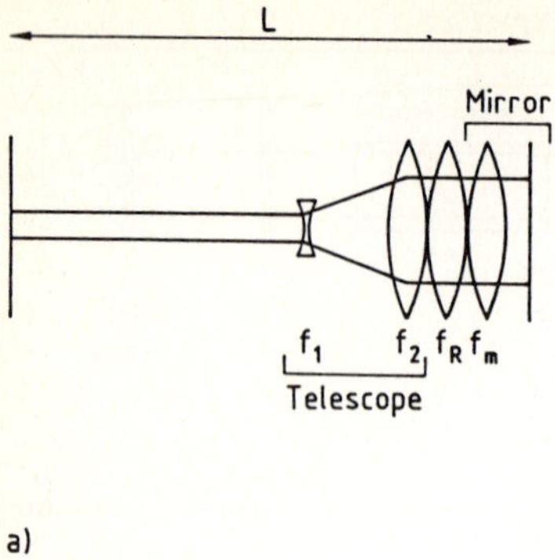

Fig. 5.29. (a) Resonator with internal telescope (f_1, f_2), focussing laser rod f_R), and curved mirror (f_M). (b) Spot size w_1 in the laser rod, and w_2 on the left mirror, as a function of telescope defocussing δ or focussing f_R of the laser rod [5.44]

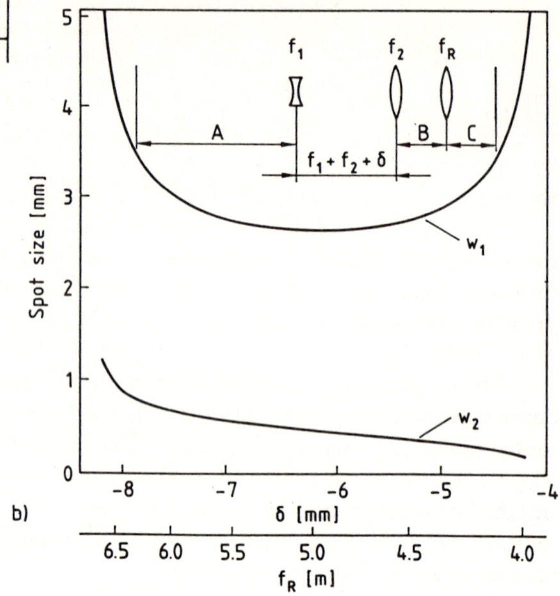

One can best understand the role of the telescope by considering, for simplicity, a short telescope of magnification M (where $f_2 = -Mf_1$) located close to the laser rod characterized by a focal length f_R, and a resonator mirror with an equivalent focal length f_{M}, as shown in Fig. 5.29a.

It can be shown that small defocusing of the telescope has two effects: it changes the spot size and it changes the wavefront curvature. Thus the telescope can be adjusted to achieve compensation of the thermal lens f_{R} of the laser rod by making

$$-\frac{1}{f_{\mathrm{T}}} = \frac{1}{f_{\mathrm{R}}} + \frac{1}{f_{\mathrm{M}}} \tag{5.42a}$$

where

$$\frac{1}{f_{\mathrm{T}}} = \frac{\delta}{f_2^2} \tag{5.42b}$$

is the focal length of the telescope for small defocusing δ. The expression on the right-hand side of (5.42a) is the optical power of the laser rod/mirror combination.

The magnification M of the telescope has to be selected for maximum insensitivity of spot size to variations of f_{R}. This is achieved for

$$\frac{1}{2M^2L} = \frac{1}{f_\mathrm{T}} + \frac{1}{f_\mathrm{R}} + \frac{1}{f_\mathrm{M}} \quad . \tag{5.42c}$$

The spot size in the laser rod is given by

$$w_1 = M(2L\lambda/\pi)^{1/2} \quad . \tag{5.42d}$$

Thus introducing the correctly adjusted telescope allows the same large mode volume in the laser rod to be maintained but with a reduction of cavity length by M^2. The main limitation of this approach is that it exposes components in the reduced beam to higher intensity and thus greater damage risk.

Hanna et al. [5.43] have presented some of their key findings in graphical form. Figure 5.29b shows spot sized w_1 (in laser rod) and w_2 (on left hand mirror) versus telescope defocusing δ, with f_R fixed (5 m), or equivalently versus f_R with δ fixed at -6.2 mm. The laser parameters are as follows: $A = 0.55$ m, $B = 0.37$ m, $C = 0.16$ m, $f_1 = -0.05$ m, $f_2 = 0.20$ m, $M = 4$.

The main feature is the broad minimum of spot size in the laser rod (upper curve), implying insensitivity of the spot size to δ. For a fixed mirror curvature and telescope setting (i.e., f_M and f_T constant) it follows from (5.42c,d) that Fig. 5.29b also represents a plot of spot size versus f_R. The minimum of the upper curve therefore implies insensitivity to fluctuations in f_R. The desired operating point is at the bottom of this minimum and the telescope must therefore be defocused by the correct amount to ensure this.

In arriving at a resonator design the main parameters to be chosen are spot size w_1, in the laser rod, resonator length L and magnification M. It is assumed that f_R is known, this parameter is usually determined by passing a He-Ne laser beam through the laser rod and measuring the beam waist at the desired pump level.

With w_1 chosen, the choice of values for L and M is made according to (5.42d) to give an acceptable compromise between a small M and hence an inconveniently large L or small L and hence large M which may then lead to excessive intensity in the contracted beam. When L (and hence M) have been chosen, and a mirror has been selected (which determines f_M) then the telescope focal length f_T is obtained from (5.42c) and finally the amount of defocusing from (5.42b).

A circular aperture to select the TEM_{00} mode is inserted and centered. In practice, it is found that the aperture diameter should be ~ 1.5 times the calculated spot diameter at the point of insertion to ensure suppression of the TEM_{01} mode.

5.2 Longitudinal Modes

5.2.1 Fabry-Perot Resonators

The Fabry-Perot resonator is not only an essential element of the laser, it is also used for mode selection and as an instument to measure laser linewidth. Formu-

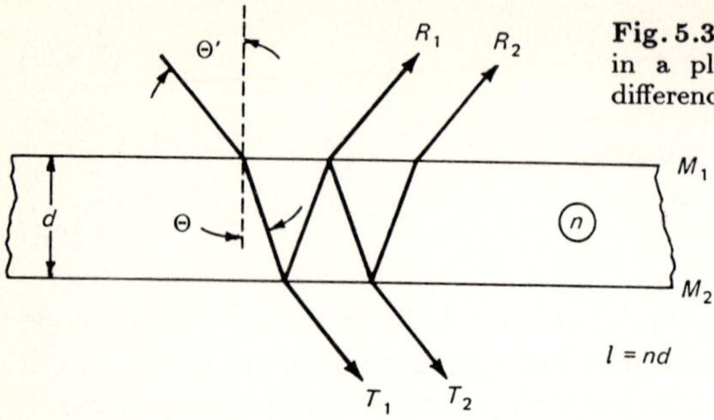

Fig. 5.30. Interference of a plane wave in a plane-parallel plate. The pathlength difference between T_1, and T_2 is $2/\cos\phi$

las needed to calculate the salient features of the different kinds of resonators employed in solid-state lasers are summarized below.

Basic Equations

A Fabry-Perot resonator consists of two plane-parallel optical surfaces; in classical optics this arrangement is known as a Fabry-Perot interferometer or as an etalon [5.46–48] (see below). Multiple reflections which occur between the surfaces cause individual components of the wave to interfere at M_1 and M_2 (Fig. 5.30). Constructive interference occurs at M_2 if all components leaving M_2 add in phase. It turns out that for such preferred directions, the components reflected from M_1 destructively interfere and that actually all the incident energy is transmitted by the etalon. For each member of either the reflected or the transmitted set of waves, the phase of the wave function differes from that of the preceding member by an amount which corresponds to a double traversal of the plate. This phase difference is

$$\delta = \left(\frac{2\pi}{\lambda}\right) 2nd \cos\Theta \quad , \tag{5.43}$$

where nd is the optical thickness between the two reflecting surfaces, $\Theta = \Theta'/n$ is the angle of refraction in the material, and λ is the wavelength. The transmission of the Fabry-Perot resonator is

$$T = \left[1 + \frac{4r}{(1-r)^2} \sin^2\left(\frac{\delta}{2}\right)\right]^{-1} \tag{5.44}$$

where r is the reflectivity of each of the two surfaces. The maximum value of the transmission, $T_{\max} = 1$, occurs in the etalon when the path length differences between the transmitted beams are multiple numbers of the wavelength:

$$2nd \cos\Theta = m\lambda : \qquad m = 1, 2, 3, \dots \quad . \tag{5.45}$$

The reflectivity of the resonator can be expressed by

$$R = \left(1 + \frac{(1-r)^2}{4r \sin^2(\delta/2)}\right)^{-1} \quad . \tag{5.46}$$

The maximum value of the reflectivity

$$R_{\max} = \frac{4r}{(1+r)^2} \tag{5.47}$$

is obtained when the path length difference of the light beam equals multiples of half-wavelength

$$2nd \cos \Theta = \frac{m\lambda}{2} \quad ; \quad m = 1, 3, 5, \ldots \quad . \tag{5.48}$$

Figure 5.31 illustrates the transmission and reflection properties of the etalon.

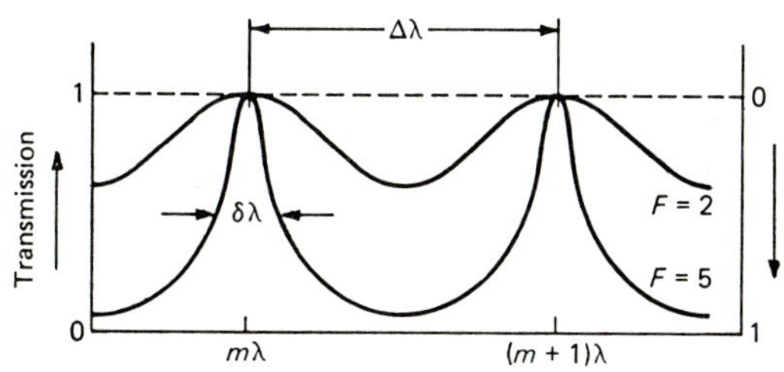

Fig. 5.31. Intensity of the reflected and transmitted beams as a function of phase difference

In the absence of absorption losses, the transmission and reflectance of an etalon are complementary in the sense that $R + T = 1$. We define the ratio of the spacing between two adjacent passbands and the passband width as finesse $F = \Delta\lambda/\delta\lambda$:

$$F = \pi \left[2\,arc\,\sin \left(\frac{2+4r}{(1-r)^2} \right)^{-1/2} \right]^{-1} \approx \frac{\pi(r)^{1/2}}{(1-r)} \quad . \tag{5.49}$$

The approximation can be used if $r > 0.5$. For small values of r the finesse approaches the value $F \approx 2$ and (5.42) is reduced to

$$R = R_{\max} \sin^2 \frac{\delta}{2} \quad . \tag{5.50}$$

We see from (5.45) that the resonance condition depends on the wavelenth λ, the optical length nd, and the angle of refraction Θ. A variation of any of these quantities will shift the etalon passband. The wavelength difference between two passbands is called the free spectral range of the etalon. The following equation gives the variations which are required to move a passband by one order of magnitude:

$$\Delta\lambda = \frac{\lambda_0^2}{2nd \cos \Theta} \quad , \quad \Delta\nu = \frac{c}{2nd \cos \Theta} \quad ,$$

$$\Delta(nd) = \frac{\lambda_0}{2 \cos \Theta} \quad , \quad \Delta\Theta = \left(\frac{\lambda}{nd} \right)^{1/2} \quad \text{at } \Theta = 0 \quad . \tag{5.51}$$

The variations required to shift the transmission or reflection spectrum by the width of one passband are

$$\delta\lambda = \frac{\Delta\lambda}{F} \quad ; \quad \delta l = \frac{\Delta l}{F}; \quad \delta\Theta = \frac{\Delta\Theta}{F} \quad . \tag{5.52}$$

Laser Resonator

If we consider for the moment only a single transverse mode, then according to (5.51) the separation of the longitudinal modes in a laser cavity is given by

$$\Delta\lambda_p = \frac{\lambda_0^2}{2L} \quad \text{or} \quad \Delta\nu = \frac{c}{2L} \tag{5.53}$$

where L is the optical length of the resonator.

For a resonator with $L = 75$ cm and $\lambda_0 = 6943$ Å one obtains $\Delta\lambda_p = 0.003$ Å. A ruby laser has a linewidth of about 0.5 Å, which means that there are approximately 160 longitudinal modes within this linewidth. If the reflectivities of the two surfaces of a resonator are different, as in a laser cavity, the reflectivity r in (5.49) is the geometric mean reflectivity of the two mirrors: $r = (R_1R_2)^{1/2}$. Typical values for a laser cavity are $R_1 = 0.5$ and $R_2 = 1.0$. Introducing these numbers into (5.49, 52), one obtains $F = 8.5$ and $\delta\lambda = 0.0004$ Å for the finesse and the spectral width of a single line in our example (Fig. 5.32).

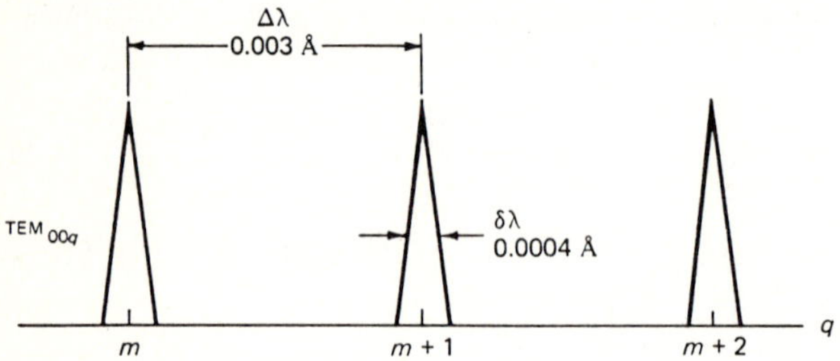

Fig. 5.32. Longitudinal modes in a laser resonator composed of a 50 % and a 100 % reflective mirror separated by 75 cm

By combining (5.49, 52, 53), the linewidth of a single longitudinal mode (assuming $R_2 = 1.0$) can be directly expressed by

$$\delta\lambda = \frac{\lambda^2[1-(R_1)^{1/2}]}{2L\pi(R_1)^{1/2}} \quad . \tag{5.54}$$

In the presence of several transverse modes, additional resonant frequencies occur in the laser cavity. The frequency separation of different TEM_{plq} modes in a laser resonator is given by [5.4]

$$\Delta\nu = \left(\frac{c}{2L}\right)\left[\Delta q + \left(\frac{1}{\pi}\right)\Delta(2p+l) arc\,\cos\left(1-\frac{L}{R}\right)\right] \quad . \tag{5.55}$$

The term on the right containing Δq gives the frequency spacing of the axial modes which belong to a single transverse mode. The term with $\Delta(2p+l)$ describes the separation of the resonance frequencies of different transverse TEM_{pl} modes. Note that the resonant frequencies depend on $(2p+l)$ and not on p and l separately. Therefore, frequency degeneracies arise when $(2p+l)$ is equivalent for different modes. By replacing $(2p+l)$ with $(m+n)$, the cavity frequencies for transverse modes expressed in Cartesian coordinates are obtained.

From (5.55) follows that the frequency spacing between transverse modes is not only a function of mirror separation, as is the case with the axial modes, but depends also on the curvature of the mirrors.

For a near plane-parallel or long-radius resonator ($L \ll R$), the second term within the brackets of (5.55) becomes small compared to Δq. In this case the resonant-mode spectrum is composed of the relatively large axial mode spacing $c/2L$, with each axial mode surrounded by a set of transverse-mode resonances, as shown in Fig. 5.33a. For example, the 75-cm-long resonator mentioned before, if terminated by two mirrors with 10-m curvatures, will have a resonance spectrum of $\Delta\nu/\text{MHZ} = 200\Delta q + 28(2p+l)$. If the mirror radii are decreased, starting from the plane-parallel configuration, the transverse-mode frequency spacing increases while the axial-mode frequency intervals remain the same, provided that the mirror separation is constant. The extreme condition is reached for the confocal resonator, where (5.55) reduces to

$$\Delta\nu = \left(\frac{c}{2L}\right)\left[\Delta q + \left(\frac{1}{2}\right)\Delta(2p+l)\right] \quad . \tag{5.56}$$

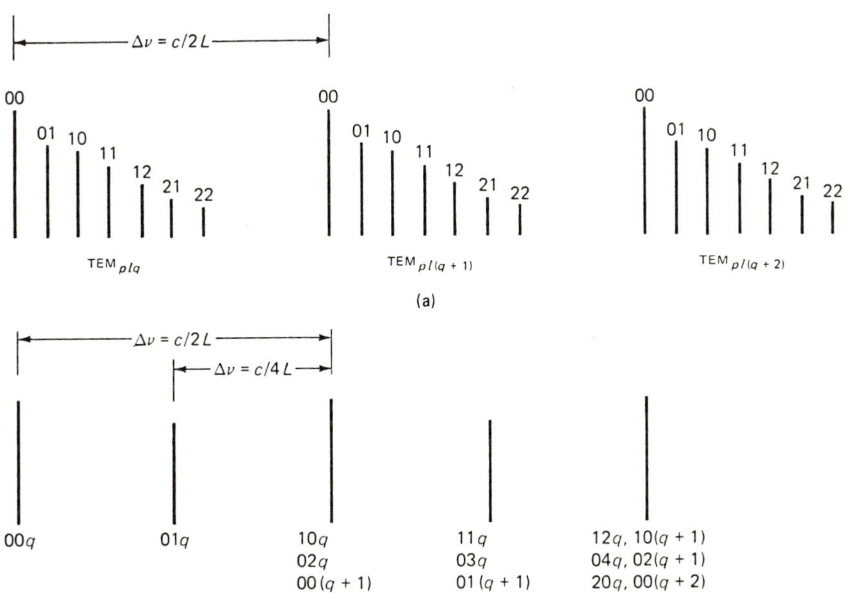

Fig. 5.33. Resonance spectrum of (a) long-radius mirror resonator and (b) confocal resonator

In the confocal resonator the resonance frequencies of the transverse modes resulting from changing p and l either coincide or fall halfway between the axial-mode frequencies (Fig. 5.33b). As the mirror curvature increases still more, the frequency interval of the transverse modes decreases and becomes zero for a concentric resonator. The frequency spacing $\Delta\nu$ expressed by (5.55) can be measured if a spectrum analyzer and photodetector of sufficiently fast response are available. The output from the detector will contain a beat-frequency signal corresponding to $\Delta\nu$ if more than one mode is oscillating.

Two techniques are widely used for obtaining spectrally narrowed output from solid-state lasers. They involve the use of either a resonant reflector or an intracavity tilted etalon. Both devices will be discussed next.

Resonant Reflectors

Fabry-Perot resonators of fixed spacing are generally referred to as etalons. If an etalon is employed in place of an output mirror in a laser cavity, it is referred to as a resonant reflector.

The normal operating region of a resonant reflector is at its maximum reflection. The value of the peak reflectivity is obtained from (5.47). If an uncoated etalon is used, the reflectivity

$$r = \left(\frac{n-1}{n+1}\right)^2 \tag{5.57}$$

must be introduced into (5.47)

As an example we will consider the properties of a single sapphire etalon employed as an output reflector in a ruby laser. If we assume a thickness $d = 3.2\,\mathrm{mm}$, an index of refraction $n = 1.76$, and an operating wavelength $\lambda = 6943\,\text{Å}$, then we obtain from the preceding equations a maximum reflectivity of $R_{\max} = 0.27$, a separation of the peaks of $\Delta\lambda = 0.43\,\text{Å}$, and a half-width of $\delta\lambda = 0.22\,\text{Å}$. As a comparison, an ordinary glass etalon with index of refraction 1.45 has a maximum reflectance of $R_{\max} \approx 0.17$. The spectral characteristics of a 2.5-mm-thick quartz etalon is shown in Fig. 5.34 (top).

The resonance peaks of multiple-resonant reflectors are sharper and more separated than is the case in single-element devices. The maximum reflectivity of a multielement resonant reflector is [5.49]

$$R_{\max} = \left(\frac{n^N-1}{n^N+1}\right)^2 \quad , \tag{5.58}$$

where n is the refractive index of the plates and N is the number of reflecting surfaces.

It will be shown that the simple theory outlined in the preceding subsection, which is applicable to single-plate etalons, can be used to predict the main features of the spectral curve of a multielement etalon. More important, the simple theory provides a quick way to see how changes in the material, plate thickness, coatings, etc., will alter the reflectivity of a given device.

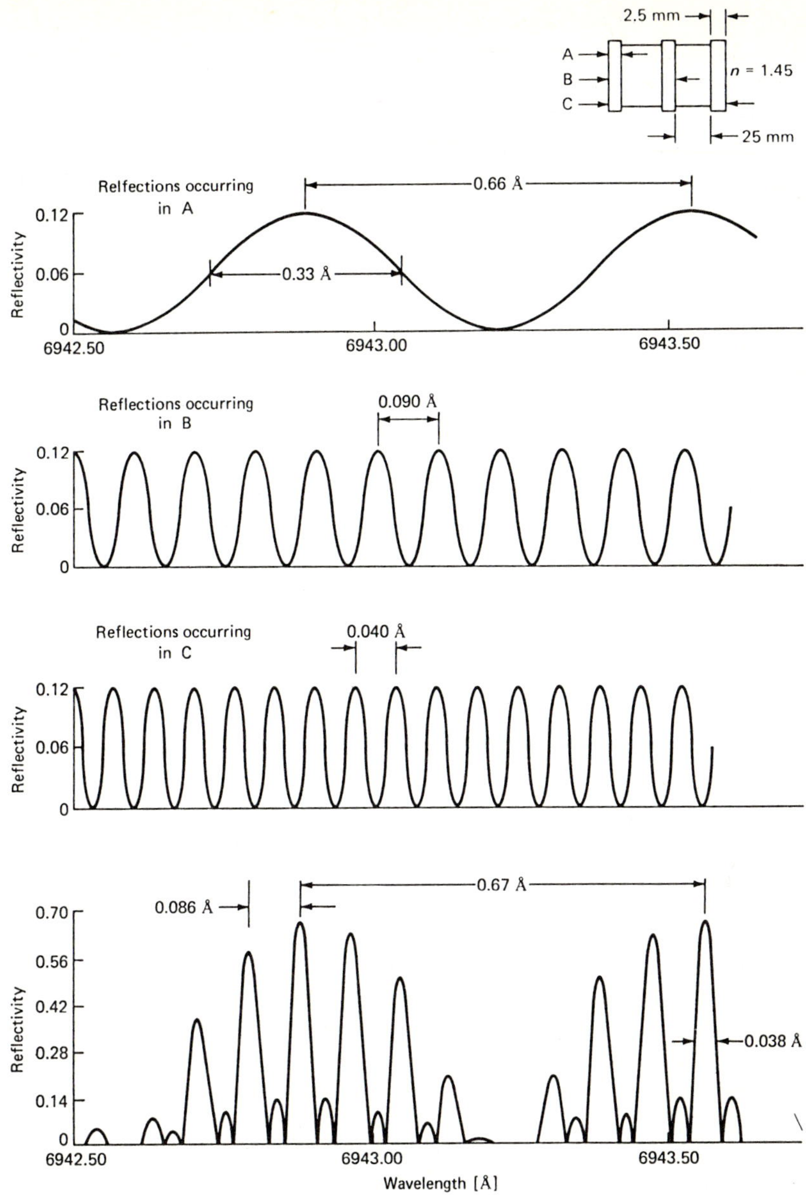

Fig. 5.34. Analysis of a three-plate resonant reflector

As an example, we will consider a three-plate etalon. The unit consists of three quartz plates with a thickness of 2.5 mm, separated by two spacers each 25 mm in length. Figure 5.34 (bottom) shows the reflectivity as a function of wavelength for this device. The curve is the result of a calculation carried out on a computer. As can be seen from this figure, the resonant reflector has a peak reflectivity of 65 % and the main peak has a half-width of 0.038 Å. The envelope of the individual resonance peaks is repeated after every 0.67 Å. There are six different resonance effects which occur in this device. With $n = 1.45$, $d_Q = 2.5$ mm, and $d_A = 25$ mm, one obtains a resonance due to: a single sapphire plate $l = 3.62$ mm, $\Delta\lambda_1 = 0.66$ Å; one plate plus one air gap $l = 28.6$ mm, $\Delta\lambda_2 = 0.084$ Å; air gap alone $l = 25$ mm, $\Delta\lambda_3 = 0.096$ Å; two plates plus air gaps $l = 32.25$ mm, $\Delta\lambda_4 = 0.074$ Å; two plates plus both air gaps $l = 57.25$ mm, $\Delta\lambda_5 = 0.042$ Å; three plates plus both air gaps $l = 60.9$ mm, $\Delta\lambda_6 = 0.040$ Å.

Comparing these results with the actually calculated curve reveals that the resonance in the single plates determines the period of the whole device. The main peaks are formed by the resonance in the air space, whereas the minor peaks are caused by resonances including both air spaces and at least two plates. If one uses sapphire instead of quartz, then with $n = 1.79$ and $N = 4$ one obtains $R_{max} = 0.67$ for a two-plate device.

Tilted Etalon

Consider an etalon of length d, refractive index n, with the surface normal inclined to the incident light beam at a small angle Θ'. The transmission of the tilted etalon is given by (5.44). The shift in the resonance wavelength due to a change in tilt angle Θ' is [5.50]

$$\Delta\lambda = \frac{-\lambda(\Delta\Theta')^2}{2n^2} \quad . \tag{5.59}$$

Fabry-Perot Interferometer

The Fabry-Perot interferometer is commonly used to measure the linewidth of the laser emission. The arrangement normally used to display the familiar Fabry-Perot rings consists of a short-focal-length negative lens, an etalon with spacings of 5 to 40 mm, a long-focal-length positive lens, and a photographic plate or screen. The first lens converts the parallel beam from the laser into a diverging cone of light.

The resonances of the etalon cause an angularly selective transmission. The light transmitted by the interferometer is collected by a lens. Light beams leaving the etalon inclined at the same angle with respect to the optical axis of the etalon form a ring in the focal plane of the lens. The corresponding number of Fabry-Perot rings are displayed by a screen or they can be photographed. In order to increase the diameter of the rings, a telescope is sometimes inserted between the interferometer and the imaging lens.

If the instrument is illuminated by a perfect monochromatic light source, one obtains a set of rings which correspond to directions for which constructive interference occurs ($2l \cos\Theta = m\lambda$). Since these interferometers are employed to measure the spectral characteristics of light beams, one is interested in the wavelength shift required to move the pattern from one ring to the next ring. This wavelength interval is called the free spectral range of the interferometer.

For an interferometer which consists of two dielectrically coated mirrors with a reflectivity of 90 % and separated by 40 mm, we obtain the following performance: The free spectral range is $\Delta\lambda_s = 0.062\,\text{Å}$ and the finesse is $F = 30$. The theoretical resolution of the instrument obtained from (5.48) is $\delta\lambda' = 0.002\,\text{Å}$. However, the practical resolution is limited by the photographic film. As a rule of thumb, one usually assumes that the resolution is about one-tenth of the free spectral range. This means that a pattern showing rings which are not wider than 10% of the ring separation indicates a laser linewidth of less than $\delta\lambda = 0.006\,\text{Å}$.

5.2.2 Spectral Characteristics of the Laser Output

If a laser is operated without any mode-selecting elements in the cavity, then the spectral output will consist of a large number of discrete frequencies determined by the transverse and axial modes. The linewidth of the laser transition limits the number of modes that have sufficient gain to oscillate. The situation is diagrammed schematically in Fig. 5.35, which shows the resonance frequencies of an optical resonator and the fluorescence line of the active material. Laser emission occurs at those wavelengths at which the product of the gain of the laser transition and the reflectivity of the mirrors exceeds unity. In the idealized example shown, the laser would oscillate at seven axial modes.

Depending on the pumping level, for ruby and Nd:YAG, one finds a linewidth of approximately 0.3 to 0.5 Å for the laser emission in the absence of mode selection. In Nd:glass, because of the wide fluorescence curve, the laser linewidth is two orders of magnitude broader.

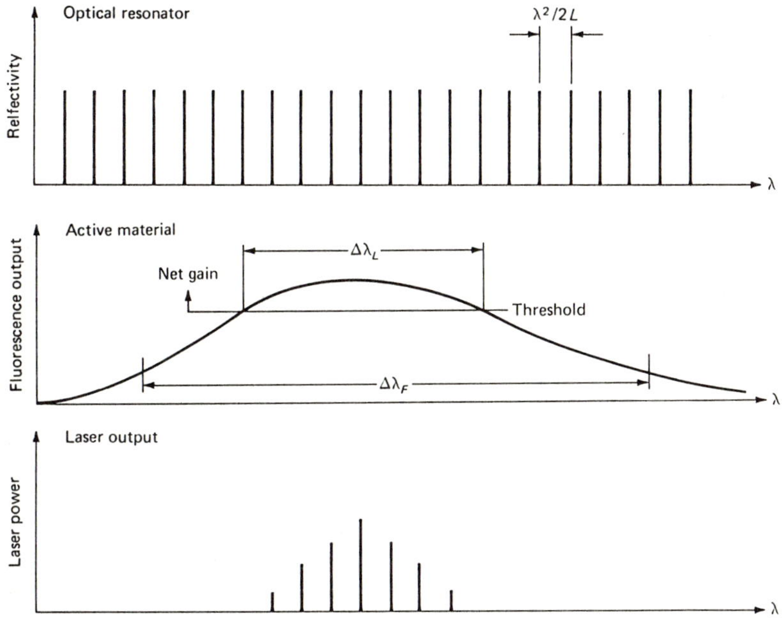

Fig. 5.35. Schematic diagram of spectral output of a laser without mode selection

The spectral characteristics of a laser are quite frequently described in terms of bandwidth, linewidth, number of axial modes, and coherence length. We will now outline the relationships among these quantities.

If the laser emission occurs at a single longitudinal and transverse mode, then the bandwidth of the laser is $\Delta\nu = c/2LF$, where F is the finesse of the optical resonator. If the laser emits K axial modes, the bandwidth between the two extreme modes is

$$\Delta\nu = \frac{(K-1)c}{2L} \quad \text{or} \quad \Delta\lambda = \frac{(K-1)\lambda^2}{2L} \quad . \tag{5.60}$$

The wave emitted from a laser which emits at a discrete number of integrally related wavelengths is strongly modulated. The situation can be illustrated by writing down the simplest case of two superimposed traveling waves whose wavelengths are specified by adjacent axial modes.

This situation is shown schematically in Fig. 5.36. The two waves interfere with one another, and produce traveling nodes which are found to be separated from one another in time by twice the cavity separation. That is, the output of such a laser is modulated at a frequency of twice the end-mirror separation ($\nu_m = c/2L$).

When lines at three integrally related frequencies are emitted, the output becomes more modulated; however, the maxima are still separated from one another by a distance of twice the mirror separation. As the number of inte-

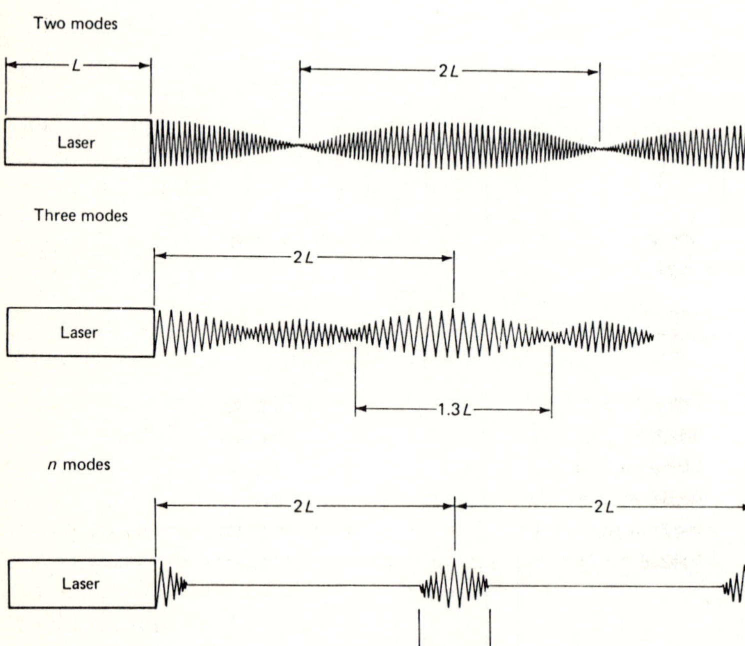

Fig. 5.36. Schematic of the output of a laser operating at two, three, and n longitudinal modes

grally related modes increases, the region of constructive interference – which is inversely proportional to the number of oscillating modes – becomes narrower.

Since Q-switched lasers have been developed for holography, the coherence length has become an important parameter in specifying the spectral characteristics of the output beam.

The temporal coherence of any spectral source is defined as the path length difference over which the radiation can still interfere with itself. Common techniques for measuring spectral bandwidth or coherence length of lasers include Fabry-Perot and Michelson interferometers, observation of beat frequencies, measurement of hologram brightness as a function of path length differences, and measurement of the fringe-free depth of field in a hologram.

In optics textbooks [5.46, 47], coherence length is defined as the path length difference for which fringe visibility in a Michelson interferometer is reduced to $1/(2)^{1/2}$. (Laser manufacturers sometimes use the 1/2 or $1/e^2$ points.)

The fringe visibility of an interferometer is defined as [5.46]

$$V = \frac{I_{\max} - I_{\min}}{I_{\max} + I_{\min}}. \tag{5.61}$$

In our subsequent discussion we will assume that the intensities of the axial modes in the laser output are equal. In this case the coherence length is directly related to the observed visibility of the interference fringes, and a simple relation between a set of longitudinal modes and temporal coherence can be obtained [5.51]. The general case of axial modes of unequal amplitudes has been treated in [5.52].

Single Axial Mode Operation of a Q-switched Laser. The reflectivity versus wavelength around the resonance points of an optical resonator can be expressed by a Gaussian distribution. Therefore, in the absence of frequency shifts or sweeps caused by the active material during laser emission, the power spectrum of the laser output is approximated by the Gaussian expression [5.52, 53].

$$P = P_0 \exp\left[-\left(2\frac{(\ln 2)^{1/2}(\nu - \nu_0)}{\delta\nu}\right)^2\right] \quad , \tag{5.62}$$

where $\delta\nu$ is the full width of the line at half-intensity. When such a pulse interacts with another identical pulse delayed by a time τ, the fringe visibility can be expressed as

$$V = \exp\left(-\frac{(\pi\tau\,\delta\nu)^2}{4\ln 2}\right) \quad . \tag{5.63}$$

If we express the transit-time difference τ in terms of an optical path-length difference $\Delta l = c\tau$, where c is the speed of light, we can write $\tau\,\delta\nu = \Delta l\,\delta\lambda/\lambda^2$. Using this substitution, (5.63) is plotted in Fig. 5.37. The fringe visibility V is reduced to $1/(2)^{1/2}$ from the peak value for $\Delta l\,\delta\lambda/\lambda^2 = 2(\ln 2)/\pi(2)^{1/2} = 0.32$. Therefore, we obtain for the total path length difference for which the fringe visibility is greater than $1/(2)^{1/2}$,

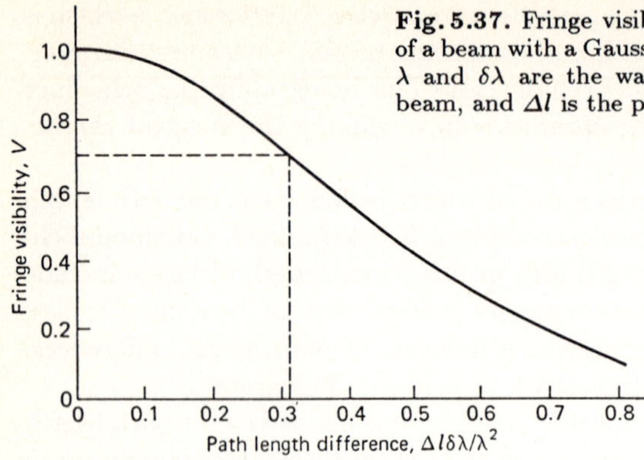

Fig. 5.37. Fringe visibility in a Michelson interferometer of a beam with a Gaussian spectral profile. The parameters λ and $\delta\lambda$ are the wavelength and the linewidth of the beam, and Δl is the path length difference

$$l_\mathrm{c} = \frac{0.64c}{\delta\nu} = \frac{0.64\lambda^2}{\delta\lambda} \quad . \tag{5.64}$$

Combining (5.64 and 54), we obtain for the coherence length L_c of a single-mode laser,

$$l_\mathrm{c} \approx \frac{4L(R_1)^{1/2}}{1-(R_1)^{1/2}} \quad , \tag{5.65}$$

where L is the resonator length and R_1 is the reflectivity of the front mirror (it is assumed that $R_2 = 1$).

The single-axial mode output pulse from a ruby oscillator having a cavity length of 75 cm and a front-mirror reflectivity of $R_1 = 0.4$ will have a coherence length of $l_\mathrm{c} = 5.2$ m. The spectral width of the single line, according to (5.64), will be $\delta\nu = 37$ MHz. The bandwidth-limited pulse length is 17 ns. If we define l_c at the $\frac{1}{2}$ or $1/e^2$ points of the visibility curve, a corresponding larger value of the coherence length is obtained.

Operation of the Laser at two Longitudinal Lines. The line-width of a single line is assumed to be very narrow compared to the mode separation. Therefore, the power spectrum of the laser is represented by two δ functions. The fringe visibility in this case is

$$V = \left| \cos\left(\frac{\pi \Delta l}{2L}\right) \right| \quad . \tag{5.66}$$

This function, which has a periodicity of $2L$, is plotted in Fig. 5.38. The condition $V \geq 1/(2)^{1/2}$ is satisfied for

$$l_\mathrm{c} = L \quad . \tag{5.67}$$

Laser Emission at N Longitudinal Modes. The power spectrum of the laser is developed into a series of δ functions with the assumption that the modes have

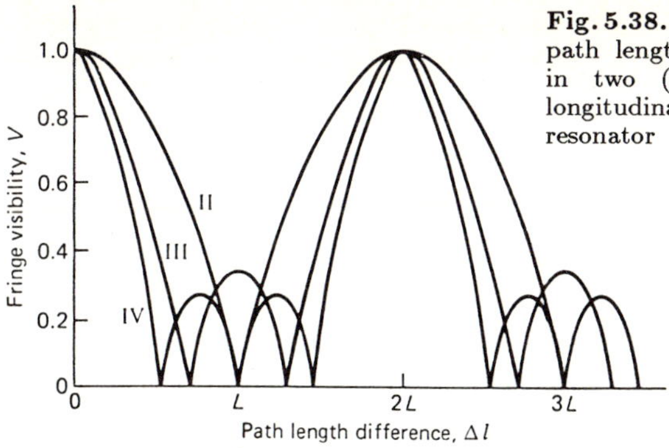

Fig. 5.38. Fringe visibility as a function of path length difference for a laser operating in two (**II**), three (**III**), and four(**IV**) longitudinal modes. L is the length of the laser resonator

equal intensity. One obtains

$$V = \left| \frac{\sin(N\pi \, \Delta l/2L)}{N \, \sin(\pi \, \Delta l/2L)} \right| \quad . \tag{5.68}$$

This function is also plotted in Fig. 5.38 for the case of a laser oscillating in three and four modes. The fringe visibility function is periodic in integer multiples of path difference $\Delta l = 2L$. The fringe visibility is equal or larger than $1/(2)^{1/2}$ for path length differences of

$$l_c = \frac{2L}{N} \quad , \qquad N \geq 2 \quad . \tag{5.69}$$

5.2.3 Axial Mode Control

A typical solid-state laser will oscillate in a band of discrete frequencies which have an overall width of about 10^{-4} of the laser frequency. Although this is a rather monochromatic light source, there are still many applications for which greater spectral purity is required.

In one of the earliest attempts to narrow the spectral width of a laser, tilted Fabry-Perot etalons were employed as mode-selecting elements [5.56]. Also the concept of axial mode selection based on an analysis of the modes of a multiple-surface resonator was introduced [5.54, 55]. In these earlier works it was shown that it should be possible to discriminate against most of the modes of a conventional resonator by adding additional reflecting surfaces. Since then, many techniques have been developed to provide a narrow spectral linewidth from solid-state lasers.

In our discussion we will distinguish between two categories of mode-selecting techniques: interferometric mode selection and enhancement of longitudinal mode selection. In the former case, a Fabry-Perot-type reflector is inserted between the two mirrors of the optical resonator. This will cause a strong amplitude modulation of the closely spaced reflectivity peaks of the basic laser resonator and thereby prevent most modes from reaching threshold.

In the second case, an inherent mode-selection process in the resonator is further enhanced by changing certain system parameters. General reviews on the subject of mode selection in lasers can be found in [5.57, 66].

Interferometric Mode Selection

The role of the resonant devices employed in interferometric mode selection is to provide high feedback for a single wavelength near the center of the fluorescence line, while at the same time discriminating against nearby wavelenths.

Resonant Reflectors. By replacing the standard dielectrically coated front mirror with a resonant reflector, the number of oscillating modes may be greatly reduced. As an example, Fig. 5.39a shows the output spectrum of a ruby laser operated without any mode-selecting elements in the cavity. The laser, operated with two dielectrically coated mirrors, emits a beam with a linewidth of about 0.5 Å at the 50 % power points. The fluorescence linewidth of ruby is 5.5 Å at room temperature; however, above laser threshold the linewidth is reduced by a factor of 10. The envelope of the 0.5-Å-wide laser line contains approximately 160 longitudinal modes for a 75-cm-long resonator. The lines are separated from each other by 0.003 Å.

The dashed line in Fig. 5.33b shows the reflectivity versus wavelength of a single sapphire etalon of 3.2-mm thickness. The curve is drawn so that the peak reflectivity concides with the maximum in the gain profile. In this case the laser will emit a beam which has a linewidth of about 0.06 Å (20 modes). Actually, if the single sapphire etalon is not temperature-controlled, the reflectivity peak

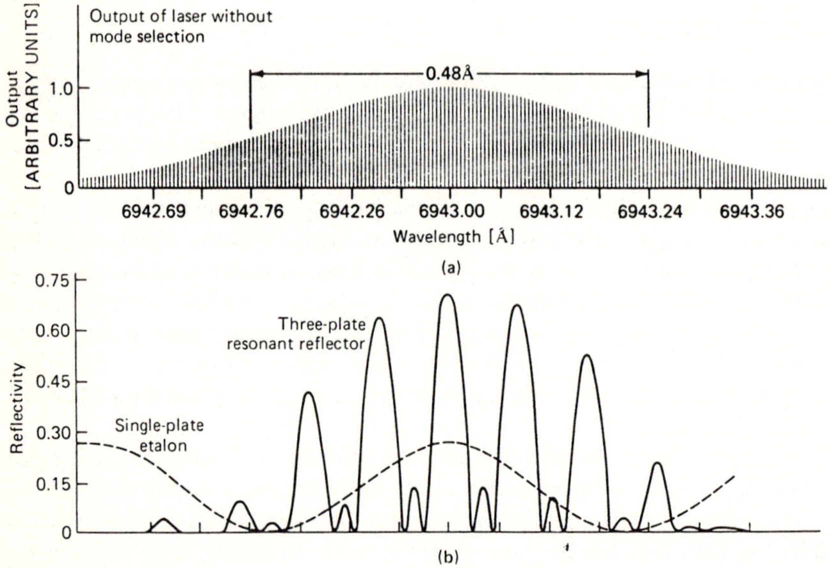

Fig. 5.39. (a) Spectral output of a ruby laser without mode selection. **(b)** Resonance curves of mode-selecting elements

can be located anywhere with respect to the gain curve. The most undesirable location would be such that the reflectivity has a minimum at the peak of the laser line. In this case, one obtains a laser output composed of two groups of lines separated by 0.45 Å.

Also shown in Fig. 5.39b is the reflectivity curve of a three-plate resonant reflector. The reflectivity peaks of this device are much narrower as compared to a single sapphire etalon; this makes such a unit a better mode selector. Experiments have shown that this type of mode selector can reduce the number of axial modes to between one and three under proper operating conditions.

Resonant reflectors featuring one, two, or three etalons are in use. The etalons are usually fabricated from quartz or sapphire. Both materials have high damage thresholds. The advantage of sapphire over quartz is that higher peak reflectivity can be achieved for the same number of surfaces. The peak reflectivities for single-, double-, and triple-plate resonant reflectors are 0.13, 0.40, and 0.66 if quartz is used; and 0.25, 0.66, and 0.87 in the case of sapphire. Etalon thickness is typically 2 to 3 mm, which assures a sufficiently large spectral separation of the reflectivity maxima within the fluorescence curve so that lasing can occur on only one peak. In multiple-plate resonators, the spacing between the etalons is 20 to 25 mm in order to achieve a narrow width of the main peak.

The design of multiple-plate resonators can be optimized with the aid of computer programs which take into account the desired reflectivity, width, and separation of the resonance peaks. *Watts* [5.49] has published procedures for the design of multiple-plate resonant reflectors which are based on the standard equations of thin-film optics. For the theoretical reflectivity, it is important that the difference in thickness of the different plates in a multiple element resonator be held to less than one-tenth of wavelength. The parallelism of the spacers between the plates should also be held to this tolerance.

Because of the relatively low peak reflectivity of resonant reflectors, these devices are used primarily with Q-switched solid-state lasers. Axial mode control with single- or multiple-resonant reflectors employed in Q-switched ruby lasers has been described in [5.58–66].

Optimum mode selection from a resonant reflector is achieved only when the reflection maximum of the device is centered at the peak of the fluorescence curve of the active material. Ideally, the gain at the adjacent reflection maxima should be insufficient to produce oscillation. Temperature tuning is the normal means of shifting the reflectivity peaks of the reflectors relative to the laser linewidth.

Let us consider the changes which take place when the temperature of either the reflector or the laser or both is varied. A change in temperature will change the thickness and the index of refraction of an etalon. These effects combine to shift the wavelength of each resonant peak by an amount

$$\frac{d\lambda}{dT} = \lambda_0 \left(\frac{1}{l}\frac{dl}{dT} + \frac{1}{n}\frac{dn}{dT} \right) \quad , \tag{5.70}$$

where n and l are the index of refraction and the thickness of the etalon,

Table 5.2. Properties of materials frequently used in the design of resonant reflectors

Quantity	Dimension	Sapphire	Quartz	BK7	Stainless steel	Air
dn/dT	$10^{-6}/\mathrm{C}$	12.6	10.3	1.86	–	−0.78
$(1/l)(dl/dT)$	$10^{-6}/\mathrm{C}$	5.8	0.55	7.0	0.9	–
$n(6943\,\text{Å})$	–	1.76	1.455	1.51	–	1.0

respectively. The first term on the right-hand side is the linear coefficient of expansion, and the second term is the thermal coefficient of the refractive index.

The temperature change necessary to cause an etalon to scan through one spectral range can be written in the form [5.59]

$$\Delta T = \frac{\lambda_0^2}{2nl(d\lambda/dT)} \quad . \tag{5.71}$$

Using the materials parameter listed in Table 5.2, we obtain for a 3-mm sapphire etalon employed in a ruby laser: $d\lambda/dT = 0.09\,\text{Å}/\mathrm{C}$ and $\Delta T = 5.0\,\mathrm{C}$. These numbers reveal that for a given wavelength the peak reflectivity of the etalon changes drastically with temperature, therefore temperature control of the resonant reflector is a necessity for stable and reproducible laser performance.

In most commercial lasers the temperature of the active material and the resonant reflector are both temperature-controlled by the cooling water circulated through the laser head and reflector housing. Ruby has a wavelength tuning rate of 0.067 Å/C. If we operate a ruby system with the above-mentioned sapphire etalon, then a 1 C rise in water temperature will shift the ruby fluorescence peak by 0.067 Å and the reflectivity peak of the etalon by 0.09 Å toward longer wavelength. The relative shift between the two peaks is 0.023 Å/C. The temperature dependence of the reflectivity of the mode selector can also be used to tune the wavelength of the laser thermally. *Schotland* [5.60] has described a frequency-tunable mode-controlled Q-switched ruby laser system that utilizes a resonant reflector. Frequency tuning is accomplished by a combination of temperature and pressure control of the reflector, which consists of two parallel flat plates separated by an air gap whose length is determined by a spacer. By varying the air pressure between the reflector plates, and hence changing the optical distance between them it is possible to tune the laser over a 0.085-Å range with a relative stability of ±0.01 Å.

Johnson and *LaGrone* [5.63] achieved frequency tuning of a ruby laser over a range of 0.3 Å with a spectral width less than ±0.01 Å using a temperature-tuned reflector. The latter reference contains a very detailed analysis of the requirements for making multiplate resonant reflectors frequency-tunable. Ruby lasers which combine both mode control and temperature tuning have also been described in [5.59, 64].

Intracavity Resonances. Mode selection can also be achieved by using the laser rod itself as an etalon, by carefully aligning the flat and uncoated end of the laser rod to a flat-cavity mirror [5.65], or by adding additional mirrors which form three- or four-mirror laser resonators [5.67].

Intracavity Tilted Etalon. The stability, simplicity, low loss, and wide adaptability of the bandwidth and free spectral range makes the tilted etalon a versatile mode selector.

The etalon is inserted at a small angle in the laser resonator. The tilt effectively decouples the internal transmission etalon from the resonator; i.e., no other resonances will be formed with other surfaces in the main resonator. If the etalon is sufficiently misaligned, it acts simply as a bandpass transmission filter. The tilted etalon has no reflection loss for frequencies corresponding to its Fabry-Perot transmission maxima. At other frequencies the reflections from this mode selector are lost from the cavity and thus constitute a frequency-dependent loss mechanism at a given angle.

A narrow region of high transmission can be obtained by using sufficiently high-reflectivity coatings on the etalon. The fact that for efficient mode selection tilted etalons must be dielectrically coated is a disadvantage in a high-power system, because of the possible damage, especially since the power density inside the etalon is approximately a factor of $(1-R)^{-1}$ higher than the power density in the main resonator. Higher resistance to damage is the main reason why resonant reflectors are preferred as mode-selecting elements in pumped-pulsed, Q-switched lasers. On the other hand, resonant reflectors cannot be used for CW-pumped Nd:YAG lasers or gas lasers because the maximum reflectivity of these devices is too low. For the latter systems, tilted etalons are used extensively as mode selectors. The application of tilted etalons for the mode control of ruby, Nd:glass, and Nd:YAG systems can be found in [5.56, 65, 68–75].

The transmission maxima can be tuned into the central region of the gain curve by changing the tilt angle or by adjusting the etalon temperature. A requirement for mode selection is that the etalon transmission drops off sharply enough to allow only one mode to oscillate. The transmission T can be expanded around its maximum, yielding

$$T(\nu)=\left[1+\frac{4R}{(1-R)^2}\left(\frac{2\pi nd\,d\nu}{c}\right)^2\right]^{-1}, \tag{5.72}$$

where $d\nu$ is the frequency deviation from the transmission maximum and R is the reflectivity of the etalon surfaces. If we assume that one mode oscillates at the center of the etalon's transmission curve ($d\nu=0$), the mode with the second largest T will be spaced at approximately $d\nu=c/2L$.

For single-mode operation, the tilted Fabry-Perot etalon must have adequate selectivity to suppress axial modes adjacent to the desired one, and its free spectral range must be large enough so that the gain of the laser is sufficiently reduced at its next resonance. The difficulty which arises from these two requirements in choosing the optimum etalon thickness can be circumvented by employing two etalons of different thickness in the resonator. *Culshaw* et al. [5.71] have obtained single-mode operation in Nd:YAG using a double Fabry-Perot fused quartz etalon arrangement. The main selective 11-mm-thick transmission etalon with 55% reflectivity discriminates agains adjacent axial modes of the particular laser cavity, while the second, less selective 2-mm-thick transmission etalon with 20% reflectivity prevents any occurrence of laser oscillations at other resonances of the main etalon. Both etalons are thermally tuned to a single axial mode near the center of the gain curve and are tilted slightly to decouple them from each other and from the flat laser mirror.

Danielmeyer [5.75] achieved single-mode operation of a Nd:YAG laser with a single 2-mm-thick quartz plate which was tilted 0.1 degree and temperature-stabilized to within 0.01 C. By slightly changing the temperature, the laser could be tuned over 40 GHz. The losses introduced by the etalon as a result of beam walk-off, imperfections of the coatings, and imperfect parallelism of the etalon surfaces are also treated in [5.75].

Enhancement of Longitudinal Mode Selection

Here we will discuss the fact that single-mode operation is easier to obtain for a narrow laser linewidth, short resonator length, a large number of round trips in a Q-switched laser, and phase modulation of the standing wave inside the active material.

Cooling of the Active Medium. The linewidth of solid-state lasers, in particular ruby, decreases for decreasing temperature. This reduces the number of axial modes which have sufficient gain to oscillate. Cooling of a ruby crystal as a means of obtaining single axial-mode operation in combination with other mode-selecting techniques was discussed in [5.65].

Shortening of the Optical Resonator. The spectral separation of adjacent axial modes is inversely proportional to the length of the resonator. Thus, in a short resonator it is easier to discriminate against unwanted modes. For example, reliable single-mode output with energies per pulse of 2 to 4 mJ has been obtained from a ruby laser with a cavity only 28 cm long [5.76]. The laser consisted of a helical flashlamp, a 7.5-cm by 0.6-cm rod, a 1.5-mm aperture, a two-plate resonant reflector, and a dye cell for Q-switching. The 1.7-mm-thick dye cell filled with cryptocyanine and methanol produced a 8-ns-long pulse.

Lengthening of the Pulse Buildup Time. It was observed very early that passive dye Q-switches tend to act as mode selectors. *Soffer* [5.77], in his first report on the production of ruby giant pulses by bleachable absorbers, observed that Fabry-Perot interferograms showed a linewidth of less than $0.02\,\mathrm{cm}^{-1}$. The same effect was found for Nd:lasers by *McFarland* et al. [5.78], who used polymethine dye to Q-switch their laser. More recently, *Bjorkholm* and *Stolen* [5.79] have reported single-mode output from a passively Q-switched ruby laser with a single-sapphire or glass flat as output mirror. The dye Q-switch consisted of a cell having a 2-mm path length filled with cryptocyanine dissolved in methanol. The Q-switch cell, which was not antireflection-coated, was placed in the cavity at an angle of several degrees. Excellent results were obtained by using chlorophyll as a mode-selecting Q-switch [5.80, 81]. Using this dye in an 80-cm-long cell, a coherence length from a ruby laser of over 10 m has been achieved [5.81].

The mode-selection property of the Q-switching dyes was explained by *Sooy* [5.82] as a result of natural selection. Longitudinal-mode selection in the laser takes place while the pulse is building up from noise. During this buildup time, modes which have a higher gain or a lower loss will increase in amplitude more rapidly than the other modes. Besides differences in gain or losses between the modes, there is one other important parameter which determines the spectral output of the laser. This parameter is the number of round trips it takes for the pulse to build up from noise. The difference in amplitude between two modes becomes larger if the number of round trips is increased. Therefore, for a given loss difference between the modes it is important for good mode selection to allow as many round trips as possible.

The development of a pulse in a dye Q-switched laser takes longer than, for example, in the case of a Pockels-cell Q-switched system [5.83]. As a result, in dye Q-switched systems, mode selectors such as etalons or resonant reflectors are more effective in discriminating against unwanted modes. In *Sooy*'s paper specific reference was made to saturable absorber Q-switching, but the analysis is equally applicable to the behavior of any Q-switch that ensures a large buildup time. There are two essential results from *Sooy*'s analysis. First, the growth with time t of the power P_n in mode n is given by

$$P_n(t) = P_{\text{on}} \exp[k_n(t - t_n)^2] \quad , \tag{5.73}$$

where P_{on} is the noise power in the mode at the start of buildup, t_n is the time at which the net gain for mode n reaches unity, k_n is equal to $(1/2T)(dg_n/dt)$, where T is the round trip time for the resonator, and g_n is the gain coefficient for mode n. Second, the ratio of the powers T_m, P_n in modes m and n, respectively, after q double passes of buildup, is given to a good approximation by

$$\frac{P_n}{P_m} = \left(\frac{1 - L_n}{1 - L_m}\right)^q (1 - L_n)^{q[(g_m/g_n)-1]} \quad , \tag{5.74}$$

where L_m, L_n are the losses per double pass for modes m and n, respectively, and g_m, g_n are the gain coefficients for these modes. The first factor in (5.74) corresponds to loss discrimination and the second factor to gain discrimination.

In most lasers, gain differences between adjacent modes are too small to play a significant role in the mode-selecting process. For mode discrimination due to different values of the reflectivity of the mode selector we obtain, from (5.74)

$$\frac{P_n}{P_m} = \left(\frac{R_n}{R_m}\right)^q \quad , \tag{5.75}$$

where $R_n = 1 - L_n, R_m = 1 - L_m$.

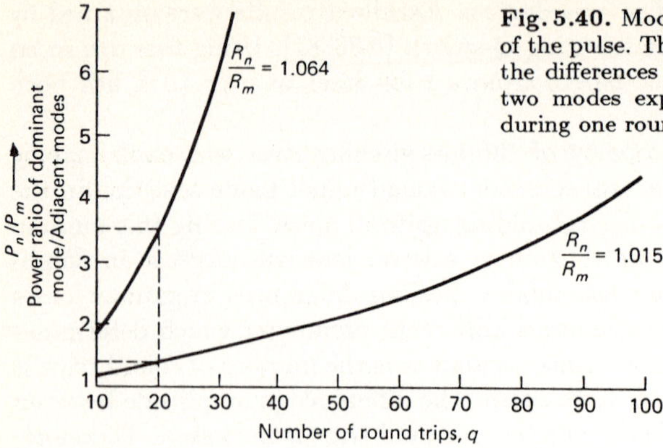

Fig. 5.40. Mode selection during the buildup of the pulse. The parameter R_n/R_m expresses the differences of the reflectivity which these two modes experience at the mode selector during one round-trip

In Fig. 5.40 the output power ratio of two modes as a function of number of round trips is plotted. Parameter is the difference in reflectivity which these modes experience at the mode selector. The reflectivity peak of a mode selector can be approximated by

$$R = R_{\max} \cos^2\left(\frac{\lambda}{\Delta\lambda_m}\frac{\pi}{2}\right) \quad , \tag{5.76}$$

where $\Delta\lambda_m$ is the FWHM of the reflectivity curve and λ is the wavelength taken from the center of the peak.

As an example, we will consider a 75-cm-long resonator and the three-plate etalon described in Sect. 5.2.1. Assuming the dominant mode right at the center of the reflecting peak, then the two adjacent modes will be shifted in wavelength by $\Delta\lambda = 0.003$ Å. With $\Delta\lambda_m = 0.038$ Å for the three-plate etalon, one obtains, from (5.72), $R/R_{\max} = 0.985$, which means that the difference in reflectivity for the two modes adjacent to the center mode is 1.5%, or $R_n/R_m = 1.015$. In order to obtain single-mode operation from this system, we would need at least 155 round trips according to (5.75) if we use as a criterion for single-mode operation that the dominant mode should be at least ten times greater in peak power than any other mode. For the cavity length chosen we require, therefore, a pulse buildup time of 775 ns.

A logical step following the foregoing analysis is to use an active Q-switch but to switch it in a way that ensures a large buildup time. This has been done by *Hanna* [5.84, 85], who operated a Pockels-cell Q-switch in a manner analogous to a saturable absorber. In this technique, initially the Pockels cell is only partially open and, therefore, presents a loss which is analogous to the low-level absorption loss or a saturable absorber. As a result of pumping, the net gain increases until it exceeds unity, and buildup of the giant pulse then starts. This buildup is monitored by a photodiode and its signal is used to trigger the Pockels cell to open completely when a preset signal level is reached. This is analogous to the bleaching of a saturable absorber at a particular intensity. Thus a long buildup time is achieved by an active Q-switch used in a way which is very closely analogous to a "passive" saturable absorber Q-switch. Utilizing this technique in conjunction with a resonant reflector, reliable single-mode operation in a Nd:$CaWO_4$ laser was obtained [5.86].

Reduction of Spatial Hole Burning. Limitations in obtaining a larger single-mode output are commonly attributed to spatial hole burning. Atoms located in the vicinity of the nodal planes of one axial mode will preferentially contribute to other modes. In gas lasers, spatial hole burning is all but prevented by the thermal motions of the atoms. In crystals, however, the amplitudes of the lattice vibrations are small compared to the light wavelength, and spatial averaging cannot take place. The efficiency of an axial mode selector can be greatly increased by providing relative motion between the atoms in the active material and the electric field of the resonator.

In principle, spatial hole burning can be eliminated with traveling-wave ring structures [5.87–95], by generating circularly polarized light in the rod [5.96], with mechanical motions [5.97], or with electro-optic phase modulations [5.98].

Traveling-wave unidirectional ring lasers have been particularly successful in achieving narrow line width from a number of solid-state lasers. As was described in Sect. 3.2, a unidirectional ring laser contains three essential elements, a polarizer, a half-wave plate and a Faraday rotator. Ring lasers can be constructed with discrete elements [5.87–93], or very recently, monolithic versions of traveling-wave ring oscillators have been developed pumped by laser diodes [5.94, 95, 99]. The first such device was the Monolithic Isolated Single mode End pumped Ring laser (MISER) developed by *Kane* et al. [5.94]. Diode pumping permits the design of very short resonators with large spacing between modes. In addition, a diode array provides a very stable pump source, with almost no amplitude fluctuations, and very little heat deposition into the lasing medium, all factors which are conducive to stable, single-line operation.

Most of the research on single-frequency, diode-laser-pumped Nd lasers has concentrated on monolithic devices where the optical cavity is formed by coated surfaces on the laser medium. While monolithic devices have significant advantages in terms of mechanical stability, they are not generally appropriate for applications requiring either rapid tuning of the laser wavelength or tuning far off from the gain peak. Therefore, diode-pumped single-frequency lasers employing external mirrors in a ring geometry appropriate for linewidth tuning requirements have been developed, too [5.93].

Single-Axial Mode Operation by Injection Locking. In the previous sub-sections, single-axial mode was described by using spectral-selective elements within the resonator cavity. In recent years, the technique of injection locking has become very successful in achieving high power single-axial mode operation.

The output from the diode-pumped narrow-linewidth ring laser, or the output from single-frequency conventional two-mirror resonator systems either diode pumped [5.100] or arc lamp pumped, can be amplified many orders of magnitude by injection seeding or locking of a powerful slave oscillator.

Diode-pumped systems have low power outputs by virtue of the output limitations of laser diodes, and in arc lamp pumped systems thermally induced birefringence and thermal distortions in the lasing medium usually limits the axial-mode selection. Although elimination of spatial hole burning by the use of quarter-wave plates at the end of the laser rod is essential for single-mode operation, the thermally induced birefringence results in residual spatial hole burning, which requires operation close to threshold.

The low power output from single-mode oscillators, leads to the concept of a narrow-line width oscillator coupled to a high-power slave oscillator.

In CW injection locking, the slave oscillator acts as a regenerative amplifier for the injected signal. If the frequency of the injected wave is detuned from the resonance of the slave oscillator, the output initially follows the injected frequency. At some detuning range, the injected wave no longer can force the slave oscillator gain below threshold for the slave oscillator's resonant modes, and multimode oscillation occurs.

On the other hand, in high-gain Q-switched lasers, the steady-state injection-locking regime is not reached. Instead, the injected wave and the initial noise signal at each axial-mode of the slave oscillator grow simultaneously on the opening of the Q switch. The axial mode that dominates in the resultant output is that mode that first saturates the gain medium, thus supressing further growth at adjacent axial modes. The mode that dominates can be selected by the injected power level and the detuned frequency range.

Practical designs of injection locking systems comprised of a low power master oscillator and a high-power slave oscillator can be found in [5.101–103].

Examples of Mode Selection in Q-Switched Ruby Lasers. Combining several of the mode-control methods discussed in this subsection will provide essentially single-mode behavior from a giant pulse ruby [5.65, 81, 104]. Publications on single-mode ruby lasers usually describe systems which are dye cell Q-switched or contain a laser rod cooled below room temperature. Since the main application of long-coherence-length ruby lasers is in holography, both techniques have severe disadvantages. Cryogenic cooling of the rod is impractical, and a dye cell Q-switch does not provide a trigger capability required for most applications.

Single-longitudinal-mode operation from Pockels-cell Q-switched ruby lasers operated at room temperature has been obtained at several laboratories. As an example in one of these systems, the rear reflector is a four-plate resonant reflector of the type shown in Fig. 5.34. The fourth plate is added to increase the peak reflectivity of the device. The front reflector is a 2.5-mm-thick single sapphire etalon. The main function of this device is to provide an optimum output coupling of 25 % and to prevent oscillation at satellite peaks (spaced 0.086 Å from the main peak) of the multiplate resonator. Both mode-selecting elements are independently temperature controlled to within ±0.2 C. The laser head contains a 10-cm-long by 0.96-cm-diameter ruby rod pumped by a helical flashlamp. Operation at the TEM_{00} mode is accomplished by inserting a 1.5-mm aperture into the 75-cm-long resonator.

Single-axial-mode operation is achieved by drastically increasing the rise time of the Pockels-cell voltage. In a standard Pockels-cell Q-switch, the rise time of the voltage pulse on the crystal is typically 20 ns, and the laser output appears normally after 50 to 100 ns. In a 75-cm-long resonator this time delay amounts to about 10 to 20 round trips for the energy to build up. In this particular case the rise time is reduced to 1 μs, and the Q-switch pulse is emitted at the end of this time period; this suggests that about 200 round trips occurred before the pulse was emitted.

The systems are all operated only slightly above threshold. Single-transverse-mode operation is obtained at output energies below 50 mJ from a single oscillator. Single-transverse- and -longitudinal-mode operation is achieved at output levels between 10 and 15 mJ In one system containing several amplifiers, single-mode output was 10 J, sufficient to obtain holograms with a scene depth of 5 m. Figure 5.41 shows Fabry-Perot interferograms and pulse-shape measurements, taken concurrently with a fast photodetector and a 519 Tektronix oscilloscope, of a ruby oscillator capable of single-axial-mode operation. The time trace of a single-mode pulse is a smooth curve resembling a Gaussian curve. If two modes are present, the beat between the modes is clearly visible as a sinusoidal modulation of the normal single-mode output pulse at a frequency $nc/2L$, where n is an integer and $c/2L$ is the frequency spacing between adja-

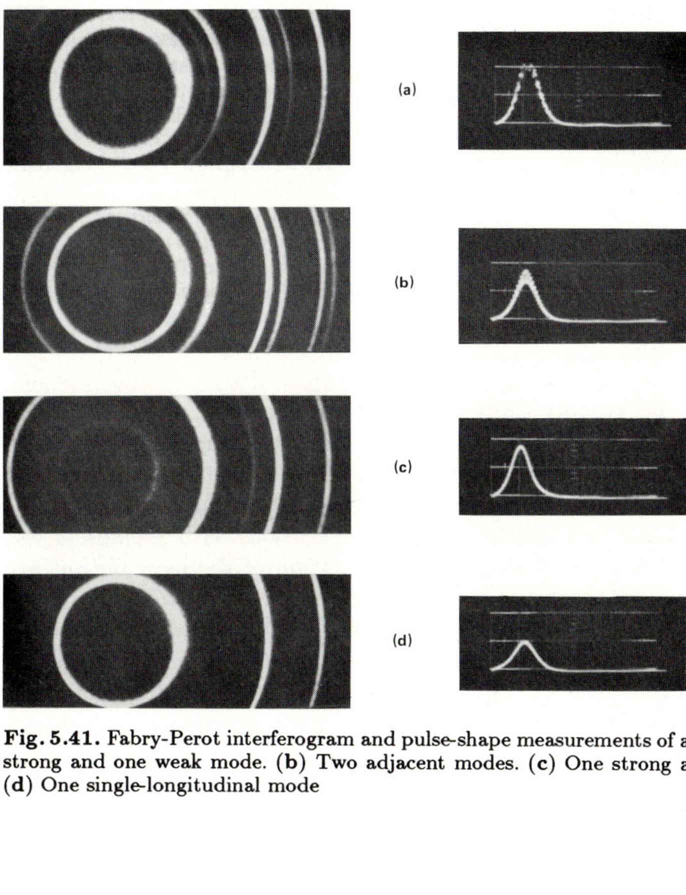

Fig. 5.41. Fabry-Perot interferogram and pulse-shape measurements of a ruby laser. (**a**) Two strong and one weak mode. (**b**) Two adjacent modes. (**c**) One strong and one weak mode. (**d**) One single-longitudinal mode

cent axial modes. Observation of the time trace is a more sensitive indicator of mode purity than is a photograph of the Fabry-Perot spectrum. Often the presence of a weak second mode is not indicated by observing the Fabry-Perot ring system, whereas it is unmistakably shown by a small degree of "ripple" on the time trace.

Figure 5.41a shows three longitudinal modes and the resulting mode beating. Figure 5.41b displays strong modulation due to the presence of two strong modes. Figure 5.41c reveals only slight evidence of resolvable mode beating; this is in agreement with the weak secondary mode in the interferogram. Figure 5.41d shows absolutely no resolvable modulation, which agrees with the clean single-mode interferogram.

The complete absence of intensity fluctuations in conjunction with the Fabry-Perot interferogram provides a positive indication of single-mode operation. At energy outputs below 12 mJ, single-mode performance is observed consistently; however, pulse-to-pulse energy fluctuations are as large as 2:1 because of the operation near threshold. Between 15 and 20 mJ the mode structures vary erratically, as shown in Fig. 5.41a–c, but never exceed a linewidth of about 0.01 Å.

The spectral characteristics of the laser were measured with a Hilger-Watts Fabry-Perot interferometer. This instrument consists of two etalons separated by 100 mm and dielectrically coated to give a reflectivity of 95 %. The free spectral range of the system is 0.025 Å, and the finesse is greater than 30. The device is capable of resolving two lines separated by 0.001 Å. A negative lens employed at the entrance of the instrument expands the 1.5-mm beam so as to just barely fill the first few orders of the Fabry-Perot. The transmitted distribution is relayed by an optical arrangement of effective focal length of 10 m and focused onto 410 Polaroid film.

5.3 Temporal and Spectral Stability

5.3.1 Amplitude Fluctuations

Fluctuations may be broadly categorized as short-term or long-term. Long-term fluctuations with periods greater than 1 s can be attributed to gradual variation in the engineering parameters of the laser, such as alignment and cleanliness of optical elements or degradation of the pump source.

Here we are concerned with short-term fluctuations with periods of less than a second. The sources of amplitude modulation of the laser output are mechanical vibrations of the optical components, thermal instabilities in the active material, variations in the pump lamp intensity, mode beating and mode hopping, relaxation oscillations, and quantum noise. The latter two phenomena, which are inherent properties of the laser oscillator, have been discussed in Chap. 3. Which of the fluctuations mentioned above are dominant in a particular situation depends on the operating mode of the laser and the frequency interval of interest.

Conventional-Mode Operation. As has been discussed in Chap. 3, in this regime relaxation oscillations dominate all other noise mechanisms.

Q-Switched Pulsed-Pumped Laser. The envelope of a Q-switched pulse can show amplitude modulation due to mode beating of axial modes. The envelope is smooth for operation at a single axial mode and appears to be smooth for a very large number of axial modes, in which case amplitude modulation due to mode beating is reduced as a result of averaging effects.

Continuous-Pumped Lasers. All the noise mechanisms listed above can be studied in a CW-pumped Nd:YAG laser. Output fluctuations at low frequencies (below 1 kHz) can be traced to mechanical and thermal instabilities and fluctuations in the pump lamp intensity. The mechanical vibrations can be originated by the environment and transmitted to the laser, or they can be generated in the laser head itself by turbulence in the cooling flow [5.105]. Any vibrations of optical components will change the losses in the resonator; as a result, the output power will fluctuate. Besides vibrations transmitted to the laser head through its support structure, vibrations from the water pump transmitted to the laser head by the cooling water and hoses as well as turbulence or air bubbles in the water surrounding the laser rod can cause mechanical vibrations of the resonator, mirrors, and laser rod [5.106]. It was found that if these mechanisms are dominant, laser performance can be stabilized by acoustically decoupling the water cooler from the laser head and reducing the flow rate.

Changes in the intensity of the pump lamps will modulate the output of the laser. In tungsten-filament lamps, pump-light variation is usually caused by voltage ripple from the power supply. A well-filtered and regulated DC power supply is, therefore, a necessity if a modulation-free output is desired. In krypton arc lamps, intensity variations are caused by arc wandering, plasma instabilities, and voltage ripple. The first two instabilities will produce random spatial and temporal variations of rod gain. Low-frequency modulation in arc lamps can be reduced by the addition of an active feedback loop to control lamp current.

Turbulence of hot air inside the laser head or dust particles in the beam can cause the optical beam to fluctuate. A bellows system which encloses the optical train is, therefore, a requirement for a low-noise laser.

Air bubbles in the cooling path of the lamps or laser rod can cause modulation of the intensity of the pump lamp. Modulation of the refractive index of the laser rod by the coolant can cause the output to fluctuate. In a typical cooling system, the laser rod is cooled by a turbulent flow of water around it. Therefore, the heat transfer between the rod and coolant changes constantly in time and space. The temperature variations within the rod create optical distortion through the dependence of the index of refraction of Nd:YAG on temperature. This mechanism becomes important if the laser rod is used as the limiting aperture for TEM_{00} operation. The use of a quartz sleeve around the laser rod to damp thermal disturbances [5.107] or an increase in the flow rate to create a more stable dynamic flow characteristic [5.108] have been suggested to overcome these problems. However, it was found [5.106] that reducing the beam diameter by means of an aperture has the same effect. The thermal inertia of the periphery of the rod provided sufficient damping to eliminate this problem.

In the 50- to 200-kHz range, depending on the particular system, output fluctuations are dominated by relaxation oscillations. Any small perturbation having a frequency component at the vicinity of the resonance frequency of the relaxation oscillation will cause large oscillations in the output power. A potential source for the excitation of relaxation oscillations is mechanical vibrations. Obviously, in order to reduce the modulation caused by relaxation oscillations, the mechanical laser structure must be extremely stable [5.109, 110].

At the lower end of the megahertz region, parasitic noise is caused by mode competition of transverse modes, whereas competition of longitudinal modes causes amplitude modulation at frequencies of several hundred megahertz. Different transverse modes occupy different regions of the laser rod; as a result of pumping nonuniformities, the gain varies for each transverse mode. Different axial modes are located at different points on the laser gain curve. Any disturbances of the laser system will encourage or discourage individual modes. Therefore, the mode pattern changes constantly. Mode selection eliminates output fluctuations associated with mode competition.

As an example, a Nd : YAG system is described which has been optimized for low parasitic noise output [5.106]. The system consists of a 3-mm by 63-mm Nd : YAG rod pumped by two tungsten-filament lamps in an elliptical cavity. The modifications which were made to the system to achieve an extremely low noise performance were as follows: The laser head was vibration isolated from the supporting structure by means of rubber pads; the laser head was acoustically decoupled from the water cooler by a ballast tank and soft plastic water hoses; the water flow was reduced from 135 to 75 cm^3/s. The laser was operated at TEM_{00} mode; this eliminated any amplitude modulation above 200 kHz. The resonator was converted from a long-radius mirror system to a confocal arrangement. After this modification, the amplitude of the relaxation oscillations was reduced by one order of magnitude. After all these modifications, the laser produced 1.2 W TEM_{00} at a lamp input power of 2 kW.

In Fig. 5.42 the depth of modulation is plotted as a function of frequency. The upper curve gives the ratio of the peak values of the noise to the average signal voltage (measured with an oscilloscope), whereas the lower curve represents the ratio of the rms value of the noise to the average signal voltage (measured with a spectrum analyzer). In the case where the fluctuations of the laser output are determined only by quantum noise, the depth of modulation is 0.8×10^{-6}. From Fig. 5.42 it follows that the spectrum of the noise has a maximum at very low frequency; the noise decreases to a minimum around 10 kHz and increases again to reach a resonant peak around 80 kHz. Above 100 kHz, a rapid decrease of amplitude modulation is observed to a level close to the theoretical limit determined by the quantum noise in the signal.

Figure 5.42 shows typical output fluctuations at the low frequency spectrum.

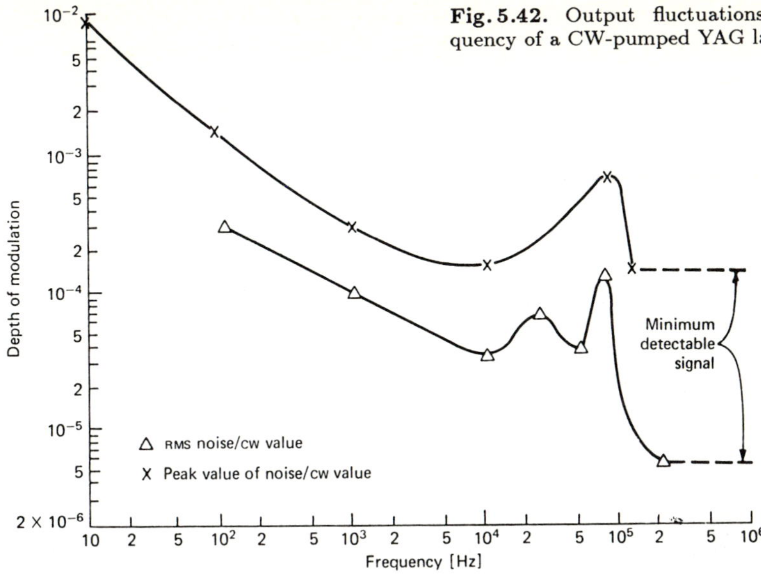

Fig. 5.42. Output fluctuations versus frequency of a CW-pumped YAG laser

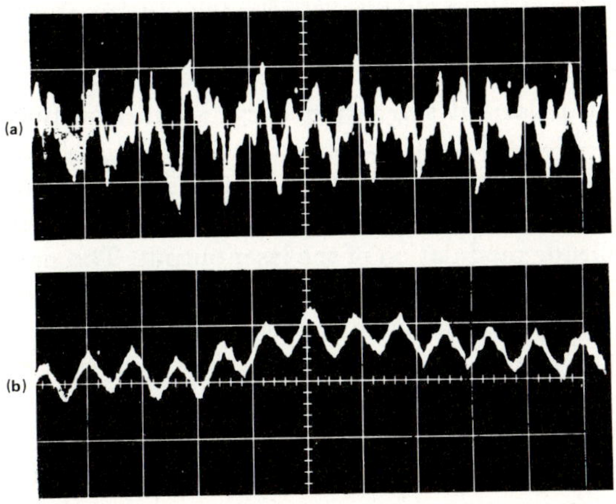

Fig. 5.43. Output fluctuations at low frequencies. Horizontal scales: (**a**) 0.5 s/div; (**b**) 10 ms/div. Vertical scale: (**a**) 2 mV/div; (**b**) 1 mV/div.

5.3.2 Frequency Changes

We will first consider the frequency changes which occur in a CW-pumped solid-state laser. The output of the laser as a function of frequency will depend on the amplitudes, frequencies, and phases of the oscillating modes. The frequency will fluctuate in a random way within a bandwidth of a few gigahertz. A drastic improvement can be achieved if mode-selecting techniques are employed which will assure single transverse and axial mode operation.

The frequency stability of a single-frequency laser is determined by its environment. Fluctuations of the optical length of the laser cavity due to thermal expansion of the support structure, temperature changes in the laser rod or other optical components, mechanical vibrations [5.111], or acoustic noises will shift the frequency. A change of the optical resonator length of half a wavelength will change the laser frequency by $c/2L$, the longitudinal mode spacing. For example, even if the laser resonator is made of a low-thermal-expansion-coefficient material, such as invar, a temperature change of 0.5 C will cause a length change of $\lambda/2$ in a 1-m-long resonator.

Therefore, in practical systems it has been found necessary to add electronic feedback devices to stabilize the frequency output of the laser. In these schemes, a discriminator measures the amount of frequency deviation from some desired center frequency and indicates the sense of the deviation. The output of the discriminator is then fed to a frequency-controlling element in the laser, for example, to an electromechanical element controlling the mirror spacing, or to an electrooptic element controlling the refractive index of a portion of the cavity [5.66]. An etalon is usually employed as the frequency discriminant. A high-Q etalon will transform a frequency modulation of the active mode into a small-intensity modulation. Phase-sensitive detection of the intensity modulation yields the error signal needed for stabilization. The two etalons required in this technique – one for mode selection and one for frequency discrimination – can also be combined into a single birefringent etalon. The birefringent etalon provides two cavities simultaneously, one for the ordinary ray, the other for the extraordinary ray. Using a birefringent intracavity etalon made from crystal quartz, a Nd:YAG laser could be stabilized to 10^{-7} in frequency [5.112]. The linearly polarized single-frequency TEM_{00} output from the laser was 150 mW.

In another scheme, the thickness of an intracavity etalon was modulated by a piezoelectric transducer [5.72]. The frequency modulation of the etalon's transmission peak causes an amplitude modulation of the laser output. The signal is detected by a photodiode mounted behind the rear mirror. The amplified error signal is used to change the length of the main resonator.

Pulsed Laser Systems

In a pulsed laser, ideally, the frequency of a single-axial-mode laser should remain constant during the pulse. In practice, it has been discovered that there can be a significant frequency modulation of the laser pulse. In ruby two effects have been observed: a time-dependent frequency shift toward the red [113–115] and an output-power-dependent blue shift [5.116–121].

The first effect represents a thermal frequency shift due to heating of the crystal as a result of the dissipation of the absorbed pump light (see also Chap. 7). As a result of changes in length and refractive index of the ruby rod during the pump cycle, frequency shifts of the order of 10 MHz/μs were observed [5.114]. The thermal effects are important if the laser is operated conventional mode or if more than one Q-switch pulse is generated during one flashlamp pulse. In the latter case, the rod has obtained a different temperature between pulses and subsequently a frequency shift will occur. In a single Q-switched laser the time duration of lasing is too short to cause noticeable thermal frequency shifts. However, the output power-dependent frequency shift is important in a single Q-switched ruby laser, since this effect broadens the bandwidth and causes a frequency sweep of the output pulse. The effect is attributed to a decrease in resonator length, as a result of a decrease in the rod's refractive index, which in turn is postulated to be directly proportional to changes in population inversion [5.117, 118]. It appears that the correct explanation for this effect is related to the fact that the Cr^{3+} ions in the ruby system slightly change their coupling to the crystal-field system with changes in state.

A change in optical length of the active material in a laser cavity, caused either by refractive index changes Δn or by a change of the physical length of the rod Δl, will cause a frequency shift $\Delta\nu$ of the reflectivity peaks of the resonator according to

$$\frac{\Delta\nu}{\nu} = \frac{\Delta n\, l_r + \Delta l\, n}{n l_r + l_0} \quad , \tag{5.77}$$

where l_r is the length of the laser rod with index n, and l_0 is the length of the air path in the resonator.

A blue shift of 85 MHz, which corresponded to a decrease in cavity length of 975 Å at a rate of 39 Å/ns, was observed in a single-mode laser operated at an energy output of about 100 mJ. The rate of frequency increase was about 3.4 MHz/ns. A linear relationship between energy output and frequency change has also been observed by *Bradley* and co-workers [5.116, 119, 120] for laser powers from 3 to 10 MW with a proportionality constant of 1.4 MHz/ns MW. Figure 5.44 shows the observed frequency drift of the Q-switched output pulse of a ruby laser as a function of time [5.121]. It can be seen from this figure that the frequency drift closely follows the temporal change in inversion. The thermally as well as the power-induced frequency shift can be partially compensated by introducing into the optical cavity an externally controlled phase shift generated, for example, by an electrooptic crystal. The linewidth and coherence length of an ideal Gaussian pulse having a linear frequency sweep is given in [5.53].

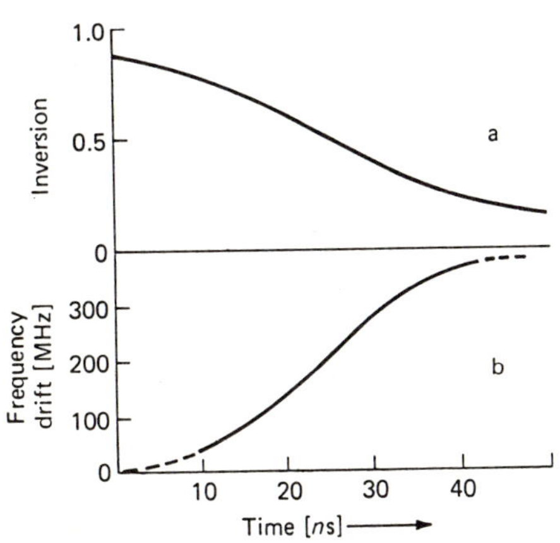

Fig. 5.44. (a) Relative inversion of a ruby rod and (b) frequency drift during a Q-switched pulse

5.4 Hardware Design

In the construction of a resonator for a laser oscillator, the important considerations are the mechanical stability and the quality of the optical components, protection of the optical train from the environment, and protection of the operator from radiation hazards.

Laser Support Structure. The key requirement of a laser resonator is that all optical components be accurately and rigidly fixed relative to each other, that the structure be insensitive to temperature variations, and that there be a provision for making angular adjustments of the mirrors. A mechanical analysis of the laser support structure has to include calculations of deformations as a result of the structure's own weight, changes in mechanical loadings, relief of internal stresses, and deformation due to temperature changes. These considerations usually result in a trade-off between size and weight on one hand and mechanical rigidity and temperature insensitivity on the other hand.

Laser systems employed for general research applications usually have an optical bench-type support structure for maximum flexibility of the experimental setup. Adjustment of the axial position of optical components is provided either by sliding carriages or by slotted holes and bolts. The benches must be sufficiently massive and well ribbed to minimize flexure. Typical support structures include "I-," "U-," "L-," and "H"-shaped beams made from aluminium extrusions. Added rigidity and long-term stability are obtained if Meehanite, a special grade of close-grain iron, is used instead of aluminium.

Lasers for industrial use are normally mounted in a fully enclosed housing for protection against a hostile environment. In most cases the optical surfaces are additionally protected by enclosing the whole optical train with rubber or Teflon bellows. Enclosures may range from a mere safety cover to a rubber-sealed dust-proof cover, all the way to a hermetically sealed cover which will maintain a positive pressure on the inside.

In military systems weight and size constraints lead to uniframe and densely packaged structures with aluminium as the choice material. Often, the whole support structure is machined from one solid block of metal.

In some applications, such as single-longitudinal-mode operation and active mode locking, the mirror separation must be kept constant to within a fraction of a wavelength over long periods of time. To achieve long-term stability of the cavity length, the lowest expansion-coefficient materials such as invar, fused quartz, or ceramics (for example, Cer-Vit) are employed. By choosing a proper geometry, incorporating bimetallic length-compensation techniques, and using flowing liquids to thermally control and stabilize the structure, at least an order of magnitude improvement in length changes can be achieved over ordinary designs. Temperature compensation of cavity-length changes can also be achieved by electronic control of the cavity length by means of piezoelectric translation.

Mirror Mounts. The requirements for a good mirror mount are an independent, orthogonal, and backlash-free tilt with sufficient resolution and good thermal and mechanical stability. Commercial mirror mounts meeting these requirements usually employ a two-axis gimbal suspension, a three-point suspension using a torsion spring, two independent hinge points formed by leaf springs, or a metal diaphragm. Rotation is achieved by micrometers or differential screws. Infinite resolution can be obtained with mounts containing piezoelectric elements. From a well-designed mirror mount a thermal sensitivity of no greater than 0.2 arc s/C can be expected.

Optical Surfaces. The polished surfaces of components, such as mirrors, laser rod, windows, lenses, doubling crystals, etc., used in a laser resonator must meet certain requirements in terms of surface quality, flatness, parallelism, and curvature. The quality of an optical surface is specified by the scratch and dig standards (MIL-13830A). According to this standard, surface quality is expressed by two numbers. The first number gives the apparent width of a scratch in microns and the second number indicates the maximum bubble and dig diameter in 10-μs steps. Components used in optical resonators should have an optical surface quality of at least 15/5. With great care a 0/0 quality can be achieved. As a comparison, commercially available lenses of standard quality have an 80/50 surface.

From the four basic techniques of shaping optical components – molding, fire polishing, felt polishing, and pitch polishing – only the two latter techniques produce an acceptable-quality surface finish for laser applications. However, whenever possible, pitch-polished rather than felt-polished components should be used. Using a polishing pitch as the lining of the polishing tool produces the finest surface quality. The pitch cold-flows and takes the shape of the work in a very short time. The actual polishing is done with a fine abrasive (rouge) in a water slurry.

Mirror blanks for the resonator are either plano-concave, plano-plano, or in special cases plano-convex, made from schlierengrade fused quartz or BSC 2. Standard curvatures are flat, 10, 5, 3, 2, 1, 0.5, and 0.25 m radius of curvature. The surface finish should be $\lambda/10$ or better. Standard mirror blanks have diameters of 25, 12.5, and 6.2 mm, and thicknesses of 6.2 and 10 mm. A parallelism of 1 arc s or better, and a flatness of at least $\lambda/20$ is required for plane-parallel blanks to be used for etalons. Low-cost general-purpose windows, used for example on dye cells, usually have a wedge of 3 to 5 minutes and a flatness of $\lambda/4$ or better.

Coatings. Evaporated dielectric films are essential parts of the resonator optics, and their properties are important for the performance of the laser. A typical laser requires a highly reflective and a partially reflective mirror coating and several antireflection coatings on the laser rod, Q-switch crystal, windows, lenses, etc.

Dielectric thin films are produced by evaporation and condensation of transparent materials in a vacuum. The dielectric thin film must adhere well to the substrate. It should be hard, impervious to cleaning, have low loss, and should exhibit a high damage threshold. Commercially available coatings are either hard coatings which will meet MIL-C-675A or semihard coatings which will pass the Scotch-tape adherence test per MIL-M-13508B.

The simplest antireflection coating is a single quarter-wave film. The reflectivity of a single dielectric layer of index n_1, which has an optical thickness of $\lambda/4$, is, at normal incidence,

$$R = \left(\frac{n_s - n_1^2}{n_s + n_1^2}\right)^2 , \qquad (5.78)$$

where n_s is the refractive index of the substrate. Zero reflection is achieved when $n_1 = (n_s)^{1/2}$.

The lowest refractive index available as stable film is MgF_2 with $n_1 = 1.38$, a value which results in a perfect antireflection coating for a substrate with $n_s = 1.90$. For $n_s = 1,80$, 1.70, and 1.5 the residual reflectance from a $\lambda/4$ MgF_2 coating is ~0.1, 0.3, and 1.4%, respectively. From Table 5.3 it follows that MgF_2 is an excellent match for ruby and Nd : YAG. In cases where the reflection from a single-layer antireflection coating is still too high, two or more dielectric-layers must be applied.

Table 5.3. Substrate characteristics

	Index of refraction	Reflectivity percent	Reflectivity with single-layer MgF_2
Fused quartz (0.63 μm)	1.46	3.5	1.5
BSC-2 (0.63 μm)	1.52	4.2	1.4
Ruby (0.69 μm)	1.76	7.6	0.15
Nd : YAG (1.06 μm)	1.82	8.4	0.1

A substrate having two layers with index values of n_1 and n_2 of optical thickness $\lambda/4$ will have a total reflectance of

$$R = \left(\frac{n_2^2 - n_1^2 n_s}{n_2^2 + n_1^2 n_s}\right)^2 . \qquad (5.79)$$

Zero reflection can be obtained if $(n_2/n_1)^2 = n_s$. If coating materials with the proper ratio n_2/n_1 are not available for a particular substrate, thicknesses which deviate from $\lambda/4$ must be used to achieve zero reflection from glass. Because the region of low reflectance of this type of coating is rather small, the coating is sometimes called "V"-coating. A very hard and durable two-layer coating frequently employed on glass substrates is the system ZrO_2-MgF_2.

High reflectivity, multilayer dielectric films can be tailored to give specific reflectance versus wavelength characteristics by the appropriate choice of the number of layers, layer thickness, and index of refraction of the materials. The simplest design of a multilayer coating is a stack of alternating films of equal optical thickness, corresponding to $\lambda/4$, but of two different refractive indices. It is most efficient to start and end with a high-index layer so that the structure will have an odd number of layers. We obtain

$$R_{\max} = \left(\frac{n_1^{l+1} - n_2^{l-1} n_s}{n_1^{l+1} + n_2^{l-1} n_s} \right)^2 , \tag{5.80}$$

where n_1 is for the high-index material, n_2 is for the low-index material, and l is the odd number of $\lambda/4$ films.

Table 5.4 lists some of the most common thin-film materials used for solid-state laser optics. Titanium oxide, cerium oxide, and zinc sulfide are used mainly as the high-index materials; magnesium fluoride and silicon dioxide are used for the low-index materials.

Table 5.4. Thin-film materials

Chemical formula	Material	Index of refraction
MgF_2	Magnesium fluoride	1.38
ThF_4	Thorium fluoride	1.50
CeO_2	Cerium oxide	2.3
ZrO_2	Zirconium oxide	2.1
SiO_2	Quartz	1.45
ZnS	Zinc sulfide	2.35
TiO_2	Titanium oxide	2.28

As follows from (5.80), the reflectance increases with increasing number of $\lambda/4$ films in discrete steps. The number of layers employed depends on the material combination and the desired reflectance value. Normally, it does not exceed 21 because scatter and absorption losses increase with the number of layers.

For commercially available mirror coatings, the maximum reflectance ranges between 99.5 and 99.8 %. Scattering and absorption losses combined are in the order of 0.5 to 0.2 %. Typical thin-film combination are ZnS-MgF_2, ZnS-ThF_4, CeO_2-MgF_2, TiO_2-SiO_2, and ZrO_2-SiO_2. The first system is very soft and must be handled with great care; the second coating is more durable and can be cleaned with alcohol and lens tissue; the last two coatings are extremely hard and durable.

A summary of optical coatings and thin-film techniques can be found in [5.122]. Optical properties of multilayer stacks and designs have been presented in the books by *Baumeister* [5.123], *Heavens* [5.124], *Macleod* [5.125], and *Vasicek* [5.126].

5.5 Unstable Resonators

The unstable resonator first described by *Siegman* [5.127, 128] has been studied extensively both theoretically and experimentally. Excellent reviews are found in [5.129, 130]. The most useful property of an unstable resonator is the attainment of a large fundamental mode volume and good spatial mode selection at high Fresnel numbers. In other words, unstable resonators can produce output beams of low divergence in a short resonator structure which has a large cross-section.

The stable resonator, whose mirror configuration corresponds to a stable periodic focusing system, has a long slender Gaussian-profile lowest-order mode whose diameter is of the order of a few times $(L\lambda)^{1/2}$. If the diameter of the laser medium is $2a$, then the area ratio of the laser medium cross section to the lowest-mode cross section is of the same order as the Fresnel number $N_F = a^2/L\lambda$ characterizing the laser system. If this Fresnel number is much larger than unity, the lowest-order mode will extract only a fraction $\approx 1/N_F$ of the energy available in the laser madium, and/or the laser must oscillate in a sizable number of higher-order modes to extract all the energy from the laser medium.

Therefore in order to produce a diffraction-limited output beam, the Fresnel number of the laser must be on the order of unity or smaller. This usually limits the diameter of the laser gain medium to a few millimeters.

The lowest-order mode in the unstable resonator, by contrast, since the unstable resonator corresponds to a divergent periodic focusing system, expands on repeated bounces to fill the entire cross section of at least one of the laser mirrors, however large it may be.

An unstable resonator may be used to obtain a nearly diffraction-limited output beam from a large-diameter gain medium which has reasonable high round-trip gain: $2G_0L_g>1.5$, where G_0 is the small-signal gain per unit length and L_g is the length of the gain medium. The light rays in an unstable resonator walk outward from the center of the laser.

The laser output is taken as a diffraction-coupled beam passing around rather than through the output mirror. An output beam from an unstable resonator usually has an annular or rectangular-annular intensity pattern in the near field.

Immediately following its invention, the significance of the unstable resonator was recognized for the extraction of diffraction-limited energy from large volume gas lasers. However, only recently have unstable resonators been applied to solid-state laser system. There are a number of reasons for this slow acceptance. The laser medium must be of high optical quality, for an unstable resonator to be effective. This requirement has limited applications of unstable resonators primarily to gas lasers because the time- and power-dependent thermal distortions occurring in solid-state lasers made this type of resonator unattractive. In addition, the output coupler of an unstable resonator, having the dimensions of a few millimeters in typical solid-state lasers, is much more expensive and difficult to fabricate in comparison to a partially transmitting mirror required for a stable resonator.

Furthermore, the alignment tolerance of an unstable resonator is more critical compared to its stable counterpart and the advantage of a large mode volume is achieved at a sacrifice of mode quality because of aperture-generated Fresnel fringes. The output from a solid-state laser is often passed through amplifier stages, or the oscillator may be followed by a harmonic generator. The near-field beam pattern of an unstable resonator which consists of a doughnut shaped beam with diffraction rings and a hot spot in the center is not very attractive in these applications.

As we will discuss later, several new designs have been successful in circumventing some of these short comings.

About 10 years after its discovery, *Byer* and co-workers [5.131, 133] applied the unstable resonator concept for the first time to a Q-switched Nd : YAG oscillator/amplifier system. They did achieve a marked improvement in Nd : YAG output energy in a diffraction-limited mode.

Despite these earlier demonstrations of the concept, the limited design flexibility brought about by the somewhat cumbersome output coupling of the unstable resonator, and its poor tolerance to mirror alignment and optical quality of the laser medium did not make it an instant favorite among solid-state laser designers. Most commercially available solid-state lasers still have stable resonators.

A stimulus for further study and research on employing unstable resonators for solid-state lasers was provided by the zig-zag slab laser. These structures, having a narrow rectangular cross section and an optical beam with very little distortions, are ideally suited for unstable resonators. Today, most flashlamp pumped and laser-diode pumped solid-state slab lasers employ unstable resonators. As a result of this activity, also several commercially available flashlamp-pumped Nd : YAG rod systems feature an unstable resonator.

The most useful form of unstable resonator is the confocal unstable resonator, introduced by *Ananev* et al. [5.132], and by *Krupke* and *Sooy* [5.134]. A primary advantage of this configuration is that it automatically produces a collimated output beam. Confocal configurations can be divided into positive and negative branches, as shown in Fig. 5.45. The negative-branch confocal configuration has significant practical advantages in the form of more easily obtainable shorter-radius mirrors and considerably easier mirror alignment tolerances. However, the positive-branch resonator seems to be universally employed in practice because the internal focal point in the negative-branch case can lead to difficulties with optical breakdown.

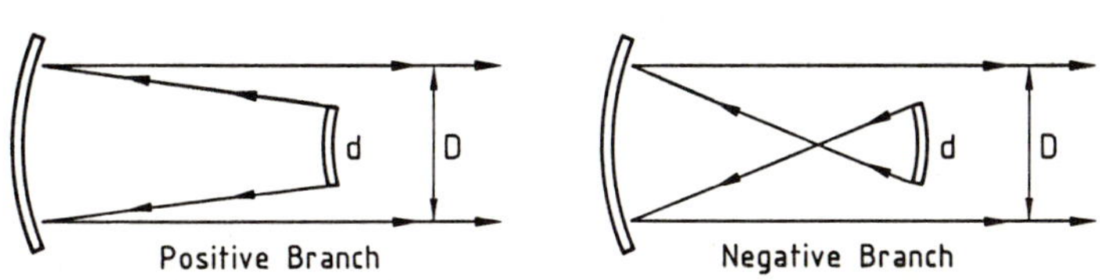

Fig. 5.45. Positive- and negative-branch confocal unstable resonators

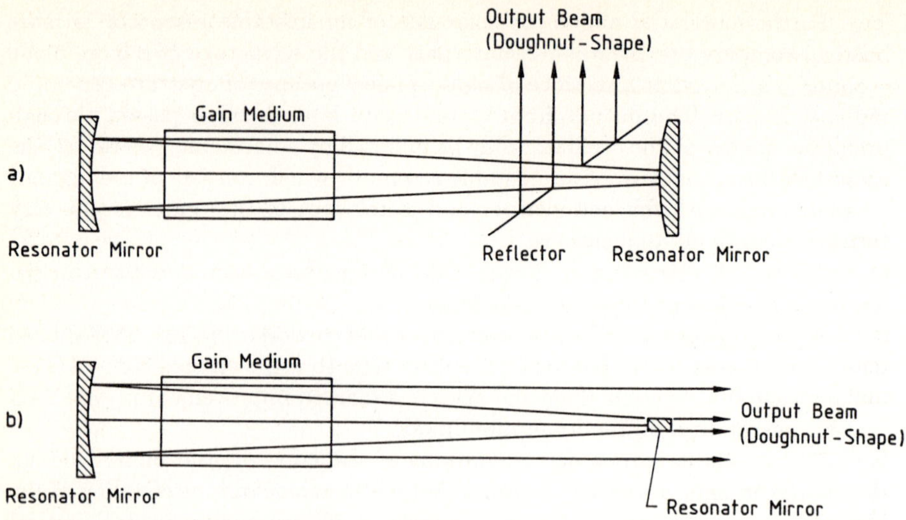

Fig. 5.46. Confocal, positive branch unstable resonators. **(a)** Scraping mirror out-coupling scheme. **(b)** Small mirror out-coupling scheme

In analogy to high-energy lasers utilizing unstable resonators the output coupling can be accomplished by means of a scraper mirror or edge coupler. Figure 5.46 shows relevant adaptions for solid-state lasers.

Figure 5.46a illustrates an unstable resonator with an output scraper mirror. It is inclined at an angle of 45° to the resonator axis and has a hole in its center which allows light to pass through it and be fed back into the resonator to sustain the lasing. Because the end mirrors and scraper are oversize, this hole determines the size and shape of the beam outcoupled from the resonator.

Figure 5.46b shows the design of a typical positive-branch, confocal unstable resonator. This design usually consists of a concave mirror M_1 and a polka-dot convex output mirror M_2, both of which are totally reflecting. The polka-dot is a small circular spot of radius d centered on a zero-power lens. The output beam is collimated as it exits the resonator around the edges of the polka-dot mirror.

While these two output coupling techniques are borrowed from high-energy lasers the designs mentioned below are more germane to the design of solid-state lasers. In an attempt to eliminate the Fresnel diffraction fringes which can cause damage to optical elements in the output of the laser, the radial birefringent element and apodized or soft aperture designs have been developed.

The radial birefringent element is based on a radial variation of phase retardation which combined with a polarizer, forms a radial intensity filter. This permits realization of Gaussian-like reflectance mirrors and soft apertures. The design of birefringent elements and their incorporation into unstable resonators was discussed in detail in [5.135–138].

In its simplest form, it is a lens of birefringent material. When a polarized optical beam passes through the lens, the radial variation in thickness of the lens creates a radial variation in the polarization of the optical beam. If the beam then passes through a polarizer, the radial variation in polarization is converted into a variation in the beam intensity. Thus the combination of a radial birefringent element and a polarizer forms a radial intensity filter which provides the capability for smoothly modulating the intensity profile of an optical beam.

Each radial birefringent element has essentially three parameters that determine the modulation profile of the beam, the center thickness, the radius of curvature, and the angle between the principal axis and a polarization axis. The first two parameters are set at the time of fabrication, and the third is varied as the need arises. Although a single-element can be designed to generate many useful profiles, multielement radial birefringent elements can generate a much larger set of modulation profiles.

A schematic of the radial birefringent-element resonator is shown in Fig. 5.47a. The reflectance profile is illustrated by Fig. 5.47b. As one can see from the curves the resonator has a flat topped intensity profile that is free from Fresnel fringes.

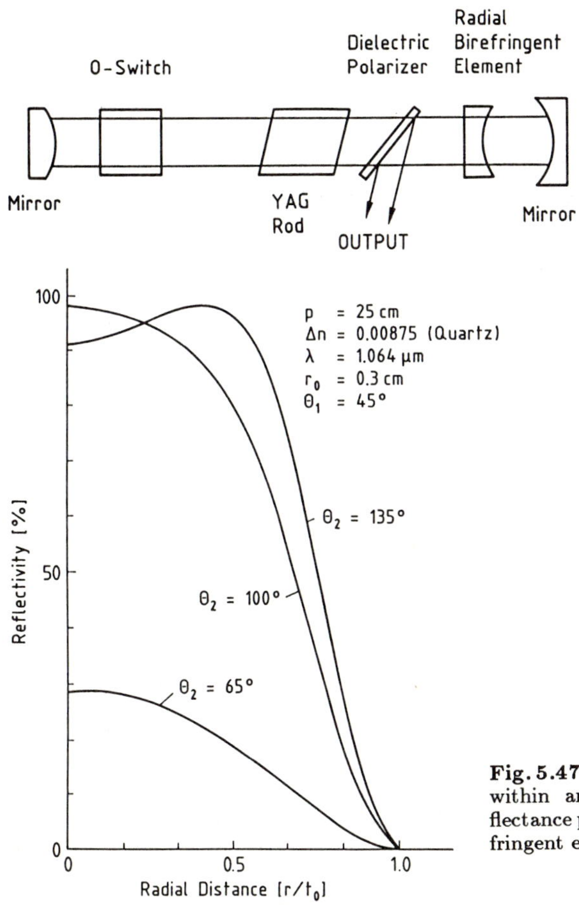

Fig. 5.47 (**a**) Radial birefringent element within an unstable resonator. (**b**) Reflectance profiles of a 2 element radial birefringent element [5.136]

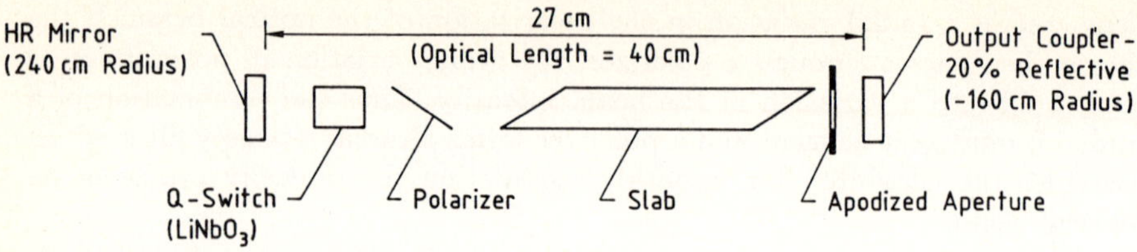

Fig. 5.48. Confocal unstable resonator for slab laser with magnification $M = 1.5$ [5.139]

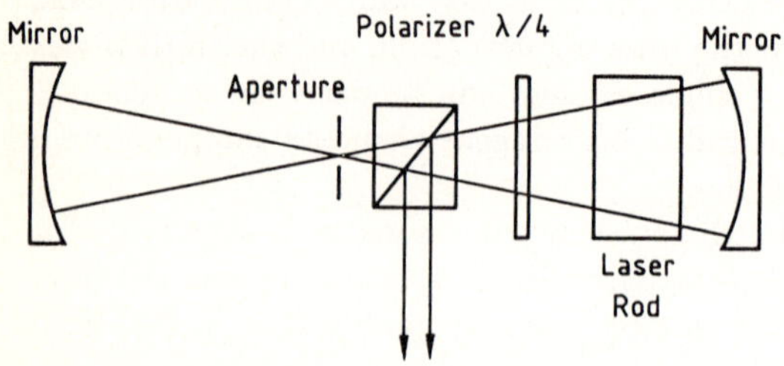

Fig. 5.49. Negative-branch unstable resonator

Figure 5.48 shows a design in which the output coupling is accomplished via a Q-switch and polarizer. An apodized aperture eliminates Fresnel diffraction rings. A different output coupler is shown in Fig. 5.49. The schematic shows a negative-branch unstable resonator featuring an internal aperture which has a diameter designed to remove diffraction rings from the near-field beam pattern. A quarter waveplate and a polarizer provide the output coupling.

Very recently, a radially variable reflectivity output coupler based on a radially variable Fabry-Perot interferometer was introduced by *De Silvestri* et al. [5.140] and applies to pulsed Nd : YAG lasers.

5.5.1 Confocal Positive-Branch Unstable Resonator

The confocal positive-branch unstable resonator is the most widely used form of the unstable resonator for solid-state lasers. It avoids a focal point in the beam and produces a collimated output beam. It has the disadvantage of a hole in the center with resultant Fresnel fringes unless either an apodized aperture or a radial birefringent element is employed as output coupler.

The design of a confocal positive-branch unstable resonator which takes into account thermal lensing of the laser rod was discussed in [5.141].

Referring to Fig. 5.45 the annular output beam has an outer diameter of D and inner diameter d, where d is also the diameter of the output coupler.

The resonator magnification

$$M = D/d \tag{5.81}$$

is the amount that the feedback beam is magnified when it travels one round trip in the resonator and becomes the output beam.

The geometrical output coupling is related to the magnification M by

$$\delta_g = 1 - \frac{1}{M^2} \quad . \tag{5.82}$$

For a confocal resonator, the mirror radii are given by

$$R_1 = \frac{-2L}{M-1} \quad , \tag{5.83}$$

$$R_2 = \frac{2ML}{M-1} \quad , \tag{5.84}$$

where L is the cavity optical length, and R_1 and R_2 are the output and back-cavity mirror curvatures. Note that the output mirror has a negative curvature and thus is convex, while the high-reflection mirror has positive curvature and is concave.

Siegman [5.129] has investigated the relationship between M and the output coupling δ for confocal unstable resonators and has shown that δ is less than the geometrically predicted value $(1 - 1/M^2)$. Resonators should be designed to operate at half-integer equivalent Fresnel numbers (N_{eq}) to obtain best mode selectivity. Equations relating M and δ under these conditions lead to the following expressions

$$\left(\frac{M^2-1}{M^2}\right) \simeq \begin{cases} 1.44\delta - 0.44\delta^2 & (N_{\mathrm{eq}} = 0.5) \\ 1.06\delta - 0.06\delta^2 & (N_{\mathrm{eq}} = 1.5) \\ \delta & (N_{\mathrm{eq}} \geq 2.5) \end{cases} \tag{5.85}$$

For positive-branch confocal unstable resonators

$$N_{\mathrm{eq}} = \frac{M-1}{2M^2} N_{\mathrm{F}} \tag{5.86}$$

where N_{F} is the conventional Fresnel number defined by the resonator length L and mode diameter d by

$$N_{\mathrm{F}} = D^2/4L\lambda \quad . \tag{5.87}$$

Physically, the half-integer equivalent Fresnel numbers correspond to Fresnel diffraction peaks centered on the output coupler leading to increased feedback into the resonator.

The design of an unstable resonator usually proceeds along the following lines: The diameter D of the laser rod and the radius R_1 of the output coupler are considered as given. The other parameters, cavity length, back-mirror curvature and output coupling, then can be calculated for selected half-integer values of N_{eq}, i.e. $N_{\mathrm{eq}} = 0.5, 1.5, 2.5, \ldots$.

For example, the resonator length is obtained on eliminating M from (5.83 and 86). One obtains

$$L = -1/2|R_1| + 1/4D[|R_1|/(\lambda N_{\mathrm{eq}})]^{1/2} \quad . \tag{5.88}$$

Once L is known M can be calculated from (5.83), and subsequently R_2 and δ_g can be determined. The actual value δ_A of the output coupling can be obtained from (5.85).

Since M is the dominant factor, determining the stability and efficiency of the system, the procedure described above usually has to be repeated several times until the right value of M is obtained. Note that the value of M has to be consistent with the gain and loss expected in the system.

As a final step one has to take into account the effect of the laser-rod focal length f, as shown schematically in Fig. 5.46. One usually chooses an available mirror curvature R_2 and calculates the rod focal length at the desired lamp input power required to achieve an effective mirror curvature $R_{2\text{eff}}$. If the mirror to rod distance is less than the rod focal length, then

$$\frac{1}{f} + \frac{1}{R_{2\text{eff}}} = \frac{1}{R_2} \quad . \tag{5.89}$$

Essentially the focusing effect of the laser rod is compensated by increasing the radius of curvature of the mirror.

Equation 5.89 is only a first-order approximation, a more rigorous treatment is found in [5.141].

5.5.2 Negative-Branch Unstable Resonators

Due to the presence of an intra-cavity focal point, the negative-branch resonator has been neglected in practical laser applications. Despite the potential problem of air breakdown this resonator merits consideration due to its unique feature of relatively large misalignment tolerances.

Ewanizky and *Craig* [5.142] found that their Q-switched Nd:YAG laser featuring negative-branch unstable resonator, was not significantly degraded with a mirror misalignment angle of as much as a few milliradians. In a similar system air breakdown was not experienced for Q-switched pulses in the order of 170 mJ and 12 ns pulse length.

Therefore, for small solid-state lasers, typical of range finders and target designators with peak powers not exceeding 5–10 MW, it is conceivable that a negative-branch unstable resonator could be employed.

The design parameters for a negative-branch resonator of the type shown in Fig. 5.50 are

$$R_1 = 2L/(M+1) \quad , \tag{5.90}$$

$$R_2 = 2ML/(M+1) \quad , \tag{5.91}$$

where L is the confocal resonator length

$$L = R_1/2 + R_2/2 \tag{5.92}$$

and M is the optical magnification

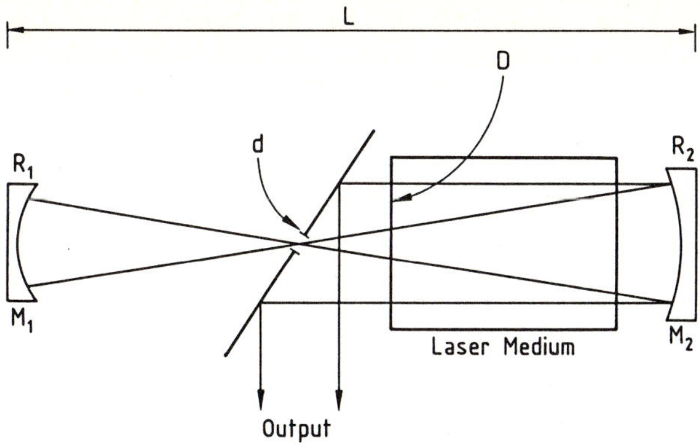

Fig. 5.50. Arrangement of a typical negative-branch unstable resonator

$$M = R_2/R_1 \quad . \tag{5.93}$$

The aperture diameter of the output mirror is

$$d = D/M \quad . \tag{5.94}$$

Recently, a variation of the negative-branch unstable resonator was described by *Gobbi* et al. [5.143]. The design is based on the proper choice for the size of the field-limiting aperture d located at the common focal plane of the mirror. If the aperture is chosen such that a plane wave incident on it is focused by mirror M_1 to an Airy disk having the same diameter d, then this results in the removal of the hot spot inside the cavity and in a smoothing of the spatial profile.

If the aperture diameter d is chosen such that

$$d = 2(0.61\lambda f_1)^{1/2} \tag{5.95}$$

where $f_1 = R_1/2$, then only the Airy disk is allowed to propagate beyond the aperture, and on reflection from the mirror M_2, it is magnified, collimated, and presented Fourier transformed at the aperture plane ready to start another similar cycle.

The radius at which the beam has zero amplitude is not determined by the geometrical magnification M, but an effective magnification imposed by diffraction.

$$M_{\text{eff}} = 1.5M = -1.5f_2/f_1 \qquad f_2 = R_2/2 \quad . \tag{5.96}$$

The diameter of the collimated beam passing through the laser is

$$D = |M_{\text{eff}}|d \quad . \tag{5.97}$$

By adding the constraint on the aperture size that it match the Airy disk, the usual hot spot in the focal plane is completely removed by diffraction. Actually the combination of aperture d and mirror M_1 acts on the resonator field as a low-pass spatial filter. This accounts for the smoothness of the field profile.

The disadvantage of this design is the limited value of D which can be achieved in practical systems. In order to fill a large active volume with diameter D, such for example a slab laser, either M_{eff} or d has to be large (5.97). In order for d to be large, it follows from (5.95), that f_1 has to be large which in turn leads to a long resonator. A large M_{eff} requires a very high gain material, for example, a Q-switched Nd:YAG oscillator. *Gobbi* et al. reported a beam diameter of $D = 4.8\,\text{mm}$ inside the laser rod. This required a resonator of 125 cm in length and a magnification of $M = 4$. In order to achieve the high gain required for this design the laser was pumped 2.5 times above threshold.

Another interesting feature of this design is the beam extraction from the resonator. Instead of a tilted scraper mirror, as shown in Fig. 5.50, beam extraction was achieved by means of a polarization coupling scheme, employing a polarizer and a quarter wave plate (Fig. 5.49).

5.6 Wavelength Selection

Vibronic lasers such as alexandrite, Ti:sapphire, Cr:GSGG have very broad gain curves, in the operation of these lasers it is necessary to use a wavelength selection technique to: (a) restrict laser action to a specified wavelength; and (b) tune the laser output. Several different methods are available (in principle) for providing the wavelength selection and tuning. These include (a) use of a prism inside the laser, (b) utilization of an adjustable optical grating within the laser, (c) use of intracavity etalons, or (d) use of one or more thin birefringent plates within the laser that are tilted at Brewster's angle.

The technique most commonly employed for the wavelength selection of tunable lasers is the birefringent filter. This technique was originated and first demonstrated by *Yarborough* and *Hobart* [5.144]. In its simplest form, the birefringent filter consists of a single thin birefringent crystal located inside the laser Fig. 5.51). For simplicity, we assume that the birefringent axes lie in the plane of the crystal, and that the crystal is tilted at Brewster's angle.

The crystal and birefringent axes, and the incident light are shown in more detail in Fig. 5.52. In this figure, α is the incidence angle between the incoming laser beam and the normal to crystal plane, and ϕ is the angle between the plane of incidence and the crystal optic axis. The phase difference δ between the ordinary and extraordinary rays emerging from the crystal is given by

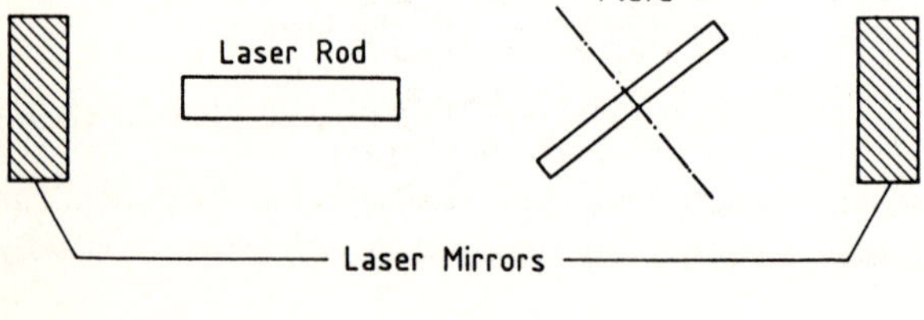

Fig. 5.51. Single-crystal wavelength selector

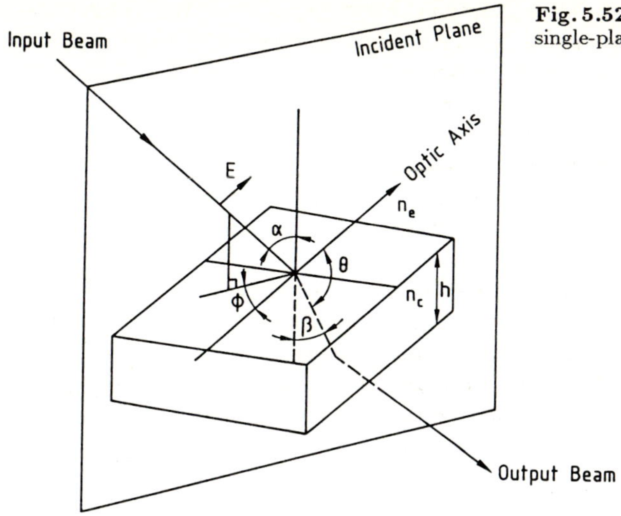

Fig. 5.52. Schematic of typical single-plate birefringent filter

$$\delta = \frac{2\pi h(n_0 - n_E)\sin^2\theta}{\lambda \sin\alpha} \tag{5.98}$$

where λ is the laser wavelength in vacuum.

An alternate representation of the single stage tuning element is given in Fig. 5.53. The input and output partial polarizers represent the front and back surfaces of the quartz crystal situated at Brewster's angle. In addition, the crystal is assumed to be rotated to $\theta = 45$ degrees in this figure.

Wavelength selection occurs with the birefringent filter because of the two different crystal indices of refraction. When the laser light has a wavelength corresponding to an integral number of full-wave retardations, the laser operates as if the filter were not present. At any other wavelength, however, the laser mode polarization is modified by the filter and suffers losses at the Brewster surfaces.

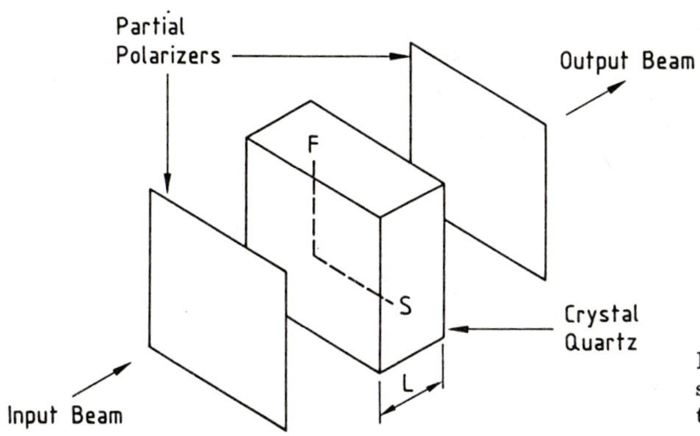

Fig. 5.53. Alternate representation of single-stage crystal filter

Tunability of the laser is achieved by rotating the birefringent crystal in its own plane. This changes the included angle ϕ between crystal optic axis and the laser axes and, hence, the effective principal refractive indices of the crystal.

The amplitude transmittance of the single-stage filter of Fig. 5.51 has been calculated by *Bloom* [5.145]. For a quartz crystal rotated to ϕ=45 degrees and tilted to Brewster's angle, the transmittance at unwanted wavelengths is about 82 %. This may or may not provide adequate suppression of unwanted wavelengths for certain lasers.

One way to lower the filter transmittance in the rejection band is to use a stack of identical crystal plates that are similarly aligned (Fig. 5.54). *Hodgkinson* and *Vukusic* [5.146] have analyzed this filter in some detail. They found, for example, that if one uses a ten-crystal filter, the resulting transmittance in the rejection band is about 15 %, which is certainly small enough to suppress unwanted laser frequencies.

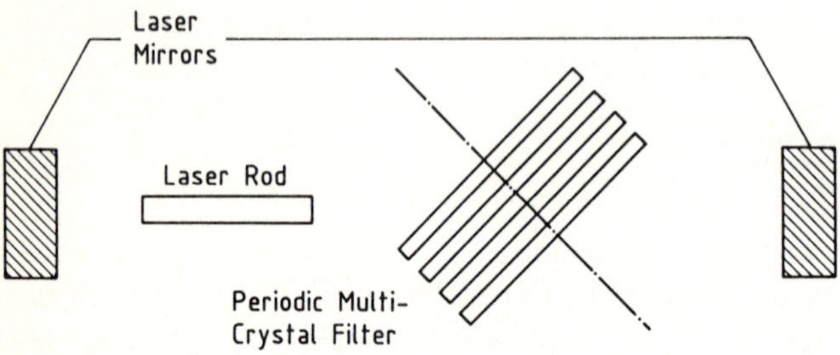

Fig. 5.54. Four-crystal wavelength selector

Another technique for lowering the filter transmittance in the rejection band is to include more Brewster's angle surfaces in the laser. This reduces the value of T_0 for the filter at the cost of increasing the width of the central passband slightly.

Still another approach for narrowing the width of the central passband of the filter is to use several crystals in series whose thicknesses vary by integer ratios. One such filter, consisting of three crystals whose thicknesses are in the ratio 1:4:16, has been analzyed by *Bloom* [5.145]. The central passband of this filter is considerably narrower than those of the previously dicussed designs. The disadvantage of this approach, however, is that numerous (unwanted) transmission spikes are now present, the largest of which has an amplitude of about 75 %.

It is obvious from these results that there are virtually an unlimited number of designs that can be tried, with corresponding tradeoffs in central passband width, stopband transmittance, presence of spikes in the stopband, and complexity of the birefringent filter. For all of these birefringent filter designs, tuning is continuous and easily implemented through rotation of the multiple crystals.

6. Optical Pump Systems

In optically pumped solid-state lasers the light source must supply the maximum possible light output in the spectral region that can be absorbed by the laser material. Electrical current, either continuous or pulsed, is supplied to the pump source and converted into optical radiation. The light source and laser material are contained in a pumping arrangement which concentrates the light from the pump source onto the laser material. In this chapter we will discuss the salient features of the three components which together comprise the optical pump system of a solid-state laser: the pump source, the power supply, and the pump cavity.

6.1 Pump Sources

In the application of light sources for pumping lasers, the primary objective is to convert electrical energy to radiation efficiently and to generate high-radiation fluxes in given spectral bands. The most efficient laser pump lamp will produce maximum emission at wavelengths which excite fluorescence in the laser material, and produce minimal emission in all spectral regions outside of the useful absorption bands. The various light sources employed to pump solid-state lasers are listed in Table 6.1.

Optical pumps for solid-state lasers can be categorized into noble gas and metal vapor discharge lamps, filament lamps, laser diodes, and pumps not based on electricity as the prime power source, such as the sun, chemical flashbulbs, and radiation obtained from detonations. Of particular interest are noble gas flashlamps, wall-stabilized cw arc lamps, and filament lamps, because these light sources are employed in the vast majority of lasers. The application of a particular pump source depends on the desired output power, the mode of operation, i.e., pulsed or continuous operation, high or low repetition rate, etc., and on the laser material which has to be pumped. Pulsed operation can be obtained from noble-gas flashlamps and laser diodes. All other light sources listed in Table 6.1 are cw or quasi-cw pump sources. A high-gain media, such as Nd : YAG, can be pumped with all of the pump sources listed in Table 6.1. However, for cw operation of Nd : YAG, the majority of systems employ either krypton-filled arc lamps or tungsten-halogen filament lamps.

For special applications, such as space communication, Nd : YAG has been pumped by metal alkali lamps, by laser diodes, and by the sun. Pulsed Nd : YAG,

Table 6.1. Optical pump sources for solid-state lasers

- Noble gas discharge lamps
  - Flashlamps
    - Linear
    - Helical
  - cw arc lamps
    - Wall stabilized
    - Vortex
- Metal vapor discharge lamps
  - Mercury arc lamps
    - Hg
    - Metal iodide doped Hg
    - Metal doped Hg
  - Alkali metal
    - Na
    - K
    - Rb
    - K-Rb
- Filament lamps
  - Tungsten-halogen
- Semiconductor diodes
  - Laser
    - GaAs
    - GaAlAs
  - Incoherent
    - GaAlAs
    - GaAsP
- Nonelectric sources
  - Sun
  - Flashbulbs
  - Detonation

ruby, and Nd : glass lasers are pumped by xenon-filled linear or helical flash-lamps. Very small Nd : YAG lasers pumped at low energy levels are sometimes pumped with krypton-filled linear flashlamps. For cw operation of ruby, high-pressure mercury arc lamps have been used, since these are the only pump sources which have sufficient brightness in the absorption bands of ruby to cause lasing.

If we compare the spectral characteristics of the pump sources listed in Table 6.1, we find that on one side we have monochromatic pump sources, such as laser diodes, and at the other extreme blackbody radiators, such as filament lamps. The discharge lamps operated at low current densities represent a compromise between a monochromatic source and a blackbody radiator. Radiation from an arc discharge lamp is made up of both line and continuum components. The output spectra of arc lamps are complicated and difficult to describe in simple terms. Nevertheless, it is useful to speak of an effective brightness temperature to relate the radiant flux density at a given wavelength to that of an ideal blackbody at that temperature. Brightness temperatures for typical pump sources are given in Table 6.2, and spectral distributions of blackbody radiators at temperatures which are achievable with typical pump sources are plotted in Fig. 6.1.

Table 6.2. Brightness temperatures of typical pump sources

Filament lamps	2,400–3,400 K
Sun	5,800 K
Pulsed arc	5,000–15,000 K
cw arc	4,000–5,500 K

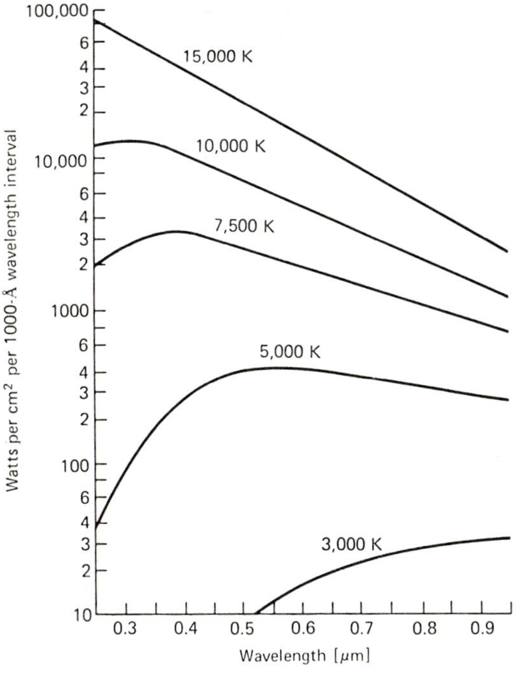

Fig. 6.1. Emission from blackbody radiators at temperatures which are achievable with typical pump sources

Table 6.3. Performance of state-of-the-art pump sources

	Krypton arc	Tungsten filament	Alkali metal	GaAlAs array
Electrical input per cm of lamp [W/cm]	800	275	160	80
Maximum total input per lamp [W]	15,000	1,500	1,000	25

In terms of average power, by far the highest rating can be obtained from a vortex stabilized cw arc lamp. However, these lamps require a closed-cycle gas circulation system and are therefore of no practical interest for commercial application of lasers. Compared to incandescent lamps, laser diodes and metal vapor lamps, the cw-pumped krypton-filled arc lamp is capable of the highest output power. All high-power commercial Nd:YAG lasers utilize this type of lamp. Table 6.3 shows a comparison of the power-handling capabilities of cw pump sources employed in Nd:YAG lasers. Photographs of typical cw pump sources and flashlamps are shown in Figs. 6.2, 3.

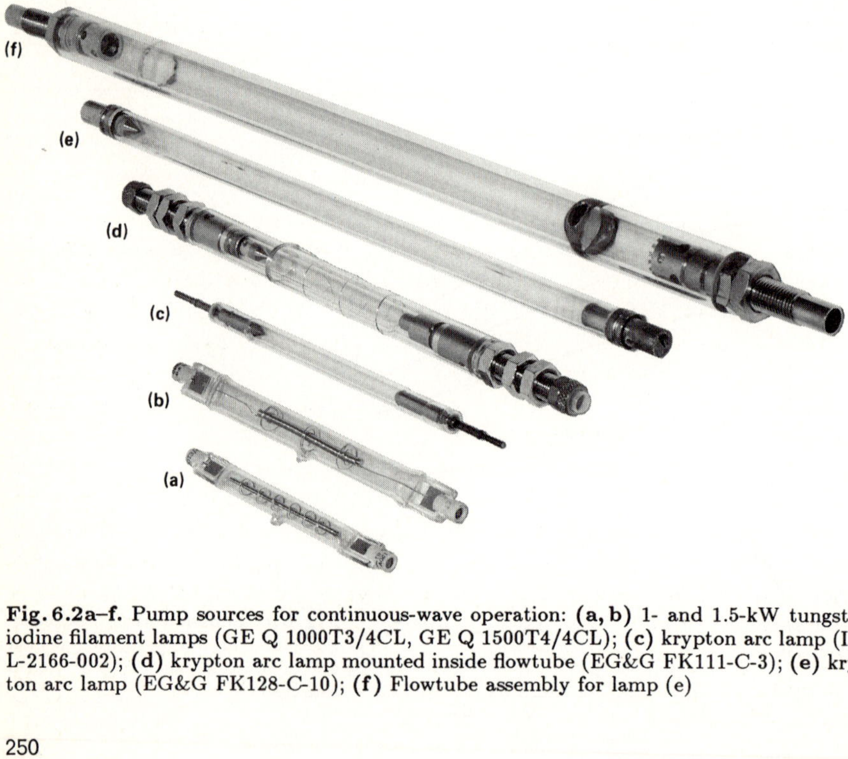

Fig. 6.2a–f. Pump sources for continuous-wave operation: **(a, b)** 1- and 1.5-kW tungsten-iodine filament lamps (GE Q 1000T3/4CL, GE Q 1500T4/4CL); **(c)** krypton arc lamp (ILC L-2166-002); **(d)** krypton arc lamp mounted inside flowtube (EG&G FK111-C-3); **(e)** krypton arc lamp (EG&G FK128-C-10); **(f)** Flowtube assembly for lamp (e)

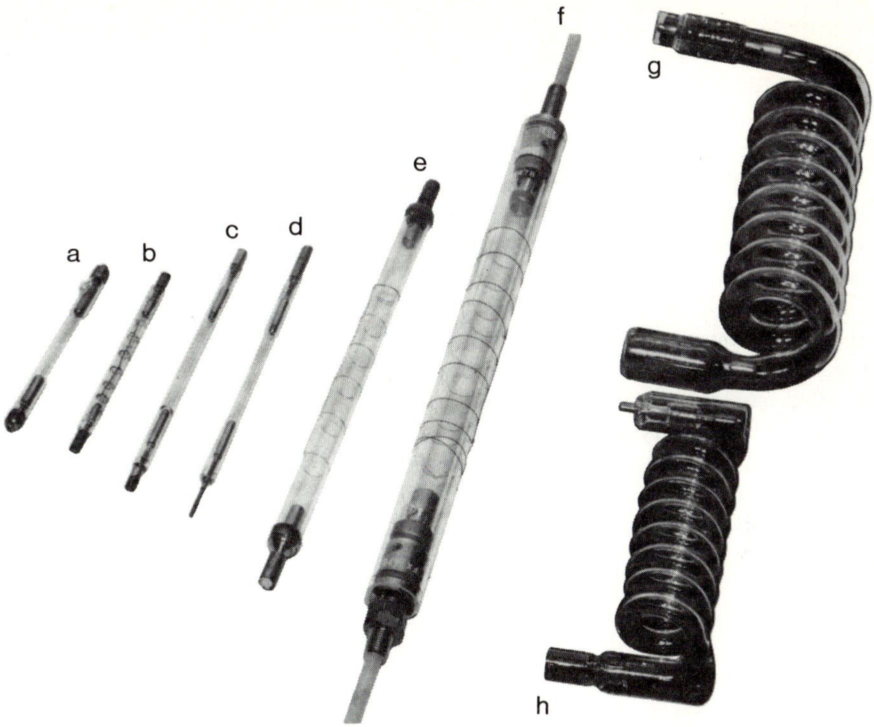

Fig. 6.3a–h. Xenon-filled helical and linear flashlamps: (**a**) Xenon Corp. Novatron 185; (**b**) ILC, Inc., L459; (**c**) ILC, Inc., 3F2; (**d**) ILC, Inc., L-210-004; (**e**) Xenon Corp. Novatron 188; (**f**) EG&G FX67B; (**g**) Spaceglass K6; (**h**) Spaceglass K1

6.1.1 Noble Gas Flashlamps

Lamp Design and Construction

Flashlamps used for laser pumping are essentially long arc devices designed so that the plasma completely fills the tube. A flashlamp consists of a linear or helical quartz tube, two electrodes which are sealed into the envelope, and a gas fill. Standard linear lamps have straight discharge tubes with wall thicknesses of 1 to 2 mm, bore diameters between 3 mm and 19 mm, and lengths from 5 cm to 1 m. Helical lamps are constructed by wrapping heated quartz tubing around mandrels. Commercial lamps have helix inner diameters as small as 1 cm and as large as 12 cm. Tube diameters and wall thicknesses are similar to those used for straight lamps. A helical lamp has a larger wall area and arc length per linear centimeter and, therefore, can deliver much higher pulse energies to a given-size laser rod than a single linear flashlamp. Helical lamps are used mainly for high-energy ruby and Nd : glass lasers. Flashlamps are typically filled with gas at a fill pressure of 300 to 700 torr at room temperature. Xenon is generally chosen as the gas fill for flashlamps because it yields a higher radiation output for a given electrical input energy than other gases. However, in a few special cases, such as small, low-energy lamps employed to pump Nd:YAG lasers, the low-atomic-weight gas krypton provides a better spectral match to the absorption bands of Nd:YAG.

Most arc lamps employ either pure tungsten for the anode and 2% thoriated tungsten for the cathode, or thoriated tungsten for the anode and a compressed pellet of porous tungsten impregnated with barium strontium aluminate for the cathode. The work function of this tungsten matrix is 1.5 eV, whereas for thoriated tungsten it is 2 eV, and for pure tungsten it is 4 eV. A cathode material with a low work function makes it easier to trigger the lamps. Thoriated tungsten is used as the anode material for the very practical reason that this alloy is easier to machine than pure tungsten. Since in a standard lamp the cathode is more emissive than the anode, flashlamps are polarized and will pass current in only one direction without damage.

There are two types of construction techniques in commercial use for sealing the electrode into the body of the flashlamp. One type, the tungsten-rod seal, uses an intermediate highly doped borsilica glass to seal the electrode to the fused quartz envelope. The glass balances the thermal stresses between fused silica, which has an extremely low coefficient of thermal expansion, and tungsten, which has a large coefficient. The second type, the solder seal, uses a copper rod, one end of which is brazed to an electrode and the other end welded to a nickel cup. The seal is made with a low-temperature indium solder between the copper-plated nickel cup and the platinum-coated end of the quartz envelope. The essential features of both types of lamps are shown in Fig. 6.4.

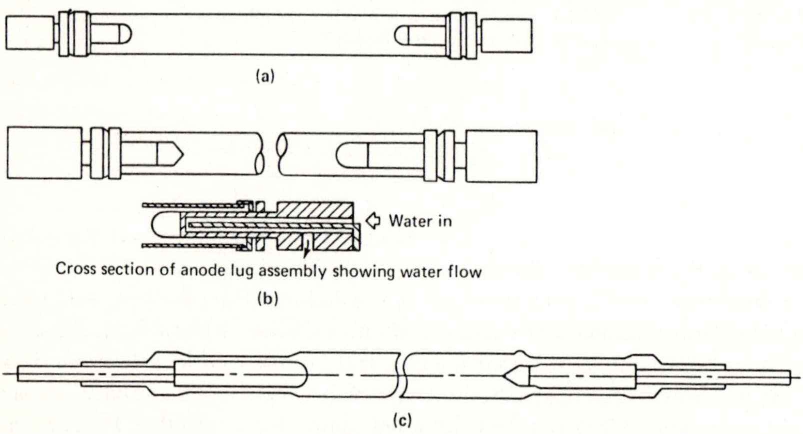

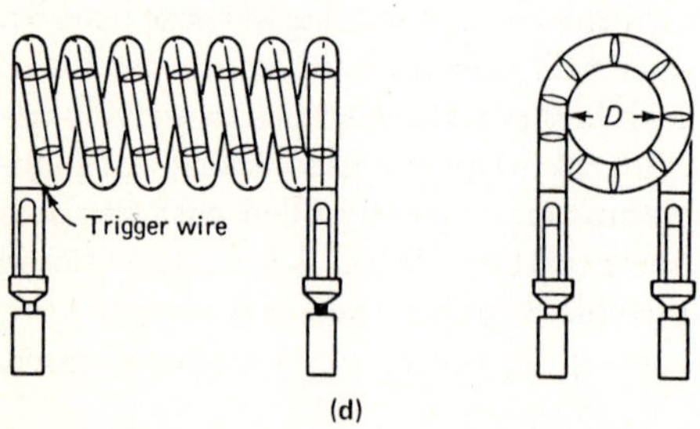

Fig. 6.4a–d. Electrode and seal construction of linear and helical flashlamps. Lamps in **(a)**, **(b)**, and **(d)** have solder seals (manufacturer EG&G); lamp in **(c)** has a graded glass seal (manufacturer ILC)

The solder seal must be operated below the softening point of the solder, which is about 180° C. A unique feature of this type of quartz-to-metal seal is that the diameter of this seal joint can be as large as the quartz tubing itself. This permits a large cross section for the electrodes and, thus, an excellent path for dissipation of heat from the electrodes. In some cases the copper stem is water-cooled internally. The lamp is filled at room temperature through the hollow copper stem. When the filling is finished, the copper tube is crushed to seal in the gas. Thus, there is no "tip-off" on this lamp and little residual stress due to formation of the seal.

The tungsten-rod seal, on the other hand, can be operated at much higher temperatures. Since the heat removal through the small-diameter tungsten stem is not very large, the quartz tubing is shrunk into close proximity to the electrode to aid cooling of the electrode in radial directions. An advantage of the rod-seal construction is the fact that during evacuation the lamps can be baked out at 1000° C, and during the filling process the electrodes can be induction-heated to a bright red for their final outgassing. The lamps are then backfilled to the appropriate gas pressure and tipped off. The tungsten-rod seal permits the fabrication of high-purity contaminant-free lamps having long shelf lives.

The envelope and the rod-seal area of flashtubes are normally cooled by free or forced air, by pressurized nitrogen, or by a liquid such as water, water-alcohol mixtures, or a fluorinated hydrocarbon. Liquid cooling permits operation of the lamps at a maximum inner-tube wall surface loading of $300\,\mathrm{W/cm^2}$ of average power. Free-air convection cooling is limited to handling about $5\,\mathrm{W/cm^2}$ of dissipation: forcing air across the flashtube envelope enables one to dissipate up to $40\,\mathrm{W/cm^2}$.

In high-average-power systems linear rather than helical lamps are normally used because with linear lamps the coolant can be more readily directed over the entire flashlamp surface. Liquid-cooled linear flashlamps are available with outer quartz jackets which permit cooling of the lamps with a highly turbulent flow. The water jacket, which is attached to the flashlamp by means of O-ring seals, is reusable after the lamp has reached its end-of-life.

Optical Characteristics

The radiation output of a gas discharge lamp is composed of several different components, each corresponding to a different physical light-emission mechanism. The relative importance of each of these mechanisms depends strongly on the power density in the lamp, and so the low-power and the high-power optical output spectra are markedly different. The total radiation is made up of both line and continuum components. The line radiation corresponds to discrete transitions between the bound energy states of the gas atoms and ions (bound-bound transitions). The continuum is made up primarily of recombination radiation from gas ions capturing electrons into bound states (free-bound transitions) and of bremsstrahlung radiation from electrons accelerated during collisions with ions (free-free transitions). The spectral distribution of the emitted light depends in complex ways on electron and ion densities and temperatures.

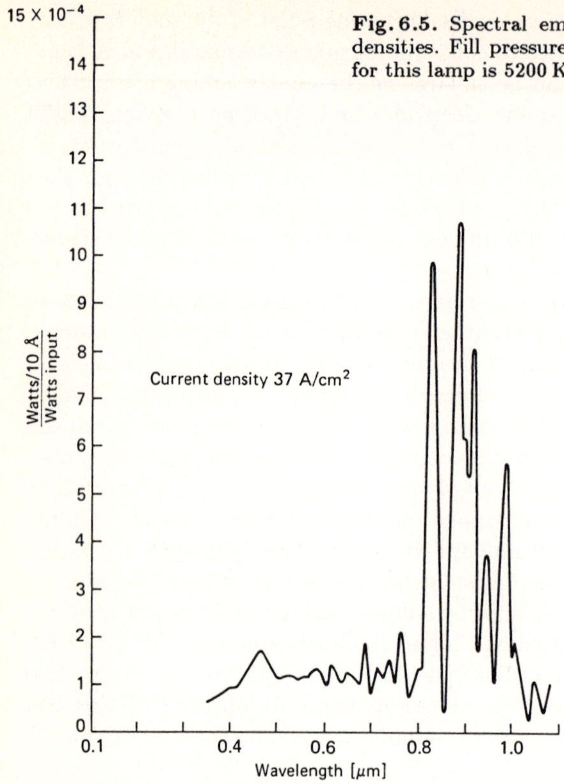

Fig. 6.5. Spectral emission from xenon at low current densities. Fill pressure 1.7 atm of Xe. Color temperature for this lamp is 5200 K [6.1]

In the continuous-power, wall-stabilized noble gas arc, current densities are such that there is a high number of bound-bound transitions and therefore the radiation spectrum is characteristic of the fill gas and is broadened by increases in pressure.

In high-current-density, pulsed laser applications, the spectral output of the lamp is dominated by continuum radiation, and the line structure is seen as a relatively minor element. Between these two cases, a pulsed-power region can exist where the pulsed power level is such that discrete line radiation is still emitted and is superimposed on a strong background pseudocontinuum. Spectral distributions over the range 0.35 to 1.1 μm were measured by *Goncz* and *Newell* [6.1] for representative pulsed and continuous xenon arc lamps. Figure 6.5 shows the output spectrum for a continuous xenon arc lamp. As can be seen, this lamp produces a very strong line structure in the infrared. In Fig. 6.6 the spectral emission of a xenon flashtube is plotted for two current densities. As a result of the high current densities, the line structure is in this case masked by a strong continuum. From Fig. 6.6 it also follows that a high current density shifts the spectral output toward the shorter wavelengths.

As we discussed at the beginning of this chapter, one can relate the radiation characteristics of a flashlamp $R(\lambda, T)$ to the characteristics of a blackbody. The departure of any practical light source from the blackbody characteristic is

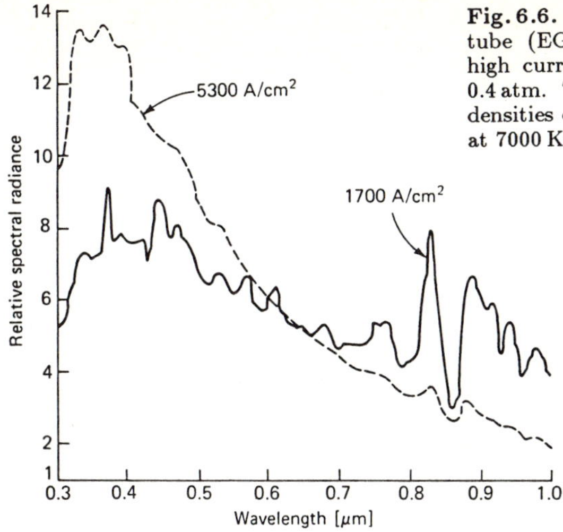

Fig. 6.6. Spectral emission of a xenon flash tube (EG&G, model FX-47A) operated at high current densities. Lamp fill pressure is 0.4 atm. The spectrum at the two current densities can be approximated by blackbodies at 7000 K and 9400 K, respectively [6.2]

accounted for in the emissivity, $\varepsilon(\lambda, T)$, which varies between zero and one, and is both wavelength and temperature dependent; thus $R(\lambda, T) = \varepsilon(\lambda, T)B(\lambda, T)$. In a flashlamp, $\varepsilon(\lambda, T)$ depends strongly on temperature and wavelength. If we assume local thermodynamic equilibrium in the plasma, the blackbody characteristic $B(\lambda, T)$ becomes the blackbody curve corresponding to the electron temperature in the plasma. In general, the emissivity of the flashlamp radiation at longer wavelengths is greater than the emissivity at shorter wavelengths. An increase in power density will result in a large increase in emissivity (and radiation) at short wavelengths, but only small changes at longer wavelengths where the emissivity is already close to unity.

The transmission properties of the flashlamp envelope will influence the spectral output. Usually the envelope is fused natural silica, which transmits light between 0.2 and 4 μm. Figure 6.7 shows the uv transmission of doped

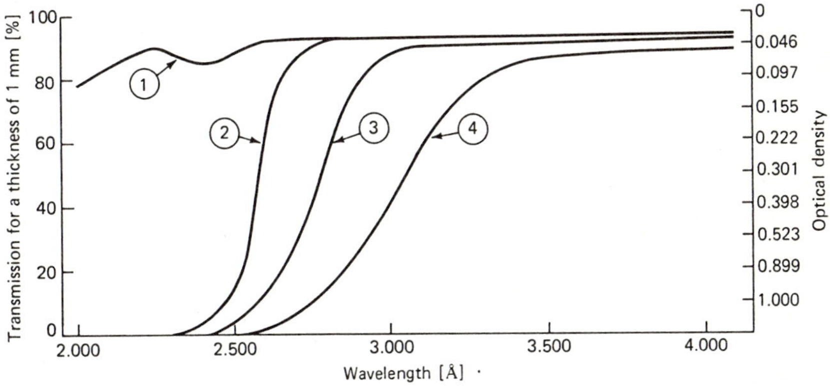

Fig. 6.7. Ultraviolet transmission (for 1-mm thickness) of different types of fused quartz. *(1)* Pursil, *(2)* Germisil, *(3)* Heliosil I, *(4)* Heliosil II

and undoped fused silica. Germisil and Heliosil, manufactured from fused silica containing a slight admixture of titanium, are sometimes employed as lamp or flow-tube envelope material to prevent the production of ozone in the air or to prevent breakdown of the flashlamp coolant, or to inhibit the formation of color centers in certain solid-state laser materials.

Lamps with cerium-doped quartz envelopes have also been designed. Quartz doped with cerium will absorb radiation below 0.31 μm and will fluoresce at wavelengths between 0.4 and0.65 μm. Spectral data [6.3] from these lamps show that a 50-mm $\times$ 4-mm bore lamp with 450 torr of pressure operated at 10 J has a 15 % greater output over the spectral range of 0.4 to 0.65 μm.

In the design of pumping geometries for high-power laser systems, it is important to know how opaque the flashtube is to its own radiation. The optical transmission of a xenon flashtube has been measured at wavelengths from 0.25 to 1.0 μm and at currents up to 400 A/cm^2 by *Emmett* and *Schawlow* [6.4]. From the experimental data plotted in Fig. 6.8 it follows that the absorption increases with current and with wavelength. Above about 0.5 μm and a current of 4000 A/cm^2, a discharge tube 1 cm in diameter is nearly opaque. At shorter wavelengths or lower currents, the discharge is fairly transparent. From Fig. 6.8 it also follows that the absorption coefficient varies linearly with the square of the current density [6.5]. Thus in a large-bore flashlamp driven at high current densities, a large percentage of the radiation emanating from the plasma core will be absorbed before it reaches the lamp surface. Hence the bulk of the radiation will come from a thin sheath near the lamp surface with a corresponding reduction in lamp efficiency.

Xenon lamps are generally used for high-power pulse applications. The lamp is a relatively efficient device as it converts 40 to 60 % of the electrical input energy into radiation in the 0.2- to 1.0-μm regions [6.1, 6]. Krypton has a radiation efficiency of 25 to 30 %. As noted in Chap. 2, principal pump bands

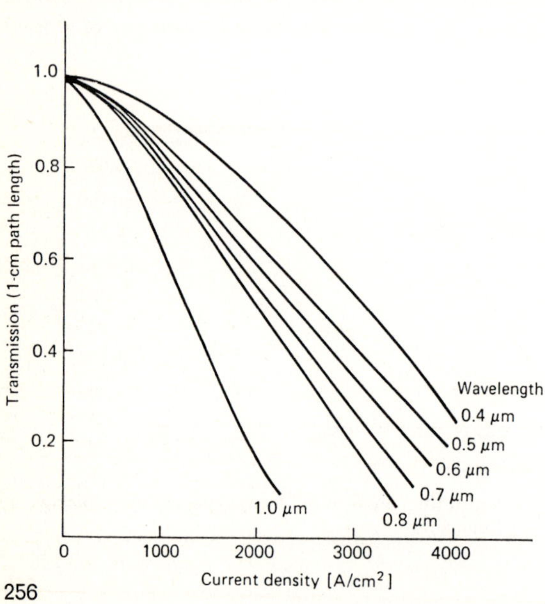

Fig. 6.8. Spectral transmission of the xenon plasma in a flashlamp as a function of current density [6.4]

of Nd:YAG are located from 0.73 to 0.76 μm and 0.79 to 0.82 μm. The xenon spectrum has no major line radiation in these bands, so pumping is primarily due to continuum radiation. However, some strong krypton lines fall within the Nd:YAG pump bands. Investigations have shown that krypton flashlamps are more efficient than xenon lamps at low power densities for pumping Nd:YAG lasers [6.6–10]. There are indications that a crossover occurs in the pumping efficiency of krypton and xenon at high peak power densities, corresponding to that power density at which the xenon continuum becomes more efficient than the krypton line structure. This crossover occurs at a power density of approximately 2×10^5 W/cm^3.

In general, the pumping efficiency of Nd:YAG of both xenon and krypton increases as the pressure is increased. In the range 450 to 3000 torr no maxima were found [6.10]. The limitation to further increases appears to be the fact that high-pressure lamps are difficult to trigger. The effect of an increase in gas fill pressure is to reduce the mean free path of the electrons and atoms in the discharge, and thus to increase their collision frequency. This leads to the production of more excited species in the discharge and the emission of more useful line radiation.

Electrical Characteristics

The impedance characteristics of a flashlamp determine the efficiency with which energy is transferred from the capacitor bank to the lamp. The impedance of the arc is a function of time and current density. Most triggering systems initiate the arc as a thin streamer which grows in diameter until it fills the whole tube. The expansion period is fast, of the order of 5 to 50 μs for tubes of bore diameter up to 1.3 cm. During the growth of the arc, lamp resistance is a decreasing function of time (Fig. 6.9). The decreasing resistance arises in part

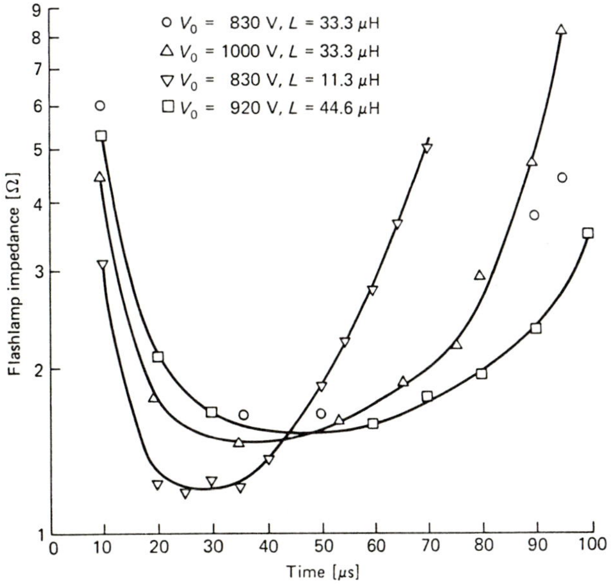

Fig. 6.9. Time dependence of flashlamp impedance for several input energies and series inductance values. The 60-mm-long and 3-mm-diameter flashlamp was filled with 450 torr of xenon. The discharge circuit consisted of a 19-μF capacitor and one inductor which was varied from 11.3 to 44.6 μH [6.81]

from the increasing ionization of the gas and from the radial expansion of the plasma. After the arc stabilizes, the voltage-current relationship is described by [6.1, 12, 13]

$$V = K_0 i^{1/2} \quad , \tag{6.1}$$

where $K_0 = kl/d$ and describes the impedance characteristics of the particular lamp. The constant k is dependent only on such parameters as gas type and gas pressure, l and d are the length and diameter of the flashlamp bore, respectively. K_0 or k is essentially the only parameter needed to describe the high-current electrical characteristics of a given flashlamp. This parameter is usually supplied by the manufacturers of the flashlamps. K_0 is found simply by flashing the lamp at some reasonable loading, while measuring the voltage and current at a convenient time during the pulse. The resistance of the flashlamp for the high-current regime with the plasma filling the bore of the tube is obtained from

$$V \cong R_{\mathrm{L}}(i) i = \left(\frac{\varrho(i) l}{A}\right) i \quad , \tag{6.2}$$

where $\varrho(i)$ is the specific resistivity of the xenon or krypton plasma, and is a function of the current density. A is the cross-sectional area of the lamp. In (6.2) the voltage drop at the electrodes, which is of the order of 20 V, has been neglected. In operating regions of interest, the flashlamp resistivity for xenon at 450 torr is related to current density j by [6.12]

$$\varrho(i) = \frac{1.13}{j^{1/2}} \quad . \tag{6.3}$$

This relationship is valid for current densities of approximately 400 to 10,000 A/cm^2 [6.14]. The lamp resistance R_{L} is obtained by introducing (6.3) into (6.2)

$$R_{\mathrm{L}}(i) = 1.27\left(\frac{l}{d}\right) i^{-1/2} \quad . \tag{6.4}$$

Comparing (6.2) with (6.1), we obtain an expression for the flashlamp parameter K_0

$$K_0 = 1.27\frac{l}{d} \quad . \tag{6.5}$$

The value of 1.27 holds for 450-torr xenon lamps. It has been found [6.3] from many flashlamp experiments that (6.5) can be generalized for other pressures and expressed as

$$K_0 = 1.27\left(\frac{P}{450}\right)^{0.2} \frac{l}{d} \quad , \tag{6.6}$$

where P is the flashlamp pressure in torr. The parameter K_0 obtained for krypton has been found to be about the same as that for xenon.

Tube Failure

The flashtube end-of-life can occur in either of two modes, and we can distinguish between catastrophic and nondestructive failures. The factors which contribute to catastrophic failure are explosion of the walls due to the shock wave in the gas when the lamp is fired, or overheating of the lamp envelope or seal and subsequent breakage due to excessive thermal loading of the lamp. The first failure mode is a function of pulse energy and pulse duration, whereas the second catastrophic-failure mode is determined by the average power dissipated in the lamp. When the flashtube is operated well below the rated maximum pulse energy and average power, the lamp usually does not fail abruptly. Rather, the tube will continue to flash with a gradual decrease in light output, which will eventually fall below a predetermined level necessary for the particular application. In the latter mode of failure, the reduced light output is caused by the erosion of the flashtube electrodes and of the quartz walls and by the gradual buildup of light-absorbing deposits within the flashtube envelope.

Flashtube Explosion

When a high-voltage trigger pulse is applied between the electrodes of a flashtube, the gas breaks down generally near the axis of the tube and a conducting filament is established. As energy is released into the channel, ohmic heating of the ambient gas causes the filament to expand radially, forming cylindrical shock waves. The shock front and its associated plasma travel radially from the tube axis to its wall. The radial velocity of the plasma discharge and the shock amplitude are proportional to the input energy. The velocity was measured to vary from about 90 to 900 m/s for inputs from 60 to 600 J into a 5-cm-long flashlamp with 6-mm bore diameter [6.11]. The cylindrical shock wave and the associated plasma heat cause stress on the inside of the tube wall which is axial in tension toward the electrodes. If the energy discharged exceeds the explosion limit for the lamp, the shock wave developed will be sufficiently intense to rupture the lamp walls and consequently destroy the lamp.

If the lamp is operated somewhat below the explosion point, catastrophic failure of the lamp is preceded by the formation of hairline cracks and a milky deposit. The microcracks, originating around minute flaws in the quartz envelope, will expand with each flash, eventually resulting in lamp failure. As a result of the high temperature of the expanding plasma, a steep temperature gradient is created near the tube surface when the plasma impinges upon the tube wall. Therefore, in addition to the progressive appearance of cracking and crazing of the inside wall, there is an associated silica deposit from evaporation and deposition [6.15].

Dugdale et al. [6.16] have investigated the effect of pulse heating on glass and ceramic surfaces. If the energy Q_0 per unit area is delivered to a material surface in time t_p, a rise in surface temperature ΔT takes place, given by

$$\Delta T = aQ_0(\pi K c_\mathrm{p} \varrho t_\mathrm{p})^{-1/2} \ , \tag{6.7}$$

where K is the thermal conductivity, c_p is the specific heat, and ϱ is the density of the material; a is a constant between 1 and 2 depending on the heat pulse waveform ($a = 1.08$ for an exponential waveform). We can relate the pulse energy per unit area to the total flashlamp electrical input energy E_0 by writing

$$Q_0 = \frac{k_1 E_0}{\pi d l} \quad , \tag{6.8}$$

where the parameter k_1 depends on the type of gas and fill pressure used. During the heat pulse, the temperature rise is confined to a depth Δx below the surface, given by

$$\Delta x = 2\left(\frac{K t_{\mathrm{p}} \varrho}{c_{\mathrm{p}}}\right)^{1/2} \quad . \tag{6.9}$$

This depth lies in the range 10^{-4} to 10^{-2} cm for glasses and ceramics subjected to heat pulses of duration 10^{-6} to 10^{-3} s. The restraint imposed by the unheated substrate may cause stresses σ_{T} to develop in the hot surface layer of magnitudes up to

$$\sigma_{\mathrm{T}} = \frac{\varepsilon \alpha \Delta T}{1 - \mu} \quad , \tag{6.10}$$

where α is the thermal expansion of the material, μ is Poisson's ratio, and ε is the Young's modulus of elasticity. The temperature gradient ΔT which will cause thermal shock damage of the envelope is obtained by equating σ_{T} with the rupture stress σ in (6.10). From (6.7, 10) follows the lamp input pulse energy which will generate stresses that exceed those of the wall material. Therefore, the explosion energy E_{ex} of thin-walled flashtubes, defined as the minimum input energy sufficient to crack the lamp catastrophically, is

$$E_{\mathrm{ex}} = k_2 L\, d\, t_{\mathrm{p}}^{1/2} \quad , \tag{6.11}$$

where k_2 depends on the type of gas and fill pressure as well as on the physical and thermal properties of the lamp envelope. The explosion energy is directly related to the inside-wall surface area of the lamp, and to the square root of the pulse duration. If L and d are measured in centimeters and t_{p} is in seconds, and if a critically damped single-mesh discharge circuit is assumed, then it has been found empirically that for xenon-filled quartz flashtubes [6.2, 17]

$$E_{\mathrm{ex}} = (1.2 \times 10^4) L\, d\, t_{\mathrm{p}}^{1/2} \quad . \tag{6.12}$$

This relation is useful for lamps with $0.5 \leq d \leq 1.5$ cm and xenon fill pressures of 300 to 450 torr, where t_{p} is defined as the time between the one-third peak light output points for a damped sinusoidal waveform and the one-half light output points for a quasi-rectangular waveform [6.18]. Equation (6.12) cannot be applied directly to helical flashtubes. With helical flashtubes, radiation from adjacent turns effectively increases the joules-per-centimeter loading on the

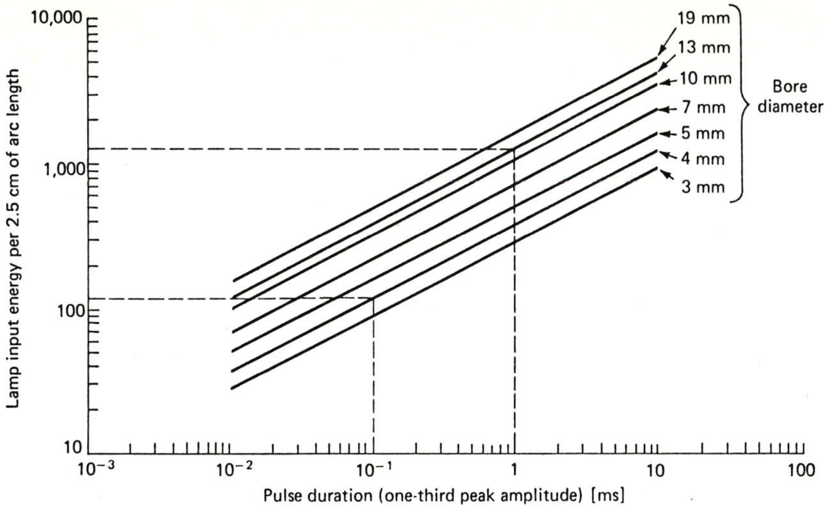

Fig. 6.10. Explosion energy of linear flashlamps (J/2.5 cm) as a function of pulse duration and bore diameter. The wall thickness of the lamps is 1 mm, except for the 19-mm-bore lamp which has a wall thickness of 1.5 mm [6.19]

quartz. E_{ex}, as calculated above, is a "free air" value which assumes that no radiation which leaves the lamp is reflected back to be reabsorbed. In an actual laser pumping cavity a certain fraction of the radiated energy – depending on the reflectivity of the cavity, its diameter, and its focal properties – is returned to the flashtube. This markedly increases the erosion and wear accompanying each discharge. In fact, the flashtube often acts as if it were being loaded with as much as 30 % more energy than is actually being dissipated in the tube. Single-pulse explosion energy as a function of pulse duration is shown in Fig. 6.10 for quartz-envelope xenon flashlamps of different bore diameters [6.19]. For a flashlamp with a specific bore size and arc length, the explosion point is a function of lamp input energy and pulse duration.

Having determined the ultimate limit for the tube, the question of the life that the tube will have in a given application has to be addressed. It has been shown that lamp life can be related to the fraction of explosion energy at which the lamp is operated. The life in flashes for a xenon-filled quartz-envelope flashlamp as a function of the single-shot explosion energy is empirically given by [6.2, 20]

$$N \cong \left(\frac{E_{ex}}{E_0}\right)^{8.5} . \tag{6.13}$$

The extremely strong dependence of flashlamp lifetime on the total input energy is a great incentive to underrate the lamp.

In Fig. 6.11, life expectancy as a function of lamp input is shown assuming that the interpulse interval is sufficiently long to permit the tube to cool between pulses [6.20]. Statistically significant data do not exist for fractions of explosion energy less than 0.2. The lifetime expressed by (6.13) is defined

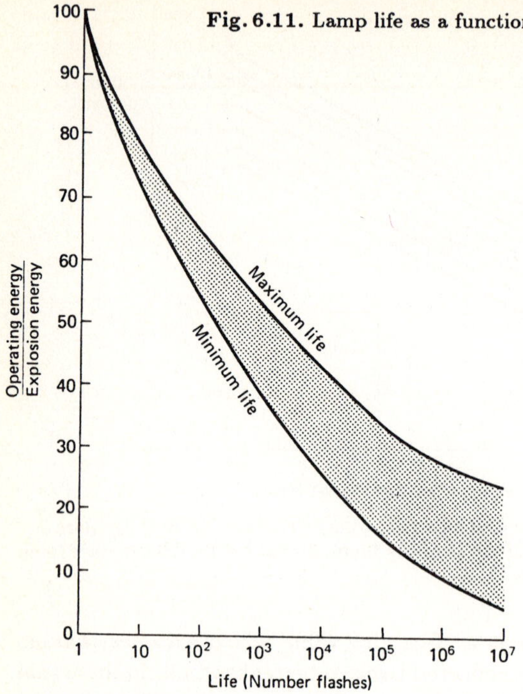

Fig. 6.11. Lamp life as a function of operating energy [%] [6.20]

as the number of shots at which the light output is reduced by 50 % of initial value. The life expectancy of a flashlamp begins to deteriorate rapidly when peak current exceeds 60 % of explosion current. Above 60 %, the lamp's life is usually limited to less than 100 discharges, even if all other parameters are optimized. At the 80 % level, explosion can occur on any discharge.

Excessive Thermal Wall Loading

Cooling of the end seals and the lamp envelope has to be adequate. The amount of average power a flashlamp can handle is determined by the quartz wall area available to dissipate the heat. The envelope of a flashlamp is under stress which is caused by the temperature and pressure differences between the inner and outer walls, and by stresses caused by the acoustical shock wave. The combined stress permissible at the envelope determines the power dissipation per unit area. The total electrical input power of a gas discharge lamp is divided between the heat dissipated by the electrodes, the quartz walls, and the emitted radiation. Typically 30 to 50 % of the lamp input power is dissipated as heat by the quartz envelope [6.21, 22]. Under high-average-power operating conditions the inner wall is close to the melting temperature of quartz, whereas the liquid-cooled outer wall surface is close to room temperature. As will be discussed in Sect. 6.1.2, the maximum heat which can be dissipated through the liquid-cooled quartz envelope of a lamp is 300 W/cm^2.

In flashlamps a temperature rise from 300 to 12,000 K is cexperienced during the pulse, which will produce a 36-fold increase in gas pressure. As a result of the additional stresses caused by the pressure increase and the shock wave, flashlamps are operated considerably below the thermal limit. Conservative numbers assuring good life expectancy are 50 to 100 W/cm^2 for water-cooled lamps.

Nondestructive Failure

Operated well below the explosion limit, flashlamp lifetime is limited by electrode erosion with the subsequent formation of opaque deposits on the lamp jacket near the lamp electrodes or by outgassing or tube leakage. Of these, cathode sputtering caused by the bombardment of the xenon ions onto the negative lamp electrode is probably the most significant. The emitted hot-cathode material deposits on the relatively cool lamp walls and effectively attenuates the amount of radiation released by the plasma. Reduction of the effect of sputtering by treatment of the cathode material with oxides and cesium, as well as better designs of the electrode support structure, including seals to promote more efficient cooling, have resulted in the development of long-life cathodes. Time-dependence measurements of sputtering in a high-repetition-rate flashtube have shown that an activated cathode markedly reduces sputtering compared to a pure tungsten cathode. *Yeamans* and *Creedon* [6.23] have observed an order of magnitude improvement in lamp life due to reduced cathode sputtering by employing a low-level, keep-alive discharge in noble gas lamps during the period between pulses.

Tube blackening which occurs in a new tube after a few shots is a sign of electrode impurity. The greatest hazard to a lamp during its final stages of processing is the introduction of water vapor from the atmosphere [6.24, 25]. If the water vapor is not entirely removed from the lamp, tungsten will react readily with water vapor traces to form WO_2 and atomic hydrogen. The tungsten oxide vaporizes from the electrode and condenses on the cooler parts of the envelope. This reaction often accounts for premature blackening of the envelope walls. Besides darkening of the envelope, the useful life of a flashlamp is limited by erratic triggering and the eventual inability to reignite; this indicates that a significant outgassing or tube leakage has occurred.

In all these cases where the input energy is a small fraction of the explosion energy, the above-listed secondary effects dominate and the life of a lamp is largely determined by the lamp process control and the precautions taken during manufacture. Therefore, (6.13) is less reliable for lifetimes above 10^6 shots. For lamps which are operated in this regime, the useful lamp life, rather than being determind by a mechanical failure, is determined by the reduction of lamp radiation below a useful level. The term "lamp life" is used loosely, since no definite standards exist with regard to the determination of the end of lamp usefulness. Lamp manufacturers usually state lamp life as the number of shots after which lamp intensity drops to 50 % of its initial value. The life expectancy is usually stated for a particular set of operating parameters and for a lamp operated in free air. In laser applications, lamp life is usually defined to end when the laser output has degraded a certain amount, for example

10%, or when the lamp fails to trigger at a specified voltage. Reliability and performance data of large xenon filled flashtubes are summarized in [6.25].

Examples of Typical Operating Conditions of Linear and Helical Lamps

In selecting a particular lamp the following steps are usually followed: (1) A flashlamp arc length and bore diameter is chosen which matches the rod size. (2) The explosion energy for the lamp chosen is calculated (Fig. 6.10). (3) An input energy is selected which is consistent with the desired lamp life (Fig. 6.11). If the lamp input required to achieve a certain laser output is not consistent with a desired minimum lamp life, the application of a multiple-lamp pump cavity has to be considered. (4) The average power loading of the lamp is calculated and compared with the limit specified by the lamp manufacturer. (5) In designing the pulse-forming network, care must be taken to avoid fast current rise times and/or reverse currents. To obtain the expected life, manufacturers recommend that the rise time of the flashlamp current be greater than 120 μs for lamps discharging more than a few hundred joules. Both electrode erosion and wall vaporization are accelerated by fast-rising peak currents.

First we will consider a typical small linear flashlamp used in military equipment such as target designators or laser illuminators. These systems operate typically at a flashlamp input of 10 J with pulse repetition rates of 10 pps. The lasers typically use 63-mm-long Nd:YAG laser rods of 6.3 mm diameter and have a Q-switched output of 100 to 200 mJ. A 4-mm-diameter and 60-mm-long lamp operated at 10 J input and a pulse width of 300 μs is operated at only 2% of the explosion energy. The wall loading of the liquid-cooled lamp operated at 10 pps is 15 W/cm^2. From the extensive data which exist for this type of lamp operated in a closed coupled laser cavity, it follows that the useful lamp life (10% drop in laser performance) is in excess of 10^6 shots. An example of a large linear lamp employed to pump ruby and Nd: glass laser rods is the FX-67B-6.5 made by EG & G. This lamp is 16.5 cm long and has a bore diameter of 13 mm. The lamp has an explosion point of 8450 J for a 1-ms-long pulse. The maximum long-term average power input for the water-cooled lamp is specified by the lamp manufacturer at 10 kW. This corresponds to an effective thermal wall loading of about 60 W/cm^2. If the lamp is operated at an input energy of 2000 J, a flashlamp life of 150×10^3 shots can be expected before the light output is decreased to 50% of its initial value.

A helical lamp with a helix inside diameter of 1.7 cm, a bore diameter of 0.8 cm, and a length of 10 cm is an example of a small lamp employed in commercial ruby and Nd: glass oscillators. Such a lamp can be pulsed with 5000 J in a 1-ms pulse at a repetition rate of 1 pps if the lamp is water-cooled. Under these operating conditions, a lamp life close to 50,000 shots can be expected. Detailed information on particular linear and helical flashtubes are contained in the data sheets issued by the major lamp manufacturers, such as EG & G, ILC, CGE, and Xenon Corp. Recent literature on flashlamp performance can be found in [6.26–30].

6.1.2 Continuous Arc Lamps

For continuous pumping of Nd : YAG lasers, the use of halogen cycle tungsten-filament lamps, noble gas arc lamps, and alkali vapor lamps has been explored. Since the discovery of the Nd : YAG laser in 1964, there has been a continual effort to find more effective means of pumping this material. One of the early devices used for pumping of the Nd : YAG laser is the tungsten-halogen filament lamp which will be discussed in Sect. 6.1.3. The pump lamps considered next were the inert gases, which are rich in line structure in the near-infrared. Xenon has the highest overall conversion efficiency and is commonly used in arc lamps. However, the infrared line spectra of xenon misses all of the Nd : YAG pump bands, which are located at 0.73 to 0.76, 0.79 to 0.82, 0.86 to 0.89, and 0.57 to 0.60 μm. It has been observed [6.21, 23, 31, 32] that the line spectrum from krypton is a better match to Nd : YAG than the line spectrum of xenon, since two of its strongest emission lines (7600 to 8110 Å) are strongly absorbed by the laser crystal (Fig. 6.12).

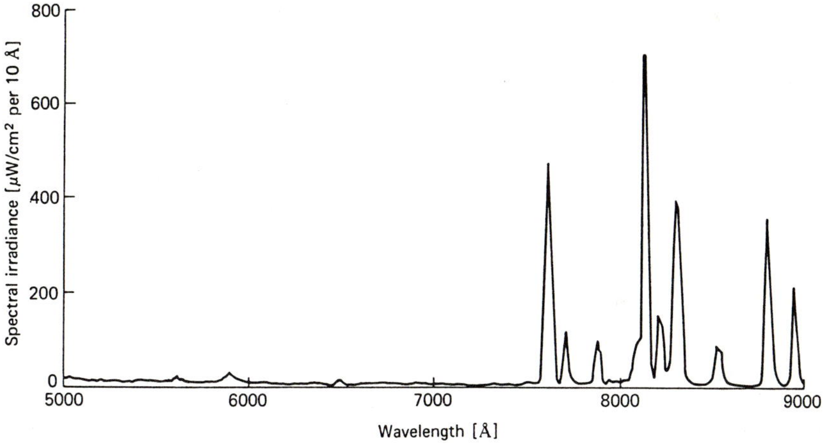

Fig. 6.12. Emission spectrum of a typical cw-pumped krypton arc lamp (6 mm bore, 50 mm arc length, 4 atm fill pressure, 1.3 kW input power). (ILC Bulletin 3533)

The laser output from a krypton-filled lamp is about twice that obtained from a xenon arc lamp operated at the same input power. Krypton arc pumped Nd : YAG lasers are currently the highest-power and the most efficient continuous solid-state lasers.

Mechanical Design. Continuous arc lamps are similar in design to linear flash-lamps (Fig. 6.4) with the exception that the cathode has a pointed tip for arc stability, whereas the anode is rounded. In the type of lamp employing a soft solder seal, additional cooling is provided to the tungsten anode cap by means of a water channel that forces some of the cooling flow through the copper cylinder and into close proximity to the tungsten. In the tungsten rod seal lamps the entire heat must be dissipated through the quartz envelope. In order to reduce thermal stresses in the walls, these lamps are often fabricated from 0.5-mm-thick tubing rather than the standard 1-mm thickness.

Cooling of the lamps is accomplished by circulating water in a flow tube surrounding the quartz envelope of the lamps. Often the flow tube will be made out of Germisil to absorb the ultraviolet content of the lamps. The design of the electrode holders and the flow tube must be such that localized boiling of the water is avoided.

Thermal Considerations. The input power capability of a cw-pumped arc lamp is determined by the permissible stresses in the quartz envelope. The stresses are caused by the temperature gradient between the inner surface of the wall and the outer surface, and by the internal gas pressure during operation of the lamp. Since thermal considerations are of utmost importance in the operation of cw lamps, we will take a closer look at the mechanisms that generate heat at the electrodes and the lamp envelope.

At both electrodes, heat is generated by bombardment from either electrons or ions, conduction of heat from the plasma, and absorption of radiation. The kinetic energy of electrons incident on the anode is very high compared to that of ions impinging on the cathode. Therefore, the anode heats up much more than the cathode. The anode also conducts more heat from the plasma, since the area of the anode that is in contact with the plasma is larger than in the case of the pointed cathode. As far as absorption of radiation is concerned, both electrodes will absorb about the same amount of power. Polished tungsten has a reflectivity of about 60 % in the visible region. Due to emission of electrons from the cathode surface, this electrode is cooled to a certain extent. The heat removed by this process is given by

$$P = \frac{\Delta E I}{e} \ , \tag{6.14}$$

where ΔE is the work function, I is the current, and e is the electron charge. With $\Delta E = 2\,\mathrm{eV}$, $I = 30\,\mathrm{A}$, and $e = 1.6 \times 10^{-19}\,\mathrm{A}$, the heat removed from the cathode is 60 W in a typical lamp.

The total electrical input power of a gas discharge lamp is divided among the heat dissipated by the electrodes, the quartz walls, and the emitted radiation. Generally, it can be assumed that 10 to 20 % of the electrical input power is dissipated as heat through the electrodes and 30 to 50 % is dissipated through the envelope [6.33]. The thermal loading of the tube envelope is caused by the loss of kinetic energy of electrons, ions, and molecules by collision with the wall of the tube. The thermal gradient in the envelope is

$$\Delta T = \frac{Qd}{K} \ , \tag{6.15}$$

where Q is the dissipated power density, d is the wall thickness, and K is the thermal conductivity. Allowing a maximum temperature difference of $\Delta T = 1800°\,\mathrm{C}$ for fused quartz between the inner and outer wall surface, then with

$K = 0.017\,\mathrm{W/°C\,cm}$ and $d = 0.1\,\mathrm{cm}$ there follows for the maximum power which can be dissipated through the quartz wall a value of $Q = 300\,\mathrm{W/cm^2}$. Absorption of radiation in a quartz envelope is negligible. However, vaporization of electrode material leads to a gradual darkening of the lamp envelope. The deposits absorb radiation, hence, the wall-temperature gradient increases during the life of the lamps. The pressure of the fill gas when hot increases by up to a factor of 10. The operating pressure depends on the cold fill pressure of the lamp, the temperature of the plasma, and the volume of dead space behind the electrodes. Thermal and pressure stresses occurring in arc lamps can be calculated using equations given by *Thouret* [6.34]. Calculations carried out on cw arc lamps [6.35] have shown that the inner wall is under compression whereas the outer wall is under tension and the total stress is largest on the outside of the quartz envelope. The rupture stress of quartz is $100\,\mathrm{kg/cm^2}$. Most lamps in commercial applications are operated at about 20 to 30 % of the ultimate tensile strength of quartz.

Electrical Characteristics. Continuous arc lamps are ignited just like flashlamps. For example, a series injection trigger transformer supplying a 30 kV spike will lower the lamp impedance to a level that is within the range of a 300-V dc supply. The operating parameters and impedance depend on current density, arc length, bore size, and fill pressure. Commercial krypton arc lamps are operated between 80 and 150 V and 20 to 50 A. Characteristic impedance values are between 2 and 10 Ω [6.33, 36]. Figure 6.13 shows typical current-voltage curves for several arc lamps filled at 4 atm of krypton. The discharge voltage and dynamic resistance increases with the increase in gas pressure [6.36].

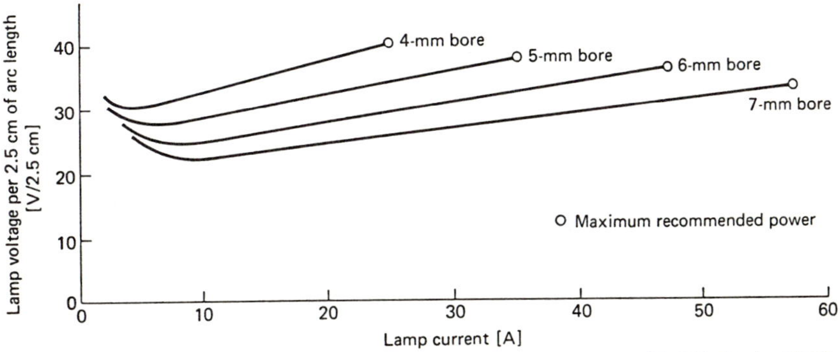

Fig. 6.13. Voltage versus current for different krypton arc lamps. (ILC Bulletin 3533)

Optical Characteristics. At current densities obtainable in standard krypton arc lamps the plasma is optically thin in the region of continuous radiation. Calorimetric measurements of the total optical output power in the spectral region 0.3 to 1.2 μm reveal a radiation efficiency of about 40 % for most Kr arc lamps. Figure 6.14 shows the dependence of the radiation efficiency on lamp input power and bore diameter. It has been found that the useful light output

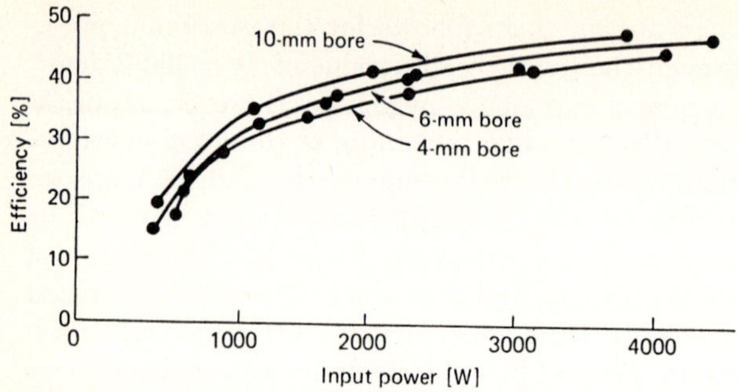

Fig. 6.14. Krypton arc lamp radiative efficiency as a function of input power; 4-, 6-, and 10-mm bore, 75-mm arc length, 4 atm of Kr fill pressure [6.33]

of krypton arc lamps for pumping Nd : YAG lasers increases with bore size, fill pressure, and input power [6.36–38]. For example, the conversion efficiency of a lamp filled with 8 atm of Kr is about 1.2 to 1.5 times as high as that of a lamp filled at 2 atm, depending on the electrical input [6.36]. A summary of the spectral data of a representative Kr arc lamp is given in Table 6.4. Note the large fraction of radiation between 0.7 and 0.9 μm.

Table 6.4. Spectral data for cw krypton arc lamps. Data are typical for lamps having a 6–13 mm bore, 7.5–25 cm arc length, 2–3 atm fill pressure, operated at 6–16 kW

Quantity	Definition	Numerical data
Radiation efficiency	Radiation output/ electrical input	0.45
Spectral output	Fraction of radiation in spectral lines/	0.40
	Fraction of radiation in continuum	0.60
Spectral power distribution	Fraction of total radiation below 0.7 μm/	0.10
	Fraction of total radiation between 0.7 and 0.9 μm	0.60
	Fraction of total radiation between 0.9 and 1.4 μm	0.30

Combining the absorption spectra of Nd : YAG with the emission spectra of the krypton lamp, the spectral utilization as a function of sample thickness has been calculated. The result is shown in Fig. 6.15. The curve illustrates rather dramatically the kind of improvement one can achieve by increasing the diameter of a Nd : YAG laser rod in a pumping cavity.

Lamp Life and Operating Characteristics of Typical cw Krypton Arc Lamps. The main degradation mechanism in these lamps is the accumulation of wall deposits. These deposits either block the radiant energy emitted from the arc or

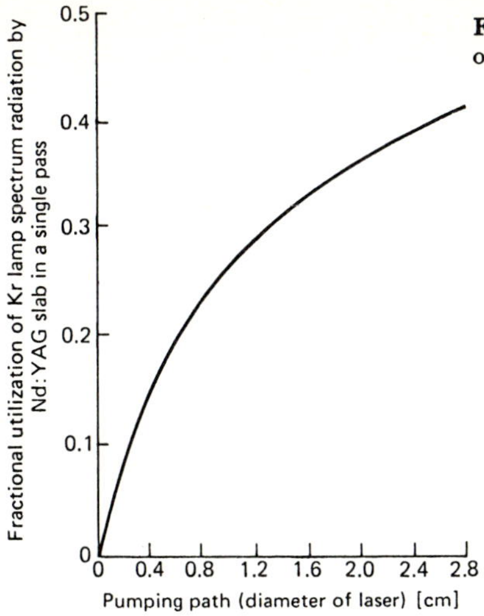

Fig. 6.15. Fractional utilization of krypton lamp output by Nd : YAG

reduce the radiance by enlarging the source diameter. The deposits arise from evaporation and sputtering from the electrodes and from the residual gaseous and high-vapor-pressure impurities left in the lamp after fabrication. The black deposit that builds up eventually completely attenuates the radiative output of the lamp. The underlying quartz becomes overheated, and catastrophic failure of the lamp occurs from excessive thermal stresses developing in the envelope walls. Black anode wall deposits are greatly reduced by the use of an internally water-cooled anode structure.

Krypton arc lamps are available from 5 to 20 cm arc length, with bore diameters from 3 to 10 mm, and are designed to handle electrical input powers of up to 15 kW [6.35]. Design parameters and performance data of typical lamps used in commercial lasers are summarized in Table 6.5. Some of the lamps listed are operated far below the manufacturer's maximum rating in order to obtain good lamp life. Typical overall efficiencies obtainable in krypton-pumped

Table 6.5. Typical operating parameters of cw pumped krypton arc lamps

Model number	FK-125-C2.75 EG&G	FK-111-C3 EG&G	5Kr2 ILC
Arc length	70 mm	75 mm	50 mm
Bore diameter	5 mm	7 mm	5 mm
Fill pressure	2 atm	2 atm	4 atm
Typical input power	3 kW	6 kW	2.5 kW
Lifetime	400–600 h	40–60 h	150–200 h
Wall loading (40 % of electrical input)	110 W/cm^2	145 W/cm^2	128 W/cm^2
Coolant flow rate	120 cm^3/s	120cm^3/s	60 cm^3/s
Electrical characteristics	100 V 30 A	112 V 56 A	84 V 30 A
Current density	150 A/cm^2	140 A/cm^2	150 A/cm^2

Nd:YAG lasers are between 2 and 3%. For example, at output levels of 100 and 250 W, 2.9 and 2.1% efficiencies were achieved with krypton lamps filled to 4 atm [6.32, 39]. Efficiencies of 3.3% have been attained at input power levels of 3 kW with a 6-mm-bore and 50-mm-long lamp filled at 8 atm of krypton [6.33].

Besides the discharge lamps discussed so far, there are a number of special designs, either noble gas filled or containing metal vapor, which have been employed in solid-state laser pumping at one time or another.

Special Noble Gas-Filled Arc Lamps. Solid-state lasers have been pumped by vortex-stabilized arc lamps, coaxial flashlamps, and rf-excited arc lamps. In a vortex-stabilized arc lamp the gas is introduced into the lamp through a ring of jets located near the lamp wall at one of the electrodes. The jets inject the gas tangentially and with an axial pitch angle. The gas passes along the length of the lamp and exits through a hole in the opposing electrode [6.40, 41]. The high-velocity swirling gas constricts the arc, serves to cool the inside wall of the quartz envelope, and carries away any vaporized electrode material. Since the arc is confined to the center of the tube and, therefore, is not in contact with the lamp walls, very high plasma temperatures can be achieved in these lamps. A vortex-stabilized plasma arc lamp filled with argon was employed to pump a cw Nd:YAG laser [6.42]. The highest power achieved was 200 W with a 3-mm × 75-mm rod and 90 kW of electrical power.

In a coaxial flashlamp the laser rod is surrounded by an annular arc area. The plasma is generated between an inner and outer quartz tube and a pair of hollow cylindrical electrodes [6.43, 44]. The laser rod, which can be liquid-cooled, is mounted at the center of the tube. The design of an electrodeless, annular, high-pressure krypton arc lamp was reported [6.45] for pumping a Nd:YAG laser crystal. The arc lamp, which was fabricated in the form of annuli with the laser rod on axis, was excited with a 27-MHz driving signal. The lamp was enclosed by a diffusely reflecting ceramic tube. At a 1.5-kW input power into the lamp, a laser output of 24 W was obtained.

Mercury Arcs. In mercury arc lamps an arc discharge takes place in an argon and mercury gas mixture. After ignition of the lamp a small amount of mercury is vaporized. This produces a high internal gas pressure and an extremely bright arc discharge. The properties of mercury arc lamps have been described in detail in [6.46]. The pressure-broadened line spectrum of mercury high-pressure tubes almost completely overlaps the absorption spectrum of ruby. The highest brightness is attained by the type A long arc tube which is particularly suitable for cw ruby lasers [6.47]. The lamp has an arc diameter of 1 mm and an arc length of between 28 and 50 mm. The maximum input power is 560 W/cm. Long arc tubes must be water-cooled in the usual way. For periodically pumped small ruby lasers, with a pulse repetition rate between 40 and 110 Hz, a type SPP1000 (Philips) mercury lamp has been used. The lamp, which is employed in movie projection systems, can handle an average power of 1 kW. A ruby, 25 mm long with 2 mm diameter, could be pumped with a pulse energy of 12 J. At the maximum repetition rate of 110 Hz an average output power of 2 W was obtained [6.48].

Doped Mercury Arcs. Mercury arcs can be doped with various elements to modify the spectral output of the lamps while maintaining the good mercury-arc characteristics [6.49, 50]. Doped Hg arc lamps are now used in almost all high-intensity lighting, such as street lights, sports arenas, etc. These lamps fall into two categories, metal-doped Hg arcs and metal iodide-doped Hg arcs.

The metal iodide-type lamps are very similar to conventional Hg arc lamps. Fused-quartz lamp jackets are used and the lamps operate at temperatures and pressures typical of the low-pressure Hg arcs. Metal iodide lamps produce a narrow line emission due to the lower pressures. Little pressure broadening and reversal occurs. Rb-I has two strong lines near 8000 Å and is a possible Nd : YAG pump lamp [6.51].

The metal-doped Hg arc lamps must be operated at temperatures above the boiling point of the metal dopant. The pressure of these lamps is also very high, and the hot metal is usually very corrosive on the lamp jacket. Therefore, single-crystal sapphire is normally used as the lamp envelope material. The output spectra of sodium-doped Hg lamps are typical of sodium, and the Hg emission is almost completely suppressed. The high operating pressure considerably broadens the sodium emission lines. Sodium also strongly reabsorbs its own radiation. This produces a "reversal" in the emitted radiation. The wings of the broad spectral line are emitted, but the center is almost completely self-absorbed. Sodium, with mercury and argon additions as buffer and starting gases, was used to pump the 5887-Å line of Nd : YAG with its self-reversed resonance radiation. A 0.7 % overall pumping efficiency was achieved.

The potassium D lines are situated almost exactly between the two Nd : YAG pump bands. As a line emitter, potassium would be very poor as a Nd : YAG pump lamp. In the high-pressure metal-doped arc, however, the broadened self-reversed K lines have two wings that cover the two Nd : YAG pump bands. Typically in a potassium-mercury-argon vapor lamp the Nd : YAG absorption bands at 0.73 and 0.76 μm and at 0.79 to 0.82 μm are matched to the broadened self-reversed resonance lines of potassium. With such a lamp, a laser output of 10 W, with a lamp input power of 420 W, was obtained [6.52]. One problem with the Na-Hg, K-Hg, and Rb-Hg lamps is that these lamps must be operated with ac current to prevent cataphoresis from separating the alkali metal and the Hg in the lamp. The resulting modulation of the laser output beam is in many applications very undesirable.

Alkali Arc Lamps. Potassium arc lamps containing no mercury were developed as pumps for Nd : YAG lasers [6.53]. This spectral emission is identical to K-Hg arc lamps without the complications arising from the presence of two components. Measurements performed on dc-operated alkali lamps indicated that the potassium-rubidium spectrum matches the excitation bands of Nd : YAG better than potassium or rubidium alone. The laser threshold with the potassium-rubidium lamp is the lowest of all lamps tested. The efficiency of the potassium-rubidium lamp is higher than that of the krypton lamp at power levels up to 1000 w, and is of the order of 3 to 3.5 %. Approximately 38 % of the input energy is radiated in the 0.7- to 1.0-μm region [6.54]. Because the alkali metals are solids at room temperature, the lamps are operated at elevated temperatures, typically between 1000 and 1200° C. Pressure and spectral regulation of the lamp are achieved by carefully controlling the alkali metal reservoir temperature. Potential applications for alkali vapor lamps include space laser communication systems, where low-threshold and low-power (100 to 200 W input) devices are required [6.40, 55].

At present, lamp technology does not permit alkali vapor lamp wall loading to exceed about 20 % of typical krypton arc lamp wall loading. The application of alkali metal vapor lamps in pulsed Nd : YAG lasers is described in [6.37].

6.1.3 Tungsten-Filament Lamps

The tungsten-filament lamp was the first and still is today a very common lamp used for pumping Nd : YAG in the cw mode of operation. Because of the development of the halogen cycle, these lamps are compact and can operate for appreciable periods near the melting point of tungsten. Tungsten-halogen lamps are very easy to operate and are low in cost. They are used in television, motion picture, and theater lighting and have been mass produced for some time. Electrically, they represent a simple resistive load and require an uncomplicated power supply. However, the lamps have a few characteristics that seriously limit their application. First, the pumping efficiency is low compared to krypton arc lamps. Second, readily available tungsten-halogen lamps are limited to 1.5 kW input power. Third, the lifetime of these lamps is typically less than 100 h when operated in a laser cavity.

The type of tungsten-halogen lamp most commonly used to pump cw Nd : YAG lasers is a 1-kW lamp with a filament length of 63 mm (General Electric Q100T3/1CL, Sylvania FCM 120 V) and a 1.5-kW lamp (General Electric Q1500T4/4CL) with a 57-mm-long filament. Both lamps have a color temperature of 3200 K and an input voltage at the rated power of 120 V. The smaller lamp has a filament diameter of 2 mm, whereas the 1.5-kW lamp has a filament diameter of 3 mm. The lamps are filled with either iodine or bromine gas. The filament, which is centered in the bulb by spiral-shaped supports, is connected to the external contacts by means of a molybdenum foil ribbon that is thin enough to deform rather than fracture the bulb.

During normal operation of any incandescent lamp, tungsten is evaporated from the hot filament and deposited on the cold inside wall of the lamp jacket. The halogen cycle chemically converts the tungsten atoms boiled off the filament into tungsten halide molecules in the low-temperature region near the lamp walls. These molecules diffuse back to the tungsten filament where the high temperature dissociates them, thereby depositing tungsten back on the filament and freeing the halogen gas (iodine or bromine) to repeat the cycle.

The halogen cycle allows operation of tungsten-filament lamps at much higher temperatures than is possible in vacuum jackets. Typically, tungsten-iodine lamps operate at 3200 K with lifetimes of 100 h compared to 2800 K temperatures for similar lifetime, vacuum-jacketed lamps. In order to keep the halogen cycle operative the bulb wall temperature must be above 250° C. Furthermore, the bulb temperature should be fairly uniform across the lamp. This, for example, necessitates that the lamps be operated in a horizontal position.

Despite their simplicity, operating the lamps inside a laser cavity requires some precautions to avoid seal failures, cracks, and overheating of the quartz envelope. The two principal modes of failure of these lamps employed inside a laser cavity are seal failure, and softening and subsequent bulging of the bulb. Overheating of the lamp envelope can be avoided by forced-air cooling of the lamp or by at least providing some circulation of air inside the laser cavity. If the seal is operated at a temperature in excess of the maximum safe limit of 340° C, the molybdenum oxidizes and expands. This can cause a failure of the seal between the lead and the quartz, permitting air to enter the lamp. Internal cracking of the quartz is usually accompanied by a series of concentric rainbow-hued rings. In most commercially available tungsten-pumped Nd : YAG lasers heat is removed from the lamp seal area by water-cooled contacts. It has also been found that allowing the lamp to warm up before the full input power is applied, reduces the frequency of seal failures. This can be attributed to the fact that thermal shocks to the seals are alleviated by a slow turn-on procedure. Other problems, such as filament sag or premature blackening of the lamp, can normally be attributed to a manufacturing problem rather than improper operating conditions. The former problem is usually caused by a tungsten wire or filament of inferior quality and the second problem is usually traced to an insufficient amount of iodine or bromine in the lamp.

Spectral Data. Tungsten-filament lamps typically operate with filament temperatures of 3200 K, which corresponds to an emission maximum occurring at 8400 Å for a blackbody radiator. Actually the tungsten-filament lamp is a graybody source having an emissivity of 0.35 at 3200 K [6.56]. Small amounts of iodine, though enough to maintain the iodine cycle, have very little effect on the spectrum [6.57]. Since the radiation from such a lamp is a broad-band continuum and the pump bands of Nd : YAG are quite narrow, only moderate laser efficiencies can be expected from filament lamps. Figure 6.16 shows the

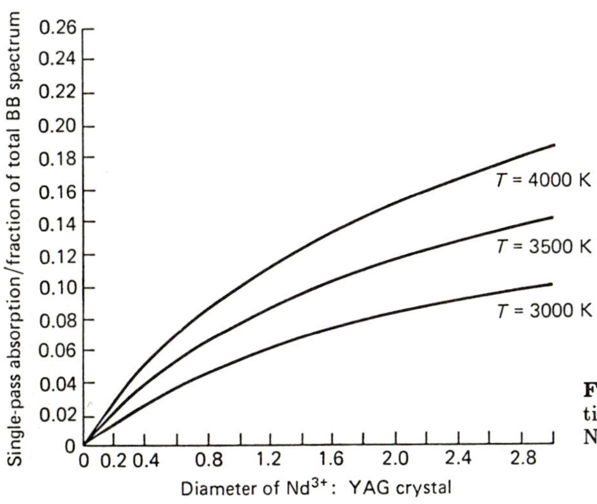

Fig. 6.16. Single-pass absorption of blackbody spectrum in Nd : YAG vs path length

result of a point-by-point computation of the absorption of Nd : YAG for blackbody spectra at various temperatures and for various thicknesses of Nd : YAG. The absorption depends strongly on the absorption path length. This is caused by the wings of the narrow absorption lines in Nd : YAG, which contribute significantly to the overall absorption if the path length is increased.

It will be noted that for a typical tungsten lamp operated at 3200 K and a Nd : YAG rod a few millimeters in diameter, the absorbed pump power in the crystal is of the order of a few percent of the total blackbody spectrum. Consistent with these considerations is the fact that the maximum observed efficiencies obtained from tungsten-pumped cw Nd : YAG lasers are of the order of 1 to 1.5 %. For example, using a spherical cavity and a filament lamp at a color temperature of 3200 K, *Church* and *Liberman* [6.58] obtained 1.0 % overall efficiency and a slope efficiency of 2.4 % at a power level of 1000 W.

6.1.4 Laser Diodes

The most promising alternative to flashlamp pumping of solid-state lasers has always been the diode laser. Throughout the last twenty years numerous laboratory devices have been assembled which incorporated single diode lasers, small laser-diode arrays or LED's for pumping of Nd : YAG, Nd : glass and a host of other Nd lasers. The low power output, low packaging density, and extremely high cost of diode lasers prevented any serious applications for laser pumping in the past. Overviews and summaries of early work in this area can be found in [6.59–62]. The digests of the Conference on Lasers and Electro-Optics (CLEO '86, CLEO '87) provide excellent surveys of current activities in laser-diode-pumped solid-state lasers [6.63].

The reason for the continued interest in this area stems from the potential dramatic increase in system efficiency and component lifetime, and reduction of thermal load of the solid-state laser material. The latter will not only reduce thermo-optic effects and therefore lead to better beam quality, but will also enable an increase in pulse repetition frequency. The attractive operating parameters combined with the low-voltage operation and the compactness of an all solid-state laser system have a potentially high payoff.

The high pumping efficiency compared to flashlamps stems from the good spectral match between the laser-diode emission and the Nd absorption bands. Actually, flashlamps have a higher radiation output to electrical input efficiency (70 %) compared to laser diodes (25–50 %), however, only a very small fraction of the blackbody spectrum is usefully absorbed by the various Nd bands.

A concomitant advantage derived from the spectral match between the diode-laser emission and the long-wavelength Nd absorption-band is a reduction in the amount of heat which is deposited in the laser material. In addition, system lifetime and reliability will be higher in laser-diode-pumped solid-state lasers as compared to flashlamp-based systems. Laser-diode arrays have exhibited lifetimes on the order of 19,000 h in cw operation and 10^9 shots in the pulsed mode. Flashlamp life is on the order of 10^7 shots, and about 200 h for cw operation. In addition, the high pump flux combined with a substantial

UV content in lamp-pumped systems causes material degradation in the pump cavity and in the coolant, which lead to systems degradation and contribute to maintenance requirements. Such problems are virtually eliminated with laser-diode-pump sources. The absence of high-voltage pulses, high temperatures and UV radiation encountered with arc lamps lead to much more benign operating features of laser-diode-pumped systems.

Laser-diode technology dates back to 1962 when laser action in GaAs diodes was first demonstrated. However, it took a decade to transform a fragile device requiring cryogenic cooling into one capable of emitting a continuous beam at room temperature. The pace of development has accelerated greatly over the past decade making diode lasers one of the fastest-moving laser technologies. In particular, device improvements such as double heterostructures, multiple-quantum-well structures, gradient-index single-quantum-well structures, monolithic phased arrays, multiple-stripe lasers which were made possible by improved manufacturing technologies such as LPE and, in particular, MOCVD have led to a dramatic reduction of threshold current, increase of slope efficiency, lifetime and output power [6.64–69].

At present, the most important application for single-diode lasers is in optical-disk recording. Fueled by strong consumer acceptance of compact-disk players, sales of 780 nm AlGaAs diode lasers grew to several million per year. Single-diode lasers utilized in these commercial applications are not practical for solid-state laser pumping due to the low power outputs. However, in parallel with the mass production of single laser diodes in Japan, went a development in the USA which is aimed at the fabrication of powerful monolithic arrays. The development of diode arrays follows two different pathways. By fabricating multiple stripes on a single GaAs substrate with small enough stripe-to-stripe spacing, one can produce an array which "inherently" phase locks. The output from a phase-locked array can be focused to an almost diffraction-limited spot. Industry motivation to develop phased arrays stems from the potential applications in read/write optical memories and laser printers.

A technological spill-over of phase-locked arrays is the development of incoherent arrays. Higher power levels, and more stripes on a single chip can be obtained if the concept of phaselocking is abandoned. Significant progress has been made recently in developing the monolithic, linear laser-diode array. Output power, slope efficiency, laser threshold and wavelength control have all been dramatically improved due to a combination of new structures and MOCVD growth techniques.

Incoherent arrays are the building blocks for solid-state laser pumps. In contrast to applications for data storage or printing, laser pumping does not require a small focusable spot size, spectral purity, and low noise. Incoherent arrays are lacking in these respects, and their main application is for laser pumping. State-of-the-art performance of linear arrays is up to 40 W/cm. Commercially, devices with 10 to 20 W/cm are available. The performance of planar arrays, produced by stacking linear arrays, has reached the impressive level of 1 kW/cm^2. Efficiencies of these devices range from 25 to 50 %. The development of diode arrays has opened up the possibility of creating practical all solid-state lasers, with potential advantages of compactness, high efficiency and long lifetime.

The prototype linear and planar arrays built to date, and the laser performance which has been achieved, have clearly demonstrated that diode pumping is no longer a question of technical feasibility, but rather a question of economic viability. High cost is the major obstacle which prevents broad acceptance of diode-pumped solid-state lasers. However, a recent study [6.62] has shown that the manufacturing, assembly and testing of diode arrays can be fully automated provided that the production volume will justify the capital investment. The change-over from the currently very labor-intensive fabrication to an automated manufacturing capability will decrease the cost per watt of output power by several orders of magnitude. There are indications that military programs will provide the required funding for this transition to occur. In this case, one can envisage gas-filled arc lamps being replaced by linear and planar laser diode arrays, just as vacuum tubes have been replaced by the transistor and later by integrated circuits.

The design and performance of laser diode arrays suitable for solid-state laser pumping will be discussed in this subsection.

Single-Diode Design

The structure of the first diode lasers was a homojunction. Such diodes typically had threshold current densities of 40,000 A/cm^2 and quantum efficiencies of 20 %. Homojunction diode lasers are made entirely of a single semiconductor compound, typically gallium arsenide, with different portions of the device having different dopings. The junction layer is the interface between n- and p-doped regions of the same material.

The development of GaAlAs-GaAs single heterojunction laser diodes offered a very significant improvement. The threshold current density decreased to 8000 A/cm^2, and the quantum efficiency increased to 40 %. In a single heterojunction diode, the active layer is sandwiched between two layers of different chemical compositions, typically GaAs and GaAlAs, which have different band gaps. A further reduction in threshold current density (1000–3000 A/cm^2) was achieved with a double-heterojunction structure. The double-heterojunction lasers have an active layer bound by two layers of different material. Usually the double-heterojunction laser is formed by a multiple-layer epitaxial process and consists of a GaAlAs-GaAs-GaAlAs structure, with a thin (about 0.5 μm) GaAs active region.

The most recent developments in high-efficiency cw pumped laser diodes are quantum-well structures (6.70–72). Quantum-well (QW) lasers are a special class of diode laser for which the active-layer thickness is less than 30 nm – the order of the de Broglie wavelength of electrons. Due to the characteristically small dimensions of QW lasers, the injected carriers are subject to quantum effects, that is, the conduction and valence bands, normally considered a continuum, become quantized. QW structures allow a better utilization of carriers for radiative transitions. As a result, QW lasers have a threshold which is sev-

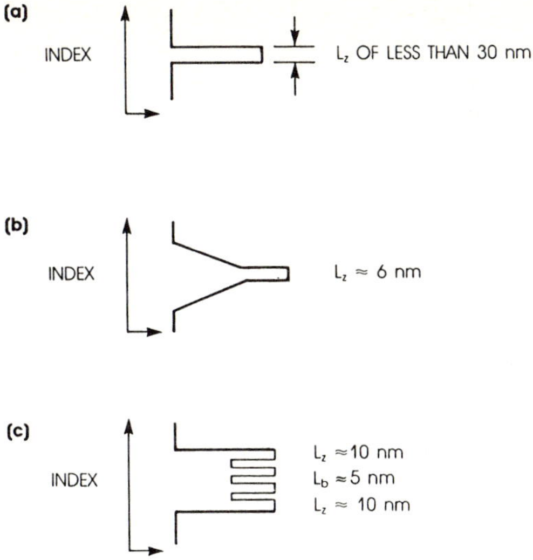

Fig. 6.17a–c. The most common types of quantum-well structures are the single QW double heterostructure **(a)**; graded-index separate confinement heterostructure **(b)**; and multiple QW structure **(c)**. L is the QW width; L_b is the barrier thickness

eral times less than that of conventional double-heterostructures (DH). The most common types of QW structures are shown in Fig. 6.17. They are the single-quantum-well (SQW) DH, the graded-index separate confinement heterostructure (GRINSCH), and the multiple-quantum-well (MQW) structure.

In an MQW structure the active layer is divided into a number of extremely thin sublayers with different band gaps. The potential wells thus formed help to improve the optical coupling between neighboring laser channels. Multiple-quantum-well structures require the ability to prepare ultra-thin GaAs and $Al_xGa_{1-x}As$-layers. Wavers with different numbers of wells, and different well and barrier thicknesses are being prepared by several companies. Commercial devices offered contain as many as 15 layers. The semiconductor layer structure consists of the substrate, cladding layer, quantum well layers, barrier layers, etc. The active zone typically consists of four 13 nm thick GaAlAs wells and three 4 nm thick barrier layers.

The ability to profile the GaAs composition of the epilayers by MBE or MOCVD techniques also made possible the preparation of a laser with graded-index waveguide and separate carrier and optical confinement heterostructure (GRINSCH). Such structures provide not only separate confinement of light and carriers to further optimize the threshold current, but also an arbitrarily graded-index profile outside the carrier confinement region. Instead of trapping electrons in a multiple-well structure, the GRIN technology is based on a graded-index structure to collect electrons and funnel them into a single quantum well. The graded-index region also provides the waveguide structure for the laser action.

The GRINSCH technology represents the state-of-the-art in cw laser-diode production. In order to produce single quantum-well structures, one needs special MOCVD equipment, which is capable of making very sharp transitions. The quantum-well structure essentially represents a monatomic layer embedded in thick layers of different compositions. The performance of laser diodes based on the graded-index, single quantum-well structure is about a factor of two better than multiple-quantum-well designs. For example, quantum efficiencies at room temperature of 50 to 60 % are routinely achieved. The output from a 60 μm stripe is as high as 1.5 W. For double heterostructures the output is typically 10 mW/μm. The damage threshold is considerably higher in single-quantum-well structures due to the lower current density than in multiple-quantum-well structures.

The GRIN structure is the latest development in the evolution of laser diodes which started with homojunction, single and double heterojunction and led to multiple- and single-quantum-well structures. During this evolutionary process quantum efficiency has been increased from a few percent to over 70 %.

The peak power from a laser diode is limited by catastrophic optical damage. The destruction of the laser facet is caused by absorption of laser radiation, which can reach intensities of several megawatts per square centimeter at the facet. Because catastrophic optical damage is determined by the power density, one can increase the total power of a laser by spreading the beam over a large area of the output facet, or by increasing the damage threshold of the facet.

Structures which widen the optical mode perpendicular to the plane of the junction include the large optical cavity, the thin active layer and the single-quantum-well/separate confinement heterostructure. In these structures the optical wave is weakly guided by a thin layer, causing the mode to spread into the cladding layers. As a result, the thickness of that region can be increased to reduce the optical flux density in the active region without reducing the external quantum efficiency. Increasing the thickness of the active region will, to some extent, lead to higher output powers. However, in order to effect order-of-magnitude improvements, the beam must be expanded laterally within the active region. In principle, the power can be spread out parallel to the junction simply by increasing the active stripe width.

The output power of a laser diode increases linearly with the stripe width. Therefore, one could consider preparing a wide-stripe laser with a width of several hundred micrometers. Such broad-area lasers do emit substantial power but, as a result of optical-field injected-charge interactions, they often lase in a multitude of incoherent filaments. Also, the large waveguide can support several transverse modes which generally change as the driving current is increased.

The relatively poor optical quality of these wide-stripe devices is not a hindrance to their use as pump sources. However, the device susceptibility to uncontrollable filamentary operation can lead to localized damage of the faces and substantially reduce the life of the diodes.

In order to achieve more stable laser operation, the broad active stripe can be divided into several nominally single-mode stripes. Multistripe structures or linear arrays, which are the building blocks of laser pumps, will be discussed below.

Linear Arrays. The high structural and compositional uniformity of LPE-, MOCVD- or MBE-grown devices allows the integration on one and the same substrate of a large number of semiconductor lasers arranged side by side. The stripes are formed by a spatial variation of injected current (gain-guided) or material composition (index-guided). The main attraction of stripe-geometry lasers is their ability to limit the number of modes in which the laser can oscillate, with the general goal being a single spatial mode and better output quality than the multimode beams produced by wide-stripe diode lasers.

Multistripe systems can be fabricated as coherent or incoherent arrays. In the coherent design each stripe operates in the fundamental transverse mode and is evanescently coupled to the adjacent stripes. Since the stripes are in close proximity to each other and since the gain-guides do not tightly confine the light laterally, the optical mode beneath each stripe couples via evanescent waves to its neighbors. Thus, the array of optical fields becomes locked in phase, and, if the phase difference between adjacent stripes is zero, the lateral radiation pattern consists of a single narrow lobe. For solid-state laser pumping, incoherent arrays are usually employed because the radiation pattern does not have to be tightly imaged into a small spot.

Incoherent arrays have an optically insulating layer between the active stripes. Therefore each stripe represents an independent layer. This has the advantage that a very large number of stripes can be built on a monolithic substrate. For example, a 25 W device produced by Spectra Diode Labs has 1000 stripes. In coherent arrays, the total width cannot exceed a certain limit, otherwise spontaneous emission and lateral lasing will occur.

Incoherent arrays can be stacked in length to any practical size. Typically, linear arrays are made from modules mounted on a common heat sink. Figure 6.18 shows a schematic diagram of a linear array comprised of 40 stripes.

The catastrophic optical damage mechanism mentioned earlier, which results from absorption of the active region at the facet, can be avoided in a new class of structures: they are the nonabsorbing facet or "window" lasers. In these devices the active layer does not extend to the facet [6.74]. Since the active layer is isolated from the surface, no absorbing regions exist and catastrophic optical damage is avoided.

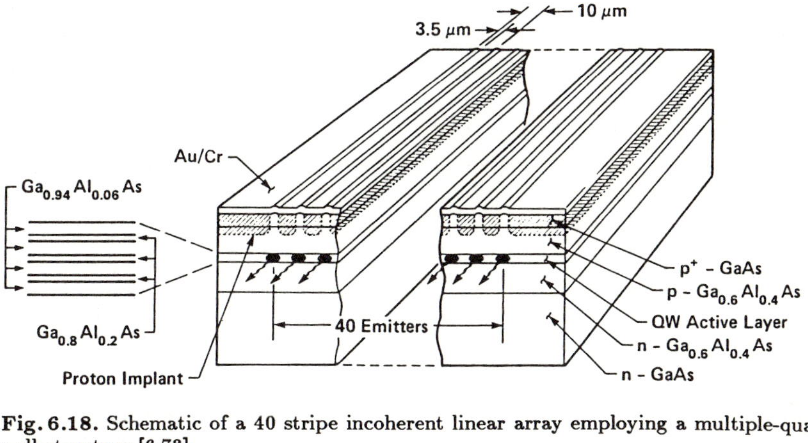

Fig. 6.18. Schematic of a 40 stripe incoherent linear array employing a multiple-quantum-well structure [6.73]

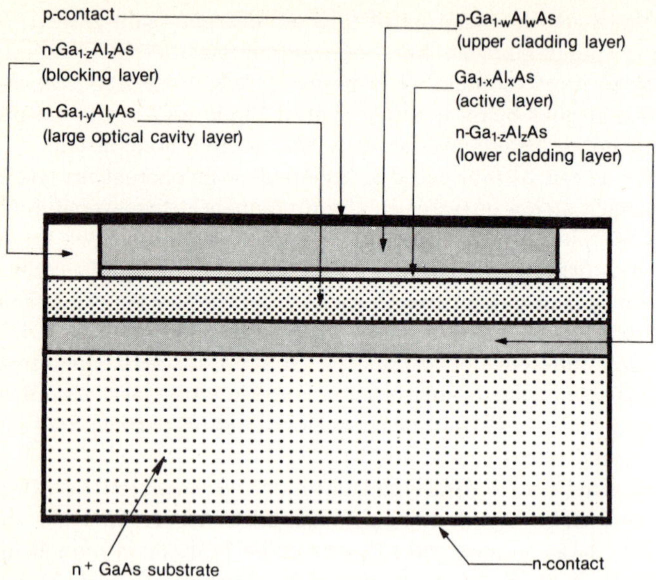

Fig. 6.19. Side view of the window large optical cavity–buried heterostructure (LOC-BH) laser [6.74]

Figure 6.19 represents a diagram of one such geometry. The interior of this structure is identical to that of a non-window heterostructure laser and consists of a thin (approximately 100 nm) n-type GaAlAs layer sandwiched between upper and lower cladding layers of p- and n-type GaAlAs, respectively. The aluminum content of each of the cladding layers is higher than that of the active layer. This double heterostructure is embedded in a current-blocking structure consisting of an n-GaAlAs layer overlying one of the p-GaAlAs, forming a reverse biased p-n junction. The aluminum concentrations of the blocking layers are similar to those of the cladding layers. The lower aluminum fraction of the active layer gives it a higher refractive index than its neighboring layers, resulting in transverse confinement for light traveling along the laser cavity. The window laser differs from the conventional buried heterostructure (BH) laser in the facet region, where the double heterostructure is replaced by AsAlAs of the same aluminum concentration as the blocking layers. At the laser's output wavelength, the absorption coefficient of this material is smaller than that of unpumped active-layer GaAlAs by several orders of magnitude. The facet can therefore be considered for all practical purposes to be transparent, and optical damage is eliminated.

The combination of monolithic multistripe laser diodes with a window structure may be the optimum configuration for solid-state laser pumping. In this way, the emitting surface is increased and the damage threshold is raised.

Planar Arrays. The only viable approach so far, demonstrated for fabricating high-power 2-D arrays of diode lasers, begins by subassembling extended linear arrays or "bars" (up to about 1 cm) of individual stripe lasers, and then stacking these linear arrays to form a 2-D array. Details of the fabrication and performance of these arrays have been described in [6.75, 76]. The process starts by growing a sequence of layers of AlGaAs on a GaAs substrate using MOCVD. These layers serve to provide carrier and optical confinement to achieve good lasing efficiency. Next the stripe-laser patterns are produced photolithographically and the contact metallizations are deposited. The wafer is then separated into 1 cm long, linear arrays by cleaving along crystal planes to form the laser cavity mirrors. Dielectric coatings are applied to the facets as soon as possible to inhibit degradation and to prevent light from being lost out the back facet. The laser bars are next soldered onto mounting plates made of a material with high thermal conductivity and with a coefficient of expansion that matches GaAs. Examples of such materials are BeO and Cu/W composites. A spacer plate is then attached to the mounting plate to provide adequate clearance between layers in the 2-D stack. The spacers can also serve as wire-bonding pads and as additional thermal conduction paths to remove heat dissipated in the laser bars. Finally, the linear subassemblies are stacked to form the 2-D array.

The basic method of stacking up linear arrays to create 2-D arrays requires a large number of fabrication processes. An intuitively appealing alternative approach is to fabricate two-dimensional arrays directly at the wafer stage, thereby substantially reducing the manufacturing complexity. At present, there is only limited research being conducted on such "surface emitting" laser structures, and the results to date have been relatively poor.

Performance of Linear and 2-D Arrays

In this subsection we will briefly discuss currently realized, as well as near-term projected, performance levels of laser diode arrays suitable for solid-state laser pumping. Parameters of particular interest are power output, efficiency, wavelength and lifetime.

Output Power. Through a combination of improved device technology, and integration of an increasing number of devices into arrays, laser diodes have achieved ever higher optical output levels in recent years. The power levels of laser-diode arrays mentioned below are for quasi-cw laser diodes, i.e., for pulse widths longer than a few microseconds. Since pumping of a Nd:YAG laser requires pulse length on the order of 200 μs, potential diode-pump sources for Nd lasers operate in the quasi-cw mode. Diode lasers capable of very high peak power have been in existence for many years, but these devices typically operate only under short pulse (10 to 100 ns) and low-duty-cycle conditions. As the pulse width of a typical GaAlAs diode laser is increased, its maximum peak power capability decreases, as shown in Fig. 6.20. The cause of this limitation is catastrophic facet degradation, a process whereby optical energy is absorbed near the facet, initiating a thermal runaway condition that raises the local temperature of the facet to the melting point. Because of the small volume

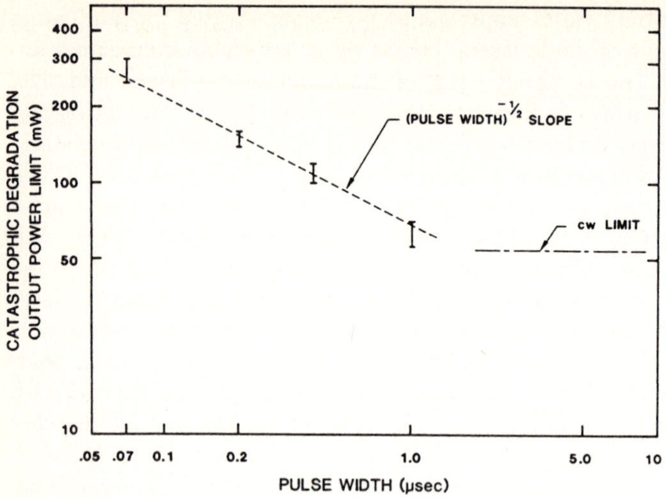

Fig. 6.20. Catastrophic degradation power vs pulse width for 6 μm wide single-stripe AlGaAs lasers [6.77]

of material involved in this process, catastrophic facet damage can occur in less than one microsecond. For pulses longer than a few microseconds, the heat generated by optical absorption diffuses out of the laser material, and the catastrophic damage limit for diode lasers operated in long-pulse mode approaches the cw damage limit. Thus, long pulse operation ($\gg 1\,\mu$s) is also called quasi-cw operation. Figure 6.21 shows the peak power versus pulse width of a commercial diode array.

We also note that the structures discussed below have incoherent outputs, because the stripes are separated by isolation regions, which prevent phase locking. By forming these laterally isolated regions, high-power laser-diode arrays can be fabricated by merely adding more stripes. Otherwise, as the width of the laser approaches and exceeds the cavity length, transverse lasing and amplified spontaneous emission compete for carriers, and the desired laser action is diminished.

A state-of-the-art array is a 1,000 stripe device with a 25 W output in a 200 μs pulse across a 1 cm bar. In this design the individual stripe produces 25 mW, which is low enough to guarantee a good lifetime. Another design

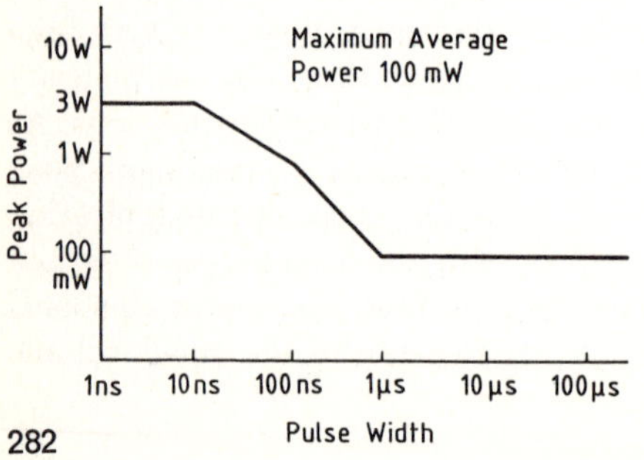

Fig. 6.21. Peak power vs pulse width for commercial diode arrays (Spectra Diode Lab SDL-2410)

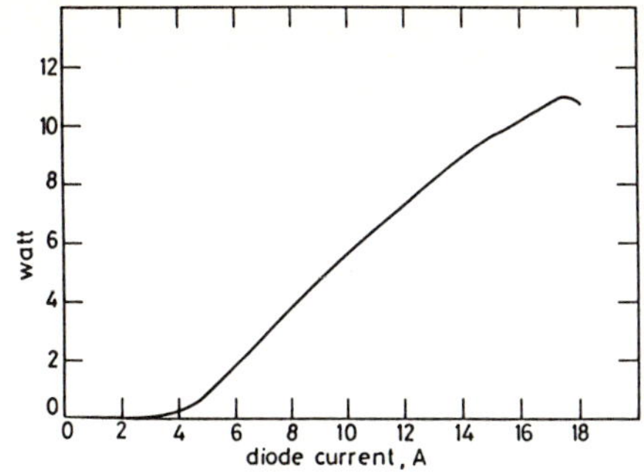

Fig. 6.22. Optical output power vs diode current for a 200 stripe monolithic array [6.78]

having a lower packaging density (20 % of the bar length) produces 11 W. This power level is generated by twenty, 10-stripe lasers. Each stripe occupies 100 μm of facet length, and the stripes are spaced on 500 μm centers. The device was operated at a repetition rate of 50 Hz by applying current pulses 150 μs long. Figure 6.22 shows a plot of output power versus diode current. The maximum peak power was 11 W at 17.3 A and the best slope efficiency was 0.96 W/A. This corresponds to a 61 % quantum efficiency. The maximum conversion efficiency of optical power output to electrical power input was 27 %.

In order to illustrate the progress made with multistripe laser diodes we have listed the progression of output powers achieved from commercial devices over the last two years.

Number of stripes	Aperture width [mm]	Output power [W]
40	0.4	2.5
140	0.8	4.0
100	1.0	5.4
200	10.0	11.5
1000	10.0	25.0

A ten-fold improvement in power was achieved within two years. From the data shown above, we see that the power output per stripe ranges from 25 to 60 mW. As was discussed before, facet damage sets the upper limit for the power emitted from one stripe. However, if the gain-guided, double-heterojunction multiple "stripe" geometry is combined with the window-laser approach, 100 mW per stripe seems quite achievable. This will lead to 1 cm bars capable of 100 W output in a 200 μs pulse. Similarly, high output powers are projected for gradient-index, single-quantum-well structures which have a higher damage threshold compared to double heterostructure devices.

Only one company, McDonnell Douglas, has currently the fabrication capability to produce stacked 2-D arrays. The company has developed the technology to produce arrays with an output power of 1 kW/cm^2 for 200 μm long pulses [6.75].

For example, one 0.5 cm^2 array produced 475 W peak power in a 200 μs long pulse. The array is made up by stacking 20 bars, each 1 cm long. Each bar produces 24 W and is fabricated from 120 μm wide, 10 stripe modules. About 80 modules are mounted on a 1 cm bar.

The largest 2-D stacked array built to date comprises seven subarrays each having an area of 0.55 cm^2. Each subarray produces a peak power of 413 W in a 216 μs long pulse, therefore a total energy per pulse of 620 mJ is achieved from the 3.85 cm^2 array. The electrical input into each subarray is 58 A at 30 V which results in an optical output to input efficiency of 24 %. The array was employed to pump a Q-switched Nd:YAG slab laser. The power density of about 1 kW/cm^2 which can be achieved from these arrays is limited by the laser-diode fabrication technology. The achievable average power is determined by the heat sink. The arrays mentioned above are typically pulsed at 50 Hz, i.e. the duty cycle is around 1 %. Therefore the average power from an array is typically 10 W/cm^2.

Utilizing advanced heat sink designs it is expected that this limit can be raised to 100–200 W/cm^2. The output levels mentioned above have been achieved with arrays operating at 20–25 % overall efficiency. Clearly if the efficiency can be increased (i.e., the window laser, GRIN structures, etc.) the heat loading will decrease accordingly and higher average powers can be achieved.

Efficiency

The energy conversion efficiency of a laser diode depends on:

a) internal quantum efficiency,
b) electrical series resistance,
c) threshold current, and the
d) operating point of the laser relative to the threshold current.

Improvements in the first three categories can be achieved by optimizing the internal structure of the laser through parameters such as layer thickness, composition and doping concentration. Item d) requires that the laser be operated at very high output power densities. This is illustrated in Fig. 6.23 with plots of power output and power conversion efficiency versus current for a 100 μm wide, single quantum well, separate confinement heterostructure laser. Threshold occurs at 1 A, and peak power of 8 W is reached at 7.2 A, at which level the laser fails catastrophically. The electrical-to-optical efficiency reaches a maximum of 45 % at approximately four times threshold. This peak value of efficiency is set primarily by the differential quantum efficiency and electrical series resistance.

The performance of commercially available laser diodes and arrays is usually expressed in terms of slope efficiency. This eliminates the need for stating specific operating points at which a particular overall efficiency is achieved.

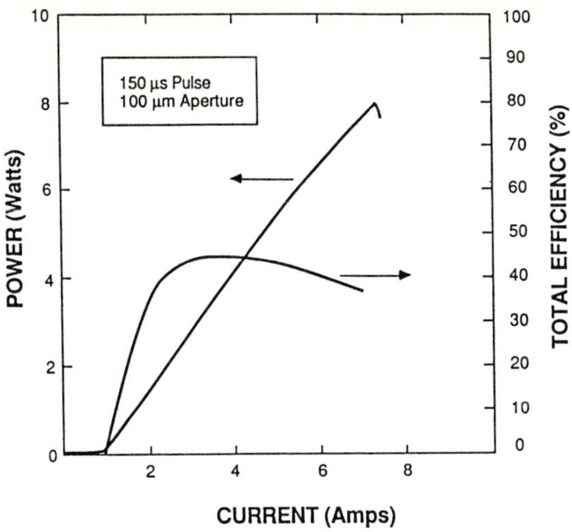

Fig. 6.23. Output power and conversion efficiency as functions of current of a high brightness laser [6.73]

With the threshold current and slope efficiency given, the overall efficiency for a particular operating point can be calculated.

Commercially available devices have slope efficiencies on the order of 25 %. Index-guided, single-quantum-well structures achieve a slope efficiency of 40–50 % on the average. Individual bars can be higher than that. Also in the window laser, a slope efficiency well over 50 % has been achieved.

Projections from several researchers indicate that slope efficiencies of 60 to 70 % will be achieved for laser diodes within the next five years.

The conversion efficiency of a diode array not only has a direct influence on the overall efficiency of the solid-state laser but is also one of the key parameters determining the design of 2-D arrays. As has been discussed in more detail in Chap. 5, heat dissipation in a 2-D array is the limiting factor with regard to average power from the laser. The average power of the laser can be increased by increasing the efficiency of the laser arrays. At present, the projected, reproducible conversion efficiency of the 2-D arrays is about 20 %. Therefore, for every watt emitted, 4 W are dissipated. On the other hand, graded-index, single-quantum-well structures have consistently exhibited conversion efficiencies of 40 % and higher. In this case, only 1.5 W are dissipated for every watt emitted – a reduction of over 60 %. Thus, the average emitted power could be increased by a factor of 2.6 without increasing the total dissipated power of the array. It is readily apparent that tremendous gains in average output power can be made by increasing the diode efficiency above 20 %. Such improvements are crucial in applications requiring high repetition rates, high average power or where high conversion efficiency itself is a primary parameter.

Lifetime Considerations. One of the several attractive features of diode lasers is their long lifetime. Room temperature values now approach 10^5 hours at cw output powers of a few milliwatts per stripe. Operating lifetimes generally range in the tens of thousands of hours for double-heterojunction GaAlAs lasers emitting about 10 to 20 mW at room temperature.

Lifetime was a serious problem with early diode lasers, but fabrication technologies have addressed the reliability issue, and as a result diode lasers have achieved the impressive operating lifetimes mentioned above. In general, the lifetime is longer for diode lasers with lower threshold currents, reflecting a general tendency for degradation to increase with operating current. As new low-threshold laser structures come into commercial use, laser lifetimes should continue to increase. Individual diodes in an array can fail for a variety of reasons, and laser diode degradation can be broadly attributed to three major causes [6.79]:

- damage to the laser mirrors (facets) which reduces their reflectivity or increases the nonradiative carrier recombination at the facets,
- ohmic contact degradation which increases the electrical and thermal resistance of the laser, and
- internal damage, which results from the formation or movement of lattice defects into the active region of the laser, decreases the internal quantum efficiency and increases the optical absorption.

The degradation modes of facet mirror damage, contact degradation and internal damage have been extensively studied by laser diode manufacturers.

As an example, Fig. 2.13 shows the results of extensive life tests performed at SDL on a multistripe laser.

Specifically, 10-stripe MQW lasers in the 810-nm region were operated at 100-mW cw output, and the drive currents were monitored as a function of time for heatsink temperatures of 30°, 70°, and 100° C.

After screening for infant mortalities at 70° C, two laser lifetimes were determined for each temperature. The lifetime for gradual degradation was obtained by extrapolating the operating current to twice its initial value. The lifetime for catastrophic failure was the actual elapsed time before sudden failure occurred. Figure 6.24 shows plots of the cumulative failures as a function of time for both lifetimes at 70° and 100° C. At 30° only the extrapolated failures due to gradual degradation are shown, because only one abrupt failure occurred at this temperature.

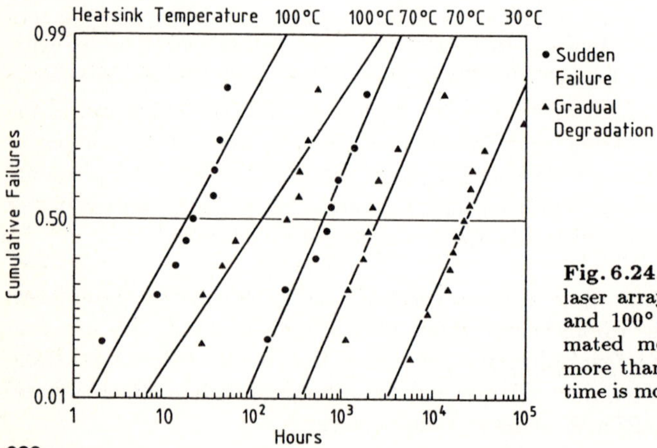

Fig. 6.24. Plot of 10-stripe diode laser array lifetimes at 30°, 70°, and 100° C. At 30° C, the estimated mean time to failure is more than 31,000 h (median lifetime is more than 22,000 h) [6.80]

From this investigation, three failure modes were identified: 1) sudden failure during burn-in (screening), which is related to the formation of dark line defects; 2) gradual degradation, which is observed at all temperatures; and 3) sudden failure related to increases in thermal resistance (attributable to failure of the bonding metallizations) at the higher temperatures. The best lasers exhibited degradation rates of 5 to 10% per thousand hours at 70° C. Below 30° the gradual degradation mechanism is the primary method of failure. At 30° C the estimated mean time to failure for the 100 mW cw 10-stripe laser is more than 31,000 h (median lifetime is more than 22,000 h).

McDonnell Douglas reported that accelerated life tests indicate mean time to failure up to 10^5 h. It has been observed that failure of an individual diode does not short out an entire array, because the bulk resistance is sufficiently high to cause a redistribution of the current. Lifetime estimates for arrays which are operated within their proper temperature limits are on the order of five to ten years' continuous operation.

Figure 6.25 shows the result of a life test. After an initial drop in power during the burn-in phase, the output did not decrease noticeably during the next 3000 hours. This time is equivalent to 4×10^{10} shots at a pulse width of 200 μs.

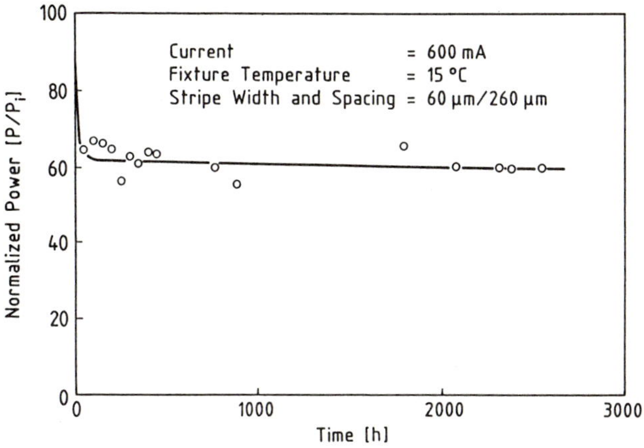

Fig. 6.25. cw lifetest of several diode bars [6.75]

Spectral Properties. The spectral properties of laser diode arrays which are most critical for the pumping of solid state lasers are the center wavelength of the emission, the spectral width of the array and the wavelength shift with temperature.

Wavelength of a diode laser depends primarily on the band gap of the material in which the electrons and holes recombine. In a binary compound like gallium arsenide the band gap has only one possible value. Where the relative proportions of different elements can vary, including ternary compounds like GaAlAs and quaternary compounds such as InGaAsP, a range of possible wavelengths can be obtained as shown.

In a number of hosts such as YAG, YLF and glass, Nd ions have substantial absorption in the vicinity of 0.807 μm, which is the emission wavelength of diode lasers with $Ga_{0.91}Al_{0.09}As$ active regions. As far as pumping of Nd lasers is concerned, the output wavelength can be tailored to the peak absorption by adjustments of the Al concentrations. Typically, a change in Al concentration of 1% results in a 10 Å change in wavelength.

More difficult to achieve is a narrow spectral width in an array. The bandwidth of the Nd:YAG absorption line at 808 nm is 20 Å for an absorption coefficient larger than $3.8\,cm^{-1}$. Individual laser diodes have a spectral width of 20 to 40 Å full width, half maximum.

Compositional changes and temperature gradients within an array lead to a much broader spectral output for the whole array as compared to a single device.

In a GaAlAs structure the peak emission changes 3 Å/°C. Therefore, in order to keep the spectral output from an array within the peak absorption region of Nd:YAG, the compositional variation has to be controlled within a fraction of 1% Al, and the temperature variation across the array has to be kept below 20° C.

The question arises, what other solid-state lasers, besides Nd, can be pumped with laser diodes?

By varying the composition of $Ga_{1-x}Al_xAl$ laser diodes, the band gap can be tuned, and output wavelengths between 770 and 900 nm can be achieved. As long as the potential lasing material has good absorption between 770 nm and 900 nm, GaAlAs laser diodes are the obvious choice.

There are two necessary material-related requirements for the creation of a room temperature cw diode laser: a direct band-gap III-V compound that emits light efficiently, and availability of a binary III-V compound substrate material with lattice-constant-matching-alloy compounds so that heterojunctions can be created.

In ternary and quaternary compounds a range of possible wavelengths can be obtained, as shown in Fig. 6.26. However, not all wavelengths depicted in Fig. 6.26 are attainable. Some compounds lack the direct band-gap energy-level structure needed for efficient production of light; these indirect band-gap materials are not suitable for use as diode lasers. Another difficulty is that ternary or quaternary compounds are grown on substrates of binary compounds; for

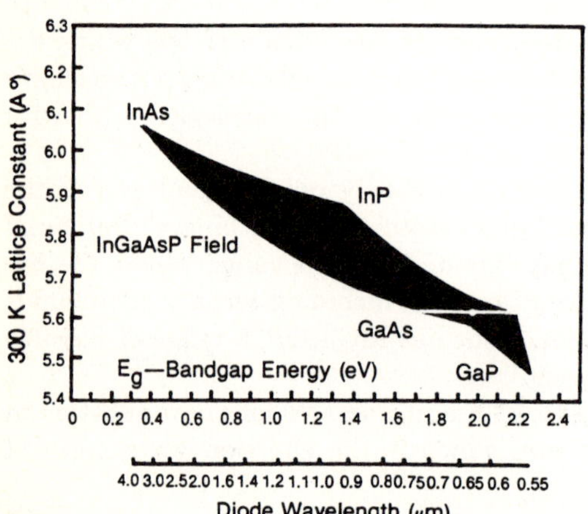

Fig. 6.26. Plot of lattice constant versus bandgap energy [6.121]

the growth to proceed properly, the spacing of atoms in the substrate must be close to that in the compound being grown.

Device-material changes require the longest development cycle, and diode lasers with good cw properties between 0.65 and 0.78 μm are still three to five years away. Emission wavelengths shorter than about 0.6 μm are not technologically straightforward, based on what we now know. At present there are no efficient (greater than 25 % internal efficiency) LED's or direct-band-gap materials with emission wavelengths much shorter than 0.65 μm, and even n- and p-type doping in such high-band-gap materials is difficult due to self-compensation.

Even if such lasers were developed, the high density of electron-hole recombination and the high energies released during nonradiative transitions in these high-band-gap materials would probably lead to fast degradation. Yellow, green, and blue diode lasers will probably not be available within the next ten years. An assessment of information dealing with laser diodes capable of operating in the 0.67 to 2.5 μm range can be found in [6.82].

6.1.5 Nonelectric Pump Sources

Besides the radiation sources discussed so far, which derive their energy from electrical power, several radiation sources have been explored for solid state laser operation which consume no electrical power.

Sun-Pumped Lasers. Considerable research has been done on sun pumping various solid-state lasers, such as Nd : glass [6.83], Nd : $CaWO_4$ [6.84], and Nd : YAG [6.85, 86]. The usual geometry of a sun-pumped laser includes a Cassegranian telescope of 60-cm to 120-cm diameter to collect the solar radiation, and suitable optics to concentrate the pump light onto one face of an end-pumped laser rod. A recently developed sun-pumped Nd : YAG laser for a space communication system has produced 5 W of cw output power [6.86]. The laser can also be pumped with a metal alkali lamp when the sun is not visible from the satellite.

Chemically Pumped Lasers. Chemical flashbulbs radiate via exothermic reactions of chemical species which liberate a large portion of the stored chemical energy as optical radiation. The advantage of this approach is the very high stored energy density of chemical reactions (tens of kilojoules stored in each gram of fuel) compared to that of electrical storage media (tenths of joules per gram of storage medium).

A laser can, for example, be built where the pump source comprises expendable chemical flashbulbs based on the combustion of oxygen and zirconium. Commercial bulbs such as Flash-Bar II manufactured by GE have an energy content in excess of 200 J/cm^3. Chemical flash pumped solid-state lasers have been studied extensively [6.87–91]. *Yang* and *Cook* [6.89] reported on the use of commercial zirconium wool-oxygen based miniature photo-flashbulbs to optically pump a Nd : glass rod. Maximum laser energy of 10.2 J was obtained by using twelve bulbs. Zirconium-oxygen mixtures have typically a flash duration of 20 to 30 ms.

Metal-shred chemical flashlamps similar to the above, except employing fluorine as oxidizer plus halocarbon accelerants can achieve burn times of less than 1 ms [6.90].

Results with quasi-cw sources operating for several seconds via the combustion of metals in flowing oxygen are given in [6.87]. A comparison of laser-output energy obtained using expendable flashbulbs filled with oxygen and various metals such Al, Zr and Hf has been given in [6.88].

Detonation Pumped Lasers. Several solid-state lasers have been built where explosives provided the necessary pump radiation. For example, *Smith* et al. [6.91] described a cyanogen-oxygen mixture reacted in a detonation chamber to pump a ruby laser.

6.2 Power Supplies

6.2.1 Operation of cw Pump Sources

Tungsten-filament lamps employed to pump Nd : YAG lasers require the least complicated supplies of all the standard pump sources used to pump solid-state lasers. In fact, since the most common filament lamps are 120 V devices, these lamps could be operated directly from the line. In practice, however, this is rarely done because this does not provide any input regulation, the laser output has a very strong 60-Hz ripple, and the lamp life is rather short as a result of the initial current surge when the lamp is turned on. For these reasons, laboratory-type power supplies for pumping Nd : YAG lasers contain at least a variable transformer, or phase-controlled SCRs, or saturable reactors for voltage control, and a rectifier and a filter.

Commercially available Nd : YAG lasers have power supplies which provide continuously variable regulated dc voltages, a start circuit which generates a slow ramp for the current rise when the lamp is turned on, and an interlock system which shuts the supply off if the laser is operated without adequate cooling, or in case of a lamp failure. A simplified schematic of a power supply which has these features is shown in Fig. 6.27. The main components of the power supply are a SCR phase-controlled rectifier bridge, a lamp voltage regulator unit and a L/C filter network. The rectified circuit consists of a bridge with phase-controlled SCRs in two of the branches. The SCRs are gated by voltage pulses from the lamp voltage regulator. The latter unit senses small voltage charges on the dc side of the supply with respect to a reference potential, amplifies these changes, and controls the phase retardation of the SCRs.

The power supply for operation of cw arc lamps is more complex because a trigger circuit is required to start the lamps, and in some cases when the lamps are filled at high pressure an additional high-voltage boost must be provided. A simplified schematic of a power supply (EMI, model EMK6000) which is employed in a large number of krypton arc pumped Nd : YAG lasers is shown in Fig. 6.28. The input is applied to a three-phase isolation transformer.

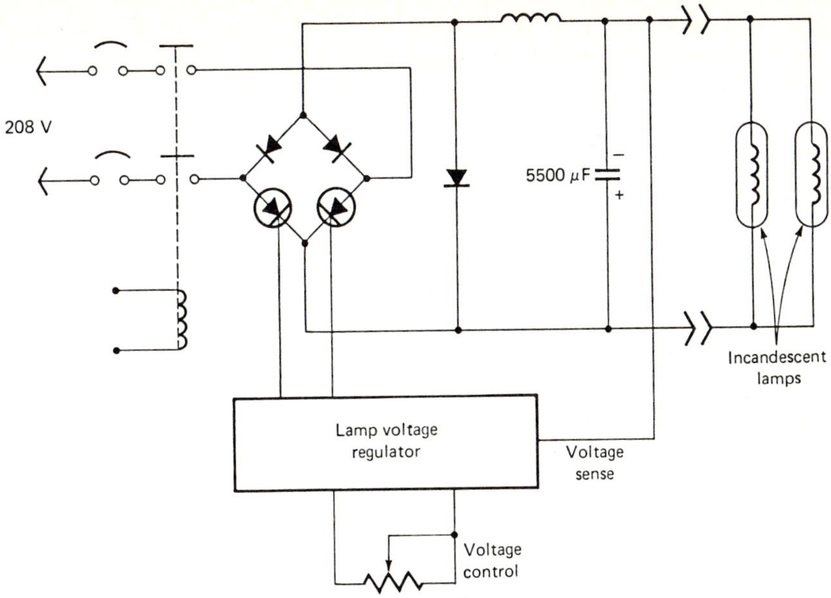

Fig. 6.27. Regulated dc power supply employed to operate filament lamps

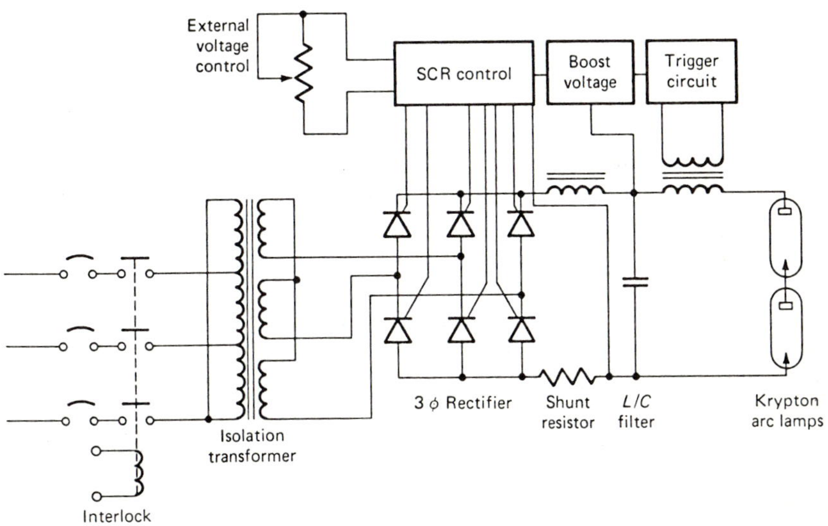

Fig. 6.28. Phase-controlled power supply for use with krypton arc lamps

Each output of the secondary of the 6 kVA power transformer is connected to the center of three pairs of series connected SCRs. This is the conventional three-phase bridge rectifier configuration, with the exception that the rectifiers are SCRs. The control unit senses the analog input from a current-sampling resistor, compares it with a reference voltage, and generates phase-controlled pulses which fire the SCRs. Control of the SCR conductance time provides regulation of the supply. The unit operates as a variable time switch for each leg of the three-phase ac input. The output of a rectifier is applied to a L/C filter network for ripple elimination. The ignitor provides the high-voltage pulse to start the lamp. This is done by discharging a capacitor with a SCR through the primary of a trigger transformer. The transformer steps up this voltage to over 30 kV, which is sufficient to ignite the lamp. For reliable lamp starting, the voltage of the main power supply must be boosted to about 600 V–1 kV during the trigger phase. This is most conveniently done by charging the filter capacitor to this voltage with a small, low-current, high-voltage supply. Power supplies of this type have been built for lamp inputs of up to 20 kW.

6.2.2 Operation of Flashlamps

The major components of a power supply employed in a flashlamp-pumped laser are a charging unit, a pulse-forming network, and a flashlamp trigger circuit.

Charging Unit

The function of the charging unit is to charge the energy storage capacitor to a selected voltage within a specified time, which depends on the desired repetition rate of the laser. The capacitor-charging source usually consists of a transformer followed by a rectifier bridge, a switching element in the primary of the transformer, a current-limiting element, a voltage sensor, and control electronics. The transformer and the rectifier bridge provide the required dc voltage for the energy-storage capacitor. In order to be able to vary this voltage and, therefore, to obtain a variable output energy from the laser, a semiconductor switch is usually included in the primary circuit of the transformer. This control element, which can be either a triac, a solid-state relay, or a pair of inverse parallel SCRs, is turned on at the beginning of the charge cycle and turned off when a preset voltage is reached in the capacitor. Control signals are derived from the capacitor voltage as in conventional dc supply designs. The charging of a capacitor presents a problem insofar as the discharged capacitor constitutes a short circuit. Without a current-limiting device in the power supply, the short circuit current is limited only by the resistance of the transformer windings. To protect the rectifier diodes and other components in the circuit, current-limiting circuits are required. Ideally, one would like a constant current supply, in which case the capacitor-charging current would be constant over the entire charging cycle. The charging current in this case would be

$$I = CV/t \quad , \tag{6.16}$$

where C is the capacitance of the capacitor, V is the final charging voltage, and t is the charging time. Figure 6.29 shows a number of circuits which are frequently employed to limit the short circuit current into a capacitive load. The most straightforward way to charge capacitors is resistance-limited charging from a constant-voltage source as shown in Fig. 6.29a. The least amount of

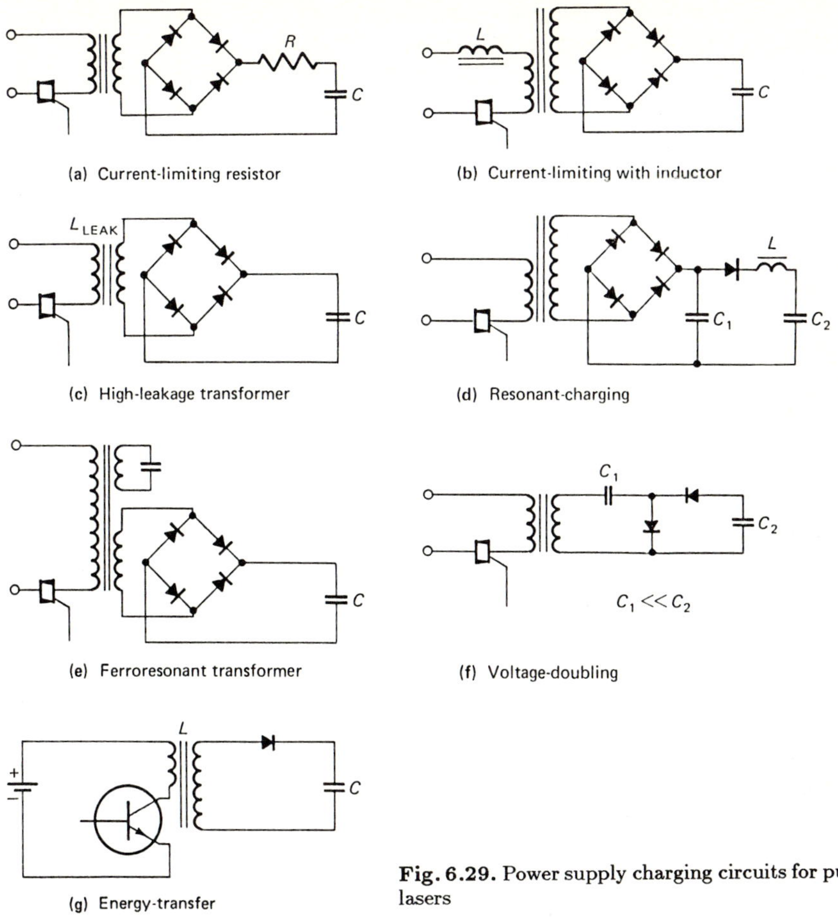

Fig. 6.29. Power supply charging circuits for pulsed lasers

power dissipated in the current-limiting resistor is equal to that stored and discharged in the capacitor if three or more time constants are used in charging. For shorter charging times, even more power is wasted in the resistor. For low-power, low-repetition-rate systems the heating losses usually can be tolerated. Since capacitor charging to high energy levels or at high repetition rates causes appreciable power waste when charging through a resistance, attempts have been made to produce power supplies without the resistance, but retaining the resistor's function of limiting initial capacitor-charging currents to safe values.

Figure 6.29b shows current limitation by an inductor in the primary of the transformer. During initial charging, when the secondary of the transformer is shorted by the discharged capacitor, the peak current is limited by the inductive reactance. The inrush current equals the primary voltage divided by the reactance. As a refinement of this technique, the inductance can be built into the transformer as primary leakage (Fig. 6.29c).

At repetition rates approaching the power-line frequency, it becomes difficult to achieve repeatable output voltage from charge to charge because of the limited number of current pulses within the charging period. For these cases it becomes necessary to use resonant charging (Fig. 6.29d). *Glascoe* [6.92] gave a complete analysis of the circuit. Current will flow from the dc source for the first half-cycle of the resonant frequency, charging the capacitor to twice the source voltage. The peak current drawn from the supply depends on the inductor, the resonant frequency of the LC circuit, and the voltage to which the capacitor is charged.

The capacitance C_1 shown in Fig. 6.29d is a filter to aid voltage regulation and is usually ten times the bank capacitance, C_2. The constant-voltage and short circuit current characteristics of ferroresonant line regulator transformers have been used to approximate a constant-current source for capacitor charging. Figure 6.29e shows the typical ferroresonant transformer with its associated components. The current-limiting characteristic of the ferroresonant transformer is attained through the use of a magnetic shunt and a resonant tank circuit [6.93]. In addition to resistive and inductive current limitations, the short-circuit current can be reduced by repetitively transferring a small amount of energy to the energy storage capacitor. In the voltage-doubling circuit shown in Fig. 6.29f, the small capacitor C_1 transfers its charge to the main capacitor C_2 during each cycle of the ac line. A circuit in which magnetically stored energy is transferred to a capacitor is shown in Fig. 6.29g.

In the following, we will describe a few typical power supplies employed in pumping solid-state lasers. Figure 6.30 shows the schematic of a power supply used to pump helical flashlamps of high-energy ruby and Nd : glass lasers. The lasers are operated at a repetition rate of a few pulses per minute. The power

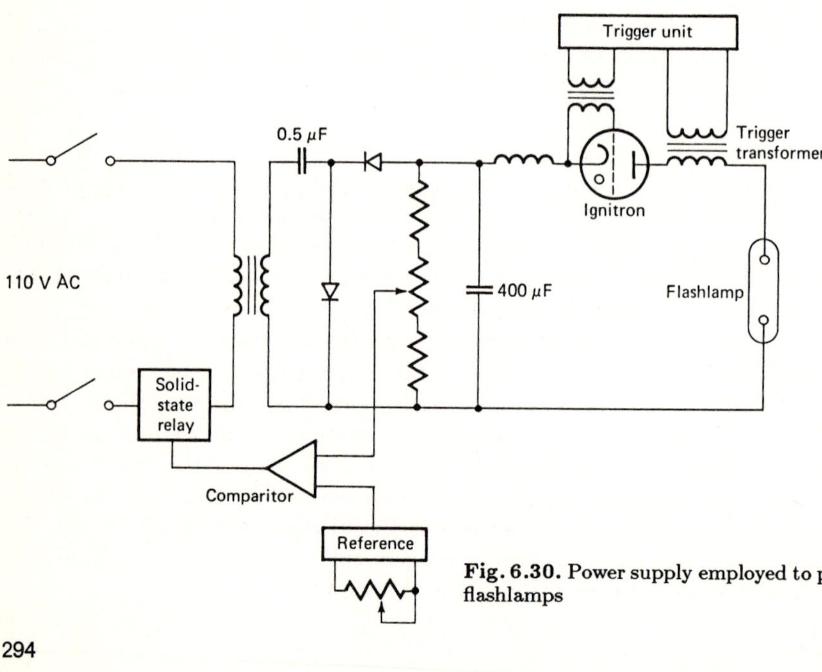

Fig. 6.30. Power supply employed to pump flashlamps

supply essentially consists of a standard power transformer with a solid-state relay in the primary, a voltage-doubling circuit, an energy storage capacitor, an ignitron, and a trigger circuit. The voltage across the 400 μF energy storage capacitor is sampled by a high-impedance voltage divider, and the voltage is fed to a comparator. If the input is equal to a reference voltage, an output from the comparator will turn off the solid-state relay. This disconnects the transformer from the line so the voltage on the capacitor remains at a constant value. A front panel knob provides prediction of the final capacitor voltage; this is done by changing the reference voltage. The small 0.5-μF capacitor provides a constant charge rate of the energy storage capacitor and, as discussed before, provides current limitation. However, many hundred charge cycles are required before the energy from the small capacitor has charged up the energy storage capacitor. Therefore, this charge-transfer technique is useful only for low-repetition-rate systems.

As we will see in the next section, large helical flashlamps require high voltages of the energy storage capacitor because of their high impedance. Since these lamps will usually self-trigger at voltages lower than the storage bank voltage, a holdoff device is required. Lamp breakdown during the charge cycle is usually prevented by an ignitron. The ignitron is a mercury-filled tube which is capable of handling very high currents. Triggering of the flashlamp is initiated by a 15- to 20-kV trigger pulse in the series injection trigger transformer and by simultaneously triggering the ignitron to allow bank discharge. In Fig. 6.30 an additional coil for shaping the flashlamp current pulse is shown. For lamps which are operated considerably below their self-triggering voltage, the ignitron may not be required. The holdoff voltage of a 7.5-cm-long linear flashtube is around 3.5 kV, whereas for a helical lamp with an arc length of 72 cm, holdoff voltage is around 6 kV.

Figure 6.31 shows a power supply which can charge a 1-kJ capacitor bank at a repetition rate of 10 pps. The unit features a high-reactance current-

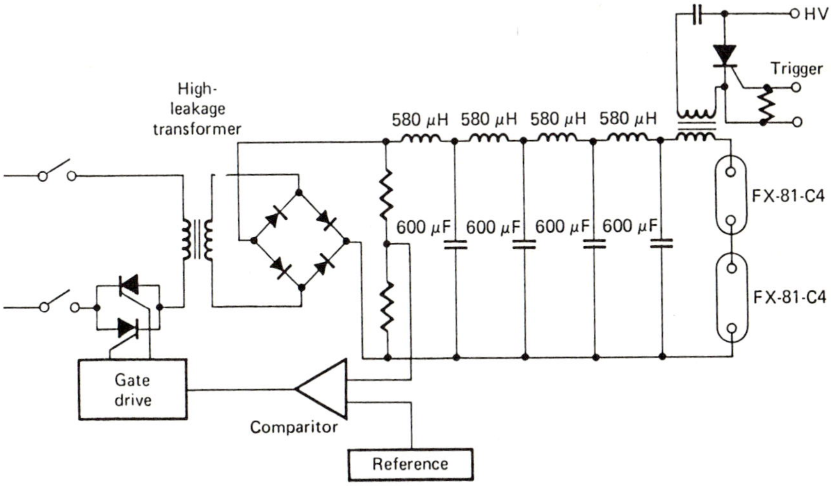

Fig. 6.31. Power supply featuring a high-leakage transformer

limiting transformer, with two inverse parallel SCRs as a switching element in the primary, and a full-wave rectifier bridge. Since this power supply is for a Nd : YAG laser welder, long pulses of the order of 5 ms are required. As we will see in the next subsection, this requires a multiple-mesh network. The trigger circuit, which we will also discuss in more detail later, consists essentially of a trigger transformer with a capacitor and a SCR in the primary. When the capacitor is discharged, a high voltage is generated at the secondary which breaks down the lamp. The power supply is designed to operate two linear flashlamps (FX81-C4). Each lamp has a bore of 10 mm and a length of 10 cm. This power supply can also be built as a three-phase supply. In this case, three separate transformers and voltage-rectifier bridges are connected in parallel at the dc output end.

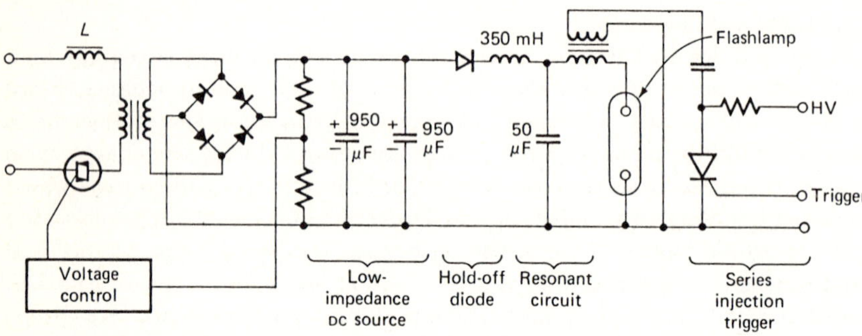

Fig. 6.32. Resonant charging of the energy storage capacitor in a high-repetition-rate system

For high repetition rates a resonant charging supply is best used. Figure 6.32 shows the simplified schematic of a power supply which is employed to pump a Nd : YAG laser at repetition rates of up to 50 pps. The laser is pumped by a single linear flashlamp with 5-mm bore diameter and 50-mm arc length at a maximum of 20 J. The dc section of the power supply represents a variable voltage source. In the primary circuit of the transformer is an inductor for current limitation and a triac for voltage control. The stepped-up voltage is rectified by a full-wave bridge and filtered by means of a bank of 950-μF electrolytic capacitors. The dc output voltage is controlled by a circuit which senses the voltage and generates trigger pulses to the gate of the triac. The resonant charging device consists of a hold-off diode, a 50-μF energy storage capacitor, and a 350-mH inductor. The flashtube trigger is generated by switching the energy from a small storage capacitor into the trigger transformer primary. The secondary of this transformer is connected in series with the pulseforming network (series injection triggering). When a flashlamp is operated at high repetition rates, the recovery characteristic of the lamp must be taken into account. An example of a recovery characteristic is shown in Fig. 6.33. The curve shows that, after cessation of current flow, the lamp remains ionized, hence at low impedance, for a time on the order of a few milliseconds.

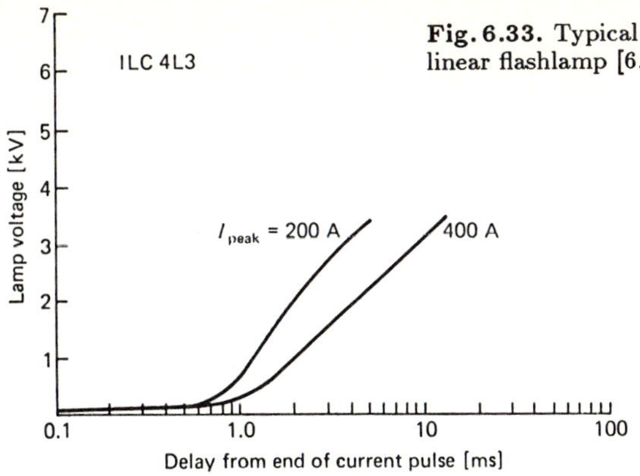

Fig. 6.33. Typical voltage holdoff recovery of a small linear flashlamp [6.20]

Power supplies used for military-type Nd : YAG lasers, such as rangefinders, target designators, or illuminators, mostly operate from batteries or on-board dc power lines. Figure 6.34 shows a simplified schematic of the power supply of an airborne target designator, which operates from 28 V dc. The unit, which consists of a dc-to-ac converter, a step-up transformer, a high-voltage rectifier, a pulse-forming network, and a parallel trigger circuit, is capable of charging the capacitor to an energy of 20 J at a repetition rate of 10 pps. The flashlamp (ILC L-213) has a bore of 4 mm and an arc length of 50 mm. Current is limited by an inductor on the high-voltage dc side of the unit. A feature of this power supply is the fact that the repetition rate of the dc-ac converter progressively increases from 1 kHz to about 5 kHz during the charge cycle in order to keep the peak-to-peak rms current drawn from the battery at a low value. This is achieved by a current sensing network which feeds a signal to the logic circuit, which in turn controls the switching frequency of the transistors in the dc-to-ac converter. At full capacitor charge, a voltage sensor turns off the driving signals to the transistors. A schematic diagram of another type of power supply used in small

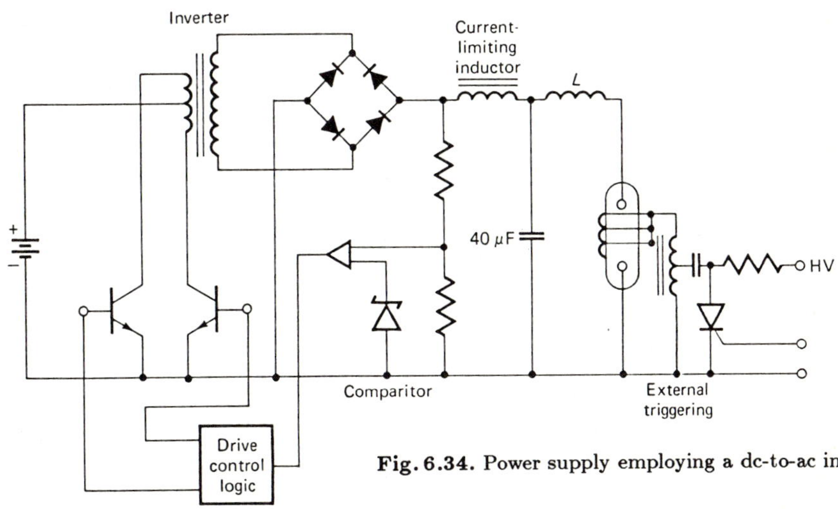

Fig. 6.34. Power supply employing a dc-to-ac inverter

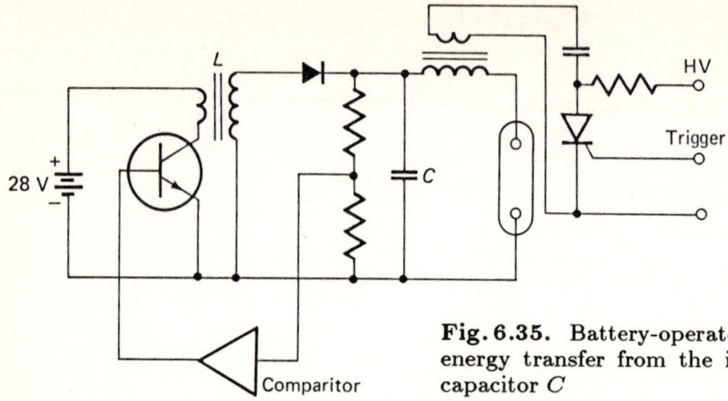

Fig. 6.35. Battery-operated power supply with energy transfer from the inductor L to the PFN capacitor C

military systems is shown in Fig. 6.35. The system operates on the principle of a flyback dc-ac converter. The transistor is turned on for a short period of time to allow current flow in the primary of the inductor. When the transistor is turned off, the magnetic energy stored in the inductor is transferred to the energy storage capacitor; at the same time the voltage is stepped up from 28 V to approximately 800 V. The device is short circuit protected since the inductive voltage kickback charges the capacitor after the transistor has turned off. The transistor is operated typically at a repetition rate of 1 to 10 kHz in order to keep the inductor small. For a 10 pps system this allows between 100 and 1000 transfers of energy from the inductor to the capacitor.

Further references on the design of power supplies employed to operate flashlamps can be found in [6.18, 94–97]. The special problems which arise in power supplies designed to charge capacitor banks with stored energies in the MJ range have been discussed in [6.97b].

Pulse-Forming Network

Flashlamps are usually operated from a single- or multiple-mesh LC network. The network stores the discharge energy and delivers it to the lamp in the desired current pulse shape. In most situations, the lamp input energy E_0, the pulse width t_p, and the lamp dimensions have been determined before the pulse-forming network is designed. The above-mentioned parameters completely describe the network's capacitance and inductance as well as the charging voltage and peak current. To familiarize the reader with the design of flashlamp energy-storage networks, we will consider first the elementary model of a pulse discharge circuit (Fig. 6.36). The capacitor C is charged to an initial voltage V_0. At time $t = 0$ the switch S is closed and the energy stored in the capacitor is delivered to the load. We assume for the moment that the load is a linear resistor R. In this case the circuit is described by the second-order differential equation

$$L\frac{d^2q}{dt^2} + R\frac{dq}{dt} + \frac{q}{C} = 0 \quad , \tag{6.17}$$

where q is the charge on the capacitor. With the initial charge on the capacitor

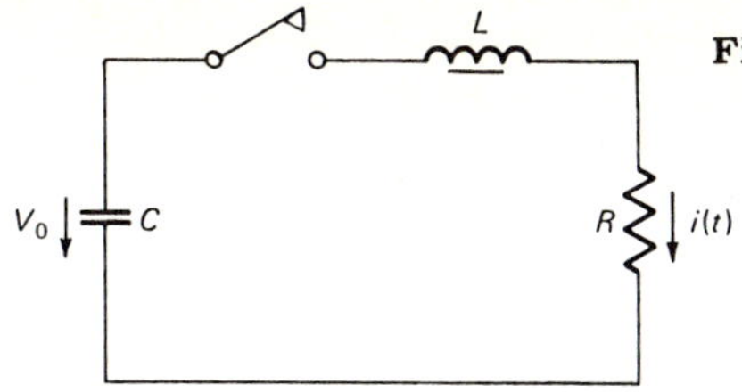

Fig. 6.36. Single-mesh discharge circuit

$q(0) = CV_0$ and the initial current $dq/dt(0) = 0$, the voltage across R is given by

$$v(t) = V_0(\alpha/\beta)\{\exp[-(\alpha-\beta)t] - \exp[-(\alpha+\beta)t]\} \quad , \tag{6.18}$$

where $\alpha = R/2L$ and $\beta = (R^2/4L^2 - 1/LC)^{1/2}$.

Depending on the relative values of R, L, and C, the voltage or current waveform across R may be as follows:

Critically Damped. In this case, $\beta = 0$ and

$$R = 2\left(\frac{L}{C}\right)^{1/2} \quad . \tag{6.19}$$

The capacitor voltage decays exponentially from V_0 to zero. If we define the rise time t_r as the time required for the voltage or current at R to rise from zero to its maximum value, we obtain

$$t_r = (LC)^{1/2} \quad . \tag{6.20}$$

The rise time measured between the 10 and 90 % points of the voltage is $0.57t_r$. The discharge current as a function of time is

$$\frac{i(t)}{i_p} = e\left(\frac{t}{t_r}\right)\exp\left(-\frac{t}{t_r}\right) \quad , \tag{6.21}$$

where the peak discharge current is

$$i_p = \frac{2V_0}{Re} \quad .$$

Overdamped. The waveform is similar to the critically damped pulse, except that the peak current is reduced and the time required to reach zero is extended. The overdamped case (β real) is defined by

(6.22)

Underdamped. The resulting pulse shape is characterized by strongly damped oscillations (β imaginary). The frequency of the oscillation depends mostly on

L and C, whereas the decaying amplitude is a function of R only. Oscillations and the associated current and voltage reversal in the discharge circuit occur if

$$R < 2\left(\frac{L}{C}\right)^{1/2} \quad . \tag{6.23}$$

Nonlinear Load. As we have seen in Sect. 6.1, the flashlamp resistance is nonlinear and can be described by $R_L(i)$ given in (6.4) or by the lamp impedance parameter K_0 defined in (6.1). The design of single-mesh circuits for driving flashlamps has been considered in [6.13] and more recently in [6.98]. In order to apply the results published in the literature, we must make the following substitutions and normalizations:

$$Z_0 = \left(\frac{L}{C}\right)^{1/2}, \quad \tau = \frac{t}{T} \quad , \quad T = (LC)^{1/2} \quad , \quad i = \frac{IV_0}{Z_0} \quad . \tag{6.24}$$

In addition, a damping factor α is defined which determines the pulse shape of the current pulse. It is

$$\alpha = K_0(V_0 Z_0)^{-1/2} \quad . \tag{6.25}$$

Solutions of the nonlinear differential equation of the lamp current for various values of α have been obtained using a computer [6.13]. The current waveform for different values of α is shown in Figs. 6.37 and 38. As can be seen from these curves, $\alpha = 0.8$ corresponds to the critically damped case. The current pulse duration t_p at the 10% points is approximately $3T$; during this time about 97% of the energy has been delivered. If we substitute

$$\alpha = 0.8 \quad \text{and} \quad t_p = 3T = 3(LC)^{1/2} \tag{6.26}$$

for a critically damped pulse into (6.24, 25), then we can determine for a given

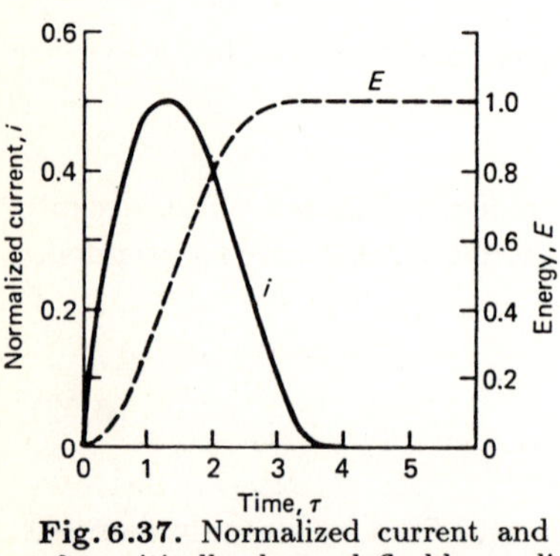

Fig. 6.37. Normalized current and energy of a critically damped flashlamp discharge circuit. Pulse shape factor α=0.8

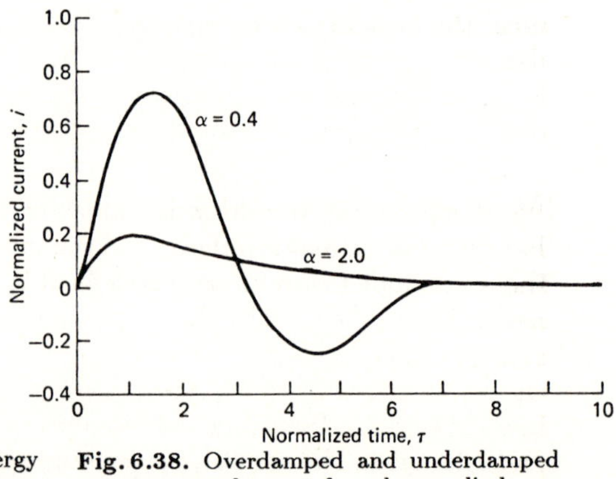

Fig. 6.38. Overdamped and underdamped current waveforms of a lamp discharge circuit

lamp type the relationship between energy input, pulse duration, pulse shape, inductance, capacitance, and operating voltage.

The energy initially stored in the capacitor is

$$E_0 = \tfrac{1}{2}CV_0^2 \quad . \tag{6.27}$$

The relation is used to eliminate V_0 from (6.25). With $\alpha = 0.8$ we obtain from (6.25) the value of the capacitor

$$C^3 = 0.09\frac{E_0 t_p^2}{K_0^4} \quad . \tag{6.28}$$

The inductance can be calculated from (6.26)

$$L = \frac{t_p^2}{9C} \quad . \tag{6.29}$$

We now have a set of three equations (6.27–29) which, given the specifications of the flashlamp parameter K_0, the desired input energy E_0, and pulse width t_p, provide explicit values of C, L, and V_0. If the lamp parameter K_0 is not specified by the lamp manufacturer, it can be calculated from (6.5). From Fig.6.37 also follows the peak current for a critically damped current pulse

$$i_p = 0.5\frac{V_0}{Z_0} \quad . \tag{6.30}$$

The rise time to reach this peak value is

$$t_r = 1.25(LC)^{1/2} \cong 0.4t_p \quad . \tag{6.31}$$

Differences between these solutions and those for the linear LCR circuit are readily apparent. For a given Z_0, V_0, and peak current, the solution of the nonlinear equation is more heavily damped than that of the linear case. The damping parameter α is dependent on V_0, the initial capacitor voltage. This is consistent with the experience that a discharge circuit, critically damped at one voltage, begins to ring as the voltage is increased.

Multiple-Mesh Networks

For more strict control of the pulse duration, and to create light pulses with flat tops, multiple-mesh networks are used. These consist of two or more LC networks in series. Switching an open-ended charged transmission line into a resistor equal to its characteristic impedance yields a rectangular pulse across the resistor. The width of the pulse equals twice the propagation time of the line. This same technique can be applied to a lumped parameter line. The principal advantage of the lumped-parameter delay line as an energy-storage pulse-forming network is its nearly constant output over the length of the pulse.

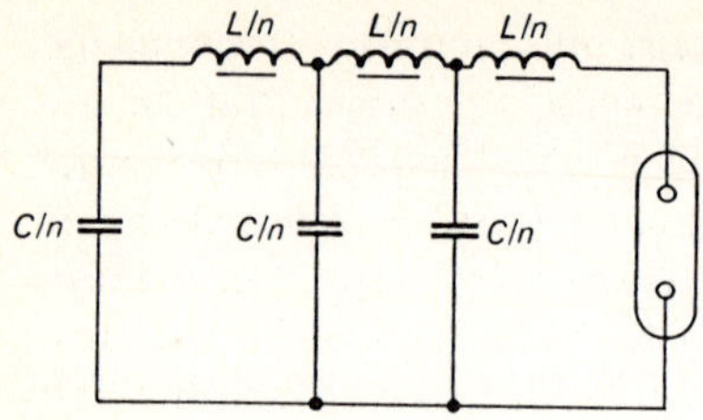

Fig. 6.39. Multiple-mesh network

This property is especially useful in producing long normal-mode laser pulses of constant power.

Most pulse-forming networks employ the "E"-type circuit of a lumped constant transmission line. In this configuration, all capacitors are of equal value and the inductance values per mesh are nearly identical (Fig. 6.39). Following are the approximate equations of a lumped-type "E" transmission line terminated by its characteristic impedance [6.92]. The characteristic impedance of the network is

$$Z_n = \left(\frac{L_\mathrm{T}}{C_\mathrm{T}}\right)^{1/2} , \tag{6.32}$$

where L_T and C_T are the total inductance and total capacitance. The pulse width at the 70 % point is

$$t_\mathrm{p} = 2(L_\mathrm{T}C_\mathrm{T})^{1/2} . \tag{6.33}$$

With the pulse duration known, the total capacitance follows from

$$C_\mathrm{T} = \frac{t_\mathrm{p}}{2Z_n} , \tag{6.34}$$

and the total inductance is

$$L_\mathrm{T} = \frac{t_\mathrm{p}Z_n}{2} . \tag{6.35}$$

In order to deliver the required energy, the capacitors of the network must be charged to a voltage

$$V = \left(\frac{2E}{C_\mathrm{T}}\right)^{1/2} . \tag{6.36}$$

The peak current during the discharge from the pulse-forming network is given by

$$i_\mathrm{p} = \frac{V_0}{2Z_n} . \tag{6.37}$$

From (6.34, 35) it follows that the network L_t and C_t are determined by the desired pulse width and the lamp impedance $R = Z_n$. The desired lamp energy input determines the charging voltage.

Once L_T, C_T, and V_0 are calculated, the desired rise time determines the number of circuit meshes in the network. Experiments have shown that current rise times of 50 μs for lamps pumped up to the 100-J level and rise times of 200 μs for lamps pumped in the kilojoule regime can cause severe electrode sputtering and crazing of the flashtube. The number of sections chosen to make up the pulse-forming network is determined by first considering the rise time and then a convenient number of components. If n is the number of meshes, then the rise time from the 10 % to the 80 % level is

$$t_r = \frac{t_P}{2n} \quad , \tag{6.38}$$

and the fall time (80 % to 10 %) is

$$t_f = 3t_r \quad . \tag{6.39}$$

The capacitance and inductance per mesh are

$$C_M = \frac{C_T}{n} \quad \text{and} \quad L_M = \frac{L_T}{n} \quad . \tag{6.40}$$

Network Components

Inductors for pulse-forming networks are usually air-core coils, except for small systems where an iron core may be used. Besides inductance, other parameters of interest are the operating voltage and the rms current which determine insulation and wire size. Energy storage capacitors ranging from 1 to 10 kV and energy levels from 10 to 5000 J are commonly used. Because space is usually at a premium, energy storage capacitors are customarily designed with a higher dielectric stress than is usually employed for conventional dc capacitors [6.99]. These low-inductance capacitors are designed for quick charge and discharge. The capacitor's dc life may be entirely different from its pulse discharge lifetime. Large discharge capacitors with storage energies in the kilojoule range are usually castor oil-impregnated paper dielectric capacitors. These capacitors are designed for energy densities of 0.04 to 0.07 J/cm^3 and 30 to 65 J/kg. In pulse discharge capacitors using high-strength dielectric materials, such as polyester or polypropylene film, energy densities of up to 0.5 J/cm^3 and 270 J/kg can be achieved. These capacitors, which are used mostly in military-type systems, have storage energies of up to a few hundred joules. Capacitor lifetime is determined by the charging voltage, the percentage of voltage reversal, and the operating temperature. In most practical situations, where a critically damped or overdamped pulse is generated and operation occurs below the peak rated voltage and at temperatures under 50° C, the capacitor life is not a serious consideration compared to other component failures. With these operating conditions, well over 10^7 discharges can be expected from the capacitor.

In Figs. 6.40 to 43 typical current waveshapes are shown which are obtained from single- and multiple-mesh networks for various types of lamps. One usually finds that the final selection of the elements of a discharge circuit depends on

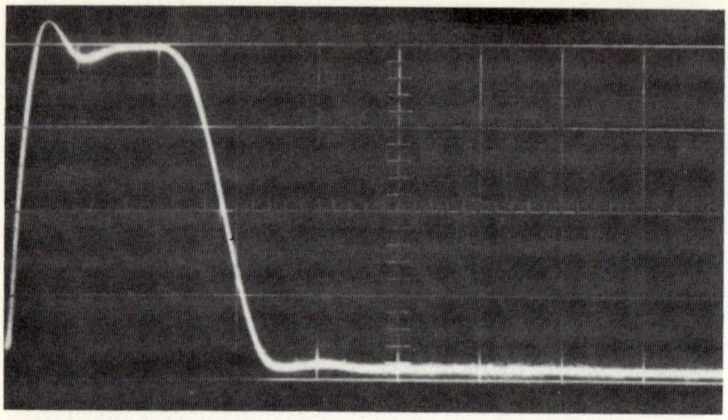

Fig. 6.40. Current waveform obtained from circuit shown in Fig. 6.31 at lamp input energy of 1 kJ and a bank voltage of 900 V. *(Horizontal)* 2 ms/div; *(vertical)* 100 A/div

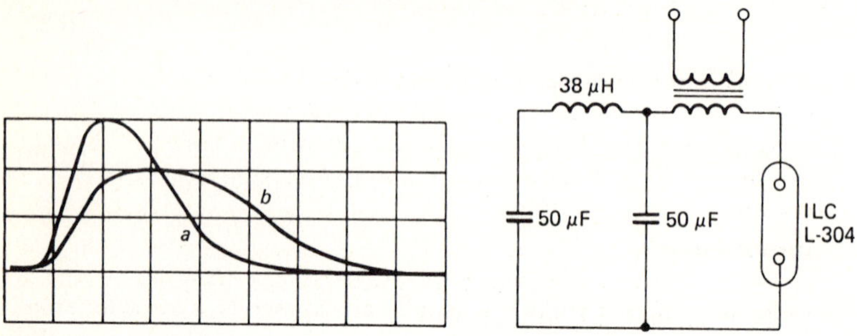

Fig. 6.41. Current pulse obtained from a *(a)* single- and *(b)* double-mesh network for 25-J input (ILC lamp model L-304). *(Horizontal)* 50 μs/div; *(vertical)* 200 A/div

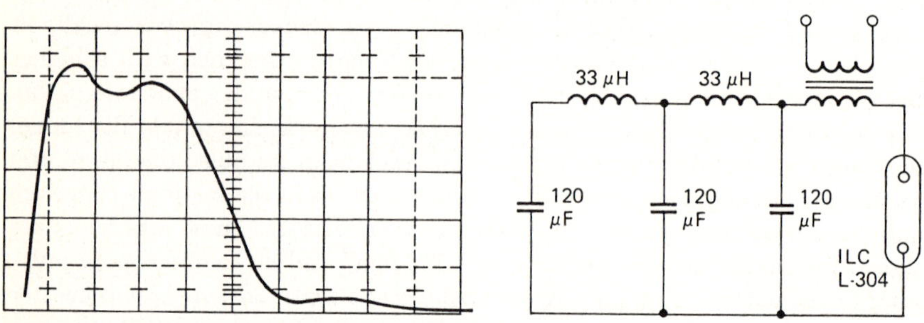

Fig. 6.42. Current pulse obtained from a three-mesh network. Input energy 300 J into ILC lamp model L-305. *(Horizontal)* 100 μs/div; *(vertical)* 200 A/div

commercially available components rather than on the exact calculated values. Furthermore, a laser system is in most cases operated over a range of input energies, and theoretically the optimum network would have to be slightly different for each input level.

We will illustrate the procedure for calculating a multiple-mesh network for the pulse shape shown in Fig. 6.40. A 5-ms-long current pulse is desired for two linear lamps connected in series and pumped at a total energy input of 1 kJ. Each of the lamps (EGG-FX81-4) has an arc length of 10 cm and a

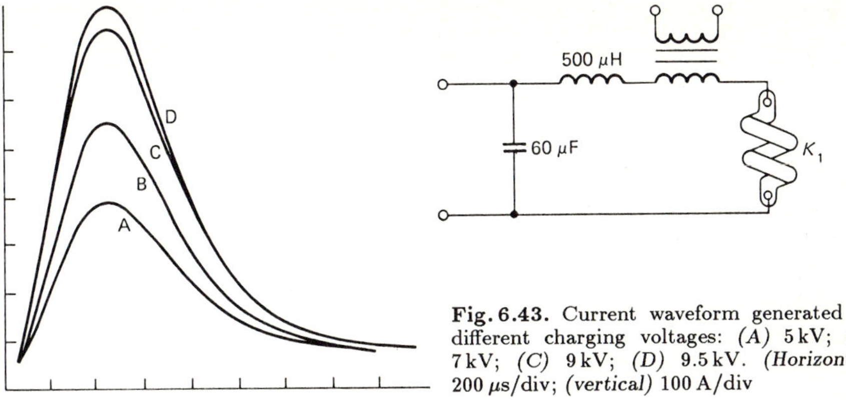

Fig. 6.43. Current waveform generated by different charging voltages: *(A)* 5 kV; *(B)* 7 kV; *(C)* 9 kV; *(D)* 9.5 kV. *(Horizontal)* 200 μs/div; *(vertical)* 100 A/div

bore of 1 cm. From (6.4) it follows that with $l = 10$ cm and $D = 1$ cm, there is a resistance of $R = 0.5\,\Omega$ for each lamp if we assume a peak current of $i_p = 650$ A. As we will see, the design of a PFN is an iterative process. First, a peak current is assumed which permits one to calculate the lamp impedance. With this parameter known the network can be designed. However, at the end of the analysis the peak current has to be rechecked. With $Z_0 = 1\,\Omega$ for the total lamp resistance, it follows from (6.34–36) that $C_T = 2500\,\mu$m, $V = 900$ V, and $L_T = 2.5$ mH. Rechecking the peak current, we find from (6.37) a value of 450 A, which means the values of C_T and L_T must be recalculated using another assumed i_p. To obtain a fairly flat 5-ms-long pulse, a four-mesh network, each section containing a 600-μF capacitor and a 580-μH inductor, is chosen. The pulse-forming network is shown in Fig. 6.31 and the current waveform is illustrated in Fig. 6.40. The measured rise time of the current pulse, which is about 0.6 ms, follows also from (6.38).

Figure 6.41 shows the current pulse of a single- and a double-mesh network. Each section consists of a 50-μF capacitor and a 38-μH inductor. For the single-mesh network the inductance is actually the saturated inductance of the trigger transformer. The flashlamp is 50 mm long and has a bore of 3 mm (ILC L-304). Total input energy is 25 J. The current pulse into a 75-mm-long, 4-mm-bore flashlamp generated by a three-mesh network is shown in Fig. 6.43. Each section of the network contains a 33-μH inductor and a 120-μF capacitor. The total stored energy is 300 J. Figure 6.43 shows the current waveform generated by a single-mesh network. The load is a helical flashlamp with a tube inside diameter of 8 mm and a total arc length of 72 cm. The lamp impedance parameter for this lamp was measured to be $K_0 = 150\,\Omega\,\mathrm{A}^{1/2}$. The values for the discharge circuit are obtained if one introduces $t_p = 500\,\mu$s and $E_0 = 2700$ J into (6.27–29). The pulses shown in Fig. 6.43 correspond to different charging voltages of 5, 7, 9, and 9.5 kV. As can be seen from this figure, despite the large range in bank voltage, input energy, and peak current, the pulses are all approximately critically damped and have pulse widths of about 500 μs.

Trigger Circuit

The discharge of the stored energy into the flashlamp is generally initiated by a high-voltage trigger pulse. The function of the trigger signal is to create an ionized spark streamer between the two electrodes so that the main discharge can occur. The initial spark streamer is formed by the creation of a voltage gradient of sufficient magnitude to ionize the gas column. The concept of a voltage gradient is important here, since it implies the existence of a stable voltage reference surface in close proximity to the flashlamp. Regardless of the triggering method used, reliable triggering cannot be achieved without this reference. Usually, an equipotential condition on the outside wall of the tube is achieved through the use of a wire wrapped around the flashlamp or by the proximity of the metal parts of the pump cavity. The ignition process can be explained by assuming that areas of the inside glass wall behave as small electrodes capacitively coupled to the reference plane. When the trigger voltage is applied, a discharge takes place between the cathode and the nearby part of the inside wall; this part will then be almost at cathode potential. Next, a discharge takes place between this part of the glass wall and a more remote area still at high potential, and so on until the anode is reached [6.100, 101].

The two most common methods of triggering flashlamps are external and series injection triggering. Other techniques are the so-called simmer and pseudosimmer triggers. In the external trigger device a wire is wrapped around the flashlamp between the electrodes and connected to the secondary of a high-voltage transformer. A high voltage is generated by discharging a capacitor through the primary of the transformer. The switching element can be either a SCR or krytron [6.102, 103]. The trigger voltages of flashtubes are between 5 and 10 kV. Figure 6.34 shows an example of external triggering. The waveform of a typical flashlamp trigger pulse is given in Fig. 6.44. The pulse shown was generated by discharging a 33-μF capacitor at a voltage of 500 V through the primary of a trigger transformer (EGG model TR132) with a turns ratio of 53 : 1.

Sometimes it is undesirable to have a wire wrapped around the flashlamp. For example, the possibility of accidental voltage break-down to the surrounding pump cavity is greatly increased by the presence of a trigger wire. In such applications, a triggering scheme which does not require an external trigger

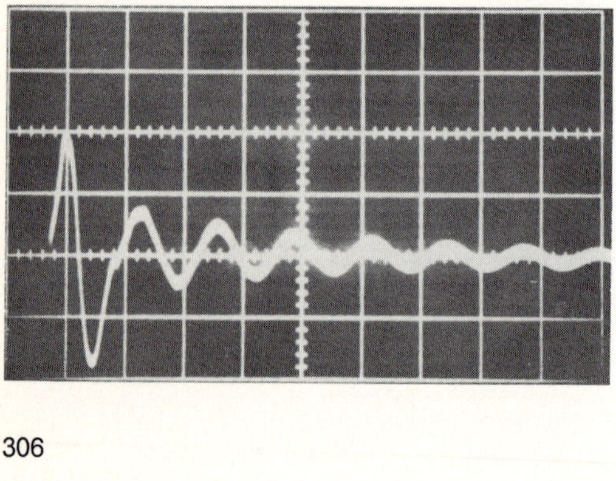

Fig. 6.44. Typical flashlamp trigger pulse. *(Horizontal)* 10 μs/div; *(vertical)* 10 kV/cm

wire is highly desirable. In the series injection method, a pulse is generated in a transformer whose secondary winding is in series with the flashlamp. The high-voltage pulse causes initial ionization of the plasma. The large lamp current which follows saturates the transformer so that its inductance is low. Examples of the series trigger technique are shown in Figs. 6.30–32, 35, and 43. Each method has its own particular advantages. External triggering provides greater design flexibility, because the secondary winding is not in the main discharge circuit. External triggering is the simplest, lightest, and most commonly used technique. Series triggering, on the other hand, is the choice when no exposed high-voltage leads are permitted. When a series trigger circuit is used, the secondary of the trigger transformer is part of the lamp discharge circuit, therefore the resistance of the secondary winding should be much less than the flashlamp resistance. This makes the series injection trigger transformer relatively bulky and heavy.

On the other hand, series triggering produces generally better pulse-to-pulse uniformity and reproducibility than external triggering. Also, in some cases longer flashtube life and more efficient pumping have been observed. These observations may possibly be explained by the fact that when a lamp is series-triggered, the core of the trigger transformer is not saturated and initial current flow is limited by the unsaturated inductance of the transformer. It appears that the reduced current growth rate during the arc expansion phase results in a more uniform arc formation and an associated smaller shock wave. An extension of this reasoning has led to the development of the so-called simmer and pseudosimmer triggering schemes. In these schemes a low-current discharge is maintained in the lamp between pulses. This keep-alive discharge is accomplished by employing a low-current dc power supply in parallel with the main supply. A typical circuit is shown in Fig. 6.45.

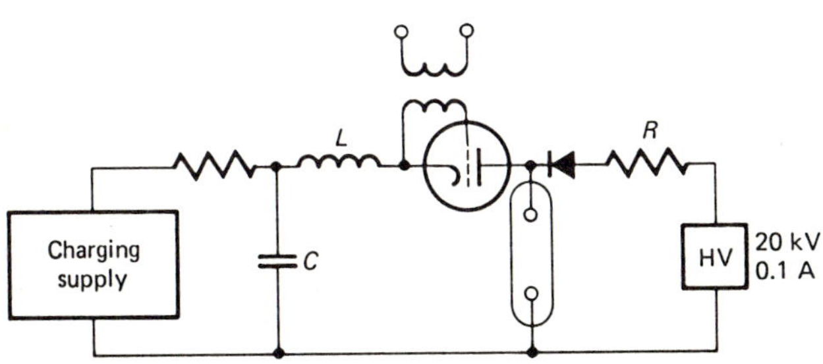

Fig. 6.45. Power supply with keep-alive discharge circuit

The simmer mode of operation requires a switching element between the lamp and the pulse-forming network. The lamp is initially ignited by the open-circuit voltage of 20 kV of the high-voltage power supply. At steady state the voltage drop across the lamp is between 500 V and 1 kV. Using this technique, considerable improvements in system performance can be achieved, including the following:

1. Increased flashlamp life: An order-of-magnitude improvement in lamp life has been observed employing a keep-alive discharge [6.23, 104]. Improvement is due mainly to a drastic decrease in cathode sputtering which is caused to a large degree by the high-voltage surge across the lamp in standard trigger schemes.
2. Reduced coolant degradation: The fluorinated hydrocarbons, such as 3M FC-104 and Dupont E-4, possess the temperature characteristics required for military equipment, but they tend to degrade from ultraviolet irradiation. A major source of this radiation is the high-voltage ignition which is eliminated in the simmer mode.
3. Reduced EMI/RFI radiation: A difficult requirement, particularly in military laser systems, is the reduction of broad-band rf radiation. For example, a laser operating at 10 pps presents three major electric field noise sources, the flashlamp igniter pulse, flashlamp current pulse, and Pockels-cell switching pulse. The first, and often the largest, source of this type of noise is eliminated in the simmer mode.
4. Increased efficiency: It has been demonstrated that in some cases the simmer mode can provide increases in Kr and Xe flashlamps' pumping efficiency for lamps operated with a pulse length of 50 to 200 μs and pulse energy levels of 5 to 20 J.
5. Lamp status control: By monitoring the keep-alive current, a broken flashlamp or a short circuit can be detected before the energy storage capacitor is discharged. This is particularly important in large systems or in systems employing many flashlamps.

The major disadvantage of this mode of flashlamp operation is the added electronics and the power consumption of the continuous discharge. Small 7.5-cm long linear lamps are operated at about 30 mA dc and consume between 5 and 10 W; large helical lamps operate at 1 kV and 100 mA, thus consuming 100 W. To reduce the power consumption of the simmer mode, a pseudosimmer was devised that combines the advantage of the light-weight external trigger transformer with the improvements achieved in a simmer-mode system [6.10]. Figure 6.46 shows a schematic block diagram of the circuit which has been used in small Nd:YAG systems. The lamp is ignited with an external trigger transformer. Lamp current initially flows through a limiting resistor, which is in parallel with the pulse-forming inductor and the SCR. Current flow through the inductor is prevented by the SCR. After an appropriate time delay, the SCR is turned on, and the high-current pulse is initiated.

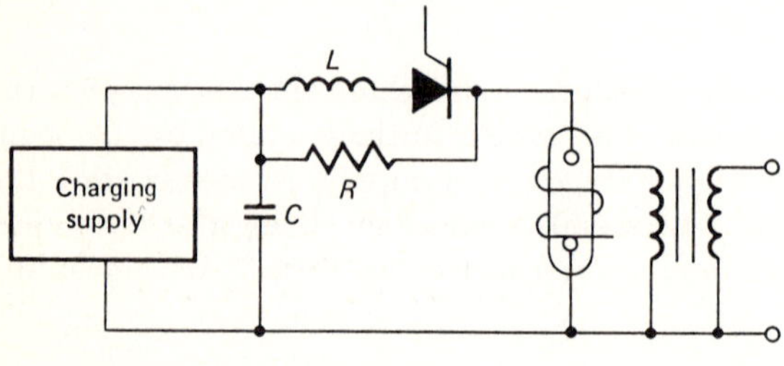

Fig. 6.46. Power supply with pseudosimmer discharge

6.3 Pump Cavities

The efficiency in the transfer of radiation from the source to the laser element determines to a large extent the overall efficiency of the laser system. The pump cavity, besides providing good coupling between the source and the absorbing active material, is also responsible for the pump density distribution in the laser element which influences the uniformity, divergence, and optical distortions of the output beam.

6.3.1 Pump Cavity Configurations

In the development of solid-state lasers, many different projection systems have been employed to transfer the radiation from a light source to the active material. Depending on the shape of the active material and the type of pump source used, one can broadly divide pumping geometries into systems in which the active material is side-pumped, end-pumped, or face-pumped.

In a side-pumped illumination system, the active material, shaped in the form of a cylindrical rod or rectangular slab, is pumped by linear, helical, or hollow lamps. Pump light strikes the active material perpendicular to the laser beam. The most common type of illumination systems fall into this category.

In a few special cases where the pump source is the sun or a light-emitting diode, the cylindrical laser rod is pumped through one of the end faces by a condenser arrangement [6.83–86, 105]. Usually a lens is employed to concentrate the light into the laser rod, which has a reflective coating on its cylindrical surface. The pump light trapped in the active material propagates along the axis of the rod similar to a light pipe. Figure 6.47 shows the pumping scheme of a sun-pumped Nd: glass laser [6.105].

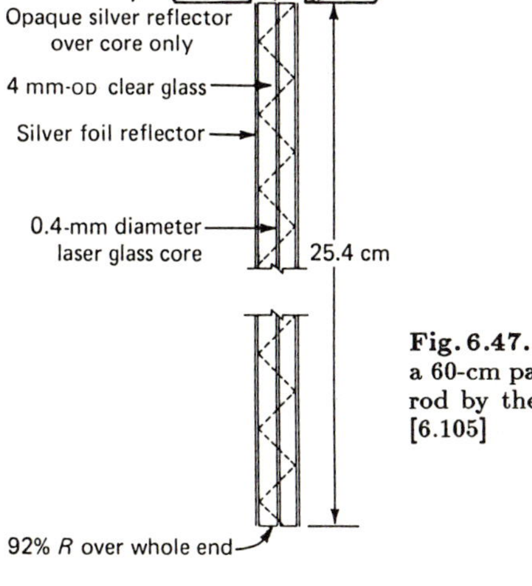

Fig. 6.47. End-pumped glass laser. Sunlight collected by a 60-cm parabola is imaged onto the end of the glass laser rod by the unity numerical aperture refractive elements [6.105]

In Chap. 7 we will discuss slab geometry designs such as the disk amplifier, the active mirror and the zig-zag laser. The first two schemes are employed in large-aperture Nd : glass lasers, whereas the zig-zag slab laser is of interest in high-average-power laser systems.

In all of these designs the flashlamp-reflector housing requires a careful design to provide uniform illumination of the slab as well as having good coupling efficiency. Figure 6.48 shows several reflector geometries which are in use

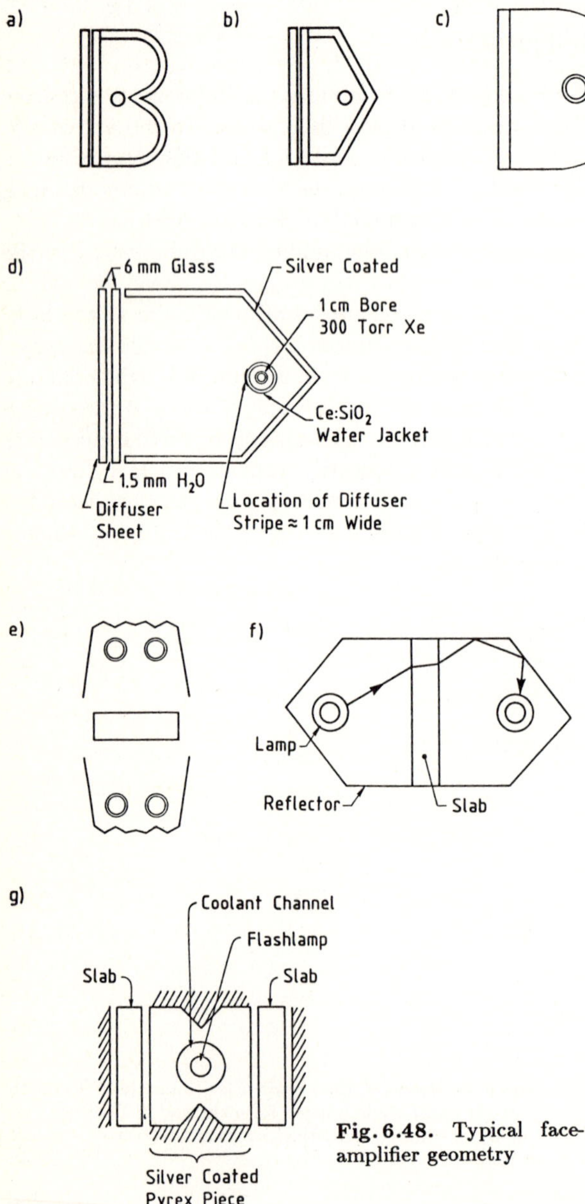

Fig. 6.48. Typical face-pumped Nd : glass disk laser amplifier geometry

for pumping slab lasers. Illustrated are the cusp-, and "V"-shaped, and semi-circular reflectors (Fig. 6.48a–c), surrounding a single lamp, as well as a "V"-shaped reflector in combination with a diffuser to spread out direct radiation and thus provide a more uniform pump intensity profile (Fig. 6.48d). A cusp-shaped reflector for a multiple lamp reflector is illustrated in Fig. 6.48e. A single slab pumped from both sides by "V"-shaped reflectors is shown in Fig. 6.48f, whereas Fig. 6.48g illustrates a reflector design employed for the pumping of two slabs by a single flashlamp. Details of the designs illustrated in Fig. 6.48 can be found in [6.151–155]. In what follows, the more common geometries for side-pumping laser rods will be discussed.

The most widely used pump cavity is a highly reflective elliptical cylinder with the laser rod and pump lamp at each focus. The elliptic configuration is based on the geometrical theorem that rays originating from one focus of an ellipse are reflected into the other focus. Therefore an elliptical cylinder transfers energy from a linear source placed at one focal line to a linear absorber placed along the second focal line. The elliptical cylinder is closed by two plane-parallel and highly reflecting end plates. This makes the cylinder optically infinitely long.

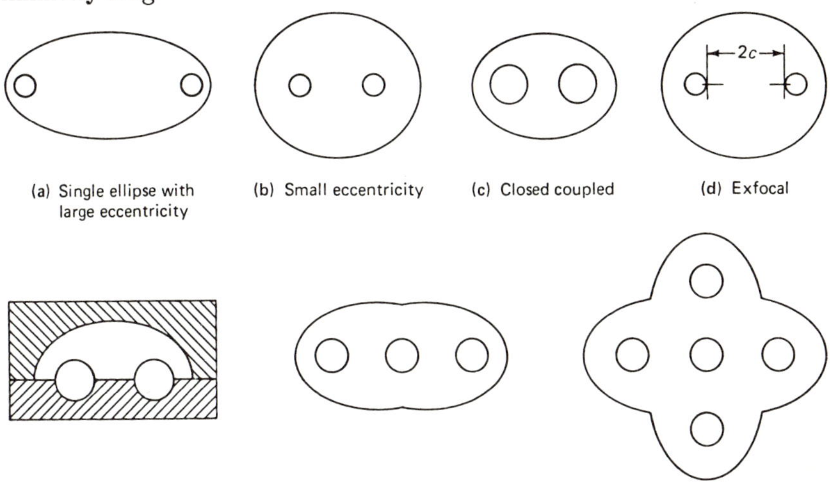

Fig. 6.49. Elliptical pump cylinders

Figure 6.49 illustrates different pump geometries based on the elliptical cylinder. A single elliptical cylinder can have a cross section with a large or small eccentricity (Figs. 6.49a and b). In the former case, the laser rod and lamp are separated by a fairly large distance, in the second case they are close together. If the elliptical cylinder closely surrounds the lamp and rod, then one speaks of a close-coupled elliptical geometry (Fig. 6.49c). As we will see, this geometry usually results in the most efficient cavity. This cavity has the further advantage that it minimizes the weight and size of the laser heads. The semi-

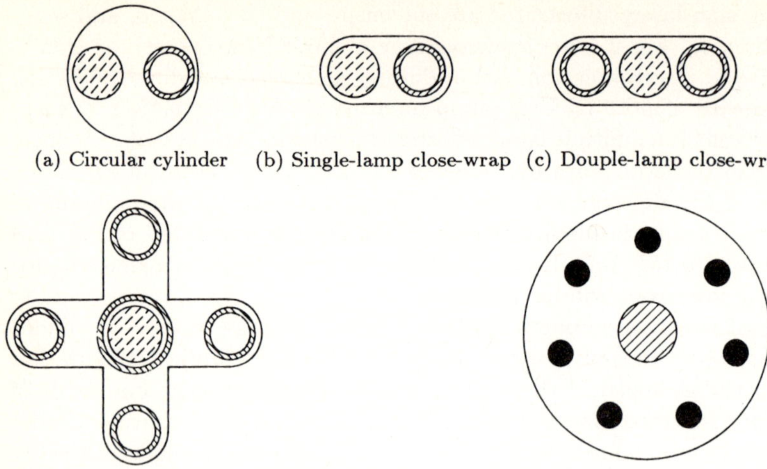

(a) Circular cylinder (b) Single-lamp close-wrap (c) Douple-lamp close-wrap

(d) Four-lamp close-wrap configuration (e) Close-coupled multiple coaxial

Fig. 6.50. Close-coupled configurations

elliptical cavity (Fig. 6.49e) is employed in cases where conduction cooling of the laser rod is desired. In an elliptical cylinder, radiation passing through the rod is quickly defocused upon further passes within the resonator and no longer reenters the rod. Focusing can be somewhat improved by locating the lamp and rod beyond the foci of the elliptical reflector [6.106]. With exfocal pumping, however, (Fig. 6.49d), the number of reflections from the cavity walls increases considerably, therefore the system efficiency is quite dependent on the wall reflectance. Since the energy delivered to a discharge lamp is limited, schemes to focus the energy from many lamps onto a single crystal are attractive. Figures 6.47f and g show two and four partial elliptical cylinders having one common axis at which the crystal is placed.

Figure 6.50 shows examples of close-coupled nonfocusing pump cavities. The lamp and rod are placed as close together as possible, and a reflector closely surrounds them. The reflector can be circular or oval in cross section (Figs. 6.50a and b). The latter type is often used in laboratory setups of low-repetition-rate pulsed lasers. The reflector is simply silver or aluminum foil wrapped around the flash lamp and laser rod. The efficiency of the closely wrapped cavity is found to be about as high as when an elliptical cylinder is used. The advantage of the pump cavity shown in Fig. 6.50b is fabrication simplicity; however, this design has a number of disadvantages: nonuniform pumping of the rod cross section and difficulties in providing adequate cooling of the rod, flashlamp, and reflector.

Multilamp close-coupled cavities provide a higher degree of pumping uniformity in the laser rod than single-lamp designs. The reflectors are used for creating the pumping density required when large-diameter rods are employed. The designs shown in Figs. 6.50c, d, and e are typically used to pump long Nd : glass laser rods.

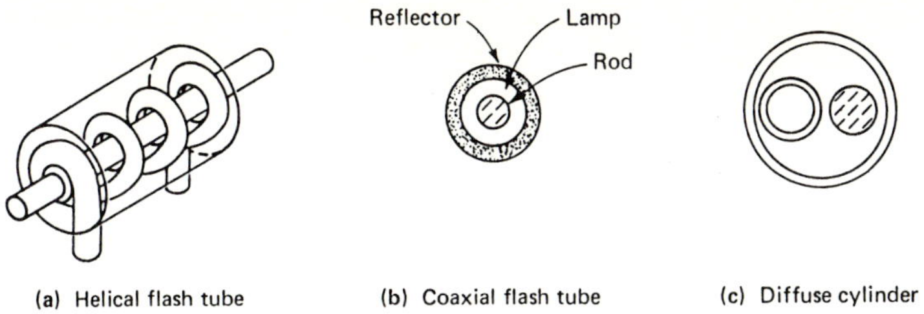

Fig. 6.51a–c. Diffuse reflecting cavities

Figure 6.51 shows designs which have diffuse-reflecting surfaces. If a helical lamp is used to pump a laser rod, the lamp is usually surrounded by a ceramic reflector, as shown in Fig. 6.51a. Occasionally, rods have been pumped by hollow lamps which were surrounded by a diffusely reflecting surface [6.45, 107, 108]. In the helical lamp and coaxial lamp pumping system shown in Figs. 6.51a and b, the efficiency is determined by the ratio of the inner lamp diameter to the rod diameter [6.108]. A very simple reflector is a ceramic cylinder with a laser rod and pump lamp in close proximity. In the design shown in Fig. 6.51c, the reflector sometimes is specular, i.e., aluminum foil or tubing. Ceramic materials have the advantage that they do not corrode or tarnish.

A high-energy transfer efficiency is achieved in pump systems with reflectors in the form of an ellipsoid of revolution [6.109] or a sphere [6.110], as shown in Fig. 6.52. In the spherical reflector the lamp and rod are positioned immediately adjacent to each other along a diameter of the reflector; in the ellipsoid of revolution the lamp and rod are positioned along the major axis between the focus and the ellipsoid surface. Somewhat similar to the ellipsoid is a reflecting cone system in which the lamp and laser rod are positioned on the axis of two cones, joined by a cylindrical reflecting surface [6.111].

The elliptical cylinders shown in Fig. 6.49 focus the pumping light in two dimensions, that is, in a plane which is the cross section of the cylinder. In contrast, the configurations illustrated in Fig. 6.52 focus the pumping light in three dimensions, thus a highly uniform rod illumination is achieved. Nevertheless, ellipsoids, spheres, and axicons have a number of disadvantages. They

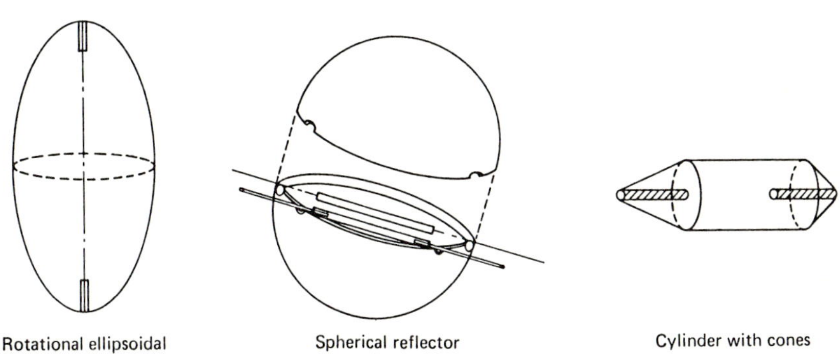

Fig. 6.52. Pump cavities with rotational symmetry

can be used for short laser rods only, and they are not very attractive from the standpoint of fabrication costs, size, and weight.

The choice of a particular pumping system usually depends on the following considerations:

1. Size of the laser rod and type of pump source: The use of small or large laser rods, single or multiple linear lamps, or helical lamps all require usually different designs.
2. Performance data: In designing a pumping system, a trade-off can be made between maximum efficiency and maximum beam uniformity by choosing either a focusing or a diffuse reflecting pump system.
3. System considerations: Cooling requirements, size and weight considerations, fabrication costs, etc., are all important factors which influence the final design.

In contrast to flashlamps or filament lamps, the emission from laser diodes is highly directional, therefore requiring different pump configurations. Arrangements for transferring pump radiation from laser diode arrays to the solid-state laser material will be discussed below.

We have to distinguish between optical systems employed for end-pumping of a laser crystal, side-pumping of a cylindrical rod, and side-pumping of a rectangular slab.

End-Pumping Configurations. The focused end-pumping configuration, if properly designed to provide matching of the pump light distribution and resonator mode, is the most efficient pump-radiation transfer scheme. Since the pump beam from the diode array is collinear with the optical resonator, the overlap between the pumped volume and the TEM_{00} mode can be very high. In addition, the coupling efficiency into the rod is high and the absorption length can be as long as the crystal. Figure 6.53 shows the experimental setup which was first used by *Sipes* [6.112]. Tightly focused end-pump geometries have become very popular in the design of small diode pumped lasers. The arrangement is employed in a number of small commercial lasers as well as for research of new materials.

For example, end pumping has been employed to obtain lasing from Nd:YAG [6.112–114], Nd:Glass [6.115], Nd:BEL [6.116], Nd:YVO_4 [6.116], Er:YLF [6.117] and Ho:AG [6.118, 120]. Also, the highest overall efficiency of any solid-state laser (12%) has been reported with the end-pump geometry [6.116].

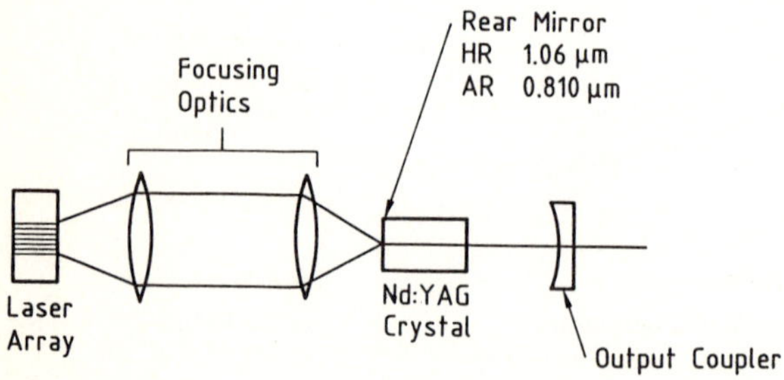

Fig. 6.53. End-pumped laser system [6.112]

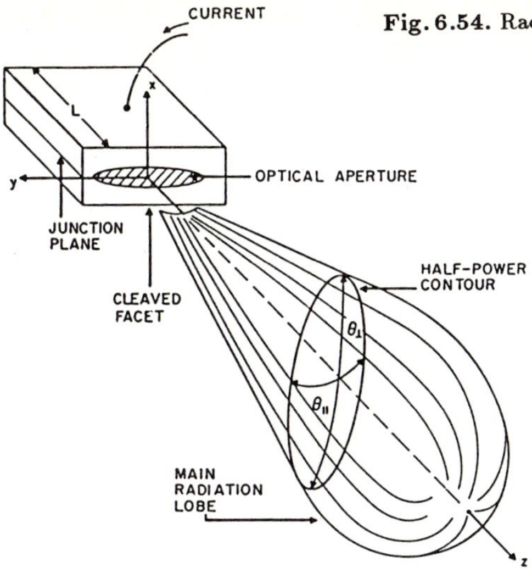

Fig. 6.54. Radiation pattern from a laser diode [6.119]

In its most basic form, end pumping involves a collimating lens with a large numerical aperture in order to collect radiation from a large cone angle, and a focusing lens to produce a small diameter spot inside the laser crystal. Since the radiation lobe emitted from a diode array has a high degree of astigmatism, as shown in Fig. 6.54, one can introduce prisms or cylindrical optics to transform the beam into a circular shape. Figure 6.55 illustrates the application of an anamorphic prism pair as part of the coupling optics between the laser diode array and the laser rod [6.113].

The purpose of the coupling optics is to shape the radiation distribution from the diode array such that the pumping volume coincides with the laser's TEM_{00} mode. Radiation from the laser diode array diverges approximately 40° in the plane perpendicular to the active layer. In the example shown in Fig. 6.55, the radiation is collected by a lens with a focal length of 6.5 mm

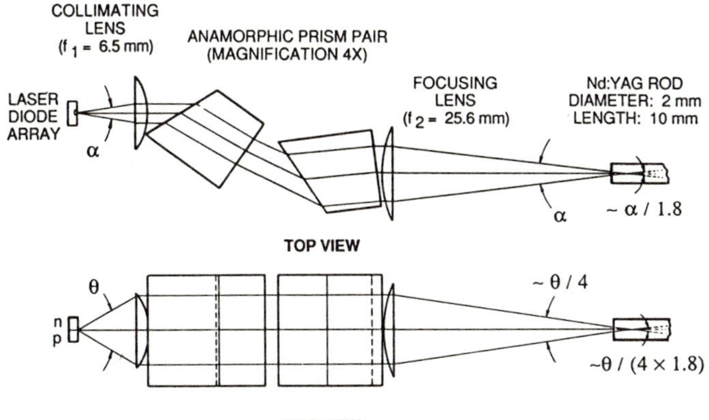

Fig. 6.55. Coupling optics between a diode array and a Nd:YAG crystal [6.113]

and a numerical aperture of 0.615. After passing through a pair of anamorphic prisms, the radiation is focused in the rod by a lens with 25.6 mm focal length. The resulting divergence inside the Nd:YAG rod is about 6°. Parallel to the junction, the low-divergence far field is collimated by lens L_1, expanded by the prism pair, and focused by lens L_2 into the rod.

In the particular example, a plano-concave configuration was chosen, with one planar rod surface, high-reflection coated at 1.06 μm and antireflection coated at 0.81 μm, for the pump light to enter the rod. The intracavity rod surface is antireflection coated at 1.06 μm and a 10 cm radius mirror was placed about 5.5 cm from the high-reflection-coated end of the 1 cm long rod.

For the cavity shown in Fig. 6.56 the maximum TEM_{00} spot size was $w_0 =$ 130 μm.

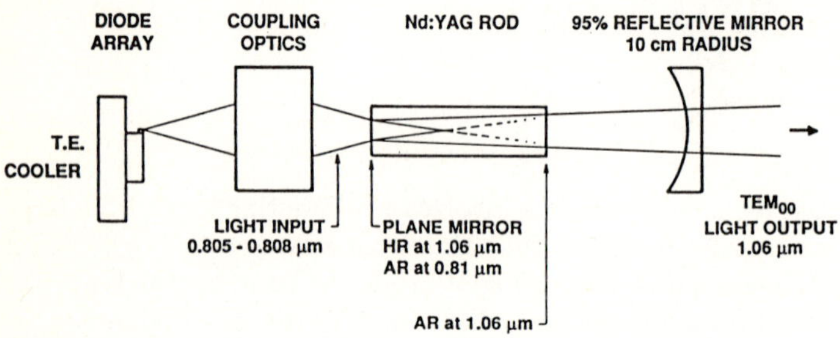

Fig. 6.56. Resonator parameters of an endpumped laser [6.113]

The equations given in Chap. 5 for the resonator modes allow the calculation of mode size and position. As we recall for a plano-concave resonator the spot size at the flat surface is given by

$$w_0^2 = (\lambda/\pi)[L(R-L)]^{1/2} \quad ,$$

where R is the concave mirror and

$$L = (l_{\rm rod}/n_{\rm rod}) + l_{\rm air}$$

where $l_{\rm rod}$ and $n_{\rm rod}$ are the length and refractive index of the rod, and $l_{\rm air}$ is the free space inside the resonator.

In general, the design of end-pumped coupling geometries usually proceeds by calculating the cross section of the TEM_{00} mode for a particular resonator structure. Then a focusing lens is chosen, which provides a spot size and divergence inside the rod of the pump radiation which closely matches the TEM_{00} mode.

End-pumping of a miniature Nd:YAG laser with laser diode arrays is an attractive means of obtaining efficient cw lasers. However, at present, the end-pump scheme is useful only for low-power lasers because the pump area is very small. In order to achieve a somewhat higher output power, two sources can be polarization-coupled to double the pump power available, or a double-ended pumping arrangement can be employed, as shown in Fig. 6.57.

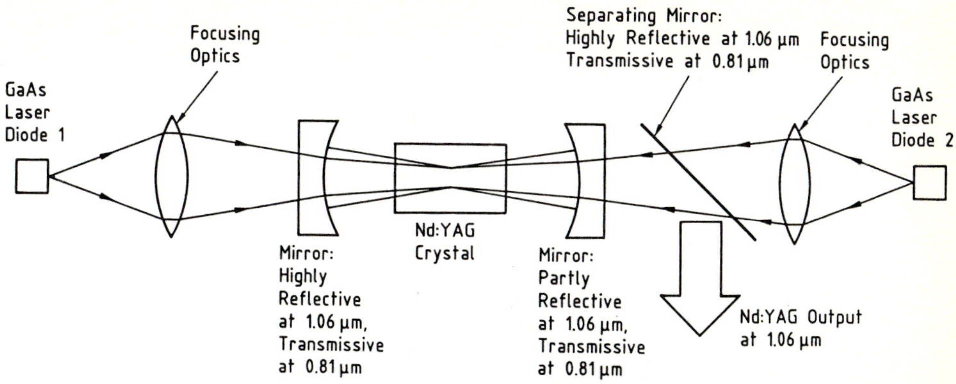

Fig. 6.57. End pumping of a Nd : YAG crystal with two GaAlAs laser diode arrays

Side Pumping of a Laser Rod. In this configuration, the diode arrays are placed along the length of the laser rod and pumped perpendicularly to the direction of propagation of the laser resonator mode. As more power is required, more diode arrays can be added along and around the laser rod. There are three practical approaches to couple the radiation emitted by the diode lasers to the rod: a) direct coupling; b) with optics between source and absorber; c) fiber-optics coupling. The direct coupling option does not allow for variations other than the placement of the diode lasers around the rod. Fiber-optics coupling is very impractical for a large number of diode lasers. Optical coupling can be achieved by using imaging optics such as lenses or elliptical and parabolic mirrors, or by non-imaging optics such as reflective or refractive flux concentrators.

Simple considerations regarding the optimum pumping configuration give high marks to the approach of using intermediate refractive optics [6.62, 122]. This is because, in this case, arrays can be spaced far apart, reducing packaging and cooling constraints. Furthermore, by using lenses, the pump distribution can be peaked at the center of the rod allowing for a better match with resonator modes. In contrast, direct pumping yields a far less optimum pump distribution and requires dense packaging.

Figures 6.58 and 59 illustrate typical designs. The laser rod is surrounded by a jacket to allow its cooling by forced convection using an axial coolant flow. The pumping cavity consists of a number of linear arrays symmetrically located around the rod. The arrays are mounted on long heat-sink structures and contain cylindrical optics. The optics allows the arrays to stand-off from the laser crystal, which is important for cooling of the laser crystal and the arrays. The output apertures of the diode lasers are imaged near the center of the rod by using a symmetrical doublet of plano-convex cylindrical lenses.

With a properly designed optical system, distribution of pump radiation can be adjusted to match the cavity mode volume. A detailed ray-trace program was developed for this geometry which permits optimization of the system. The laser arrays can either be arranged symmetrically around the laser crystal, or they can be distributed in a half cylinder. In the latter case, the laser crystal is

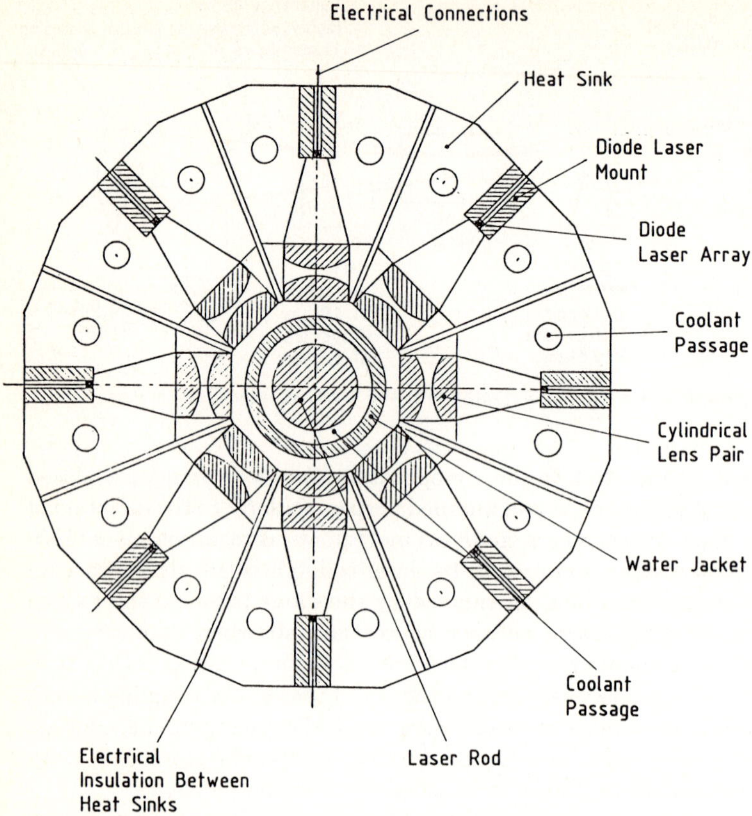

Fig. 6.58. Side pumping of a laser rod in a symmetrical pump configuration

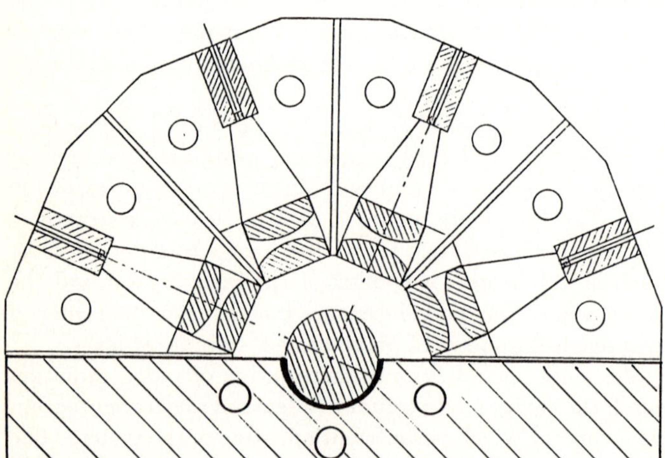

Fig. 6.59. Side pumping of a cylindrical laser rod employing a hemispherical pump geometry

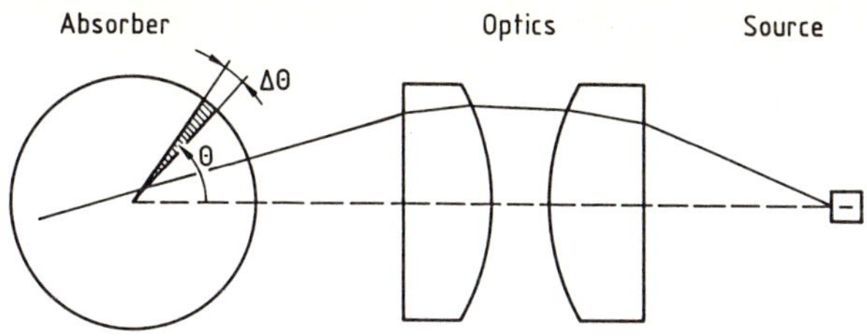

Fig. 6.60. Definition of optical path parameters for a side-pumped laser

mounted in a heat sink containing a back reflector. Figure 6.60 shows the optical path of a typical ray in the side-pumped configuration. Figure 6.61 exhibits the results of a Monte-Carlo program written for this configuration. The curves give the distribution of pump energy in the laser rod at specific angles, and also the average power density is given by integrating over all angles. Table 6.6 summarizes the pump efficiency which can be achieved with this design. The radiation transfer efficiency in a side-pumped geometry depends very strongly on the optical density (1/absorption coefficient α times optical pathlength l) of the laser crystal. In the case of Nd : YAG, the absorption coefficient is high over only a narrow linewidth on the order of 30 Å. Therefore, if the diode array has a much larger effective linewidth, due to variations in material composition or thermal gradients, the absorption increases, and efficiency suffers.

Table 6.6. Efficiency of pump radiation absorption in Nd : YAG as a function of rod diameter and diode-laser wavelength-distribution width. (Center wavelength of the distribution: λ_c=807 nm) Reflection losses are neglected

Bandwidth [Å]	Rod diameter [mm]	6	8
200		63 %	79 %
40		91 %	96 %
20		93 %	98 %

The side-pump geometry is not as efficient as the end-pump design, however, it permits scaling of the laser to large power levels. As a matter of fact, the side-pump geometry becomes more efficient for higher-power systems. Inefficiencies in side-pump geometries are normally result of a small absorption length, low pumping density, and wasted pump energy due to resonator-mode and pump-distribution mismatch. Large lasers require large rod diameters, thus increasing the pump efficiency; they also require densely packaged, high-power diode arrays which provide gains over the whole cross section of the rod at sufficient intensity to permit the use of unstable resonators. All these factors contribute to a more efficient utilization of the pump radiation, as compared to small systems.

Side-Pumped Slab Laser. Side-pumped slab lasers are usually pumped by 2-D arrays or densely-packed linear arrays. In a slab laser, the face of the crystal

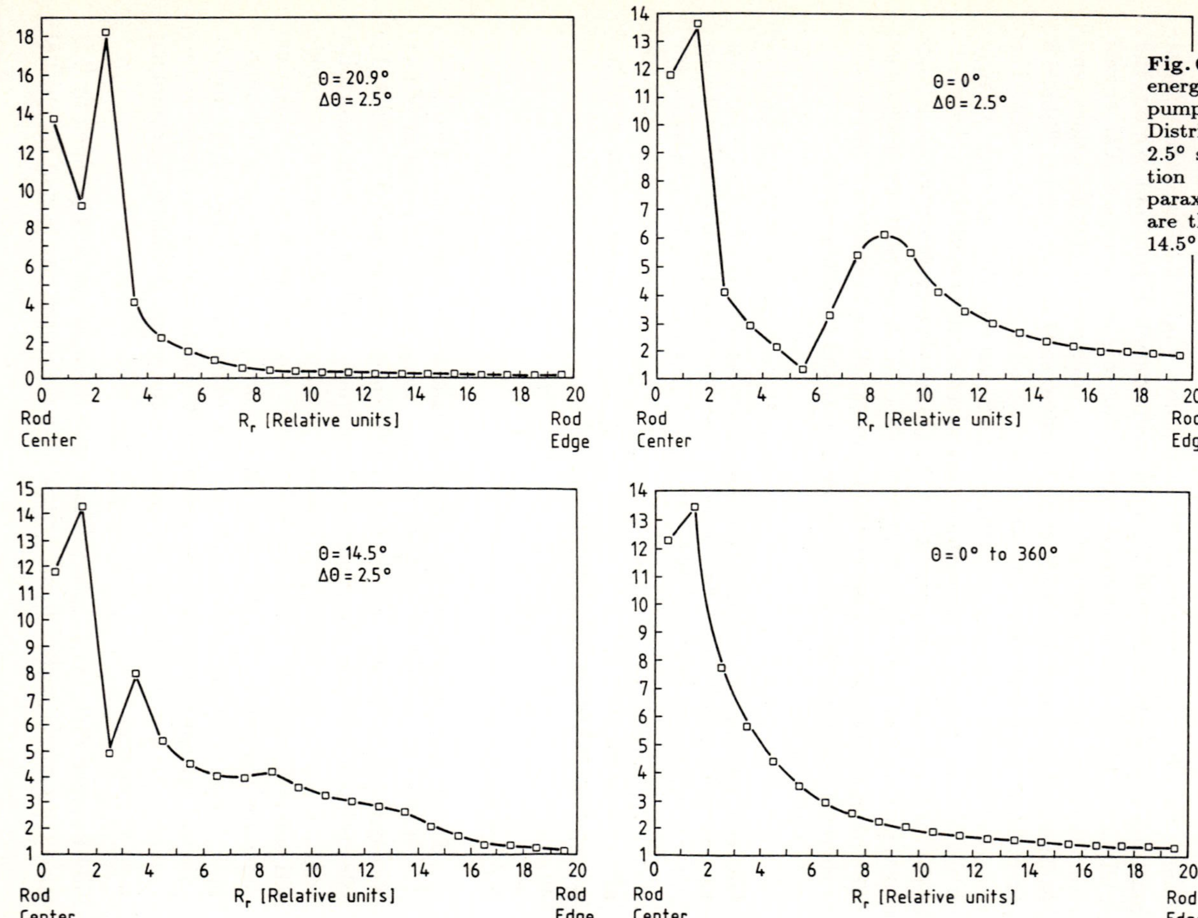

Fig. 6.61. Distribution of the energy density in a side-pumped Nd:YAG rod [6.62]. Distribution is calculated for a 2.5° slice of the rod cross section oriented at Θ from the paraxial direction. Illustrated are the cases $\Theta = 20.9°$, 0°, 14.5° and the angular average.

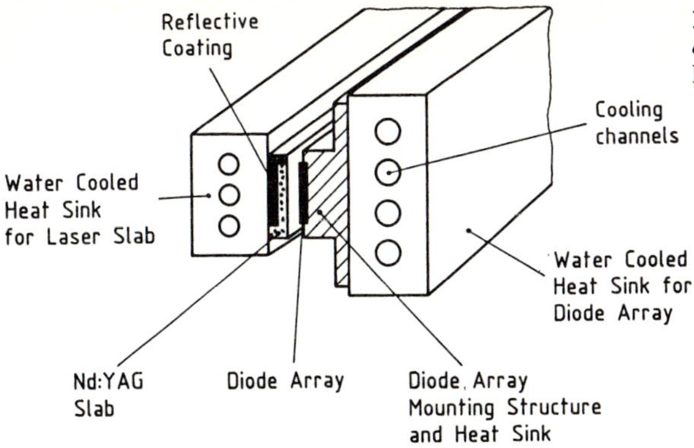

Fig. 6.62. Typical geometry of a laser-diode-pumped slab laser

and the emitting surface of the laser diodes are in close proximity, and no optics is employed. Figure 6.62 shows a typical design.

The Nd:YAG slab is pumped from one face only. The opposite face is bonded to a copper heat sink containing a reflective coating to return unused pump radiation back into the slab for a second pass. An antireflection coating on the pump face is used to reduce coupling losses (the diode array is not in contact with the YAG). Liquid cooling is employed to remove heat from the YAG and diode heat sink.

Lasers of the type above have been built [6.75] with 3.85 cm^2 of pump area, containing seven planar arrays each with an emitting area of 0.55 cm^2. On the average, each array produced a peak power of 413 W, in a 216 μs pulse. Electrical input to each array was 58 A and 30 V, which resulted in an optical output/electrical input efficiency of 24.2 %. The power density achieved at the arrays corresponds to 750 $\mathrm{W/cm}^2$. The total output energy derived by the seven arrays corresponds to 624 mJ. The Nd:YAG slab had a pump length of 8.9 cm, a height of 0.69 cm and a thickness of 0.67 cm.

6.3.2 Energy Transfer Characteristics

Theoretical Coupling Efficiencies

The amount of pump energy which is transferred from the source to the laser rod can be approximated by

$$\eta = \eta_{\mathrm{ge}} \times \eta_{\mathrm{op}} \quad , \tag{6.41}$$

where η_{ge} is the geometrical cavity transfer coefficient. It is the calculated fraction of rays leaving the source which reach the laser material either directly or after reflection from the walls. The parameter η_{op} expresses the optical efficiency of the cavity and essentially includes all the losses in the system. This parameter can be expressed as

$$\eta_{\mathrm{op}} = r_{\mathrm{w}}(1 - r_{\mathrm{r}})(1 - a)(1 - f) \quad , \tag{6.42}$$

where r_w is the reflectivity of the cavity walls at the pump bands, r_r is the reflection losses at the laser-rod surface or at the glass envelopes of the cooling jackets and Fresnel losses of any filters inserted in the cavity, a is the absorption loss in the optical medium between the lamp and the laser rod such as coolant liquid, filters, etc., and f is the ratio of the nonreflecting area of the cavity to the total inside area. This factor accounts for losses due to openings in the reflector which are required, for example, to insert the laser rod and pump lamp. Equations (6.41, 42) are rough approximations based on the assumption that all the lamp radiation undergoes just one reflection. If one were to include direct radiation and multiple reflections from the cavity walls, these equations would have to be developed in series expressions.

The elliptical cavity has been most extensively discussed in the literature [6.123–130]. As mentioned before, in this configuration a linear lamp and a laser rod, possibly with different radii, are placed at the foci of an elliptical cylinder. The pump source is usually believed to be a cylindrical radiator having a Lambertian radiation pattern. This implies that it appears as a source having constant brightness across its diameter when viewed from any point.

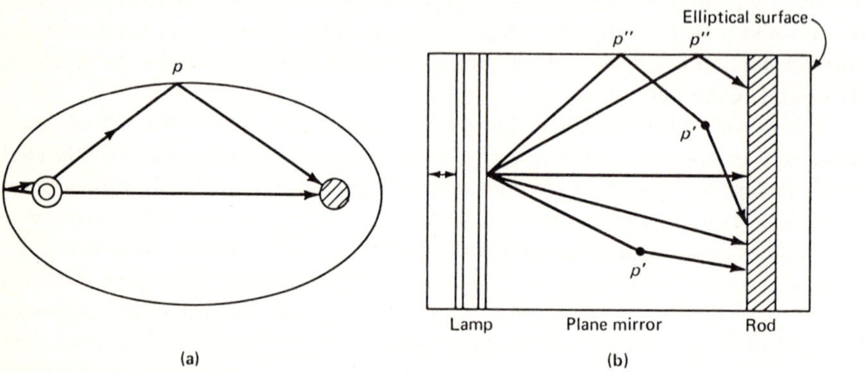

Fig. 6.63a, b. Trajectories of photons emitted from the pump source. **(a)** Cross section. **(b)** Top view of elliptical pump cavity

In Fig. 6.63 trajectories of photons are shown, which originate from a volume element dv of the source. In Fig. 6.63a the rays leave normal to the surface of the source and, therefore, remain in a cross-sectional plane. Figure 6.63b shows trajectories of photons which leave the source at an angle with respect to a cross-sectional plane of the ellipse. In this case the photons can be reflected off the end-plate reflectors as well as undergo a reflection at the elliptical cylinder. In an elliptical pump cavity, all rays emanating from one point of the source are transformed into parallel lines at the laser rod. Each line corresponds to rays leaving the source inclined at the same angle with respect to the major axis of the ellipse, but at different angles with respect to the cross-sectional plane.

Image formation, in its usual sense, is meaningless for the elliptical cylinder, since rays emanating in different directions from a point on the source

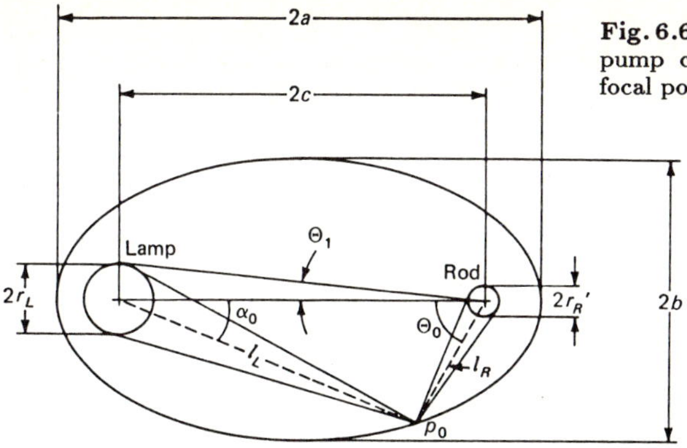

Fig. 6.64. Cross section of elliptical pump cavity. Eccentricity $e = c/a$; focal point separation $c = (a^2 - b^2)^{1/2}$

converge after reflection from the cavity at altogether different points. In a pump cavity, all that is actually desired is the transfer of radiant energy from the source to the laser rod. From the foregoing considerations, it is obvious that the condition of an infinitely long elliptical pump cavity can be satisfied by enclosing the ends of the cylinder with highly reflective plane mirrors. Therefore, in the analysis of this arrangement the two-dimensional case can be treated by considering the light distribution in a plane perpendicular to the longitudinal axes of the cylinder.

In the theoretical expressions obtained for the efficiency of elliptical configurations, the pump system is usually characterized by the ratio of the rod and lamp radii r_R/r_L, the ratio of the lamp radius to the semimajor axis of the ellipse r_L/a, and the eccentricity $e = c/a$, where $2c$ is the focal separation. Consider any point P on the surface of the cavity with a distance l_R from the crystal, and distance l_L from the lamp as shown in Fig. 6.64. Suppose that the lamp has a radius r_L, then, as a consequence of the preservation of angles, upon reflection the image will have the radius $r'_L = r_L l_R / l_L$. This means that the portion of the elliptical reflector nearer the lamp forms a magnified image at the laser rod while that portion nearer the crystal forms a reduced image of the lamp. A point P_0, with corresponding angles α_0 and Θ_0 measured from the lamp and rod axis, respectively, may be defined dividing these two regions. At this point, the ellipse generates an image of the lamp which exactly fills the crystal diameter. We must allow for this effect of magnification and demagnification when determining how much energy is captured by the crystal. From the properties of an ellipse and noting that at P_0, $l_R/l_L = r_R/r_L$, we obtain

$$\cos\alpha_0 = \frac{1}{e}\left[1 - \frac{1-e^2}{2}\left(1 + \frac{r_R}{r_L}\right)\right] \quad , \quad \text{and} \tag{6.43}$$

$$\sin\Theta_0 = \left(\frac{r_L}{r_R}\right)\sin\alpha_0 \quad . \tag{6.44}$$

The geometrical cavity transfer coefficient can be calculated by considering what fraction of the energy radiated by the lamp into an angle d_α is trapped by the crystal. Integration over all angles leads to [6.123, 128]

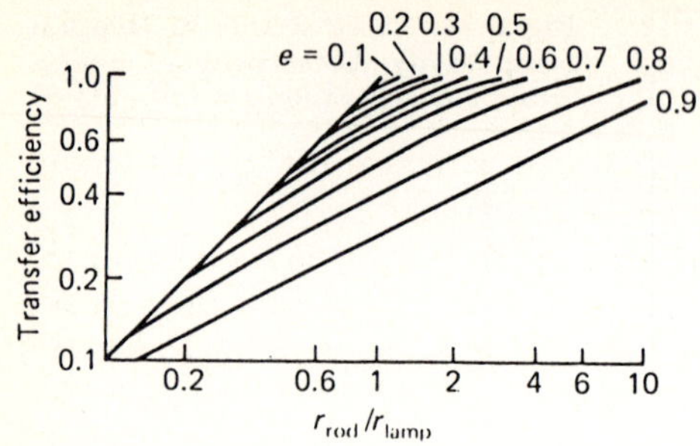

Fig. 6.65. Dependence of transfer efficiency of an elliptical reflector on the quantity r_R/r_L and eccentricity e [6.131]

$$\eta_{ge} = \frac{1}{\pi}\left[\alpha_0 + \left(\frac{r_R}{r_L}\right)\Theta_0\right] \quad . \tag{6.45}$$

This expression is plotted in Fig. 6.65. A certain portion of the reflecting surface behind the lamp is screened from the crystal by the lamp itself. In filament lamps and in pulsed flashlamps, which resemble a blackbody source, the radiation reflected back into the lamp will be absorbed. On the other hand, the plasma in cw arc lamps is optically thin, and the reflected pump radiation is transmitted through the plasma. If we assume the radiation reflected back to the pump source to be lost, we must reduce the angle Θ_0 in the above formula by Θ_1, where

$$\sin\Theta_1 = \frac{r_L}{4ae} \quad , \quad \text{and} \tag{6.46}$$

$$\eta'_{ge} = \frac{1}{\pi}\left[\alpha_0 + \left(\frac{r_R}{r_L}\right)(\Theta_0 - \Theta_1)\right] \quad . \tag{6.47}$$

This transfer coefficient is obtained from Fig. 6.66 by subtracting the lower set of curves from the upper curves.

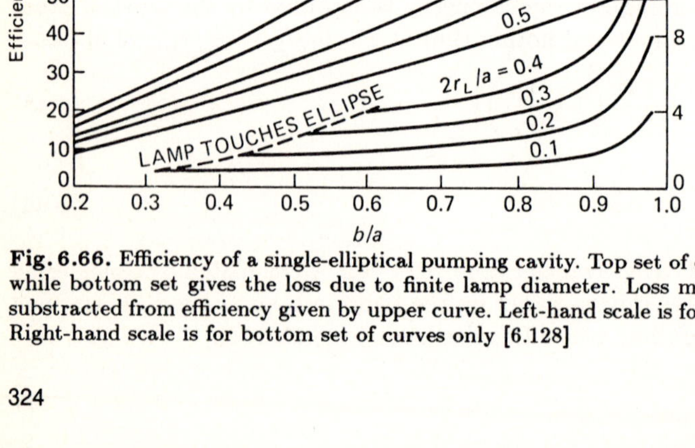

Fig. 6.66. Efficiency of a single-elliptical pumping cavity. Top set of curves is for $2r_L/a = 0$, while bottom set gives the loss due to finite lamp diameter. Loss multiplied by r_R/r_L and substracted from efficiency given by upper curve. Left-hand scale is for top set of curves only. Right-hand scale is for bottom set of curves only [6.128]

It is seen from Figs. 6.65, 66 that the efficiency increases with an increase in the ratio r_R/r_L and with a decrease in the eccentricity e. This result stems from the fact that the magnification of the pump source increases with the eccentricity of the ellipse. A superior efficiency is therefore obtained for an almost circular cavity and as small a pump source as possible.

It should be noted that in these calculations the crystal is assumed to absorb the radiation which falls upon it; no second traverses of light through the crystal are considered. Furthermore, no allowance was made for the reflection loss at the cavity walls or losses due to Fresnel reflection at the surface of the laser rod. Multiple reflections can become significant if the absorption coefficient of the crystal is small compared to the inverse of the crystal diameter [6.124, 127].

Figure 6.67 shows a three-dimensional illustration of the dependence of the transfer efficiency in an elliptical cylinder on the reflectivity γ of the end plates, the ratios r_L/r_R, and $2a/r_R$ [6.126]. For the elliptical portion of the cavity a reflectivity of one is assumed. The curves are calculated for a length of the cylinder of $l = 20r_R$ and a distance between the rod and the lamp of $2c = 8r_R$. Therefore, a large value of the variable $2a/r_R$ indicates that the eccentricity of the ellipse $e = c/a$ is small, and vice versa. As may be seen from Fig. 6.67, the efficiency increases if one decreases the radius of the lamp. It should be noted that for $\gamma = 1$ the cavity should have an eccentricity as small as possible in agreement with other theoretical calculations. However, when $\gamma < 1$, the optimum eccentricity increases with decreasing values of the reflectivity. For example, for $r_L/r_R = 1$ and a reflectivity of $\gamma = 0.60$, the optimum ellipse has a major axis of $2a/r_R = 16$ or an eccentricity of 0.5. For a

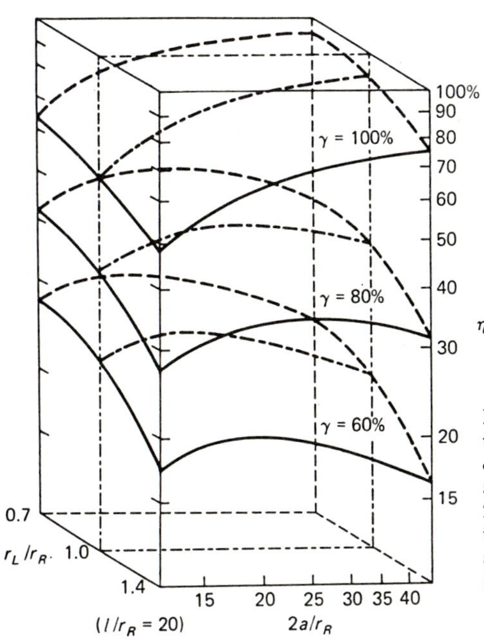

Fig. 6.67. Three-dimensional illustrations of the dependence of total transfer efficiency η upon three parameters that are reflectivity γ of mirrors, r_L/r_R, and $2a/r_R$, where r_L and r_R are the radii of the lamp and the laser rod, respectively, and $2a$ is the major axis of the ellipse ($2c/r_R = 8$) [6.126]

given focal point separation $2c$, an ellipse with a large eccentricity has a small cross-sectional area (Figs. 6.49a and b), which tends to reduce the losses caused by the end mirrors.

The analytical methods developed for the design of pumping configurations imply idealized systems in which reflection and refraction losses are usually ignored and only a single reflection from the walls is considered. In order to analyze pumping systems with fewer idealizations, statistical sampling methods, such as the Monte-Carlo method, have been employed. The Monte-Carlo method consists essentially of tracing, in detail, the progress of large numbers of randomly selected rays over many events (i.e., absorptions, reflections, and refractions) throughout the pumping cavity. The average transfer efficiency of these rays can then be taken as representative of all possible rays in the cavity. Monte-Carlo calculations for a variety of geometrical configurations are reported in [6.132–135]. For the single-elliptical pump cavity [6.132], the adverse effect of a small value of the quantity r_R/r_L was confirmed. The analysis also showed the efficiency to be strongly dependent on the reflector reflectance. On the other hand, the calculated efficiency values appear to be less dependent on the eccentricity than indicated by the results of earlier work [6.123, 128], in which multiple passes were ignored.

The results of calculations performed for a single-elliptical cavity containing a ruby rod pumped by a xenon flashlamp are shown in Table 6.7. Shown is the absorption of pump radiation by the ruby rod after one and three reflections, and the fraction of radiation returned to the lamp and absorbed by the cavity walls. In the case of perfectly reflecting walls ($\gamma = 1$), the only two absorbers in the cavity are the ruby rod and the lamp. For small ellipses (large r_R/a) and ellipses of low eccentricity, a significant fraction of the light is incident on the rod which has not undergone a reflection on the cavity walls. Also, in some cases, a large fraction of pump radiation is returned to the source. For ellipses with eccentricities between 0.2 and 0.6, the transfer efficiency is reduced by about 35 % if reflectivity drops from $\gamma = 1$ to $\gamma = 0.6$. For ellipses with very large eccentricities, the decrease is about 50 %. In these cavities there is less direct pumping, and the light has to undergo more reflections before beeing absorbed by the rod.

Multiple-elliptical laser pump cavities generally consist of a number of partial-elliptical cylinders having a common axis at which the crystal is placed. The results of calculations performed to obtain the efficiency of various multiple cavities with varying sources and crystal diameters show that, for equal diameters of the source and the crystal, the single-elliptical cavity has the highest efficiency. However, a cavity with two or more partial-elliptical cavities will allow higher input power, at some sacrifice in overall efficiency, if a high-power output laser is required. If the diameters of the source and crystal are different, these conclusions must be modified. It should also be pointed out that the multi-ellipse cavity results in more uniform pumping of the laser crystal.

Referring back to the case of a single-elliptical cylinder, (6.45) gives the efficiency for the case when the lamp does not block radiation. For multi-elliptical pump cavities one obtains the expression

Table 6.7. Transfer of pump radiation in a single-elliptical pump cavity [6.132]. Values are for $r_R = r_L$

Eccentricity e	Lamp radius/semimajor axis r_R/a	Wall reflectivity γ	Efficiencies: Direct radiation	After 1 reflection	After 3 reflections	Radiation returned to lamp	Radiation absorbed by walls
		1.0	0.18	0.56	0.64	0.36	0
	0.395	0.8	0.16	0.51	0.58	0.25	0.17
		0.6	0.16	0.38	0.41	0.23	0.35
0.4		1.0	0.05	0.67	0.74	0.25	0.18
	0.160	0.8	0.04	0.52	0.63	0.19	0.18
		0.6	0.04	0.45	0.49	0.12	0.39
		1.0	0.10	0.52	0.62	0.37	0
	0.395	0.8	0.12	0.45	0.51	0.26	0.23
		0.6	0.11	0.39	0.41	0.21	0.38
0.6		1.0	0.02	0.55	0.76	0.22	0
	0.160	0.8	0.02	0.47	0.61	0.12	0.27
		0.6	0.02	0.37	0.46	0.11	0.43
		1.0	0.17	0.71	0.73	0.26	0
	0.195	0.8	0.17	0.62	0.63	0.20	0.17
0.2		0.6	0.17	0.48	0.49	0.15	0.36
		1.0	0.07	0.77	0.82	0.16	0
	0.12	0.8	0.07	0.69	0.70	0.10	0.20
		0.6	0.08	0.54	0.55	0.09	0.36
		1.0	0.02	0.40	0.69	0.29	0
	0.195	0.8	0.02	0.30	0.47	0.23	0.30
0.8		0.6	0.03	0.25	0.34	0.21	0.45
		1.0	0.02	0.42	0.81	0.17	0
	0.12	0.8	0.02	0.30	0.54	0.14	0.32
		0.6	0.02	0.29	0.40	0.09	0.51

$$\eta_{eg} = \frac{1}{\pi}\left[(\alpha_0 - \alpha_2) + \left(\frac{r_R}{r_L}\right)\Theta_0\right] \quad . \tag{6.48}$$

Here, α_0 has been reduced by an angle α_2 which represents that portion of the reflecting wall that was cut away. Light falling in this region is assumed to be lost. For a double-elliptical pumping cavity,

$$\cos\alpha_2 = \frac{2e}{1+e^2} \quad . \tag{6.49}$$

Equation (6.48) is plotted in Fig. 6.68.

We will briefly discuss pumping systems other than elliptical cylinders. The analysis of a system with a circular reflective cylinder is analogous to that for an elliptical system. The lamp and laser rod are symmetrically positioned at the same distance from the cylinder axis. The system efficiency increases with an increase in the ratio r_R/r_L and decreases for increases in the ratios of F/R_0, where F is the distance from the cylinder axis to the axes of the lamp or rod [6.136].

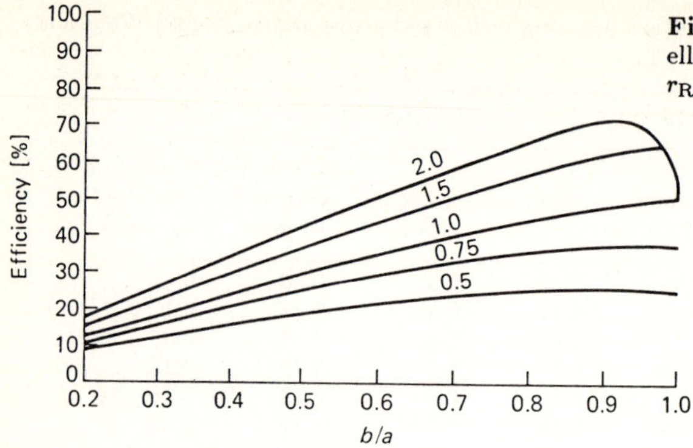

Fig. 6.68. Efficiency of a double-elliptical cavity. Parameters are for r_R/r_L [6.128]

The transfer efficiency of spherical and ellipsoidal cavities has been calculated in [6.137], and a review of most common pumping systems has been given by *Kalinin* and *Mak* [6.131].

The transfer efficiency of laser pumping cavities that have diffusely reflecting walls can be estimated if we assume that the light inside the cavity is isotropic. Since all surfaces will be equally illuminated, the light absorption at each surface will be the product of the area and the fractional absorption. Since all radiation must eventually be absorbed somewhere, the proportion of light absorbed at a surface area S_1 having a fractional absorption A_1 will be [6.134]

$$\eta = \frac{S_1 A_1}{\sum_{i=1}^{n} S_i A_i} , \tag{6.50}$$

where the summation covers the entire internal surface of the cavity. If S_1 is the exposed surface area of the laser crystal and A_1 is the effective absorption over this surface, the transfer efficiency of the laser configuration is given by

$$\eta = \frac{S_1 A_1}{S_1 A_1 + S_2 A_2 + S_3 A_3 + S_4} , \tag{6.51}$$

where S_2 and A_2 are the diffusely reflective wall area and absorption coefficient of the cavity, S_3 is the surface area of the lamps, and A_3 is the absorption of pump radiation by the lamp. S_4 is the area covered by holes in the cavity wall; these surfaces are treated with unit absorption, i.e., $A_4 = 1$.

An important parameter in the formula for transfer efficiency is A_1, the fraction of photons which strike the laser rod surface and are absorbed by the laser material. This parameter, which has been called the capture efficiency, depends on the absorption coefficient, the refractive index, and the radius of the laser rod.

Figure 6.69 shows the capture efficiency A_1 of a cylindrical crystal exposed to isotropic light [6.134]. Radiation incident on the rod which is not absorbed is either reflected off the surface or passes through the laser material. For large values of αr_R the capture efficiency is limited only by Fresnel reflection losses

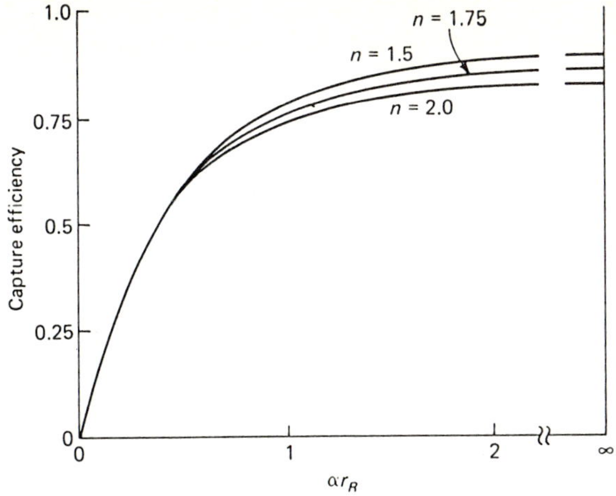

Fig. 6.69. Capture efficiency of a cylindrical laser rod exposed to isotropic light. The parameters n, r, α are the refractive index, radius, and absorption coefficient of the rod [6.134]

on the crystal surface. Using this grossly simplified analysis, no significant discrepancy between it and the Monte-Carlo calculation is usually found.

With the aid of (6.51) and Fig. 6.69, we will calculate the transfer efficiency of the diffusely reflecting cavities shown in Figs. 6.65 and 3.5. Both cavities are 10 cm long and contain a 1-cm-diameter ruby crystal. In the first case, the crystal is pumped by a linear flashlamp and, in the second case, by a helical flashlamp. For ruby with an index of refraction of 1.76 and an average absorption coefficient in the pump bands of $1.0\,\mathrm{cm}^{-1}$, we obtain a capture efficiency of $A_1 = 0.60$ from Fig. 6.69; the surface area is $S_1 = 31.4\,\mathrm{cm}^2$. For both lamps we assume that 10 % of the pump radiation is reflected back to the lamp and is reabsorbed by the plasma ; that is, $A_3 = 0.1$. The linear lamp is of the same size as the laser rod ($S_3 = 31.4\,\mathrm{cm}^2$). The helical lamp is approximated by an annular region of isotropically radiating material with an inner diameter of 1.8 cm. The reflectivity of the diffusely reflecting walls is assumed to be 95 %. The cavity diameter is 4 cm, from which we obtain $S_2 = 150\,\mathrm{cm}^2$, if we include the end walls. The area covered by holes and apertures is about $S_4 = 0.1 S_2$. Introducing these parameters into (6.51), we obtain $\eta = 0.42$ and $\eta' = 0.40$ for the linear and helical lamp-pumped ruby system.

Empirical Data

Theoretical considerations of the efficiency of pump cavities are of utility primarily in their qualitative aspects and in their illustration of the gross features of the energy transfer in the various reflector configurations. Even under very idealized assumptions, calculation of the amount of energy absorbed in the laser rod is very tedious since the spectral properties of the pumping source, the absorption spectrum of the laser material, and all possible path lengths of the elementary rays through the laser rod, determined by the specific reflector configuration, must be taken into consideration. Not even the most thorough computations can take into account the effect of mechanical imperfections in the fabrication of the reflectors, optical obstructions caused by seams and gaps, aberrations by flowtubes, losses due to holes and apertures in the end plates, etc.

Summarizing the experimental work carried out on pump-cavity design, we may draw the following conclusions:

- For maximum efficiency and relatively short rods (up to 10 cm in length) a small single-elliptical cylinder, an ellipsoid of revolution, or a spherical cavity is best. For larger rods, the closed-wrapped geometry (Fig. 6.50b) is about as efficient.
- Small elliptical cavities with low major axis-to-rod diameter ratios are more efficient than large cavities. In a small elliptical cavity the fraction of direct radiation is high, and most of the pump radiation is incident on the rod after a single reflection on the cavity walls. Therefore, in this geometry, a reflectivity of less than unity, imperfections in the geometry, and obstruction of the light path are less detrimental to the efficiency than is the case in a large elliptical cylinder. Another strong argument for making elliptical cavities as small as possible is the increased probability of the lost radiation being redirected to the laser rod if the ratio of cavity volume to laser rod volume is small.
- In agreement with theory, the highest efficiency in an elliptical cylinder is obtained for a large r_R/r_L ratio. However, lamp life considerations demand that in most cases a ratio of one is chosen.
- A high reflectivity is extremely important, whereas mechanical tolerances of the elliptical shape are not very critical in small ellipses. This is not surprising because in a close-coupled ellipse the cross sections of lamp and rod are fairly large with respect to the dimensions of the ellipse, which reduces the requirement for accurate imaging.
- The length of elliptical-cylinder cavities should be as great as possible in relation to the diameter, but the laser rod, pump lamp, and cavity should all be the same length.
- Optimum small elliptical-cylinder pump cavities are equally as good as ellipsoid-of-revolution pump cavities.
- Multiple cylindrical ellipse cavities may be utilized to increase the number of pump lamps and to produce symmetrical pumping. If the cavity size is kept small, this can be achieved with a minimum sacrifice in efficiency. Figure 6.70 shows the reflection characteristics of a double-elliptical cavity.

In designing a close-coupled single-elliptical cylinder, the following procedure can be followed: First, the smallest separation of rod and pump source is determined according to

$$2c = s_1 + \frac{d_r + d_L}{2} \quad , \tag{6.52}$$

where d_r and d_L are the clearances required for the rod and lamp assembly, respectively. These two dimensions are normally several times larger than the

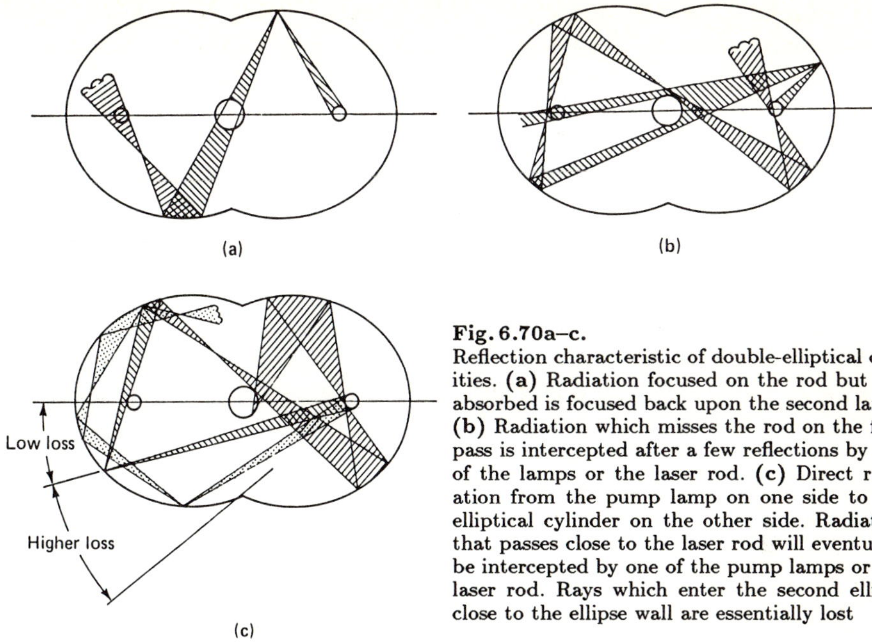

Fig. 6.70a–c.
Reflection characteristic of double-elliptical cavities. **(a)** Radiation focused on the rod but not absorbed is focused back upon the second lamp. **(b)** Radiation which misses the rod on the first pass is intercepted after a few reflections by one of the lamps or the laser rod. **(c)** Direct radiation from the pump lamp on one side to the elliptical cylinder on the other side. Radiation that passes close to the laser rod will eventually be intercepted by one of the pump lamps or the laser rod. Rays which enter the second ellipse close to the ellipse wall are essentially lost

physical diameter of the rod and lamp in order to allow space for flow tubes, O-rings, rod and lamp supports, rod holders, lamp terminals, cooling channels, etc. s_1 is the separation between the rod and lamp assembly; it is a safety factor to allow for tolerance buildup.

After the separation of the focal points has been determined, we can calculate the major axis of the ellipse:

$$2a = s_1 + 2s_2 + d_r + d_L \quad , \tag{6.53}$$

where s_2 is the clearance of the rod and lamp assembly from the elliptical surfaces. We see from (6.52, 53) that in the extreme, $s_1 = s_2 = 0$, $a = 2c$, and the eccentricity of the ellipse is $e = 0.5$. In practical cases, e will be somewhat lower.

The minor axis of the ellipse follows from

$$b = (a^2 - c^2)^{1/2} \quad . \tag{6.54}$$

Figure 6.78 shows the cross section of a double-elliptical cavity of a cw-pumped Nd:YAG laser designed to yield a close separation between laser rod and lamps. Figure 6.79 shows the outside configuration of the laser head.

Pump-Light Distribution in the Laser Rod

The pump-light distribution over the cross section of the active material is the result of a combination of three effects: the illumination properties of the pump cavity, refractive focusing occurring in the rod itself, and nonuniform absorption of pump radiation. How dominant these individual effects are depends on the pumping geometry, i.e., focusing or diffuse reflecting, on the treatment of the cylindrical surface of the rod, i.e., rough or polished, and on the optical thickness, i.e., the product of absorption coefficient and radius.

We can superficially distinguish between strongly focused, uniform, and peripheral pump light distribution. A nonuniform distribution of pump energy in the active medium leads to a nonuniform temperature and gain coefficient distribution. Saturation effects, threshold energy, and possible laser rod damage depend on the gain distribution across the rod. The distribution of pump light in a cylindrical laser rod will be considered for two cases of practical interest: a strongly focusing elliptical cylinder and a diffuse reflecting cavity.

Considering first the case of a laser rod in a diffuse reflecting cavity, we find that if a laser rod with a polished surface is exposed to isotropic pump radiation, appreciable focusing results from the refraction of the pump light at the cylindrical surface. Rays which strike the rod periphery tangentially will pass the center of the rod at a radius [6.138]

$$r_{\mathrm{i}} = \frac{r_{\mathrm{R}}}{n} \quad . \tag{6.55}$$

Furthermore, all radiation incident on the surface passes through a core bound by the diameter $2r_{\mathrm{i}}$. The pump light intensity in this region is higher than in the boundary region $r_{\mathrm{R}} \geq r \geq r_{\mathrm{i}}$. The focusing action from the refraction of pump light at the surface of a polished rod is modified by absorption [6.131, 139]. The pump intensity as a function of path length x in the rod is given by

$$\frac{I(\nu)}{I_0(\nu)} = \exp[-\alpha(\nu)x] \quad . \tag{6.56}$$

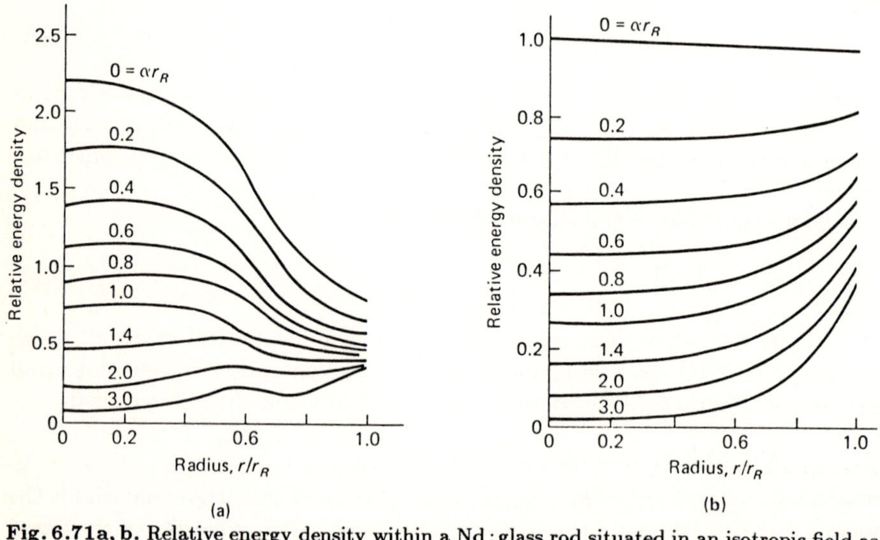

Fig. 6.71a, b. Relative energy density within a Nd:glass rod situated in an isotropic field as a function of normalized radius for different absorption values. **(a)** With polished, **(b)** with frosted lateral surface [6.131]

Pump radiation propagating to the center of the rod is attenuated by absorption at the periphery. The compensating effects of refractive focusing and pump light absorption can be seen in Fig. 6.71a. Plotted is the relative energy density within a Nd: glass rod situated in an isotropic field as a function of normalized radius for a number of absorption values [6.131]. It is seen that for an optically thin rod, refractive focusing is predominant and the pump density is highest in the central region ($r \ll 0.63 r_R$) of the rod. On the other hand, in rods which are optically dense, the pump intensity is highest on the periphery.

Several solutions exist to improve the pumping uniformity of laser rods located in diffuse cavities. The internal focusing effect can be reduced by roughening the surface of the rod, immersing the rod in a liquid (ideally an index-matching fluid), and cladding the rod. In the latter case the active core is surrounded by a transparent layer, such as sapphire for ruby rods and glass for Nd: glass rods. This makes the collecting cross section larger than the absorbing cross section.

Figure 6.71b shows the pump-light distribution in a rod with a frosted lateral surface. No focusing effect is observed and the rod exhibits a peripheral pump light distribution which is st[Arictly determined by absorption. At low absorption values αr_R a fairly uniform pump light distribution is obtained in the laser material. Generally, the most uniform pump distribution is obtained in a laser rod which has a roughened side wall, is optically thin, and is pumped by a helical flashlamp.

If the laser rod is in a focusing pump cavity, the light source is imaged on the laser rod. When the region of maximum energy density in free space is smaller than the cross section of the laser rod, a highly illuminated core results, whose diameter is no longer given by the outside diameter of the rod but by the image-forming properties of the pump system and the index of refraction of the active material.

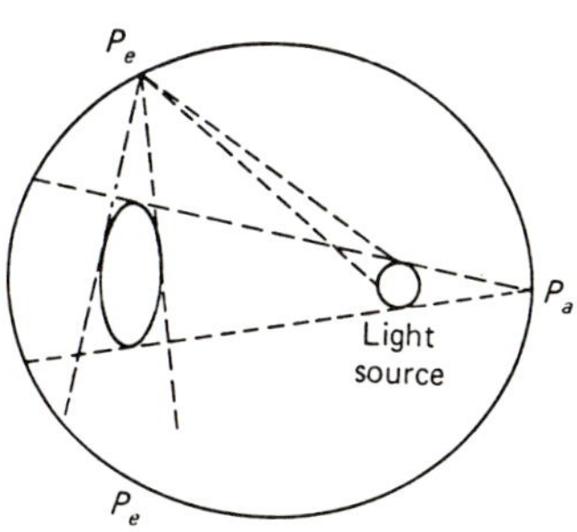

Fig. 6.72. Reflections of different light cones emerging from a circular light source located inside an elliptical reflector

Figure 6.72 shows the reflection of different light cones emerging from a circular light source of an elliptical reflector. Reflection at point P_a gives the beam of greatest diameter, perpendicular to the major axis. Reflection at P_e gives the beam of least diameter parallel to the major axis. The three cones which are reflected at points P_a and P_e define the region through which the total flux of the light source passes. This oval region has a long axis (ϱ_a) parallel to the minor axis of the ellipse, and a short axis (ϱ_b) parallel to the major axis of the ellipse. We have [6.140]

$$\varrho_a = r_1 \frac{a/e+1}{a/e-1} \quad \text{and} \quad \varrho_b = r_1 \frac{(a/e)^2-1}{(a/e)^2+1} \ , \tag{6.57}$$

where r_1 is the radius of the light source. Note that ϱ_a and ϱ_b are the boundaries in free space, inside a laser rod these dimensions are reduced by the ratio $1/n$.

From these considerations it follows that, while the emission of the light source is axially symmetrical, the energy density distribution at the site of the laser is no longer axially symmetrical in an elliptical cavity. There is a preferred axis perpendicular to the long axis of the ellipse. The region of maximum energy density is of interest for laser operation, since it is here that the material first reaches the threshold inversion. The focusing action is strongest in a laser rod with a polished surface located in an elliptical pumping cavity.

The distribution of the pump energy within the laser rod can be studied by taking near-field pictures of the pattern of fluorescence emission. A typical picture obtained in this way from a ruby rod pumped by a flashlamp in a single elliptical cavity is shown in Fig. 6.73 [6.142]. The pump light is mostly concentrated in a central core approximately elliptical in shape with its major axis orthogonal to the major axis of the elliptical cylinder, which is in agreement with the aforementioned theoretical considerations. Facing the flashlamp, a wedge-shaped bright region, which is caused by direct radiation from the flashlamp, is present in the rod.

Figure 6.74 shows the result of a computer calculation carried out to determine the pump radiation distribution in a ruby rod pumped by a xenon flashlamp in a single-elliptical cavity [6.132]. Indicated in Fig. 6.74 are the ar-

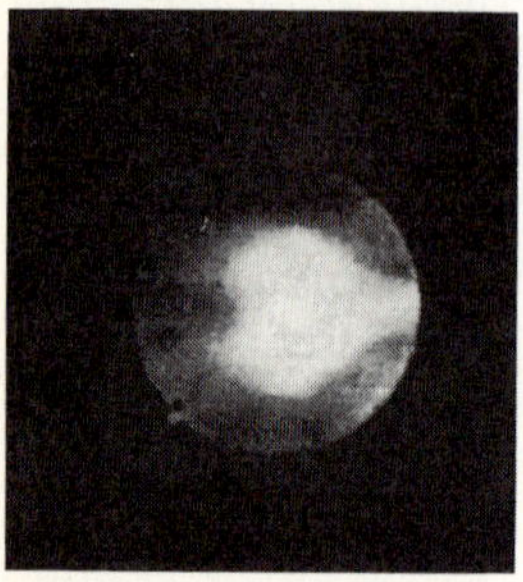

Fig. 6.73. Distribution of fluorescence in a polished ruby rod as a result of illumination in an elliptical reflector [6.141]

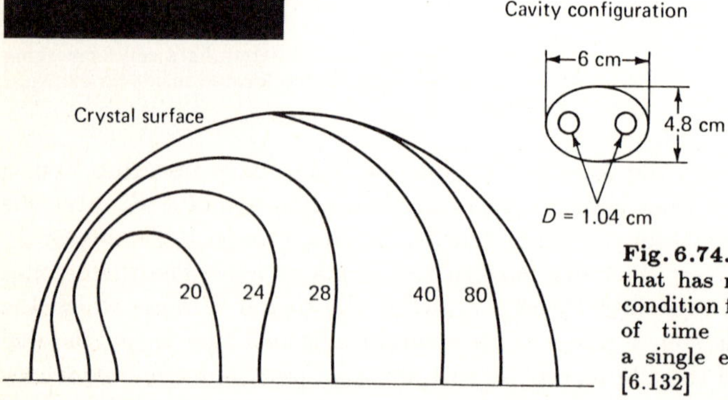

Fig. 6.74. Area of ruby crystal that has reached the threshold condition for lasing as a function of time (arbitrary units) in a single elliptical pump cavity [6.132]

eas of the ruby crystal which reach threshold as a function of time. The pump nonuniformity in the crystal is again a combination of the focusing action of the ellipse and direct radiation from the pump source.

In order for strong focusing to occur, it is necessary that the rod be fairly transparent to the pump radiation. Standard-doped ruby, Nd: glass, and Nd: YAG rods of up to 1 cm diameter are sufficiently transparent for strong focusing effects. With three-level systems, where a substantial change in the ground-state population will occur during pumping, a rod that is initially opaque may become transparent in the course of the pump pulse. The pump radiation can "bleach" the rod, so to speak, or "burn" its way in. The bleaching of ruby rods can be very substantial when the rods are used as storage devices (amplifiers, giant pulse lasers). Here, no regeneration is present during the pump phase of the laser cycle and the ground state can be almost completely depleted.

One of the consequences of focusing is that the threshold for oscillation is lowered in the focal region. Concentration of pump light at the center of the rod can be desirable in systems which operate at the TEM_{00} mode, since operation in this mode requires maximum gain in the center.

If the lateral surface of the rod is frosted, then the pump radiation entering the rod remains diffuse; in this case radiation from the entire surface arrives at each point within the rod. Frosting of the lateral surface of the rod provides for a considerably more uniform gain distribution, and a larger cross section of the rod is utilized. However, an asymmetric gain distribution remains, since the rod is pumped from one side. To eliminate this asymmetry, a dual flashtube and close-coupled geometry can be utilized. This configuration provides for a relatively uniform gain distribution. If the laser rod is surrounded by a flowtube for water cooling, the glass tube can be ground to achieve a homogeneous pump light distribution in the laser rod [6.142].

6.3.3 Mechanical Design

In the construction of a pump cavity, several critical design areas can be identified; these are efficient cooling of the laser rod, lamps and reflector; the design of the various O-ring seals; the selection of the reflector base material; polishing and plating procedure; and prevention of arcing. In addition, human engineering aspects have to be considered, such as, for example, ease of lamp replacement. In the following we will address these potential problem areas.

The mechanical design of a pump cavity is determined mainly by two considerations: the geometry chosen for efficient energy transfer from the pump source to the laser material, and the provisions required for extracting the heat generated by the pump source. Optically pumped lasers have efficiencies of a few percent, therefore, almost all the electrical energy supplied to the lamp will have to be removed as heat from the pump cavity. The causes of this low efficiency can be divided between the poor conversion of electrical energy to energy absorbed by the lasing material (typically 5 to 10 %) and the poor utilization of this absorbed energy in contributing to the output of the laser (approximately 20 to 30 %).

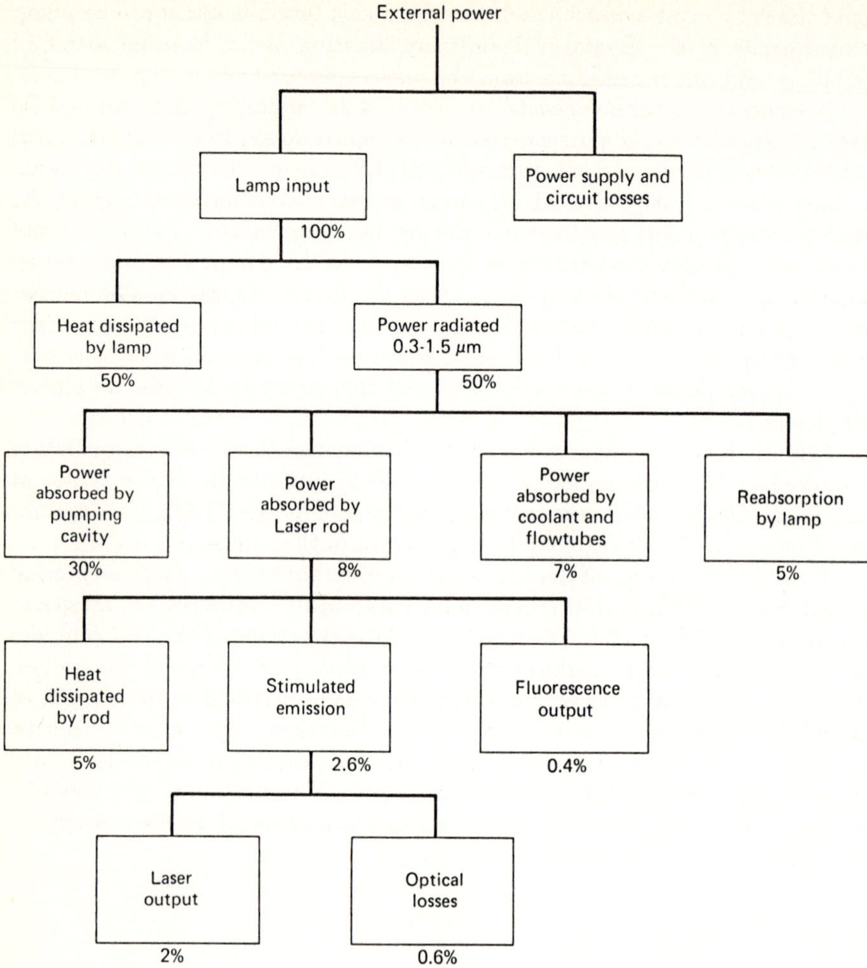

Fig. 6.75. Energy balance in an optically pumped solid-state laser system. (The percentages are fractions of electrical energy supplied to the lamp)

Thermal Load of Cavity Components

Let us consider the radiant energy-transfer processes in the pump cavity of a laser cavity. Figure 6.75 shows in a simplified way the energy balance in a laser system. The electrical input power supplied to the lamp is either dissipated as heat by the lamp envelope and electrodes or emitted as radiation. A portion of the radiation will be absorbed by the metal surfaces of the pump cavity. The radiation reflected from the walls will either be absorbed by the lasing medium or will return to the lamp. The light which is absorbed by the lamp will add energy to the radiation process in the same way as the electrical power does, and the returned light will be radiated with the same efficiency as the power supplied electrically. One consequence of the reabsorption is that a lamp, when enclosed in the pumping cavity, is operated under a higher thermal loading, resulting in shorter lamp life than when operated in the open for the same electrical input power. Since most laser cavities are liquid-cooled, a distinction is made (Fig. 6.57) between the radiation actually absorbed by the rod and the radiation absorbed by the surrounding cooling liquid, flowtubes, filters, etc. The pump power absorbed by the laser rod causes stimulated emission and fluorescence at the laser wavelength and other main emission bands. The remainder is dissipated as heat by the laser material. The percentages which appear in Fig. 6.75 are based on the electrical input power to the lamp. The numbers are typical for cw and pulsed pumped Nd : YAG lasers employing noble gas arc lamps.

The specific percentage values of Fig. 6.75 were measured on a Nd : YAG laser pumped by two cw krypton arc lamps contained in a gold-plated double-elliptical pump cavity. The radiation efficiency of the lamps was measured calorimetrically by surrounding a lamp with an opaque liquid calorimeter. The energy balance inside the cavity was also measured calorimetrically by measuring separately the heat removed by the lamps, laser rod, and cavity cooling loops. The transfer efficiency of the pump cavity and the power absorbed by the rod itself was determined by measuring the heat extracted from the laser rod cooling loop if a black-anodized copper rod, a Nd : YAG laser crystal, and a quartz rod were installed in the cavity [6.143]. With a black absorber installed in the cavity in place of the laser crystal, 43 % of the available pump radiation was absorbed. This number represents the cavity transfer efficiency. Table 6.8 summarizes the energy balance in the laser cavity. As a comparison, Table 6.9 shows the results of an energy-transfer analysis performed on a large Nd : glass disk amplifier [6.144a]. Useful information for the calculation of heat load and energy balance in pump cavities can be found in [6.144b,c].

Table 6.8. Energy transfer in a cw krypton arc lamp, pumped Nd : YAG laser

Heat dissipation of lamps	53 %
Heat dissipation of laser rod assembly	14 %
Heat dissipation of pump reflector	30 %
Radiation output	3 %
Electrical input to lamps	100 %

Table 6.9. Energy transfer in a Nd : glass disk amplifier

Circuit losses	8 %
Lamp heat	50 %
Heating of pump cavity walls	30 %
Ultraviolet absorption	7 %
Heating of glass disks	2 %
Fluorescence decay	2 %
Useful laser energy	1 %
Electrical input to lamps	100 %

Spectral Properties of the Materials Employed in the Design of Pump Cavities

Since the reflectivity of the metal surfaces as well as the transmission of the cooling fluid in the cavity are wavelength dependent, the spectrum of the pump light incident on the laser rod is different from the source emission spectrum. Ideally, in the transmission of the radiation from the source to the laser rod, one would like to have minimum optical losses in the pump bands of the laser material, and total absorption of all pump energy in spectral regions which do not contribute to the laser output. In this way the thermal heat load and the associated optical distortions in the active material would be kept at a minimum. Particularly undesirable is the ultraviolet content of the pump light, because it is detrimental to ruby- and neodymium-laser output and causes solarization in most materials [6.145, 146]. Furthermore, ultraviolet radiation leads to rapid deterioration of any organic materials in the pump cavity, such as silicon O-rings. Also undesirable is the formation of ozone by the ultraviolet radiation, because it leads to corrosion of metal parts in the cavity. Pump radiation, which has a longer wavelength than the stimulated emission, does not contribute to the laser output but does heat up the laser crystal and leads to optical distortions. The intensity and spectral content of the pump radiation reaching the laser rod depend on the reflectivity of the cavity walls, spectral filters placed inside the pump cavity, and the cooling medium.

The cavity walls can consist of specular-reflecting metal surfaces, diffuse-reflecting surfaces from ceramics, and compressed powders of inorganic materials or, in special cases, the reflector can be lined with a dielectric thin-film

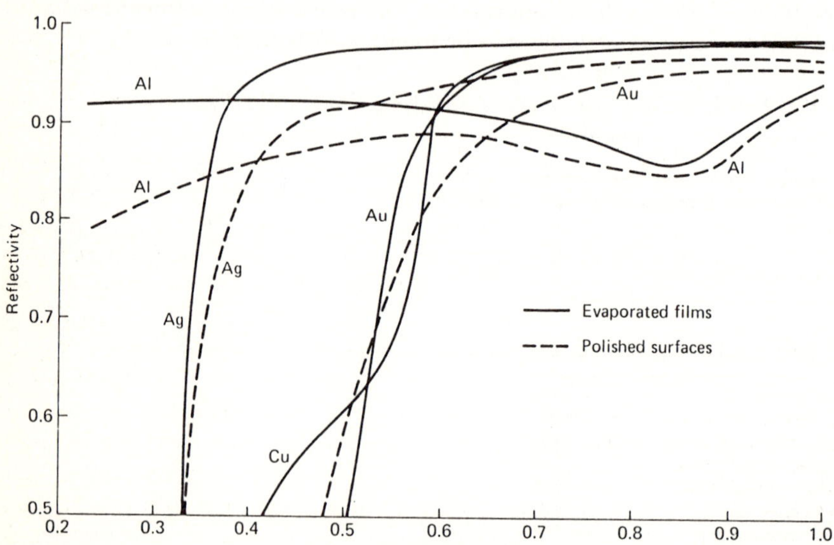

Fig. 6.76. Reflectivity versus wavelength for metals commonly used in the design of laser pump cavities

coating. The metals most commonly employed to obtain specular-reflecting surfaces in laser cavities are aluminum, silver, and gold. The reflectance versus wavelength of these materials is shown in Fig. 6.76. The reflective metal surfaces are usually obtained by evaporation, sputtering, polishing, or electroplating. The reflectance of a good evaporated coating is always higher than that of a polished or electroplated surface.

The cavity walls must have a high reflectivity at the absorption bands of the laser material. Therefore, for pumping ruby with the absorption bands located between 4100 and 5600 Å, only aluminum and silver can be employed, since gold has a low reflectivity at these short wavelengths. Compared to silver, aluminum has a higher reflectivity for wavelengths shorter than about 0.35 μm, whereas for all longer wavelengths the reflectivity of silver is higher. Maintaining the high reflectivity of silver presents a real problem. During the aging of silver in air, a layer of silver sulfide forms on the surface, causing the reflectance to drop. This tarnishing may be prevented by coating the silver with a thin protective layer of a transparent material, such as SiO_2. Alternatively, if the silver reflectors are immersed in an inert cooling fluid or operated in a dry nitrogen atmosphere, they will maintain their high reflectance for long periods of time. Because of the problems associated with silver, the most commonly used metal in pump cavities containing ruby rods is aluminum. For pumping Nd : YAG or Nd : glass, the situation is different. As can be seen from Fig. 6.76, aluminum has a minimum in its reflectance curve in the near-infrared. Therefore, both silver and gold have higher reflectances in the main pump bands of neodymium lasers. For cw-pumped Nd : YAG lasers, where most of the pumping occurs in the wavelength region between 0.7 and 0.9 μm, gold is used exclusively because, in contrast to silver, it does not tarnish. In high-power pumped-pulsed neodymium lasers, a considerable amount of the flashlamp radiation is in the Nd pump bands located around 0.53 and 0.58 μm. In these systems silver-coated reflectors are usually employed.

In focusing geometries the base material for the reflector is either aluminum, copper, or stainless steel. Aluminum is usually employed for light-weight systems. If used in ruby systems, aluminum, when highly polished, does not require plating. However, aluminum reflectors in Nd : YAG lasers require silver plating for pulsed-pumped systems and gold plating for cw-pumped systems. Usually, aluminum is plated with nickel first. Nickel provides a hard surface which polishes very easily. After polishing, a flash of either silver or gold is applied. If weight considerations are not too stringent, a better choice for the reflector base material is copper, since copper has a lower thermal expansion, higher thermal conductivity compared to aluminum, and nickel plating on copper is much more durable than on aluminum. Kanigen-plated copper can easily be polished to very good tolerances. Both aluminum and copper must be nickel-plated before polishing if a good surface is to be obtained. Polishing nickel-plated copper or aluminum reflectors can represent a problem in some cases because of the danger of polishing through the nickel. In this case stainless steel offers an attractive alternative despite its thermal conductivity, which is a factor 10 lower than copper. Stainless steel can be polished to the highest optical finish. A thin layer of gold or silver is electroplated or evaporated onto the surface after the polishing process.

A highly polished and reflective pump cavity is paramount in focusing geometries for the attainment of good efficiency. A poor electroplating quality showing a haze or orange peel on the reflector surface, flaking, pits, and cracks, is of great concern to the designer of laser systems. The improvement in laser output obtained by gold plating a highly polished stainless steel reflector is between a factor of 2 and 3 for a cw-pumped Nd:YAG system. A comparison of the laser output of a pulsed ruby system from a highly polished and an unpolished aluminum reflector showed an improvement of 50%.

A diffuse reflector is usually fabricated from ceramics or compressed powder of MgO or $BaSO_4$. The powder is usually contained in the inner space between two concentric quartz tubes. The reflection from these materials is in the order of 90 to 98% and fairly independent of wavelength for the region of interest. A reflector surrounding a helical lamp can be made from ceramic, aluminum, stainless steel, etc. Diffuse reflecting, molded ceramic inserts do not tarnish or corrode. Some ceramic materials containing impurities, however, show discoloration after prolonged exposure to flashlamp radiation. The difference in output due to different diffuse reflectors in a system containing a helical flashlamp and a ruby crystal is illustrated in Fig. 6.77.

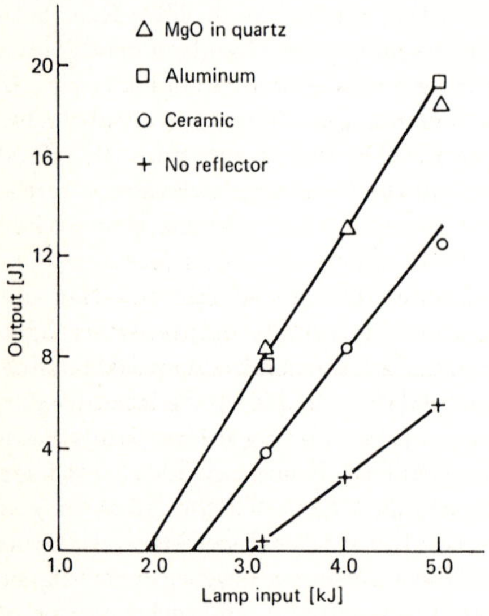

Fig. 6.77. Output from a ruby laser pumped by a helical flashlamp for different reflector materials. Laser was operated in conventional mode

In special cases, where it is important to minimize the heat load of a laser rod, dichroic thin-film coatings can be employed as reflective surfaces. These coatings, which are applied on either glass or metal surfaces, are designed such that they reflect pump radiation but transmit all unwanted radiation. The transmitted radiation is absorbed by the metal walls of the cavity or by an absorber surrounding the glass reflector [6.147a].

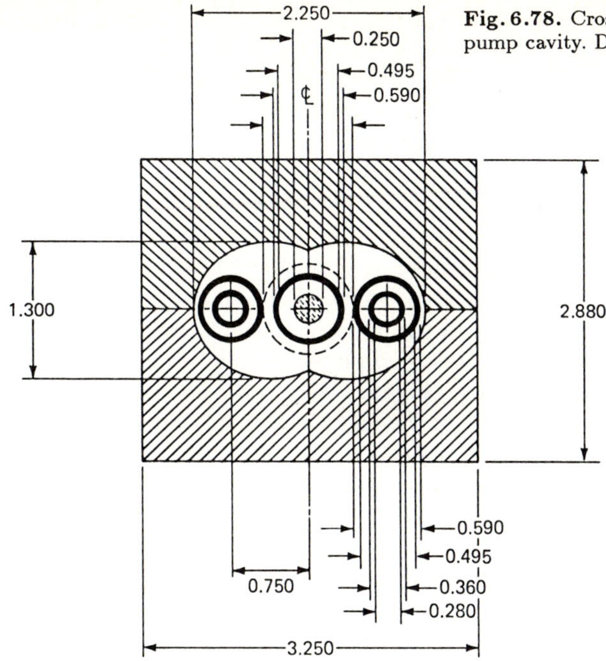

Fig. 6.78. Cross section of a double-elliptical pump cavity. Dimensions are in inches

The purpose of spectral filters placed inside the pump cavity is to provide adequate absorption of the intense ultraviolet radiation from xenon flashlamps. One way to accomplish this is to surround the laser rod or flashlamp with a tube made of Nonex, Pyrex, titanium-doped quartz, samarium-doped glass ED-5 and ED-6, or colored filter glass with a sharp cutoff. Instead of a tube as filter, a flat plate can be inserted, for example, in an elliptical-pump cylinder separating the rod and lamp (Fig. 6.78). Other approaches include the use of ultraviolet-free lamps which are manufactured by employing an ultraviolet-absorbing or reflecting envelope [6.147b]. In glass rods, ultraviolet radiation can be prevented from reaching the active material by cladding the rod with samarium-doped glass.

The spectral properties of the cooling fluid can be utilized to remove some of the unwanted pump radiation. Water, if used as a coolant, is very effective in absorbing radiation at wavelengths longer than 1.3 μm. Where absorption of ultraviolet radiation by the laser material must be held to a minimum, potassium chromate, potassium dichromate, or sodium nitrite can be added to the cooling water.

More recently, the efficiency of Nd : glass and Nd : YAG lasers pumped by xenon flashlamps was improved by circulating fluorescent dyes around the laser rod [6.148–150]. The dyes absorb in the spectral region in which the Nd ions do not absorb, and become fluorescent at one of the pump bands of Nd^{3+}. Rhodamine 6G dissolved in ethanol was identified as the best dye, giving better than 50 % improvement in the laser output.

Cooling Techniques

In elliptical-pump cavities, the laser rod and lamp are often liquid-cooled by circulating the coolant in flowtubes which surround these elements. The inside of the pumping chamber itself is dry. However, in most cases the body of the reflector contains cooling chambers through which the coolant fluid passes.

The pumping cavity usually consists of two parts that separate along the plane of the major axis. The end-plate reflectors, in addition to serving their optical function, can be used to mount and index the elliptical sections and to provide precision mounts for the laser rod assembly, as shown in Fig. 6.79. The rod assembly, consisting of the laser rod mounted in rod holders (Fig. 6.80a) can slide into precision-bored holes in these end plates. O-rings around the rod holder will seal this unit from the cavity. Figure 6.80b shows an example of a large Brewster-angle ruby rod installed in the end plate of an elliptical cavity without the use of rod holders. The drawing also shows the cooling channel and plenum chamber employed to force the cooling water into the annular cooling path defined by the ruby rod and the flowtube.

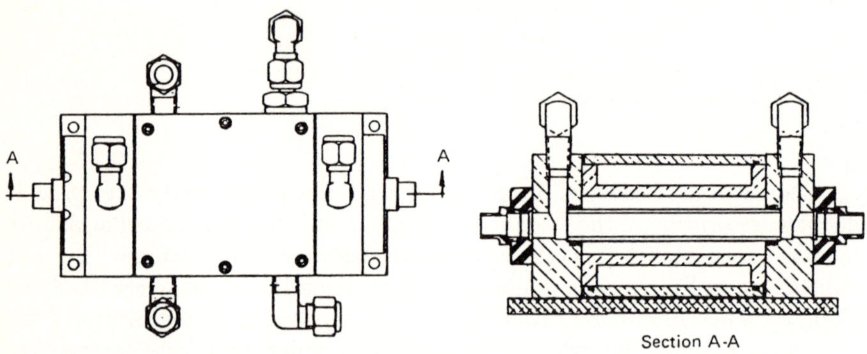

Fig. 6.79. Outside configuration of a double-elliptical pump cavity

Figure 6.81 shows a drawing of a double-elliptical pump cavity featuring flowtubes. Note the elaborately designed cooling channels, which are machined close to the reflector surface. This design provides a large cooling surface and high-velocity flow for efficient heat exchange. The 25-cm-long double-elliptical cavity is capable of dissipating 40 kW of lamp input power.

In the so-called flooded cavity approach, the whole inside of the pump cavity is immersed in cooling fluid. The absence of flowtubes and separate cooling chambers for the reflector makes this type of cavity very compact and simple in design. For example, only one inlet and outlet are required for the cooling loop, whereas in an elliptical cavity with flowtubes, one pair of coolant ports with the associated fittings, tubing, etc., is required for each reflector half, lamp, and laser rod. Also, in this design, lamp and laser rod can be brought very close together, and no reflection losses from additional glass surfaces are encountered. Figure 6.82 shows a photograph of a commercial cw-pumped Nd:YAG laser featuring a single elliptical reflector in a flooded cavity.

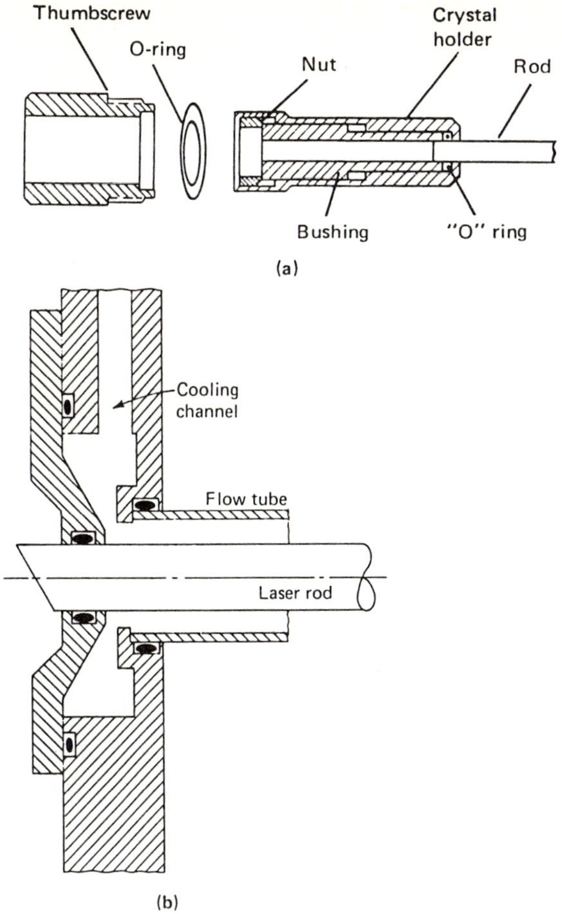

Fig. 6.80a, b. Laser rod mounting techniques **(a)** with and **(b)** without separate rod holder

Most liquid-cooled military-type Nd:YAG lasers and most commercial cw-pumped Nd:YAG lasers feature this design, because of its compactness and simplicity. Figure 6.83 shows a photograph of a liquid-immersed pump cavity used for a high-repetition-rate military Nd:YAG laser. The cavity is sealed by one large O-ring in the top cover. The reflector inserts are machined from aluminum which is nickel-plated, polished, and silver-plated. The laser head is machined from aluminum which is hard-anodized (Mil. Spec. A-8625) to prevent corrosion.

Figure 6.84 shows an exploded view of a single- and double-elliptical pump cylinder using immersion cooling. The laser head is machined from a solid block of acrylic. This material alleviates the problem of arcing and eliminates the need to feed the lamp anode through the laser head. The reflector inserts are machined from copper and are gold-plated.

Great care has to be taken in the design of liquid-immersed cavities of the type displayed in Figs. 6.82–84 to achieve symmetrical high-velocity flow along

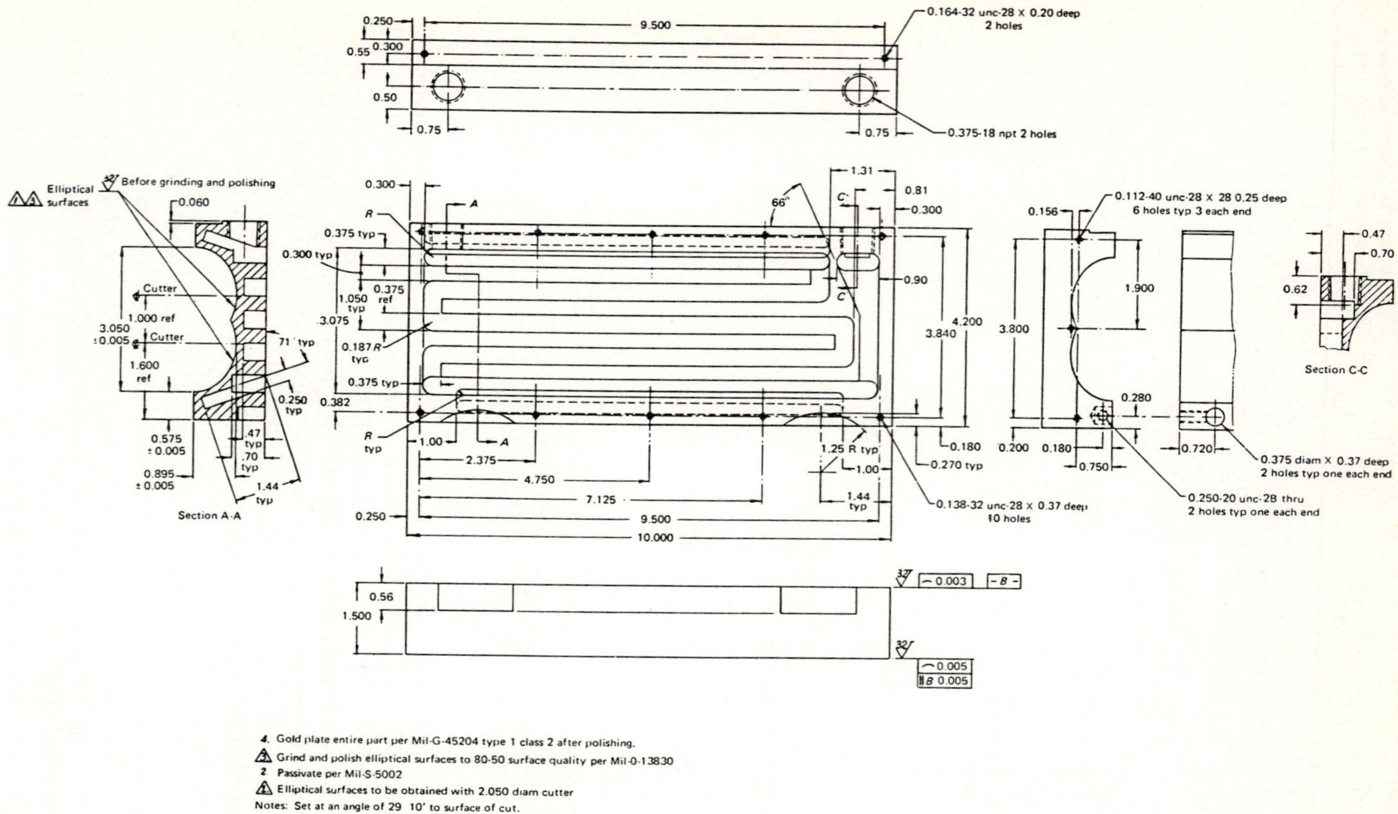

Fig. 6.81. Machine drawing of a double-elliptical pump cavity. Dimensions are in inches

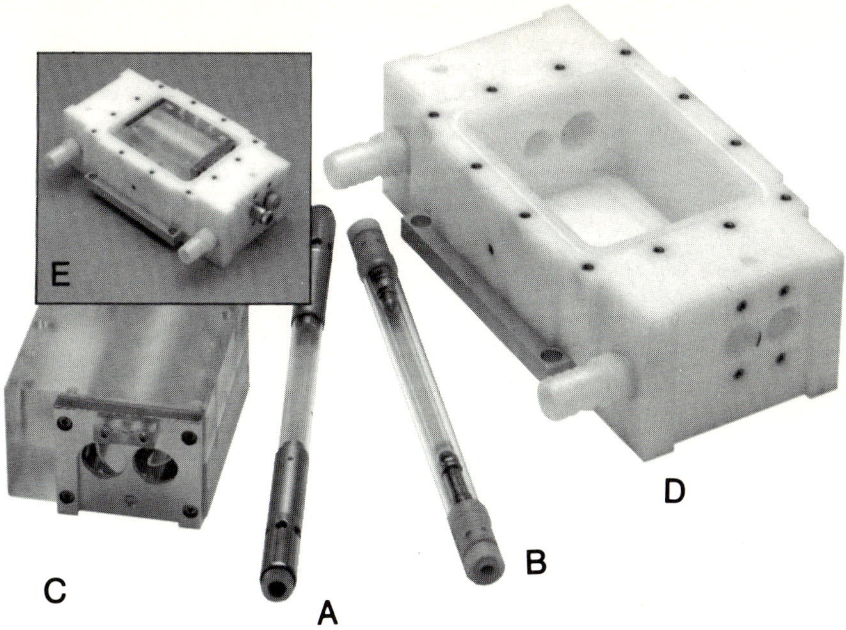

Fig. 6.82. Commercial cw-pumped Nd: YAG laser featuring a flooded pump cavity design. The photo shows the laser rod assembly (A), krypton arc lamp assembly (B), single elliptical reflector (C), and pump housing (D). The insert (E) shows the assembled pump head with top removed. (Quantronix Corp. Model 114)

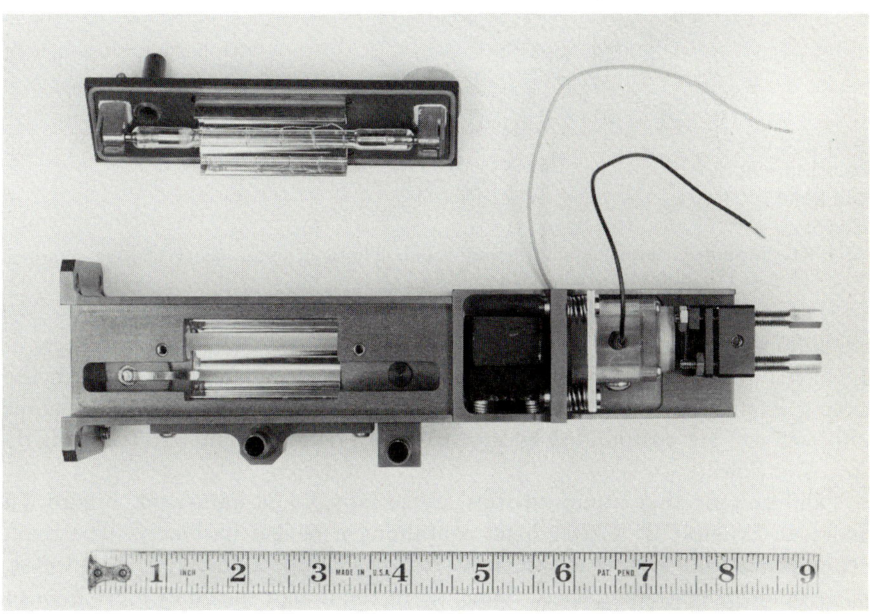

Fig. 6.83. Laser head of a military Nd: YAG laser

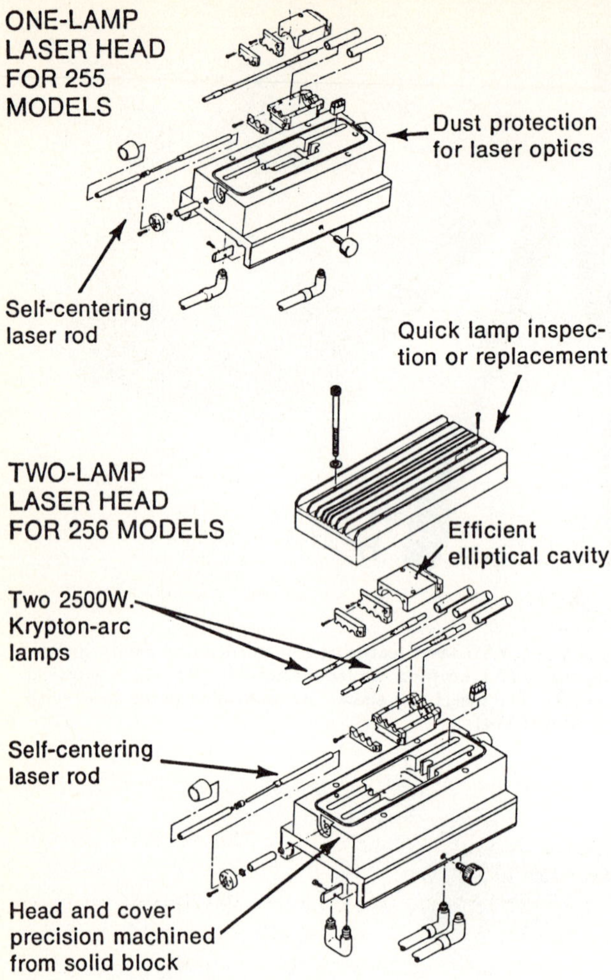

Fig. 6.84. Exploded view of a single- and double-elliptical pump cavity of a cw-pumped Nd:YAG laser (Holobeam models 255 and 256)

the lamp and laser rod. Otherwise, the different cross sections and pockets in the cavity result in low-velocity and, in extreme cases, stagnant areas in the cooling loop. Note that in the design shown in Figs. 6.82 and 84 the lamps and laser rod are surrounded by flowtubes to provide symmetrical and highly turbulent flow.

Diffuse-reflecting pump cavities are usually liquid-immersed. Figure 3.5 shows an example of a laser head containing a helical flashlamp. The main elements of such a system are the laser rod, flashlamp, a closely fitted reflector, and a housing containing these elements. In the design shown in Fig. 3.5, laser rod, flashlamp, and reflector are all immersed in water. In helical-pump lamp systems the laser material is sometimes surrounded by a flowtube to define a

Fig. 6.85. Example of a diffuse-reflecting pump cavity

cooling channel which forces water first through an annulus between rod and flowtube and then back over the helical lamp.

Figure 6.85 shows a laser head employed to pump laser rods up to 15 cm in length by a single linear flashlamp. Lamp and rod are contained in a diffuse reflector consisting of barium sulfate which is compressed between two quartz tubes. Inside, the pumping chamber is filled with water. The body of the laser head is made from acrylic.

If the heat load in a cavity is not excessive, the laser rod can be attached to a heat sink. Of the cavities pictured in Figs. 6.49 and 50 the semi-elliptical and close-coupled designs are particularly suitable for this type of cooling. A typical design of a conduction-cooled laser rod (Fig. 6.49e) consists of a reflector with semi-elliptical top half and a flat bottom half which contains two longitudinal grooves for carrying the laser rod and flashlamp. Thermal conduction between the bottom part, which acts as heat sink, and the laser rod can be obtained by mechanically clamping the rod into the groove. Sometimes a thin gold foil is inserted between the rod and the heat sink which will yield and conform to the irregularities of the laser rod surface under the clamping force. An example of an air-cooled laser head is shown in Fig. 6.86. The pump cavity of this Nd:YAG laser, which is pulsed at a repetition rate of 10 pps, is cooled by forced air generated by an axial blower. The Nd:YAG rod, 25 mm by 3 mm, is cemented to a copper heat sink with silver epoxy.

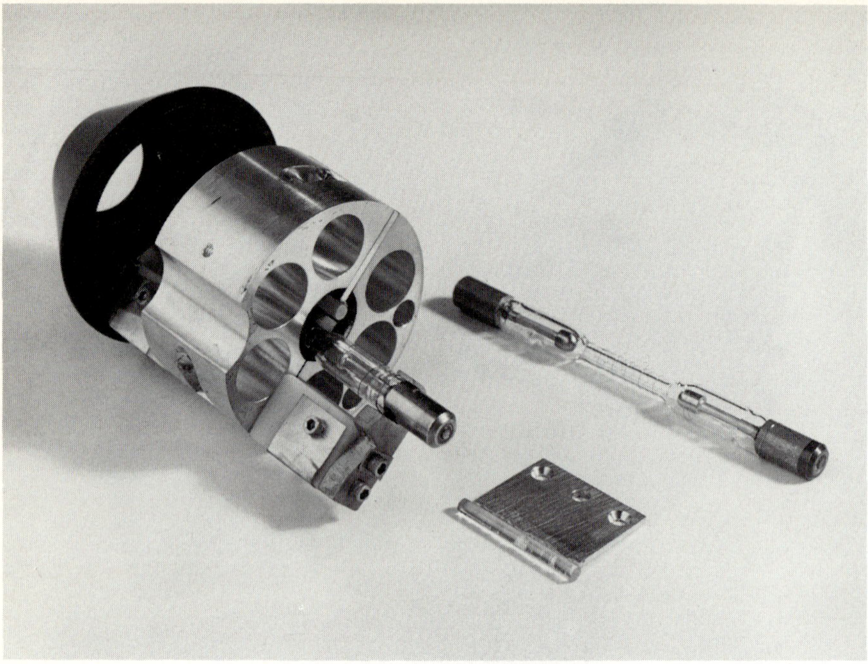

Fig. 6.86. Laser head of a small air-cooled Nd : YAG laser

The optical elements of a laser cavity can also be cooled by forced air or gas generated by axial blowers, fans, or pumps. This technique is usually employed in very low-power or low-repetition-rate systems. Forced-air cooling of tungsten-filament lamps is a common technique in Nd : YAG lasers. However, the laser rod and cavity are liquid-cooled. In some military Nd : YAG lasers a closed-cycle, high-pressure nitrogen gas loop is employed; this will be discussed in more detail in Chap. 7.

Sealing Techniques

An O-ring seal, if properly designed, is very reliable and does not present a problem. Seal areas within the pump cavity are considerably more critical because organic materials exposed to the pump radiation will quickly deteriorate. If O-rings made from an organic material are used to seal the laser rod, flowtubes, or lamps, they should be well shielded from radiation. Figure 6.80 shows several examples of O-ring locations which provide maximum protection from pump light. White silicon O-rings are usually found to be best suited for laser applications.

Besides O-rings, several other techniques are used to seal laser rods, in particular, ruby rods. The first technique involves the use of metal O-rings. For example, a very satisfactory method of sealing laser rods is achieved by extruding tin rings in a sleeve crystal holder (similar to Fig. 6.80a) into spaces between the laser rod and rod holder. Another technique involves the use of a mechanical press-fit between the ruby rod and a polycrystalline sapphire sleeve, both of which have a small taper. This technique, although expensive, is ideal from an optical point of view because it allows pumping of the ruby over its full length. Unlike Nd:YAG and Nd:glass, unpumped ruby absorbs strongly at the laser wavelength, therefore it is desirable in this material to minimize unpumped regions of the rod.

7. Heat Removal

Any solid-state laser operating in either cw or pulsed mode of operation must dissipate an appreciable amount of heat. The heat arises because:

1. The energy difference between the pump bands and the fluorescence energy levels is lost to the host lattice through radiationless transitions;
2. the quantum efficiency of the fluorescence processes involved in the laser transitions is less than unity, therefore some of the photon lose their total energy to the host lattice;
3. the spectral distribution of the pump light is such that there is considerable absorption by the host directly, mainly in the ultraviolet and infrared bands, in which case all energy in those spectral regions is converted into heat.

7.1 Thermal Effects in Laser Rods

Thermal effects in laser rods are brought about by a combination of heat generation due to absorption of pump radiation and heat flow due to cooling processes. Heating and cooling of the laser material leads to a nonuniform temperature of the rod, which results in a distortion of the laser beam due to a temperature- and stress-dependent variation of the index of refraction. The type of optical distortions which occur in a laser rod, as a result of a nonuniform temperature distribution, are thermal lensing and thermally induced berefringence due to the photoelastic effect of thermal strains. In addition, optical distortions can arise as a result of an elongation and bending of the rod. The particular temperature profile which exists in the laser material depends to a large degree on the mode of operation, i.e., cw pumped, single shot, or repetitively pulse pumped.

In the case of cw operation, a long cylindrical laser rod with uniform internal heat generation and constant surface temperature assumes a quadratic radial temperature dependence. This leads to a similar dependence in both the index of refraction and the thermal strain distribution. In a pulse-pumped system, laser action occurs only during the pump pulse or shortly thereafter in the case of Q-switching, therefore the main interest is centered around the time interval of the pumping pulse. Theoretical and experimental investigations have shown that heat transport during the pump pulse, which usually has a duration between 0.2 and 5 ms, can be neglected. Therefore in single-shot operation the optical distortions are solely a result of the existence of thermal gradients generated by non-uniform pump-light absorption. In repetitively pulse-pumped systems, distortions will occur from the cumulative effects of non-uniform pump processes and thermal gradients due to cooling. Which effects dominate depends, as we will see, on the ratio of the pulse interval time to the thermal relaxation time constant of the rod. At repetition intervals, which are short compared to the thermal relaxation time of the laser rod, a quasi-thermal steady-state will be reached where the distortions from pumping become secondary to the distortions produced by the removal of heat from the laser material.

7.1.1 cw Operation

Temperature Distribution

We consider the case where the heat generated within the laser rod by pump light absorption is removed by a coolant flowing along the cylindrical rod surface. With the assumption of uniform internal heat generation and cooling along the cylindrical surface of an infinitely long rod, the heat flow is strictly radial, and end effects and the small variation of coolant temperature in axial direction can be neglected. The radial temperature distribution in a cylindrical rod with thermal conductivity K, in which heat is uniformly generated at a rate Q per unit volume, is obtained from the one-dimensional heat conduction equation [7.1]

$$\frac{d^2T}{dr^2} + \left(\frac{1}{r}\right)\left(\frac{dT}{dr}\right) + \frac{Q}{K} = 0 \quad . \tag{7.1}$$

The solution of this differential equation gives the steady-state temperature at any point along a radius of length r. With the boundary condition $T(r_0)$ for $r = r_0$, where $T(r_0)$ is the temperature at the rod surface and r_0 is the radius of the rod, it follows that

$$T(r) = T(r_0) + \left(\frac{Q}{4K}\right)(r_0^2 - r^2) \quad . \tag{7.2}$$

The temperature profile is parabolic, with the highest temperature at the center of the rod. The temperature gradients inside the rod are not a function of the surface temperature $T(r_0)$ of the rod. The heat generated per unit volume can be expressed as

$$Q = \frac{P_a}{\pi r_0^2 L} \quad , \tag{7.3}$$

where P_a is the total heat dissipated by the rod and L is the length of the rod. The temperature difference between the rod surface and the center is

$$T(0) - T(r_0) = \frac{P_a}{4\pi KL} \quad . \tag{7.4}$$

The transfer of heat between the rod and the flowing liquid creates a temperature difference between the rod surface and the coolant. A steady state will be reached when the internal dissipation P_a is equal to the heat removed from the surface by the coolant

$$P_a = 2\pi r_0 Lh[T(r_0) - T_F] \quad , \tag{7.5}$$

where h is the surface heat transfer coefficient and T_F is the coolant temperature. With $F = 2\pi r_0 L$ being the surface area of the rod, it follows that

$$T(r_0) - T_F = \frac{P_a}{Fh} \quad . \tag{7.6}$$

Combining (7.4 and 7.6), one obtains for the temperature at the center of the rod

$$T(0) = T_F + P_a\left(\frac{1}{4\pi KL} + \frac{1}{Fh}\right) \quad . \tag{7.7}$$

Thus, from the geometry and the appropriate system and materials parameters, the thermal profile of the crystal can be determined, except that h must be

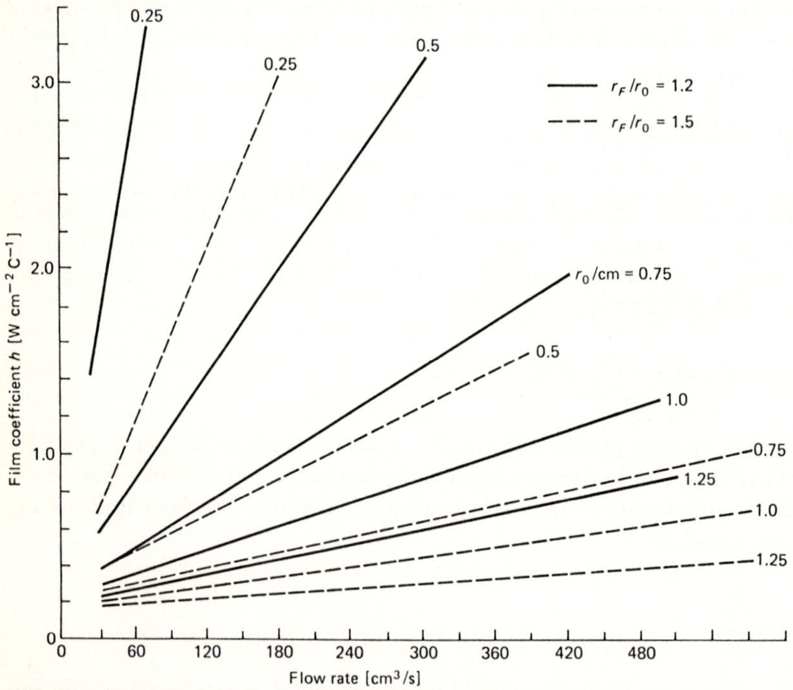

Fig. 7.1. Surface heat transfer coefficient h as a function of flow rate. Parameters are the rod radius r_0 and the flowtube inner radius r_F. [7.4]

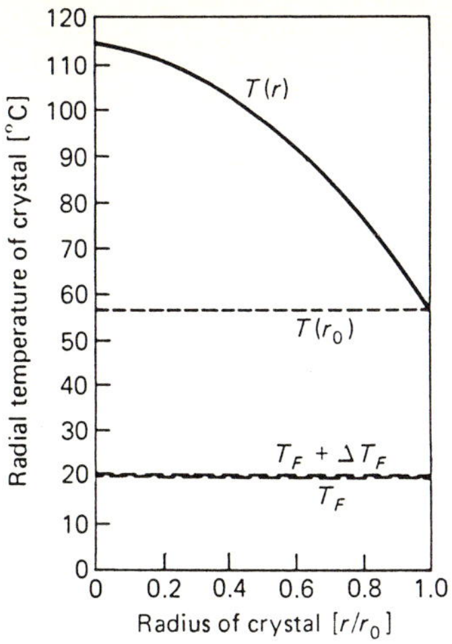

Fig. 7.2. Radial temperature distribution within a Nd : YAG crystal as a function of radius. T_F is the temperature of coolant entering flowtube assembly; ΔT_F is the axial temperature gradient; and $T(r_0)$ is the rod surface temperature. [7.3]

evaluated. This coefficient is obtained from a rather complex expression involving the thermal properties of the coolant, the mass flow rate of the coolant, the Reynolds, Prandtl, and Grashof numbers, and the geometry [7.2–4]. In Fig. 7.1 the heat transfer coefficient for water is plotted as a function of flow rate for cases of practical interest. Parameter is the rod radius r_0 and the inside radius r_F of a circular cooling channel.

Figure 7.2 shows as an example the radial temperature profile in a Nd : YAG rod calculated from (7.7). The laser, which delivered between 200 to 250 W of output, was pumped at 12 kW of input power. The following parameters have been used in the numerical calculations of the temperature profile in the crystal: rod length $L = 7.5$ cm; rod radius $r_0 = 0.32$ cm; flowtube inside radius $r_F = 0.7$ cm; power dissipated by the Nd : YAG rod $P_a = 600$ W; mass flow rate of the coolant $m^* = 142$ g/s; fluid temperature entering the cavity $T_F = 20°$ C. As can be seen from this figure, the maximum temperature of the crystal occurring at the center is 114° C. The large temperature gradient of 57° C between the center of the crystal and the surface is responsible for the high stresses present in the material.

Thermal Stresses

The temperature gradients generate mechanical stresses in the laser rod, since the hotter inside area is constrained from expansion by the cooler outer zone. The stresses in a cylindrical rod, caused by a temperature distribution $T(r)$, can be calculated from the equations given by *Timoshenko* and *Goodier* [7.5]. Equations (7.8–10) describe the radial σ_r, tangential σ_ϕ, and axial σ_z stress in an isotropic rod with free ends and a temperature distribution according to (7.2)

$$\sigma_r(r) = QS(r^2 - r_0^2) \quad , \tag{7.8}$$

$$\sigma_\phi(r) = QS(3r^2 - r_0^2) \quad , \tag{7.9}$$

$$\sigma_z(z) = 2QS(2r^2 - r_0^2) \quad , \tag{7.10}$$

where the factor $S = \alpha E[16K(1-\nu)]^{-1}$ contains the materials parameters; E is Young's modulus; ν is Poisson's ratio; and α is the thermal coefficient of expansion. The stress components, σ_r, σ_ϕ, σ_z represent compression of the material when they are negative and tension when they are positive. We notice that the stress distributions also have a parabolic dependence on r. The center of the rod is under compression. The radial component of the stress goes to zero at the rod surface, but the tangential and axial components are in tension on the rod surface by virtue of the greater bulk expansion in the rod's center compared to the circumference.

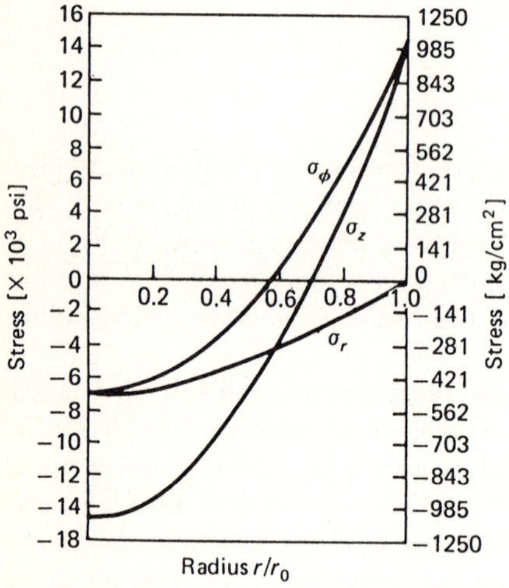

Fig. 7.3. Radial (σ_r), tangential (σ_ϕ), and axial (σ_z) stress components within a Nd:YAG crystal as a function of radius. [7.3]

Figure 7.3 gives the stresses as a function of radius inside the Nd:YAG rod whose temperature profile was shown in Fig. 7.2. From these curves it follows that the highest stresses occur at the center and at the surface. Since the tensile strength of Nd:YAG is 1800 to 2100 kg/cm^2, the rod is stressed up to 75% of its ultimate strength. As the power dissipation in the rod is increased, the tension on the rod surface increases and may exceed the tensile strength of the rod, thereby causing laser rod fracture. It is of interest to determine at what power level this will occur. Using (7.9, 10) we find, after adding the two stress components $\sigma_z(r_0)$, $\sigma_\phi(r_0)$ vectorially,

$$\sigma_{\max} = \frac{\alpha E}{8\pi K(1-\nu)} \frac{P_a}{L} \quad . \tag{7.11a}$$

We note from (7.11) that the tension on the surface of a laser rod depends on the physical constants of the laser material and on the power dissipated per unit length of the material, but does not depend on the cross section of the rod. Upon substitution of $\sigma_{\max}$ with the value of the tensile strength of Nd:YAG, we find that with 115 W dissipated as heat per centimeter length of Nd:YAG rod, the tension on the surface of the rod equals the tensile strength of the material. The actual rupture stress of a laser rod is very much a function of the surface finish of the rod, and can vary by almost a factor of three in Nd:YAG [7.6].

Stress Fracture Limit

The mechanical properties of the laser host material determine the maximum surface stress that can be tolerated prior to fracture. If there were no other constraints, such as stress induced focusing and birefringence, the thermal loading and thus average output power of a rod laser would be increased until stress fracture occurred. If we let σ_{max} be the maximum surface stress at which fracture occurs, then we can rewrite (7.11a) as follows

$$\frac{P_a}{L} = 8\pi R \quad \text{where} \quad R = \frac{K(1-\nu)}{\alpha E}\sigma_{max} \tag{7.11b}$$

is a "thermal shock parameter". A larger R indicates a higher permissible thermal loading before fracture occurs. The table below lists typical values for a number of laser materials if we assume a standard surface treatment.

Material	Glass	GSGG	YAG	Al_2O_3
thermal shock parameter R [W/cm]	1	6.5	7.9	100

Photoelastic Effects

The stresses calculated in the previous subsection generate thermal strains in the rod, which in turn produce refractive index variations via the photoelastic effect. The refractive index of a medium is specified by the indicatrix, which in its most general case is an ellipsoid. A change of refractive index due to strain is given by a small change in shape, size, and orientation of the indicatrix [7.7]. The change is specified by small changes in the coefficients B_{ij} [7.8],

$$B_{ij} = P_{ijkl}\varepsilon_{kl} \quad (i,j,k,l = 1,2,3) \tag{7.12}$$

where P_{ijkl} is a fourth-rank tensor giving the photoelastic effect. The elements of this tensor are the elastooptical coefficients. ε_{kl} is a second-rank strain tensor.

We will confine our calculation to the case of a Nd:YAG laser rod. The method, however, is applicable to any material if the proper photoelastic matrix is used. Since Nd:YAG is a cubic crystal, the indicatrix is a sphere. Under stress the indicatrix becomes an ellipsoid. Nd:YAG rods are grown with the cylindrical axes along the [111] direction. The light propagates in this direction, and thus the change of the refractive index along the [111] is of interest.

Since the transverse stresses are in radial and tangential direction – relative to the coordinate system shown in Fig. 7.4 – the local indicatrix also orients

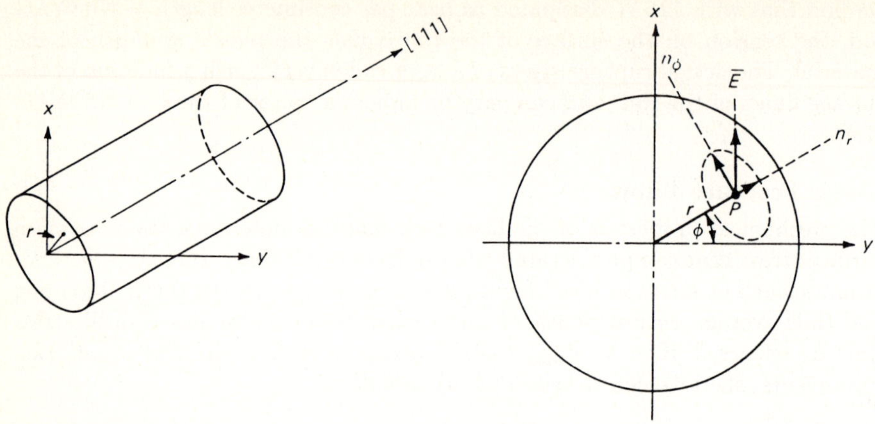

Fig. 7.4. Crystal orientation for a Nd:YAG rod *(left)* and orientation of indicatrix of the thermally stressed Nd:YAG rod in a plane perpendicular to the rod axis *(right)*

its axis in these directions. In a cylindrical coordinate system the photoelastic changes in the refractive index for the r and ϕ polarizations are given by [7.8]

$$\Delta n_r = -\tfrac{1}{2} n_0^3 \Delta B_r \quad \text{and} \tag{7.13}$$

$$\Delta n_\phi = -\tfrac{1}{2} n_0^3 \Delta B_\phi \quad . \tag{7.14}$$

A considerable amount of tensor calculation is required to determine the coefficients ΔB_r and ΔB_ϕ in a plane perpendicular to the [111] direction of the Nd:YAG crystal. The technique used in these calculations consists of introducing the elastooptic coefficients for Nd:YAG into the matrix form P_{mn} of the photoelastic tensor. The published values of these coefficients are given in a [100]-oriented Cartesian coordinate frame [7.9]: $P_{11} = -0.029$; $P_{12} = +0.0091$; and $P_{44} = -0.0615$. The strain tensor is obtained from the stresses calculated in (7.8–10). After a coordinate transformation to bring P_{mn} and ε_{kl} into the same coordinate system, the tensor operation according to (7.12) can be performed. Introducing the expression for ΔB_r and ΔB_ϕ into (7.13, 14), the refractive index changes are given by [7.10, 11]

$$\Delta n_r = -\frac{1}{2} n_0^3 \frac{\alpha Q}{K} C_r r^2 \quad , \tag{7.15}$$

$$\Delta n_\phi = -\frac{1}{2} n_0^3 \frac{\alpha Q}{K} C_\phi r^2 \quad , \tag{7.16}$$

where C_r and C_ϕ are functions of the elastooptical coefficients of Nd:YAG.

$$C_r = \frac{(17\nu - 7)P_{11} + (31\nu - 17)P_{12} + 8(\nu + 1)P_{44}}{48(\nu - 1)} \quad ,$$

$$C_\phi = \frac{(10\nu - 6)P_{11} + 2(11\nu - 5)P_{12}}{32(\nu - 1)} \quad .$$

The induced birefringence is determined from (7.15, 16)

$$\Delta n_r - \Delta n_\phi = n_0^3 \frac{\alpha Q}{K} C_B r^2 \quad , \quad \text{where} \tag{7.17}$$

$$C_B = \frac{1+\nu}{48(1-\nu)}(P_{11} - P_{12} + 4P_{44}) \quad .$$

Inserting the values of the photoelastic coefficients and the material parameters of Nd:YAG, $\alpha = 7.9 \times 10^{-6}/^\circ\mathrm{C}$, $K = 0.11\,\mathrm{W/cm}^\circ\mathrm{C}$, $\nu = 0.3$, $n_0 = 1.82$, into (7.15–17), one obtains $C_r = 0.017$, $C_\phi = -0.0025$, $C_B = -0.0097$, and

$$\Delta n_r = (-2.8 \times 10^{-6})Qr^2 \quad ,$$

$$\Delta n_\phi = (+0.4 \times 10^{-6})Qr^2 \quad ,$$

$$\Delta n_r - \Delta n_\phi = (-3.2 \times 10^{-6})Qr^2 \quad ,$$

where Q has the dimension watts per cubic centimeter and r is measured in centimeters.

Thermal Lensing

Having explored the stresses in the laser rod, we now turn to the optical distortions which are a result of both the temperature gradients and the stresses. The change of the refractive index can be separated into a temperature- and a stress-dependent variation. Hence

$$n(r) = n_0 + \Delta n(r)_\mathrm{T} + \Delta n(r)_\varepsilon \quad , \tag{7.18}$$

where $n(r)$ is the radial variation of the refractive index; n_0 is the refractive index at the center of the rod; and $\Delta n(r)_\mathrm{T}$, $\Delta n(r)_\varepsilon$ are the temperature- and stress-dependent changes of the refractive index, respectively.

The temperature-dependent change of refractive index can be expressed as

$$\Delta n(r)_\mathrm{T} = [T(r) - T(0)]\left(\frac{dn}{dT}\right) \quad . \tag{7.19}$$

With the aid of (7.2, 4) we obtain

$$\Delta n(r)_\mathrm{T} = -\frac{Q}{4K}\frac{dn}{dT}r^2 \quad . \tag{7.20}$$

As can be seen from (7.15, 16 and 20), the refractive index in a laser rod shows a quadratic variation with radius. An optical beam propagating along the rod axis suffers a quadratic spatial phase variation. This perturbation is equivalent to the effect of a spherical lens. The focal length of a lens-like medium where the refractive index is assumed to vary according to

$$n(r) = n_0\left(1 - \frac{2r^2}{b^2}\right) \tag{7.21}$$

is given by [7.12]

$$f \cong \frac{b^2}{4n_0 L} \quad . \tag{7.22}$$

This expression is an approximation where it was assumed that the focal length is very long in comparison to the rod length. The distance f is measured from the end of the rod to the focal point.

The total variatio of the refractive index is obtained by introducing (7.15, 16, and 20) into (7.18):

$$n(r) = n_0\left[1 - \frac{Q}{2K}\left(\frac{1}{2n_0}\frac{dn}{dT} + n_0^2\alpha C_{r,\phi}\right)r^2\right] \quad . \tag{7.23}$$

As was discussed in the previous subsection, the change of refractive index due to thermal strain is dependent on the polarization of light, therefore the photoelastic coefficient $C_{r,\phi}$ has two values, one for the radial and one for the tangential component of the polarized light. Comparing (7.23) with (7.21) yields

$$f' = \frac{K}{QL}\left(\frac{1}{2}\frac{dn}{dT} + \alpha C_{r,\phi} n_0^3\right)^{-1} \quad . \tag{7.24}$$

In our final expression for the focal length of a Nd:YAG rod, we will include the contributions caused by end effects. Perturbations of the principal thermal distortion pattern occur in laser rods near the ends, where the free surface alters the stress character. The so-called end effects account for the physical distortion of the flatness of the rod ends. Self-equilibrating stresses causing a distortion of flatness were found to occur within a region of approximately one diameter from the ends of Nd:glass [7.13] and one radius from the end for Nd:YAG [7.14]. The deviation from flatness of the rod ends is obtained from

$$l(r) = \alpha_0 l_0[T(r) - T(0)] \quad , \tag{7.25}$$

where l_0 is the length of the end section of the rod over which expansion occurs. With $l_0 = r_0$ and (7.2), we obtain

$$l(r) = -\alpha_0 r_0 \frac{Qr^2}{4K} \quad . \tag{7.26}$$

The focal length of the rod caused by an end-face curvature is obtained from the thick-lens formula of geometric optics [7.7]

$$f'' = \frac{R}{2(n_0 - 1)} \quad , \tag{7.27}$$

where the radius of the end-face curvature is $R = -(d^2l/dr^2)^{-1}$. From these expressions follows the focal length of the rod caused by a physical distortion of the flat ends:

$$f'' = K[\alpha Q r_0 (n_0 - 1)]^{-1} \quad . \tag{7.28}$$

The combined effects of the temperature- and stress-dependent variation of the refractive index and the distortion of the end-face curvature of the rod lead to the following expression for the focal length:

$$f = \frac{KA}{P_a}\left(\frac{1}{2}\frac{dn}{dT} + \alpha C_{r,\phi} n_0^3 + \frac{\alpha r_0 (n_0 - 1)}{L}\right)^{-1} \quad , \tag{7.29}$$

where A is the rod cross-sectional area and P_a is the total heat dissipated in the rod. If one introduces the appropriate materials parameters for Nd : YAG into (7.29), then one finds that the temperature-dependent variation of the refractive index constitutes the major contribution of the thermal lensing. The stress-dependent variation of the refractive index modifies the focal length about 20 %. The effect of end-face curvature caused by an elongation of the rod is less than 6 %.

Ignoring the end effects, we notice that the focal length is proportional to a material constant and the cross section A of the rod and is inversely proportional to the power P_a dissipated as heat in the rod. At first, it may be surprising that the length of the rod does not enter the equations. However, in a longer rod, for example, the reduction in power dissipation per unit length is offset by a longer path length.

We see from (7.29) that we have little flexibility in influencing the focal length. The material constants are determined when we choose the laser material: the dissipated power P_a is determined by the application (even though we may be able to reduce the heat load by avoiding unusable pump radiation); thus the only remaining design parameter is the rod cross section. The focal length can be increased by inreasing A, but this is usually not a practical way of solving the problem.

According to (7.29), the focal length of a cylindrical laser rod, where heat is generated uniformly within the bulk material, can be written as

$$f = M P_{\text{in}}^{-1} \quad , \tag{7.30}$$

where M contains all the material parameters of the laser rod and an efficiency factor η which relates the electrical input power to the power dissipated as heat in the rod ($P_a = \eta P_{\text{in}}$).

We can introduce a laser rod sensitivity defined as

$$\frac{d(1/f)}{dP_{\text{in}}} = M^{-1} \quad . \tag{7.31}$$

The sensitivity factor desribes how much the optical power $1/f$ of a laser rod changes with a change in input energy to the lamp. To understand the importance of this factor, it is necessary to go back to the resonator theory. With the laser rod operating at a given power level, the designer chooses the resonator optics to provide the desired output beam pattern. In doing so, he takes into account the lensing of the laser rod. However, no system works at a constant power level. Power-supply fluctuations, lamp aging, general system deterioration, just to mention a few parameters, change the heat input to the rod, and thereby, the optical focusing power of the rod. The sensitivity factor tells how sensitive a laser rod is to these changes. The designer must ensure that his resonator design is capable of maintainig the output beam within specifications in spite of these fluctuations. For a Nd:YAG rod 0.63 cm in diameter and assuming that 5 % of the electrical input power to the lamp is dissipated as heat, we obtain a change of focusing power of 0.5×10^{-3} diopters per watt of lamp input variation.

Returning now to (7.29), we can see that the rod acts as a bifocal lens with different focal lengths for light with radial and tangential polarization. Since a linear polarized wave or a nonpolarized wave incident on the crystal will always have components in radial and tangential direction, two focal points are obtained. For Nd:YAG one finds a theoretical value of $f_\phi/f_r = 1.2$, whereas measurements show ratios varying from 1.35 to 1.5 [7.15]. A difference in focal length between different polarizations means that a resonator designed to compensate for the rod lensing for radial polarization cannot also compensate for the lensing of tangentially polarized light.

In Fig. 7.5, theoretical and measured thermally induced back focal lengths of various laser rods are plotted as a function of lamp input. Experimentally, the rod focal length is usually determined by projecting a HeNe laser beam through the rod and measuring the position of beam diameter minimum. Figure 7.5 also shows the thermally induced focusing in a Nd:LaSOAP rod operated at 20 pps [7.16]. At this high repetition rate it was shown that the dynamic optical distortions are purely a function of input power, just as is the case in cw operation. The data illustrate the very strong thermal lensing which occurs in Nd:LaSOAP as a result of the seven-times-lower conductivity of this material as compared to Nd:YAG.

Comparing the expermental results with (7.29), we find that the focal length does not always vary exactly as the inverse of lamp input power. For example, the curve in Fig. 7.5 for the 10-cm-long rod can be approximated by $f \propto P_{\text{in}}^{-1.5}$, whereas for the curve of Nd:LaSOAP we obtain $f \propto P_{\text{in}}^{-1.2}$. Applying the sensitivity factor of (7.31) to the measured curves, we find that Nd:LaSOAP is more sensitive to pump fluctuation than Nd:YAG and Nd:YALO. Typical values for Nd:YAG and Nd:YALO are $0.5\text{–}1.0 \times 10^{-3}$ diopters/W, and $2.5\text{–}4 \times 10^{-3}$ diopters/W for Nd:LaSOAP.

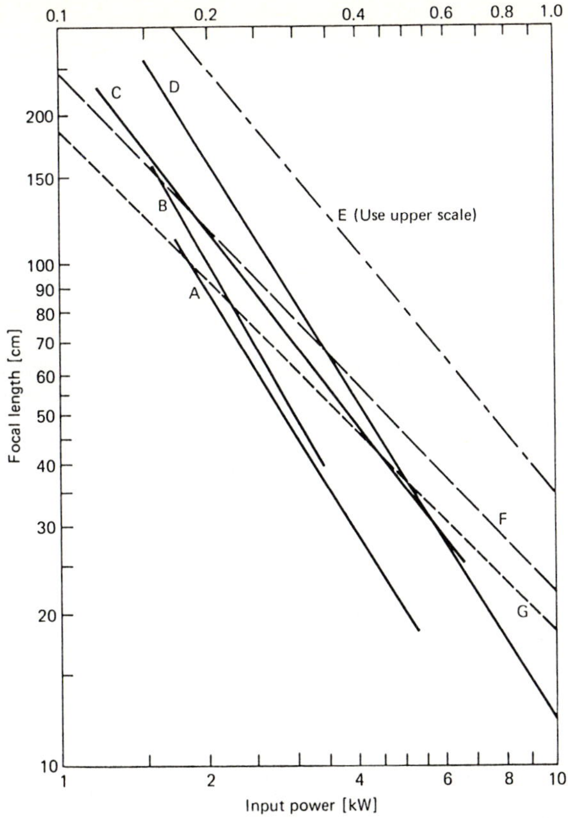

Fig. 7.5. Thermally induced back focal length as a function of lamp input power for a variety of solid-state lasers: *(A-E)* Measurements of the average focal length $(f_r + f_\phi)/2$ of *(A)* *B*-axis Nd:YAlO$_3$ rod, 7.5 × 0.62 cm; *(B)* Nd:YAG rod, 10 × 0.62 cm; *(C,D)* Nd:YAG rods, 7.5 × 0.62 cm, curves are for different rods and pump cavities; *(E)* Nd:LaSOAP rod, 7.5 × 0.6 cm (use upper scale for input power). *(F,G)* Theoretical back focal length of a Nd:YAG rod assuming that 5% of the electrical input power will be dissipated as heat in the 7.5 × 0.63 cm diameter crystal. Shown is the focal length for the radially polarized *(G)* and tangentially polarized *(F)* beam components. The value for M, see (7.30), is $M_\phi = 22 \times 10^4$ W cm, and $M_r = 18 \times 10^4$ W cm.

Stress Birefringence

We will now investigate the influence of thermally induced birefringence on the performance of a solid-state laser. Taking Nd:YAG as an example, it was shown in (7.15, 16) that the principal axes of the induced birefringence are radially and tangentially directed at each point in the rod cross section and that the magnitude of the birefringence increases quadratically with radius r. As a consequence, a linearly polarized beam passing through the laser rod will experience a substantial depolarization. We refer to Fig. 7.4, where we have chosen a point $P(r, \phi)$ in a plane perpendicular to the rod axes. At this point we have a radial refractive index component n_r, which is inclined at an angle

ϕ with respect to the y axis and a tangential component n_ϕ perpendicular to n_r. Assume that $\overline{E}$ is the polarization vector for incident radiation. Radiation incident at point P must be resolved into two components, one parallel to n_r, and the other parallel to n_ϕ. Since $\Delta n_r \neq \Delta n_\phi$, there will be a phase difference between the two components and the light will emerge elliptically polarized. This will occur for all points of the rod cross section with the exception for points located along the x and y axes in Fig. 7.4. Radiation incident along the y axis will see only one refractive index, n_ϕ, while along the x axis, n_r will be the only refractive index.

Birefringence effects in pumped laser rods can be studied in a polariscopic arrangement in which the expanded and collimated light beam from a HeNe laser serves as an illuminator for the observation of the rod between crossed polarizers. Because of thermally induced birefringence, the probe light suffers depolarization and is partially transmitted by the analyzer. The transmitted light forms the so-called isogyres, which display the geometrical loci of constant phase difference. Photographs of conoscopic patterns for various pump powers of a Nd:LaSOAP rod between crossed polarizers are shown in Fig. 7.6.

The isogyre pattern exhibits a cross and ring structure, where the arms of the cross are parallel and orthogonal to the incident polarization. As mentioned before, the crosses correspond to those regions of the crystal where an

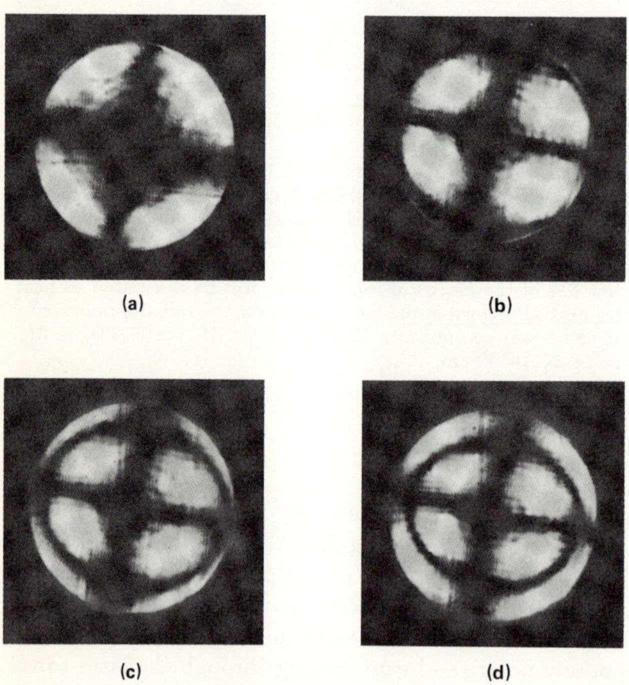

Fig. 7.6a–d. Thermal stresses in a 7.5-cm long and 0.63-cm-diameter Nd:LaSOAP crystal. The rod was pumped at a repetition rate of 40 pps by a single xenon flashlamp in an elliptical pump cavity. Input power (**a**) 115 W, (**b**) 450 W, (**c**) 590 W, (**d**) 880 W [7.16]

induced (radial or tangential) axis is along a polarizer axis, so that the induced birefringence results only in a phase delay and not in a polarization rotation. The dark rings correspond to an integral number of full waves of retardation.

A number of operations of solid-state lasers, such as, for example, electro-optical Q-switching, frequency doubling, and external modulation of the beam, require a linearly polarized beam. An optically isotropic material, such as Nd:YAG, must be forced to emit a linearly polarized beam by the introduction of a polarizer in the resonant cavity. In the absence of birefringence, no loss in output power would be expected. However, the thermally induced birefringence that does occur causes a significant decrease in output power and a marked change in beam shape.

For a system with intracavity polarizer, the effect of depolarization involves two phenomena: coupling of power into the orthogonal state of polarization followed by subsequent removal of that component by the polarizer, and modification of the main beam by the depolarization process leading to a distortion of beam shape. When a birefringent crystal is placed between a polarizer and analyzer that are parallel, the transmitted intensity is given by [7.7]

$$\frac{I_{\mathrm{out}}}{I_{\mathrm{in}}} = 1 - \sin^2(2\phi)\,\sin^2\left(\frac{\delta}{2}\right) \quad , \tag{7.32}$$

where ϕ is the angle between the polarizer and one of the principal birefringence axes, and δ is the polarization phase shift of the light emerging from the crystal. The index difference, $\Delta n_\phi - \Delta n_r$, leads to a phase difference

$$\delta = \frac{2\pi}{\lambda} L(\Delta n_\phi - \Delta n_r) \quad . \tag{7.33}$$

To illustrate the magnitude of stress birefringence in a Nd:YAG rod, the difference in optical path length

$$\delta_l = \frac{L(\Delta n_\phi - \Delta n_r)}{\lambda} \tag{7.34}$$

is plotted in Fig. 7.7 as a function of pump power. The following constants were used: $P_a = 0.05 P_{\mathrm{in}}$, $L = 7.5\,\mathrm{cm}$, $r_0 = 0.31\,\mathrm{cm}$, and $\lambda = 1.06 \times 10^{-4}\,\mathrm{cm}$. As can be seen from this figure, at maximum lamp input power of 12 kW, the path-length difference is approximately six wavelengths. The data plotted in Fig. 7.7 are in excellent agreement with the values obtained from conoscopic interference patterns [7.10, 11, 17]. Assuming a plane wave, we can calculate the total transmitted intensity from (7.32, 33) by integrating over the cross-sectional area of the rod. The following integral must be evaluated,

$$\left(\frac{I_{\mathrm{out}}}{I_{\mathrm{in}}}\right)_{\mathrm{T}} = \left(\frac{1}{\pi r_0^2}\right) \int\limits_{\phi=0}^{2\pi} \int\limits_{r=0}^{r_0} \left[1 - \sin^2(2\phi)\,\sin^2\frac{\delta}{2}\right] r\,dr\,d\phi \quad . \tag{7.35}$$

with $\delta = C_{\mathrm{T}} P_a (r/r_0)^2$, where $C_{\mathrm{T}} = 2n_0^3 \alpha C_B / \lambda K$. Integration of (7.35) yields

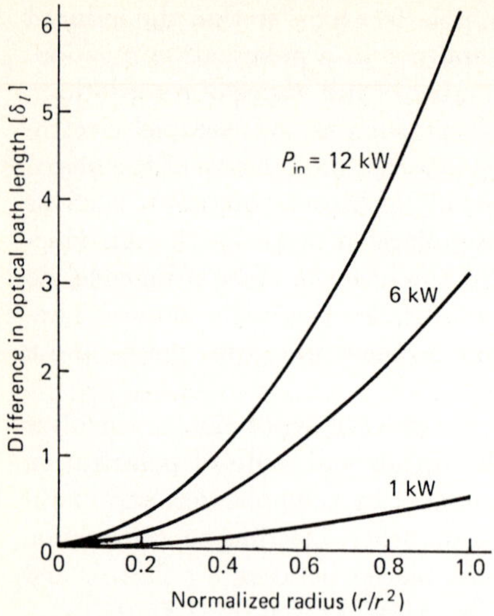

Fig. 7.7. Differences in optical path length as a function of normalized rod radius in a Nd:YAG rod. Parameter is the lamp input power

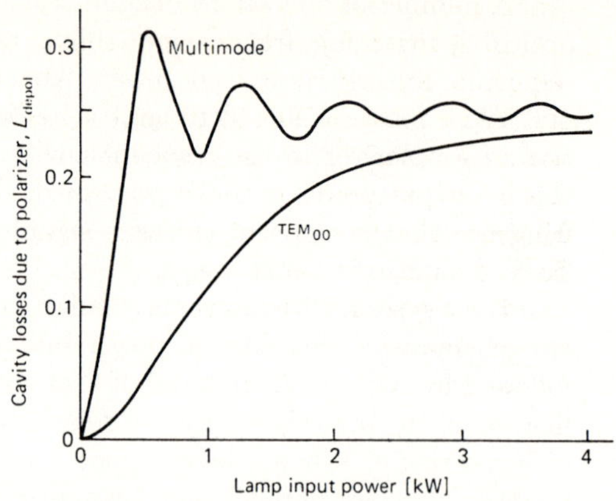

Fig. 7.8. Calculated resonator loss caused by the combination of thermally induced birefringence in a Nd:YAG rod and an intracavity polarizer

$$\frac{I_{out}}{I_{in}} = \frac{3}{4} + \frac{\sin\,(C_T P_a)}{4 C_T P_a} \quad . \tag{7.36}$$

A laser resonator containing a polarizing element is optically equivalent to a laser rod of twice the actual length located between a polarizer and analyzer. Subtracting I_{out}/I_{in} from unity and multiplying the phase difference δ by a factor of two yields the fraction of the circulating power which is polarized orthogonal to the intracavity polarizer. This beam, caused entirely by birefringence, will actually be ejected from the cavity and represents the depolarization loss of the resonator:

$$L_{depol} = 0.25[1 - \mathrm{sinc}(2C_T P_a)] \quad . \tag{7.37}$$

This loss is plotted in Fig. 7.8, where the same constants as for Fig. 7.7 have been used. Note that this curve represents the round-trip loss in the cavity for a plane-parallel beam which was used as an approximation for a highly multimode laser beam.

Also shown is the depolarization loss for a TEM_{00} mode for which it was assumed that the beam radius w is $w = r_0/2$. A similar calculation as the one carried out for a plane wave yields a loss factor [7.18]

$$L_{depol} = 0.25\left(1 + \frac{16}{C_T^2 P_a^2}\right)^{-1} \quad . \tag{7.38}$$

For the same lamp input power the losses for the TEM_{00} mode are less than for a highly multimode beam. This is expected since the energy in the TEM_{00} mode is concentrated nearer the center of the rod, where the induced birefringence is smaller.

Equations (7.37, 38) express the depolarization losses if a polarizer with a very high extinction ratio, such as a Glan-type polarizer, is inserted in the cavity. In cw-pumped Nd:YAG lasers, very often a single quartz plate orientated at the Brewster angle is utilized to obtain a linearly polarized output beam. The reflectivity of fused silica for the "S" polarization at the Brewster angle is 12.6 % per surface at 1.06 μm, which results in a round-trip loss in the cavity of 58 %. In case of a Brewster-plate polarizer in the cavity, the round-trip losses expressed by (7.37, 38) must be multiplied by the factor 0.58 [7.18].

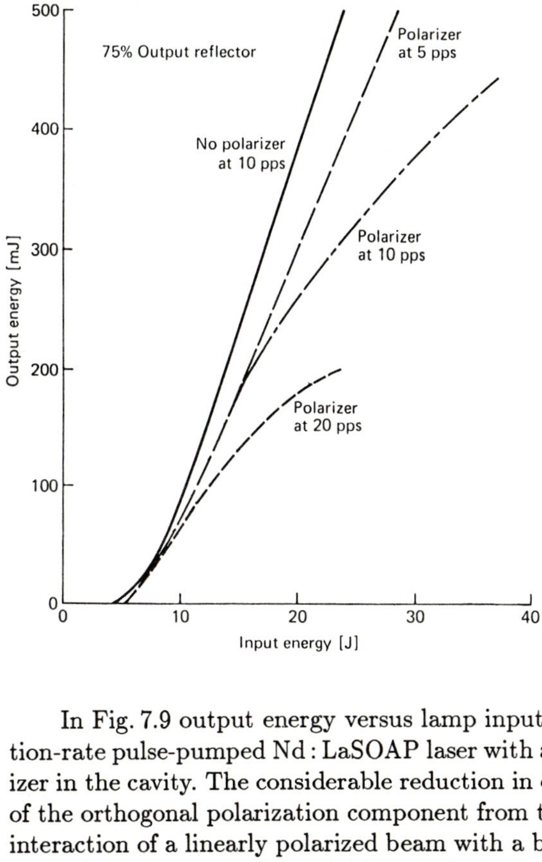

Fig. 7.9. Conventional-mode output from a Nd:LaSOAP laser with Glan-type polarizer in the resonator [7.16]

In Fig. 7.9 output energy versus lamp input is measured for a high-repetition-rate pulse-pumped Nd:LaSOAP laser with and without a Glan-type polarizer in the cavity. The considerable reduction in output resulting from rejection of the orthogonal polarization component from the cavity is quite evident. The interaction of a linearly polarized beam with a birefringent rod and a polarizer not only leads to a substantial loss in lasing efficiency, but also a severe distortion of the beam shape occurs. Figure 7.10 shows the output beam shape obtained from a cw-pumped Nd:YAG laser with and without an intracavity polarizer. The output is obtained in form of a cross with a bright center. As discussed before, the depolarization losses are smallest in the rod center and in directions parallel and orthogonal to the preferred direction of the polarizer.

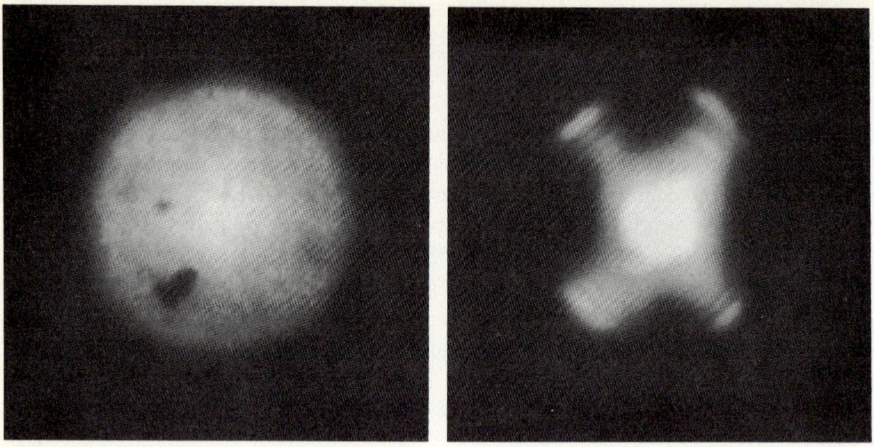

Fig. 7.10. Output beam pattern fo high-power cw Nd:YAG laser without *(left)* and with *(right)* a Brewster plate in the cavity

Areas of high depolarization losses are removed from the cavity by the polarizer and are, therefore, missing in the output beam.

Compensation of Thermally Induced Optical Distortion in Cylindrical Laser Rods

Various methods of compensation for spherical and nonspherical lensing effects and birefringence produced by the laser rod have been suggested and tried to date.

Thermal Lensing. Of all the optical distortions produced in the laser rod, the spherical component of the lensing effect is the easiest to compensate for. In a uniformly heated cylindrical rod which is uniformly cooled around its outer periphery, stress is radially symmetric. This is the situation most desirable in a laser rod, since it is indicative of a radially symmetric refractive index gradient which is required for optimum optical compensation of thermal lensing in the laser rod. Compensation of the spherical component of the thermal focusing effect can be accomplished by grinding concave spherical surfaces on one or both ends of the laser rod, by introducing a negative single-element lens into the resonator [7.19] or by suitably designing the resonator configuration. The latter approach has been discussed in detail in Chap. 5. All three techniques lead to a substantial reduction of multimode beam divergence of a cw-pumped laser and to an enhancement of the TEM_{00} mode power.

Compensation of Thermal Birefringence. The majority of schemes involve the insertion of devices into the laser cavity which will cause an intra-cavity polarization rotation. The objective of this approach is to achieve equal path length at each point of the rod cross section for radially and tangentially polarized light. The best approach of the problem of birefringence compensation appears to be a 90° rotation with a quartz rotator located between two laser rods. Both rods are inside the laser cavity, and the part of a mode that is radially polarized in the first rod is tangentially polarized in the second rod. Since each part of the beam passes through nearly identical regions of the two rods, the birefringence induced by one rod is cancelled by the other. With this technique, linearly polarized outputs with negligible power loss and TEM_{00} output of 50 to 70 % of the multimode power have been achieved [7.20]. In principle, a 90° intracavity rotation could also be achieved by placing a $\frac{1}{8}$-wave Faraday rotator between the rod and the rear mirror in a single-rod oscillator.

The aforementioned techniques can provide a compensation for birefringence and bifocusing, but in long rods the complex interaction between birefringence and bifocusing is difficult to correct in a lumped element. In addition, insertion losses for compensation devices can be significant. These difficulties could be avoided if the principal axes of the laser rod birefringence were in Cartesian coordinates instead of polar coordinates. One might accomplish this in cubic or isotropic materials by altering the shape of the rod cross section such that it will produce rectangular isotherms. Geometries of the active material other than cylindrical will be discussed at the end of this chapter.

A reduction of birefringence in Nd : YAG lasers has been achieved by employing rods which have their axis along a crystallographic axis where the thermally induced birefringence of the rod is at a minimum [7.21]. An analysis of birefringence as a function of the orientation of the laser rod axis with respect to the crystallographic coordinate system indicates that the thermally induced stress birefringence in Nd : YAG laser rods is dependent on growth direction. For all rod-axis orientations other than in the [111] crystal direction, the birefringence is angular-dependent. However, for high-power systems, the losses are essentially constant regardless of growth direction and/or orientation of the polarizer in the laser cavity. For systems in which the dissipated power in the Nd : YAG rod is small (50 W or less), the losses can be minimized by choosing the proper rod-axis orientation with respect to crystallographic direction and the optimium orientation of the polarizer in the laser resonator. The optimum rod-axis direction is in the [001] crystal direction or equivalent.

By replacing the cubic YAG host with the orthorhombic yttrium orthoaluminate ($YAlO_3$), the effects of thermal birefringence can be reduced to negligible levels. In optically isotropic laser crystals, the indicatrix patterns are radially oriented and all linearly polarized waves undergo depolarization. If the host material, however, is an optically anisotropic crystal, such as $YAlO_3$, thermally induced stresses cause a slight modification of the uniform anisotropy. This perturbation is nonuniform but small, since the index difference due to the crystal anisotropy is 10^2 to 10^3 times as large as the index variation due to thermal stresses. Since the indicatrix essentially retains its orientation, linearly polarized waves may propagate through the crystal without suffering depolarization. As was mentioned in Chap. 2, the exploration of Nd : YALO as a laser material has been hampered by the unavailability of good quality crystals.

Nonsperhical Aberrations. In cw-pumped Nd : YAG lasers, optically thin, small-diameter rods located in focusing cavities are usually employed. In these situations the pump light is concentrated on the rod axis, and our assumption of a uniformly heated cylindrical rod is no longer valid. If the fine-ground surface of the laser rod were a perfect Lambertian diffuser, then the pump light intensity at the rod axis would be lower than at the surface because of the Nd absorption. In fact, the surface seems to have distinctly nonuniform forward scattering properties, with the result that pump light is more intense on the rod axis than at the edges. Since the focal length of a given zone in the rod is inversely proportional to the intensity of the absorbed light at that zone, it is easy to undestand that the focal length of the center of the rod is less than that at the edges. The pump-induced aberration in the rod is illustrated in Fig. 7.11. The amount of spherical aberration depends on the surface finish of the rod, the type of pumping cavity used, and the diameter of the pump lamp. For a ray passing through the rod at a zone of radius r, the focal length f_{r} is always greater than at the center and the difference $f_{\mathrm{r}} - f_{\mathrm{c}}$ is proportional to r^2. This effect is positive spherical aberration. A positive lens with spherical surfaces produces a similar effect of opposite sign. Thus in applications where the laser beam is passed through a positive lens, spherical aberration can be partially compensated.

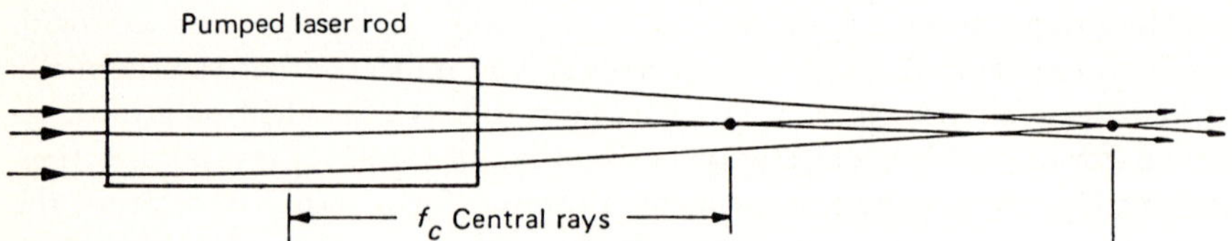

Fig. 7.11. Spherical aberration in a laser rod

Astigmatism in the laser rod can be caused by nonsymmetrical pumping or by grown-in rod imperfections. Astigmatism can be corrected by passing the beam through a spherical lens suitably tilted away from normal incidence. For a thin plano-convex lens of radius R tilted at an angle Θ with respect to the normal, the focal lengths (f_{T} and f_{Θ}) for rays lying in the plane of incidence (tangential plane) and perpendicular to the plane of incidence (sagittal plane) are

$$f_{\mathrm{T}} = \frac{R\cos^2\theta}{(n^2-\sin^2\theta)^{1/2}-\cos\theta}\ , \quad f_{\mathrm{s}} = \frac{R}{(n^2-\sin^2\theta)^{1/2}-\cos\theta}\ , \tag{7.39}$$

where n is the refractive index. We see from (7.39) that by introducing a tilt Θ, the ratio $f_{\mathrm{T}}/f_{\mathrm{s}}$ can be varied, thus introducing an easily controlled amount of astigmatism into the system to cancel out effects produced in the laser rod.

7.1.2 Single-Shot Operation

If a laser rod is pumped by a single isolated pump pulse, a transient thermal profile will be established which is the result of a fast pump-introduced heating process followed by a slow recovery of the rod to thermal equilibrium. Figure 7.12 shows schematically the transient temperature distribution in an optically pumped cylindrical laser rod. During the flashlamp pulse, energy is absorbed by the laser crystal. At the end of the pump pulse, the temperature of the rod has increased unformly by ΔT above the coolant temperature if we assume uniform absorption of pump radiation. Since only the surface of the rod is in contact with the coolant, a radial heat flow develops at the end of the pump pulse. The initially uniform temperature distribution changes to a parabolic temperature profile which decays with a certain time constant. Since lasing action occurs during or at the end of the pump cycle, the optical distortion in a single-shot laser arises from the thermal gradients produced during the pump cycle and not from heat flow as a result of cooling.

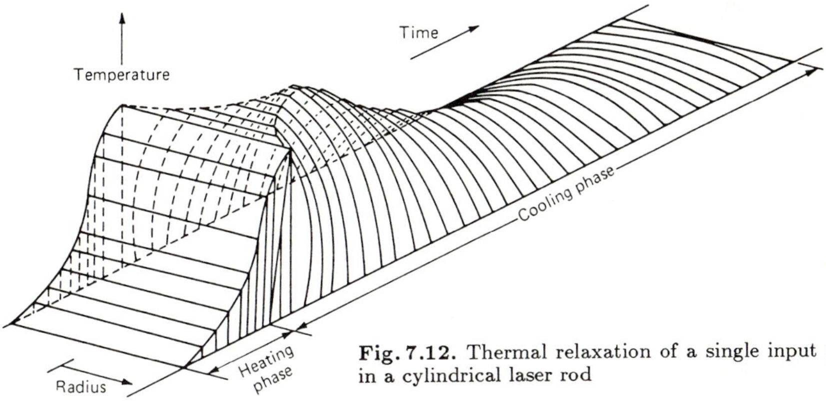

Fig. 7.12. Thermal relaxation of a single input pulse in a cylindrical laser rod

If the temperature of the laser rod increases uniformly over the rod cross section, the bulk temperature rise at the end of the pump pulse is given by

$$\Delta T = \frac{Q}{cV\varrho} \quad , \tag{7.40}$$

where Q is heat deposited in the rod, c is the specific heat, V is the volume, and ϱ is the density of the laser material. The bulk temperature rise as a function of time depends on the shape of the flashlamp pulse. The energy deposited in the rod is

$$Q = \int_0^\infty P(t)dt \quad , \tag{7.41}$$

where $P(t)$ is the pulse shape of the pump power. The time dependence of the temperature during the pump pulse can be obtained from

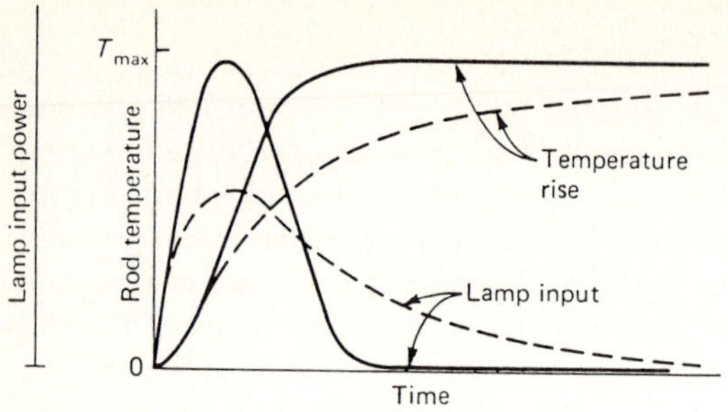

Fig. 7.13. Temperature rise in a laser rod for different lamp input pulse shapes

$$T(t) = (cV\varrho)^{-1} \int_0^t P(t)dt \quad . \tag{7.42}$$

In Fig. 7.13 the temperature rise for several common pulse shapes is illustrated. Measurements have shown that the temperature increases without any observable delay with respect to the energy deposited in the rod. In air-cooled laser heads, a considerable amount of rod heating can take place after the pump pulse because of conductive and convective heat transfer from the hot flashtube. In liquid-cooled systems, most of this long IR radiation is absorbed and carried away by the coolant.

If the temperature rises uniformly in the active material, thermal effects in an oscillator are limited to a frequency shift of the emitted radiation and an increase in optical length of the resonator. Nonuniform pumping will also lead to thermal lensing and birefringence. If one assumes a pump-induced thermal profile $T(r)$ at a given time t_1 during the pump cycle, then thermal lensing and birefringence for this particular instant in time can be calculated as was discussed in Sect. 7.1.1 for the cw case.

The most common technique for observing pump-induced thermal effects is to insert the laser rod in one arm of an interferometer and to observe the time-dependent fringe pattern with a high-speed framing camera [7.22–27]. Another interferometric technique consists of counting the transient fringes with a fast photodetector mounted behind a pinhole [7.28]. The transient response of the rod to the pump pulse has also been studied by photographing the induced stress birefringence by means of a plane polariscope [7.29] or by measuring thermal lensing [7.30] or beam divergence [7.31] during the pump pulse.

Table 7.1 summarizes the experimentally determined temperature excursions and path-length changes occurring at the end of the pump pulse for a variety of laser rods and pumping arrangements. The temperature listed in this table are usually inferred from an interferometric measurement of the optical path-length changes. If we ignor changes in refractive index resulting from thermal stress, then optical path length and temperature changes are related according to

$$\Delta(n_0 l_0) = n_0 l_0 (\alpha_l + \alpha_n) \Delta T \quad , \tag{7.43}$$

where $n_0 l_0$ is the undistorted optical length, $\alpha_l = (1/l_0)(dl/dT)$ is the ther-

Table 7.1. Temperature rise at end of pump cycle

Material	Rod size length × diameter [cm]	Input energy [J]	Temp. rise [C]	Temp. profile	Pump geometry	Increase in optical lenght [λ]	Ref.
Nd:glass	15 × 1.5	7,000	6.4	Concave negative lens	Focusing linear lamp	10	[7.22]
Ruby	15 × 1.5	7,000	15.3	Concave	Focusing linear lamp	32	[7.22]
Ruby	7.5 × 0.6	1,250	4.0	Convex	Helical lamp	10	[7.23]
Ruby	6.3 × 0.63	800	3.5	–	Helical flashlamp	8	[7.28]
Nd:glass	6.3 × 0.63	800	4.9	–	Helical flashtube	7	[7.28]
Ruby	5 × 0.69	5,000	6.7	Convex	Helical lamp polished rod	–	[7.26]
Nd:glass	7.5 × 1.0	11,500	15	Concave	Helical lamp twice as long as rod	9	[7.25]

mal coefficient of linear expansion, and $\alpha_n = (1/n_0)(dn/dT)$ is the thermal coefficient of refractive index.

Most measurements reveal that the optical path-length change is actually concave and the rod behaves like a negative lens. The implication is that the absorption of pump energy is higher at the outer edge of the rod than in the center. Figure 7.14 shows an example of a concave temperature profile

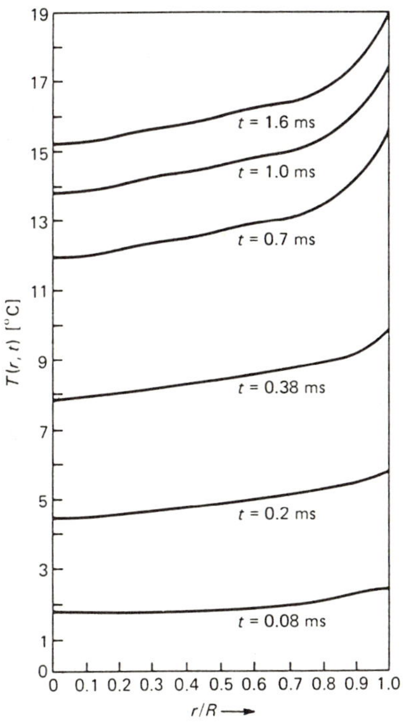

Fig. 7.14. Temperature rise in a 7.5-cm-long and 1.0-cm-diameter Nd:glass rod pumped by a 15-cm-long helical flashlamp operated at 11,500 J input. [7.25]

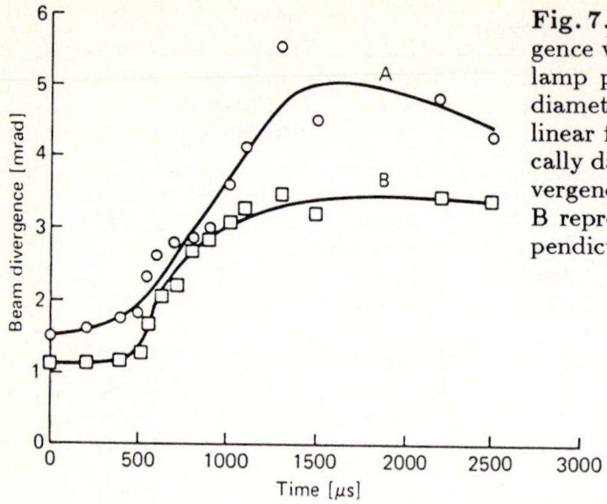

Fig. 7.15. Ruby laser amplifier beam divergence versus time after the inition of the flashlamp pump. The ruby laser rod, 12.7 mm in diameter and 7.6 cm long, was pumped by two linear flashlamps with an 800-μs, 5000-J critically damped pulse. Curve A represents the divergence in the plane of the flashlamps; curve B represents the divergence in the plane perpendicular to it. [7.30]

calculated from the fringe pattern of an interferometric measurement [7.25]. The total path-length change in the center of this rod was nine wavelengths. The increase in beam divergence in a ruby amplifier rod during the pump cycle is presented in Fig. 7.15 [7.30]. Maximum gain was obtained in the rod after 750 μs, at which point the beam divergence was twice the initial divergence.

7.1.3 Repetitively Pulsed Lasers

In the previous section we discussed pump-induced thermal distortions which occur for single-shot lasers operated either in normal pulse or Q-switched operation. In this regime the pulse input intervals are long with respect to the thermal relaxation of the laser rod and the active material has returned to ambient temperature before the onset of the next pump pulse. If the laser is operated at a repetition rate at which there exists a residual temperature distribution from the previous pulse, a temperature buildup inside the material will occur. The residual temperature will accumulate for subsequent pulses until a steady-state condition is reached. Figure 7.16 shows the temperature profile of a laser operated at a fairly low repetition rate. The residual temperature has not completely decayed between pulses, and the initial uniform temperature rise caused by the pump source is modified to a slightly parabolic temperature profile.

Figure 7.17 shows a thermal buildup at a pulse interval smaller than the thermal relaxation time. Under these conditions, the laser rod will soon establish a steady-state operating condition where the radial flow of heat out of the rod will be equal to the total heat being put into the rod. During the transition period the cylindrical rod passes through several regions of radial thermal profiles as a steady-state equilibrium is approached. In the extreme case, when the pulse interval time becomes negligibly small as compared to the thermal relaxation time, the thermal profile in the laser rod approaches the cw condition. In this mode of operation the thermal effects, which have been calculated

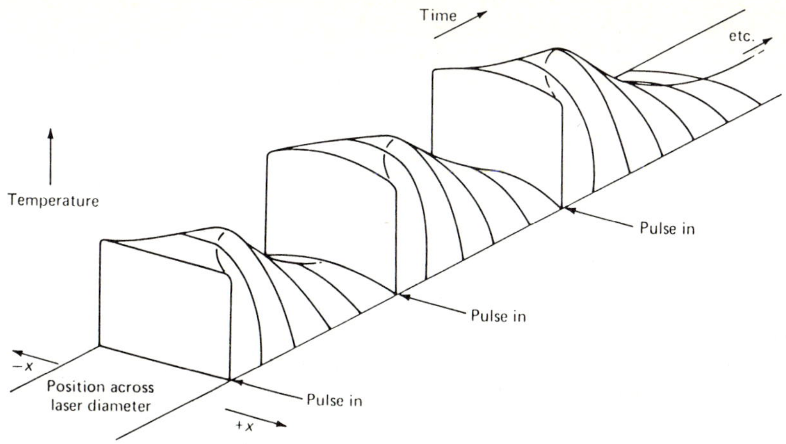

Fig. 7.16. Thermal relaxation in a solid cylinder with pulse input intervals near the thermal relaxation time

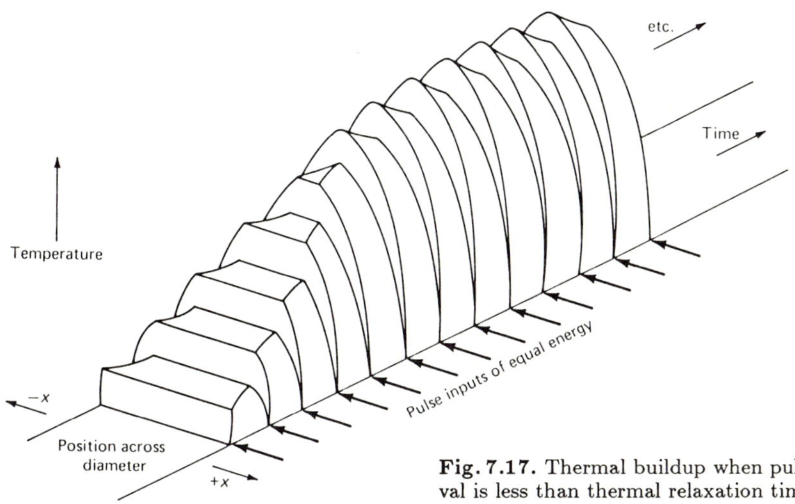

Fig. 7.17. Thermal buildup when pulse interval is less than thermal relaxation time

for Nd : glass [7.32–34] and Nd : YAG [7.10, 11, 14], depend only on the average input power.

The transient temperature distribution in a laser rod, as illustrated, for example, in Figs. 7.16 and 17, has an effect of large proportions on energy output and beam divergence of the laser. Calculations treating the thermal relaxation of a cylindrical laser rod which is suddenly heated by a pump pulse can be found in [7.35–38]. A general theory describing the transient thermal profile in a pulse-pumped rod for a large variety of operating parameters has been presented by *Koechner* [7.4]. The calculations were performed by solving the heat diffusion equation for a cylindrical rod for radial heat flow. The general solution of this equation is given by *Carslaw* and *Jaeger* for an infinite circular cylinder [7.1].

The general expression of the radial temperature profile $T(r,t)$ in a solid rod as a function of time can be expressed in a series of Bessel functions of the first kind and zero-, first-, and second-order (J_0, J_1, J_2)

$$\frac{T(r,t)}{\Delta T} = \sum J_{0,1,2}\left(\frac{t}{\tau}, \frac{r}{r_0}, \frac{\Delta t}{\tau}, M, g, A\right) \quad , \tag{7.44}$$

where ΔT is the initial temperature rise, t/τ is the normalized time, r/r_0 is the normalized rod radius, $\Delta t/\tau$ is the normalized pulse repetition rate (Δt is the pulse interval time), M is the number of consecutive pulses ($M \to \infty$ at steady state), g specifies the temperature distribution in the rod at the end of the first pump pulse, and A is a cooling parameter which specifies the cooling condition of the rod. The full expression of (7.44) is very complicated and must be evaluated by computer. Figures 7.18–21 illustrate some of the results of the analysis. Shown are thermal distributions in laser rods for a variety of system parameters. Since all quantities are normalized, the results are applicable to any kind of laser host material, rod size, cooling flow, pump energy, etc. In order to be able to apply the results to practical cases, we have to discuss briefly the normalization parameters τ, ΔT, and A. The thermal time constant of the rod is

$$\tau = r_0^2/k \quad , \tag{7.45}$$

where r_0 is the rod radius and k is the thermal diffusivity, which is related to the materials parameters by

$$k = K/c\gamma \quad , \tag{7.46}$$

where K is the thermal conductivity, γ is the mass density, and c is the specific heat. The initial temperature distribution at the end of the pump cycle is approximated by a parabolic function

$$T(r,0) = \Delta T\left[1 + g\left(\frac{r}{r_0}\right)^2\right] \quad , \tag{7.47}$$

where ΔT is the temperature rise in the center of the rod and g is a measure of the pumping nonuniformity; for example, $g = 0$ indicates a uniform pump distribution across the rod, $g > 0$ represents a pumping nonuniformity leading to a negative lensing effect, and focusing of pump radiation in the rod can be expressed by $g < 0$. The amount of heat deposited in the laser rod is given by

$$Q = 2\pi l_0 c\gamma \int_0^{r_0} T(r,0) r \, dr \quad . \tag{7.48}$$

From (7.47, 48) it follows that the temperature rise ΔT in the center of the rod at the end of the first pump cycle is

$$\Delta T = \frac{Q}{Vc\gamma(1 + g/2)} \quad . \tag{7.49}$$

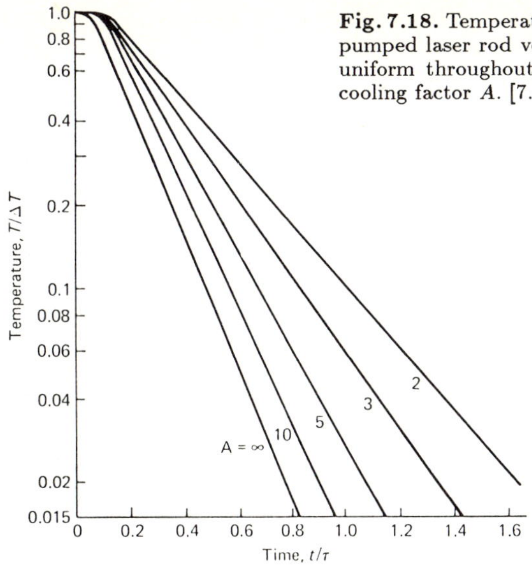

Fig. 7.18. Temperature decay in the center of an optically pumped laser rod versus time. Initial temperature rise is uniform throughout the rod ($g = 0$). Parameter is the cooling factor A. [7.4]

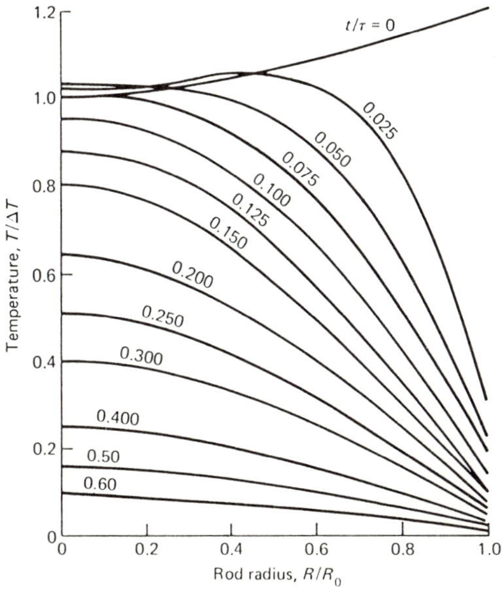

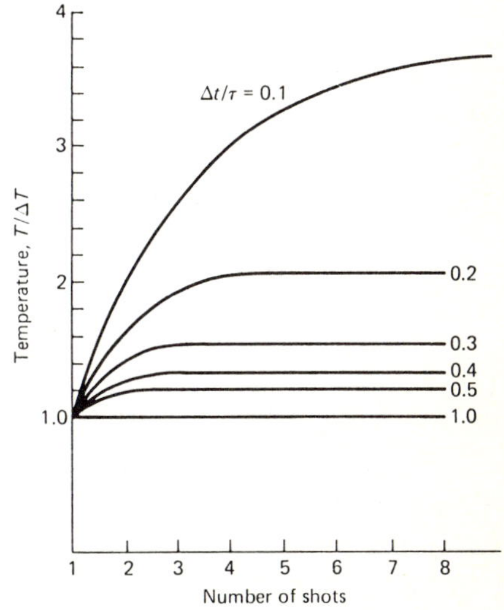

Fig. 7.19. Temperature profile of a solid laser rod for single-shot operation. Pumping coefficient $g = 0.2$, cooling factor $A = 10$. [7.4]

Fig. 7.20. Temperature buildup in the center of a repetitively pumped laser rod versus number of shots. Parameter is the normalized repetition rate. Cooling factor $A = 10$, pumping coefficient $g = 0$. [7.4]

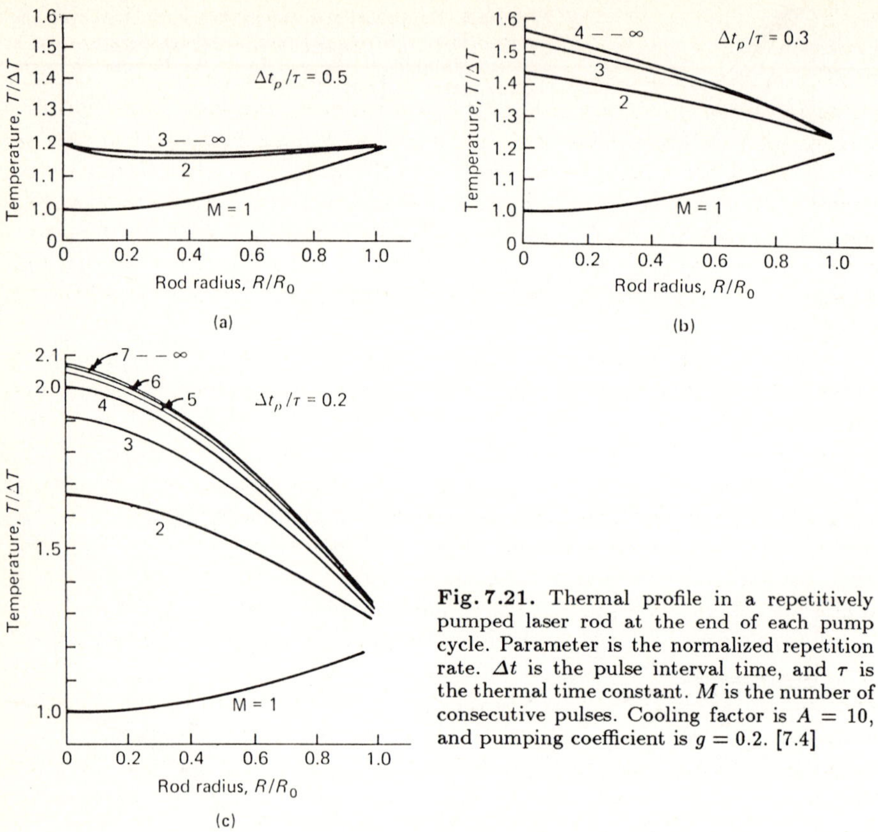

Fig. 7.21. Thermal profile in a repetitively pumped laser rod at the end of each pump cycle. Parameter is the normalized repetition rate. Δt is the pulse interval time, and τ is the thermal time constant. M is the number of consecutive pulses. Cooling factor is $A = 10$, and pumping coefficient is $g = 0.2$. [7.4]

The cooling condition of the rod is specified by a dimensionless parameter

$$A = r_0 h/K \quad , \tag{7.50}$$

where h is the surface heat transfer coefficient. Typical values for h were presented in Fig. 7.1.

Figure 7.18 illustrates the thermal relaxation of a cylindrical laser rod pumped by an isolated pulse. The temperature is plotted for the center of the rod as a function of normalized time. The value $A = \infty$ represents the case of infinitely good thermal contact between the laser crystal and the cooling fluid; i.e., the boundary surface of the rod is held at a constant temperature during the heat removal process. With this idealized assumption, the decay of thermal effects depends only on the thermal diffusivity and rod diameter. A thermal relaxation time

$$\tau' = r_0^2/4k \tag{7.51}$$

can be defined at which the temperature in the center of the rod has decayed to $1/e$ of the initial value ΔT.

In actual cases, however, it is the temperature of the coolant fluid which is constant and the surface of the rod which varies in temperature. As a result of this assumption, the temperature decay in a rod depends on the material parameters and the conduction of heat across the flow boundary. This condition is expressed by a finite value of A. For most liquid cooled laser rods the value of the rod-surface cooling parameter A is between 2 and 10. Therefore, in practical situations the thermal relaxation processes are considerably slower than in the case where a zero thermal impedance between the rod surface and the cooling fluid is assumed. This explains the discrepancy which existed between the theoretical and measured decay times of thermal effects in laser rods [7.39].

Figure 7.19 shows the temperature profile in a solid rod as a function of time for a cooling factor of $A = 10$. It is assumed that the initial temperature distribution at the end of the pump pulse ($t/\tau = 0$) is parabolic, with the temperature of the cylindrical surface 20% higher than in the center of the rod ($g = 0.2$). As one can see from this figure, at the water-cooled cylindrical surface the temperature decreases very rapidly, causing large temperature gradients. The temperature in the center changes very little for about 0.075 time constants. Actually, the temperature in the center shows a slight increase due to heat flow toward the center immediately after the flashlamp pulse.

If a laser is repetitively pumped, the thermal buildup in the rod depends on the ratio of the pulse interval time to the thermal time constant. In Fig. 7.20 the temperature buildup at the center of a cylindrical rod is plotted as a function of number of pump pulses. The parameter is the ratio $\Delta t/\tau$. The values for the temperature buildup are obtained at the end of each pump pulse. During the pulse interval time the temperature decreases, of course. The cooling factor was chosen $A = 10$ and the pumping coefficient $g = 0$. In the case of $\Delta t = \tau$, any distortion introduced by the pumping process has a chance to relax out before the onset of the next pump pulse. If the pulse interval time Δt is shorter than the thermal relaxation time τ, a temperature buildup occurs. For example, for $\Delta t/\tau = 0.2$, the incremental temperature rise at the center of the rod is more than twice as high as compared to the single-pulse operation.

Figure 7.21 shows temperature profiles at the end of each pump pulse for different repetition rates. As one can see from this figure, the initially concave temperature profile is changed to a convex profile. The absolute temperature as well as the temperature gradient at each point of the rod cross section increase for the higher repetition rates. Again, $A = 10$ and $g = 0.2$ were chosen for the cooling factor and the pumping coefficient, respectively. After the first pulse, the outside of the rod is at a higher temperature than the center as a result of the assumed nonuniform pump light absorption. In Fig. 7.21a the negative focusing effect created by nonuniform pumping is partially compensated by the interpulse cooling process. In Figs. 7.21b and c the initial diverging condition of the rod changes to a strongly focusing condition after quasi-steady state is reached.

The cooling condition of a laser rod can be improved by drilling a hole in the center of the rod. This is done occasionally in large ruby amplifier systems which must operate at high repetition rates. The calculations show, as one would expect, that the improvement is most dramatic for a hollow cylinder cooled on the inner and outer surface [7.4, 37]. However, because of practical difficulties which arise in channeling the coolant flow across the emitting faces of the rod, this is rarely done. But even for a rod with an uncooled core, a substantial improvement can be achieved compared to a solid rod. The improvement stems from a reduction of the path length for the heat flow to the cylindrical surfaces. For example, in a rod with a hole diameter of $R_1 = 0.3R_0$, the path for heat transfer is reduced by 30 %, whereas the cross-sectional area, which determines the maximum laser output power, is reduced by only 9 %. If the bore of the hollow cylinder is uncooled, then the hottest region of the rod is right at the inner boundary surface.

The thermal relaxation in a laser rod can be determined experimentally by measuring the thermally induced birefringence [7.40] or lensing [7.36, 41, 42] or by observing an interference fringe pattern [7.39, 43]. Figure 7.22 shows the temperature decay in a ruby rod, 1 cm in diameter and 12 cm long [7.39]. The temperature was calculated from the interference pattern which was recorded with a movie camera. According to (7.45), the thermal time constant of this rod is about 2 s. The temperature in the center of the rod has decayed to 0.37 of the initial value after 1.1 s. Comparing these data with Fig. 7.18 reveals a cooling parameter of this water-cooled rod of $A = 2$ or a surface heat transfer coefficient of $h = 1.68\,\mathrm{W/cm^2\,K}$.

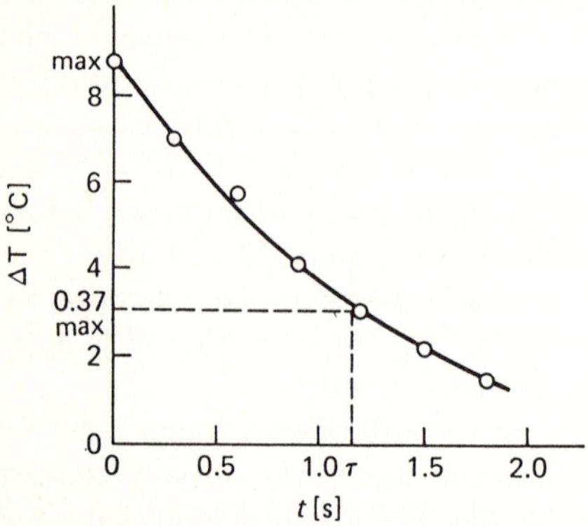

Fig. 7.22. Temperature difference ΔT between center and edge of a ruby rod after the pump pulse. [7.39]

The experimental results obtained from birefringence measurements can be divided into three regimes: inherent birefringence present in the laser rod; induced birefringence during the pump cycle; and variations in birefringence after the pump pulse due to thermal conduction within the laser rod.

Figure 7.23 shows the light pattern transmitted through the crossed polarizers for a Nd : glass rod pumped by a single flashlamp pulse [7.40]. A 16-mm movie camera was used to record the pattern on the screen as a function of time. The first photograph is a measure of the static birefringence in the laser rod. The distorted nature of the second pattern indicates that the heat disposition did not possess perfect cylindrical symmetry. However, as the rod thermalizes

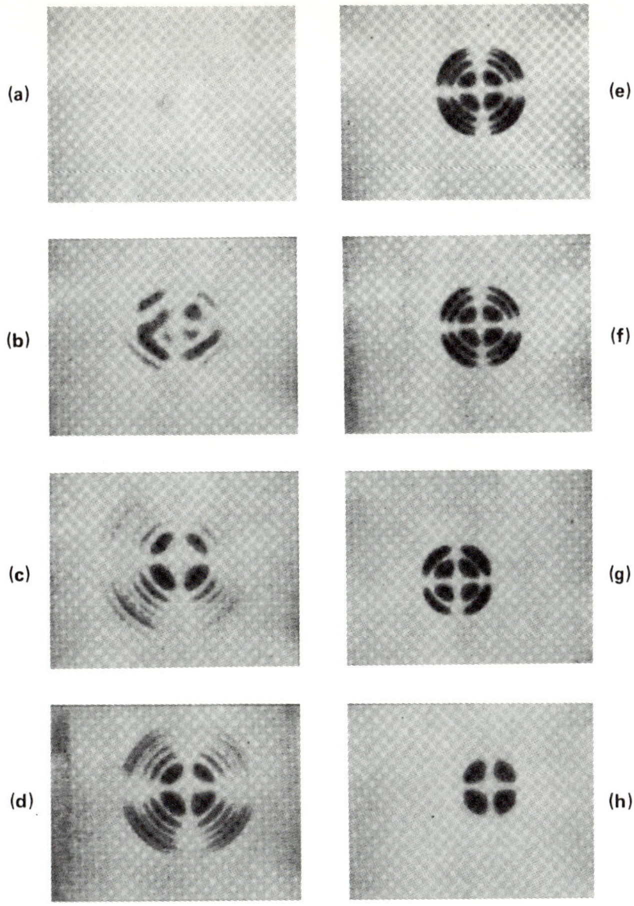

Fig. 7.23a–h. Thermal birefringence in a 15.2–cm-long by 1.5-cm-diameter Nd : glass rod pumped by two flashlamps in a double-elliptical pump cavity. Total pump energy was 7000 J. **(a)** $t < 0$; **(b)** $t = 0.1$ s: **(c)** $t = 2.4$ s; **(d)** $t = 3.3$ s; **(e)** $t = 8.7$ s; **(f)** $t = 9.5$ s; **(g)** $t = 13.1$ s; **(h)** $t = 18.6$ s. [7.40]

during cooling, the symmetry does clearly develop. It is of interest to note that even after 18 s, considerable birefringence still remains in the rod. This is not surprising since the thermal time constant for a glass rod of this size is $\tau = 140$ s. From Fig. 7.18 follows, if we assume a cooling factor of $A = 10$, that it takes at least 80 s before the temperature elevation at the center has decayed to less than 10 % of its initial value.

The thermally induced lens-like behavior of a water-cooled ruby rod (0.48 cm in diameter) was experimentally measured by passing a collimated HeNe laser probe beam through the rod and observing the behavior of the probe beam during and after the lamp pulse [7.36]. It was determined that during the 400-μs lamp pulse, the ruby rod behaves as a negative lens and about 120 ms later, it behaves as a positive lens that increases in strength and reaches a maxi-

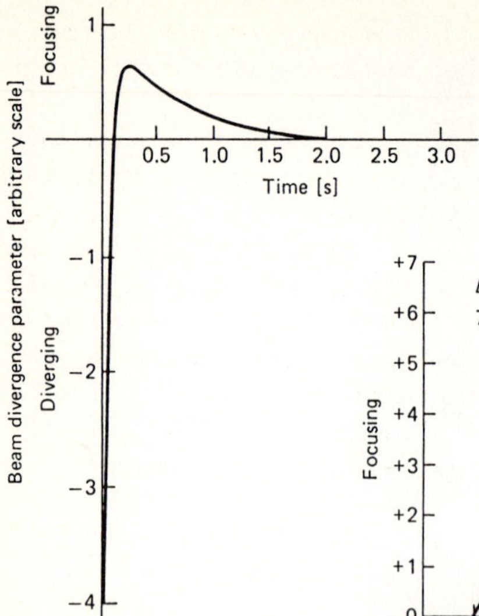

Fig. 7.24. Calculated time-dependent thermal lens effect of a 0.48-cm-diameter ruby rod. [7.36]

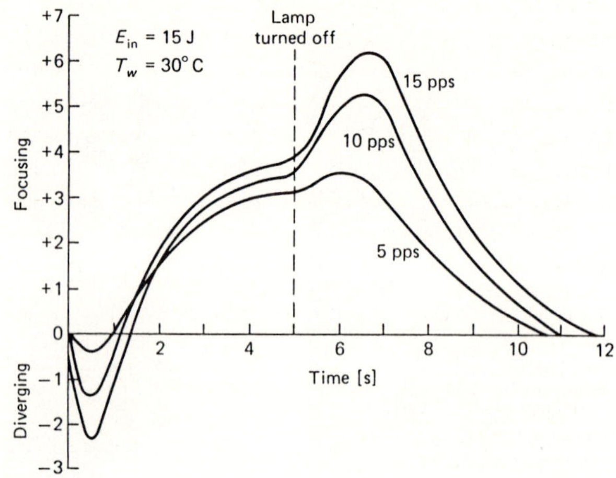

Fig. 7.25. Thermal lensing in a Nd : YAG laser rod as a function of repetition rate. [7.41]

mum 250 ms after the lamp pulse, and then completely decays approximately 1.8 s after the lamp pulse. Figure 7.24 represents a calculated curve which best describes the experimentally observed focusing action of the rod. The coefficient of surface heat transfer was determined to be $h = 0.27\,\mathrm{W/cm^2\,K}$. From (7.45, 50) it follows that $\tau = 0.45$ s and $A = 0.15$. Comparing the values with the measured data, one finds that because of the small rod-surface cooling parameter it takes four thermal time constants before the rod has returned to thermal equilibrium.

Figure 7.25 shows thermal lensing in a repetitively pumped Nd : YAG rod. A transient defocusing and a quasi-steady-state equilibrium with positive focusing are identified. At a repetition rate of 5 pps the rod reaches steady state after about 5 s or 25 pulses. For the higher repetition rates steady state had not been reached after 5 s, at which time the system was turned off. From Fig. 7.25 it follows that it takes about 5 to 7 s for the rod to return to equilibrium after the flashlamp has been turned off. The laser rod used in these experiments measured 7.5 cm by 0.63 cm and was pumped by a single linear flashlamp. The surface heat transfer coefficient was caculated to be $0.79\,\mathrm{W/cm^2\,K}$. From these parameters we calcualte $A = 1.9$, $\tau = 2.1$ s, and $\Delta t_p/\tau = 0.1$. With these values and the graphs presented in [7.4], good agreement between the calculated and measured transient periods is obtained. If this laser rod is operated above 5 pps, the time between pulses is so short compared to th thermal time constant that essentially no temperature decay takes place between pulses. The temperature profile in the rod is then identical to the one obtained at cw operation for the same input power. This is the regime where, for example, thermal lensing is found to be insensitive to pulse repetition rate as long as the average power is constant [7.42].

The transient optical distortions in a large Nd:glass rod for single-shot and repetitive pumping were reported in [7.43]. A 2.5-cm diameter by 30-cm-long rond was pumped by a helical flashlamp in a 0.8-ms-long pulse. The time necessary for the water-cooled Nd:glass rod to return after firing to initial no-strain condition was on the order of 10 min. Introducing the appropriate materials parameters into (7.45), we obtain for the thermal time constant $\tau =$ 6 min. Since about 1.7 time constants are required for the thermal distortions to completely decay, we can estimate a surface cooling parameter of $A = 2$ to 3 from Fig. 7.18. If this rod is pulsed every 25 s (corresponding to $\Delta t/\tau = 0.07$), it takes about 25 shots before the rod has reached steady-state condition.

Table 7.2. Beam divergence as a function of repetition rate in a ruby amplifier

Time between firings	3 s or more	2.5 s	2 s
$\Delta t/\tau$	0.38	0.31	0.25
Residual temperature $T/\Delta T$	0.32	0.47	0.55
Uncorrectable beam divergence [mrad]	0.25	0.33	1.0

As our last example of thermal distortions in repetitively pumped lasers, we have listed in Table 7.2 the increase of uncorrectible beam divergence with repetition rate in a large ruby amplifier rod [7.44]. The amplifier was followed by a telescope which was adjusted at each repetition rate to minimize beam divergence. This allows one to compensate for spherical aberrations introduced by the thermal profile. The ruby rod had a diameter of 2 cm with a 0.6-cm core drilled out in the center. The rod length was 20 cm. The strongly turbulent coolant flow resulted in a surface heat transfer coefficient of 2.4 W/cm^2 K. From these data it follows $A \cong 6$ and $\tau = 8$ s. Little difference in the magnitude of uncorrectible beam divergence was noticed from single-shot operation to a pulse interval time as long as 3 s. However, at a pulse interval of 2.5 s a noticeable increase occurred, which became rather dramatic at pulse intervals of 2 s. Comparing the pulse interval time with the thermal time constant, we obtain from Fig. 7.18 the residual temperature present in the rod at the onset of the second pulse. Up to repetition rates of 20 ppm, the residual temperature in the center of the rod was less than 32 % of its initial value; this percentage increased to 55 % at the repetition rate of 30 ppm.

7.2 Cooling Techniques

7.2.1 Liquid Cooling

The primary purpose of the liquid is to remove the heat generated in the laser rod, pump source, and laser cavity. Sometimes the coolant serves additional functions, such as index matching, thereby reducing internal reflections which could lead to depumping modes, or as a filter to remove undesirable pump radiation. The coolant is forced under pressure to flow over the rod and lamp surfaces. These elements are located either inside flowtubes or in cooling chambers machined out of the main body of the laser head. The temperature difference between the part to be cooled and the liquid is a function of the velocity and the cooling properties of the flowing liquid. At low velocities, the flow is laminar and most of the temperature drop is due to pure conduction across a stationary boundary layer at the liquid interface. For higher velocities, the flow becomes turbulent, leading to a more efficient heat transfer process with a subsequent lower temperature drop. Turbulent flow requires a greater pressure differential for the same volume flow, but the necessary differential usually is still small compared to the total pressure difference associated with the complete cooling system.

The temperature increase of the coolant, as it passes through the laser cavity, is given by

$$\Delta T = \frac{Q}{C_p m} \quad , \tag{7.52}$$

where Q is the extracted heat, C_p is the specific heat of the coolant, and m is the mass flow rate. Water is preferably used as a coolant for solid-state lasers, with the exception of military lasers which have to operate over an extremely wide temperature range. A standard specification for airborne military systems is MIL-E-5400K, class 1, which requires operation of electronic equipment over the range of $-54°$ to $+55°$ C.

The major difficulty in finding coolants for use in systems which have to be stored and operated at freezing temperatures is the strong ultraviolet radiation within the laser cavity. This causes most of the fluids, which would otherwise have suitable optical and physical parameters, to decompose. Extensive tests have indicated that under these conditions, ethylene glycol and water mixtures and fluorinated hydrocarbons are the best choices for laser coolants. Interesting properties of these and other commonly used coolants of solid-state lasers are given in Table 7.3.

From purely heat transfer considerations, water is by far the best fluid. As compared to the other coolants, it has the highest specific heat and thermal conductivity and the lowest viscosity. If we express the mass flow rate by the volumetric flow $f_{\mathrm{V}} = m/\varrho$, where ϱ is the density of the fluid, then after introducing appropriate parameters from Table 7.3 into (7.52), we obtain the temperature rise in water:

Table 7.3. Room-temperature properties of various coolants

Parameter		Water	Water 60 %, methyl alcohol 40 %	FC-104[a]	E-4[b]	Ethylene glycol 50 %, water 50 %
Specific heat	[cal/g° C]	1.0	0.84	0.24	0.24	0.79
Viscosity (poise)	[g/cm s]	1×10^{-2}	0.8×10^{-2}	1.4×10^{-2}	4.1×10^{-2}	3.0×10^{-2}
Thermal conductivity	[cal/cm s K]	1.36×10^{-3}	0.91×10^{-3}	0.33×10^{-3}	0.16×10^{-3}	1.01×10^{-3}
Density	[g/cm^3]	1.0	0.905	1.79	1.76	1.06
Prandtl number N_{Pr}	–	7.4	7.4	10.2	61.5	23.5
Volumetric thermal expansion coefficient	[° C^{-1}]	0.643×10^{-4}	4.14×10^{-4}	9.0×10^{-4}	6.4×10^{-4}	5.7×10^{-4}
Boiling point	[° C]	100	65	104	194	110
Freezing point	[° C]	0	−29	−62	−94	−36

[a] Manufacturer 3 M Company, fluorocarbon "FC" series.
[b] Manufacturer Du Pont, Freon "E" series.

$$\Delta T[\mathrm{C}] = \frac{0.24P[\mathrm{kW}]}{f_\mathrm{v}[\mathrm{ltr/s}]} , \tag{7.53}$$

where P is the heat carried away by the water. As a comparison, if FC-104 is employed as a coolant, we obtain a temperature rise

$$\Delta T[\mathrm{C}] = \frac{0.56P[\mathrm{kW}]}{f_\mathrm{v}[\mathrm{ltr/s}]} . \tag{7.54}$$

For the same heat extraction P and flow rate f_v, the temperature rise in FC-104 is over twice the temperature increase of water. The lower viscosity of water compared to the other coolants results in a smaller pressure drop in the cooling lines.

Water has the additional advantage over all other coolants that it is chemically stable under intensive ultraviolet radiation. Water-alcohol mixtures or fluorinated hydrocarbons have C–H or F–H bonds which break if the energy from the UV radiation exceeds the bond energy. The coolants become acidic, precipitates develop and the optical transmission decreases. The degradation of these coolants can be related to their UV absorbance cutoff wavelength, which shifts toward longer wavelengths as the coolant degrades [7.45]. Dissociation followed by complete carbonization of the coolants listed in Table 7.3 have led to the development of UV-free flashlamps. The envelope of these lamps is usually made from ultraviolet-absorbing titanium-doped quartz. Even though most military systems employ flashlamps with UV-absorbing quartz envelopes, the amount of UV remaining may still cause coolant degradation. Once an ethylene glycol and water mixture has been exposed to UV radiation, it continues to build up acidity even when not in use. Stainless steel plumbing has been used with ethylene glycol and water systems to preclude corrosive buildups. When the fluorinated hydrocarbons are exposed to UV radiation, hydrofluoric acid is formed. This acid reacts with silicon. Hence 356 aluminum and other similar alloys cannot be used in fluorinated hydrocarbon coolant systems. Cloudiness that formerly developed in fluorinated hydrocarbon coolants after continued exposure to UV radiation has been eliminated by proper processing and control of the fluid to ensure that all impurities have been removed.

Cooling Equipment

With the exception of lasers cooled directly by tap water, a closed-loop cooling system is employed which consists in its most basic form of at least a liquid pump, a heat exchanger, and a reservoir. Commercially available coolers contain, in addition to these components, a particle filter, a demineralizer, gauges, and sensors for monitoring flow, temperature and pressure. If common tap water is used, periodic cleaning is necessary to remove deposition of organic and mineral deposits. In closed-loop systems, if demineralization and filtering are employed, the need to clean surfaces exposed to the cooling fluid is essentially eliminated.

The heat exchanger removes heat from the closed-loop system by thermal coupling to an outside heat sink. This can be accomplished in several ways: In a liquid-to-air heat exchanger the coolant is passed through an array of fins through which air is blown by a fan similar to a car radiator. In a liquid-to-liquid heat exchanger the heat generated within the closed loop is exhausted to external water. Figure 7.26 shows the plumbing diagram of a typical cooler with a liquid-to-liquid heat exchanger. The water flows from the reservoir to a centrifugal pump, through a heat exchanger, into the laser head, and back again into the reservoir. This sequence of components minimizes the static pressure in the laser head. The temperature of the closed-loop water is regulated by a control valve in the external supply line. The valve probe is located in the reser-

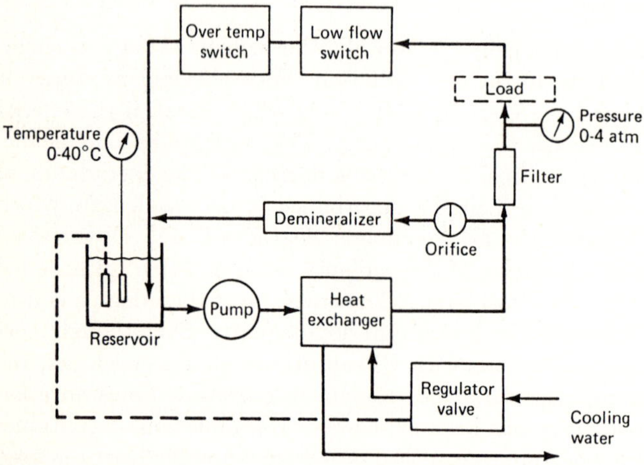

Fig. 7.26. Schematic of a water cooler containing a water-to-water heat exchanger

voir. As the temperature at the reservoir increases, the valve is opened, thereby allowing more external cooling water to flow through the heat exchanger. The system contains an in-line honeycomb filter to remove particulate matter and a bypass demineralizer which will maintain low electrical conductivity in the water and minimize corrosion. The return line is monitored by a low-flow interlock and an over-temperature switch. If flow falls below a preset value or the discharge temperature exceeds a certain limit, the interlock turns off the power supply. The system also contains gauges to display temperature in the return line and the discharge pressure.

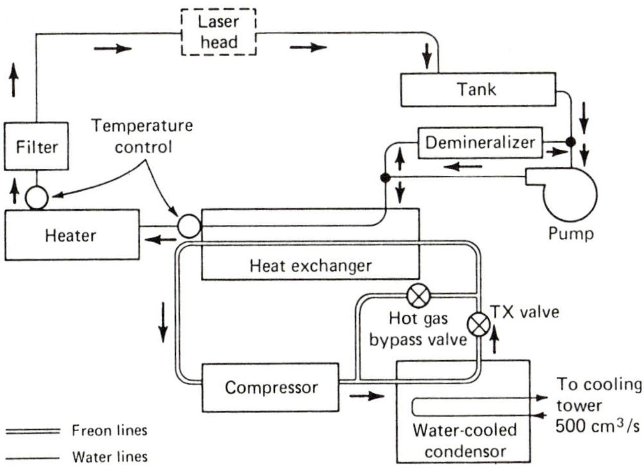

Fig. 7.27. Schematic of a water cooler employing a refrigeration unit

In both types of cooler we have discussed so far, the temperature of the closed loop can be regulated only over a relatively narrow temperature range. Furthermore, the temperature of the cooling loop is always above ambient air temperature or the temperature of the external cooling water. Figure 7.27 shows a diagram of a cooler which maintains the cooling water at a precise, reproducible temperature which is independent of the ambient air or water temperature. This cooler contains a thermostatically controlled refrigeration stage between the heat exchanger and the external water supply. Freon is cycled in a close loop between the heat exchanger and a water-cooled condensor by means of a compressor. Any changes in the heat dissipation of the load, or temperature variations in the external lines, are compensated by a hot gas bypass valve which regulates the amount of refrigeration. Smaller units contain aircooled condensors, thus eliminating the need for an external water source.

7.2.2 Air or Gas Cooling

In low average power lasers, especially portable systems, forced air is sometimes used to cool the laser rod and flashlamp. Air flow is generated by employing miniature axial or centrifugal blowers or fans which have been designed for air cooling of electronic equipment. The air flow required for cooling the laser head is calculated from the dissipated heat and the maximum temperature difference along the air stream. For standard air (20° C, 1 atm) we obtain

$$f_\mathrm{v}[\mathrm{ltr/min}] = \frac{49P[\mathrm{W}]}{\Delta T[\mathrm{C}]} \quad , \tag{7.55}$$

where f_v is the air flow and ΔT is the temperature difference between the inlet air and the exhausted air. The numerical factor contains the thermal properties of air. The air flow which can be obtained from a given fan depends on the static pressure the fan must work against. Small fans which are used for cooling lasers usually provide an air flow of 300 to 600 ltr/min at a static pressure of 30 torr. Allowing a temperature rise of 10° C of the exhausted air, heat removal capacity will be in the 100- to 200-W range. Figure 6.86 shows an example of an air-cooled Nd:YAG laser. A vane axial fan located upstream generates an air flow which passes through the pump cavity and over the rod and flashlamp. The air is exhausted through a nozzle in front of the unit. Besides convection cooling, the laser rod is also cooled by conduction into a copper heat sink to which it is mounted.

In lasers with high power dissipation, difficulties arise if one attempts to force low-pressure air into a small cavity while maintaining a high heat transfer coefficient. It is soon found that the power required to pump atmospheric pressure air through the cavity is so large as to make the system not feasible. As the density of air is increased, for instance by a factor of 20, the pressure drop required to move the same mass flow of air through the cavity is reduced to 1/20th and the pump power required is reduced to 1/400th.

A cooling system which has been employed very successfully in small military laser systems is based on the use of compressed dry nitrogen as the coolant medium to transfer the heat generated in the laser cavity to the ambient air. The compressed nitrogen is circulated through the laser pumping cavity by means of an axial flow blower. Nitrogen exhausted from the cavity is then ducted through fins of a heat exchanger, where it gives up the energy picked up in the laser cavity. The cooled nitrogen is then ducted back through the cavity again to complete the nitrogen cooling loop. A fan provides the required air flow through the heat exchanger. Figure 7.28 shows a typical laser-cooler module which contains a laser crystal, flashtube, trigger transformer, and cooling system. The unit is charged to a pressure of 20 atm; with dry nitrogen at this high pressure the heat transfer capability of compressed nitrogen and FC-104 for example, shows that nitrogen at 20 atm has not only the same specific heat value but also has twice the value of heat transfer coefficient as FC-104 for equal mass flow rates.

7.2.3 Conductive Cooling

In a variety of commercial and military systems the laser rod is mounted directly to a heat sink, as shown schematically in Fig. 7.29. Good conduction

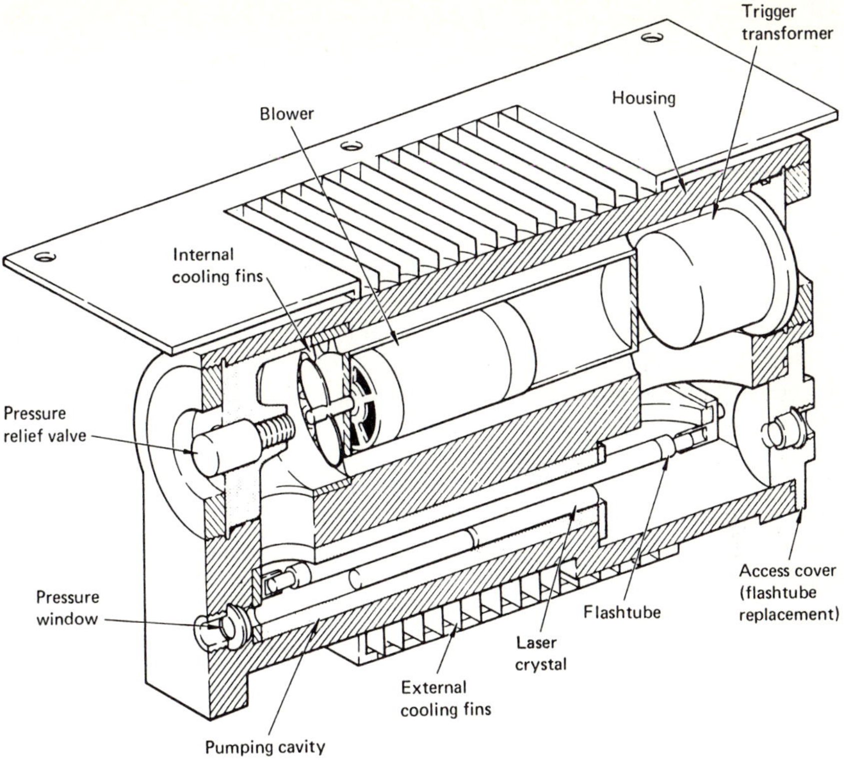

Fig. 7.28. Laser head cooled by pressurized nitrogen. (Courtesy Hughes Aircraft Company.)

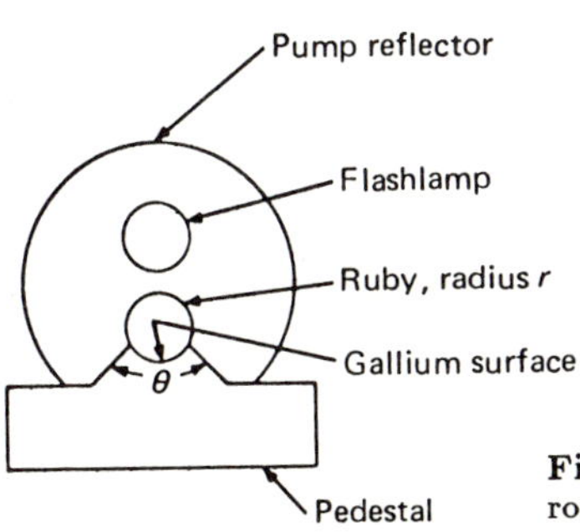

Fig. 7.29. Typical geometry of a conductively cooled laser rod

cooling of the laser element requires intimate thermal contact between the laser rod and the heat sink. The laser rod can be mechanically clamped, soldered, or bonded to the heat sink. If the laser rod is mechanically clamped to a heat sink, a temperature gradient across the rod-clamp interface will develop. This is

$$\Delta T_1 = Q/hA \quad . \tag{7.56}$$

Experimental results have shown that between a ruby rod and an aluminum heat sink the heat transfer coefficient can be as low as $h = 1\,\mathrm{W/cm^2\,{}^\circ C}$. Usually, the rod is mounted in a carefully machined saddle of the heat sink with a wetting agent such as indium or gallium at the interface.

In a space-born Nd : YAG laser system the rod was soldered to a mounting structure of pure niobium, having a thermal coefficient of expansion which closely matches that of Nd : YAG. The mounting surface for the rod is a groove to provide intimate contact with the rod over 90° of its periphery. The rod was soldered to the heat sink with indium solder after the contact surfaces were gold-plated. The heat was removed from the rod mounting structure by a heat pipe. A fin-type radiator at the condensor end of the heat pipe radiated the heat into space [7.46].

In another technique, a small Nd : YAG rod was bonded to a copper heat sink by means of a silver-filled epoxy adhesive. In this case, ultraviolet-free flashlamps must be employed to avoid decomposition of the epoxy. The temperature distribution in a conductively cooled rod has been calcualted in [7.47] with the assumption that heat is uniformly produced by absorption of radiation and no losses occur except by conduction through the contact surface of the heat sink.

7.3 Noncylindrical Laser Elements

Practical limitations arise in the operation of any solid-rod laser due to the thermal gradients required to dissipate heat from the rod. The ratio of heat load to output power is typically in the range from 2 : 1 to 10 : 1. As was discussed in Sect. 7.1, the thermally loaded cylindrical laser medium exhibits optical distortions which include thermal focusing, stress induced biaxial focusing, and stress induced birefringence. These thermally induced effects severely degrade the optical quality of the laser beam and eventually limit the laser output power, either because of an unacceptably poor beam pattern or because of thermal stress induced breakage of the rod. An incremental improvement can be achieved by employing a hollow cylindrical rod with gas or liquid cooling, or by dividing a large-diameter rod into a bundle of small-diameter parallel rods.

The limitations imposed by the rod geometry have long been recognized and a number of designs have emerged which became known as axial-gradient laser, slab-geometry laser, disk and active mirror laser.

Axial Gradient Laser

In this design, a solid rod is segmented into disks which are perpendicular or at the Brewster angle to the optics axis (similar to stacked coins). The individual disks are face-cooled by forcing a suitable cooling fluid through the spaces between the disks. With this configuration the heat flow paths are essentially parallel to the optical axis, therefore radial distortions should be minimized. If the disks are arranged perpendicular to the optic axis, either an index-matching fluid or antireflective coatings have to be employed. Depending on the geometry and holding structure used, the disks can be circular, elliptical, square, or rectangular. Disk lasers have been built with Nd:glass [7.48,49], Nd:YAG [7.50], and ruby [7.51,52] as active material. Designs featuring up to 50 disks have been constructed. The disks are typically 10 to 20 mm in diameter, 5 to 10 mm thick, and separated by 0.5 to 1 mm. Although the axial-gradient laser appeared to offer a great potential improvement over solid rods in average power output and beam divergence, because of its better cooling capabilities, the experimental results were very disappointing. Problem areas which are inherent to this type of device are stresses and optical distortions in the disks due to edge cooling effects, optical losses due to surface scatter and attenuation in the coolant, and mechanical problems associated with the holding structure and cooling manifold. Since these engineering problems could not be resolved work stopped on this type of device in the early 1970s.

Rectangular-Slab Laser

The rectangular-slab laser provides a larger cooling surface and essentially a one-dimensional temperature gradient across the thickness of the slab. Figure 7.30 shows the geometry of a rectangular slab laser. The z axis coincides with the optical axis of the slab.

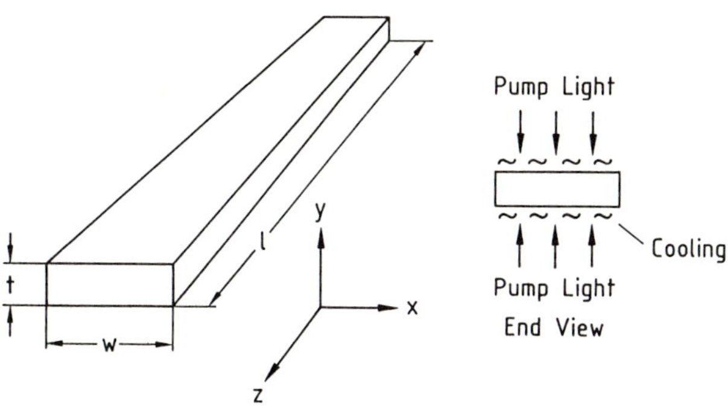

Fig. 7.30. Geometry of a rectangular slab laser

The slab has a thickness t and a width w. The upper and lower surfaces are maintained at a constant temperature by water-cooling, and the sides are uncooled. Provided the slab is uniformally pumped through the top and bottom surfaces the thermal gradients are negligible in the x and z directions and the thermal analysis is reduced to a one-dimensional case, i.e. temperature and stress are a function of y only. This, of course, is only true for an infinitely large plate in x and y, and uniform pumping and cooling. Under these conditions we find that the temperature assumes a parabolic profile (Fig. 7.31a).

The maximum temperature which occurs between the surface and the center of the slab ($Y = t/2$) is given by

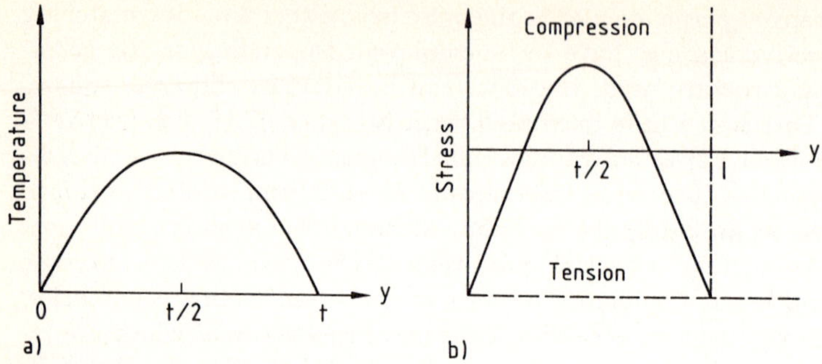

Fig. 7.31a,b. Temperature profile (a) and stress (b) in a rectangular slab

$$\Delta T = \frac{t^2}{8K} Q \tag{7.57}$$

where Q is the heat deposition, t is the thickness and K the thermal conductivity of the slab. For example, a heat deposition of $2\,\mathrm{W/cm^3}$, in a 1 cm thick glass slab ($K = 0.01\,\mathrm{W/cm^3}$) will create a temperature difference between the cooled surfaces and the center of 25° C.

The temperature rise causes stress in the slab according to

$$\sigma_\mathrm{s} = \frac{2\alpha E}{3(1-\nu)} \Delta T = \frac{\alpha E t^2}{12(1-\nu)K} Q \tag{7.58}$$

where σ_s is the surface stress for the slab. The surfaces are in tension and the center is under compression as shown in Fig. 7.31b.

If we introduce the "thermal-shock parameter" from Sect. 7.1 we can calculate the maximum temperature difference allowed between the surface and the center before thermal fracture occurs; it is

$$\Delta T_\mathrm{max} = \frac{3R}{2K} \quad . \tag{7.59}$$

With $R = 1\,\mathrm{W/cm}$ for Nd : glass, one obtains

$$\Delta T_\mathrm{max} = 150°\,\mathrm{C} \quad .$$

Stress fracture at the surface limits the total thermal power absorbed by the slab per unit of face area. For slabs of finite width W, the power per unit length at the stress fracture limit is given by

$$\frac{P_a}{L} = 12\sigma_\mathrm{max} \frac{(1-\nu)K}{\alpha E} \left(\frac{W}{t}\right) \tag{7.60}$$

where W/t is the aspect ratio of a finite slab. It is interesting to compare the surface stress of a rod and slab for the same thermal power absorbed per unit length. From (7.11, 60) it follows

$$\frac{(P_a/L)_{\text{rod}}}{(P_a/L)_{\text{slab}}} = \frac{2\pi}{3}\left(\frac{t}{W}\right) \quad . \tag{7.61}$$

Thus, for superior power handling capability relative to a rod, the aspect ratio of the slab must be greater than 2.

The temperature and stress profile leads to a birefringent cylindrical lens. The focal lengths of the birefringent lens are [7.53]

$$f_x = \frac{N - S_1}{2L} \quad ; \quad f_y = \frac{N - S_2}{2L} \tag{7.62}$$

for x and y polarized light, respectively. The parameter N is the contribution from thermal focusing, i.e.,

$$N = \frac{dn}{dT}\left(\frac{Q}{2K}\right) \quad , \tag{7.63}$$

and the parameters S_1 and S_2 are related to stress induced focusing

$$S_1 = \frac{Q\alpha E}{2(1-\nu)K}(B_\perp + B_\parallel) \tag{7.64}$$

$$S_2 = \frac{Q\alpha E}{(1-\nu)K}B_\perp \quad . \tag{7.65}$$

A comparison of the focal length of a rod with that of a rectangular slab for the same heat deposition Q shows, that the slab has approximately twice the optical power of the rod. In a slab, however, for incident radiation polarized along either the x or y directions the stress induced depolarization is zero. In a rectangular rod, the main axes of the resulting index ellipse are oriented parallel to the x and y axes of the rod, and a beam polarized along either one of these axes can propagate in a direction parallel to the z axis without being depolarized. This is, as mentioned earlier, in contrast to the situation in a cylindrical rod, where a plane-polarized beam does suffer depolarization because the direction of the main axes of the index ellipsoid vary from point to point within the cross section of the rod. The advantage of the rectangular geometry is obtained at some cost: The focal lengths are shorter than for the cylindrical rod, and cylindrical rather than spherical compensating optics are needed. The idea of the zig-zag path in a slab which will be discussed next is to eliminate to a first-order focusing in a slab.

Slab Laser with Zig-Zag Optical Path

In the previous section we considered propagation straight through an infinite slab. Due to the rectilinear pumping and cooling geometry of the lasing medium, thermal gradients and thermally induced stresses are present only in the y direction. Therefore, for light polarized in either the x or y directions, stress-induced biaxial focusing and depolarization losses are eliminated. How-

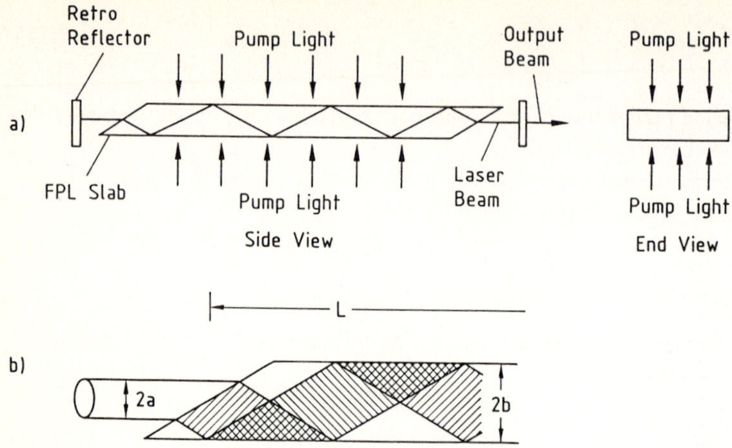

Fig. 7.32a,b. Schematic of a zig-zag laser with Brewster angle faces. **(a)** Schematic, **(b)** Detail of Brewster-angle entrance. The coordinate system is the same, as chosen in Fig. 7.30

ever, the slab still behaves as a thin cylindrical lens with a focal length shorter by a factor of two relative to that of a rod.

The cylindrical focusing in the slab can be eliminated by choosing propagation along a zig-zag optical path. In the zig-zag geometry, the optical beam does not travel parallel to the z axis, as was the case for the straight-through optical path. Instead the beam traverses the slab at an angle with respect to the x-z plane using total internal reflection from the slab y faces. This geometry is shown in Fig. 7.32.

The laser beam is introduced into the slab with suitable entrance/exit optics, usually through surfaces at Brewster's angle, as shown in Fig. 7.32b. In order to maintain total internal reflection for the desired path length the two opposing faces have to be highly polished and have to be fabricated with a high degree of parallelism. The same two optical faces are also employed for pumping and cooling of the slab.

In the ideal case, this geometry results in a one-dimensional temperature gradient perpendicular to the faces and a thermal stress parallel to the faces. Since the thermal profile is symmetrical relative to the center plane of the slab the thermal stress averaged from one slab surface to the other is zero. Thus, for a beam traversing from one slab surface to the other, the stress-optic distortion is compensated to a high degree. Also, since all parts of a beam wavefront pass through the same temperature gradients in a surface-to-surface transit, no distortion results from the variation of refractive index with temperature. Thus, in this geometry, thermal distortion effects are fully compensated within the host material in the ideal case. In the practical application of the arrangement, thermal distortions arise only as a 2nd- or higher-order effect with pumping power, due to departure from the ideal.

The advantage of this configuration is the combination of two ideas: the elimination of stress induced birefringence by virtue of the rectangular geom-

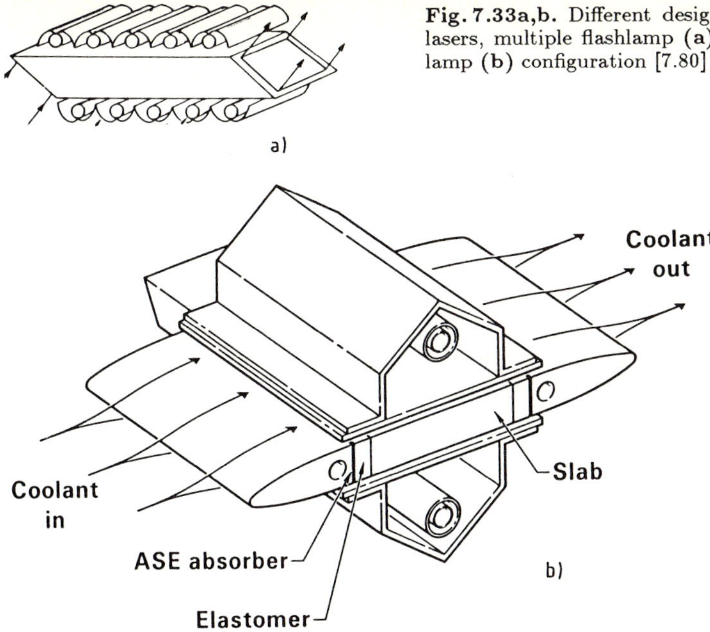

Fig. 7.33a,b. Different designs of zig-zag slab lasers, multiple flashlamp (**a**) and single flashlamp (**b**) configuration [7.80]

etry, and the elimination of thermal and stress induced focusing by optical propagation along a zig-zag path.

The concept of a slab-laser geometry with a zig-zag optical path confined in the slab by total internal reflection was first proposed by *Martin* and *Chernoch* [7.54, 55]. Figure 7.33 shows two different designs of a zig-zag slab laser. In Fig. 7.33a a multitude of flashlamps are oriented transverse to the optical axis of the slab, and in Fig. 7.33b a single lamp is placed along the axis of the slab on each side. Appropriate reflectors are employed to distribute the pump light evenly over the slab surface area. Uniform pumping and cooling of the optical surfaces is absolutely crucial for good performance of a zig-zag slab laser, as will be discussed later.

Despite the elegance with which the zig-zag slab design addresses the problem of thermal distortions inherent in all solid-state lasers, the performance from these systems has been noticeably poorer than predicted.

In practice, the slab laser approximates only an ideal infinite slab thus far assumed. Pump or cooling induced gradients across the slab width (normal to the plane of reflection), as well as end effects, are often the reason why the slab configuration shows a disappointing optical performance. Distortions can only be eliminated in a slab of infinite extent, which is uniformly pumped and cooled. For a slab of finite width and length, edge and end effects give rise to distortions, and practical solutions of flashlamp pumping and cooling of the slab lead to unavoidable nonuniformities.

For example, gradients in the z-direction are encountered in the transition region between the pumped area and the unpumped ends of the slab. Additional gradients along the z and x axes are caused by nonuniform pumping and cooling, as well as by thermal boundary conditions at the top and bottom of the slab. The major source of gradients in the x direction is a nonuniform illumination of the slab by the flashlamps in conjunction with the reflectors, and cooling of the slab at the edges outside the pumping boundary.

Since there is no compensation in the y and z direction, thermal gradients in these directions must be minimized.

In order to reduce end effects, a smooth temperature profile in the transition region is usually attempted. Figure 7.34 illustrates a particular design whereby a plenum provides for the mixing of flowing with stagnant water to produce a gradually decreasing surface heat transfer coefficient.

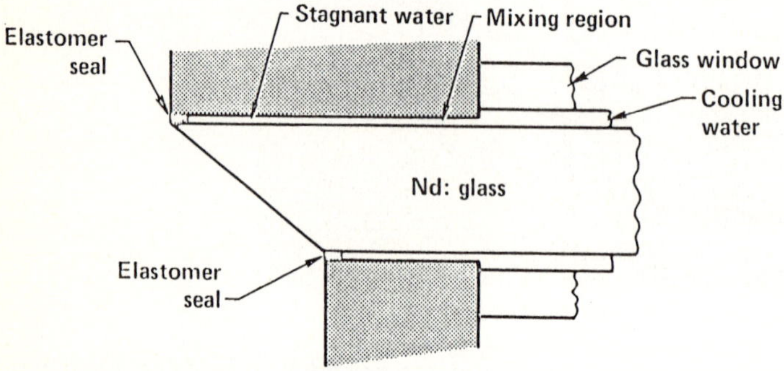

Fig. 7.34. Cooling and sealing of the ends of a slab laser [7.80]

The severeness of surface deformation, and bulging of the entrance and exit faces of a slab under thermal loading is illustrated in Fig. 7.35. These deformations in conjunction with index gradients lead to focusing effects of the slab laser. A detailed evaluation of the limitations of zig-zag slab lasers due to thermal distortions can be found in [7.53].

Zig-zag slab lasers have been built with Nd: glass [7.55–59], Nd: SOAP [7.60], Nd: YAG [7.59, 61–63], and Nd: GGG [7.62, 64] as the lasing medium. In addition to these flashlamp-pumped systems, several laser-diode pumped Nd: YAG slab lasers have been constructed [7.65, 66]. Earlier work at General Electric [7.56] resulted in Nd: glass Q-switched oscillator-amplifier system sharing a common slab. Typically, a one-pass oscillator yielded an output of about 250 mJ/pulse, then several passes through the same slab, provided the gain and energy extraction required to increase the output up to 2.5 J per pulse at repetition rates of 3 to 10 pps. A diagram of the beam folding arrangement for a typical oscillator-amplifier in a single slab is shown in Fig. 7.36 [7.67]. The same group also built a Nd: YAG system with a $8 \times 16 \times 128\,\mathrm{mm}^3$ slab which yielded 150 W of cw output, 5 J at 20 pps, and 250 mJ/pulse at 320 pps In the Q-switched mode [7.61].

Eggleston et al. [7.58] constructed a dual-slab Nd: glass zig-zag laser with single-side pumping. A single slab $6.3 \times 56 \times 350\,\mathrm{mm}^3$ produced 80 W of output power in an oscillator with only three fringes of distortion. The choice

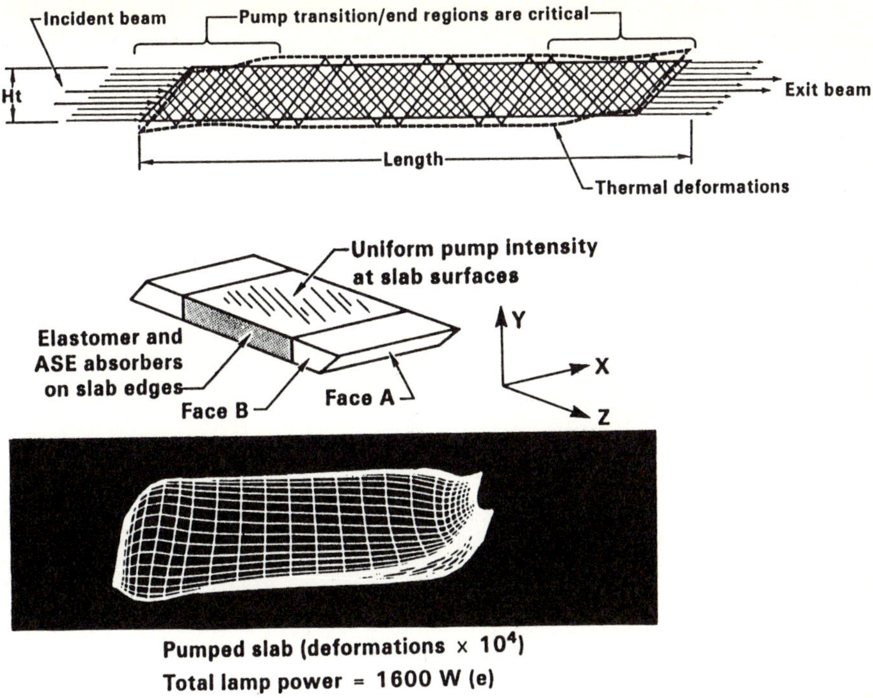

Fig. 7.35a,b. Deformations in a thermally loaded slab laser **(a)** Beam Trajectories, **(b)** 3-D deformations computed in pumped glass slabs. [7.80]

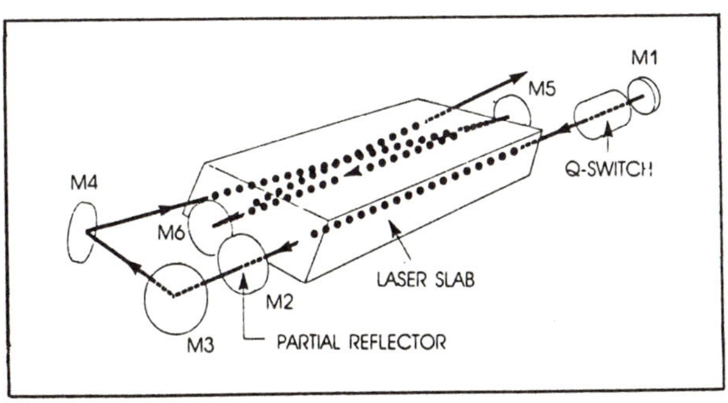

Fig. 7.36. Example of a multipass zig-zag slab laser. Shown is an oscillator and a 3 pass amplifier

of a dual slab, as shown in Fig. 7.37, was implemented to develop a compact oscillator-amplifier structure. Two slabs were placed on either side of a row of 16 flashlamps. The slabs were cooled with an ethylene glycol/water mixture. Two reflectors on the outside of each slab recycle pump light not absorbed during the first pass. The performance of the oscillator is summarized in Fig. 7.38. Significant distortions are encountered near 11.5 kW average input power, which is close to the fracture limit. Both slabs made from Nd:LHG5 fractured at an input power of 13 kW.

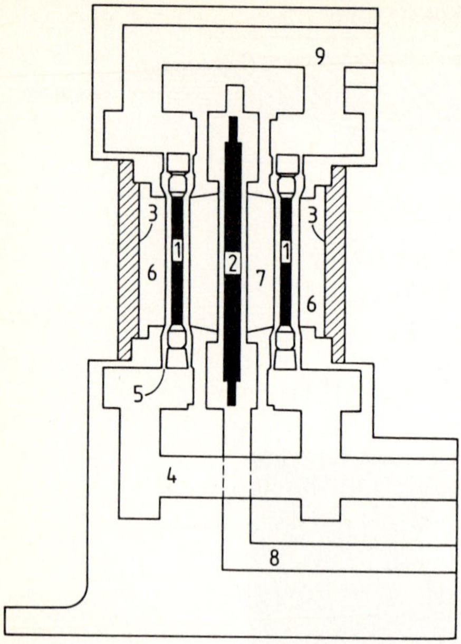

Fig. 7.37. Layout of a dual-slab laser. Labelled items are *(1)* glass slabs, *(2)* flashlamps, *(3)* back reflectors, *(4)* slab coolant inlet plenum, *(5)* injection nozzles, *(6)* reflector covers, *(7)* central pyrex body housing lamps, *(8)* lamp coolant inlet, *(9)* slab coolant outlet plenum. [7.58]

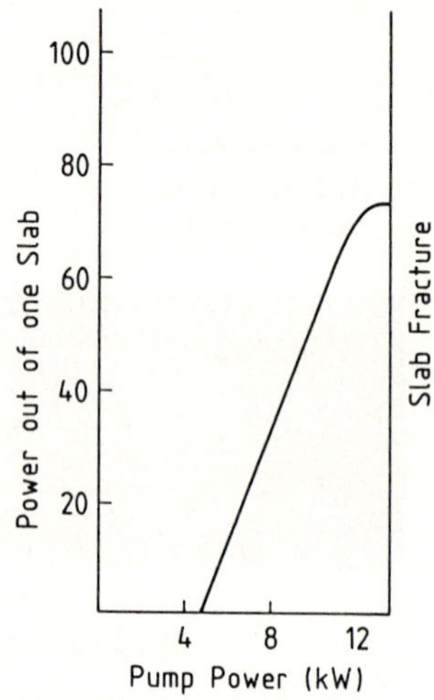

Fig. 7.38. Output power versus input power at 10 Hz operation. Both slabs fractured at a power input of 13 kW. [7.58]

The construction of a high average power Nd : YAG slab laser has been described in [7.64]. The laser which contained a $6 \times 20 \times 150\,\mathrm{mm}^3$ large Nd : YAG slab produced 70 W average output at 100 Hz. At the 0.7 J per pulse output, the input was 150 J per pulse. The free running oscillator had an overall efficiency of 0.5 %. Design details are shown in Fig. 7.39. The example illustrates the care which must be taken to achieve uniform pumping and a cooling geometry that maintains a nearly constant surface temperature of the slab. The former is achieved with suitably shaped reflectors, and the latter required a row of coolant injection nozzles and an inlet and outlet plenum.

A maximum average power of 61 W and a slope efficiency of 0.9 % was obtained at 7 pps from a Nd : GGG slab laser [7.64]. The slab was $9 \times 35 \times 77\,\mathrm{mm}^3$ cut from a GGG boule containing 2 at. % Nd. The slab was designed to have 12 internal reflections in a single pass. A cross-section of the pumping configuration is displayed in Fig. 7.40. The insulators attached to the sides of the slab are employed to reduce thermal edge effects.

In order to increase the average power from a zig-zag Nd : glass slab laser it was suggested to move the slab between the pumping lamps to distribute the thermal loading over the area of the glass while maintaining high gain in the pumped volume [7.68].

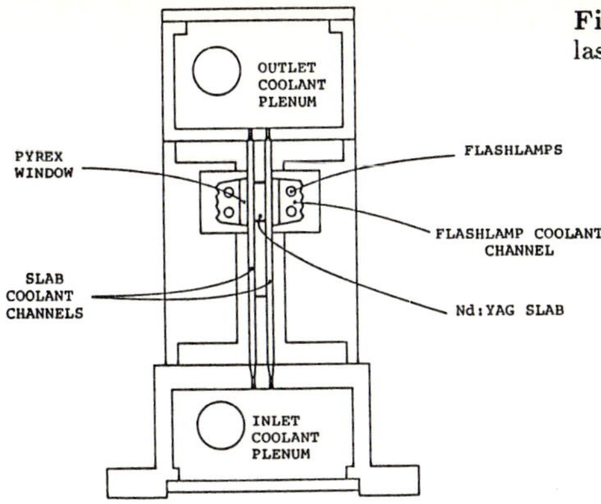

Fig. 7.39. Cross-section of a Nd:YAG slab laser [7.64]

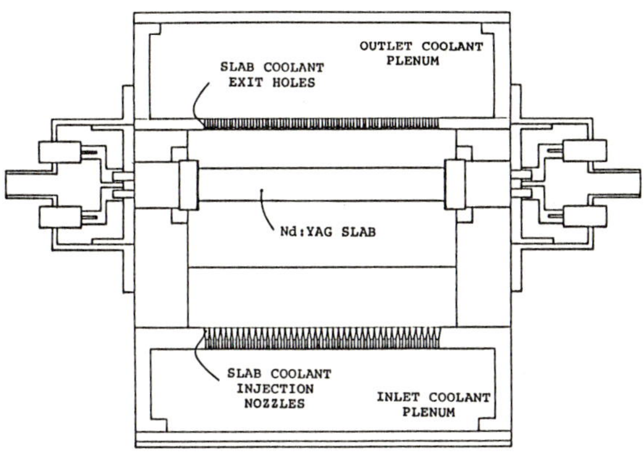

Fig. 7.40. Design of a Nd:GGG slab laser [7.64]

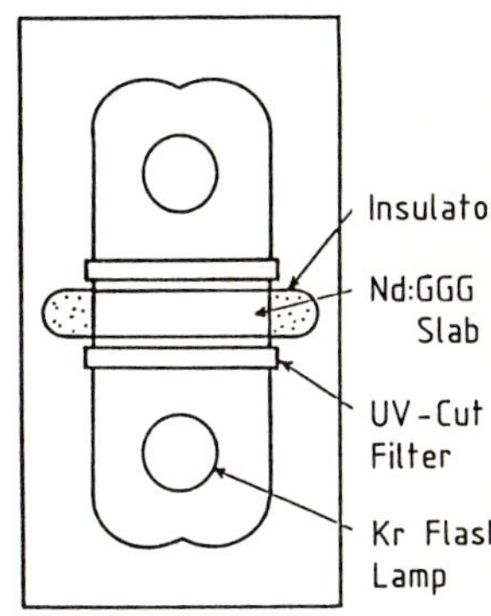

After 15 years of development work at General Electric [7.54–56, 63, 67] and more recently intensive efforts at Stanford University [7.53, 62, 57, 59] and at a number of industrial laboratories [7.58, 64, 69] the slab laser has still not found general acceptance by the laser community.

The reasons for that are a number of practical engineering problems which have prevented so far the realization of the potential advantages of the zig-zag slab laser over the rod geometry. a) *Low Efficiency.* The awkward rectangular beam geometry which requires conversion to a circular beam in useful optical applications, combined with the fact that the full width of the slab can usually not be used because of distortions at the edges, results in a poor utilization of the pumped volume. Therefore all slab lasers built to date exhibit rather modest efficiencies. b) *Residual Distortions.* Deformation of the ends and pump faces of the slab due to thermal strain have resulted in beam qualities considerably below expectations. c) *High Fabrication Cost.* The high optical fabrication costs of the slab, combined with the very demanding and therefore expensive laser head design caused by the complicated mechanical mounting and sealing geometry, and stringent pumping and cooling requirements, have also prevented wider application of this technology.

Active Mirror Amplifier

The active-mirror concept was invented at General Electric [7.70] and extensively developed at the Laboratory for Laser Energetics, University of Rochester [7.71–73]. In this approach, a single circular or rectangular disk is pumped and cooled from the backside, and laser radiation to be amplified enters from the front, as shown in Fig. 7.41a. The front face of the disk has an anti-reflection

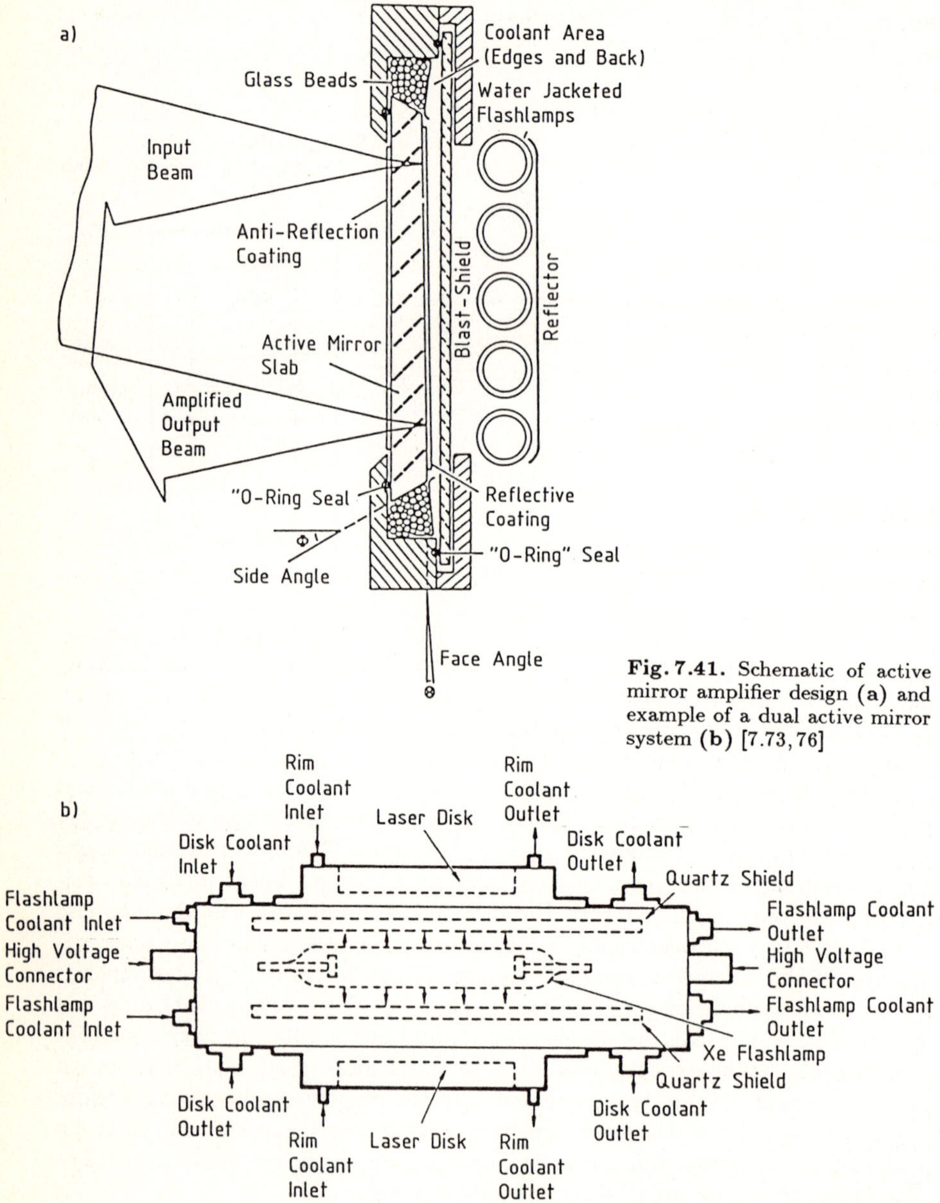

Fig. 7.41. Schematic of active mirror amplifier design **(a)** and example of a dual active mirror system **(b)** [7.73, 76]

coating for the laser radiation, whereas the backside has a thin-film coating which is highly reflective for the laser radiation, and transparent for the flash-lamp pump radiation. The laser radiation passes twice through the disk thus improving extraction efficiency. The advantage of this type of amplifier is that it can be pumped uniformly and efficiently; the disadvantage is that the slab can be pumped only through one face, and this unsymmetrical pumping leads to some thermal distortion. Originally designed for Nd : glass systems requiring large apertures such as needed for fusion research, active-mirror amplifiers have recently also been developed with Nd : YAG [7.74] and Nd : Cr : GSGG [7.75] as the lasing medium.

Active mirrors were originally developed for single-shot operation with the Omega laser system at the University of Rochester. More recently, designs have been developed for high-average-power operation [7.74, 77]. For example, Fig. 7.41b shows a schematic of an active mirror amplifier comprised of two disks pumped by a single row of flashlamps. Clear fused quartz windows are mounted between the flashlamp array and the laser disks to form separate cooling plenums. Parasitic oscillation and amplified spontaneous emission suppression are provided by a liquid layer directly adjacent to the laser disk rim (typically a mixture of ethyleneglycol and water) and by an absorbant black coating located on the disk holder. Cooling of the disk rim, important in controlling birefringence, is accomplished by slow circulation of the coolant mixture. The amplifier has been operated at up to 4 Hz repetition rate.

Disk Amplifiers

Solid-state glass disk amplifiers were considered very early in the development of high-brightness pulsed laser systems in order to solve problems of cooling, aperture size, gain uniformity, and pumping efficiency. In a disk laser amplifier the surfaces are set at Brewsters angle, as shown in Fig. 7.42. The slab faces have a high-quality optical finish to minimize scattering loss. The beam is linearly polarized in the *p*-plane to avoid reflection loss. Flashlamps are used for pumping, with lamp radiation incident on the disk faces and transverse to the beam direction. Thus, the disks are said to be face pumped. Nd : glass amplifiers containing glass disks were first built in the late 1960s at General Electric, University of Rochester, Los Alamos Laboratory and at the Naval Research Laboratory. Over the years this type of amplifier has been highly developed at laboratories engaged in laser-fusion studies, and particularly at the Lawrence Livermore National Laboratory [7.78, 79].

A thorough treatment of all optical, mechanical and electrical aspects of disk laser amplifiers can be found in the Lawrence Livermore National Laboratory Laser Program Annual Reports. Also the book by *Brown* [7.73] provides a detailed discussion of the various engineering issues related to disk lasers. In the earlier designs, the disks were elliptical in shape and mounted in a cylindrical pump geometry as indicated in Fig. 7.42a. The amplifiers for the more recent fusion systems such as NOVA have a rectangular pump-geometry because of their greater pump efficiency (Fig. 7.42b). Disk amplifier with apertures as large as 74 cm have been built and operated.

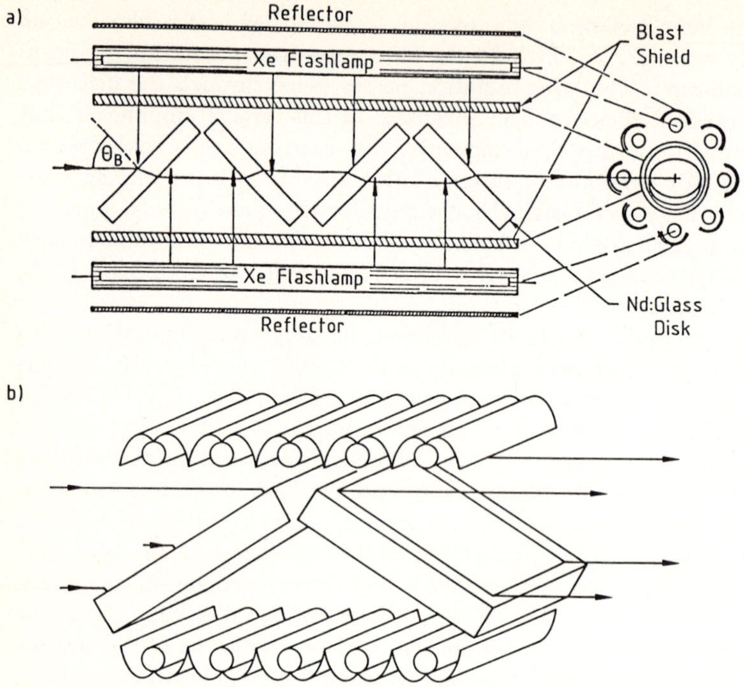

Fig. 7.42a,b. Schematic diagram of disk amplifiers (**a**) cylindrical geometry, (**b**) rectangular geometry

Generally, an even number of disks is used in an amplifier since beam translation occurs due to refraction in a single disk. Nonlinear beam steering in disks can be minimized if the disks are placed in an alternate rather than parallel arrangement. The entire periphery of each disk is generally surrounded by a metallic sheath or holder. A blast shield is normally placed between the flashlamp array and the disks, this makes it possible to completely encase the disk assembly. It also protects the disks in case of a flashlamp explosion.

The concept of face pumping, as opposed to pumping through the edges of a disk or slab, is important in that the pump radiation can be distributed, in an easily controlled manner, uniformly over the faces of the slab. Pumping in this way yields uniform gain over the aperture for a laser beam, but more importantly, the slab heating is uniform laterally, thus allowing transverse temperature gradients to be minimized. Such temperature gradients give rise to distortion in the slab just as in the rod geometry. In contrast to the rod geometry, however, the transverse gradients in the slab geometry arise only from spurious effects in pumping and cooling, whereas in the rod geometry, transverse temperature gradients arise from the principal cooling mechanism.

The Brewster's-angle disk lasers have been primarily developed for very high peak-power and relatively low average-power operation, with the slabs and coolant all approaching thermal equilibrium in the interval between pulses. To

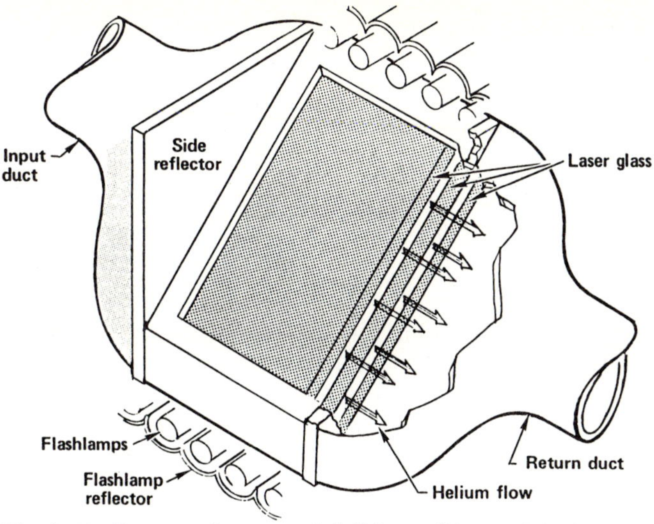

Fig. 7.43. Concept of a gas cooled disk amplifier module designed for high average power operation. [7.80]

operate the Brewster's-angle slab laser at high average power requires adequate cooling, as, for example, with forced convection over the faces. In addition, since the laser beam must pass through the coolant, the problem of thermal distortion in the coolant must be dealt with. This operating mode requires that the slabs be sufficiently thin so that the cooling time is several times less than the laser repetition period. Figure 7.43 shows a configuration of a gas cooled disk module [7.80] which could potentially be developed for high average power operation, particularly if the Nd : glass slabs are replaced with crystalline laser host materials.

8. Q-Switches and External Switching Devices

The most important device to be discussed in this chapter is the laser Q-switch. This device, located inside the laser resonator, drastically changes the power and temporal characteristic of the beam obtained from a laser oscillator. Other switches employed outside the resonator include optical gates, which are used to select one pulse from a train of pulses or to slice a small portion from a larger pulse, and optical isolators, which are employed to eliminate feedback between amplifier stages.

With the exception of mechanical devices causing a rotation, oscillation, or translation of optical components and the passive dye cell, which changes its transmission depending on the intensity of the beam, most devices rely upon interaction of the light beam with electric, magnetic, and acoustic fields. Devices are in use which utilize the electrooptic Kerr and Pockels effects, the magnetooptic Faraday effect, and the acoustooptic effect.

8.1 Pulse-Reflection Mode Q-Switches

A mode of laser operation extensively employed for the generation of high pulse power is known as Q-spoiling or Q-switching. It has been so designated because the optical Q of the resonant cavity is altered when this technique is used [8.1,2]. As was discussed in Chap. 3, the quality factor Q is defined as the ratio of the energy stored in the cavity to the energy loss per cycle. Consequently, the higher the quality factor, the lower the losses.

In the technique of Q-switching, energy is stored in the amplifying medium by optical pumping while the cavity Q is lowered to prevent the onset of laser emission. Although the energy stored and the gain in the rod are high, the cavity losses are also high, lasing action is prohibited, and the population inversion reaches a level far above the threshold for normal lasing action. The time for which the energy may be stored is on the order of τ_f, the lifetime of the upper level of the laser transition. When a high cavity Q is restored, the stored energy is suddenly released in the form of a very short pulse of light. Because of the high gain created by the stored energy in the active material, the excess excitation is discharged in an extremely short time. The peak power of the resulting pulse exceeds that obtainable from an ordinary long pulse by several orders of magnitude. Because of its extremely high power, the pulse so produced is called a giant pulse.

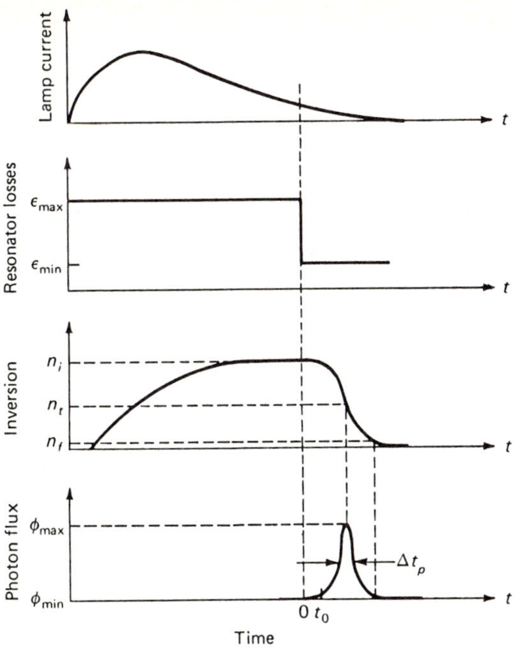

Fig. 8.1. Development of a Q-switched laser pulse. Shown is the flashlamp output, resonator loss, population inversion, and photon flux as a function of time

Figure 8.1 shows a typical time sequence of the generation of a Q-switched pulse. Lasing action is disabled in the cavity by a low Q of the cavity. Toward the end of the flashlamp pulse, when the inversion has reached its peak value, the Q of the resonator is switched to some high value. At this point a photon flux starts to build up in the cavity, and shortly afterwards a Q-switch pulse is emitted.

8.1.1 Q-Switch Theory

A number of important features of a Q-switched pulse, such as energy content, peak power, pulsewidth, rise and fall times, and pulse formation time, can be obtained from the rate equations discussed in Chap. 1. In all cases of interest the Q-switched pulse duration is so short that we can neglect both spontaneous emission and optical pumping in writing the rate equations.

From (1.61, 58), it follows that

$$\frac{\partial \phi}{\partial t} = \phi\left(c\sigma n\frac{l}{l'} - \frac{\varepsilon}{t_R}\right) \quad \text{and} \tag{8.1}$$

$$\frac{\partial n}{\partial t} = -\gamma n \phi \sigma c \quad . \tag{8.2}$$

In (8.1) we expressed the photon lifetime τ_c by the round-trip time $t_R = 2l'/c$, and the fractional loss ε per round trip according to (3.7). Also, a distinction is made between the length of the active material l and the length of the resonator l'. Q-switching is accomplished by making ε an explicit function of time (e.g., rotating mirror or Pockels cell Q-switches) or a function of the photon density (e.g., saturable absorber Q-switching). The losses in a cavity can be represented by

$$\varepsilon = -\ln R_1 + L + \zeta(t) \quad , \tag{8.3}$$

where the first term represents the coupling losses, the second term contains all the incidental losses such as scattering, diffraction, and absorption, and $\zeta(t)$ represents the cavity loss introduced by the Q-switch. For a particular explicit form of $\zeta(t, \phi)$, the coupled rate equations can be solved numerically with the boundary condition $\zeta(t < 0) = \zeta_{\max}$; $\zeta(t \geq 0) = 0$. In many instances Q-switches are so fast that no significant change of population inversion takes place during the switching process; in these cases ζ can be approximated by a step function.

In the ideal case, where the transition from low Q to high Q is made instantaneously, the solution to the rate equations is particularly simple [8.3, 4]. In this case we assume that at $t = 0$ the laser has an initial population inversion n_i, and the radiation of the cavity has some small but finite photon density ϕ_i. The solution of the rate equations is illustrated in Fig. 8.1. Initially, the photon density is low while the laser is being pumped and the cavity losses are $\varepsilon_{\max} = -\ln R_1 + L + \zeta_{\max}$. The loses are suddenly reduced to $\varepsilon_{\min} = -\ln R_1 + L$. The photon density rises from ϕ_i, reaches a peak $\phi_{\max}$ many orders of magnitude higher than ϕ_i, and then declines to zero. The population inversion is a monotone decreasing function of time starting at the initial inversion n_i and ending at the final inversion n_f. We note that the value for n_f is below the threshold inversion n_t for normal lasing operation. At n_t the photon flux is maximum and the rate of change of the inversion dn/dt is still large and negative, so that n falls below the threshold value n_t and finally reaches the value n_f. If n_i is not too far above n_t, that is, initial gain is close to threshold, then the final inversion n_f is about the same amount below threshold as n_i is above and the output pulse is symmetric. On the other hand, if the active material is pumped considerably above threshold, the gain drops quickly in a few cavity transit times τ_R to where it equalizes the losses. After the maximum peak power is reached at n_t, there are enough photons left inside the laser cavity to erase the remaining population excess and drive it quickly to zero. In this case the major portion of the decay proceeds with a characteristic time constant τ_c, which is the cavity time constant.

We proceed with the rate equations by taking the quotient of (8.1) and (8.2) in order to eliminate time as a parameter:

$$\frac{d\phi}{dn} = -\frac{l}{l'\gamma}\left(1 - \frac{n_\mathrm{t}}{n}\right) \quad , \tag{8.4}$$

where $n_\mathrm{t} = \varepsilon/2\sigma l$ is the inversion population at threshold obtained from (3.7, 15). According to (8.1), peak output occurs when $d\phi/dt = 0$, or $n = \varepsilon/2\sigma l$.

Since this last expression is identical to the threshold inversion, maximum photon density and, therefore, peak power occurs when the inversion has fallen to the threshold value n_t. Equation (8.4) can be integrated to give

$$\phi(t) = \frac{l}{l'\gamma}\left[n_\mathrm{t}\ln\left(\frac{n(t)}{n_\mathrm{i}}\right) + n_\mathrm{i} - n(t)\right] \quad . \tag{8.5}$$

With $n(t) = n_\mathrm{t}$, we obtain $\phi(t) = \phi_\mathrm{max}$ from (8.5). Since the photons decay with a lifetime $\tau_\mathrm{c} = t_\mathrm{R}/\varepsilon$, the peak power is $P_\mathrm{p} \approx \phi_\mathrm{max} Alh\nu/\tau_\mathrm{c}$. Only the fraction $\ln(1/R_1)/\varepsilon$ represents useful output, therefore

$$P_\mathrm{p} = \frac{Vh\nu\ln(1/R_1)}{\frac{1}{2}\gamma t_\mathrm{R}}\left\{n_\mathrm{i} - n_\mathrm{t}\left[1 + \ln\left(\frac{n_\mathrm{i}}{n_\mathrm{t}}\right)\right]\right\} \quad . \tag{8.6}$$

This expression shows that the peak power from a Q-switched laser is inversely proportional to the cavity round-trip time t_R, in agreement with experimental observations.

The residual inversion remaining in the active material after the emission of a Q-switched pulse can be determined from (8.5) by setting the left-hand side of the equation equal to zero. The result is a transcendental expression for the final inversion n_f:

$$n_\mathrm{f} - n_\mathrm{i} = n_\mathrm{t}\ln\left(\frac{n_\mathrm{f}}{n_\mathrm{i}}\right) \quad . \tag{8.7}$$

Figure 8.2 shows a graphical solution of this equation [8.3]. As can be seen from Fig. 8.2, increasing the pumping level (larger $n_\mathrm{i}/n_\mathrm{t}$) will result in the removal

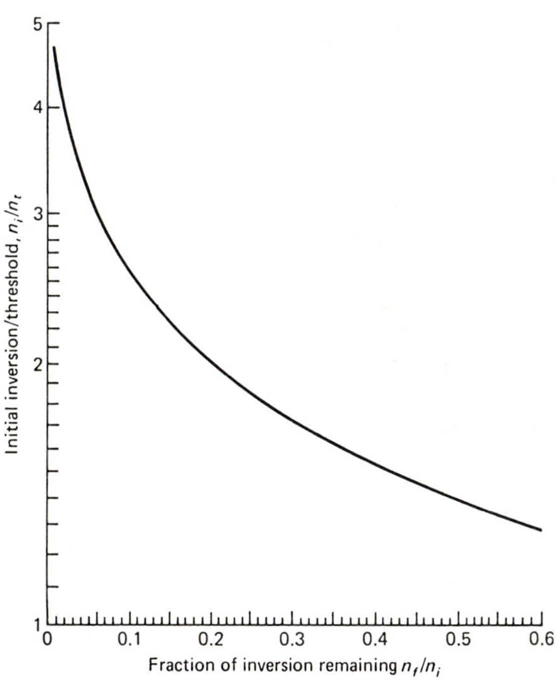

Fig. 8.2. Fraction of initial inversion n_i remaining in the active materal after emission of a Q-switched pulse. The parameters n_t and n_f are the inversion at threshold and the final inversion, respectively. [8.3]

of a larger fraction of the stored energy (smaller n_f/n_i). Once both n_i and n_f are known, the total energy E in the Q-switched pulse can be determined:

$$E \approx \left(\frac{1}{\gamma}\right) Vh\nu(n_i - n_f)\frac{\ln(1/R)}{\ln(1/R) + L} \quad . \tag{8.8}$$

If we now define the pulse duration by $\Delta t_p = 2E/P_{max}$, we have

$$\Delta t_p = \tau_c \frac{n_i - n_f}{n_i - n_t[1 + \ln(n_i/n_t)]} \quad . \tag{8.9}$$

The rise time of a Q-switched pulse has been calculated in [8.3].

As illustrated in Fig. 8.1, the emission of the Q-switched laser pulse does not occur until after an appreciable delay, during which time the radiation density builds up exponentially from noise. During this time the inversion (and hence the gain) of the amplifier remains virtually unchanged, so that until the radiation density becomes a significant fraction of the peak radiation density, we may make the reasonable approximation that ϕ grows according to

$$\phi(t) \approx \phi_i \exp(2g_0 l - \varepsilon)\frac{t}{t_R} \quad . \tag{8.10}$$

Thus, the "buildup" time t_D required for the evolution of the Q-switched pulse, let us say up to $\phi = \phi_{max}/20$, is given approximately by

$$t_D = \frac{t_R \ln(\phi_{max}/20\phi_i)}{2g_0 l - \varepsilon} \quad . \tag{8.11}$$

To use (8.11) one needs to know ϕ_i, the initial photon density (although only to within an order of magnitude or so). To a good approximation ϕ_i can be obtained by considering the photons which are spontaneously emitted in the direction of the laser beam,

$$\phi_i \approx n_2 \frac{\Omega}{4\pi}\frac{\tau_c}{\tau_f} \quad , \tag{8.12}$$

where Ω is the solid angle of the laser beam.

We will illustrate the use of (8.6–12) by giving a numerical example of a typical ruby laser. We assume a ruby rod of $l = 10$ cm length and $A = 0.75\,\mathrm{cm}^2$ cross section located in a resonator with $l' = 50$ cm. In a typical situation, the output mirror has a reflectivity of 50 % ($R_1 = 0.5$) and the total optical losses are 20 % ($L = 0.2$). Furthermore, we assume that the ruby crystal is pumped to a level which yields a gain of $g_0 = 0.1\,\mathrm{cm}^{-1}$. From these data, it follows from Fig. 4.2 that $n_i = 0.22 n_{tot}$, and from (3.13) that the gain at threshold $g_{th} = 0.045\,\mathrm{cm}^{-1}$ or $n_t = 0.10 n_{tot}$. With $n_i/n_t = 2.2$, it follows then from Fig. 8.2 that the final inversion $n_f = 0.04 n_{tot}$. If we introduce these values into (8.6, 8, 10, and 12), we find that the ruby oscillator emits a Q-switch pulse with a peak power of $P_{max} = 290$ MW and a total energy of $E_{out} = 2.3$ J. The pulse width is $\Delta t_p = 16$ ns, and the pulse buildup time is $t_D = 58$ ns. An analysis for optimizing the energy output of Q-switched lasers is presented in [8.5].

Slow Q-Switching

Instead of a step function, we will now consider the case of a resonator loss that varies in time [8.6, 7].

A form of Q-switch commonly used in practice is a rotating prism driven by a high-speed motor. A typical spinning rate is 24,000 rpm. At this rate of rotation it takes about 40 ns to sweep over an angle of 0.1 mrad. The loss rate variation for this case can be assumed to be of the form

$$\varepsilon = -\ln R_1 + L + B \cos(\omega t) \quad . \tag{8.13}$$

In a slow Q-switch the development of a Q-switch pulse depends on the ratio of pulse delay time t_D to the switching time. This is illustrated in Fig. 8.3 for three extreme cases. In Fig. 8.3a, the pulse delay time is longer than the switching time of the Q-switch. The photon flux starts to increase exponentially at $t = 0$, and at the time ϕ_max is reached and a pulse is emitted the Q-switch has already passed the point of ε_min. The pulse energy is relatively small because the cavity losses are not at a minimum at the time of maximum photon density.

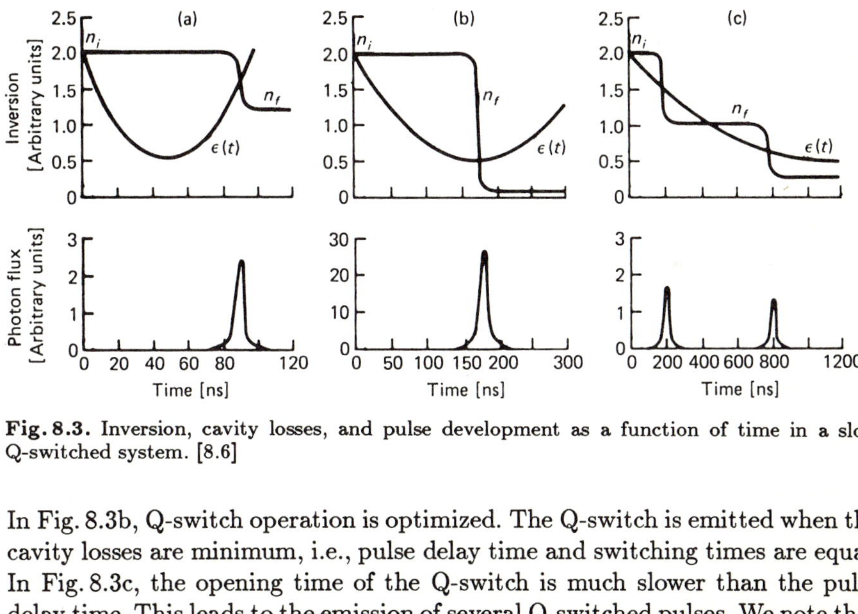

Fig. 8.3. Inversion, cavity losses, and pulse development as a function of time in a slow Q-switched system. [8.6]

In Fig. 8.3b, Q-switch operation is optimized. The Q-switch is emitted when the cavity losses are minimum, i.e., pulse delay time and switching times are equal. In Fig. 8.3c, the opening time of the Q-switch is much slower than the pulse delay time. This leads to the emission of several Q-switched pulses. We note that the first pulse is larger than the second and occurs at a larger value of ΔN; the second pulse occurs near the minimum loss level and is much lower in energy. We can explain the occurrence of multiple pulsing as follows: At $t = 0$ the resonator losses are low enough for the photon density to grow exponentially. After 200 ns the photon density has built up to its maximum level and a pulse is emitted. After the pulse is emitted the inversion reaches the level n_f. Because n_f is below threshold, the photon density ϕ is very small and a steady state is reached until the slowly changing loss rate term ε decreases enough so that the condition for photon density buildup occurs. The cycle then starts over, but because it now begins with a lower initial value of the inversion a small pulse is produced.

It should be noted that for this multiple pulsing to occur, the first pulse must be emitted at the point of a relatively high loss rate so that another pulse can build up and be emitted when the output loss rate factor is lower and probably near its minimum.

Multiple pulsing can be avoided either by shortening the switching time of the Q-switch (higher rotational speed) or by increasing the pulse delay time, for example, by increasing the cavity length. From the foregoing consideration it follows that in a laser with fixed pumping level and mirror separation, maximum output is obtained only at one particular speed.

Continuously Pumped, Repetitively Q-Switched Systems

A very important class of laser systems, employed extensively in micromachining applications, is the cw-pumped, repetitively Q-switched Nd:YAG laser. In these laser systems the population inversion undergoes a cyclic variation, as shown in Fig. 8.4. Between Q-switches the population inversion rises from a value n_f to a value n_i. The buildup of the inversion under the influence of a continuous pumping rate and spontaneous decay is described as a function of time by

$$n(t) = n_\infty - (n_\infty - n_\mathrm{f}) \exp\left(-\frac{t}{\tau_\mathrm{f}}\right) \quad , \tag{8.14}$$

where n_∞ is the asymptotic value approached as t becomes large compared to the spontaneous decay time τ_f. The value n_∞, which depends on the pump input power, is reached only at repetition rates small compared to $1/\tau_\mathrm{f}$. For repetition rates larger than $1/\tau_\mathrm{f}$, the curves representing the buildup of the population are shorter segments of the same exponential curve followed for the lower repetition rates.

During the emission of a Q-switch pulse, the inversion falls from n_i to n_f. Figure 8.5 shows the development of the Q-switched pulse on an expanded time

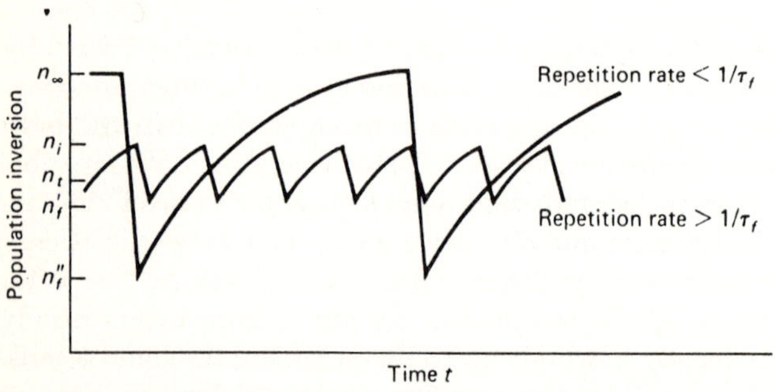

Fig. 8.4. Population inversion versus time in a continuously pumped Q-switched laser. Shown is the inversion for two different repetition rates. At repetition rates less than $1/\tau_\mathrm{f}$, the inversion approaches the asymptotic value n_∞

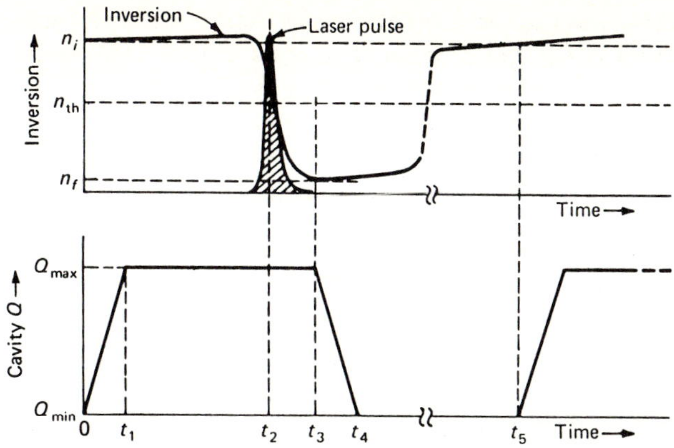

Fig. 8.5. Development of a Q-switched pulse in a cw-pumped system. [8.8]

scale [8.8]. At $t = 0$ the cavity Q factor starts to increase until it reaches its maximum value $Q_{\max}$ at $t = t_1$. Pulse formation ensues until the full pulse output is achieved at $t = t_2$. Stimulated emission ceases at $t = t_3$; at this time continued pumping causes the inversion to start to increase. At the point where the inversion begins to increase, $t = t_3$, the cavity Q begins to decrease, reaching its minimum value at $t = t_4$. During the time period t_5–t_3, the inversion is allowed to build up to its inital value $n_{\rm i}$.

The theory of *Wagner* and *Lengyel* [8.3], summarized before for the case of single Q-switched pulses, can also be applied to the case of repetitive Q-switches, with some modifications that take into account the effects of continuous pumping. Equations (8.1, 2) are applicable, however, we will set $\gamma = 1$ and $n = n_2$ because all repetitively Q-switched lasers of practical use are four-level systems. The inversion levels $n_{\rm f}$ and $n_{\rm i}$ are connected by (8.7). During the low-Q portion of the cycle, the inversion n_2 is described by the differential equation

$$\frac{dn_2}{dt} = w_{\rm p}(n_{\rm tot} - n_2) - \frac{n_2}{\tau_{\rm f}} \quad . \tag{8.15}$$

With the assumption that $n_2 \ll n_{\rm tot}$ and with

$$n_\infty = w_{\rm p}\tau_{\rm f}n_{\rm tot} \tag{8.16}$$

we obtain (8.14) as a solution. For repetitive Q-switching at a repetition rate f, the maximum time available for the inversion to build up between pulses is $t = 1/f$. Therefore,

$$n_{\rm i} = n_\infty - (n_\infty - n_{\rm f})\exp\left(-\frac{1}{\tau_{\rm f} f}\right) \tag{8.17}$$

in order to have the inversion return to its original value after each Q-switch cycle.

During each cycle a total energy $(n_i - n_f)$ enters the coherent electromagnetic field. Of this a fraction, $T/(T+L)$, appears as laser output. Therefore, the Q-switched average power P_{av} at a repetition rate f is given by

$$P_{av} = \frac{Tf}{T+L}(n_i - n_f)h\nu V \quad . \tag{8.18}$$

The peak power P_p of the Q-switched pulses is obtained from (8.6). The effective pulse width Δt_p can be calculated from the peak and average powers according to $\Delta t_p = P_{av}/P_p f$. It is convenient to calculate the ratios P_p/P_{cw} and P_{av}/P_{cw}, where P_p is the Q-switched peak power, and P_{cw} is the cw power from the laser at the same pumping level. For cw operation the time derivatives in (8.1 and 2) are zero, and we have

$$P_{cw} = \frac{T}{T+L}\left(\frac{n_\infty - n_t}{\tau_f}\right)h\nu V \quad . \tag{8.19}$$

Because of the transcendential functions expressed by (8.6, 7, and 17), the ratios P_p/P_{cw} and P_{av}/P_{cw} cannot be expressed in closed form. The result of numerical calculations is shown in Figs. 8.6 and 7 [8.9].

In Fig. 8.6 the ratio of Q-switched peak power output to the maximum cw output is plotted versus repetition rate for a Nd : YAG laser. For repetition rates below approximately 800 Hz ($\tau_f f \approx 0.20$) the peak power is independent of repetition rate. At these low repetition rates there is sufficient time between

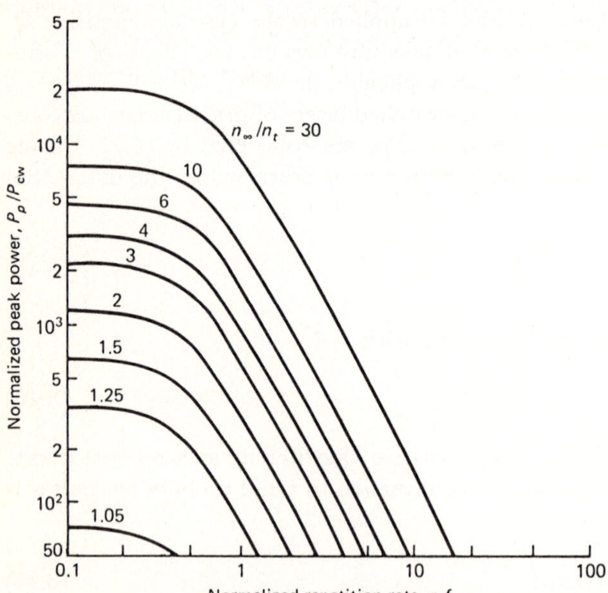

Fig. 8.6. Ratio of peak power to cw power as a function of repetition rate for a typical cw-pumped Nd : YAG laser. The parameter n_∞/n_t expresses the fractional inversion above threshold. [8.9]

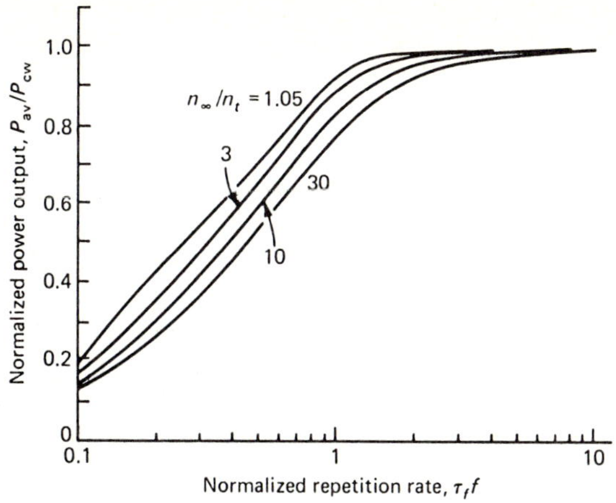

Fig. 8.7. Ratio of Q-switched average power to maximum cw power as a function of normalized repetition rate f. [8.9]

pulses for the inversion to reach the maximum value n_∞. In the transition region between 0.8 and 3 kHz, peak power starts to decrease as the repetition rate is increased. Above 3 kHz, the peak power decreases very rapidly for higher repetition rates.

Figure 8.7 shows the ratio of Q-switched average power to cw power as a function of repetition rate. Above a repetition rate of approximately 10 kHz, the Q-switch average power approaches the cw power. At low repetition rates the average power is proportional to the repetition rate. In Fig. 8.8 the experimentally determined peak power, average power pulse width, and pulse buildup time is plotted as a function of repetition rate for a Nd:YAG laser. In accordance with theory, for higher repetition rates the pulse width and the pulse buildup time increase as a result of the reduction of gain.

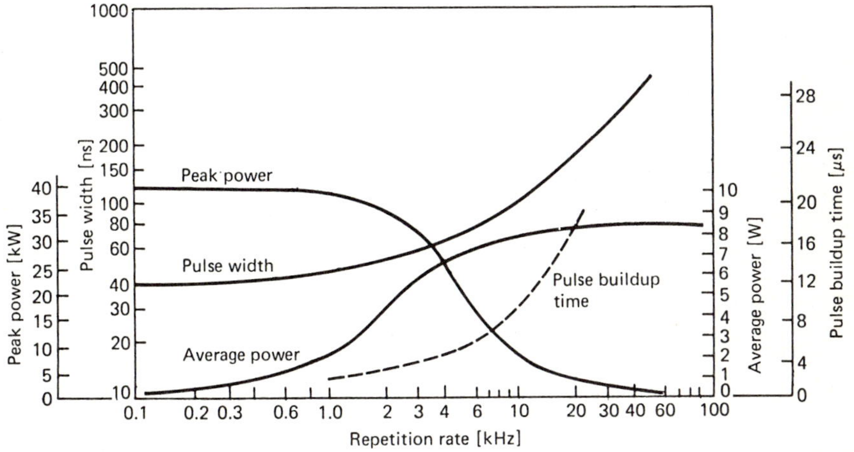

Fig. 8.8. Performance of a cw-pumped Nd:YAG laser system. The laser, containing a 7.5-cm by 0.5-cm rod doped with 0.8 % Nd, delivered 0.5 W at the TEM_{00} mode. Lamp input power was 5.5 kW. Plotted is the peak power, average power, pulse buildup time, and pulse width as a function of repetition rate

In the following sections we will describe and compare different Q-switch techniques.

8.1.2 Mechanical Devices

Q-switches have been designed based upon rotational, oscillatory, or translational motion of optical components. These techniques have in common that they inhibit laser action during the pump cycle by either blocking the light path, causing a mirror misalignment, or reducing the reflectivity of one of the resonator mirrors. Near the end of the flashlamp pulse, when maximum energy has been stored in the laser rod, a high Q-condition is established and a giant pulse is emitted from the laser.

The first mechanical Q-switch consisted of nothing more than a rotating disc containing an aperture [8.10]. This method was soon abandoned in favor of rotating mirrors or prisms which allow much faster switching times [8.11]. In some cases one of the resonator mirrors was attached to one end of a torsional rod driven at its mechanical resonance frequency [8.12]. Minute translations of one mirror have been utilized to change the Q of the cavity. One technique consists of replacing the output mirror of the laser oscillator by a two-plate Fabry-Perot resonator. By modulating the spacing between the two plates, for example, with a piezoelectric transducer, the device can be shifted from a transmission peak to a reflection peak [8.13]. Another technique makes use of frustrated total internal reflection [8.14]. When a material of matching index of refraction is brought to within a fraction of a wavelength of the surface of a total internally reflecting roof prism, the total internal reflection is destroyed. Consequently, the light incident on the prism is no longer reflected.

Of these devices, only the rotating mirror and prism Q-switches are still employed in contemporary laser systems. The spinning reflector technique for the generation of Q-switched pulses, as shown in Fig. 8.9, involves simply rotating one of the two resonant cavity reflectors so that parallelism of the reflectors occurs for only a brief instant in time.

If a plane mirror is employed as the rotating element, the axis of rotation must be aligned to within a fraction of a milliradian parallel to the face of the opposing reflector. This difficulty can be overcome, and usually is, by using a roof prism as the rotating element. In this case critical prealignment of the resonator mirror can be completely dispensed with. If the roof of the prism

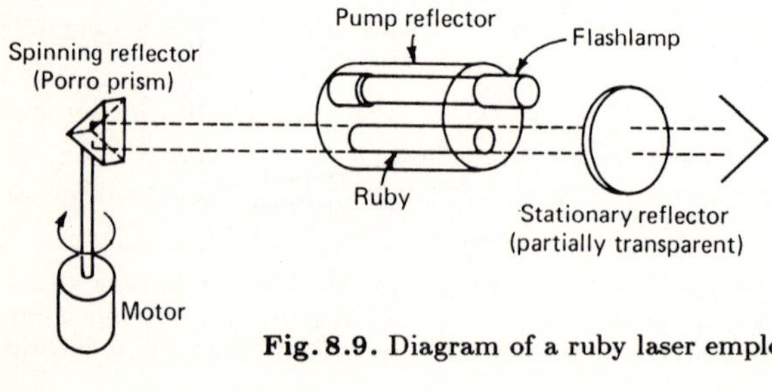

Fig. 8.9. Diagram of a ruby laser employing a spinning prism Q-switch

is perpendicular to the axis of rotation, then the retroreflecting nature of the prism assures alignment in one direction, while the rotation of the prism brings it into alignment in the other direction. In the case of a ruby laser, the roof of the prism should be either parallel or perpendicular to the polarization of the laser emission, so that the reflected light remains plane-polarized. Experiments have shown that an angular accuracy of $\pm 3°$ is sufficient to assure optimum performance.

Synchronization of the flashlamp with the mirror is usually obtained by the use of a magnetic pickup which senses the position of the rotating shaft and generates an electric signal to trigger the flashlamp at the appropriate time. The time interval between the generation of the magnetic head pulses and optical alignment is adjustable by controlling the mechanical location of the magnetic head. The time delay is adjusted so that at maximum population inversion the rotating reflector is parallel to the fixed mirror. For example, for a prism rotating at 24,000 rpm, the flashlamp has to be triggered 115° prior to optical alignment if a delay of 800 μs is desired. A second magnetic head can be used to generate a synchronization pulse just prior to the laser pulse which can be used to trigger external events. For a trigger pulse which precedes the Q-switch pulse by, for example, 1 μs, the magnetic head has to be affixed to the rotating prism at a position 2.25° ahead of alignment for the above-mentioned rotational speed. For the typical laser resonator, 30 or 40 cm in length, a rotational speed on the order of 20,000 rpm is needed to ensure that the laser output contains just a single pulse. As the mirror speed is increased up to about 40,000 rpm, some further reduction in the pulse duration can be obtained. For example, the angular speed of a prism rotating at 24,000 rpm is $\omega = d\alpha/dt = 2.5\,\text{mrad}/\mu\text{s}$. Experiments have shown that laser action is inhibited in common laser geometries for a mirror misalignment on the order of 1 mrad. This angle, in conjunction with the rotational speed mentioned above, results in a switching time of 400 ns.

Mechanical Q-switches are essentially slow Q-switches which tend to generate multiple pulses, as explained in Sect. 8.1.1. There are optical techniques for achieving effective multiplication of the actual rotational speed of the spinning reflector and any number of additional stationary reflectors [8.11, 15].

For mechanically Q-switching, cw-pumped Nd : YAG lasers, multisided rotating prisms rather than single prisms or mirrors are employed for the attainment of high repetition rates [8.16,–19]. As we have seen from Fig. 8.7, in a Nd : YAG laser the average power increases with repetition rate up to about 8 kHz. A repetition rate of 5 kHz was obtained from a Nd : YAG laser which was Q-switched by a 12-sided prism rotating at 25,000 rpm [8.17]. Cw-pumped repetitively Q-switched Nd : YAG lasers contain either a hexagonal or octagonal prism driven by an ac or dc motor at speeds around 12,000 rpm. Precise alignment of the prism faces at the Q-switching position is necessary to ensure high output peak power. Misalignment due to such causes as motor axis precession, pyramidal error, and tilt between the rotation axis and the prism faces must be prevented.

Rotating-mirror devices are simple and inexpensive. They are insensitive to polarization and, therefore, birefringence effects. Hence, more energy from the laser can be extracted under certain conditions as compared to electrooptic Q-switches. However, the mechanical Q-switches suffer from the tendency to emit multiple pulses. The devices are also very noisy, and they require frequent maintenance because of the relatively short lifetime of the bearings. Due to these disadvantages the rotating prism Q-switch has been replaced by the acousto-optic Q-switch in commercially available cw-pumped Nd:YAG lasers.

8.1.3 Electrooptical Q-Switches

Very fast electronically controlled optical shutters can be made by using the electrooptic effect in crystals or liquids. The key element in such a shutter is an electrooptic element which becomes birefringent under the influence of an external field. Birefringence in a medium is characterized by two orthogonal directions, called the "fast" and "slow" axes, which have different indices of refraction. An optical beam, initially plane-polarized at 45° to these axes and directed normal to their plane, will split into two orthogonal components, traveling along the same path but at different velocities. Hence, the electrooptic effect causes a phase difference between the two beams. After traversing the medium, the combination of the two components results, depending on the voltage applied, in either an elliptical, circular, or linearly polarized beam [8.20]. For Q-switch operation only two particular voltages leading to a quarter-wave and half-wave retardation are of interest. In the first case, the incident linearly polarized light is circular polarized after passing the cell, and in the second case the output beam is linearly polarized; however, the plane of polarization has been rotated 90°.

The two most common arrangements of components for Q-switching are shown in Fig. 8.10. In Fig. 8.10a the electrooptic cell is located between a polarizer and the rear mirror. The inclusion of the polarizer is not essential if the laser radiation is polarized, such as, for example, in ruby. The sequence of operation is as follows: During the flashlamp pulse, a voltage $V_{1/4}$ is applied to the electrooptic cell such that the linearly polarized light passed through the polarizer is circularly polarized. After being reflected at the mirror, the radiation again passes through the electrooptic cell and undergoes another $\lambda/4$ retardation, becoming linearly polarized but at 90° to its original direction. This radiation is ejected from the laser cavity by the polarizer, thus preventing optical feedback. Towards the end of the flashlamp pulse the voltage on the cell is switched off, permitting the polarizer-cell combination to pass a linearly polarized beam without loss. Oscillation within the cavity will build up, and after a short delay a Q-switch pulse will be emitted from the cavity.

In the arrangement in Fig. 8.10b, an electric voltage must first be applied to the cell to transmit the beam. In this so-called pulse-on Q-switch, the cell is located between two crossed polarizers. As before, polarizer P_1, located between the laser rod and the cell, is not required if the active medium emits a polarized beam. During the flashlamp pulse, with no voltage applied to the cell, the cavity

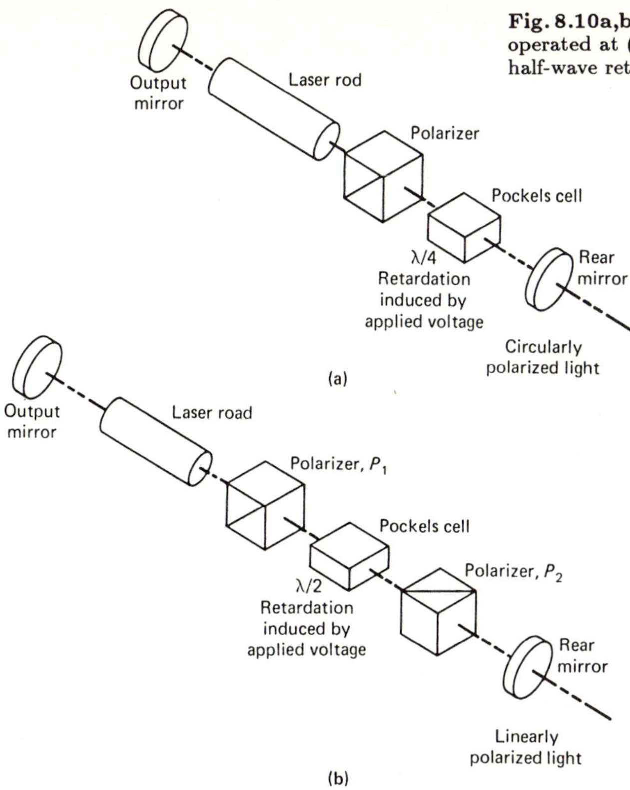

Fig. 8.10a,b. Electrooptic Q-switch operated at (**a**) quarter-wave and (**b**) half-wave retardation voltage

Q is at a minimum due to the crossed polarizers. After a period of time a voltage $V_{1/2}$ is applied to the cell which causes a 90° rotation of the incoming beam. The light is therefore transmitted by the second polarizer P_2. Upon reflection at the mirror the light passes again through polarizer P_2 and the cell, where it experiences another 90° rotation. Light traveling toward the polarizer P_1 has experienced a 180° rotation and is therefore transmitted through P_1.

Two types of electrooptic effects have been utilized in laser Q-switches: the Pockels effect, which occurs in crystals which lack a center of point symmetry, and the Kerr effect, which occurs in certain liquids. Pockels cells, which require a factor of 5 to 10 lower voltage than Kerr cells, are the most widely used active devices for Q-switching pulsed lasers.

Pockels Cell Q-Switch

Q-switches are classified into two types: cells in which the electric field is applied along the direction of the optical beam, and cells in which the electric field is perpendicular to the direction of the optical beam [8.21, 22].

Pockels Cells with Longitudinal Field. In this class of Q-switches the electric field is applied parallel to the crystal optical axis and in the same direction as the incident light. The dependence of the index of refraction on the electric

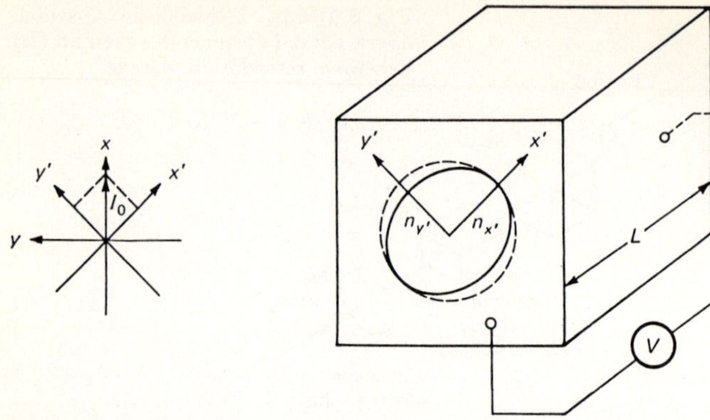

Fig. 8.11. Change of the index ellipsoid in a KDP crystal when an electric field is applied parallel to the z axis, I_0 is an incident wave polarized in the x direction; x and y are the crystallographic axes; x' and y' are the electrically induced axes

field can be described in terms of a change in orientation and dimensions of the index ellipsoid. The crystals used for longitudinal electrooptic Q-switches are uniaxial in the absence of an electric field; that is, there is only one value of refractive index in the direction of light propagation [8.23–27]. The index ellipsoid is an ellipse of revolution about the optic (z) axis. As indicated in Fig. 8.11, the index ellipsoid projects as a circle on a plane perpendicular to the optic axis. The circle indicates that the crystal is not birefringent in the direction of the optic axis. When an electric field is applied parallel to the crystal optic axis; the cross section of the ellipsoid becomes an ellipse with axis x' and y', making a 45° angle with the x and y crystallographic axes. This angle is independent of the magnitude of the electric field. The length of the ellipse axes in the x' and y' directions are proportional to the reciprocals of the indices of refraction in these two directions.

We will now express the phase shift between the orthogonal components corresponding to a wave polarized in the x' and y' directions. Changes of the refractive index Δn are related by the electrooptic tensor r_{ij} of rank 3 to the applied field.

$$\Delta\left(\frac{1}{n_i^2}\right) = \sum_{j=1}^{3} r_{ij}E_j \quad , \tag{8.20}$$

where $i = 1, \ldots, 6$ and $j = 1, \ldots, 3$.

Generally there exist 18 linear electrooptic coefficients r_{ij}. However, in crystals of high symmetry, many of these vanish. For phosphates of the KDP family, r_{63} is the only independent electrooptic coefficient which describes the changes in the ellipsoid when a longitudinal field is applied to the crystal. The change of refractive index in the x' and y' directions is [8.25]

$$n_{x'} = n_0 + \tfrac{1}{2}n_0^3 r_{63}E_z \quad , \quad n_{y'} = n_0 - \tfrac{1}{2}n_0^3 r_{63}E_z \quad , \tag{8.21}$$

where n_0 is the ordinary index of refraction and E_z is the electric field in the z direction. The difference in the index of refraction for the two orthogonal components is then

$$\Delta n = n_0^3 r_{63} E_z \quad , \tag{8.22}$$

For a crystal of length l, this leads to a path-length difference Δnl and a phase difference of $\delta = (2\pi/\lambda)\Delta nl$.

The phase difference δ in a crystal of length l is related to the voltage $V_z = E_z l$ applied across the faces by

$$\delta = \frac{2\pi}{\lambda} n_0^3 r_{63} V_z \quad . \tag{8.23}$$

It should be noted that δ is a linear function of voltage and is independent of the crystal dimensions. If linearly polarized light is propagated through the crystal with the direction of polarization parallel to the x or y axis, as shown in Fig. 8.11, the components of this vector parallel to the electrically induced axes x' and y' will suffer a relative phase shift δ. In general, orthogonal components undergoing a relative phase shift produce elliptically polarized waves. Thus, the application of voltage in this configuration changes linearly polarized light to elliptically polarized light. If the light then passes through a polarizer, the resulting light intensity will be a function of the ellipticity and therefore the voltage applied to the crystal. A simple derivation shows that, with the analyzer axis oriented at right angles to the input polarization direction, the voltage and the transmitted light intensity I are related by [8.20]

$$I = I_0 \sin^2 \frac{\delta}{2} \quad , \tag{8.24}$$

where I_0 is the input light intensity.

For Q-switch operation, two particular values of phase shift are of interest; these are the $\lambda/4$ and $\lambda/2$ wave retardations which correspond to a phase shift of $\frac{\pi}{2}$ and π. With linearly polarized light being applied, for example, in the x direction, as shown in Fig. 8.11, the output from the crystal is circularly polarized if $\delta = \frac{\pi}{2}$. For $\delta = \pi$ the output beam is linearly polarized, but the plane of polarization has been rotated 90°.

From (8.23) it follows that the voltage required to produce a retardation of π is

$$V_{1/2} = \frac{\lambda}{2 n_0^3 r_{63}} \quad . \tag{8.25}$$

Deviations from this voltage will change the transmission $T = I/I_0$ of the Pockels cell according to

$$T = \sin^2\left(\frac{\pi}{2}\frac{V}{V_{1/2}}\right) \quad . \tag{8.26}$$

This equation is obtained by combining (8.23–25).

A number of crystals which have been used in longitudinal modulators are listed in Table 8.1. The most useful materials are KD*P and CD*A be-

Table 8.1. Electrooptic parameters of $\bar{4}2m$-type crystals

Material	Index of refraction at 0.55 μm	Electrooptic constant, r_{63} [μm/V $\times$ 10^{-6}]	Typical half-wave voltage [kV] at 0.55 μm
Ammonium dihydrogen phosphate (ADP)	1.53	8.5	9.2
Potassium dihydrogen phosphate (KDP)	1.51	10.5	7.5
Deuterated KDP	1.51	26.4	3.0
Rubidium dihydrogen phosphate (RDP)	1.51	15.5	5.1
Ammonium dihydrogen arsenate (ADA)	1.58	9.2	7.2
Potassium dihydrogen arsenate (KDA)	1.57	10.9	6.5
Rubidium dihydrogen arsenate (RDA)	1.56	14.8	4.9
Cesium dihydrogen arsenate (CDA)	1.57	18.6	3.8
Deuterated CDA	1.57	36.6	2.0

cause they possess the lowest half-wave voltages. The latter material, which is relatively new, has a resistivity three orders of magnitude lower than KD*P, leading to a high loss tangent. For this reason almost all commercially available Pockels cell Q-switches with longitudinal field employ KD*P crystals. KD*P and its isomorphs are the most widely known electrooptic crystals; their general properties are reviewed in [8.21, 8.27–33]. They are grown at room temperature from a water solution and are free of the strains often found in crystals grown at high temperature. Excellent crystals as large as 5 cm in any dimension can be obtained commercially at nominal cost. Although the crystals are water-soluble and fragile, they can be handled, cut, and polished without difficulty. The crystals have a high optical transmission in the range 0.22 to 1.6 μm, and they possess relatively large electrooptic coefficients. In KDP the half-wave voltage is of the order of 7.5 kV. By using KD*P this figure is reduced by a factor of more than 2. The half-wave voltage can be further reduced by a factor of 2 by utilizing two crystals optically in series and electrically in parallel.

All crystals listed in Table 8.1 are hygroscopic and must be protected from atmospheric water. This protection is typically provided by enclosing the crystal in a cell which is hermetically sealed or filled with index-matching fluid. As a result, six surfaces are encountered by a laser beam on a single transit through the cell. Transmission losses can be reduced by the use of antireflection coatings on the cell windows. While hard, damage-resistant AR coatings cannot be applied to KDP and its isomorphs, losses at the crystal itself can be very much reduced by the use of an index-matching liquid such as 3M's FC77. Other techniques employed to minimize insertion losses of Pockels cells include a Brewster-angle design [8.34] and uncoated crystals with plane-parallel faces accurately aligned to the cavity mirrors [8.35].

Transmission of typical electrooptic shutters with liquid-immersed crystals and AR-coated windows is about 90%. The electric field is usually applied to the crystal by means of a pair of metal electrodes containing apertures which are either bonded or evaporated onto the square ends of the crystal or simply held in place by compression. The main drawback of end-plate electrodes is their geometry, which gives rise to a nonuniform electric field across the clear aperture. The field strength in the aperture varies from a maximum around the inner edge of the rings to a minimum at the geometric center [8.36]. Fringing necessitates operation at considerably higher voltages than if the field were uniformly applied. Partial compensation is attained by making the crystal length roughly 30% greater than the clear aperture diameter.

Typical commercially available Pockels cells employing KD*P have quarter-wave voltages between 3.5 and 4. kV at 0.69 μm and 5 to 6 kV at 1.06 μm. Apertures range from 9 to 25 mm, and the maximum safe operating level for most KD*P Pockels cell Q-switches is around 200 MW/cm^2.

A considerable improvement in fringe uniformity can be achieved by using cylindrical band electrodes applied to the end of the barrel of cylindrical-shaped crystals [8.37], as shown in Fig. 8.12. Band electrodes allow an optical transmission uniformity to within a few percent across the clear aperture. This

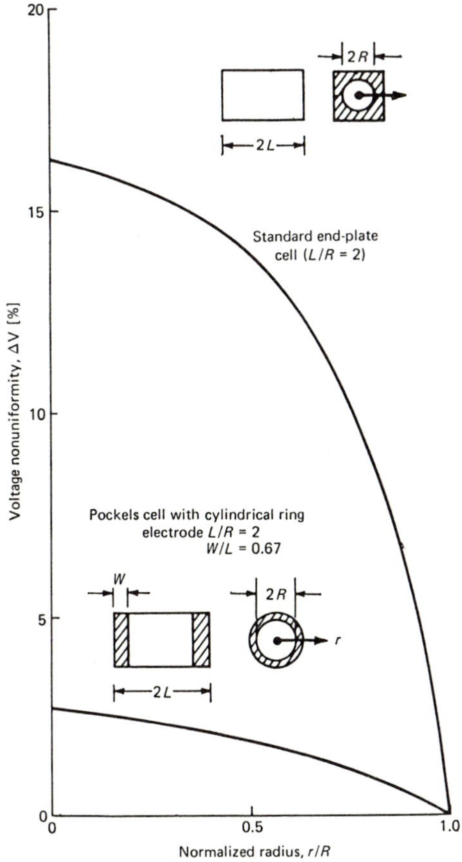

Fig. 8.12. Voltage nonuniformity ΔV versus normalized radius r/R for a cylindrical ring electrode (*lower curve*) and a standard end-plate cell. [8.37]

represents an order-of-magnitude improvement compared to standard cells containing rectangular crystals and end-plate electrodes. The Fig. 8.12 also shows computer-calculated curves of the voltage nonuniformity ΔV as a function of the normalized radius r/R for a conventional endplate cell and a cylindrical ring electrode cell [8.37]. The value of ΔV is the amount by which the voltage on the electrodes must be raised to obtain full half-wave radiation at a point r/R as compared to $r = R$. The more uniform electric field distribution within the crystal having cylindrical band electrodes results in higher extinction ratios and lower half-wave voltages. By use of low-reactance connecting leads, rise times in Pockels cells on the order of 500 ps and lower have been achieved.

Pockels Cells with Transverse Fields. In this configuration, the electric field is perpendicular to the direction of the beam. For this geometry, the half-wave voltage depends on the ratio of thickness to length of the crystal, which has the advantage that by proper choice of the crystal geometry a considerably lower voltage is required as compared to longitudinally applied fields. In Fig. 8.13a a KDP crystal is shown where the optical propagation direction is at 45° to the x and y axes while the field is applied along the z axis. The half-wave voltage $V_{1/2}$ for this case is given by

$$V_{1/2} = \frac{\lambda d}{n_o^3 r_{63} l} , \tag{8.27}$$

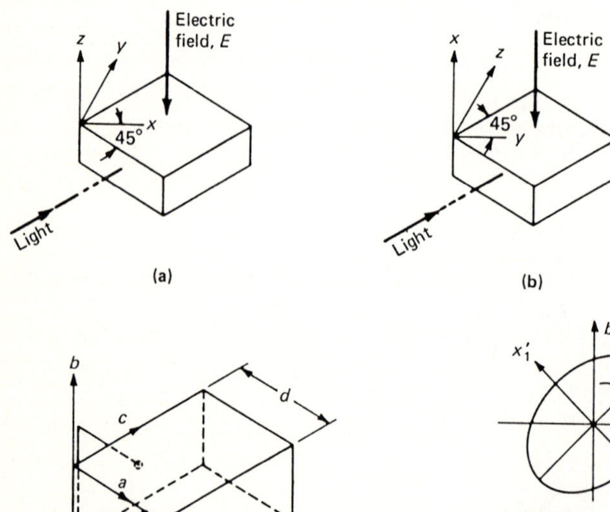

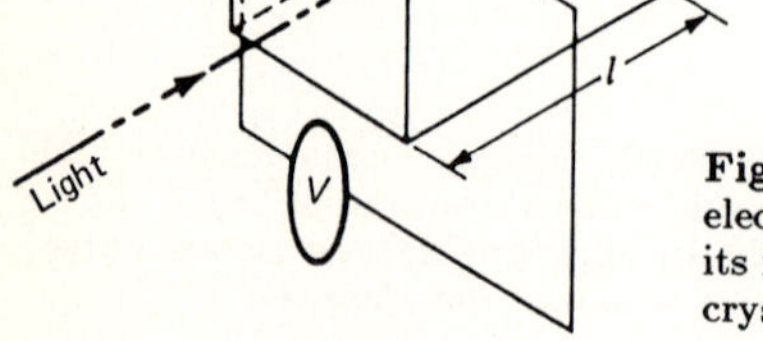

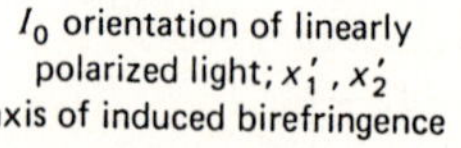

Fig. 8.13a–c. Electrooptic Q-switches with transverse electric fields: (a, b) give the configuration for KDP and its isomorphs; (**c**) shows the orientation for a $LiNbO_3$ crystal

where l and d are the crystal length and thickness, respectively. The light is linearly polarized at 45° to the z axis. In another arrangement, the optical propagation direction in the KDP crystal is at an angle of 45° to the y and z axes while the electric field is applied along the x axis, as shown in Fig. 8.13b. If the light is polarized at 45° to the x axis, then the half-wave voltage is given by [8.20]

$$V_{1/2} = \left(\frac{1}{n_o^2} + \frac{1}{n_e^2}\right)^{3/2} \frac{\lambda}{r_{41}} \frac{1}{2(2)^{1/2}} \frac{d}{l} \quad , \tag{8.28}$$

where n_o and n_e are the ordinary and extraordinary indices of refraction, respectively.

Since propagation in the 45° z cut or 45° y cut KDP and its isomorph is at an angle other than parallel or normal to the optic axis, the extraordinary and ordinary ray propagate in different directions through the crystal. To compensate for this walk-off as well as for temperature effects, two crystals of suitable orientation are usually employed [8.38–43]. Because of the need to compensate for the angular and temperature dependence of birefringence in KDP-type crystals, transverse field devices have not become popular as Q-switches in this class of materials.

However, the material $LiNbO_3$ has found widespread applications as a Q-switch [8.44–48]. This crystal, unlike KD*P, is not hygroscopic and can be operated in a transverse electrode orientation with the light propagation along the optic axis. It does not require the sealed-cell assemblies necessary for KD*P crystals, therefore transmission of an AR-coated $LiNbO_3$ crystal can be as high as 98 %. In an arrangement as shown in Fig. 8.13c, light propagates along the c axis of the crystal. If the light is polarized parallel to the a axis and an electric field is applied parallel to the a axis, the half-wave retardation is

$$V_{1/2} = \frac{\lambda d}{2 r_{22} n_o^3 l} \quad , \tag{8.29}$$

where l is the length of the crystal in the c direction, d is the distance between the electrodes along the a axis, and r_{22} is the electrooptic coefficient.

The half-wave voltage is directly proportional to the distance between the electrodes and inversely proportional to the path length. At a wavelength of $\lambda = 1.064\,\mu\mathrm{m}$, the linear electrooptic coefficient is $r_{22} = 5.61 \times 10^{-6}\,\mu\mathrm{m/V}$ and the refractive index of the ordinary ray is $n_o = 2.237$ in $LiNbO_3$. The theoretical half-wave voltage obtained from (8.29) for a typical crystal size of 9 mm × 9 mm × 25 mm is 3025 V. This is the value for a dc bias voltage; for pulsed Q-switch operation a 30 to 40 % higher voltage is required [8.45]. In pulsed operations, the value of r_{22} is smaller than for dc operation, which is explained by the fact that for static fields the electrooptic effect is equal to the sum of the intrinsic electrooptic effect and that induced by the piezooptic effect [8.49, 50]. $r_{22} = r'_{22} + p_{2k} d_{2k}$, where p_{2k} and d_{2k} are the elastooptic and piezoelectric constants of the crystal. When the bias voltage is turned on or off rapidly, the piezooptic effect is absent, since deformation of the crystal cannot occur during the fast switching times typical for Q-switch operation.

Since most military applications involve operation over a large temperature range, lithium niobate is also often chosen for its smaller variation in electrooptic coefficients. This feature minimizes the requirements for temperature and optical feedback stabilization controls. At the present time many military rangefinders and target designators utilize lithium niobate Q-switches in a high repetition rate (10 to 60 pps), low power density (5 to 10 MW/cm^2) mode. In these applications crystal temperatures may exceed 80° C. In contrast, while KD*P Q-switches can be used at 70° C, they should never be used above 55° C when the electric field is applied continuously, because this leads to a fogging effect on the crystal surfaces. The major drawback of lithium niobate is that it is not suitable for high-power-density applications and users place a limit of between 10 to 50 MW/cm^2 on its peak optical power-handling capabilities. KD*P Q-switches are usually specified where large apertures, as large as 40 mm, and high peak power densities are involved. Q-switches employing $LiNbO_3$ crystals have apertures up to 9 mm and require half-wave voltages between 2.5 and 6 kV at 1.06 μm depending on the crystal geometry.

Prelasing and Postlasing Occurring with Electrooptical Q-Switches. Theoretically, an infinite loss is introduced into the cavity by a Pockels cell operated at a proper $\lambda/4$ or $\lambda/2$ voltage. In practice, field strength nonuniformities, inhomogeneities in the crystal, birefringence introduced by mechanically clamping the crystal, and nonperfect polarizers limit the extinction ratio of an electrooptic Q-switch to a few hundred. The extinction ratio is defined as the ratio of the maximum to the minimum transmitted intensity obtained between crossed polarizers as the applied voltage is varied through a half-wave voltage.

If the gain of the laser exceeds the loss produced by the Q-switch, normal lasing will occur before the instant of Q-switching. This phenomenon, which is termed "prelasing", is due to the Pockels cell and polarizer combination not acting as perfect shutter, so that there is still some feedback from the 100 % mirror. Prelasing is most likely to occur first just prior to the time of Q-switching since the population inversion, the stored energy, and the gain are largest at that time.

Because of the high gain of Nd:YAG, depumping of the stored energy by prelasing is a particular problem with this material. Tests have also shown that prelasing may be a major cause of damage to the Pockels cell crystal or polarizers. Prelasing, if it occurs, allows a pulse buildup from a "seed" pulse in a small region of the laser rod. In the area of the laser rod where prelasing occurs, the Q-switched pulse develops more rapidly compared to the rest of the pulse, which must develop its output from the spontaneous emission noise. As a result, a very high peak power density will occur in this small part of the rod.

Appropriate design precautions must be undertaken to ensure that prelasing does not occur. The first major requirement is to operate the modulator at its optimum extinction ratio. This requires that the correct dc bias voltage be applied and that the Pockels cell crystal and polarizers be aligned properly.

Typically the optic c axis of the Pockels cell crystal must be parallel to the laser beam direction to within 10 arc min or less. Alignment of the c axis is best performed by centering the optic-axis figure (the Maltese-cross pattern) on the resonator optical axis. This is done by illuminating the crystal with a diffuse light source and observing the crystal between crossed polarizers. A pattern of a cross surrounded by a series of circles appears. The line connecting the center of the cross to the point at which the observation is made is exactly parallel to the c axis. Furthermore, to establish the proper condition for Q-switching, the crystal must be aligned so that either the a or b axis is parallel to the polarization direction of the laser ($LiNbO_3$ crystals and longitudinal field KD^*P). Figure 8.14 shows the extinction ratio versus bias voltage for a 9 mm × 9 mm × 25 mm $LiNbO_3$ crystal [8.51].

In practice, the main Q-switched pulse may be followed by one or more pulses of lower amplitude (postlasing). The second pulse may occur from several hundred nanoseconds to several tens of microseconds after the main pulse. Postlasing results from piezooptic effects in the electrooptic crystal. The piezoelectric action of the applied voltage compresses the crystal, and when that voltage is removed the crystal remains compressed for some definite interval of time. This compression generates a retardation of the optical wave by means of a strain-birefringence effect, thus creating a loss in the cavity which becomes smaller with time as the compression relaxes. Figure 8.15a shows a waveform picture of the voltage conventionally applied to a Pockels cell, and Fig. 8.15b shows the loss versus time for an actual Pockels cell switch. It was found [8.46] that the loss drops to about 25 % and then decays to zero in approximately

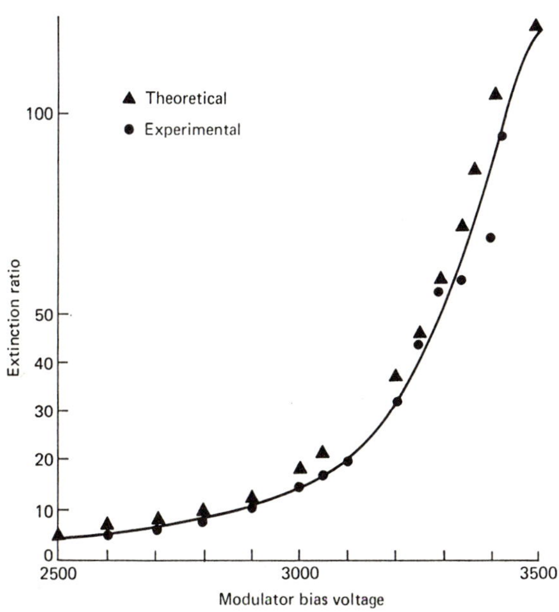

Fig. 8.14. Extinction ratio versus applied voltage of a $LiNbO_3$ Pockels cell-calcite prism assembly. [8.50]

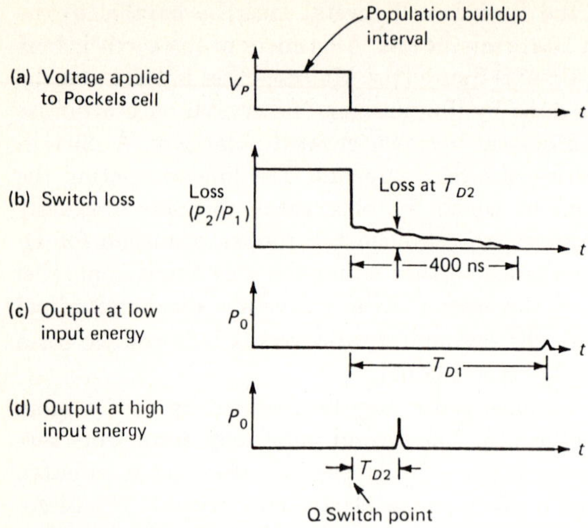

Fig. 8.15. Piezooptic effect in $LiNbO_3$. [8.46]

400 ns. If the resulting decay time of the loss is longer than the output pulse buildup time, the main laser pulse will be complete before the loss reaches its minimum. Thus some energy will remain in the rod after the first output pulse. This residual energy may produce a second pulse when the cavity loss reaches its minimum. By creating this time-dependent loss, the elastooptic effect seriously affects the efficiency of a Q-switched laser.

At low input energy levels there is a long time delay between the switching time and the time the output pulse actually appears, as shown in Fig. 8.15c, and thus there is negligible loss in efficiency. However, at higher input energy levels the time delay t_D becomes shorter and thus the laser suffers a considerable output loss due to the switch loss, as shown in Fig. 8.15d. This, besides prelasing, accounts for the roll-off in output efficiency in the Q-switched laser at higher input energy levels.

The elastooptic effect, which is quite pronounced in $LiNbO_3$ but is also observed in KD^*P, can be minimized by switching the bias voltage to a negative value rather than to zero.

Examples of Pockels Cell Q-Switched Lasers. Output energies between 100 and 250 mJ are obtained from Nd : YAG oscillators. The energy extraction is limited by the high gain of this material, which leads to prelasing and subsequently to a depopulation of the inversion.

Ruby and Nd : glass laser oscillators produce, depending on size, between 0.1 and 10 J in the Q-switched mode. For example, a ruby rod 75 mm long and 6 mm in diameter pumped by a linear flashlamp at about 150 J will produce around 100 mJ of Q-switched output. Ruby and Nd : glass rods, 100 mm long and 10 mm in diameter pumped at 2000 J in a 500-μs flashlamp pulse, will typically generate pulses with an energy content of 2 J. Large rods, for example 150 mm by 20 mm, pumped around 5 kJ, will produce Q-switched pulses up to 10 J in energy. The width of the Q-switched pulses is usually between 10 and

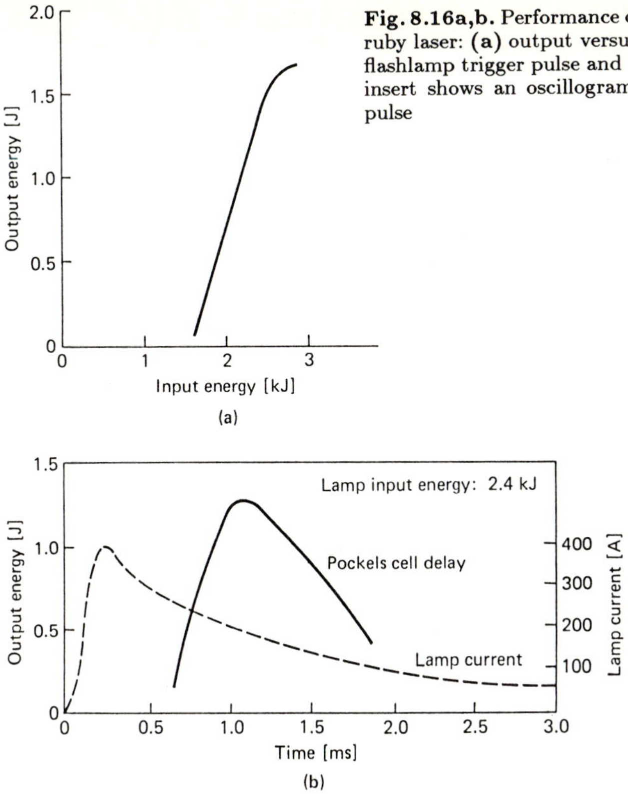

Fig. 8.16a,b. Performance of a Pockels cell Q-switched ruby laser: **(a)** output versus input; **(b)** delay between flashlamp trigger pulse and Pockels cell switching. The insert shows an oscillogram of the flashlamp current pulse

25 ns, and the pulse buildup time t_D is between 50 and 100 ns in typical systems. In the larger systems, the energy output is actually limited by the maximum permissible power density on the optical components. In order to avoid optical damage, most commercial solid-state laser oscillators are operated at power densities around 150 MW/cm^2. Therefore, from an oscillator containing a 1-cm-diameter Nd: glass or ruby rod a peak power $\sim$100 MW can be expected and $\sim$500 MW from a large, 2-cm-diameter rod.

Figure 8.16a shows a plot of output versus input energy of a typical Q-switched ruby laser containing a 10-cm by 1-cm rod pumped by a helical flashlamp. In Fig. 8.16b the output from the laser at fixed lamp input is plotted as a function of the delay between the flashlamp trigger pulse and the opening of the Pockels cell. The maximum energy extraction is obtained for a delay of 1.1 ms. This time corresponds to a point near the end of the flashlamp pulse at which the lamp current has dropped to 50 % of its initial value. Prior to this time the stored energy in the rod increases because of the pumping action of the lamp, and after the optimum time of 1.1 ms, population inversion decreases because of the diminishing pump power and the increase of spontaneous emission losses.

Figure 8.17 shows the dependence of the pulse width of a Q-switched ruby laser on the energy output, with cavity length and output reflectivity as a parameter. According to (8.9), one would expect that the pulse width Δt_p

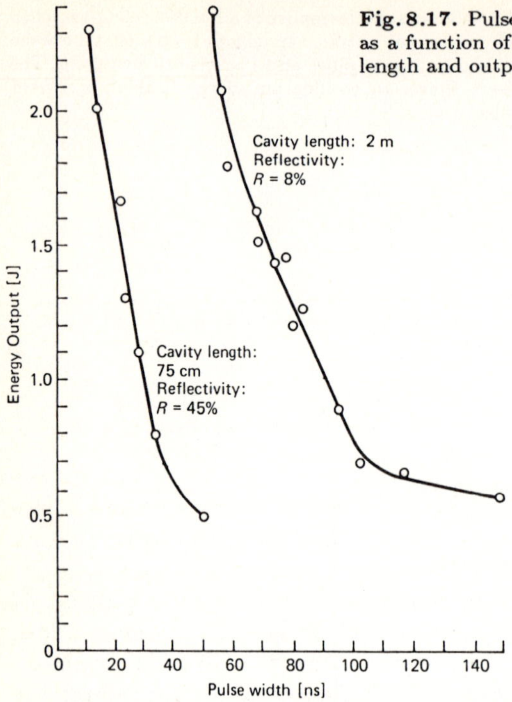

Fig. 8.17. Pulse width from a Q-switched ruby laser as a function of energy output. Parameters are cavity length and output reflectivity

increases for a system with a long cavity (large τ_c), a low gain (small n_i) and a low output reflectivity (high n_t). The experimental results illustrate this dependence of the pulse width on the system parameters. Decreasing the output energy, achieved by reducing the input power, will result in a low gain and low initial inversion n_i. By combining the effects of low gain, high threshold, and long cavity, pulses up to 150 ns can be produced from Q-switched lasers.

Employing a feedback loop to control the switching of a Pockels cell, pulse durations of up to 600 ns have been obtained from a Q-switched ruby laser [8.52] and up to 1.4 μs in Nd:YAG lasers [8.53, 54]. The principle of electrooptical feedback control is depicted in Fig. 8.18. The resonator incorporates a Pockels cell as a Q-switching element. Negative feedback on the circulating power is applied through the action of photo-detector-derived voltage on the Pockels cell. In this way the laser rod stored energy is released at a controlled rate.

Optimization of Pockels Cell Q-Switch Designs in the Presence of Thermally Induced Birefringence. In low-repetition-rate systems, or at low input powers depolarization losses can be ignored, but in higher-average-power systems depolarization losses become significant.

The polarizer, required for Q-switch operation employing a Pockels cell, rejects any radiation not polarized in the proper plane of polarization. This can lead to large depolarization losses in the presence of thermally-induced birefringence, as may occur in repetition-rated systems. Several techniques have been developed in conjunction with Pockels-cell Q-switches which act on both

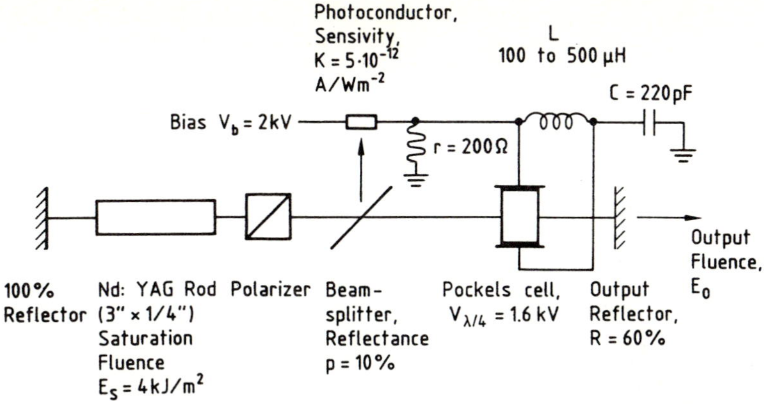

Fig. 8.18. Electrooptic feedback control for stretching of Q-switched pulses. [8.53]

polarizations. In general, the techniques are based on the use of a calcite polarizer in the resonator which separates the two orthogonally polarized beams. The methods differ in the ways these two beams are treated in the resonator.

One advantage of a variable-reflectivity resonator is that the depolarized component of the beam contributes to the output rather than acting as a loss, as in a conventional resonator. The outcoupling of laser radiation in a variable-reflectivity resonator is accomplished with a polarizer and phase retarder by inducing a phase shift in a linearly polarized beam. This is done in a conventional resoator with two total end reflectors by a phase-retardation element such as a quarter-wave plate. In the design by *Chun* et al. [8.55], the two resonator mirrors comprised a pair of crossed prisms. In the Porro-prism resonator the phase shift can be induced without a retardation plate; the proper combination of prism-induced phase shift and prism azimuth angle results in the desired laser outcoupling. Figure 8.19 is a schematic diagram of an electrooptic, Q-switched, Porro-prism resonator. A calcite polarizer and electrooptic Pock-

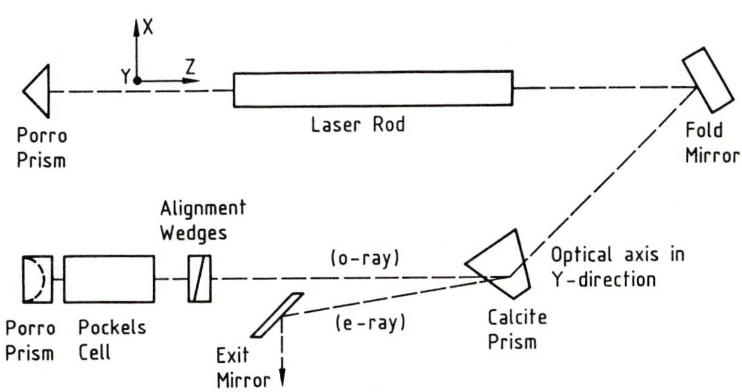

Fig. 8.19. Variable reflectivity resonator which allows extraction of the orthogonally polarized beam

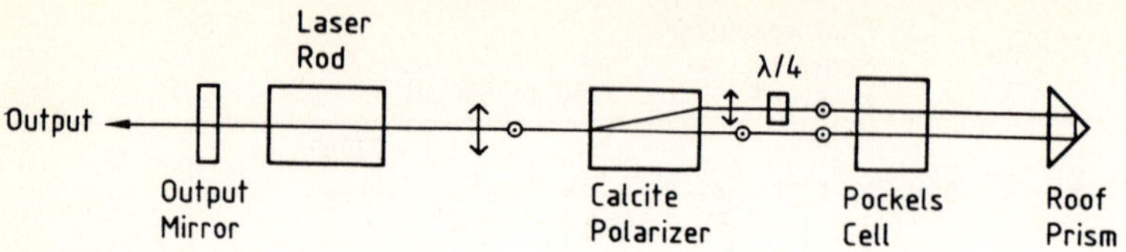

Fig. 8.20. Polarization insensitive Q-switch based on spatial separation of orthogonally polarized beams

els cell are located within the laser resonator. The calcite polarizer generates polarized laser-cavity resonant mode and allows the extraction of the orthogonally polarized component. Therefore the depolarized component of the beam contributes to the output.

Rankin et al. [8.56] used a calcite polarizer which separates the two orthogonal polarization components into two parallel beams. A quarter-wave plate is inserted into one of the beams, as shown in Fig. 8.20 before both beams are passed through a large aperture Pockels cell. A roof prism returns the beams back to the Pockels cell/polarizer assembly.

More recently *Richards* [8.57] proposed a design which exploits directional differences between the two orthogonal beam components after passing through a birefringent prism. A schematic diagram of the laser is shown in Fig. 8.21.

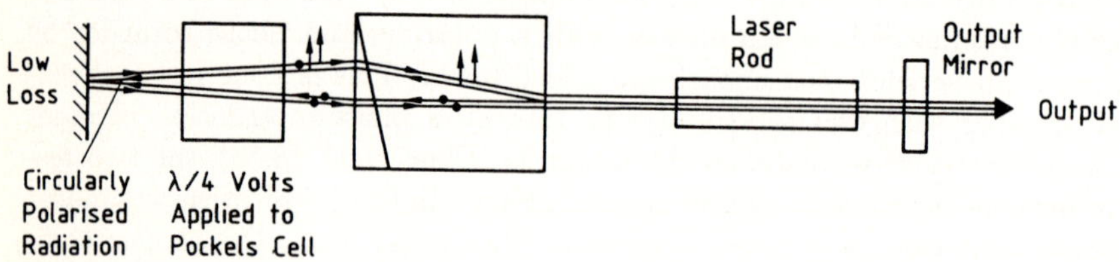

Fig. 8.21. Polarization insensitive Q-switch based on angular separation of orthogonally polarized beams

It is very similar to a conventional Q-switched laser and contains a mirror at each end of the optical cavity, one completely reflecting and the other partially transmitting to provide output. Other components include a pumped laser rod, in this case Nd : YAG, a Pockels cell, and a birefringent prism that replaces the linear polarizer commonly used in Q-switched lasers. After passing through the birefringent prism, radiation is split up into two orthogonally polarized components, the extraodinary (e) and the ordinary (o) rays, that propagate in two slightly different directions. When these two rays are reflected by mirror M_2 back through the prism, a further separation of the two rays will occur. In this case the losses in the cavity will be very high and laser action will be suppressed. However, if a quarter-wave voltage is applied to the Pockels cell, the e and o rays returning to the prism from M_2 will be interchanged. In this case, the walk off occurring in the first pass will be canceled by an equal and opposite walk off during the return pass, hence the beam returning to the laser rod will be low, allowing a Q-switched pulse to develop.

Kerr Cell Q-Switch

Like the Pockels cell, the Kerr cell is a device which can produce a controllable birefringence by the application of a voltage to a cell. In this case, the cell contains a liquid, usually nitrobenzene, instead of a crystalline solid. In the Kerr cell the birefringence is proportional to the square of the applied voltage. The difference between the index of refraction for light polarized parallel to the direction of the inducing field and the index of refraction for light polarized at right angles to the direction of the inducing field is given by

$$\Delta n = \lambda B E^2 \quad , \tag{8.30}$$

where B is the Kerr constant and E is the transverse field strength. From (8.30) it follows that the phase difference is

$$\delta = 2\pi l B E^2 \quad , \tag{8.31}$$

where l is the length of the region in which the transverse field exists. Since $E = V/d$ for parallel plates, where d is the electrode separation and $l = 2L$ because the light in a resonator travels twice through the cell of length L, we obtain for the half-wave retardation voltage ($\delta = \pi$)

$$V_{1/2} = \frac{d}{2(BL)^{1/2}} \quad . \tag{8.32}$$

For nitrobenzene, $B = 26 \times 10^{-6}$ cgs at 6943 Å and the half-wave voltage becomes

$$V_{1/2} = \frac{30d}{(L)^{1/2}} \quad , \tag{8.33}$$

where d and L are in centimeters and V is in kilovolts. For typical values of d and L, V must be 10 to 40 kV. The axes of induced birefringence in a Kerr cell are at 45° with respect to the applied field. Figure 8.22 shows

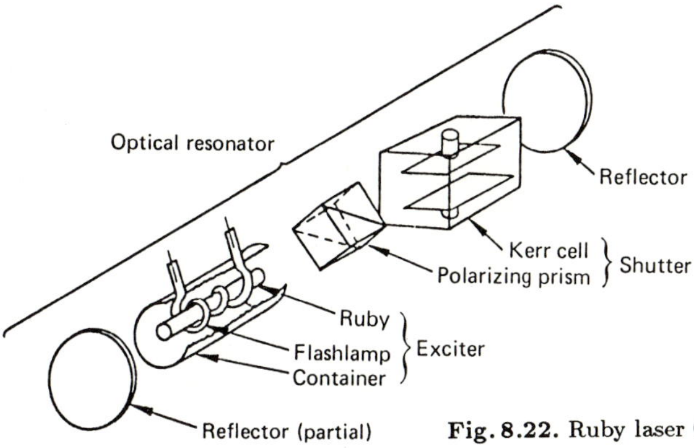

Fig. 8.22. Ruby laser Q-switched with a Kerr cell

a Kerr cell Q-switch operated at a quarter-wave retardation voltage. Laser systems featuring Kerr cells are described in [8.58, 59]. As mentioned earlier, because of the higher voltage requirement of Kerr cells as compared to Pockels cells, problems associated with high leakage currents due to impurities in the nitrobenzene, and difficulties in constructing leakproof cells, this type of Q-switch is not used very often anymore. However, more recent experiments in which the electric field of a high-powered laser beam was employed to operate a Kerr cell have generated renewed interest in this device. In these experiments the Kerr cell was actually operated as a camera shutter employed to photograph a frequency-doubled mode-locked pulse [8.60].

Drivers for Electrooptic Q-Switches

In the operation of an electrooptically Q-switched laser, it is necessary to switch voltages electrically in the range between 3 kV and 15 kV depending on the crystal material, geometry, and optical wavelength. The driver for the Pockels cell must be a high-speed, high-voltage switch which also must deliver a sizeable current. The cell has a few tens of picofarads capacitance which is charged (or discharged) to several kilovolts in a few nanoseconds. The resulting current is of the order of 10 to 20 A. Common switching techniques include the use of vacuum tubes, cold cathode tubes, thyratrons, SCRs, and avalanche transistors.

Vacuum tubes are usually applied if a rapid sequence of Q-switch pulses must be generated for holographic applications, for example; or if a particular waveform of the electronic signal is desired, for example, to achieve longitudinal mode selection [8.61]. A cold cathode gas tube, such as the EG&G KN-6 krytron, is an attractive device, being of reasonable size and having a very fast switching time. Nevertheless, it is a gas tube with a limited lifetime, and the operating voltage is restricted to values below 8 kV. A typical circuit diagram for a Pockels cell driver using a krytron is shown in Figure 8.23a. The $LiNbO_3$ crystal operated at a quarter-wave retardation voltage of 3.3 kV is switched to a negative bias voltage of 800 V to compensate for the piezoelectric effect. The bias voltage is of the same polarity as that of the holdoff voltage but applied to the opposite electrode. The net result is that when the holdoff voltage is removed (at the instant of Q-switching), the bias across the modulator becomes negative. In systems which must operate over a large temperature range, such as military systems, a circuit is usually incorporated into the system which controls the quarter-wave voltage level at the krytron high-voltage supply.

Hydrogen thyratrons can be operated at higher voltages than krytrons. A circuit for operation of a cell at the pulse-on mode is shown in Fig. 8.23b. When the thyratron conducts, it short-circuits the HV cable between the thyratron and Pockels cell, inducing a negative high-voltage equivalent to the half-wave voltage across the very-high-impedance Pockels cell. The voltage remains nearly constant for a time equal to twice the cable transit time. The voltage across the cell then decays exponentially.

Semiconductor switches, such as SCRs or avalanche transistors stacked in series, have been successfully employed to drive Q-switches. The approach here

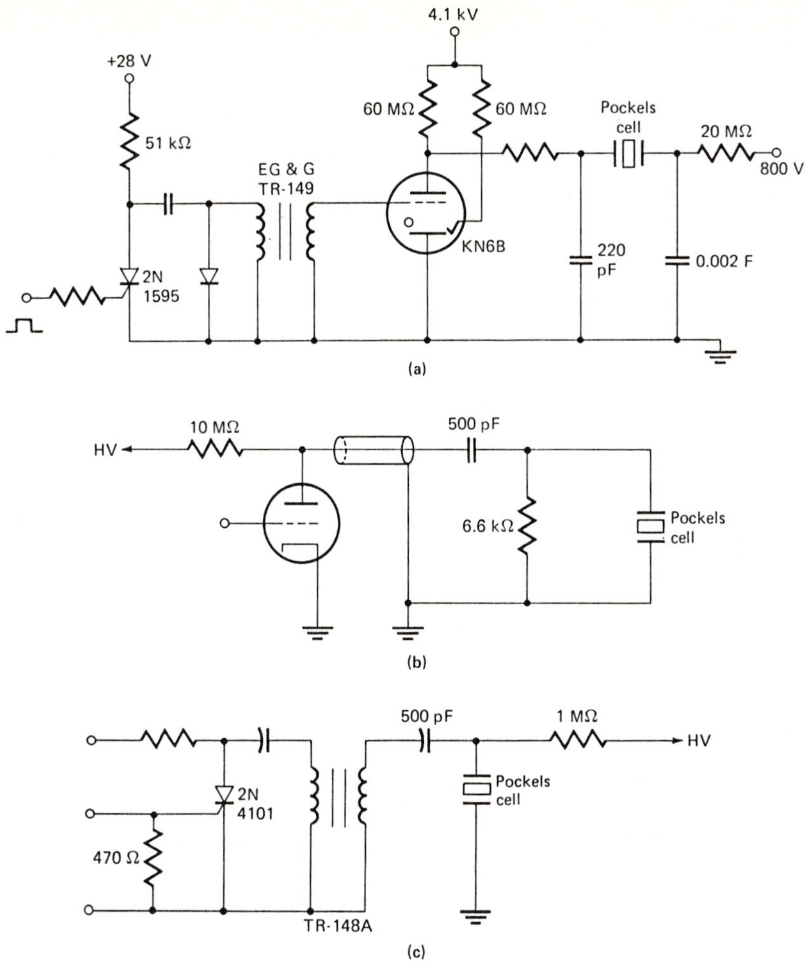

Fig. 8.23a–c. Circuit diagram of Pockels cell drive electronics using **(a)** a krytron; **(b)** a hydrogen thyratron; **(c)** an SCR

is to use a voltage of several hundred volts and step it up to several kilovolts by a high frequency step-up transformer. A transformer-driven Pockels cell switched by a SCR at a rise time of 250 ns is shown in Fig. 8.23c [8.48].

8.1.4 Acoustooptic Q-Switches

In acoustooptic Q-switches, an ultrasonic wave is launched into a block of transparent optical material, usually fused silica. A transparent material acts like an optical phase grating when an ultrasonic wave passes through it. This is due to the photoelastic effect, which couples the modulating strain field of the ultrasonic wave to the optical index of refraction [8.62–67]. The resultant grating

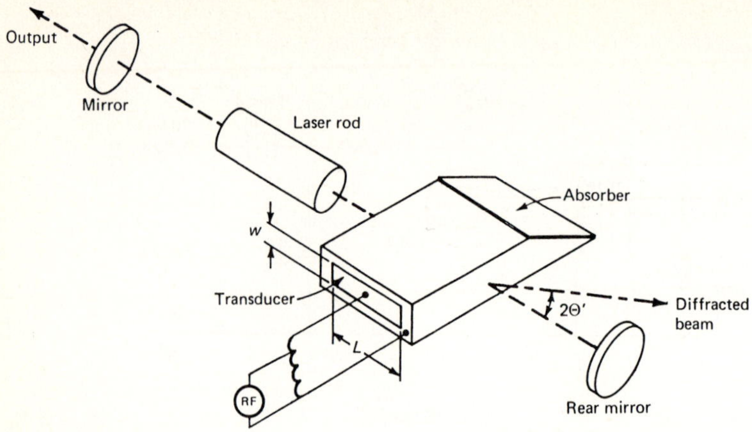

Fig. 8.24. Acoustooptic Q-switch employed in a cw-pumped Nd:YAG laser. The major components of the Q-switch are a very thin quartz crystal transducer having a thickness of a half acoustic wavelength, a fused silica block to which the transducer is epoxy-bonded, an inductive impedance-matching network, an acoustical absorber at the side opposite to the transducer, a water-cooling jacket system to conduct away generated heat, a thermal interlock cut-out which automatically turns off the rf power if the temperature in the modulator rises beyond allowable limits, a Bragg angle adjustment, a 50 MHz rf driver containing a pulse generator, and logic which allows adjustment of the repetition rate from 0 to 50 kHz

has a period equal to the acoustic wavelength and an amplitude proportional to the sound amplitude.

If a light beam is incident upon this grating, a portion of the intensity will be diffracted out of the beam into one or more discrete directions (Fig. 8.24). By properly choosing the parameters, the diffracted beam can be deflected out of the laser resonant cavity, thereby providing an energy loss which is sufficient to Q-spoil the cavity.

The ultrasonic wave is typically launched into the Q-switch block by a piezoelectric transducer which converts incident electromagnetic energy into ultrasonic energy. The laser is returned to the high Q-state by switching off the driving voltage to the transducer. With no ultrasonic wave propagating through it, the fused silica block returns to its usual state of high optical transmission, the deflected beam disappears, and a giant laser pulse is emitted [8.9, 68–70]. two different types of diffraction effects are observed depending on the optical and acoustic wavelengths λ and Δ, and the distance l over which the light interacts with the acoustic beam.

Raman-Nath Scattering

Raman-Nath scattering occurs when either the interaction path is very short or when the ultrasonic frequency is very low, i.e.,

$$l\lambda \ll \Delta^2 \quad . \tag{8.34}$$

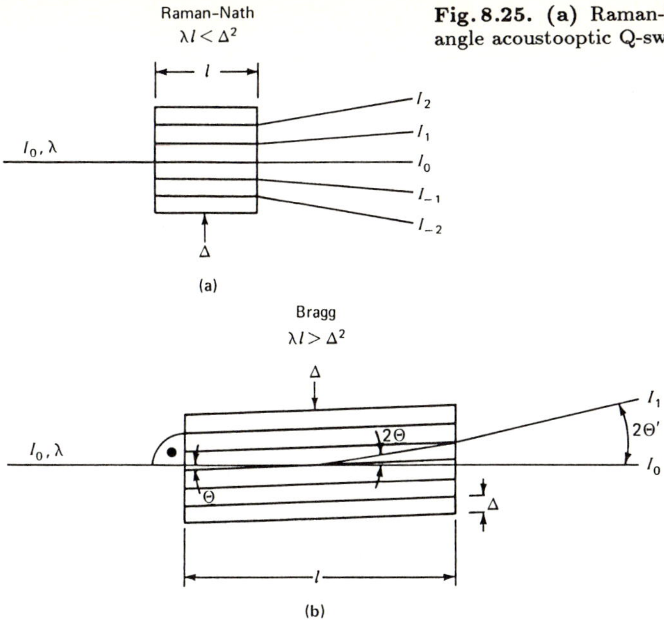

Fig. 8.25. (a) Raman-Nath and (b) Bragg angle acoustooptic Q-switches

Maximum light scattering is observed if the light beam and acoustical wave are perpendicular with respect to each other. The light beam is scattered symmetrically in many higher diffraction orders, as shown in Fig. 8.25a. The intensity of the individual orders is [8.66]

$$\frac{I_n}{I_0} = \mathrm{J}_n^2(\Delta\phi) \quad , \tag{8.35}$$

where I_n is the intensity of the nth order, I_0 is the intensity of the incident light, J_n is the Bessel function of nth order, and $\Delta\phi = 2\pi\Delta n l/\lambda$ is the amplitude of the phase grating. The amplitude $\Delta\phi$ has been shown to be [8.63, 71]

$$\Delta\phi = \pi\left(\frac{2}{\lambda_o^2}\frac{l}{w}M_2 P_{ac}\right)^{1/2} \quad , \tag{8.36}$$

where λ_o is the optical wavelength, P_{ac} is the acoustic power, l and w are the dimensions of the flat rectangular transducer, and M_2 includes a group of materials parameters known as the acoustooptic figure of merit,

$$M_2 = \frac{n^6 p^2}{\varrho v^3} \quad , \tag{8.37}$$

where n is the index of refraction, p is the appropriate photoelastic coefficient, ϱ is the density, and v is the acoustic velocity.

Bragg Scattering

When the frequency of the acoustical wave is raised and the interaction path is lengthened, higher-order diffraction is eliminated and only two light beams of zero and first order become predominant. The condition for Bragg scattering to occur is

$$l\lambda \gg \Delta^2 \quad . \tag{8.38}$$

In the Bragg regime the optical and acoustical beams are offset slightly from normal incidence to interact at the Bragg angle

$$\sin\Theta = \frac{\lambda}{2\Delta} \quad . \tag{8.39}$$

It should be noted that λ, Δ, and Θ are measured inside the medium (that is, $\lambda = \lambda_o/n$). As shown in Fig. 8.25b, if measured externally the angle between the incident light and the acoustic wave is $\Theta' = n\Theta$ and the scattering angle is twice that, i.e.,

$$2\Theta' = 2n\Theta \approx \frac{\lambda_o}{\Delta} \quad . \tag{8.40}$$

The intensity I_1 of the scattered beam is [8.71]

$$\frac{I_1}{I_0} = \sin^2\left(\frac{\Delta\phi}{2}\right) \quad , \tag{8.41}$$

where the phase amplitude is the same as defined in (8.36).

Depth of Modulation

It is apparent from (8.35–41) that the amount of diffracted power depends on the materials parameters expressed by M_2, the ratio of length to width of the interaction path, and the acoustical power P_{ac}. In a given material such as, for example, fused silica, the value of the photoelastic coefficient p in (8.37) depends on the plane of polarization of the light beam with respect to the ultrasonic propagation direction and on the type of ultrasonic wave, i.e., longitudinal or shear wave. With shear wave generation the particle motion is transverse to the direction of the acoustic wave propagation direction. In this case the dynamic optical loss is independent of polarization in isotropic materials such as fused quartz [8.65, 72]. Table 8.2 lists the pertinent material parameters for an acoustooptic Q-switch fabricated from fused silica.

Since at a small depth of modulation the sine function in (8.41) can be taken equal to its argument, the diffracted power is proportional to the figure of merit M_2. From Table 8.2 it follows, then, that in a Bragg angle device employing longitudinal-mode ultrasound, light polarized perpendicular to the acoustic wave vector is deflected five times stronger than light polarized parallel to this direction. The shear wave device, which operates independent of the

Table 8.2. Material parameters of acoustooptic Q-switches employing fused silica (Bragg and Raman-Nath devices)

Acoustic wave	p coefficient	Polarization of optical beam with respect to acoustic wave vector	Velocity of sound $\times 10^5$ [cm/s]	Figure of merit $\times 10^{-18}$ [s^3/g]	Acoustical power P_{ac} [W] for 1 % deflection ($l/w = 10$)
Shear wave	$p_{44} = 0.075$	Independent	3.76	0.47	0.42
Longitudinal	$p_{11} = 0.121$	Parallel	5.95	0.30	0.67
Longitudinal	$p_{12} = 0.270$	Perpendicular	5.95	1.51	0.13

light polarization, has a higher diffraction efficiency than parallel-polarized light in a longitudinal device, but a substantially lower efficiency compared to the perpendicular-polarized light in a longitudinal modulator.

The fraction of incident light which is scattered by the acoustooptic modulator determines whether the Q-switch can hold off laser action. In order to Q-switch an unpolarized laser system, such as Nd:YAG, the lower of the efficiency factors of a longitudinal device determines the extinction ratio of the device. Therefore, unpolarized lasers are usually Q-switched with shear wave devices, whereas for polarized lasers, such as Nd:$YAlO_3$ or Nd:YAG, which contain polarizing elements in the cavity, longitudinal modulators are employed. Because a longitudinal Q-switch in which the large p_{12} coefficient is utilized is very attractive, since substantially lower rf powers are required for the same diffraction efficiency compared to a shear wave device, techniques were devised to Q-switch also unpolarized lasers with this modulator. One commercially available Q-switch, for example, contains two longitudinal modulators orientated at 90° with respect to each other [8.73]. In another device the modulator is sandwiched between two quarter-wave plates which provide a 90° rotation of the plane of polarization of the light beam after each reflection from a resonator mirror [8.74].

The amount of acoustical power required to achieve a certain diffraction efficiency can be calculated from (8.35–41). For the practical case of a longitudinal Bragg angle device having a length of $l = 50$ mm and a transducer width of $w = 3$ mm, we find from the values given in Table 8.2 that a scattered fraction of 2.1 %/W of acoustic power is theoretically obtained for light polarized in the direction of the acoustic wave propagation. Typically, one measures a scattered fraction of 0.8 to 1 % per watt of electrical power, which indicates a conversion efficiency of the order of 40 to 50 %. For light polarized perpendicular to the acoustic beam vector, a value of 4 to 5 % per watt is typically obtained. For example, at an application of about 20 W to the transducer, a 20 % single-pass loss is achievable for parallel-polarized light. At this level the dynamic optical loss rises up to 70 % for perpendicularly polarized light. At 70 % the deflection process is noticeably saturated, and further increase in the loss is obtained only at a much higher expenditure of driving power. High-powered cw-pumped Nd:YAG lasers usually require a single-pass dynamic loss in excess of 20 %. For example, in a Bragg angle shear wave device, experimentally achievable diffraction efficiencies are of the order of 1 % per watt of electrical power. Commercially available units of this type are operated up to 60 W of rf power to the transducer, which results in a 40 % diffraction efficiency. The dynamic resonator loss introduced by the Q-switch is $L = \eta(2-\eta)$, where η is the deflection efficiency. In our example, where $\eta = 0.4$, we obtain a double-pass insertion loss of 64 %, which is sufficient to Q-switch most high-power Nd : YAG systems.

Design Features of Acoustooptic Q-Switches

We will calculate the parameters of a typical shear mode Bragg angle device employed for Q-switching a cw-pumped Nd : YAG laser. Essentially all commercially available Q-switches consist of a fused silica block to which a crystalline quartz or a $LiNbO_3$ transducer is bonded. Both the transducer and the fused silica contain vacuum-deposited electrodes to allow for electrical connections. An inductive impedance-matching network usually couples the signal of the rf generator to the quartz transducer. Virtually all acoustooptic Q-switches are single-pass devices; i.e., the acoustic wave generated by the transducer is absorbed after traveling across the interaction medium. The absorber, consisting of a piece of lead attached to the tapered end of the quartz block, prevents reflected acoustical waves from interfering with the incident light beam.

However, Q-switches have been built in which an ultrasonic standing wave was allowed to build up resonantly by reflection from the parallel face opposite to the transducer. This has the advantage that intensities at least ten times higher can be achieved than would exist in single-pass devices [8.75]. The disadvantage of this technique is that the laser can be Q-switched at only one fixed repetition rate.

Typical acoustooptic Q-switches can be operated in five different modes: internally driven Q-switch operation, externally gated pulse operation, an externally gated cw operation, a single-shot mode, and normal cw operation (with no rf modulation on the Q-switch).

Although the figure of merit M_2 of fused quartz is quite low, its optical high quality, low optical absorption, and high damage threshold make it superior to other, more efficient acoustooptic materials, such as lithium niobate ($LiNbO_3$), lead molybdate ($PbMoO_4$), tellurium dioxide (TeO_2), and dense flint glass. These materials are usually employed in low-power light modulators and optical scanners [8.66, 71, 76]. For an optical wavelength of 1.064 μm, a transducer drive frequency of 50 MHz, an index of refraction $n = 1.45$, and an acoustical velocity of $v = 3.76 \times 10^5$ cm/s for a shear wave in quartz, one obtains an acoustical wavelength of $\Lambda = 75\,\mu$m, a Bragg angle of $\Theta \approx 17$ arc min, and a scatter angle of $2\Theta' \approx 49$ arc min. For the light and acoustic waves to intersect at the Bragg angle, a parallelogram angle of the fused quartz block of $89° 43'$ must be chosen (Fig. 8.25b). For a cell 50 mm long, the Bragg condition (8.38) is satisfied. The width of the transducer perpendicular to the acoustic and optical propagation direction is typically 3 mm, which is about twice the beam diameter for TEM_{00}-mode Nd : YAG lasers.

To be able to deflect the beam out of the cavity, the frequency of the rf signal driving the transducer is in the 20- to 50-MHz range. For these frequencies we obtain from (8.40) scattering angles in silica between 0.3 and 0.8° for an optical wavelength of 1.06 μm. The Q-switch must be able to switch from the high-loss to the low-loss state in less than the time required for the laser pulse to build up if maximum output energy is to be achieved. If this condition is met there will be no appreciable loss of laser output energy due to scattering by the switch. The overall turnoff time of an acoustooptical Q-switch is dominated not by electronic switching time but by the transit time of the sound wave across the beam diameter. Because the ultrasound is traveling at an acoustic velocity v in the block, which is typically of the range of 5 mm/μs, the transit time of the sound wave across the beam diameter D is D/v and will be of the order of 200 ns/mm of optical beam diameter. This time is short compared to the giant pulse evolution time in many laser systems, but may be too long for some high-gain lasers. The low-gain characteristics of cw-pumped solid-state lasers do not require very high contrast for Q-switching but do demand exceptionally low insertion loss. Since the best optical-quality fused silica with antireflection coatings can be used as the active medium in the acoustooptical Q-switch, the overall insertion loss of the inactive Q-switch can be reduced to less than 0.5 % per pass. The low insertion loss of the acoustooptic Q-switch offers the convenience of converting from Q-switched to cw operation simply by removing the rf drive power. Performance data of an acoustooptic Q-switched Nd : YAG laser are shown in Fig. 8.8.

8.1.5 Dye Q-Switch

The operation of passively Q-switched lasers has been of intense interest since their successful operation in 1964 [8.77–79]. The dye Q-switch is a very simple device. It consists of a dye cell which is placed inside the optical resonator, preferably between the laser medium and the rear mirror, as shown in Fig. 8.26a. The dye initially absorbs the laser rod fluorescent emission to the degree that the rear reflector is optically isolated from the remainder of the laser cavity. When the dye suddenly bleaches, the laser radiation can reach the rear reflector and laser oscillation occurs.

The generic name for the type of material which is used in the dye cell is saturable absorber. In these liquids the absorption coefficient decreases readily with increasing light intensity, as shown in Fig. 8.26b. Thus, the material becomes more transparent as the light becomes more intense. The increase in transparency of the dye is also frequently called “bleaching” of the dye. The decrease of the absorption coefficient for a saturable absorber or dye is related to the incident light intensity in the following way:

$$\gamma = \frac{\gamma_0}{1-(I/I_s)} \quad , \qquad (8.42)$$

where γ_0 is the absorption coefficient at zero intensity, and I_s is the saturation

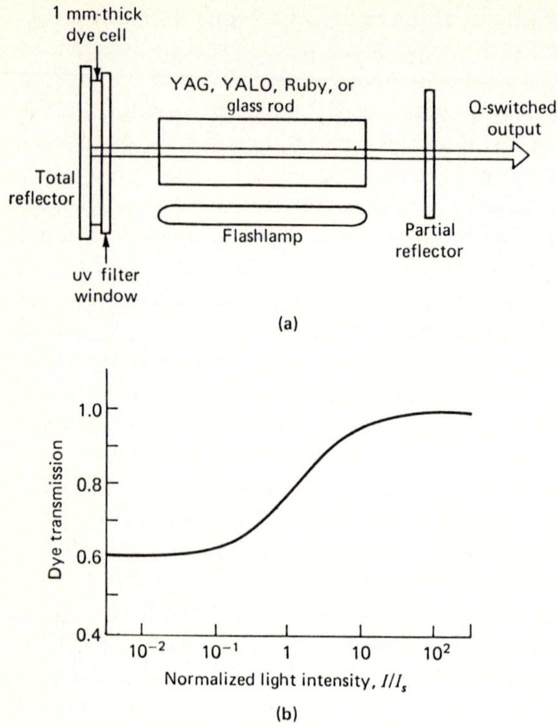

Fig. 8.26. **(a)** Laser system Q-switched with a saturable absorber. **(b)** Nonlinear transmission of a saturable absorber versus light intensity in units of the saturation intensity I_s of the dye

intensity, the intensity at which γ decreases by one-half. The bleaching process in a saturable absorber is based on saturation of a spectral transition. The dye molecules in the solution absorb photons and are transferred to an excited state from which they return at the end of an average lifetime τ_S to the ground level. When the dye is inserted into the laser cavity it will look opaque to the laser radiation until the photon flux is large enough to depopulate the ground level. If a sufficient number of molecules are excited, the dye becomes completely transparent to the laser radiation, a situation which is similar to a three-level laser material pumped to a zero inversion level. The saturation intensity I_S of an optically thin saturable absorber can be expressed as

$$I_S = \frac{h\nu}{\sigma_S \tau_S} \quad , \tag{8.43}$$

where σ_S is the cross section of the absorbing centers, and their effective lifetime in an excited state is τ_S. The response time of a saturable absorber to a step function in light intensity of amplitude I is

$$\tau_R = \frac{\tau_S}{1 + I/I_S} \quad . \tag{8.44}$$

From these expressions, we see that if a saturable absorber is to be appreciably bleached at reasonable light levels (between a few kilowatts and a few megawatts per square centimeter), and at the same time respond quickly, it must have a large cross section and a reasonably short lifetime. For this reason, organic dyes have been the logical choice for saturable absorbers, particularly cryptocyanine [8.78, 79], chlorophyll [8.80], and many phthalocyanines [8.78, 81] for Q-switching ruby lasers. Cryptocyanine with methanol as a solvent has an absorption bandwidth of 370 Å peaked at 7040 Å [8.78]. As mentioned before, for a dye to be effective for Q-switching, the dye cross section σ_s must be much larger than the laser cross section. Cryptocyanine [8.78] has a value of $\sigma_s = 8.1 \times 10^{-16}\,\mathrm{cm}^2$, as compared to the absorption cross section of ruby which is $\sigma = 1.2 \times 10^{-20}\,\mathrm{cm}^2$. The saturation power density of cryptocyanine in methanol is about $I_s = 5\,\mathrm{MW/cm}^2$. The same dye in acetonitrile has yielded quite satisfactory results at power levels up to gigawatts per square centimeter in cells between 1 and 2 mm in thickness. At low peak power levels, less than a few megawatts per square centimeter, chloroaluminum phthalocyanine is sometimes used because of its low saturation density of approximately $0.1\,\mathrm{MW/cm}^2$ [8.81].

Although a wide variety of dyes for use at 6943 Å have been studied, only a few dyes have been reported for use with neodymium lasers. They are manufactured by the Eastman Kodak Company and designated as Q-switch solutions #9860, 9740, 19 152, and 14 015.

The saturation power density for dyes diluted with 1,2-dichloroethane or chlorobenzene is of the order of $50\,\mathrm{MW/cm}^2$. Eastman 14 015, with 1,2-dichloroethane as a solvent, appears to be more stable than the first two dyes mentioned above.

Eastman Kodak 14 015 (licensed by IBM Corp) contains transition metals and relaxes more slowly compared to the other dyes in order to avoid mode-locking effects.

The temporal and spectral characteristics of the output pulse of a dye-switched laser depend strongly on the dye relaxation time. If the dye has a very short relaxation time and can follow fast oscillations in the intensity, it will attempt to mode-lock the pulse instead of Q-switching it. In this case high peak power pulses in the subnanosecond range are obtained. If, on the other hand, the dye's relaxation time is of the order of a cavity transit time or longer, it will give very narrow spectral emission. Dyes which exhibit very short overall relaxation times are cryptocyanine and Kodak's #9860 and #9740 solutions. Dyes which relax more slowly usually contain transition metals such as, for example, Kodak 14 015.

The unsaturated transmission of typical Q-switches is of the order of 40 to 60 %. The dye concentration or the path length may be varied to achieve various transmissions, thus changing the laser output power and pulse width. If it is convenient to measure experimentally the single-pass gain, $G = \exp(g_0 l)$, of the laser medium when it is fully pumped, then the dye cell transmission T_s is given by

$$T_s \geq \frac{1}{G(R_1)^{1/2}} \quad , \tag{8.45}$$

where R_1 is the reflectivity of the output mirror.

Since cw-pumped solid-state lasers will not generally be able to saturate or bleach the dye, the operation of the dye Q-switch is practically limited to pulsed solid-state lasers.

There are a number of distinct disadvantages associated with the passive saturable Q-switch:

1. *Large jitter.* The time between the triggering of the flashlamp and the emission of a Q-switched pulse is associated with a jitter which is typically of the order of 10 to 100 μs.
2. *Degradation of the dye.* Damage to a dye Q-switch can happen two ways. First, ultraviolet light from the flashlamp pumping the laser, particularly below 3500 Å, will break down the long-chain dye molecules. A simple solution is to place a uv filter with sufficient absorption to reduce this kind of degradation in front of the dye cell. Howeve, dye Q-switches do require a certain amount of maintenance, since the optical density of the dyes decreases even while the dyes are kept in the dark. Second, the dye solvents change their index of refraction with an increase in light intensity, and the core of the dye through which the laser light passes has a higher index of refraction than the outside. Hence, a lens effect, called self-trapping, occurs which can greatly increase the light power density in the dye. This effect can be minimized by working with a fairly thin dye cell ($\sim$1 mm).
3. *Absorption losses.* Residual absorption in the dye cell due to incomplete bleaching or absorption at the laser wavelength originating in an excited state of the saturable absorber results in a lower energy output than would be obtained from either a rotating mirror or electrooptic shutter Q-switch.

Advantages of passive dye Q-switches include economy, simplicity of operation, and the emission of the output pulse in a narrow linewidth [8.81, 82]. A typical linewidth of a dye Q-switched ruby laser is of the order of 0.02 Å.

Passive dye Q-switches usually consist of two glass windows separated by an annular spacer. Common dye cells have path lengths ranging from 1 to 10 mm. The windows are antireflection-coated to minimize losses. Very often the 100 % reflective rear mirror is made part of the dye cell, in which case the coating is immersed in the dye solution. Dye Q-switches are either simple cells containing a sealed fill hole for insertion of the dye or they contain a closed-cycle liquid loop. The latter systems consist of a dye cell, a liquid pump, and a large-volume reservoir. The pump and nonglass components of the liquid loop must be carefully selected for use with the corrosive solvents common to passive dyes. Compared to static dye cells, considerably better system performance is achieved by circulating the dye. Because fresh dye is circulated into the optical cavity for successive shots, flowing dye-cell Q-switches have a much improved shot-to-shot reproducibility and are capable of higher-repetition-rate operation. Furthermore, because of the large volume of dye employed in flowing systems, maintenance intervals, which include density adjustments of the dye, are greatly increased.

The particular dye concentration will depend on the lamp input energy, resonator length, and ambient temperature. For example, if the lamp input is raised, the dye concentration should be increased, and vice versa. Adjustment of the passive dye to the correct optical density is usually performed by observing the laser output pulse with a fast photodiode connected to an oscilloscope. If one starts with a very low dye concentration which is slowly increased, then one observes first conventional lasing which changes to multiple Q-switch pulses until at the optimum concentration a single Q-switch pulse is obtained. If the concentration is further increased, the laser energy will decrease and the oscillator will finally cease to lase.

A variation of the dye Q-switch is the plastic Q-switch. These Q-switches, made from transparent plastic material, are impregnated with dye solutions [8.83]. These Q-switches, which are most suitable for giant pulses up to 1 to 3 MW, can be used repeatably before damage occurs. Commercially available Q-switches have transmittance values ranging from 50 to 70 % [8.83]. Low maintenance transmittance combined with simplicity, space and weight savings have resulted in the application of plastic Q-switches in certain low-repetition-rate military and scientific systems.

Reliable Q-switch operation of a Nd:YAG laser operating at 20 Hz and producing 190 mJ for more than 10^6 shots has been reported in [8.84]. The plastic Q-switch consisted of a Kodak polyester sheet (#15 064) glued between two glass or sapphire optical surfaces in order to better dissipate the absorbed energy and minimize the distortions due to its poor optical quality. *Michelangeli* et al. [8.84] reported that the best performance was achieved with a configuration as shown in Fig. 8.27. The design employs a polarizer-output-coupler cavity. By rotating the quarter-wave plate it is possible to vary the output coupler reflectivity and optimize the laser intensity impinging on the plastic Q-switch. Also in the configuration shown, the laser beam passes through the saturable absorber four times for each round trip. This allowed the use of a plastic dye switch with a low optical density (OD= 0.42). The Q-switch pulse had a width of 6 ns and the energy mentioned above was achieved with a lamp input of 14 J.

Recently a new type of passive Q-switch emerged. Researchers in several countries have discovered that F_2^- color centers in LiF can achieve Q-switched Nd:YAG laser action [8.85–87]. The color centers are induced in the crystal by irradiation with gamma sources or electron beams. Repeatable Q-switch performance for over 10^6 shots at 20 Hz at room temperature has been reported

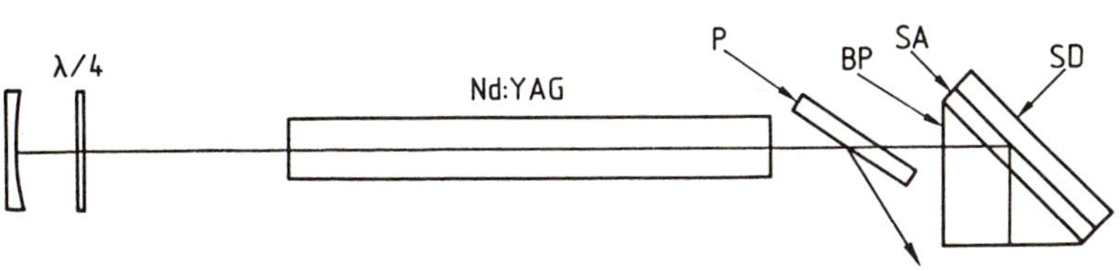

Fig. 8.27. Schematics of the laser cavity configurations (*SA:* polyester saturable absorber, *P:* Brewster-type dielectric polarizer, *BP:* BK7 glass prism having a total reflecting surface, *SD:* sapphire disk having a total reflecting surface)

[8.87]. Passive Q-switching was observed using a 5 cm long undoped F_2:LiF crystal in a Nd:YAG oscillator. The optical density was 0.37 at low intensities, and the Q-switch bleached at intensities in excess of $0.5\,\mathrm{MW/cm^2}$. At a flashlamp input of 11 J an output of 21 mJ was obtained in a 30 ns long pulse.

8.2 Pulse-Transmission Mode Q-Switches

A means for generating extremely short Q-switched laser pulses involves Q-switching the laser with 100 % mirrors on both ends of the cavity and then, at the peak of the circulating power, rapidly switching the output mirror from 100 to 0 % reflection. This leads to a rapid dumping of the entire optical energy from within the cavity. This particular mode of operation, first proposed by *Vuylsteke* [8.88], is termed pulse-transmission mode (PTM). One of the advantages of this technique is the production of Q-switched pulses whose width is primarily a function of the oscillator cavity length, rather than the gain characteristics of the laser cavity as in the pulse-reflection mode (PRM). Specifically, the laser pulse width at the half-power points will be equivalent to the round-trip transit time in the cavity, with the condition that the Q-switch employed be switched within this same time period. Thus, based on allowable cavity dimensions, pulse widths in the range of 2 to 5 ns are feasible for oscillators whose typical PRM pulse widths are of the order of 10 to 20 ns.

Figure 8.28 shows the optical layout of a ruby oscillator employed to generate short pulses by the pulse-transmission mode. We will explain the operation of the system by assuming that the ruby *c* plane is perpendicular to the plane of the paper. When the flashlamp is fired, the horizontally polarized ruby fluorescence is transmitted by the thin-film or calcite polarizer, thereby preventing regeneration. Upon reaching peak-energy storage in the ruby, the Pockels cell is biased to its half-wave retardation voltage. The resulting vertically polarized

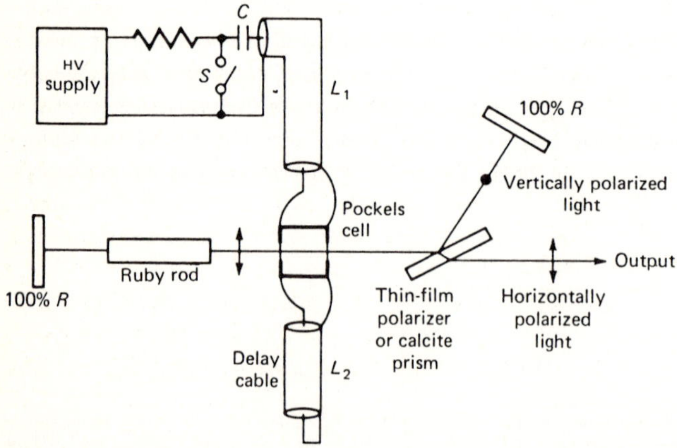

Fig. 8.28. Optical layout of a ruby laser employing PTM Q-switching

light is reflected by the polarizer to the off-axis mirror, and regeneration occurs in the cavity. When the giant pulse power in the cavity reaches its peak value, the bias is removed from the Pockels cell in a time period of about 2 ns. The cavity energy then literally drains out of the cavity in the time required for the radiation to travel one round trip in the optical cavity. The combination of the polarizer, Pockels cell, and 100 % mirror comprises what amounts to a high-speed voltage-variable mirror whose reflectivity is changed from 0 during the pumping cycle to 100 % during the pulse buildup, and back to 0 during the cavity dumping phase. For this reason, the PTM technique is sometimes referred to as time variable reflector technique (TVR).

To illustrate the practical realization of a PTM oscillator, we will consider a typical ruby oscillator consisting of a 10-cm by 1-cm ruby rod, a Pockels cell, a thin-film polarizer, and two 99 % mirrors. If we assume a 75-cm-long cavity, we obtain a round-trip transit time of 5 ns. The ruby rod is pumped by a 1-ms-long flashlamp pulse, and the Pockels cell is switched the first time after about 0.8 ms, as would be the case in normal PRM Q-switching. The time delay between switching the Pockels cell and the occurrence of peak power in the cavity is typically 60 ns. In order to extract the stored cavity energy, the bias on the Pockels cell is reduced to zero after this time delay. This can be accomplished by means of the circuit shown in Fig. 8.28. In this arrangement, the Pockels cell is connected in-line between coaxial cables L_1 and L_2. Closing the switch S will discharge the capacitor C into the transmission line L_1. When the voltage pulse reaches the Pockels cell, the optical beam will experience a 90° polarization rotating and the Q-switch pulse will start to build up from noise. Assuming a perfect 50-Ω impedance of the Pockels cell, no reflection will occur at the cell and the voltage pulse will travel to the end of the shorted transmission line L_2. At that point the pulse will be reflected with a 180° phase shift. When the reflection reaches the cell, the voltage on the crystal will be zero. Therefore the length of cable L_2 determines for how long the voltage is applied to the Pockels cell.

The performance of a ruby oscillator having the above-mentioned system parameters is illustrated in Fig. 8.29. Shown is the power inside the resonator if the energy is not dumped (Fig. 8.29a). The measurement was made by monitoring the leakage radiation through one of the 99 % mirrors with a fast detector and oscilloscope. The peak power and energy inside the resonator were determined to be 190 MW and 1 J, respectively. Figure 8.29b shows the circulating power monitored through the same mirror if the energy is dumped and the system is operated as a PTM oscillator. The internal cavity power reaches almost the same value as in Fig. 8.29a, then falls in about 5 ns to almost zero. This shows that all but a small percentage of the available energy has been dumped from the cavity. Figure 8.29c reveals that the dumped pulse is triangular, with a 10 to 90 % rise time of 3.0 ns and a pulse width of 5.3 ns. This width compares exactly with the cavity round-trip transit time within experimental error. The total energy of this pulse was measured to be 0.75 J. The peak power of this pulse is thus 1.4×10^8 W. The rise time of the output pulse is determined by the switching speed of the hydrogen thyratron which was used to discharge

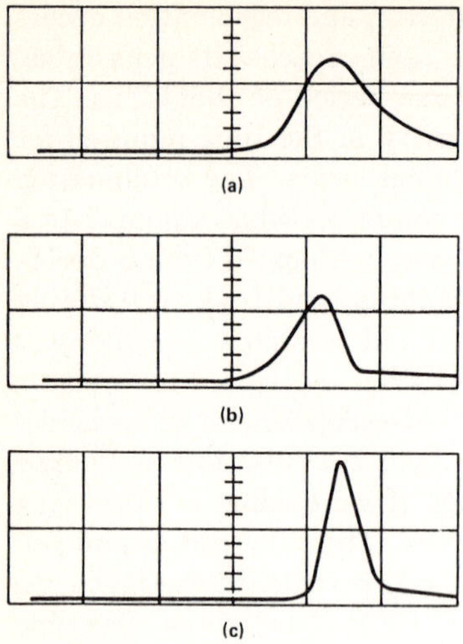

Fig. 8.29a–c. Performance of a ruby oscillator with a PTM switch. Photo scales 10 ns/div. **(a)** Internal cavity power at 99 % mirror without dumping. Pulse rise time 7 ns, pulse width 12 ns, peak power 190 MW. **(b)** Internal power, with cavity dumping, drops from 180 MW to almost zero in 5 ns. **(c)** Power dumped from cavity; rise time 3.0 ns, pulse width 5.3 ns, energy 0.74 J, and peak power 140 MW

capacitor C. The experimental data reveal that 75 % of the stored energy was extracted from the resonator.

In practical situations the design of PTM lasers is completely dominated by the requirement of keeping the power density within the cavity below the damage level. The usual calcite polarizer, being the component with the lowest damage threshold, has been replaced in contemporary oscillators by highly damage-resistant thin-film polarizers. These components permit PTM oscillators to operate at power densities up to $300\,\mathrm{MW/cm^2}$.

It is not necessary in a PTM oscillator to use the same Pockels cell for both the Q-switch initiation and cavity dumping. Earlier systems employed two Pockels cells for these functions [8.89–91]. Also, instead of a fixed-delay transmission line a more precise synchronization between the peak power in the cavity and the switching of the Pockels cell can be achieved if the cavity radiation is monitored by a detector mounted behind one of the 99 % mirrors [8.92]. Other variations of PTM oscillators are described in [8.93–95]. PTM systems can be built with any solid-state laser; for example, Nd : glass oscillators have produced 3-ns pulses with energies up to 180 mJ [8.95]. Cavity dumping is also possible with cw-pumped lasers; this will be discussed next.

Cavity Dumping of cw-Pumped Lasers

Cavity dumping can be compared with the Q-switching of a continuously pumped laser. In both cases energy is assumed to be discharged from the laser in the form of a repetitive train of light pulses. However, energy accumulation and storage between output pulses are primarily in the optical field for cavity dumping, and primarily in the atomic population inversion for Q-switching.

The finite buildup time of the field inside the laser cavity and the time required to repump the inversion set an upper limit to the repetition rate available from Q-switched lasers. This maximum value of repetition rate for Q-switched

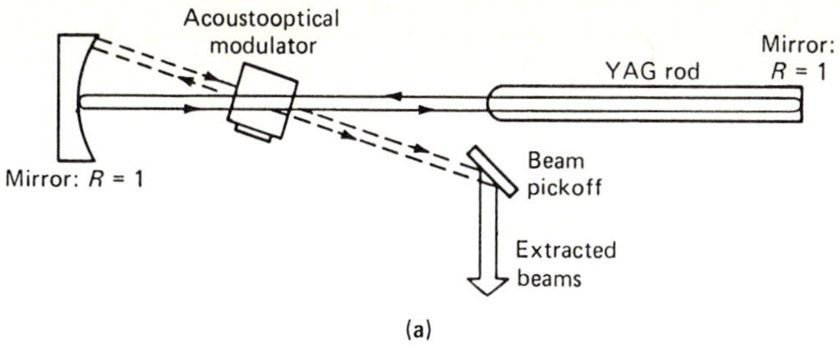

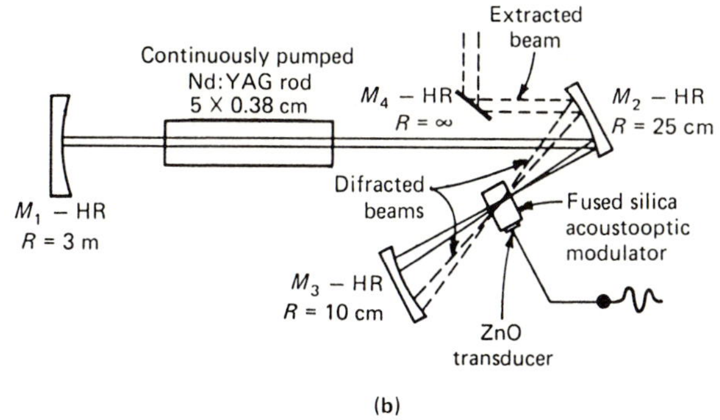

Fig. 8.30. Common arrangements for cavity dumping of cw-pumped solid-state lasers. The broken lines indicate the beams which are diffracted by the modulator

Nd:YAG lasers is of the order of 50 to 100 kHz. Cavity dumping of continuously pumped lasers is a way to obtain pulsed output at higher repetition rates than are available by Q-switching. Repetition rates from 125 kHz up to several megahertz for cavity dumping were achieved with cw-pumped Nd:YAG lasers [8.96, 97].

Figure 8.30 shows two common arrangements employed for cavity dumping of cw-pumped solid-state lasers [8.96, 97]. Essentially all systems of this type employ acoustooptic modulators as the switching element. In order to obtain fast switching action, the incident beam must be focused to a narrow waist inside the modulator. The two oscillator designs differ in the way the optical beam is focused into the modulator.

In Fig. 8.30a the modulator is located at a beam waist created by a concave mirror and by the thermal lens properties of the laser rod. The acoustic wave in the fused silica causes Bragg scattering of the forward- and backward-traveling light beam in the resonator. The two diffracted beams which are obtained from the cavity-dumped oscillator are initially traveling in opposite directions, therefore their frequencies are shifted to a value of $\omega + \Omega$ and $\omega - \Omega$, where ω is the frequency of the incident beam and Ω is the frequency of the acoustic wave [8.98, 99]. The two diffracted beams are extracted from the cavity as a single beam and deflected out of the system by a mirror. In Fig. 8.30b the cavity is formed by three high-reflectivity mirrors M_1, M_2, and M_3. The mirror curvature and the distance between M_2 and M_3 are chosen such that the light beam between M_2 and M_3 is focused to a small diameter at the center of curvature of M_3. The modulator is inserted at the waist of the optical beam.

Acoustooptic modulators employed for cavity dumping differ from their counterparts used in Q-switch applications in several respects:

1. Compared to Q-switching, the cavity-dump mode requires much faster switching speeds. The rise time in an acoustooptic modulator is approximately given by the beam diameter divided by the velocity of the acoustic wave. In order to obtain rise times around 5 ns, a value which is required for efficient cavity dumping, the incident beam must be focused to a diameter of approximately 50 μm.
2. For efficient operation in the cavity-dump mode, it is important that esentially all the circulating power be diffracted into the first diffraction order. In a Bragg device the diffraction efficiency increases with the rf carrier frequency, therefore modulators employed in cavity dumpers operate at considerably higher frequencies, i.e., 200 to 500 MHz as compared to acoustooptic Q-switches.
3. In order to generate an output pulse in the cavity-dump mode, a short rf pulse is applied to the modulator, whereas in an acoustooptic Q-switch the rf carrier is turned off for the generation of an output pulse.
4. The cavity is never kept below threshold condition as in the Q-switched mode of operation. If the cavity is dumped of all its energy, the field has to build itself from the noise level. Repetitive cavity dumping was observed to become unstable in this case. If the repetition rate is lowered, the laser material is pumped higher above threshold between pulses, therefore, according to the theory of *Wagner* and *Lengyel* [8.3], a larger fraction of the stored energy is extracted from the system. The lower limit of the dumping repetition rate is reached when the internal laser energy decreases to one photon immediately after dumping. The upper limit of the cavity dumping repetition rate is set by the switching speed of the modulator. Repetition rates as high as 10 MHz have been reported. From a cw-pumped Nd:YAG laser capable of 10 W of cw power, peak powers of 570 W with a pulse duration of 25 ns have been obtained at a 2 MHz repetition rate [8.100]. For high-data-rate communications systems, cavity dumping of cw-pumped lasers is sometimes combined with mode locking [8.101, 102].

8.3 Optical Gates External to Resonator

There are three types of frequently occurring requirements in pulsed laser technology which call for an optical transmission gate. One requirement is to select a single pulse from a train of mode-locked pulses, generally for further optical amplification in laser amplifiers. Such pulses are typically several tens of picoseconds wide, and successive pulses in the train can be adjusted between a 5- to 10-ns interval. The envelope of such a train is roughly Gaussian. The gate, sometimes called a "pulse extraction switch" or shutter, when used in this manner, is set to threshold at an energy level corresponding to the energy achieved one or two pulses preceding the maximum, which then triggers the gate long enough to pass the next pulse.

The second requirement is to transmit a small portion, several nanoseconds wide, of a Q-switch laser pulse which is typically several tens of nanoseconds wide. The gate, when used in this manner, is frequently called a "slicer" or "shutter".

The third requirement is to block the optical path at various points within a chain of high-power laser amplifiers in the absence of the laser pulse. This is often necessary to prevent self-oscillation of an amplifier chain due to reflections off the end face of each stage and to protect the optical components of the laser from light reflected back from a target. Reflection of laser light occurs when, for example, laser radiation is focused on a solid target to produce a plasma.

Electrooptic Shutters

In contemporary laser systems the selection of a single pulse in a mode-locked pulse train or the narrowing of the width of a Q-switched pulse is accomplished by means of a fast-switching Pockels cell. This device is inserted between the laser oscillator and the first preamplifier. The principal problem is to synchronize the opening of the shutter with the occurrence of the laser pulse. This problem is solved by either electronically or optically triggering the Pockels cell. In the former case a delay cable between the Q-switch and the external gate provides a fixed delay between these two devices. In the second case, the Pockels cell is triggered by the laser pulse itself, either by a fast detector or by a laser-triggered spark gap.

The schematic of a typical optical gate employing a Pockels cell as a shutter and an optically triggered spark gap driver is shown in Fig. 8.31. The system consists of a spark gap assembly, a Pockels cell, a Glan-Thompson polarizer and analyzer, a high-voltage power supply, and a shorted coaxial transmission line which determines the pulse duration. The optical gate shown in Fig. 8.31 can perform either of the two functions mentioned before, i.e., tailoring the pulse emitted by a Q-switched laser in order to adapt its characteristics to specific applications requiring, for example, short rise time or short pulse duration, or selection of a single picosecond pulse from a train of pulses.

We will explain the function of the spark gap triggered gate shown in Fig. 8.31 by considering a semipolarized train of pulses produced by a mode-locked oscillator. This train is incident upon the first Glan-Thompson polarizer, which increases the degree of polarization to approximately 100 : 1. The transmitted part of the pulse passes through the KD^*P crystal unaffected and is then rejected by the second Glan-Thompson polarizer. The rejected energy is focused by a convex lens onto one electrode of the spark gap. By adjustment of the spark gap separation and gas fill pressure and suitable optical filtering, it is possible to arrange that one of the higher-energy mode-locked pulses causes

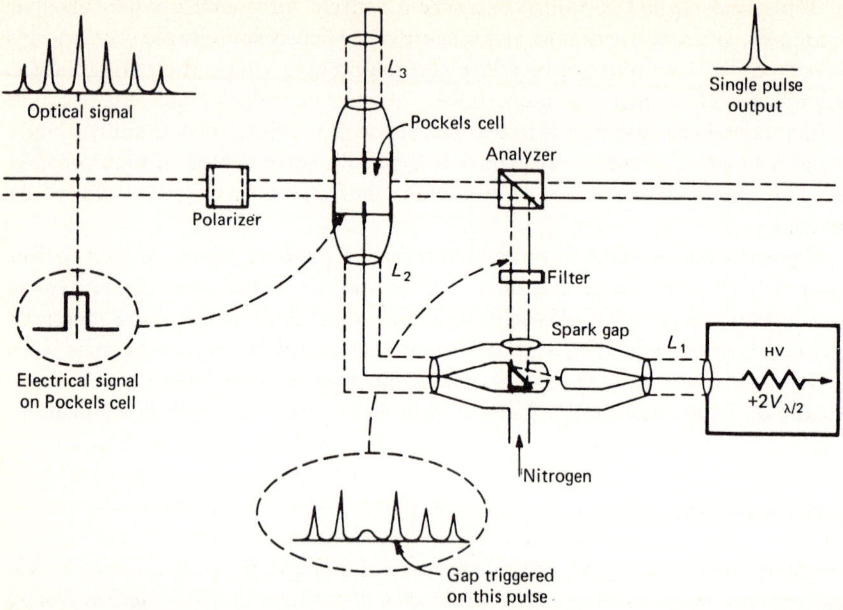

Fig. 8.31. Experimental arrangement for extracting subnanosecond pulses with a fast Pockels cell. Polarizer and analyzer are crossed 90° to each other, so that the analyzer passes light only when the Pockels cell receives a 10-kV pulse from the spark gap (which is triggered by the laser pulse) and briefly rotates a component of the beam by 90°

the gap to avalanche and conduct the current from the high-voltage supply to the electrode of the KD*P crystal. As long as this bias remains on the crystal, the plane of polarization of radiation transmitted through the polarizer will be rotated by 90°. The analyzer is oriented to transmit this polarization, thus allowing radiation to be transmitted from the system. A reflection cable in the electrical circuit then turns the high-voltage pulse off after the proper time interval, returning the system to its original state. This can be adjusted to occur from 3 to 15 ns after the spark gap breakdown occurs, thus allowing only one pulse in the mode-locked train to be transmitted. The initial time delay between the laser trigger pulse and the opening of the cell is determined mainly by the length of the coaxial cable L_2, whereas the opening time of the switch is determined by the length of the coaxial cable L_3. The coaxial line L_1 is initially charged from a voltage source at twice the half-wave retardation voltage $V_{1/2}$. After triggering of the spark gap, a voltage step of amplitude $V_{1/2}$ starts traveling into the line L_2 connected to the Pockels cell. The shutter is therefore switched on, after a time delay which can be adjusted through the line-matched Pockels cell, the voltage step propagated into the transmission line L_3 is back reflected at its shorted end with a sign change, and shuts off the cell after a time delay proportional to the length of L_3. To modify the time duration of the pulse, it is therefore sufficient to modify the length of the transmission line L_3.

By inserting an appropriate neutral-density filter in front of the spark gap, the switch can be used as a slicer whose transmission time is controlled by the length of the transmission line L_3. The optical filters are adjusted so that the gap breaks down somewhere on the leading edge of the 20-ns giant pulse incident on the focused spark gap.

Laser-triggered spark gaps are usually constructed with two electrodes: one solid with a tungsten insert, the other sectioned and hollow, containing a small prism or a lens. In most designs one of the electrodes can be screwed in or out to adjust the gap spacing. The spark gaps are pressurized with dry nitrogen at pressures ranging from 2 to 15 atm. The gap separation and nitrogen pressure are adjusted so that the proper voltage is applied to the Pockels cell when the gap breaks down. Typical Pockels cell voltages range from 8 to 15 kV. The power supply must deliver twice that voltage. The switch shown in Fig. 8.31 contains two Pockels cells operated optically in series. This allows an improvement in dynamic contrast and the use of a lower bias voltage. Typical rise times of fast-switching optical gates are of the order of 1 to 3 ns. A small delay and low jitter of the laser-triggered spark gap requires the use of high gas pressures and small gap spacing. The construction and operation characteristics of pressurized laser-triggered spark gaps are described in [8.103–110]. Optical gates which employ photodiodes instead of spark gaps to trigger the Pockels cell switch are described in [8.111, 112].

Isolators

The purpose of isolators is to avoid the initiation of parasitic oscillations in an amplifier chain and to protect the preceding amplifiers from the energy reflected from a target. Ideally, an isolator should be a gate which opens only during the passage of the pulse. Devices which are employed for this purpose include dye cells, electrooptic shutters of the type described before, and Faraday rotators [8.113–117]. Dye cells are the least expensive and simplest of the isolators mentioned above. However, their extinction ratio is only of the order of 2 : 1 or 3 : 1. Electrooptic shutters, because of their limited apertures, are usually employed in the preamplifier section of large laser systems. The main function of dye cells and electrooptic shutters placed between amplifier stages is to prevent self-oscillations of a high-gain amplifier chain. For large Nd : glass amplifier systems, a typical rule of thumb is to place a dye cell behind each amplifier stage and an electrooptic isolator at points where the small-signal gain exceeds 30 dB. Protection of the amplifier chain from laser radiation reflected back from a target is achieved in large Nd : glass laser systems by placing a Faraday rotator between the last amplifier and the target. Since these devices can be built with large apertures from high-damage-threshold glasses, Faraday isolators can be placed before or after the last stage of amplification where the power transmitted is the greatest.

The Faraday isolator transmits light in the forward direction and attenuates it strongly in the reverse direction. This nonreciprocal action, produced by means of the Faraday effect, is observed when a plane-polarized beam of light is passed through a piece of glass subject to a strong magnetic field applied

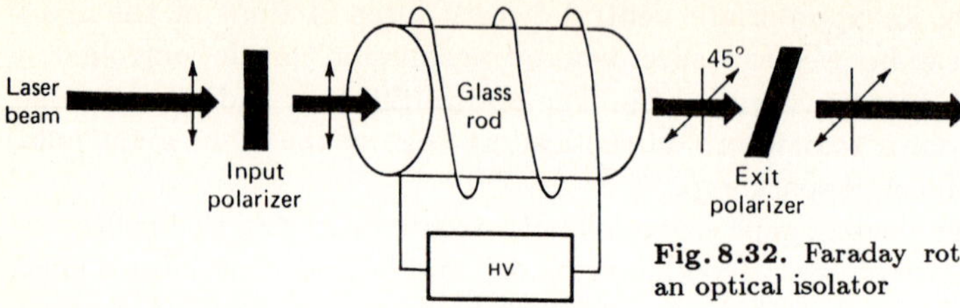

Fig. 8.32. Faraday rotator employed as an optical isolator

parallel to the direction of the light beam. Under these conditions, rotation of the plane of polarization of the beam takes place. The angle of rotation is given by the relation

$$\Theta = VH_x l \quad , \tag{8.46}$$

where V is the Verdet constant [arc min/Gauss cm], H_x is the component of the magnetic field in the direction of light propagation [Gauss], and l is the length of the light path in glass [cm].

Figure 8.32 shows the use of the Faraday effect in an optical isolator. A polarizer allows a plane-polarized beam of monochromatic light to pass through a glass rod which is placed along the axis of a solenoid. The current in the solenoid produces a magnetic field sufficient to give the plane of polarization of the beam a rotation of 45°. A second polarizer is mounted with its plane of polarization at 45° relative to the first polarizer and, apart from absorption and losses in the glass rod and second polarizer, the plane-polarized light beam will emerge from the second polarizer unattenuated. A beam of light passing through the second polarizer in the reverse direction will experience rotation of its plane of polarization in the glass rod in a direction determined by the magnetic field direction. The beam emerging from the glass rod in the reverse direction will, therefore, be polarized at 90° to the transmission plane of the first polarizer and will be almost entirely extinguished, depending on the quality of the polarizers.

The types of glass used in laser systems must have a high Verdet's constant, a low absorption at the laser wavelength, and a high threshold value with respect to damage by intense laser radiation. Materials which are used in Faraday rotators include the glasses Hoya FR-4, FR-5, Owens Illinois EY-1, and Soveril B-1664. The Verdet constant for these glasses at 1.06 μm is between 0.08 and 0.007 arc min/G cm.

As an example of such a device we will briefly consider a commercially available unit manufactured by CILAS [8.115]. The Faraday isolator (rf 70) consists of a 64-mm-diameter by 135-mm-long glass rod in a solenoid magnet. A stack of eight glass plates in front and behind the glass rod serve as polarizer and analyzer. A 45° rotation of the polarization of the incident beam is achieved by discharging two 140 μF capacitors initially charged to 10 kV into the coil, and thereby generating a peak current of 2100 A. The ratio of forward to backward transmission is about 15 dB. Besides pulsed Faraday rotators, there are also cw systems in use which employ superconducting magnet coils operated at the temperature of liquid helium [8.116].

9. Mode Locking

As we have seen in the previous chapter, the minimum pulse width obtainable from a reflection-mode Q-switched laser is of the order of 10 ns because of the required pulse buildup time. With the pulse-transmission technique, the pulse width can be reduced to a minimum of 1 to 2 ns. The limitation here is the length of the cavity, which determines the pulse length. Ultrashort pulses with pulse widths in the picosecond regime are obtained from solid-state lasers by mode locking. Employing this technique, which phase-locks the longitudinal modes emitted by the laser, the pulse width is inversely related to the bandwidth of the laser emission.

In a free-running laser, both longitudinal and transverse modes oscillate simultaneously without fixed mode-to-mode amplitude and phase relationships. The resulting laser output is a sort of time-averaged statistical mean value. Restricting oscillation of the laser to the TEM_{00} mode for the moment, we recall from Chap. 5 that in a typical laser cavity there are perhaps a few hundred axial modes which fall within the frequency region where the gain of the laser medium exceeds the losses of the resonator.

In Fig. 9.1 the spectral and temporal structure of the radiation inside a laser cavity are shown for a non-mode-locked laser. In the frequency domain,

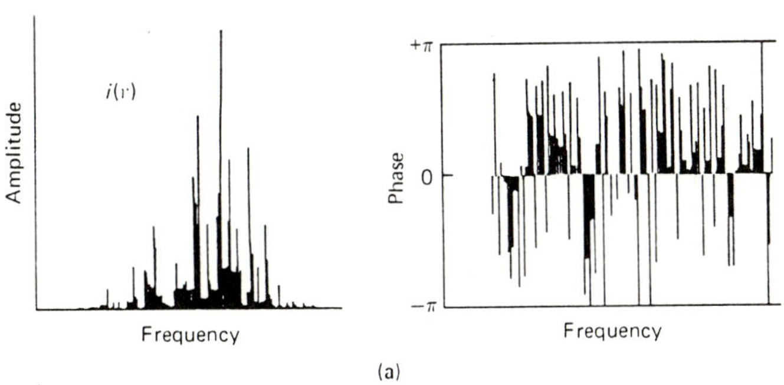

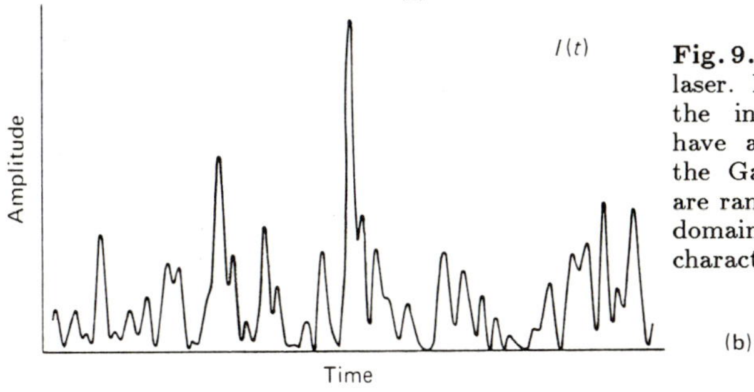

Fig. 9.1. Signal of a non-mode-locked laser. In the frequency domain **(a)** the intensities $i(\nu)$ of the modes have a Rayleigh distribution about the Gaussian mean and the phases are randomly distributed. In the time domain **(b)** the intensity has the characteristic of thermal noise

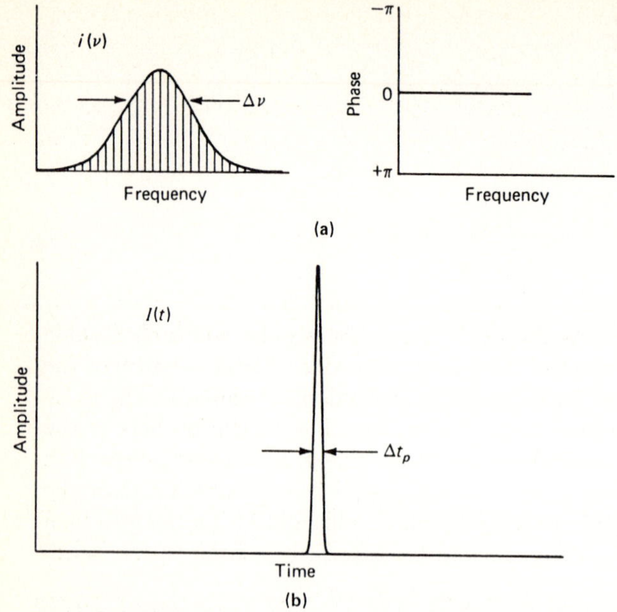

Fig. 9.2. Signal structure of an ideally mode-locked laser. The spectral intensities (**a**) have a Gaussian distribution, while the spectral phases are identically zero. In the time domain (**b**) the signal is a transform-limited Gaussian pulse

the radiation consists of a large number of discrete spectral lines spaced by the axial mode interval $c/2L$. Each mode oscillates independent of the others, and the phases are randomly distributed in the range $-\pi$ to $+\pi$. In the time domain, the field consists of an intensity distribution which has the characteristic of thermal noise.

If the oscillating modes are forced to maintain a fixed phase relationship to each other, the output as a function of time will vary in a well-defined manner. The laser is then said to be "mode-locked" or "phase-locked". Figure 9.2 shows the output signal of an ideally mode-locked laser. The spectral intensities have a Gaussian distribution, while the spectral phases are identical zero. In the time domain the signal is a single Gaussian pulse. As can be seen from this figure, mode locking corresponds to correlating the spectral amplitudes and phases. When all the initial randomness has been removed, the correlation of the modes is complete and the radiation is localized in space in the form of a single pulse.

Because the intensity profiles $I(t)$ shown in Figs. 9.1b and 2b circulate around inside the cavity with a repetition rate determined by the round-trip transit time, these signals will repeat themselves and appear in the laser output at a rate of $c/2L$. Therefore, mode locking results in a train of pulses whose repetition period is twice the cavity transit time.

If we make some simplifying assumptions, we can obtain a general idea about the pulse width and peak power of the mode-locked output pulses. The structure of an optical pulse is completely defined by a phase and an intensity. Whether these refer to time or to frequency is immaterial since, if the description in one domain is complete, the profiles in the other are obtained from the Fourier transform. However, there is no one-to-one correspondence between the two intensity profiles $I(t)$ and $i(\nu)$, since each depends not only on the other but also on the associated phase function. The only general relationship between the two is

$$\Delta\nu\Delta t \geq K \quad , \tag{9.1}$$

where Δt and $\Delta\nu$ are the full width of half-height of $I(t)$ and $i(\nu)$, respectively, and K is a constant of the order of unity. In particular, the shortest pulse obtainable for a given spectral bandwidth is said to be transform-limited; its duration is

$$t_\mathrm{p} = \frac{K}{\Delta\nu} \quad . \tag{9.2}$$

For mode-locking purposes $\Delta\nu$ corresponds to the gain bandwidth $\Delta\nu_\mathrm{L}$ of the laser, although sometimes in the literature $\Delta\nu$ is more loosely interpreted as being the atomic linewidth of the laser medium. The number of axial modes which are contained within the oscillating bandwidth is $N = \Delta\nu_\mathrm{L} t_\mathrm{R}$, where $t_\mathrm{R} = 2L/c$ is the round-trip time in the optical resonator. The width of the individual mode-locked pulses is therefore

$$t_\mathrm{p} \approx \frac{1}{\Delta\nu_\mathrm{L}} \approx \frac{t_\mathrm{R}}{N} \quad . \tag{9.3}$$

Equation (9.3) expresses the well-known result from Fourier's theorem that the narrower the pulse width t_p, the larger the bandwidth required to generate the pulse. From (9.3) follows also the interesting fact that the pulse width of the mode-locked pulses roughly equals the cavity round-trip time divided by the number of mode-locked pulses.

For simplicity, the spectral intensity in Fig. 9.2a was chosen to be a Gaussian function and hence the temporal profile has also a Gaussian distribution. The most general Gaussian optical pulse is given by

$$E(t) = \left(\frac{E_0}{2}\right) \exp\left(-\alpha t^2\right) \exp\left[\mathrm{j}(\omega_\mathrm{p} t + \beta t^2)\right] \quad . \tag{9.4}$$

The term α determines the Gaussian envelope of the pulse, and the term $\mathrm{j}\beta t$ is a linear frequency shift during the pulse (chirp). From (9.4) follows for the pulse width at the half-intensity points

$$t_\mathrm{p} = \left(\frac{2\ln 2}{\alpha}\right)^{1/2} \quad , \tag{9.5}$$

and a bandwidth again taken at the half-power points of the pulse spectrum [9.1]

$$\Delta\nu_\mathrm{p} = \frac{1}{\pi}\left[2\ln 2\left(\frac{\alpha^2+\beta^2}{\alpha}\right)\right]^{1/2} \quad . \tag{9.6}$$

Note how the frequency chirp contributes to the total bandwidth. The pulse width-bandwidth product is a parameter often used to characterize pulses. For Gaussian pulses, the pulse width-bandwidth product is given by

$$t_p \Delta\nu_p = \left(\frac{2\ln 2}{\pi}\right)\left[1+\left(\frac{\beta}{\alpha}\right)^2\right]^{1/2} . \tag{9.7}$$

For the important special case $\beta = 0$ (i.e., no frequency chirp), one obtains $K = t_p \Delta\nu_p \approx 0.441$.

Mode-locked lasers can have high peak powers, because the power contained in the entire output of the uncoupled laser is now contained within the more intense ultrafast pulses. From (9.3) follows for the ratio of pulse-on-to-pulse-off, i.e., the duty cycle, a value of $\Delta t_p/t_R = 1/N$, so that the peak power of the pulse is N times the average power resulting from incoherent phasing of the axial modes. For a Nd:glass laser, N can be as high as 10^4.

The generation of mode-locked pulses from a laser requires that the longitudinal modes be coupled together. This is achieved experimentally by placing inside the laser cavity either an externally driven loss or phase modulator, or a passive device whih exhibits saturable absorption.

Active mode-locking is the technique employed in cw-pumped lasers. A continuous train of mode-locked pulses at a relatively high repetition rate (several hundred megahertz) is obtained, as shown in Fig. 9.3. Saturable absorbers are the preferred way of obtaining mode-locked output from the pulse-pumped laser. Saturable dyes mode-lock and simultaneously Q-switch the laser output. Therefore, from a dye mode-locked laser a relatively short train of mode-locked pulses is generated, as illustrated in Fig. 9.4. The envelope of the train of pulses

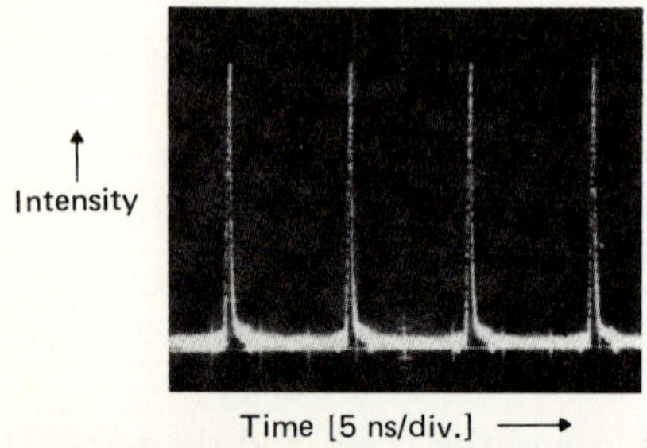

Fig. 9.3. Oscilloscope trace of a train of mode-locked pulses from an actively mode-locked cw-pumped Nd:YAG laser

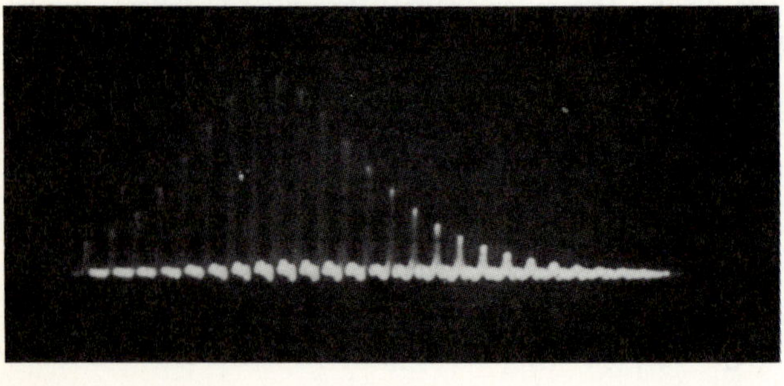

Fig. 9.4. Oscilloscope trace of a typical passively mode-locked Nd:glass laser. Horizontal 20 ns/div. Rise and fall times and pulse width are all detector-limited in this photograph and also in Fig. 9.3

has approximately the duration of a regular Q-switched pulse. Details of the generation of mode-locked pulses with active and passive devices will be discussed in the following sections. Exxcellent tutorial discussions of the work on mode-locked lasers with extensive references can be found in [9.2–7].

9.1 Passive Mode Locking

In 1965 the nonlinear absorption of saturable absorbers was first successfully used both to Q-switch and mode-lock solid-state lasers [9.8, 9]. Figure 9.5 shows various configurations of mode-locked oscillators whose three essential compo-

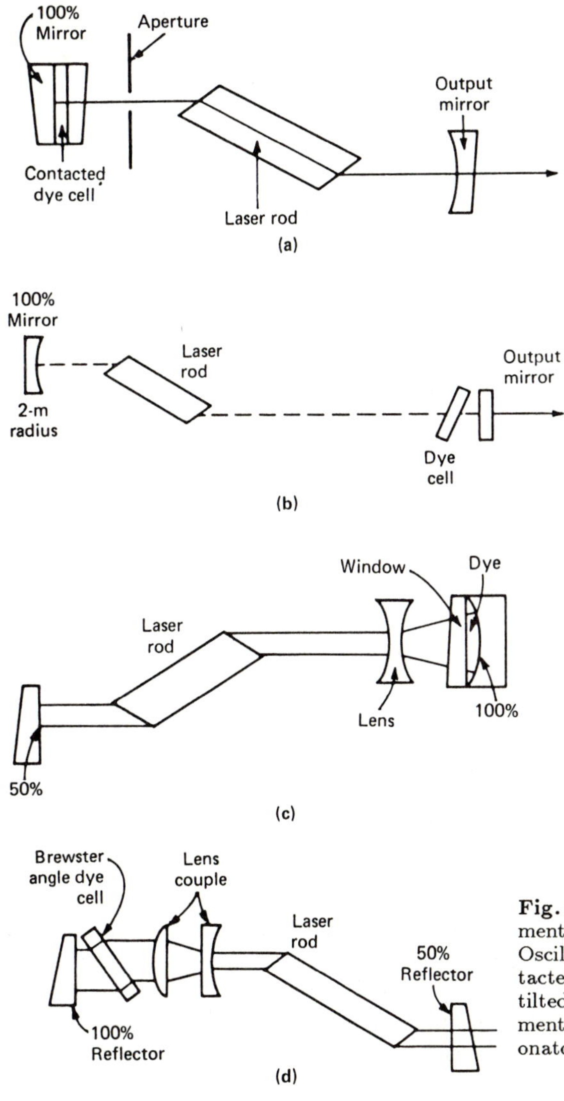

Fig. 9.5a–d. Experimental arrangements for passively mode-locked lasers. Oscillator cavity with optically contacted dye cell (a) and with dye cell tilted at Brewster angle (b) arrangements, including telescopes in the resonator, are shown in (c) and (d)

nents are the optical cavity, the laser rod, and the dye cell containing the nonlinear absorber. Concave mirrors reduce the stringent requirements of stability of the laser cavity and facilitate optical alignment. A small aperture discriminates against higher-order transverse cavity modes and achieves pure TEM_{00}-mode radiation. On the other hand, longitudinal mode selection due to etalon effects must be carefully avoided by employing Brewster angles at the laser rod and wedged plates for the dye cell and laser mirrors.

We will now discuss the pulse formation in a passively mode-locked laser. A computer simulation of the evolution of a mode-locked pulse train from noise is shown in Fig. 9.6 [9.10]. This figure shows the transformation of irregular pulses into a single mode-locked pulse. In Figs. 9.6a–c the noise-like fluctuations are linearly amplified, however, a smoothening and broadening of the pulse structure can be seen. In Figs. 9.6d–f the peak-to-peak excursions of the fluctuations have increased and, in particular, the amplitude of the strongest pulse has been selectively emphasized. In Fig. 9.6f the background pulses have been completely suppressed.

The successive stages in the development of mode-locked pulses illustrated in Fig. 9.6 have been experimentally observed in ruby and glass lasers [9.11–13]. In these experiments electrooptic streak cameras were employed which clearly showed the transformation of irregular fluctuating pulses into a periodic succession of mode-locked pulses. The understanding of passive mode locking in solid-state systems was greatly improved after a time-domain description of mode locking was introduced by *Letokhov* [9.14]. According to this model, which has been adopted by many researchers [9.4, 10–13, 15], the simultaneous Q-switched and mode-locked pulse evolution from noise can be divided into three stages.

Linear Amplification Stage. In the initial stage, population inversion of the laser transition and the transmission of the bleachable dye are not affected by the radiation. As a result, both the amplification and absorption processes can be considered to be linear. The intensity pattern is that of spontaneous emission with a spectral content about equal to the fluorescence bandwidth. Interference of the laser modes with random phase relations leads to fluctuations of the light intensity. The total number of fluctuations is large, being of the order of the number of cavity modes, but there is a small number of intensity peaks exceeding the average intensity significantly. The chaotic sequence of fluctuations shown in Fig. 9.6a represents the laser radiation in the early stage of pulse generation.

As time goes on, the gain increases above threshold and the noise-like signal is amplified. During the linear amplification a natural mode selection takes place because the frequency-dependent gain favors cavity modes in the center of the fluorescence line. As a result of the spectral narrowing caused by the amplification process, a smoothening and broadening of the amplitude fluctuations occurs, as shown in Figs. 9.6b and c.

The following numbers are typical for the linear amplification process of the Nd : glass, a large number of longitudinal modes is initially excited. For

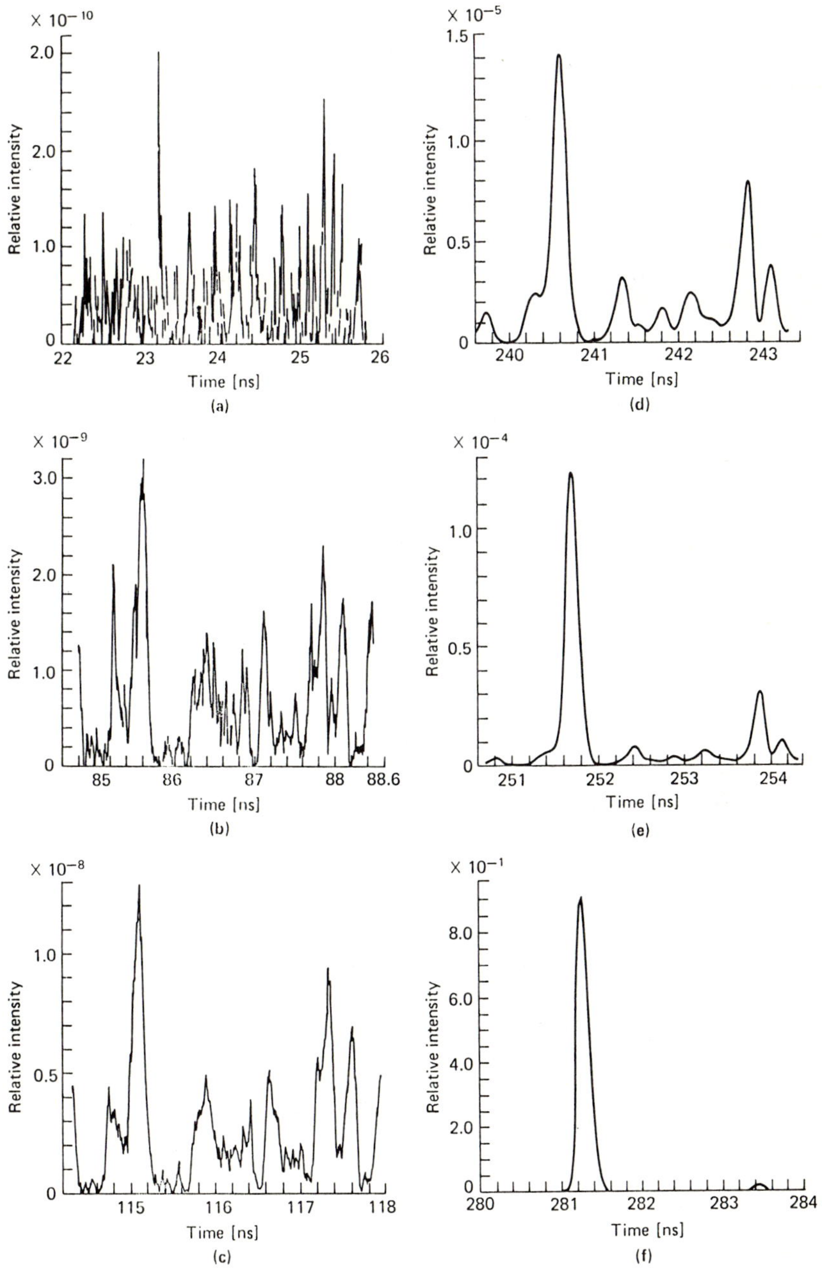

Fig. 9.6a–f. Computer simulation of the evaluation of a mode-locked pulse from noise. **(a–c)** regime of linear amplification and linear dye absorption, **(d–e)** nonlinear absorption in the dye cell, **(f)** regime of nonlinear amplification, dye completely bleached. [9.10]

a typical cavity length of $\sim 1\,\mathrm{m}$, one calculates $\sim 4 \times 10^4$ cavity modes in Nd:glass with $\delta\nu_F \approx 7500\,\mathrm{GHz}$. This bandwidth also implies a duration of the fluorescence peaks of $10^{-13}\,\mathrm{s}$ at the beginning of the linear amplification. Assuming a net gain of several percent per cavity round trip at the center of the laser line, the linear stage comprises $\sim 10^3$ cavity transits corresponding to a buildup time of $10^{-5}\,\mathrm{s}$. The light intensity rises by many orders of magnitude to approximately $10^7\,\mathrm{W/cm^2}$, and the fluorescence peaks increase in duration to about $10^{-11}\,\mathrm{s}$.

Nonlinear Absorption. In this second phase of pulse evolution, the gain is still linear but the absorption of the dye cell becomes nonlinear because the intensity peaks in the laser cavity approach values of the saturation intensity I_s of the dye (Fig. 8.22b). In the nonlinear regime of the mode-locked laser we note two significant processes acting together: first, there is a selection of one peak fluctuation or at least a small number. The most intense fluctuations which were built up during the linear amplification stage preferentially bleach the dye and grow quickly in intensity. The large number of smaller fluctuations, on the contrary, encounter larger absorption in the dye cell and are effectively suppressed. The second effect is a narrowing of the existing pulses in time, which broadens the frequency spectrum. The shapes of the pulses are affected by the nonlinearity of the dye because the wings of the pulse are more strongly absorbed than the peak. The second phase ends when the absorbing transition in the dye cell is completely saturated. Under favorable conditions the final transmission is close to one; i.e., the dye is transparent. The nonlinear action of the absorber at the intermediate power regime was illustrated in Figs. 9.6d and e.

From the foregoing considerations follows that a saturable absorber employed for mode locking must have (1) an absorption line at the laser wavelength, (2) a linewidth equal to or greater than the laser line width, and (3) a recovery time shorter than the round-trip time of a pulse circulating between the two mirrors forming the laser cavity. In Table 9.1 the saturation parameter and the recovery time are listed for four dyes commonly used in mode-locked lasers [9.5].

Nonlinear Amplification. The final phase of the pulse evolution occurs when the intensity is sufficiently high for complete saturation of the absorber transition to take place and for the amplification to be nonlinear. This is the regime

Table 9.1. Saturation density I_s and recovery time τ_D of various dyes employed for mode-locking

Dye	Eastman No. 9740	Eastman No. 9860	DDI	Cryptocyanine
I_s [W cm^{-2}]	4×10^7	5.6×10^7	$\sim 2 \times 10^7$	5×10^6
τ_D [ps]	8.3	9.3	14	22
Laser	Nd	Nd	Ruby	Ruby

of high peak power. During the nonlinear stage the pulse intensity quickly rises within ~ 50 cavity round trips to a value of several gigawatts per square centimeter. As was shown in Fig. 9.6f, at this point the background pulses have been almost completely suppressed. Successive passages of the high-intensity radiation pulse through the resonator result in a pulse train appearing at the laser output. Finally, the population inversion is depleted and the pulse decays. It would be expected that the nonlinear amplification would further broaden the spectrum and narrow the temporal width of the pulse. However, this does not happen, as can be seen by comparing Figs. 9.6e and f. This is due mostly to the onset of new processes, which cause the high-intensity pulse to evolve into a very complicated temporal and spectral structure.

The output pulses from a mode-locked laser are far more complex than would be predicted by a simple theory because at these high light intensities various nonlinear effects occur which strongly affect the generated pulse. The pulse-width measurement consistently yields a duration longer than the reciprocal of the bandwidth. *Treacy* [9.16] observed that pulses were chirped and, hence, could be compressed by a dispersive delay line outside the laser cavity. Subpicosecond structures were observed by several groups [9.7, 17–19], and more recently, several researchers reported the observation of spectral broadening and modulation that were typical of self-phase modulation [9.20–22].

It is generally believed now that during the high-intensity stage of pulse development, the characteristics of the output pulses are largely determined by the intensity-dependent index of refraction change in the laser rod and dye solvent. Under different conditions or at different times, self-phase modulation can produce either a positive frequency sweep or the breakup of a pulse into a burst with the character of random noise. As the pulse grows, the nonlinear index of refraction causes the more intense portions to experience longer optical path lengths, expanding the leading edge of the pulse and compressing the trailing edge to produce a positive frequency sweep. If there is limited self-focusing, self-phase modulation will continue and the spectral shifts will become significant compared to the laser gain profile. The pulse shape will be distorted by the reduced gain for the shifted components. Normal dispersion will then cause the different frequency components of the phase-modulated, distorted pulse to interfere. In Nd : glass, an irregular sequence of subpicosecond light bursts of total duration of 10 to 15 ps finally results. Measurements of individual pulses as a function of position within the train clearly showed the change of pulse properties as a function of the number of cavity round trips [9.23–25]. It was shown [9.25, 26] that pulses positioned several hundred nanoseconds before the maximum of the pulse train display optimum properties. Because of the small intensity (several $10^8\,\mathrm{W\,cm^{-2}}$) one finds frequency modulation and dispersion effects negligible.

The results of extensive experimental studies of the time and spectral structure of mode-locked laser radiation has been summarized in [9.25–29]. Mode-locked pulses are usually detected with the help of a fast photodiode and a traveling-wave oscilloscope. For example, the Tektronix model 519 traveling-wave oscilloscope has approximately a 1-GHz bandwidth. A combination of this oscilloscope and an ITT F4014 photodiode will have a rise time of about 0.5 ns. An oscilloscope display of the mode-locked laser output is an indispensible diagnostic tool for performing the alignment of the components, adjusting input power and dye concentration, etc., and checking the overall performance of the system such as number of mode-locked pulses, duration of the envelope, presence of background radiation, etc. However, the photodiode-oscilloscope combination does not provide adequate time resolution for direct measurement of the pulse duration of individual mode-locked pulses. Although image-tube streak cameras [9.30] and ultrahigh-speed cameras with Kerr cell shutters [9.31] having time resolutions of a few picoseconds have been built, so few now exist that most workers, instead, employ nonlinear optical effcts, such as harmonic generators [9.32, 33] and two or three photon absorption methods for pulsewidth measurements [9.34].

9.1.1 Design and Performance Characteristics of Passively Mode-Locked Solid-State Lasers

Most passively mode-locked lasers are similar in design to the oscillator configurations shown in Fig. 9.5. Usually, the oscillator is followed by an external optical gate which transmits only one pulse from the pulse train emitted by the laser. Despite the relatively simple construction of a passively mode-locked laser oscillator, the output will be very unpredictable unless dye concentration, optical pumping intensity, and resonator alignment are carefully adjusted. It is not uncommon to find that the average pulse duration from one train of picosecond pulses to the next changes significantly and that the pulse train envelopes are not reproducible [9.35]. For instance, one may find that, in ~ 10 shots, optimum mode locking occurred only once. With the proper saturable absorber and with a judiciously chosen optical system, the probability of obtaining clean mode-locked pulse trains which are free of subsidiary pulses is typically 0.6 to 0.7 in passively mode-locked solid-state lasers. With extreme care, which involves the optimization of each individual component, systems have been reported which consistently mode-locked 80 to 90 % of the time [9.36–38].

The design elements of particular interest in a mode-locked system are the resonator configuration, the type of dye and solvent used, the concentration and thickness of the dye, and the construction of the dye cell. In the following we will discuss these elements and their influence on the oscillator output. The performance parameters of prime concern are the pulse energy and width, the signal-to-noise ratio, and the shot-to-shot reproducibility.

Resonator Configuration. One major requirement in the resonator design of a mode-locked system is the complete elimination of reflections which can occur from components located between the two cavity mirrors. This is accomplished by employing laser rods with Brewster's angle at the ends, placing the dye cell at Brewster's angle in the resonator, and by using cavity mirrors which are wedged. Reflection from an optical surface which is parallel to the cavity mirrors will create a secondary resonator. The mode-locked pulse will be split into several pulses which will circulate inside the resonators with different round-trip times. The result is a very erratic output usually consisting of several superpositioned pulse trains or containing subsidiary pulses in the train [9.2, 35, 39]. With all optical surfaces inside the resonator, either at Brewster angle or antireflection-coated and tilted away from the resonator axis, coupled resonator structures can be avoided and the occurrence of satellite pulses is minimized. Similar attention must be paid to avoid back-reflection into the cavity from external components.

In order to reduce the number of reflective surfaces in the laser cavity and, therefore, minimize the possibility of secondary reflections, the dye cell and rear mirror are often combined. As shown in Fig. 9.5a, the 100 % mirror takes the place of the rear dye-cell window. A saturable absorber, where the dye is in contact with the cavity mirror, not only provides the most reliable mode-locking operation but also yields the shortest pulses [9.2, 40]. Typically, dye cells vary in length from 1 cm to 0.1 cm; however, it was found that a thin dye layer is preferable for mode locking [9.40, 41]. The resonator length is usually chosen between 1 and 1.5 m; in this case the pulse separation is of the order of 10 n, which makes the selection of a single pulse with an external gate relatively simple. The reflectivity of the front mirror is typically between 50 and 60 %. The output reproducibility from a mode-locked laser is drastically improved by using at least one curved mirror instead of two plane mirrors [9.38]. Very often a mode-selecting aperture is inserted into the resonator of a mode-locked oscillator because in multitransverse-mode systems the power density is so large in localized areas that component damage frequently occurs. Sometimes a Galilean telescope or a single lens is included in the laser cavity to increase the beam diameter, which reduces the optical power density inside the dye and at the rear mirror. Examples of oscillators containing beam-expanding optics in the resonator are shown in Figs. 9.5c and d [9.37, 42].

Saturable Dyes Suitable for Mode-Locking. As was mentioned before, a dye employed to mode-lock a laser has to have a fast recovery time which is of the order of the duration of the mode-locked pulse. Generally speaking, saturable absorbers with long relaxation times tend to narrow the spectral width, while saturable absorbers with fast relaxation times tend to broaden the spectral width of lasers. This is the reason why for longitudinal-mode selection of passive Q-switched lasers, dyes with very long recovery times are desirable, as discussed in Chap. 5.

Nd : glass and Nd : YAG lasers are mode-locked with Eastman 9740 or 9860 dye suitably diluted with 1,2-dichloroethane or chlorobenzene. The dye concentration is usually chosen to produce a linear transmission between 50 and 80 % at 1.06 μm through the cell. Experimental data concerning the properties of these dyes were published in [9.43, 44].

Practically all dyes useful in the generation of mode-locked pulses decompose when exposed to ultraviolet light. In particular, Eastman 9740 and 9860 are extremely sensitive to light in the uv region. To eliminate dye breakdown due to uv radiation from flashlamp and ambient light, the dye cell should be well light-shielded with only a small aperture exposed for the laser beam. Uv-absorbing glass and quartz can be used for the dye cell windows, and special uv-free fluorescent lighting around the laser area is usually installed. A suitable fluorescent lamp is the General Electric Series F96 T12/Gold. Otherwise, while handling these dyes, fluorescent lights should be turned off and a tungsten-filament lamp should be used for illumination.

A new dye – Eastman 14015 – was introduced for mode-locking neodymium lasers. According to preliminary data, it appears that this dye is much more stable and yields a longer performance life than the dyes mentioned above. Eastman 14015 has the interesting property that the recovery time is drastically affected by the solvent [9.45]. It was found that the dye dissolved in 1,2-dichloroethane produces Q-switching, but it does not mode-lock the Nd: glass laser. This was attributed to the relatively long lifetime of the dye in its triplet state. Other solvents were found to shorten the recovery time, and mode locking was achieved with iodoethane, methyl sulfoxide, and ethyl sulfide as solvents. Ruby lasers are usually mode-locked with either cryptocyanine or 1, 1′-diethyl-2, 2′-dicarbocyanine iodide (DDI) [9.42]. The former dye can be diluted with nitrobenzene, acetone, ethanol, or methanol. The most consistent mode-locked operation is achieved with acetone, mainly because the absorption peak for cryptocyanine in acetone exactly coincides with the ruby line, while in the other solvents it is displaced by more than 100 Å compared to the ruby wavelength. Shorter mode-locked pulses and a more reliable operation is achieved with DDI diluted in methanol or ethanol. The absorption peak of the DDI-methanol combination occurs at 7060 Å. By some substitutions on the polymethine chain of DDI a new dye, DDI*, was developed which, diluted with methanol, has an absorption peak at the ruby wavelength [9.46]. It was also found that in DDI* a temperature change from 3 to 40° C resulted in only a 3 % shift of the absorption, and during a 16-week period a 4 % absorption shift occurred compared to cryptocyanine with a 16 % shift. Highly reproducible ultrashort mode-locked pulses with duration of the order of 5 ps were obtained by adding rhodamine 6G to a solution of DDI in methanol [9.47]. In the same laser, DDI in methanol yielded pulses of 10 ps in length and cryptocyanine in acetone resulted in pulse widths around 20 ps. Other related dyes, such as DDCI and DTCDI, have also been used to mode-lock ruby lasers and have resulted in pulse widths of 20 to 100 ps, respectively [9.48].

Saturable absorbers should be replenished with fresh solution periodically. By far the most reliable performance from mode-locked dye systems is obtained if the dye is circulated through the cell from a large reservoir. The pumping action assures uniform mixing of the dye; because of the large volume, the dye's concentration remains constant over a long period of time and fresh dye is exposed to each laser pulse [9.49, 50]. For a discussion of the structure and properties of organic dyes the reader is referred to [9.51].

Performance Data of Mode-locked Solid-state Lasers. The wide-spread interest in the Nd: glass laser arose because this material possesses a much broader linewidth than other solid-state lasers and would, therefore, be expected to produce the shortest mode-locked pulses. Typical commercially available mode-locked Nd : glass lasers produce pulse trains about 50 to 200 ns wide, containing pulses of 5 to 20 ps in duration. The Brewster-ended rods have dimensions ranging from 8 to 20 cm in length and up to 1.5 cm in diameter. Mode locking is achieved with dye cells having a path length of several millimeters. Mirrors or windows are wedged at least 30 arc/min so that they do not act as mode selectors. The total energy output can vary from 25 mJ to several hundred millijoules, and the energy content of the pulses in the middle of the train is between 1 to 10 mJ. Engineering aspects of mode-locked Nd : glass systems have been discussed in [9.36, 38].

From ruby lasers of approximately similar geometry as the glass lasers mentioned above, pulse durations that have been quoted range from 5 ps to 30 ps [9.41, 42, 46, 52–55]. Thicknesses of dye cells range typically from several millimeters to a fraction of a millimeter. Optimum absorber cell transmission for most systems is between 0.6 and 0.8. For a typical ruby rod, 10 cm long and 1 cm in diameter, the total energy in the mode-locked pulse train is around 50 mJ and the energy per single pulse is about 2 to 5 mJ in the center of the train.

Mode-locked Nd : YAG lasers typically employ laser rods ranging from 3×5 mm to 6×75 mm in size. With 1-mm-thick dye cells pulses between 20 and 40 ps in duration have been produced [9.49, 56–58]. Depending on dye concentration and flashlamp energy, the pulse trains are usually 10 to 80 ns long. Figure 9.7a shows a typical mode-locked train generated by a Nd : YAG laser. A portion of the pulse in the center was switched out in order to measure the pulse width. Figures 9.7b and c show an oscillogram and a two-photon absorption measurement of this pulse.

It has been found that for maximum reliability of operation it is necessary to temperature-control the dye solution. A change in absorption of 1 % per degree was found in the diluted Eastman dyes 9740 and 9860 [9.58]. Very stable and reproducible performance was achieved with a flowing dye system with the dye temperature controlled to ± 0.1 C.

Pulsed mode-locked Nd : YAG systems are usually operated at a pulse repetition rate of up to 5 pps and in extreme cases up to 30 pps [9.59]. The systems usually incorporate a large pump and temperature-controlled dye reservoir, micropore filter, and a laminar flow dye cell. Materials which come into contact with the dye solution must be limited to stainless steel, teflon, and glass.

Mode-locked pulses ranging from 5 to 15 ps were obtained by mode-locking Nd : CaLaSOAP with Eastman dye 9740 in dichlorethane [9.60]. A 0.25-mm-thick dye was formed by the 65 % mirror and a wedged fused silica window. The train contained typically 15 to 20 pulses.

Generation of Bandwidth-limited Pulses. In order to generate extremely short bandwidth-limited mode-locked pulses from solid-state lasers, the following procedures must be followed:

1) Complete elimination of parasitic mode selection which causes a re-

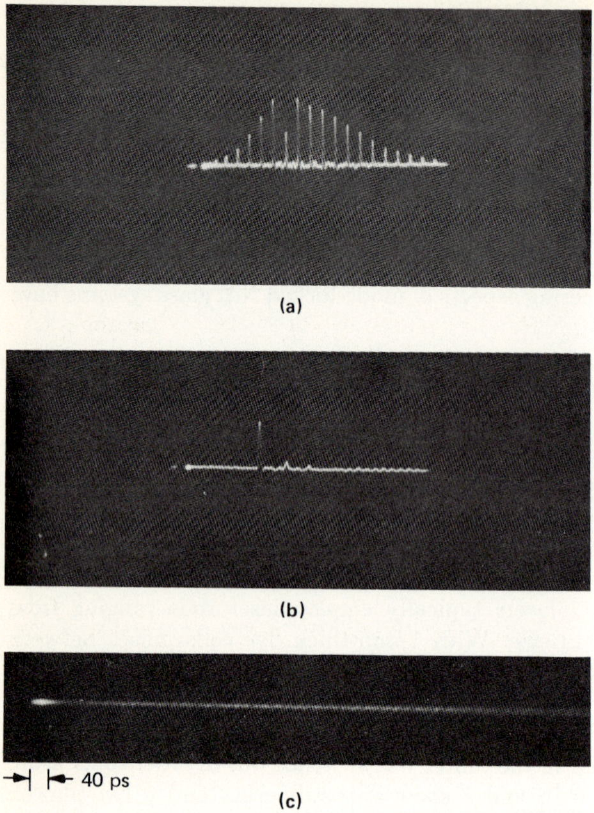

Fig. 9.7a–c. Passively mode-locked Nd:YAG laser. Oscillograms of **(a)** the pulse train, **(b)** a single pulse, and **(c)** a two-photon absorption measurement

duction in the number of excited modes. From (9.3) it follows that purposely reducing the number of longitudinal modes leads to a longer pulse duration. On the other hand, flattening of the gain-frequency profile and therefore increasing the number of modes helps to generate shorter light pulses [9.61, 62].

2) Use of a narrow-path dye cell. It has been found that the width of the individual pulses in a mode-locked train have a direct relationship to the optical path length of the dye cell. The most reliable mode locking and the shortest pulses are obtained when the saturable absorber is placed in contact with one of the dielectric mirrors and the dye thickness l satisfies the condition $l \leq c\tau/2n$, where τ and n are the relaxation time and refractive index of the dye, respectively, and c is the speed of light. For Eastman dyes A9740 and A9860, $l \approx 1\,\mathrm{mm}$. The relationship between dye thickness and pulse duration has been investigated by *Bradley* et al. [9.18, 63], who measured their shortest pulses for $l \approx 30\,\mu\mathrm{m}$.

3) Selection of a single pulse from the early portion of the train. Measurements with picosecond streak cameras have shown that the shortest pulses occur in the initial part of the train [9.63–65]. As was explained earlier, during the high-intensity-portion of the pulse development, self-phase modulation causes frequency broadening of the pulses. *Bradley* and *Sibbett* [9.63] mea-

sured pulses of 3 ps duration at the beginning of a pulse train generated from a Nd : glass laser, mode-locked with a 50-μm-long dye cell, increasing to about 10 ps at the end of the train.

4) Operation close to threshold. Since the pulse duration $t_p \approx 1/N$, where N is the number of locked modes, one might be tempted to increase the pump power in an effort to excite more modes. However, if the pump energy is too high, it is possible for more than one train of mode-locked pulses to develop during the pump cycle. Experience shows that pumping the laser just above threshold is essential for reproducible mode-locking operation.

A mode-locked pulse can also be further shortened by several temporal compression techniques. The approaches mentioned in the literature include single pass or multiple pass pulse compression employing saturable absorbing dyes [9.66]. Particularly effective are regenerative pulse compression schemes, where a mode-locked pulse from a master oscillator is injected into a regenerative amplifier containing an intracavity saturable dye. As the pulse is amplified through the nonlinear transmission range of the dye, its leading and trailing edges experience more loss than its peak, effectively compressing it in time [9.67–69].

Variable Pulse Width from Passively Mode-locked Systems. Several applications of mode-locked laser systems, such as laser-produced plasma experiments, require the generation of relatively long mode-locked pulses with durations of several hundred picoseconds. By combining mode locking with longitudinal mode selection, it is possible to extend the pulse width from the picosecond to the nanosecond range. A reduction of the number of axial modes and therefore an expansion of the pulse width can be achieved by inserting frequency-selective elements into the cavity. Employing tilted transmission etalons or diffraction gratings or a combination of both, variable pulse widths were obtained from Nd : glass, Nd : YAG, and ruby lasers [9.65, 70]. Figure 9.8 shows typical results for a saturable absorber mode-locked and Q-switched Nd : YAG system [9.71].

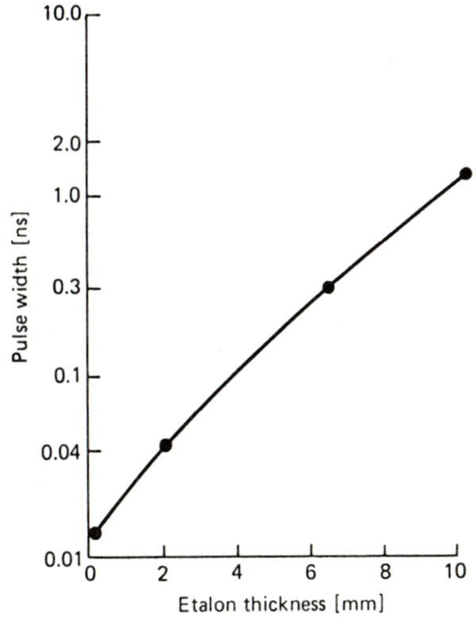

Fig. 9.8. Pulse width versus etalon thickness for a mode-locked Nd : YAG laser. [9.71]

With no etalon the pulse width, as determined from two-photon fluorescence, was 20 ps. When tilted etalons of different thicknesses were inserted in the cavity and adjusted such that the transmission maximum coincided with the peak fluorescence of the 1.06-μm line, stable operation was obtained at pulse widths up to 1.25 ns. Agreement between the experimentally determined pulse duration and the spectral characteristic of the cavity shows that mode locking with spectral mode selection is understood well enough so that lasers can be designed to produce pulses having durations anywhere in the range from 10^{-11} to 10^{-9} s.

Signal-to-Noise Ratio. One important characteristic of the emission from a mode-locked laser is the ratio between the peak intensity of the pulse and the background radiation between the principal pulses. Especially when single mode-locked pulses are amplified to high energy levels, any radiation preceding the main pulse is very undesirable. For oscillators employed in multiple amplifier systems, a signal-to-noise ratio in excess of 20 dB is usually required. This performance level can be achieved with carefully designed systems which are operated close to threshold. The signal-to-noise ratio from an oscillator can be improved by passing the radiation through a fast-relaxing saturable dye. Having the dye cell located external to the oscillator will attenuate the low-intensity background radiation more than the high peak power mode-locked pulse [9.72].

9.2 Active Mode Locking

By placing inside a laser cavity either a phase (FM) modulator or an amplitude (AM) modulator driven at exactly the frequency separation of the axial modes, one can cause the laser to generate a train of mode-locked pulses with a pulse repetition rate of $f_\mathrm{m} = c/2L$. Active mode locking, performed on cw-pumped lasers such as the Nd : YAG system, is achieved by inserting into the resonator an electrooptic or acoustooptic modulator.

Loss Modulation. From a frequency-domain viewpoint, introducing a time-varying transmission $T(t)$ through an amplitude modulator inside the laser resonator creates sidebands on each oscillating axial mode which overlap with adjoining axial modes. The operation can best be described as follows: assume that the mode with the frequency ν_0, nearest the peak of the laser gain profile, will begin to oscillate first. If a loss modulator operating at a frequency f_m is inserted into the resonator, the carrier frequency ν_0 will develop sidebands at $\pm f_\mathrm{m}$. If the modulating frequency is chosen to be commensurate with the axial mode frequency separation $f_\mathrm{m} = c/2L$, the coincidence of the upper $(\nu_0 + f_\mathrm{m})$ and the lower $(\nu_0 - f_\mathrm{m})$ sidebands with the adjacent axial mode resonances will couple the $\nu_0 - f_\mathrm{m}$, ν_0, and $\nu_0 + f_\mathrm{m}$ modes with a well-defined amplitude and phase. As the $\nu_0 + f_\mathrm{m}$ and $\nu_0 - f_\mathrm{m}$ oscillations pass through the modulator, they will also become modulated and their sidebands will couple the $\nu_0 \pm 2f_\mathrm{m}$

modes to the previous three modes. This process will continue until all axial modes falling within the laser linewidth are coupled.

Viewed in the time domain, the same intracavity modulating element, with its modulation period equal to the round-trip transit time $2L/c$, can reshape the internal circulating field distribution repeatedly on each successive round trip inside the cavity. For example, light incident at the modulator during a certain part of the modulation cycle will be again incident at the same point of the next cycle after one round trip in the laser resonator. Light suffering a loss at one time will again suffer a loss on the next round trip. Thus, all the light in the resonator will experience loss except that light which passes through the modulator when the modulator loss is zero (Fig. 9.9a). Light will tend to build up in narrow pulses in these low-loss time positions. In a general way we can see that these pulses will have a width given by the reciprocal of the gain bandwidth, since wider pulses will experience more loss in the modulator, and narrower pulses will experience less gain because their frequency spectrum will be wider than the gain bandwidth.

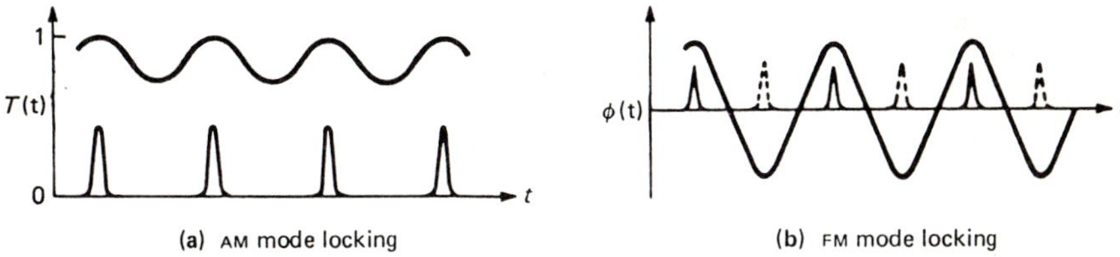

Fig. 9.9a, b. Mode-coupling behavior in the case of (a) AM mode locking and (b) FM mode locking

Siegman and *Kuizenga* [9.73, 74] developed an elementary analysis of mode locking in homogeneous lasers by following a single mode-locked pulse through one round trip around the laser cavity. For steady-state mode locking the pulse shape should be unchanged after a complete roundtrip. The self-consistent solution carried out for a Gaussian pulse leads to a simple expression which shows the dependence of the mode-locked pulse width on linewidth, modulation frequency, depth of modulation, and saturated gain.

For an acoustooptic AM modulator operating in the Bragg regime, as well as for electrooptic AM modulators, the single-pass transmission function is

$$T(t) \approx \cos(\delta_{\mathrm{AM}} \sin \omega_{\mathrm{m}} t) \quad , \tag{9.8}$$

where δ_{AM} is the modulation index and $\omega_{\mathrm{m}} = 2\pi f_{\mathrm{m}}$ is the angular frequency of the modulation. In the ideal mode-locking case, the pulse passes through the modulator at the instant of maximum transmission. This occurs twice in every period of the modulation signal ω_{m}, and hence one drives these modulators at a modulation frequency equal to half the axial mode spacing of the laser. Expanding (9.8) at the transmission maximum gives for the round-trip modulation function

$$T(t) = \exp(-\delta_{\mathrm{AM}}^2 \omega_{\mathrm{m}}^2 t^2) \quad , \tag{9.9}$$

which results in a pulse width for the AM mode-locked laser [9.75] of

$$t_{\mathrm{p}}(\mathrm{AM}) \approx \frac{(gl)^{1/4}}{2\delta_{\mathrm{AM}}^{1/2}}\left(\frac{1}{f_{\mathrm{m}}\Delta\nu}\right)^{1/2} , \qquad (9.10)$$

where g is the saturated gain coefficient at the line center, $\Delta\nu$ is the gain bandwidth of the laser, and l is the length of the active medium. The AM mode-locked pulses have no frequency "chirp"; i.e., $\beta_{\mathrm{AM}} = 0$ in (9.4).

Phase Modulation. Light passing through an electrooptic phase modulator will be up- or down-shifted in frequency unless it passes through at the time when the intracavity phase modulation $\delta(t)$ is stationary at either of its extrema. The recirculating energy passing through the FM modulator at any other time receives a Doppler shift proportional to $d\delta/dt$, and the repeated Doppler shifts on successive passes through the modulator eventually push this energy outside the frequency band over which gain is available from the laser medium. The interaction of the spectrally widened circulating power with the narrow laser linewidth leads to a reduction in gain for most frequency components. Thus, the effect of the phase modulator is similar to the loss modulator, and the previous discussion of loss modulation also applies here. As shown in Fig. 9.9b, the existence of two phase extrema per period creates a phase uncertainty in the mode-locked pulse position, since the pulse can occur at either of two equally probable phases relative to the modulating signal. The quadratic variation of $\delta(t)$ about the pulse arrival time also produces frequency "chirp" within the short mode-locked pulses.

In the FM case, the internal phase modulator introduces a sinusoidally varying phase perturbation $\delta(t)$ such that the round-trip transmission through the modulator is given approximately by

$$T(t) \approx \exp\left(\pm \mathrm{j}\delta_{\mathrm{FM}}\omega_{\mathrm{m}}^2 t^2\right) , \qquad (9.11)$$

where δ_{FM} is the peak phase retardation through the modulator. The $\pm$ sign corresponds to the two possible phase positions at which the pulse can pass through the modulator, as mentioned earlier. With these parameters the pulse width of phase mode-locked pulses is given by [9.73]

$$t_{\mathrm{p}}(FM) \approx \frac{(gl)^{1/4}}{2(\delta_{\mathrm{FM}})^{1/4}(f_{\mathrm{m}}\Delta\nu)^{1/2}} . \qquad (9.12)$$

In an electro-optic phase modulator the phase retardation is proportional to the modulating voltage, hence $\delta_{\mathrm{FM}} \propto P_{\mathrm{m}}^{1/2}$, where P_{m} is the drive power into the modulator. Therefore, we obtain from (9.12) for the pulse width $t_{\mathrm{p}}(\mathrm{FM}) \propto P_{\mathrm{m}}^{-1/8}$, which indicates that the pulses shorten very slowly with increased modulator drive. More effective in shortening the pulses is an increase of the modulationl frequency. Since $f_{\mathrm{m}} = c/2L$, the pulse width will be proportional to the square root of the cavity length.

In order to calculate the pulse width from (9.12), we can calculate the saturated gain coefficient g by equating the loop gain with the loss in the resonator

$$2gl \approx \ln\left(\frac{1}{R}\right) \quad , \tag{9.13}$$

where R is the effective reflection of the output mirror and includes all losses. For a typical Nd:YAG laser with 10 % round-trip loss, that is, $R = 0.9$, a resonator length of 60 cm, and a linewidth of 120 GHz, the pulse length is given by $t_p(\mathrm{FM}) = 39(1/\delta_{\mathrm{FM}})^{1/4}$. For $\delta_{\mathrm{FM}} = 1$ rad, which is easily obtainable, pulses of 39 ps can be generated. The mode-locked pulses obtained through FM modulation show a linear frequency shift

$$\beta_{\mathrm{FM}} = \pi^2 \left(\frac{\delta_{\mathrm{FM}}}{4gl}\right)^{1/2} \Delta\nu f_{\mathrm{m}} \quad , \tag{9.14}$$

where β_{FM} was defined in (9.4).

Dynamics of Mode Locking. If a laser cavity is mechanically or optically disturbed, the mode-locked pulse train may be upset sufficiently to cause a momentary, substantial broadening of the pulses themselves. In such a case, the cavity must recover under the influence of the modulator. A transient description of the buildup of mode locking has been developed for the extreme case where the laser cavity field starts out from noise. For an AM modulator with a modulation function according to (9.8), a transient solution of the mode-locking process yields [9.76, 77]

$$t_p = \frac{t_{p0}}{[\tanh(M/M_0)]^{1/2}} \quad , \tag{9.15}$$

where t_{p0} is the steady-state pulse width given by (9.10), M is the number of round trips from the time the modulator is turned on, and the constant M_0 is given by

$$M_0 = \frac{\Delta\nu}{4(2gl)^{1/2}\delta_{\mathrm{AM}} f_{\mathrm{m}}} \quad . \tag{9.16}$$

According to (9.15) it requires $M \approx 1.5M_0$ for the pulse width to approach 5 % of the final steady-state value. For a typical cw-pumped Nd:YAG laser with $\Delta\nu = 120$ GHz, $f_{\mathrm{m}} = 100$ MHz, $2gl = 0.2$, and $\delta_{\mathrm{AM}} = 0.3$, almost 3000 transits or 15 μs are required to bring the pulsewidth down near its steady-state value.

9.2.1 Design of Actively Mode-Locked Laser Systems

As is the case with passive mode-locked systems, etalon effects due to intracavity elements will reduce the bandwidth of the system and broaden the mode-locked pulses. In a cw-pumped Nd:YAG laser, even a weak etalon effect due to the modulator or rod surfaces can decrease the effective value of $\Delta\nu$ by a large amount. Therefore, one of the most important considerations in a practical mode-locking system is the elimination of residual reflections and optical interference effects in the laser cavity. On the other hand, it is possible to use a tilted etalon inside a mode-locked laser to deliberately lengthen the pulse width. With uncoated quartz etalons of thickness between 1 and 10 mm, good control of the pulse width can be achieved [9.75].

Mode locking of cw-pumped Nd:YAG lasers has been accomplished with acoustooptic loss modulators [7.78, 79] and electro-optic phase or loss modulators [9.80–82]. Detailed design information of stable and reliable actively mode-locked Nd:YAG and Nd:YLF systems can be found in [9.83, 84]. The crystals used in electro-optic modulators are usually lithium niobate or barium sodium niobate. We will consider $LiNbO_3$ as an example to illustrate typical operating parameters of an electro-optic modulator employed in active mode locking.

For $LiNbO_3$ operated as a phase modulator, two cases are of particular interest. First, suppose that the direction of light propagation is in the x direction and that the modulating field is applied in the z direction. A pure phase retardation can be obtained by polarizing the light either along the y axis or the z axis. One obtains

$$\Delta n_y = -\tfrac{1}{2} r_{13} n_o^3 E_z \quad \text{or} \quad \Delta n_z = -\tfrac{1}{2} r_{33} n_e^3 E_z \quad , \tag{9.17}$$

where n_o and n_e are the ordinary and extraordinary index of refraction, and r_{13}, r_{33} are the electro-optic coefficients of $LiNbO_3$. Because r_{33} is larger than r_{13} in $LiNbO_3$, optimum phase modulation is obtained by polarizing the incident light in the z direction. The total phase change in the crystal is $\delta = 2\pi\,\Delta n a/\lambda_0$, where a is the length of the crystal in the x direction. If a voltage $V = V_0 \cos \omega_m t$ is applied across the crystal in the z direction, the peak single-pass phase retardation of the modulator is

$$\delta'_{FM} = \frac{\pi r_{33} n_e^3 V_0 a}{\lambda_0 d} \quad , \tag{9.18}$$

where d is the dimension of the crystal in the z direction. The effective value of δ_{FM} is obtained by multiplying (9.18) with the mode-coupling coefficient of the crystal, [9.85]:

$$\delta_{FM} = \left[\cos\left(\frac{\pi Z_0}{L}\right) \text{sinc}\left(\frac{\pi a}{2L}\right)\right] \delta'_{FM} \tag{9.19}$$

where Z_0 is the distance of the modulator from the end mirror and L is the optical length of the laser cavity. The term in brackets, which varies between 0 and 1, represents the degree of synchronism between the optical and modulating fields. If the $LiNbO_3$ modulator is operated with the light polarized in the y direction, the term $r_{33} n_e^3$ in (9.18) must be replaced by $r_{13} n_o^3$. For $LiNbO_3$ these parameters are $n_o = 2.24$, and $n_e = 2.16$ at $1.06\,\mu$m, and $r_{13} = 8.6 \times 10^{-10}$ cm/V and $r_{33} = 30.8 \times 10^{-10}$ cm/V. Typically, a phase retardation of approximately $\delta_m = 1$ rad results with 300 V across a 5-mm × 5-mm × 20-mm crystal.

In order to produce a pure loss modulation in $LiNbO_3$ the light propagates along the z axis (optic axis) and the modulating field is applied in the x direction. If the incident beam is polarized in the x direction, a single pass through the crystal will cause a polarization rotation of $\pi r_{22} n_o^3 a E_x / \lambda_0$ radians. For a sinusoidally varying voltage $V_x = V_0 \cos \omega_m t$, the peak single-pass power loss to the incident polarization is

$$\delta_{AM} = \sin^2 \left(\frac{\pi r_{22} n_o^3 V_0}{\lambda_0} \frac{a}{d} \right) \quad . \tag{9.20}$$

Hence, when used with a Brewster plate internal to the laser, this crystal orientation acts as a pure loss modulation. From these considerations follows that in order to convert from phase to loss modulation it is necessary to substitute an x axis $LiNbO_3$ for a z-axis one and to reduce the frequency by a factor of 2.

Electrically, the modulator consists of a pair of electrodes between which the crystal is placed. An inductive loop and a tuning capacitor are shunted across these electrodes and are used to achieve the desired resonant frequency.

In a Nd : YAG laser, the average power obtained in mode-locked operation is very nearly the same as the average power from the same laser without mode locking [9.86]. We do not consider here the reduction in average power that may occur in a laser simply from inserting the modulator element in the cavity as a result of optical losses in the modulator element. For example, in a Nd : YAG laser phase modulated with a 2-cm-long $LiNbO_3$ crystal at a frequency of 388 MHz, an average power of 1.2 W TEM_{00} mode was achieved [9.80]. The laser produced a train of pulses 30 ps in duration and with peak powers over 100 W. The rf drive power was 4.8 W. The laser consisted of a 63-mm-long by 3-mm-diameter laser rod pumped by two tungsten lamps in a double-elliptical pump cavity. The cw output power of the laser with the modulator inserted in the cavity was 1.3 W. In a similar arrangement, mode-locked pulses varying from 40 ps to 200 ps were generated by inserting etalons into the cavity [9.74]. In a commercial Nd : YAG laser, phase modulation is achieved with a barium sodium niobate crystal having the size of a 5-mm cube which was driven with a 5-W rf source at 200 MHz [9.81].

Electronic Feedback Loops. Active mode locking of a laser can be achieved by using a tunable oscillator and adjusting the driving frequency to agree with the cavity length; or, alternatively, selecting a fixed frequency and adjusting the mirror spacing to correspond. However, slow shifts in oscillator frequency or, more commonly, changes in effective cavity length due to thermal effects in the laser rod and laser support structure will cause a gradual loss of resonance. In the loss-modulated laser, this change causes the pulse to go through the modulator at a time when the polarization rotation is nonzero and, hence, increases the average loss introduced by the modulator. In the phase-modulated laser, it causes the pulse to pass through the modulator at a time when the instantaneous phase retardation is changing with time. Thus, a Doppler shift of the train results, which shifts the pulse frequencies toward the wings of the line, which reduces the output power of the laser. Actively mode-locked lasers are very sensitive to even very small frequency detuning between the applied modulation frequency f_m and the exact round-trip repetition frequency $c/2L$ of the laser cavity. For the pulse-modulation effects to be properly cumulative over many pulse transits, the two frequencies must be very accurately synchronized. Modulation frequency tolerances less than ~ 1 kHz for a modulation frequency of several hundred megahertz are typical for cw mode-locked Nd:YAG lasers. In order to maintain stable mode-locked operation, it is necessary to use some form of electronic feedback. The primary role of the electronic system is to maintain the crystal driving frequency in exact resonance with the round-trip cavity frequency or equivalently with the cavity intermode spacing.

Next, we will consider several types of feedback loops that have been used in mode-locked lasers for long-term operation.

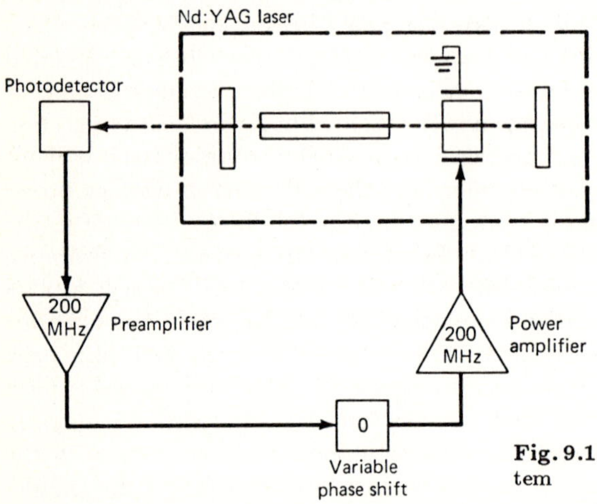

Fig. 9.10. Direct-drive mode-locking system

Oscillator Loop. A simple mode-locking system can be built by allowing the laser to determine its own drive frequency, as shown in Fig. 9.10. This can be accomplished by using a high-speed photodetector to sense the first beat frequency $c/2L$ of the oscillator. This signal is amplified, phase-shifted a variable amount, and then applied to the intracavity modulator. The adjustable phase shifter compensates for the delay in the feedback loop. When the phase of the electric signal on the modulator and the signal derived from the laser equal an integer times π, the loop goes into regeneration and the laser is mode-locked. The resultant mode-locking system is a closed-loop oscillator using the laser cavity as the basic reference. Such a system will automatically track changes in cavity geometry [9.87].

In a phase-modulation system there is an ambiguity in the phase relationship between the laser pulse train and the modulator driving signal, as the pulses may pass through the modulator at either of the two extreme voltage points. Because of this phase ambiguity, two possible pulse trains can be obtained – one at 0° phase with respect to the modulator drive and one at 180°. The laser will always operate with only one of these pulse trains running at a time, because to do otherwise would result in a larger net loss in the laser cavity. However, small perturbations to the laser can cause random shifting from one phase to the other, the phase change requiring several microseconds. An automatic phase-modulating mode-locked system, which operates by synchronizing the modulator drive signal to the first-order mode beat of the laser, suffers from sporadic loss of lock as the laser hops from one stable phase to the other.

A loss-modulation system does not suffer from this ambiguity. However, one of the inherent problems with a loss modulator is the tendency to excite the natural relaxation oscillations in the cw laser. These relaxation oscillations cause fluctuations in amplitude, accompanied by damped oscillations at frequencies in the 30- to 150-kHz range. For this reason, most commercially built active mode-locked lasers employ phase modulation.

In a phase-modulated laser we can avoid the failure mode which occurs due to a spontaneous switch of 180° in the phase of the output pulse train by utilizing the second-order beat frequency. A mode-locking control system that has proven to be quite stable is shown in schematic form in Fig. 9.11 [9.81]. The oscillator is phase-modulated at the fundamental beat frequency of 200 MHz by a modulator driven from a voltage-controlled oscillator (VCO). A photodetector mounted behind the rear mirror of the oscillator samples a portion of the beam. A bandpass filter selects the second harmonic of the cavity beat frequency. The resultant amplified 400-MHz signal as compared in a phase detector with a 400-MHz comparison signal derived by multiplying the output of the VCO by 2. The phase-detected output is amplified and used to control the VCO. Initial mode locking is performed manually by adjusting the bias of the VCO.

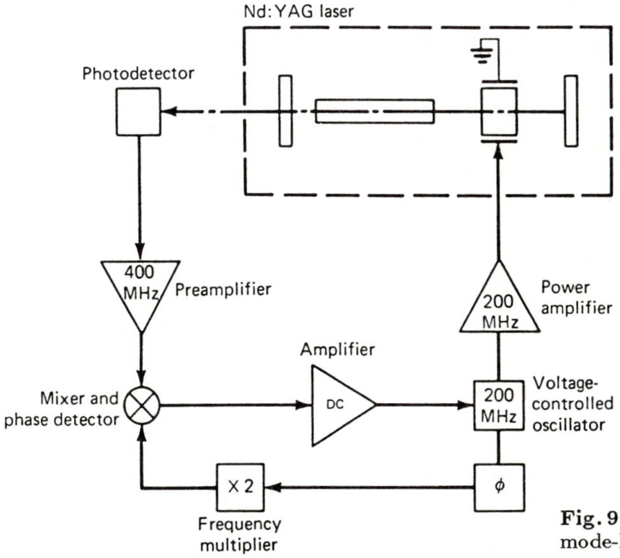

Fig. 9.11. Automaticclosed-loop mode-locking system. [9.81]

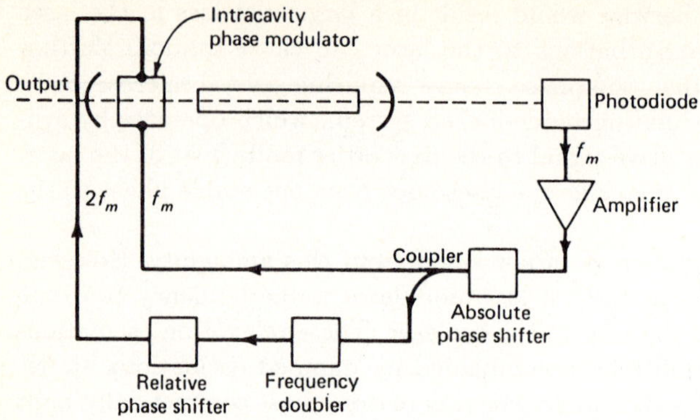

Fig. 9.12. An oscillator loop corrected for phase switching. [9.89]

In another technique two drive frequencies, the fundamental and harmonic beat frequency, are both applied to the modulator to suppress the 180° phase switching of the optical pulse train [9.89]. An oscillator loop that incorporates this feature is shown in Fig. 9.12. Here the fundamental beat note f_{m} is detected, amplified, and applied to the intracavity modulator. Part of this signal is coupled out, doubled in frequency, and applied to the crystal that is in a doubly resonant circuit. There are two phase-shift adjustments: an absolute phase shift and a relative phase shift. The relative phase shifter is adjusted in the open loop as described previously to suppress the pulse train phase switching. The absolute phase shifter is then adjusted to provide stable loop operation.

Servo Loops. It is possible to construct another type of feedback loop called a servo loop where variations in the laser generate an error signal which can be used to correct the frequency of the drive oscillator or to correct the cavity length. A simple technique for doing this takes advantage of the fact that as $c/2L$ changes with respect to f_{m}, the phase of the mode-locked pulse train relative to the modulation drive signal also changes. Both the loss- and phase-modulated lasers provide such a discriminant. Experimental curves of the phase of the pulse train as a function of the detuning frequency for a Nd:YAG laser are shown in Fig. 9.13. Using this discriminant, an error signal proportional to Δf_{m} is derived which, for example, in the scheme illustrated in Fig. 9.14, is used to adjust the length of the cavity [9.80, -81]. In Fig. 9.14 the pulses are detected with a photodiode whose output consists of beat notes at all multiples of the modulation frequency. To avoid the ambiguity which occurs in phase-modulated systems, the first harmonic of the beat frequency is filtered out and amplified, after which both this signal and a portion of the signal driving the intracavity modulator (whose phase is used for comparison purposes) are beat down to an intermediate frequency. The two rf signals are then mixed and the dc output of the mixer is amplified. The resultant dc signal is then proportional to the phase difference between the pulse train and the rf modulator drive signal, and can be used to stabilize the phase of the pulse train.

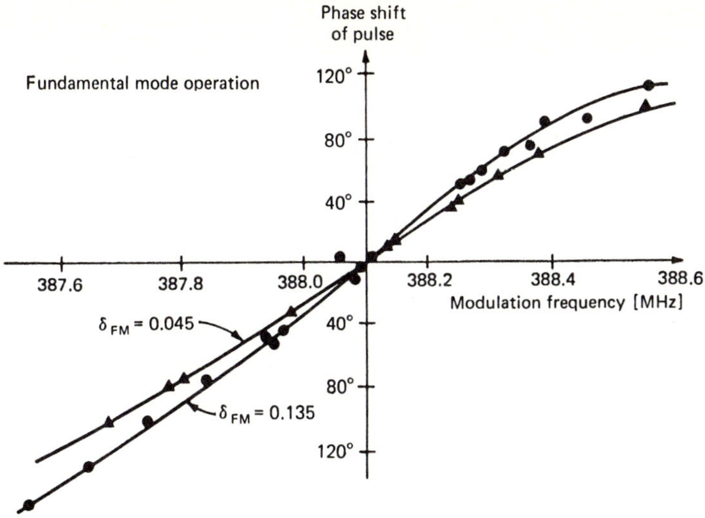

Fig. 9.13. Experimental curves of the phase of the pulse train relative to the modulation drive signal as a function of detuning the modulation frequency from the axial mode interval. [9.85]

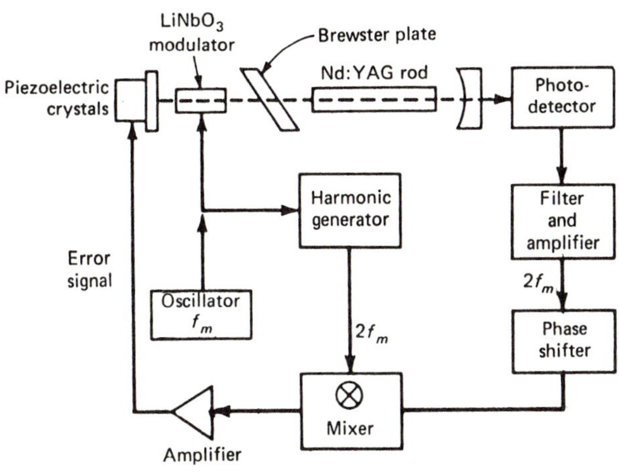

Fig. 9.14. Mode-locking system with phase-detection technique. [9.85]

Extensive lists of references on actively mode-locked solid-state lasers can be found in several excellent reviews [9.75, 86, 88]. Besides in Nd:YAG, active mode locking has also been achieved in ruby [9.90, 91], Nd:glass [9.92–94] and Nd:YLF [9.84, 95, 96].

At the conclusion of this chapter a few topics related to active mode locking will be briefly mentioned. Mode locking at harmonic multiples of the basic repetition frequency $c/2L$ has been obtained by driving the intracavity modulator at the appropriate modulation frequency $f_m = Nc/2L$, where $N > 1$. Intracavity phase modulation at harmonics up to $N = 5$ has been achieved in a Nd:YAG laser [9.97–99]. These techniques are of interest for high-data-rate communications systems. Several authors describe mode locking and second-harmonic generation inside the laser resonator using the same nonlinear crystal [9.81, 100, 101] or a separate nonlinear crystal [9.79].

9.3 Active-Passive Mode Locking

Due to the statistical nature of passive mode locking, the pulses generated, frequently include satellite pulses and have large shot-to-shot variations in pulsewidth and energy. Purely active Q-switching and mode locking of a cw Nd: YAG laser can produce stable pulses but at very low energy. Therefore several authors [9.102–108] have developed active-passive Nd systems comprised of an acousto-optic modelocker and a saturable dye. In these systems, the pulse train amplitude stability and pulse contrast is improved with respect to purely passive systems. Perhaps the simplest experimental scheme is the Nd.: YAG colliding pulse method reported by *Vanherzeele* et al. [9.104]. More complex regenerative amplifier systems have been reported for both Nd: YAG and Nd: glass [9.107, 108]. Simple, effective and low-cost systems employing an acousto-optic modulator have been described by a number of researchers [9.102, 103, 105, 106].

10. Nonlinear Devices

Nonlinear optical devices, such as harmonic generators and parametric oscillators, provide a means of extending the frequency range of available laser sources. In 1961, *Franken* [10.1] and coworkers detected ultraviolet light at twice the frequency of a ruby laser beam when this beam was propagated through a quartz crystal. This experiment marked the beginning of an intense investigation into the realm of the nonlinear optical properties of matter.

Frequency conversion is a useful technique for extending the utility of high-power lasers. It utilizes the nonlinear optical response of an optical medium in intense radiation fields to generate new frequencies. It includes both elastic (optical-energy-conserving) processes, such as harmonic generation, and inelastic processes (which deposit some energy in the medium), such as stimulated Raman or Brillouin scattering.

There are several commonly used elastic processes. Frequency doubling, tripling, and quadrupling generate a single harmonic from a given fundamental high-power source. The closely related processes of sum- and difference-frequency generation also produce a single new wavelength, but require two high-power sources. These processes have been used to generate high-power radiation in all spectral regions, from the ultraviolet to the far infrared. Optical parametric oscillators and amplifiers generate two waves of lower frequency. They are capable of generating a range of wavelengths from a single frequency source, in some cases spanning the entire visible and near-infrared regions.

As far as inelastic processes are concerned, the Raman process can be utilized in solid-state lasers for the generation of additional spectral output lines. The strongest interaction is for the output shifted towards a longer wavelength (first Stokes shift), but at sufficiently high pump intensities additional lines at longer as well as shorter wavelengths with respect to the pump wavelength will appear. (Stokes and anti-Stokes lines.)

Although it produces a small wavelength shift, stimulated Brillouin scattering is mainly of interest for the realization of phase-conjugating mirrors. The application of phase conjugation, or wavefront reversal, via stimulated Brillouin scattering offers the possibility of minimizing thermally-induced optical distortions [10.2] which occur in solid-state laser amplifiers.

Nonlinear optical effects are analyzed by considering the response of the dielectric material at the atomic level to the electric fields of an intense light beam. The propagation of a wave through a material produces changes in the spatial and temporal distribution of electrical charges as the electrons and atoms react to the electro-magnetic fields of the wave. The main effect of the forces exerted by the fields on the charged particles is a displacement of the valence electrons from their normal orbits. This perturbation creates electric dipoles whose macroscopic manifestation is the polarization. For small field strength this polarization is proportional to the electric field E. In the nonlinear case, the reradiation comes from dipoles whose amplitdues do not faithfully reproduce the sinusoidal electric field that generates them. As a result, the distorted reradiated wave contains different frequencies from that of the original wave.

In a given material, the magnitude of the induced polarization per unit volume $\boldsymbol{P}$ will depend on the magnitude of the applied electric field $\boldsymbol{E}$. We can therefore expand $\boldsymbol{P}$ in a series of powers of $\boldsymbol{E}$ and write:

$$\begin{aligned} \boldsymbol{P}_l(\omega_j) = {} & \boldsymbol{X}^{(1)}_{lm}\boldsymbol{E}_m(\omega_j) + \ldots + \boldsymbol{X}^{(3)}_{lmn}\boldsymbol{E}_m(\omega_r)\boldsymbol{E}_n(\omega_s) \\ & + \boldsymbol{X}^{(4)}_{lmnp}\boldsymbol{E}_m(\omega_r)\boldsymbol{E}_n(\omega_s)\boldsymbol{E}_p(\omega_t) + \ldots \\ & + \boldsymbol{X}^{(6)}_{lmn}(-i\omega_r)\boldsymbol{E}_m(\omega_r)\boldsymbol{B}_n(\omega_s) + \ldots \end{aligned} \quad (10.1)$$

where $\boldsymbol{P}$ and $\boldsymbol{E}$ are vectors linked by tensors of second ($\boldsymbol{X}^{(1)}$), third ($\boldsymbol{X}^{(3)}, \boldsymbol{X}^{(6)}$), and fourth ($\boldsymbol{X}^{(4)}$) rank. The values of the tensor coefficients are functions of frequency and temperature. The subscripts j, r, s, and t denote different frequency components, and l, m, n, and p are Cartesian indices that run from 1 to 3. In (10.1) only those terms are listed which give rise to optical phenomena treated in this book.

For small field strength the polarization is proportional to the electric field $\boldsymbol{E}$ and is accounted for by the polarizability tensor $\boldsymbol{X}^{(1)}_{lm}$. Linear optics encompass all the interaction of light and dielectrics where the first term of (10.1) is a valid approximation. In linear optics the index of refraction is given by

$$n = (1 + 4\pi\boldsymbol{X}^{(1)})^{1/2} = \varepsilon^{1/2} \quad (10.2)$$

where ε is the dielectric constant of the material.

The $\boldsymbol{X}^{(3)}$ term is responsible for second-harmonic generation, optical mixing, and the Pockels effect. The nonlinear polarization tensor $\boldsymbol{X}^{(3)}$ vanishes in crystals that have a center of symmetry. In these crystals second-harmonic generation is not possible. The third-rank tensor $\boldsymbol{X}^{(3)}$ in general has 27 components. As a result of crystal symmetry, many of the components of $\boldsymbol{X}^{(3)}$ will be zero or equal to other components of the tensor. Furthermore, for those crystals of main interest to us, there is usually one predominant coefficient associated with a single light propagation direction which yields maximum harmonic power. If $\omega_r = \omega_s$ we obtain the relationship of second-harmonic generation

$$\boldsymbol{P}_l(2\omega) = \boldsymbol{X}^{(3)}_{lmn}\boldsymbol{E}_m(\omega)\boldsymbol{E}_n(\omega) \quad . \quad (10.3)$$

The generation of harmonics is a special case of optical mixing in nonlinear materials. The simultaneous application of two fields with frequencies ω_r and ω_s

produces a polarization at the sum and difference frequencies. The polarizations produced are of the form

$$P_l\begin{pmatrix}\omega_r+\omega_s\\ \omega_r-\omega_s\end{pmatrix}=X^{(3)}_{lmn}E_m(\omega_r)E_n(\omega_s)\quad . \tag{10.4}$$

In general, if three waves ω_r, ω_s, and $\omega_i = \omega_r + \omega_s$ are superimposed in a nonlinear medium, each wave is coupled to the other two through polarization waves. This is called the parametric interaction of three waves. Therefore, it is possible to convert energy into radiation at the sum and difference frequencies. From the point of view of parametric amplification, the second-harmonic generation is a special case of interaction between two waves with a common frequency.

The Pockels effect is obtained if one of the electric fields is taken to be a dc field applied across a suitable crystal. For $\omega_s = 0$ and $E_n = E_{\rm dc}$ one obtains $\omega_j = \omega_r$, and the index of refraction becomes a function of $E_{\rm dc}$, i.e., $P_l(\omega_j) = [X_{lmn}E_{\rm dc}]E_m(\omega_j)$.

The $X^{(4)}$ term of (10.1) couples the nonlinear polarization to three electric field vectors. If $\omega_r = \omega_s = \omega_t$ is the fundamental frequency, then $\omega_j = 3\omega$ and $P_l(3\omega)$ will generate the third harmonic.

For $E_n = E_p$ and $\omega_s = -\omega_t$ the $X^{(4)}$ term transforms to

$$P_l(\omega_j) = [X^{(4)}_{lmnn}E^2_{\rm av}]E_m(\omega_j) \tag{10.5}$$

where $E_{\rm av}$ is the average electric field strength. In this case the index of refraction becomes a function of the light intensity, which leads to nonlinear processes, such as the Kerr effect, and electrostriction.

With the application of a dc magnetic field ($\omega_s = 0$) the $X^{(6)}$ term describes the Faraday effect.

10.1 Harmonic Generation

In this section, we will review the basic theory and discuss system parameters and materials properties which affect harmonic generation.

10.1.1 Basic Equations of Second-Harmonic Generation

The process of harmonic generation by an incident wave of frequency ω_1 must be viewed as a two-step process: First, a polarization wave at the second harmonic $2\omega_1$ is produced which has a phase velocity and wavelength in the medium which are determined by n_1, the index of refraction for the fundamental wave, that is, $\lambda_p = c/2\nu_1 n_1$. The second step is the transfer of energy from the polarization wave to an electromagnetic (em) wave at frequency $2\omega_1$. The phase velocity and the wavelength of this em wave are determined by n_2, the index of refraction for the doubled frequency, that is, $\lambda_2 = c/2\nu_1 n_2$. For efficient energy transfer it is necessary that the two waves remain in phase, which implies that $n_1 = n_2$. Since almost all materials have normal dispersion in the optical region, the radiation will generally lag behind the polarization wave. The phase mismatch between the polarization wave and the em wave for collinear beams is usually expressed as the difference in wave number

$$\Delta k = \frac{4\pi}{\lambda_1}(n_1 - n_2) \quad . \tag{10.6}$$

If Maxwell's equations are solved for a coupled fundamental and second-harmonic wave propagating in a nonlinear medium, then the ratio of the power generated at the second-harmonic frequency to that incident at the fundamental is given by [10.3]

$$\frac{P_{2\omega}}{P_\omega} = \tanh^2\left[lK^{1/2}\left(\frac{P_\omega}{A}\right)^{1/2} \frac{\sin \Delta kl/2}{\delta kl/2}\right] \quad , \tag{10.7}$$

where

$$K = 2\eta^3\omega_1^2 d_{\text{eff}}^2 \quad , \tag{10.8}$$

l is the length of the nonlinear crystal, A is the area of the fundamental beam, η is the plane-wave impedance $\eta = \sqrt{\mu_0/\varepsilon_0\varepsilon} = 377/n_0[\Omega]$, ω_1 is the frequency of the fundamental beam, and d_{eff} is the effective nonlinear coefficient of the nonlinear polarizability tensor $\boldsymbol{X}^{(3)}$ in (10.1). The dimension of d_{eff} in (10.8) is given in the MKS system and includes ε_0, the permittivity of free space, thus d_{eff} [As/V^2]. Some authors exclude ε_0 from the d coefficient, in this case $d\,[\text{As/V}^2] = 8.855 \times 10^{-12} d\,[\text{m/V}]$. The conversion from the cgs system to MKS units becomes $d\,[\text{As/V}^2] = 3.68 \times 10^{-15} d\,[\text{esu}]$.

For low conversion efficiencies, (10.7) may be approximated by

$$\frac{P_{2\omega}}{P_\omega} = l^2 K \frac{P_\omega}{A} \frac{\sin^2 (\Delta kl/2)}{(\Delta kl/2)^2} \quad . \tag{10.9}$$

For a given wavelength and a given nonlinear material, K is a constant. The conversion efficiency, therefore, depends on the length of the crystal, the power density, and the phase mismatch. For a crystal of fixed length, the second-harmonic power generation is strongly dependent on the phase mismatch expressed by the sinc^2 function, as is illustrated in Fig. 10.1. In this case, a variation of Δk was obtained by changing the crystal temperature. The harmonic power is at maximum when $\Delta k = 0$, that is, at the exact phase-matching temperature. For a fixed Δk, the second-harmonic power as a function of distance l along the crystal grows and decays with a period of $\Delta kl/2 = \pi$. Half of this distance has been termed the coherence length l_c. It is the distance from the entrance face of the crystal to the point at which the second-harmonic power will be at its maximum value. The reader is warned not to confuse this parameter with the coherence length l_c of the laser beam, which was defined in Chap. 5.

For normal incidence the coherence length is given by

$$l_c = \frac{\lambda_1}{4(n_2 - n_1)} \quad . \tag{10.10}$$

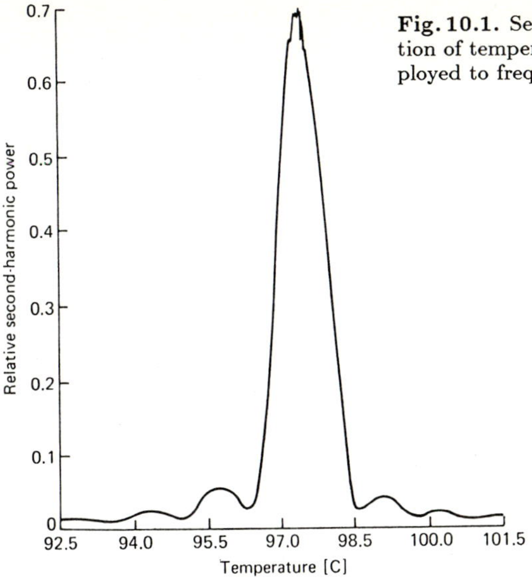

Fig. 10.1. Second-harmonic generation as a function of temperature in a $Ba_2NaNb_5O_{15}$ crystal employed to frequency-double a Nd:YAG laser

Expressing the phase mismatch Δk in terms of coherence length in (10.9), one obtains

$$\frac{P_{2\omega}}{P_\omega} = l_c^2 K \frac{4}{\pi^2} \frac{P_\omega}{A} \sin^2\left(\frac{\pi l}{2 l_c}\right) \quad . \tag{10.11}$$

The oscillatory behavior of (10.11) is shown in Fig. 10.2 for several values of l_c. For the ideal case $l_c = \infty$, the second-harmonic conversion efficiency is proportional to the square of the crystal length, at least in the small-signal approximation

$$\frac{P_{2\omega}}{P_\omega} = l^2 K \frac{P_\omega}{A} \quad . \tag{10.12}$$

Clearly, if the crystal is not perfectly phase-matched ($l_c = \infty$), the highest second-harmonic power we can expect to generate will be the signal obtained

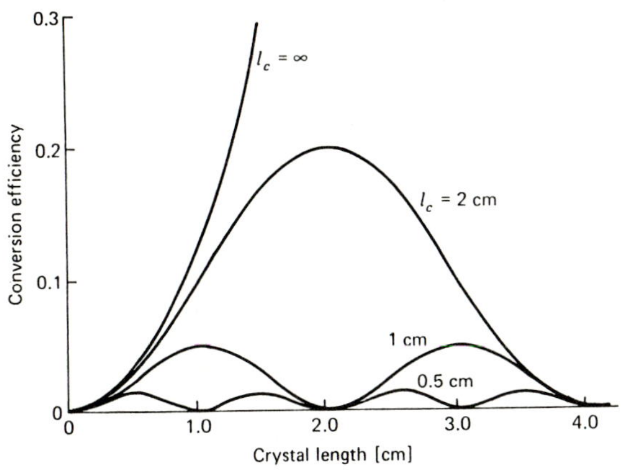

Fig. 10.2. Second-harmonic power conversion efficiency as a function of distance l from the entrance surface of a CDA crystal. Parameter is the coherence length l_c ($K = 1.3 \times 10^{-9}\,W^{-1}$; $\lambda = 1.06\,\mu m$; $I = 100\,MW/cm^2$)

after the beam propagates one coherence length, no matter how long the crystal is. The decrease in harmonic power, for example, between l_c and $2l_c$, is explained by a reversal of the power flow. Instead of power being coupled from the polarization wave into the em wave, it is coupled from the em wave into the polarization wave: i.e., the power is coupled back into the input beam. Thus we find that the power oscillates back and forth between the harmonic and the fundamental wave.

Index Matching

With typical dispersion values in the visible region, the coherence length in most crystals is limited to about 10 μm. For this reason the intensity of second-harmonic power is small. Only if n_1 can be made substantially equal to n_2 will relatively high efficiencies of frequency doubled power be obtained.

An effective method of providing equal phase velocities for the fundamental and second-harmonic waves in the nonlinear medium utilizes the fact that dispersion can be offset by using the natural birefringence of uniaxial or biaxial crystals [10.4, 5]. These crystals have two refractive indices for a given direction of propagation, corresponding to the two allowed orthogonally polarized modes; by an appropriate choice of polarization and direction of propagation it is often possible to obtain $\Delta k = 0$. This is termed phase matching or index matching.

We shall restrict our discussion to the most frequently encountered case of birefringence, that of uniaxial crystals. These crystals, to which the very important nonlinear crystals KDP and its isomorphs and $LiNbO_3$ belong, have an indicatrix which is an ellipsoid of revolution with the optic axis being the axis of rotation, as shown in Fig. 10.3. The two allowed directions of polarization and the indices for these directions are found as follows: We draw a line through the center of the ellipsoid in the direction of beam propagation (line $0P$ in Fig. 10.3). Then we draw a plane perpendicular to the direction of propagation. The intersection of this plane with the ellipsoid is an ellipse. The two axes of this ellipse are parallel to the two directions of polarization and the length of each semi-axis is equal to the refractive index in that direction.

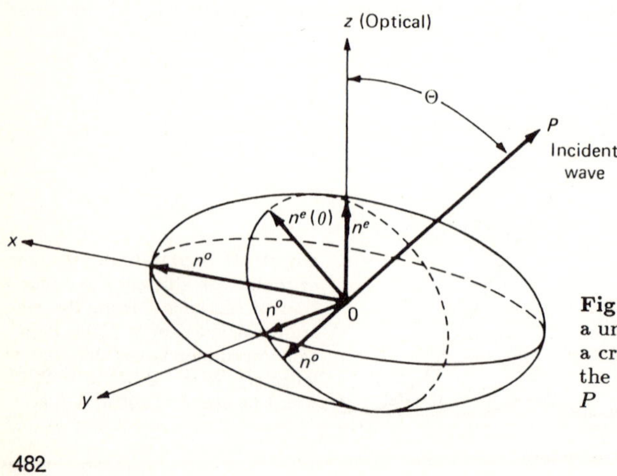

Fig. 10.3. Indicatrix ellipsoid of a uniaxial crystal. Shown is also a cross section perpendicular to the light propagation direction P

We now examine how the indices of refraction vary when the direction of propagation is changed. We notice that for the direction of polarization perpendicular to the optic axis, known as the ordinary direction, the refractive index is independent of the direction of propagation. For the other direction of polarization, known as the extraordinary direction, the index changes between the value of the ordinary index n_0 when $0P$ is parallel to z, and the extraordinary index n_e when $0P$ is perpendicular to z. When the wave propagation is in a direction Θ to the optic axis, the refractive index for the extraordinary wave is given by [10.6]

$$n^e(\Theta) = \frac{n^o n^e}{[(n^o)^2 \sin^2 \Theta + (n^e)^2 \cos^2 \Theta]^{1/2}} , \tag{10.13}$$

where the superscripts "o" and "e" refer ot the ordinary and the extraordinary rays.

Changing the point of view, we look now at the shape of the wavefronts for these two rays instead of their direction and polarization. Assuming that the input is a monochromatic point source at 0, the expanding wavefront for the o ray is spherical, whereas the spreading wavefront for the e ray is an ellipsoid. This property of crystals is described by the index surface, which has the property that the distance of the surface from the origin along the direction of the wave vector is the refractive index. For an uniaxial crystal this surface has two sheets – a sphere for ordinary waves polarized perpendicular to the optic axis with index n^o, and an ellipsoid for extraordinary waves with index $n^e(\Theta)$. By definition the optic axis is that direction at which the o and e rays propagate with the same velocity. If the value of $n^e - n^o$ is larger than zero, the birefringence is said to be positive and for $n^e - n^o$ smaller than zero the birefringence is negative; the corresponding crystals are called positive or negative uniaxial. Figure 10.4 shows a cross section of the index surface of a

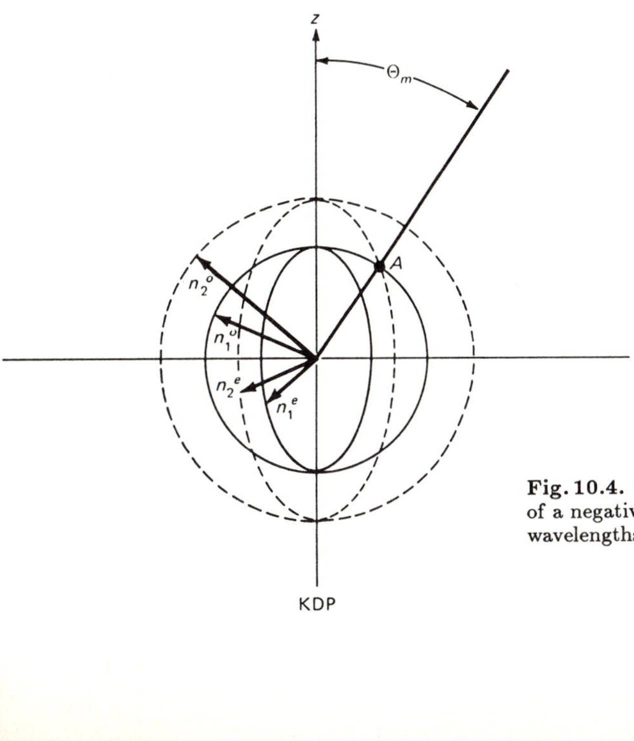

Fig. 10.4. Cross section of the index surface of a negative uniaxial crystal for two different wavelengths

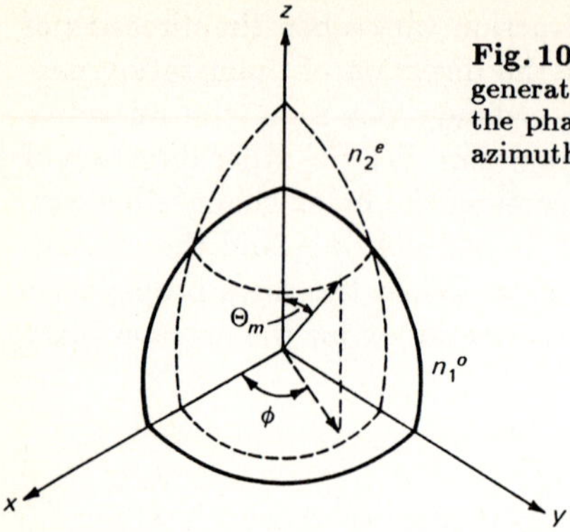

Fig. 10.5. Direction for phase-matched second-harmonic generation (type I) in a uniaxial crystal, where Θ_m is the phase-matching angle measured from z, and ϕ is the azimuth angle measured from x

negative uniaxial crystal (for the moment we consider only the solid lines n_1^{o} and n_1^{e}). The complete surfaces are generated by rotating the given sections about the z axis. The wavefront velocity v and the refractive index n are related by $v = c/n$, where c is the velocity of light.

Both refractive indices n^{o} and n^{e} are a function of wavelength. Figures 10.4 and 10.5 illustrate how the dependence of the refractive index on beam direction, wavelength, and polarization can be utilized to achieve angle-tuned phase matching. The dashed lines in these figures show the cross section of the index surfaces n_2 at the harmonic frequency. As can be seen, the negative uniaxial crystal has sufficient birefringence to offset dispersion, and the matching condition can be satisfied for a beam deviating from the z axis by the angle Θ_{m}.

The directions for phase-matched second-harmonic generation are obtained by considering the intersection of the index surfaces at the fundamental and harmonic frequencies. As was mentioned earlier, frequency doubling may be considered as a special case, where two incident waves with electric fields $\boldsymbol{E}_m$ and $\boldsymbol{E}_n$ are identical wave forms. There are two types of processes in harmonic generation, depending on the two possible orientations for the linear polarization vectors of the incident beams. In the type I process both polarization vectors are parallel: in the type II process the polarization vectors are orthogonal.

In a negative uniaxial crystal there are two loci where the index surfaces intersect and $\Delta k = 0$ [10.10]

$$
\begin{aligned}
&n_2^{\mathrm{e}}(\Theta_{\mathrm{m}}) = n_1^{\mathrm{o}} \quad \text{type I} \\
&n_2^{\mathrm{e}}(\Theta_{\mathrm{m}}) = \tfrac{1}{2}[n_1^{\mathrm{e}}(\Theta_{\mathrm{m}}) + n_1^{\mathrm{o}}] \quad \text{type II} \quad ;
\end{aligned}
\tag{10.14}
$$

first, in a symmetrical cone at Θ_{m} (type I) about the optic axis, where two o-rays at ω are matched to an e-ray at 2ω; second, in a cone at Θ_{m} (type II), where an o-ray and an e-ray at ω are matched to an e-ray at 2ω.

The harmonic power is not independent of the azimuthal angle of the phase matched direction (Fig. 10.5). In general, d_{eff} is a combination of one or several coefficients of $\boldsymbol{X}^{(3)}$ and the angles Θ and ϕ which define the direction of the wave propagation vector. For example, for KDP and its isomorphs and type I index matching, one obtains [10.7, 8]

$$d_{\text{eff}} = d_{14} \sin 2\phi \sin \Theta_{\text{m}} \quad .$$

The phase matching angle Θ_{m} is obtained by combining (10.14 and 13)

$$\sin^2 \Theta_{\text{m}} = \frac{(n_1^{\text{o}})^{-2} - (n_2^{\text{o}})^{-2}}{(n_1^{\text{e}})^{-2} - (n_2^{\text{o}})^{-2}} \quad .$$

Maximum interaction is achieved when a single beam incident on the crystal is equally divided into two fundamental beams propagating along the crystal's x and y axis. Therefore, one obtaines $\phi = 45°$ and $\boldsymbol{E}_m = \boldsymbol{E}_n$ in (10.3). In type I phase matching the fundamental beams are polarized perpendicular to the optical axis and the harmonic beam is polarized parallel to the optical axis of the crystal.

For type II phase matching one obtains

$$d_{\text{eff}} = d_{14} \cos 2\phi \sin 2\Theta_{\text{m}} \quad .$$

In this case the beam propagation vector is orientated at 90° with respect to the x axis ($\phi = 90°$) for maximum interaction. A single linearly polarized fundamental beam incident on the crystal may be equally divided into o and e rays by orienting the polarization vector at 45° with respect to the x axis. The polarization vector of the harmonic beam is rotated 45° from the fundamental beam with the polarization parallel to the optical axis.

Comparing type I and II phase matching, we find that type I is more favorable when Θ_{m} is near 90°, whereas type II leads to a higher d_{eff} when Θ_{m} lies near 45°. For $LiNbO_3$ and type I phase matching we obtain

$$d_{\text{eff}} = d_{31} \sin \Theta_{\text{m}} + d_{22} \cos \Theta_{\text{m}}(4 \sin^3 \phi - 3 \sin \phi) \quad .$$

Critical Phase Matching

If phase matching is accomplished at an angle Θ_{m} other than 90° with respect to the optic axis of a uniaxial crystal, there will be a small angle between the direction of power flow (Poynting vector) of the fundamental and second harmonic.

For a negative uniaxial crystal and type I phase matching, this angle is given by [10.9]

$$\tan \varrho = \frac{(n_1^{\text{o}})^2}{2} \left(\frac{1}{(n_2^{\text{e}})^2} - \frac{1}{(n_2^{\text{o}})^2} \right) \sin 2\Theta \quad . \tag{10.15}$$

The angle ϱ has the effect of limiting the effective crystal volume over which harmonic generation can take place. The beams separate or walk off in a distance of order

$$l_a \approx \frac{a}{\varrho} \tag{10.16}$$

called the aperture length, where a is the beam diameter. This effect, although present, is not too serious in large-diameter beams. For example, in KDP one obtains $\varrho = 1.8°$ which results in $l_a = 15.8$ cm for a beam of 5 mm diameter.

Another, much more serious, limitation of angular phase matching is due to the divergence of the interacting beams. For second-harmonic generation in a negative crystal we find from (10.6) that $\Delta k = 0$ if $n_1^o = n_2^e(\Theta_m)$, which is exactly true only at $\Theta = \Theta_m$. In Sect. 10.1.2 it will be shown that for a small deviation $\delta\Theta$ from the phase-matching angle one obtains $\Delta k \approx \delta\Theta \sin 2\Theta_m$ [10.7, 8, 10, 11]. This linear variation of Δk with Θ can be large enough to cause practical difficulties in real devices. Consider as an example second-harmonic generation in KDP at 1.065 μm with $\Theta_m = 42°$. The linear change in Δk with Θ is sufficiently great to restrict the divergence from the phase-matched direction to approximately 1 mrad if the coherence length is to be greater than 1 cm. Phase matching under these unfavorable conditions is termed "critical phase matching".

Noncritical Phase Matching

If the refractive indices can be adjusted so that $\Theta_m = 90°$, by variation of a parameter such as the temperature or chemical composition of the crystal, the linear change in Δk with Θ is eliminated and the restriction on beam divergence is due to a much smaller quadratic term $\Delta k \sim (\delta\Theta)^2$ [10.7, 8, 10, 11]. In addition, at $\Theta_m = 90°$ there are no walk-off effects due to double refraction. For example, second-harmonic generation in $LiNbO_3$ occurs at 90° for 1.064-μm radiation provided the crystal is at a temperature of $\sim$47° C. The allowable divergence from the phase-matched direction is greater than 10 mrad if the coherence length is to be 1 cm. For these reasons, 90° phase matching is often called noncritical phase matching. Provided that n_1^o and n_2^e are nearly equal and $d(n_1^o - n_2^e)/dT \neq 0$, noncritical phase matching can be achieved by temperature tuning the crystal.

Because of its practical importance, in this section we have emphasized phase matching in negative uniaxial crystals. The reader interested in a systematic review of phase matching in uniaxial and biaxial crystals is referred to the paper by *Hobden* [10.10]. Frequency doubling with focused Gaussian beams has been discussed in [10.12, 13]. The subject of nonlinear optics has been treated in several books [10.3, 11, 14–16] and tutorial review articles [10.16, 17]. For a general introduction to crystal optics the reader is referred to standard texts [10.6, 19, 20].

10.1.2 Parameters Affecting the Doubling Efficiency

High-efficiency second-harmonic conversion depends on parameters which are related to the laser source, such as power density, beam divergence, and spectral linewidth, and parameters associated with the harmonic generator, such as the value of the nonlinear coefficient, crystal length, angular and thermal deviation from the optimum operating point, absorption, and inhomogeneities in the crystal. The dependence of the second-harmonic power on these parameters will be examined in this subsection.

The Dependence of Harmonic Generation on Laser Parameters

Power Density. From (10.9) it follows that the conversion efficiency is proportional to the power density of the fundamental beam, whereas the harmonic power itself is proportional to the product of fundamental power and power density. At conversion efficiencies above 20 %, the second-harmonic generation starts to deviate markedly from the linear relationship of (10.9) because of depletion of the fundamental beam power. At these high efficiencies, (10.7) should be used. Figure 10.6 shows a plot of conversion efficiency versus power density for a CDA crystal employed to frequency-double a Nd laser. The parameter is the coherence length l_c. The curve for $l_c = \infty$ gives the theoretically maximum achievable conversion efficiency as a function of power density. Figure 10.6 illustrates the strong dependence of conversion efficiency on the coherence

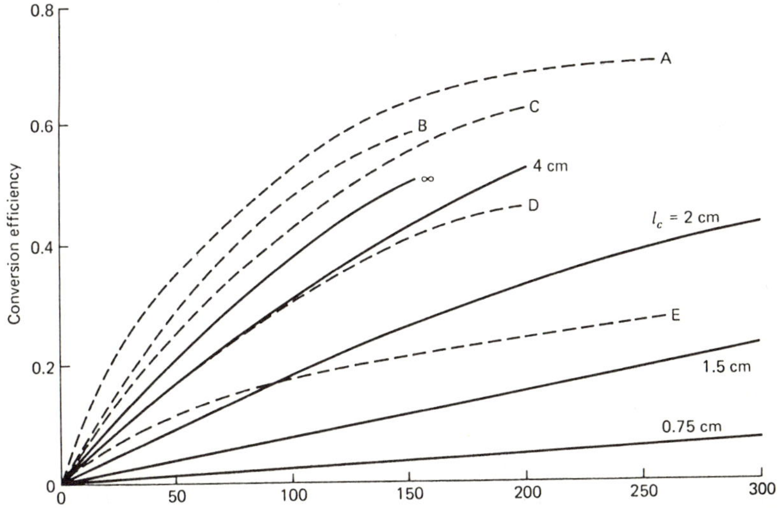

Fig. 10.6. Second-harmonic conversion efficiency versus power density. Solid lines: Theoretical values for a 2-cm-long CDA crystal with coherence length l_c as parameter ($\lambda = 1.06\,\mu$m, $n = 1.55$, $K = 1.3 \times 10^{-9}\,\mathrm{W}^{-1}$). Dashed lines: Experimental values. A, Nd : glass laser, 2.5-cm-long KDP crystal [10.21]; B, ruby laser, 1.45-cm-long RDA crystal [10.22]; C, Nd : YAG laser, 1.75-cm-long CDA crystal [10.23]; D, Nd : YAG laser, 1.35-cm-long CD*A crystal [10.23]; E, Nd : YAG laser, CDA crystal [10.24]

length l_c. In almost all practical cases the coherence length is limited by the beam divergence and the bandwidth of the laser beam and by angular and thermal deviations of the crystal from the phase-matching angle and temperature. Also plotted in Fig. 10.6 are experimentally achieved conversion efficiencies as a function of power density for a variety of crystals.

Beam Divergence. When collinear phase-matched second-harmonic generation is used, the light waves will have a small but finite divergence. It is necessary to consider the mismatch Δk of the wave vector for small deviations $\delta\Theta_m$ from the phase-matched direction. An expansion for $n_1^o - n_2^e(\Theta)$ taken for a direction close to the perfect phase-matching direction Θ_m yields

$$n_1^o - n_2^e(\Theta) = \frac{\partial[n_1^o - n_2^e(\Theta)]}{\partial\Theta}(\Theta - \Theta_m) \quad . \tag{10.17}$$

The expression given in (10.13) for the dependence of n_2^e on the angular direction in the crystal can be very well approximated by [10.21]

$$n_2^e(\Theta) = n_2^o - (n_2^o - n_2^e)\sin^2\Theta \quad . \tag{10.18}$$

Introducing (10.18) into (10.17) gives

$$n_1^o - n_2^e(\Theta) = \delta\Theta(n_2^o - n_2^e)\sin 2\Theta_m \quad , \tag{10.19}$$

where we note that $\partial n_1^o/\partial\Theta = 0$, and $\partial n_2^o/\partial\Theta = 0$. From (10.9) follows that the conversion efficiency will be reduced to one-half of its peak value for $\mathrm{sinc}^2(\Delta kl/2) = 0.5$ or

$$\frac{\Delta kl}{2} = 1.39 \quad . \tag{10.20}$$

After introducing (10.19, 20) into (10.6) and multiplying the resulting expression by n_1^o we obtain

$$\Delta\Theta = \frac{0.44\lambda_1 n_1^o/l}{(n_2^o - n_2^e)\sin 2\Theta_m} \quad , \tag{10.21}$$

where $\Delta\Theta = 2n_1\delta\Theta$ is the full angle at the half-intensity points of a curve representing second-harmonic power generation versus the angular deviation from the phase matching direction. The angle $\Delta\Theta$ can also be interpreted as the beam divergence of the laser at which conversion efficiency drops to one-half of its peak value. Under noncritical phase-matching conditions ($\Theta_m = 90°$), we can make the approximation $\sin 2(90° + \delta\Theta) \approx 2\delta\Theta$, and instead of (10.21) we obtain

$$\Delta\Theta = \left(\frac{0.44\lambda_1/l}{(n_2^o - n_2^e)}\right)^{1/2} n_1 \quad . \tag{10.22}$$

Spectral Linewidth. Expanding $n_1^{\mathrm{o}} - n_2^{\mathrm{e}}(\Theta)$ for small wavelength changes around the central wavelength λ_0 at which phase matching occurs results in [10.25]

$$n_1^{\mathrm{o}} - n_2^{\mathrm{e}}(\Theta) = \left(\frac{\partial n_1^{\mathrm{o}}}{\partial \lambda_1} - \frac{1}{2}\frac{\partial n_2^{\mathrm{e}}(\Theta)}{\partial \lambda_2}\right)(\lambda - \lambda_0) \quad . \tag{10.23}$$

From (10.6, 20, 23) we obtain an expression for $\Delta\lambda = 2\delta\lambda$, where $\delta\lambda = \lambda - \lambda_0$ is the deviation from the phase-matching wavelength at which the doubling efficiency drops to one-half

$$\Delta\lambda = 0.44\lambda_1/l\left(\frac{\partial n_1^{\mathrm{o}}}{\partial \lambda_1} - \frac{1}{2}\frac{\partial n_2^{\mathrm{e}}(\Theta)}{\partial \lambda_2}\right) \quad . \tag{10.24}$$

Spectral Brightness. From the foregoing considerations it becomes clear that in order to achieve maximum second-harmonic power, the laser source should have a high power density, small beam divergence, and narrow linewidth. These properties of the laser can also be expressed by a single parameter, namely the spectral brightness [W/cm^2 sterad Å]. For high-efficiency second-harmonic generation the laser must exhibit a high spectral brightness, which can be achieved by transverse and longitudinal mode selection. Experimental data showing the dependence of second-harmonic generation on the mode structure of the laser can be found in [10.26, 27]. In general, one finds that ruby and Nd : YAG lasers have sufficiently narrow linewidths for efficient harmonic generation. In these lasers attention is focused mainly on obtaining a diffraction limited beam, i.e., TEM_{00}-mode operation. In Nd : glass, however, the broad linewidth of the laser without axial mode selection can present a severe limitation in obtaining high conversion efficiencies.

If a laser has many randomly phased longitudinal modes, account must be taken of the intensity fluctuations in second-harmonic generation. These fluctuations enhance the average second-harmonic output by a factor $(2 - 1/n)$ when n randomly phased longitudinal modes are present [10.3, 21, 27, 28]. This is true only for low conversion efficiencies, where second-harmonic generation follows a square-law dependence. For high conversion efficiencies, where second-harmonic generation can be approximated by a linear dependence, this factor approaches one.

Very high conversion efficiency for the entire pulse requires that all the radiation incident on the nonlinear crystal be converted efficiently, regardless of its intensity, bandwidth, polarization, or amplitude and phase variations. The range of intensity over which the efficiency is high is especially important. The intensity is necessarily nonuniform because it drops to zero at the spatial and temporal edges of the beam, and there may also be amplitude nonuniformities. The standard technique of frequency conversion uses a single crystal of a nonlinear material for frequency doubling. With a single nonlinear crystal, the dynamic range is necessarily limited to moderate values. *Eimerl* [10.29] found that using two crystals for each conversion step, the dynamic range can be much larger. The two crystals are arranged so that the output wave generated in the first one is polarized orthogonally to that generated in the second. The two harmonics therefore add in quadrature, which gives the technique its name quadratic frequency doubling. The conversion efficiency of these quadrature arrangements is much less-sensitive to laser pulse nonuniformities than that in single-crystal methods. Consequently, very high conversion efficiency is possible for typically nonuniform laser pulses.

Parameters of the Harmonic Generator which Affect Doubling Efficiency

Temperature. The indices of refraction of the crystal at the phase-matching temperature T_0 can be expanded in a Taylor series for a nearby temperature [10.30]

$$n_1 - n_2 = (T - T_0)\frac{d(n_2 - n_1)}{dT} \quad . \tag{10.25}$$

With the aid of (10.6, 20) we obtain an expression of the temperature sensitivity of the doubling crystal with respect to second-harmonic generation,

$$\Delta T = \frac{0.44\lambda_1}{l\, d(n_2^e - n_1^o)/dT} \quad , \tag{10.26}$$

where ΔT is the full width at half-maximum of the temperature range over which second-harmonic generation is possible in a particular crystal (Fig. 10.1). Temperature changes of the doubling crystal may be the result of ambient temperature variations, or they may be caused by absorption losses in the crystal. Experimentally determined thermal tuning ranges for several crystals of practical interest are shown in Fig. 10.7.

Phase-Matching Angle. The sensitivity required to angular-adjust a doubling crystal and maintain its alignment is given by (10.21). The angle $\Delta\Theta$ now becomes the full acceptance angle over which the crystal can be tilted before

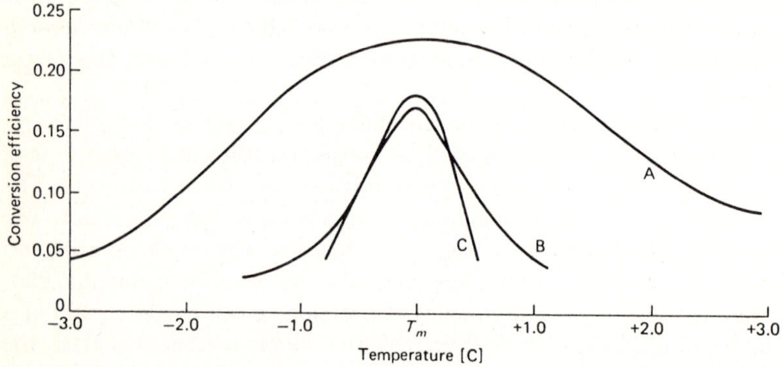

Fig. 10.7. Thermal tuning curves of several nonlinear crystals. A, CDA, 1.06 μm, $I = 100$ MW/cm^2, 1.8-cm-long crystal; B, ADP, 0.53 μm, $I = 10$ MW/cm^2, 4-cm-long crystal; C, RDA, 0.69 μm, $I = 100$ MW/cm^2, 1.5-cm-long crystal

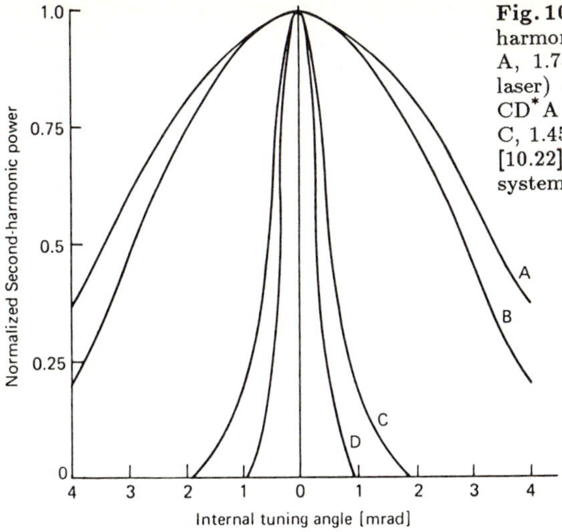

Fig. 10.8. Angular dependence of second-harmonic power in nonlinear crystals. A, 1.75-cm-long CDA crystal (Nd:YAG laser) at 20° C [10.23]; B, 1.35-cm-long CD*A (Nd:YAG laser) at 20° C [10.23]; C, 1.45-cm-long RDA crystal (ruby laser) [10.22] D, 2.5-cm-long KDP (Nd:glass system), [10.21]

the second-harmonic power decreases to less than half. Figure 10.8 illustrates the angular tuning range for several crystals of practical interest.

Absorption. Absorption in the crystal will lead to thermal gradients as well as thermally induced stresses. The associated refractive index nonuniformities severely restrict the crystal volume over which phase matching can be achieved [10.31, 32]. Heating of the nonlinear crystal due to absorption is particularly a problem in doubling experiments involving high-power cw beams. A self-induced thermal distribution in the nonlinear material can be caused by absorption of the fundamental or harmonic beams or by multiphoton absorption processes [10.24, 33].

Optical Homogeneity. Index inhomogeneities in nonlinear optical crystals are an important limiting factor in second-harmonic generation, since the phase-matching condition cannot be satisfied over the whole interaction path of the beams. Effects of crystal inhomogeneities, such as striae, have been discussed by several authors [10.30, 34].

Nonlinear Coefficient. From purely theoretical considerations it would seem that the crystal with the highest nonlinear coefficient would be the most desirable material to use in a doubling experiment. However, in practical situations it turns out that damage threshold, optical quality, angular and thermal tuning range, acceptance angle, etc., are equally important parameters. From (10.11) it follows that it is possible to trade off the nonlinear coefficient d for the interaction length l_c. A material with a low nonlinear coefficient but with properties which allow a long interaction length can be as efficient as a short crystal with a high nonlinear coefficient.

Figure of Merit. The parameters which affect the doubling efficiency can be combined in a figure of merit which characterizes either the laser source, the crystal, or both. For example, *Eimerl* [10.35] has shown that the performance of a nonlinear frequency doubler is uniquely determined by the output brightness of the pump laser and the figure of merit of the nonlinear material. Furthermore, the analysis revealed that the maximum conversion efficiency is only dependent on the optical properties of the material, and the brightness of the pump laser, but is independent of the physical size of the nonlinear crystal. The size is adjusted to avoid optical damage in the material. The constraint is that the material under consideration must be available in large enough size to fit the requirements of the laser system. The nonlinear frequency doubling crystal is characterized by the nonlinear index, and the parameters defining the tolerance to angular, wavelength, and temperature detuning from phase-matched conditions.

In the analysis mentioned above second-harmonic conversion is characterized by two parameters relating the conversion efficiency, the pump intensity and the nonlinear properties of the material. These are the nonlinear "drive" which is the source term for generation of the electric field at the second harmonic, and the detuning which is the phase mismatch between second-harmonic waves at the exit and entrance planes of the crystal. The drive is given by

$$\eta_0 = C^2 I l^2 \quad , \tag{10.27}$$

where C is proportional to the nonlinear coefficient of the material, I is the pump intensity, and l is the crystal length. The detuning is given by

$$\delta = 0.5\,\Delta k l \quad , \tag{10.28}$$

where Δk is the wavevector mismatch related to the angular sensitivity β of the crystal by $\Delta k = \beta_\theta \Delta\theta$. The divergence of the laser is $\Delta\theta$. The criteria for efficient harmonic generation are that

$$l < \eta_0 < 5 \quad , \quad \text{and} \quad \delta \ll \pi/10 \quad . \tag{10.29}$$

10.1.3 Properties of Nonlinear Crystals

Survey of Materials

The nonlinear crystals can be classified into two groups according to their physical properties. Crystals grown from water solutions are fragile, hygroscopic, and sensitive to thermal shock. The crystals of this group, to which KDP and its isomorphs belong, are somewhat difficult to handle because the crystals are soft, and the polished faces may be fogged if they are held with bare hands or exposed to a humid atmosphere. On the other hand, the crystals are easy to grow, they are available in large sizes, and they are of excellent optical quality. Crystals grown from the melt are relatively hard, nonhygroscopic and less sensitive to thermal shock. Important members of this group of crystals are $LiNbO_3$ and KTP. The optical quality is usually inferior to water-grown crystals because of refractive index nonuniformities associated with the crystal growth condition. Table 10.1 lists the primary nonlinear materials which are used in conjunction with solid-state lasers.

KDP and Its Isomorphs. The crystals of this family have proven to be the most important group of useful second-harmonic generators. The crystals which are all negative uniaxial, belong to point group $\bar{4}2$ m and, thus, have a tetragonal symmetry. The crystals are grown at room temperature from a water solution which makes large, distortion-free, single crystals available. Transparency exists from 0.22 to 1.6 μm for the phosphates, and from about 0.26 to 1.6 μm for the arsenates. Deuteration increases the infrared limit to about 1.9 μm. The greatest attributes of this family of crystals as a nonlinear device material are their resistance to laser damage and their high optical quality. Opposing these advantages, there are several disadvantages. The materials have fairly low refractive indices, typically 1.50 to 1.55, so that they also have small nonlinear coefficients. All of the KDP isomorphs are water-soluble and have a maximum safe operating temperature of about 100° C. The crystals are sensitive to thermal shock, and should be heated slowly at a rate of less than about 5° C/min.

Table 10.1. Nonlinear optical coefficients for primary nonlinear materials

Material	Symbol	Formula	Relative nonlinear coefficient d/d_{36} (KDP)	Nonlinear coefficient [a] $[10^{-24}\ \mathrm{As/V^2}]$	Index of refraction n^0 (1.06 μm)	Reference
Ammonium dihydrogen phosphate	ADP	$NH_4H_2PO_4$	1.2	$d_{36} = 4.96$	1.50	[10.27]
Potassium dihydrogen phosphate	KDP	KH_2PO_4	1.0	$d_{36} = 4.16$	1.49	[10.36]
Potassium dideuterium phosphate	KD*P	KD_2PO_4	1.06	$d_{36} = 4.43$	1.47	[10.36]
Rubidium dihydrogen phosphate	RDP	RbH_2PO_4	0.92	$d_{36} = 3.81$	1.49	[10.37, 38]
Rubidium dihydrogen arsenate	RDA	RbH_2AsO_4	0.64	$d_{36} = 2.66$	1.55	[10.22]
Cesium dihydrogen arsenate	CDA	CsH_2AsO_4	0.92	$d_{36} = 3.81$	1.55	[10.23, 36]
Cesium dideuterium arsenate	CD*A	CsD_2AsO_4	0.92	$d_{36} = 3.81$	1.55	[10.23, 36]
Lithium iodate	–	$LiIO_3$	14.0	$d_{31} = 58.4$	1.86	[10.39]
Lithium niobate	–	$LiNbO_3$	13.4	$d_{31} = 55.8$	2.23	[10.40]
Barium sodium niobate	–	$Ba_2NaNb_5O_{15}$	38.0	$d_{32} = 159.4$	2.26	[10.41]
Potassium titanyl phosphate	KTP	$KTiOPO_4$	29.2	$d_{33} = 121.3$	1.74	[10.42–45]
Beta-barium borate	BBO	β-BaB_2O_4	4.1	$d_{22} = 17.1$	1.66	[10.46, 47]
Urea	–	–	2.98	$d_{14} = 12.4$	1.48	[10.48]
Silver gallium sulfide	–	$AgGaS_2$	46.8	$d_{36} = 194.8$	2.46	[10.49]
Silver gallium selenide	–	$AgGaSe_2$	83.0	$d_{36} = 345.3$	2.49	[10.50]

[a] Absolute value of nonlinear coefficients is based on $d_{36}(\mathrm{KDP}) = 1.15 \times 10^{-9}$ [esu]. Conversion: $d[\mathrm{As/V^2}] = 8.855 \times 10^{-12} d[\mathrm{m/v}] = 3.68 \times 10^{-15} d$ [esu]

ADP and KDP were the first crystals used for the demonstration of phase-matched second-harmonic generation. Isomorphs of these materials have similarly been used in nonlinear optics, the most widely known isomorphs being deuterated KDP, which is normally designated as KD*P. The most complete compilation of data on KDP, KD*P, and ADP can be found in [10.51]; other references of interest are [10.27, 52, 53]. Some of the other isomorphs have been used because the temperature dependence of their refractive index allows 90° phase matching for particular interactions. CDA and CD*A 90° phase-match the important 1.06-μm transition of Nd:YAG and Nd:glass [10.23, 24, 54]. RDP and RDA are mainly employed to frequency-double the output from ruby lasers [10.22, 37, 55].

Calculations of various basic properties of KDP related to the generation of the second, third, and fourth harmonics of 1 μm laser radiation for five different conversion processes can be found in [10.56]. The paper contains a detailed tabulation of phase matching angles, sensitivity to angular mismatch, wavelength, and temperature.

The range of wavelength and temperature over which KDP isomorphs 90° phase-match is shown in Fig. 10.9. Figure 10.10 shows the crystal and electric vector orientation for harmonic generation in KDP and its isomorphs. KDP and KD*P crystals are commercially available for either type I or type II angle-tuned phase matching. Type II is more efficient than type I, but places more restrictions on input beam quality. The crystals are usually supplied in cells

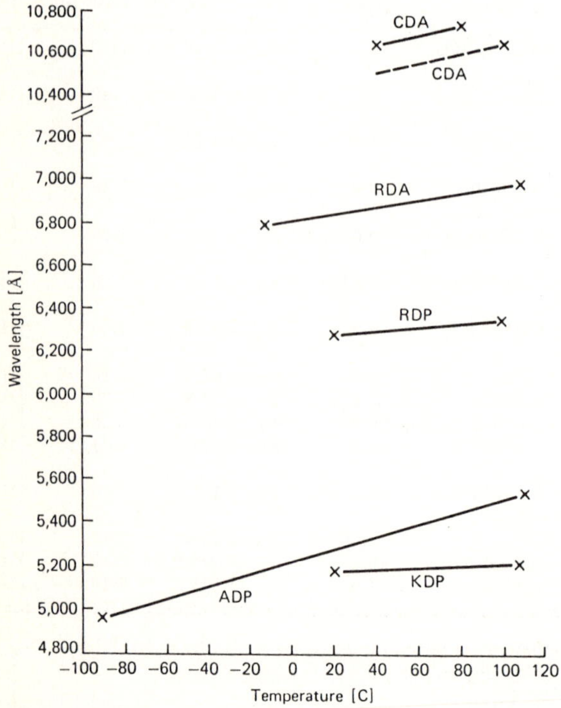

Fig. 10.9. Fundamental wavelength versus temperature in KDP-type crystals for 90° phase matching [10.53]

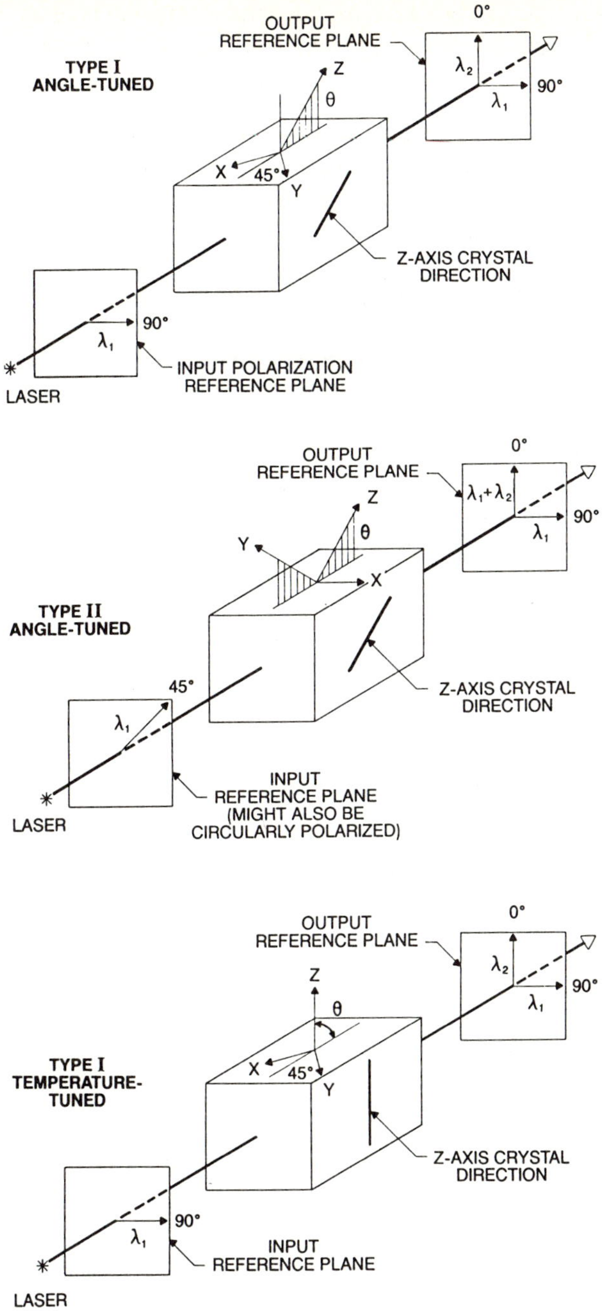

Fig. 10.10. Crystal and electric vector orientation for harmonic generation in KDP and its isomorphs [10.57]

sealed in dry inert atmosphere or index matching fluid. ADP, CDA and RDA crystals are also available mounted in heated cells for 90° phase matching.

$LiNbO_3$. This material is nonhygroscopic and hard, taking a good polish readily. The uniaxial crystal, belonging to the trigonal point group 3 m, has a large nonlinear coefficient relative to KDP. The crystals of lithium niobate are transparent in the region 0.42 to 4.2 μm. Temperature sensitivity of birefringence is such that, by varying the temperature, phase matching can be achieved at 90° to the optical axis.

The temperature of Li_xNbO_3 for noncritical phase matching depends strongly on the composition of the crystal. However, only crystals of the congruent composition ($x = 0.972$) can be pulled from the melt with good optical homogeneity. For this composition the phase matching temperature is about 4° C for second-harmonic generation of Nd lasers. Unfortunately, $LiNbO_3$ is particularly susceptible to photorefractive damage from propagation of visible or uv radiation. This optical-index damage is reversible if the ambient temperature of the crystal is increased to about 170° C which is above the annealing temperature for the photorefractive damage. So-called "hot" $LiNbO_3$ is prepared by increasing the proportion of the Li_2O or MgO in the melt. The nonstoichiometrically grown "hot" crystals suffer from optical inhomogeneities in the form of striations normal to the growth direction. Despite the high nonlinear coefficient, and the ease of handling and polishing, the poor optical quality of $LiNbO_3$, the need for operation at elevated temperatures combined with a relatively low damage threshold have severely limited practical applications for this material.

The results of the extensive studies of the growth and crystal structure of lithium niobate have been published in the literature [10.58, 59]. Other papers have highlighted some of the practical problems that arise in the use of lithium iobate [10.40, 60–64]. Investigations of the radiation-induced index inhomogeneities which are either caused by ion impurities or hydrogen ions (protons) in the lithium niobate can be found in [10.65–67].

Barium Sodium Niobate. This crystal is biaxial; however, since n_x is approximately equal to n_y to a first approximation, at least insofar as the gross properties of phase matching are concerned, the crystal can be assumed to be uniaxial. This crystal is physically similar to $LiNbO_3$. However, it possesses higher nonlinear coefficients and is not readily susceptible to (reversible) optical index damage. At about 100° C noncritical phase matching is possible for 1.06 μm radiation. The large nonlinear coefficient combined with 90° phase matching, has made possible very efficient second-harmonic generation from a continuous-wave Nd: YAG laser, providing an intense source of 5300-Å radiation [10.41, 68, 69]. However, the use of $Ba_2NaNb_5O_{15}$, which transmits light in the region 0.46 to 1.1 μm, presents other problems. Great difficulties are involved in preparing high optical quality crystals. Commercially available crystals suffer from striation, scatter centers, and built-in strain. The most persistent defects are growth striations which are periodic variations of refractive index normal to the growth direction. Light is diffracted and refracted by these index variations, making good phase matching very difficult to achieve. Because of these difficulties the crystal is no longer offered commercially.

Lithium Iodate and Lithium Formate Monohydrate. Lithium iodate crystals are transparent in the region 0.32 to 4.0 μm, while formate crystals are transparent in the 0.25- to 1.2-μm range. Both crystals are grown from water solutions. Lithium iodate has nonlinear coefficients which are comparable to those of lithium niobate, although it does not suffer from the refractive index damage problems that plague the niobate. Also, $LiIO_3$ possesses an optical quality which is far superior to $Ba_2NaNb_5O_{15}$. Consequently, lithium iodate has found application in efficient intracavity second-harmonic generation in Nd:YAG lasers. Since the refractive indices are very stable with respect to temperature, phase matching is achieved by angle tuning [10.39, 70–72]. Interesting features of the formate crystals are a near noncritical phase matching for Nd laser radiation and an insensitivity of the phase-match direction to temperature changes [10.73, 74].

Potassium Titanyl Phosphate

The recently developed crystal $KTiOPO_4$ (KTP) has been shown to possess excellent qualities for uses as a nonlinear optical material. It offers nonlinear optical coefficients comparable to $Ba_2NaNb_5O_{15}$, can be phase matched using either a type-I or type-II interaction, has a high damage threshold, and is inert chemically with a high mechanical stability.

Using type-II interactions KTP exhibits unique combination of properties. In this configuration it has large angular and temperature bandwidths as well as high nonlinear coefficients and damage threshold. KTP crystallizes in the orthorhombic point group mm2 [10.75]. The crystal structure, refractive indices, and phase match parameters have been reported in [10.42, 43–45, 76]. KTP is the best nonlinear material for Nd : lasers to emerge in recent years. At the present time, the major drawback is the difficult growth process required to

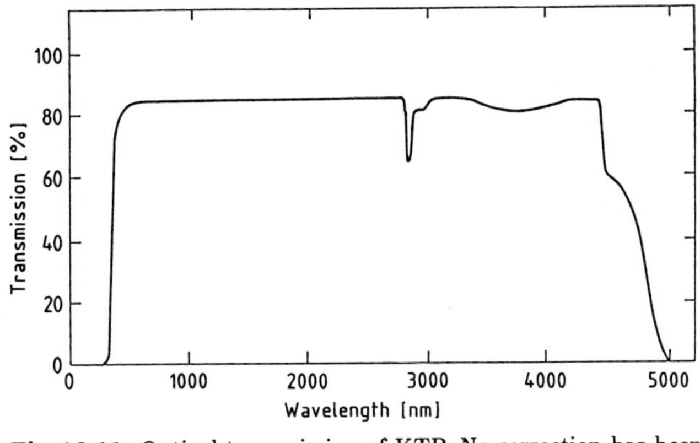

Fig. 10.11. Optical transmission of KTP. No correction has been made for reflection losses [10.44]

produce these crystals, which leads to high cost and small size crystals. Until recently, crystals produced by the flux or hydrothermal growth method, have been limited to a few millimeters in size. However, currently crystals up to 1 cm^3 are commercially available. Figure 10.11 shows the transmission curve for KTP. The material is transparent from 0.35 μm to about 4.5 μm. The optical spectrum in structure-free except for traces of OH^- absorption bands observed at 2.8 and 3.8 μm.

Figure 10.12 shows the crystal orientations for phase match condition for type-II interaction at 1.06 μm. It should be noted, that the phase match angle ϕ measured from the x-axis in the xy plane is different for flux and hydrothermally grown crystals.

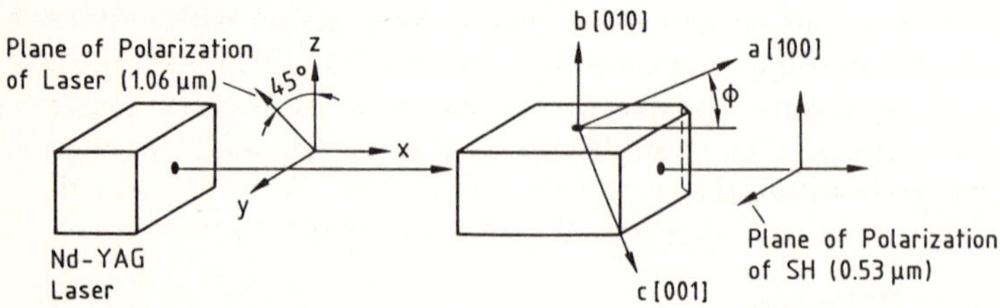

Fig. 10.12. Orientation of KTP for Type-II interaction at 1.06 μm. $\phi = 26°$ for hydrothermally and $\phi = 21°$ for flux-grown material [10.43]

Beta-Barium Borate. The material β-BaB_2O_4 (BBO) is a new nonlinear optical crystal which possesses many excellent properties for nonlinear frequency conversion in a spectral range that extends from the ultraviolet to the mid-infrared [10.77]. The salient features of the BBO crystal are a wide transparency and an angle-tuning range for phase matching, a very high damage threshold and a wide temperature acceptance width. The nonlinear coefficient of BBO is about four times higher than the comparable value of KDP. Crystal sizes up to 1 cm^3 are currently available. Beta-barium borate is a very new material and the data base is limited at the present time. Relevant data can be found in [10.46, 47, 77–80]. An interesting feature of BBO is the fact that it is transparent out to 1890 Å, second-harmonic generation down to 2048 Å has been demonstrated [10.80].

Organic Materials. Efficient harmonic frequency conversion has been reported in a number of organic crystals. Interest in organic materials for applications in nonlinear optics stems from the extremely large nonlinearities, high damage thresholds and transparency in the uv, which some of these crystals have shown, although these properties are not all found in the same crystal. For an organic molecule to exhibit a high optical nonlinearity it must possess a conjugated system of bonds which give rise to a strong π-electron delocalization. By adding donor and acceptor groups at the opposite ends of the conjugated system this delocalization can be further increased. The strong charge transfer between such groups leads to a resonance enhancement of the nolinear susceptibility.

Finally, the molecule must be able to crystallize in a structure lacking a center of inversion. This can be achieved by adding to it suitable substituents or combining it with another crystal.

A large number of organic crystals have been selected, synthesized and investigated for possible applications as nonlinear optical materials [10.81–84].

The best-known nonlinear organic crystal is urea [10.83, 84] which has been employed in a number of parametric oscillators in the wavelength range of 500 to 1125 nm [10.48, 85, 86]. Other organic crystals which are currently being explored for nonlinear applications include MNA [10.87–89], DAN [10.88, 89], MAP [10.83], COANP, PAN and MBANP (see, for example, [10.90] for comparison of these last three materials).

The majority of the experiments with these nonlinear organic crystals has been carried out at 1.06 μm radiation for phase-matched second-harmonic generation or for parametric frequency conversion. The crystals are usually used in the form of thin films or platelets ranging in thickness from 15 μm to a few millimeters. Work on nonlinear frequency conversion of laser radiation with organic crystals is still in the early R&D phases and only a few materials, such as urea, for example, are commercially available at the present time.

Critical Issues of Material Selection. For the design of a frequency converter the following properties of the nonlinear crystal are of key importance: value of the nonlinear coefficient, damage threshold, phase-matching and transparency range, available crystal size and optical homogeneity, and chemical and mechanical stability.

These parameters and properties will be reviewed for the important nonlinear materials.

Nonlinear Coefficient. A large nonlinear coefficient is a necessary, although not a sufficient condition for a crystal to yield a high conversion efficiency. KDP has become the reference against which other nonlinear materials are compared.

Lithium iodate and lithium niobate have nonlinear coefficients which are about equal and an order of magnitude higher than KDP and its isomorphs. If the optical nonlinearity of a material were the only parameter to consider, then barium sodium niobate would be far superior compared to the other materials listed in Table 10.1 because it possesses a nonlinear factor three times larger than $LiNbO_3$ and $LiIO_3$. One of the most promising new crystals is KTP which has almost as high a nonlinear coefficient as barium sodium niobate.

Damage Threshold. The selection of a particular nonlinear material for use in a solid-state laser is mainly predicted upon high damage threshold, with secondary emphasis placed on the magnitude of the nonlinear coefficients. Table 10.2 summarizes damage threshold levels and absorption losses of commercially available crystals. The damage threshold for KDP and its isomorphs is quite high, with somewhat higher values for the phosphates than for the arsenates. Very high damage thresholds have also been reported for KTP and BBO.

Table 10.2. Typical damage threshold level and absorption for nonlinear crystals

Material	Absorption [cm^{-1}] 1.06 μm	0.53 μm	0.26 μm	Damage threshold [MW/cm^2]	Reference
KDP	0.07	0.01	0.20	400	[10.21, 24, 63]
KD*P	0.006	0.006	0.02	400	[10.24]
ADP	0.006	0.01	0.01	400	[10.24, 91]
CDA	0.04	0.01	–	300	[10.24, 38]
CD*A	0.01	0.009	–	300	[10.24, 38]
RDP	0.04	0.01	0.02	300	[10.37]
RDA	–	–	–	300	[10.22, 38]
$Ba_2NaNb_5O_{15}$	0.002	0.04	–	10 –25	[10.24, 92]
$LiIO_3$	0.002	–	–	10	[10.24, 93]
$LiNbO_3$	–	0.002	–	6–40	[10.58, 94]
KTP	0.001	0.001	–	250	[10.42, 44]
BBO	–	–	–	4600	[10.78]

From Table 10.2 it is clear that for high-power lasers, such as pumped pulsed, Q-switched systems, our choice of nonlinear materials is limited to crystals belonging to the KDP family, or to KTP and BBO if the small crystal size of these new materials is acceptable. The other nonlinear materials, such as $LiNbO_3$, $Ba_2NaNb_5O_{15}$, and $LiIO_3$, have such low damage thresholds that their application has been restricted to harmonic generation of cw-pumped Nd:YAG lasers.

Phase-Matching and Transparency Range. In Table 10.3 the phase-matching conditions of commonly used nonlinear crystals are listed for wavelengths of major interest. The dependence of the phase-matching condition on temperature, angle, and wavelength variations from the ideal condition is expressed by ΔT, $\Delta\Theta$, and $\Delta\lambda$. These parameters, which were given in (10.21, 24, and

Table 10.3. Phase-matching parameters of important nonlinear materials

Crystal	Phase-matching condition: Wavelength [μm]	Temp [$^\circ C^{-1}$]	Angle [deg]	Temperature tolerance [$^\circ$C cm]	Angular tolerance [a] [mrad cm]* [mrad $(cm)^{1/2}$]	Bandwidth tolerance [Å cm]
KDP	0.69	25	50.6	3.4	1.0*	~6
KD*P	1.06	25	40.5	6.7	1.7*	~65
ADP	0.53	46	90	0.8	32	~1.2
RDA	0.69	97	90	3.3	40	–
CDA	1.06	42	90	5.8	69	–
CD*A	1.06	102	90	5.9	72	–
$Ba_2NaNb_5O_{15}$	1.06	105	90	0.5	43	–
$LiNbO_3$	1.06	165	90	0.6	47	~2.3
$LiIO_3$	1.06	20	29.4	–	0.7*	–
KTP	1.06	25	90	25	15	5.6
BBO	1.06	25	23	55	1.5	–

[a] Measured external to crystal

26), define the interval between the half-power points of the curve of second-harmonic power against temperature, angle, and wavelength, respectively.

If birefringence is a slow function of T, then the crystal has a wide central peak, making it easier to maintain peak output in the presence of small temperature fluctuations. In this regard, KTP far outperforms any other nonlinear material.

Crystals which have a birefringence which is not strongly dependent on wavelength allow efficient doubling of laser beams having a broad spectral bandwidth.

As was explained earlier, in a 90° phase-matching condition the effect of beam divergence on the interaction length is minimized. For example, as shown in Table 10.3, crystals operated at the 90° phase-matching angle have acceptance angles of the order of degrees, whereas crystals which are critically phase-matched, have an acceptance angle of around 1 mrad.

It is also important, that the crystal has a low absorption coefficient at both the fundamental and harmonic wavelength. Particularly critical are absorption losses in intracavity doubling experiments and in high average-power harmonic generation.

In the visible regime, the absorption is around 1 % per cm for most crystals of the KDP family, as shown in Table 10.2. Crystals grown from the melt such as $LiNbO_3$ usually have higher absorption losses; although extremely low losses have been reported for KTP as indicated in the table. In the infrared region the absorption edge is shifted toward longer wavelengths in the deuterated crystals of the KDP family. This explains the lower absorption of ADP, KD^*P, and CD^*A as compared to KDP and CDA at 1.06 μm. For doubling experiments involving second-harmonic generation in the ultraviolet, ADP is usually the best choice because of this crystal's low absorption in this region.

Crystal Size. KDP is superior to any other nonlinear crystal with regard to availability in large sizes combined with excellent optical quality. Crystals with diameters as large as 27 cm have been fabricated. On the other hand, crystals grown from the melt such as KTP, BBO and $LiNbO_3$ are limited in size to about 1 cm^3.

Stability. As mentioned before, crystals grown from the melt such as $LiNbO_3$, KTP, BBO, are generally hard, chemically stable and can easily be polished and coated. Solution grown crystals such as KDP and its isomorphs are soft and hygroscopic. These crystals need to be protected from the atmosphere.

Material Selection. The ideal nonlinear optical material must combine several properties to be suited for use in a practical device. All materials which are now commonly used for nonlinear optical devices have some deficiency which limits their utility. $Ba_2NaNb_5O_{15}$ for example has one of the highest nonlinear optical coefficients known for a material transparent in the visible; but it has a damage threshold several orders of magnitude less than the best optical materials. Ideally one would like to combine the high nonlinear optical coefficient of

Table 10.4. Performance of externally frequency-doubled laser systems

Laser	Nonlinear material	Wavelength [μm]	Conversion efficiency %	Power density [MW/cm^2]
Nd:$YAlO_3$	CDA	1.08–0.54	22	42
Nd:YAG	CDA	1.06–0.53	33	125
Nd:YAG	CDA	1.06–0.53	45	250
Nd:YAG	CD^*A	1.06–0.53	37	250
Nd:YAG	RDP	1.06–0.53	36	250
Nd:YAG	$Ba_2NaNb_5O_{15}$	1.06 –0.53	35	20
Nd:YAG	$LiNbO_3$	1.06–0.53	9.2	44
Nd:YAG	CD^*A	1.06–0.53	30	120
Nd:YAG	CDA	1.06–0.53	16	80
Nd:glass	KDP	1.06–0.53	51	
Nd:glass	KDP	1.06–0.53	50	
Nd:glass	KDP	1.06–0.53	30	400
Nd:glass	ADP	0.53–0.26	30	150
Ruby	RDA	0.69–0.35	37	180
Ruby	$LiIO_3$	0.69–0.35	18	10
Ruby	ADP	0.69–0.35	18	177
Nd:YAG	CD^*A	1.06–0.53	57	
Nd:YAG	KD^*P	1.06–0.53	45	80
Nd:glass	KD^*P	1.05–0.53	83	9500
Nd:YAG	BBO	1.06–0.53	70	140

System parameters (fundamental wavelength)	Doubler parameters	Reference
100 mJ, 6 mm, 40 pps, 8 ns, 12 MW	$T = 42°$ C $\alpha = 90°$ $l = 3.5$ cm	[10.97]
25 MW, 5 mm, 18 ns, 1.2 mr, TEM_{00}	$T = 40.3°$ C $\alpha = 90°$ $l = 2.0$ cm	[10.54]
50 MW, 5 mm, 12 ns, 1.1 mr, TEM_{00}, 20 pps	$T = 48°$ C $\alpha = 90°$ $l = 1.75$ cm	[10.23]
50 MW, 5 mm, 12 ns, 1.1 mr, TEM_{00}	$T = 112.3°$ C $\alpha = 90°$ $l = 1.35$ cm	[10.23]
50 MW, 5 mm, 12 ns, TEM_{00}, 1.1 mr, 10 pps, 2–4 axial modes	$T = 25°$ C $\alpha = 51°$ $l = 1.53$ cm	[10.55]
60 mJ, 12 ns, 5 mm, 10 pps	$T = 112°$ C $\alpha = 90°$ $l = 0.5$ cm	[10.33]
1 mJ, 7 ns, TEM_{00}, 10 pps	$T = 168°$ C $\alpha = 90°$ $l = 1.5$ cm	[10.94]
500 mJ, 14 ns, 3.5 mr, 6 mm		[10.98]
180 mJ, 20 MW, 5 mm, 4 mr, 10 pps	$T = 39.9°$ C $\alpha = 90°$ $l = 1$ cm	[10.24]
29 J, 19 Å, TEM_{00}, 40 μrad, 20 ns	$T = 30°$ C $\alpha = 45°$ $l = 2.5$ cm	[10.21]
20 J, 1 ns, 0.05 Å, 45 mm	$T \approx 30°$ C $\alpha \approx 45°$ $l = 5$ cm	[10.99]
10 J, 20 ns, TEM_{00}, 0.2 mr, 0.5 Å, YAG oscillator	$T \approx 30°$ C $\alpha \approx 45°$ $l = 3.8$ cm	[10.100]
3 J, 20 ns, TEM_{00}, 0.5 Å, 0.2 mr	$T = 30°$ C $\alpha \approx 85°$ $l = 2.5$ cm	[10.100]
23 MW, 10 ns, 4 mm, 0.7 mr, TEM_{00}, single long. mode	$T = 96.5°$ C $\alpha = 90°$ $l = 1.45$ cm	[10.22]
20 ns, 3 mm, 1.5 mr, 10 MW/cm^2	$T = 25°$ C $\alpha = 52°$ $l = 1.1$ cm	[10.71]
5 J, 18 ns, 14 mm, 1.5 mr 277 MW	$T = 25°$ C $\alpha = 52°$ $l = 1.6$ cm	[10.101]
0.5 J, 15 ns 20 pps	$T = 102°$ C $\alpha = 90°$ $l = 1.4$ cm	[10.102]
2.7 J, 17 ns, 0.4 mr 12.5 mm	$T = 25°$ C $\alpha \approx 40°$ $l = 3$ cm	[10.103]
420 J, 0.6 ns, 10 cm	$T = 25°$ C $\alpha = 57°$ $l = 1$ cm	[10.104]
400 mJ, 10 ns, 6 mm	$T = 25°$ C $l = 0.68$ cm	[10.78]

$Ba_2NaNb_5O_{15}$, the 90 degree phase matchability of CD*A, the transparency range and thermal conductivity of sapphire, and the low absorption and damage threshold of the phosphate glasses. In addition, one would like the ideal material to be chemically, mechanically and thermally stable.

At the present time, the leading candidate material for nonlinear experiments with solid-state lasers is KTP. Although KTP does not equal the best qualities of any of the materials mentioned above, its combination of these properties makes it an exceptionally good material for frequency doubling a Nd laser. KTP has a large nonlinear coefficient, high damage threshold and large angular and temperature acceptance range. Its major drawbacks are the limited size (about $1\,\mathrm{cm}^3$ at present) and the high cost associated with the difficult growth process.

Other important alternatives are CD*A, $LiIO_3$, BBO and $LiNbO_3$. For frequency doubling of Nd lasers, both $LiNbO_3$ and CD*A, have the disadvantage of requiring a thermally controlled oven for noncritical phase matching. The complexity and power requirement of the temperature controller and oven are a major drawback for most applications. $LiIO_3$ has a low damage threshold, restricting the material to cw pumped solid-state lasers. BBO and a number of other materials are still in the developmental stages.

KDP is the standard of the industry and together with its isomorphs is the only crystal available for laser systems requiring apertures larger than 1 cm. Despite the lower nonlinear coefficient compared to the other materials discussed in this section, conversion efficiencies as high as 80 % have been obtained from these crystals. Since a high conversion efficiency is actually the result of material properties as well pump source characteristics, the high peak power, narrow spectral bandwidth, small beam divergence and clean spatial and temporal beam profiles obtained from Nd : glass lasers employed in fusion research, make possible such high harmonic conversions even in crystals with modest nonlinearity.

Recent comparisons of different nonlinear materials can be found in [10.95, 96].

A summary of the performance of state-of-the-art frequency-doubled pulsed lasers is given in Table 10.4.

10.1.4 Intracavity Frequency Doubling

In the previous section we discussed frequency doubling by placing a nonlinear crystal in the output beam of the laser system. Frequency doubling a cw-pumped laser system in this manner would result in an unacceptable low harmonic power because large conversion efficiencies require power densities which are not available from a cw-pumped laser. One obvious solution to this problem is to place the nonlinear crystal inside the laser resonator, where the circulating power is approximately a factor $(1/T)$ larger than the output power. The power is coupled out of the resonator at the second-harmonic wavelength by replacing the output mirror with transmission T by one which is 100 % reflective at the fundamental and totally transmitting at the second harmonic. Functionally, the second-harmonic crystal acts as an output coupler in a manner analogous to the transmitting mirror of a normal laser. In the normal laser the transmitting mirror couples out power at the laser frequency, whereas the nonlinear crystal inside the laser couples out power at twice the laser frequency. Because advantage is taken of the high power density inside the laser cavity, it is only necessary to achieve a conversion efficiency equal to the optimum mirror transmission to convert completely the available output at the fundamental to the harmonic. For example, for a cw-pumped Nd:YAG laser with an optimum output coupling of $T = 0.1$, an intracavity conversion efficiency of 10 % will produce an external conversion of 100 %, in the sense that the total 0.53-μm power generated in both directions by the nonlinear crystal is equal to the maximum 1.06-μm power which could be extracted from the cavity without the nonlinear crystal.

There are several disadvantages associated with intracavity doubling. A nonlinear crystal of poor optical quality will drastically degrade the performance of the laser. Amplitude fluctuations are strongly magnified by the combination of the nonlinear process and the gain of the active material. The harmonic power is generated in two directions, which requires an additional dichroic mirror for combining the two beams. The other alternative to intracavity doubling of a cw laser is external doubling with a strongly focused beam. However, since the amount of second harmonic produced is proportional to the product of fundamental power and fundamental power density, the reduced power in an external-doubling configuration requires a much higher power density in order to produce the same harmonic output as could be produced by doubling inside the cavity. Since a greater power density is required for external frequency doubling, crystal damage is more likely to occur in this configuration. In addition, crystal acceptance angles are frequently too narrow to permit tight focusing of the beam onto the crystal. For flashlamp-pumped operation, on the other hand, the fraction of circulating power that is coupled out is so high that there is little advantage in placing the crystal inside the laser.

The choice of nonlinear materials for intracavity frequency doubling of the Nd:YAG and Nd:$YAlO_3$ laser are KTP, BBO, $LiNbO_3$, $LiIO_3$ and CD*A. Compared to flashlamp-pumped systems, in internally frequency-doubled cw lasers the average power is usually high and the peak power low. Therefore, a high nonlinear coefficient, small absorption losses, and good optical quality are the determining factors for the selection of a particular crystal. In order to increase the conversion efficiency of cw pumped systems, the power density can be increased by employing an acousto-optic Q-switch or modelocker in the resonator.

Cavity Configuration

There are three primary factors which affect the choice of a laser cavity configuration. In order to achieve efficient harmonic generation it is important to obtain a high spatial power density inside the nonlinear crystal. Concurrently, the intracavity beam must be large enough inside the laser rod to utilize the

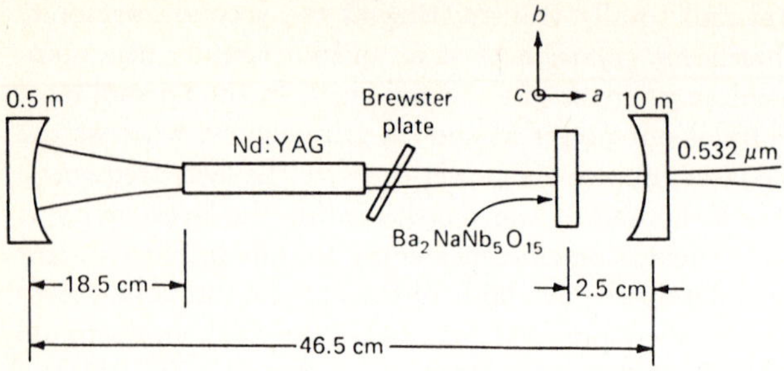

Fig. 10.13. Internal second-harmonic generation [10.105]

maximum rod volume which can contribute to TEM_{00}-mode oscillation. This generally requires that the beam cross-sectional area be at least an order of magnitude larger inside the laser rod than inside the nonlinear crystal. Finally, since intracavity harmonic generation produces a beam of harmonic power in each of two directions, it is desirable to select a cavity configuration which permits recovery of both beams. A resonator configuration which allows for a large TEM_{00}-mode volume in the laser rod and provides a high power density in the nonlinear crystal is shown in Fig. 10.13. The cavity, formed with one 10-m and one $\frac{1}{2}$-m radius-of-curvature mirror, represents a near-hemispherical configuration. Both mirrors have high-reflection coatings for the 1.06-μm radiation and the front mirror is transparent for the 0.53-μm radiation. Only the green light, emitted toward the flat mirror, leaves the cavity; the green light traveling in the opposite direction is absorbed in the Nd : YAG crystal. The nonlinear crystal is located at the position of minimum spot size. The $Ba_2NaNb_5O_{15}$ is oriented in a temperature-controlled oven so that phase-matched second-harmonic generation can be achieved along the crystallographic a axis utilizing the nonlinear coefficient d_{32}. The Brewster-angle plate in the cavity serves to polarize the laser in the proper plane. Also, the position of the laser rod in the cavity was chosen so that self-aperturing by the laser rod permitted the laser to operate only in the TEM_{00} mode. With this arrangement a total cw power of 1.1 W at 0.532 μm has been produced, which equals the available 1.064-μm TEM_{00} output of the basic Nd : YAG laser [10.105].

If it is desired to obtain all the harmonic power in a single output beam, it is necessary to employ a dichroic mirror to reflect one of the beams back in the same direction as the other. Figure 10.14 shows a technique which allows the extraction into a single output beam of the second-harmonic power which is generated in both directions by the frequency-doubling crystal [10.106]. The frequency-doubled beam has a polarization which is rotated 90° with respect to the polarization of the fundamental beam. The dichroic mirror M_1 is designed to reflect the 1.06-μm beam completely and transmit virtually all the orthogonally polarized 0.53-μm beam. Mirror M_2 is a 100 % reflector for both 1.06-μm and 0.53-μm mirrors. In this way the forward and reverse green beams are combined into one.

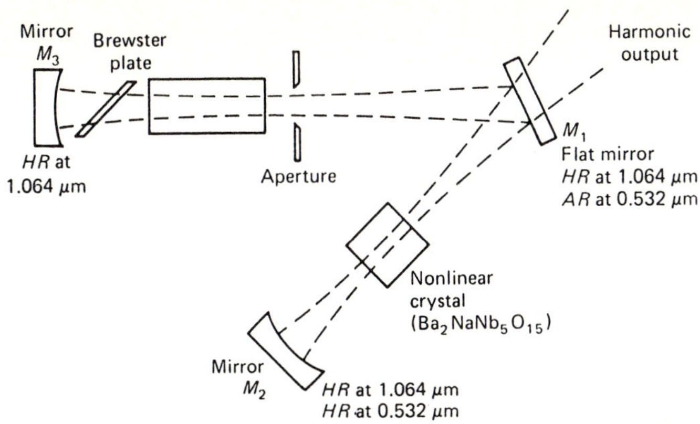

Fig. 10.14. Three-mirror, folded cavity configuration allows harmonic power generated in two directions to be combined into a single output beam [10.106]

To prevent any destructive interference effects, the phase shifts due to dispersion between the 1.06- and 0.53-μm beams in the air path between the doubling crystal and the dielectric coating of mirror M_2 must be adjusted so that they are in phase as they reenter the doubling crystal after reflection from M_2. This is accomplished by providing a translation adjustment for the mirror M_2 to move it toward or away from the doubling crystal, such that the net phase shift between the two waves is a multiple of 2π. The computed phase shift, based on the measured 1.06–0.53 μm dispersion of air of 3.6×10^{-6}, is approximately 29.0°/cm. For doubling of the 1.06-μm line of Nd:YAG, this requires a mirror movement of no more than 3.28 cm [10.107], and the accuracy required to correct for the phase shift is of the order of millimeters.

It is of interest to note that the arguments presented here are valid only if both the fundamental and the second harmonic are reflected by the same mirror. If they do not share a common mirror, then the phase shift between the fundamental and the second harmonic will depend on the optical spacing between the mirrors used to reflect the two wavelengths. Thus, a displacement of only an optical wavelength will produce a full 360° relative phase shift [10.106, 107]. When the phase shift between the second-harmonic power generated to the right and the harmonic power reflected from mirror M_2 is ϕ, the total second harmonic power P_{SH2} is

$$P_{SH2} = 2(1 + \cos \phi) P_{SH1} \quad , \tag{10.30}$$

where P_{SH1} is the single-pass second harmonic generated. The oscillating behavior predicted by (10.30) has been verified experimentally [10.107], and for $\phi = 0$ the predicted second-harmonic power enhancement of a factor 4 was measured. The fourfold increase in output power is due to the fact that for $\phi = 0$ the electric field strength is doubled and the harmonic power increases with the square of the field. Thus the incorporation of a harmonic reflecting mirror not only directs the harmonic in one direction but permits a possible increase of a factor of 2 in total second-harmonic generation.

The beam diameter at various locations inside a laser cavity of the type shown in Fig. 10.14 is a function of mirror curvature, laser rod size and face curvature, nonlinear crystal orientation, and the placement and spacing of all these and any other intracavity elements (such as Brewster plates, intracavity lenses, or apertures), as well as of the thermal focusing present in the rod. A computer program which takes into account all of the factors mentioned above in determining the beam diameter at various intracavity locations is presented in [10.108].

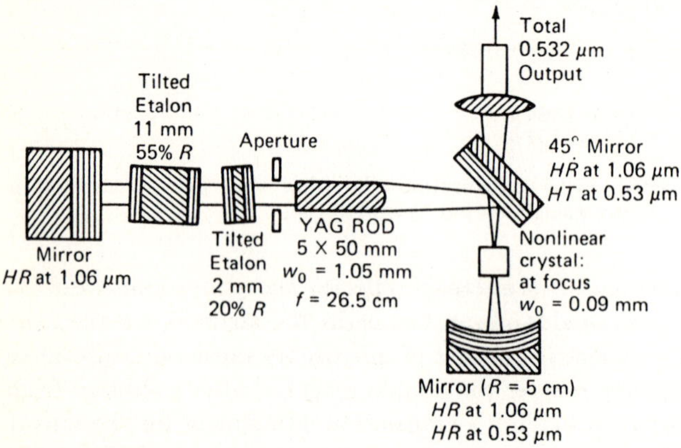

Fig. 10.15. Single-ended output by use of three-mirror, folded cavity configuration (similar to Fig. 10.11). The intracavity 45° mirror is highly reflecting at 1.06 μm but transparent at 0.53 μm. The short-radius curved mirror is highly reflecting at both 1.06 and 0.53 μm [10.109]

Intracavity frequency doubling of a laser operating in many longitudinal modes will generate fluctuations in the instantaneous as well as the average harmonic power. The nonlinear process in the resonator actually magnifies the fluctuations occurring due to mode beating at the fundamental. Single-longitudinal-mode operation or mode locking avoids this problem. Figure 10.15 shows an intracavity frequency-doubled Nd:YAG laser containing two quartz etalons in the resonator to obtain single-frequency operation [10.109]. The cavity is designed to obtain an adequate TEM_{00}-mode volume in the laser rod and a small waist at the nonlinear crystal. The beam waist is generated by a short-radius mirror and the Nd:YAG rod with one end curved to act as an internal focusing element. Frequency-doubled output in a single direction is achieved with a three-mirror cavity similar to Fig. 10.14. The multiple-dielectric-layer 45° mirror favors *s*-polarized laser oscillation at 1.06 μm and very effectively transmits the second-harmonic *p*-polarized radiation. The system shown in Fig. 10.15 produces unidirectional power outputs of 0.3 to 0.5 W at 0.53 μm.

An analysis of optical second-harmonic generation internal to the laser cavity has been presented in [10.105, 110]. The steady-state condition for intracavity doubling can be determined if we equate the round-trip saturated gain of the laser to the sum of the linear and nonlinear losses

$$\frac{2g_0 l^*}{1+I/I_s} = L + K'I \quad , \tag{10.31}$$

where g_0 is the unsaturated gain coefficient, l^* is the length of the laser medium, I is the circulating power density in the laser rod, and I_s is the saturation power density of the active material. All linear losses occurring at the fundamental frequency are lumped together into the parameter L; the quantity $K'I$ is the nonlinear loss. The nonlinear coupling factor K', defined by

$$I(2\nu) = K'I^2(\nu)$$

is related to K in (10.12); it is $K' = \kappa l^2 K$, where κ accounts for different power densities in the laser rod and nonlinear crystal in the case of focused beams; that is, $\kappa = I_{\text{crystal}}/I_{\text{rod}}$.

From the theoretical treatment of intracavity doubling, it follows that a maximum value of second-harmonic power is found when

$$K'_{\max} = \frac{L}{I_s} \quad . \tag{10.32}$$

The magnitude of the nonlinearity required for optimum second-harmonic production is porportional to the loss, inversely proportional to the saturation density, and independent of the gain. Thus, for a given loss, optimum coupling is achieved for all values of gain and, hence, all power levels. As we recall from Chap. 3, in a laser operating at the fundamental wavelength the optimum output coupling of the front mirror is gain-dependent. The usefulness of (10.31) stems from the fact that by introducing the materials parameters for a particular crystal into (10.8) and combining with (10.31), the optimum length of the nonlinear crystal is obtained. For example, from (10.8) we obtain $K' = 1.5 \times 10^{-6} l^2 \kappa$ for $Ba_2NaNb_5O_{15}$. With (10.31) and $I_s \approx 10^3\,\mathrm{W/cm^2}$ for Nd:YAG, the optimum crystal length is $l^2 = 660\,L/\kappa$ [cm^2]. Assuming a round-trip loss of $L = 0.02$ and a power-density enhancement of $\kappa = 50$, a crystal of $l = 0.5$ cm is sufficient to optimally couple the second harmonic from the laser.

From the theory presented in [10.110], it follows that if the loss due to the inserted second-harmonic crystal is small compared to the total internal loss, the value of K' is the same for fundamental and second-harmonic output coupling. Also, the maximum second-harmonic power equals the fundamental power obtainable from the same laser.

Intracavity frequency doubling of a cw-pumped laser can be combined with Q-switching or mode locking. In a typical system of the first type the laser contains, in addition to the laser rod and harmonic generator, an acoustooptic Q-switch in the resonator. In the case of an internally doubled Q-switched laser, the harmonic power generated cannot be simply related to the harmonic coupling coefficient, since the effects of pump depletion and pulse stretching must be considered. *Murray* and *Harris* [10.111] have made a computer study of this problem, and from their data the harmonic power extracted from the laser can be related to the pumping rate above threshold, and to the normalized harmonic coupling coefficient. In Q-switched systems, it is generally found that the coupling strength is two to three orders smaller than in cw systems

[10.112, 113]. This is due to the fact that fo the higher peak powers in a Q-switched system the optimum second-harmonic output occurs at a lower value of the nonlinear parameter. The low K' value permits the use of a mirror configuration which is well within the region of stable oscillation and yields a fairly large beam waist. The configuration is thus relatively insensitive to slight misalignment and motion of the resonator elements. In the case of the cw system the optimum second-harmonic output is obtained for a resonator operating with an extremely small beam waist (and thus larger K') which is sensitive to both transverse and axial motions of the resonator elements.

Repetitively Q-switched Nd:YAG lasers have been internally frequency-doubled with $Ba_2NaNb_5O_{15}$ [10.112, 114], $LiIO_3$ [10.113], $LiNbO_3$ [10.97, 115], CD^*A [10.116] and KTP [10.117, 118]. With KTP as the nonlinear material output powers of up to 20 W have been achieved, and with CD^*A harmonic powers as high as 35 W were generated. In addition to Nd:YAG, other Nd lasers such as Nd:YLF [10.118] and Nd:$YALO_3$ [10.97, 115] have been internally frequency doubled.

In applications requiring high amplitude stability, one may wish to frequency-double the output from a cw-pumped mode-locked laser. With proper temperature control of the nonlinear crystal, it is even possible to perform both the electrooptic modulation and optical harmonic generation in a single intracavity nonlinear crystal [10.119–123]. Since the harmonic conversion efficiency is greatest at the peak of the mode-locked pulses, the harmonic conversion process tends to flatten and broaden the pulses, opposing the modulator's efforts to sharpen the pulses [10.124].

The pulse-lengthening mechanism can be understood by considering the losses introduced to the resonator by the mode-locking modulator and the harmonic generator. The loss due to second-harmonic generation is caused by the transmitted output power; this loss decreases with longer pulses because peak power and, hence, harmonic conversion efficiency decreases with longer pulses. The loss introduced by the mode-locking modulator, on the other hand, increases with longer pulses, since this loss results from the tails of the pulse passing through the modulator under less than optimum conditions. The pulse length of the mode-locked laser is affected by both these losses. To a first approximation, the pulse length will be that which minimizes the sum of these two losses, as is illustrated in Fig. 10.16 for a typical Nd:YAG laser [10.119]. The pulse-lengthening effect decreases the peak power as well as the average power of the laser. Therefore, the enhancement of second-harmonic power due to mode locking is not as large as one would expect. Theoretical expressions for the enhancement of second-harmonic generation to be expected from mode locking have been given in [10.125, 126].

Combined mode locking and second-harmonic generation is useful for obtaining high conversion efficiencies with low-loss nonlinear crystals, which are either too short or have too low a nonlinear coefficient to achieve a high conversion by cw operation of the laser. In order to minimize cavity losses, usually a single crystal, such as $Ba_2NaNb_5O_{15}$ or off-stoichiometric $LiNbO_3$, is employed to perform both the mode locking and the second-harmonic generation. This

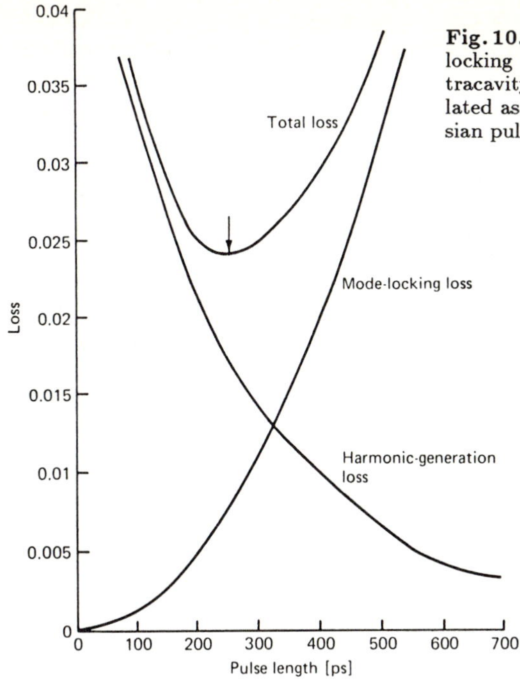

Fig. 10.16. Loss introduced by the mode-locking modulator and loss introduced by intracavity second-harmonic generation, calculated as a function of pulse length for a Gaussian pulse in a typical Nd : YAG laser [10.119]

requires an oven fixture capable of heating the nonlinear crystal as uniformly as possible and of coupling the rf power to two faces of the crystal. Internally frequency-doubled and mode-locked Nd : YAG lasers have produced 0.5 to 1.5 W of output power at 0.53 μm [10.119, 123]. Pulse widths are typically 250 to 330 ps.

10.1.5 Third-Harmonic Generation

Frequency up-conversion to the third harmonic has recently attracted considerable attention for Nd : glass laser beams for fusion applications. Third-harmonic generation is a two-step process involving two harmonic generators. In the first crystal ("doubler") some fraction of the fundamental radiation is converted to the second harmonic, followed by a second crystal ("tripler"), in which unconverted fundamental radiation is mixed with the second harmonic to produce the third harmonic. The equations governing frequency mixing in nonlinear crstals have been given in [10.127]:

$$dE_1/dZ = -iK_1 E_3 E_2^* \exp(-\mathrm{i}\Delta k \cdot Z) - \tfrac{1}{2}\gamma_1 E_1 \ , \tag{10.33a}$$

$$dE_2/dZ = -iK_2 E_3 E_1^* \exp(-\mathrm{i}\Delta k \cdot Z) - \tfrac{1}{2}\gamma_2 E_2 \ , \tag{10.33b}$$

$$dE_3/dZ = -iK_3 E_1 E_2 \exp(\mathrm{i}\Delta k \cdot Z) - \tfrac{1}{2}\gamma_3 E_3 \ . \tag{10.33c}$$

Here the E_j's are the complex electric vectors of waves propagating in the Z direction with frequencies ω_j, where $\omega_3 = \omega_1 + \omega_2$. The electric field of wave j is the real part of $E_j \exp(\mathrm{i}\omega_j t - ik_j Z)$, and the phase mismatch $\Delta k = k_3 - (k_1 + k_2)$ is proportional to the deviation $\Delta\theta$ of the beam path from the phase-matching direction. The γ_j's are absorption coefficients. For tripling, $\omega_2 = 2\omega_1$, $\omega_3 = 3\omega_1$, $K_2 \simeq 2K_1$, and $K_3 \simeq 3K_1$.

Equations (10.33a, b) describe single rays as well as complete beams [10.128–130], and solutions are presented in Fig. 10.17 for the phase-matched case ($\Delta k = 0$). The efficiency refers to the combined fundamental and second-harmonic input power to the tripler.

The parameter M is the ratio of second-harmonic power to total power in the tripler:

$$M = P_{2\omega}/P_\omega + P_{2\omega} \quad . \tag{10.34}$$

If the input photons at ω and 2ω are matched 1:1, then $P_{2\omega} + 2P_\omega$ and $M = 0.67$. In principle, complete conversion of the input beams to the third harmonic can be achieved. For other values of M, after a ray has propagated an optimum distance Z_{opt} into the crystal, the number of photons in one of the components depletes to zero, and as z increases beyond Z_{opt} the mixing process reverses and the third-harmonic radiation reconverts.

Successful tripling is made difficult because the distance Z_{opt} is a sensitive function of both M and intensity; for example, Fig. 10.17 shows that input at $4\,\mathrm{GW/cm^2}$ will convert with an efficiency of 25% if $M = 50\%$ but 80% if $M = 60\%$.

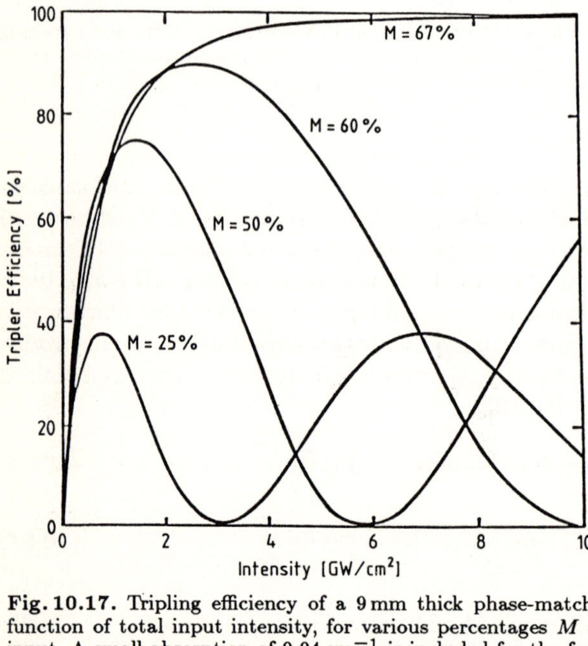

Fig. 10.17. Tripling efficiency of a 9 mm thick phase-matched KDP Type-II crystal as a function of total input intensity, for various percentages M of the second harmonic in the input. A small absorption of $0.04\,\mathrm{cm^{-1}}$ is included for the fundamental [10.128]

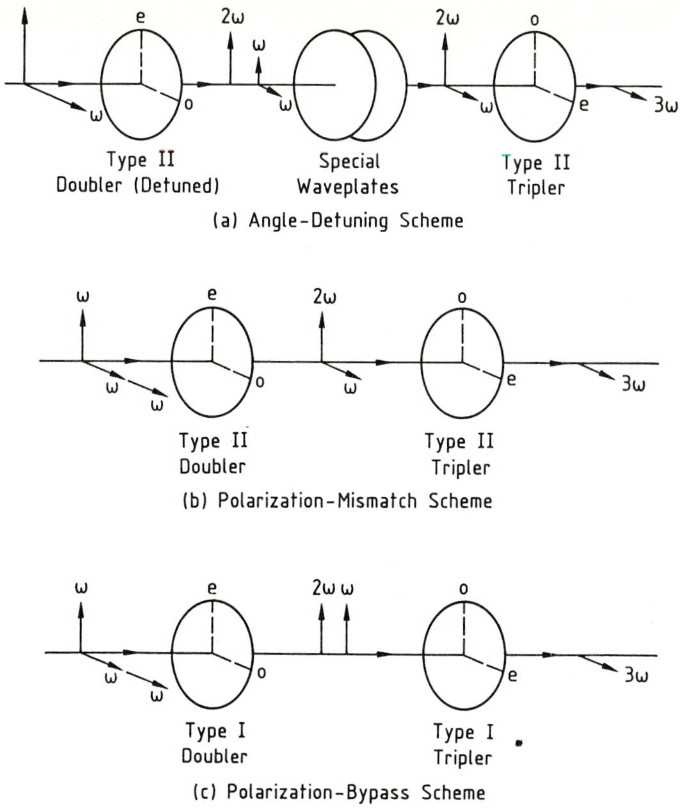

Fig. 10.18. Different tripling schemes. Fundamental wavelength emerges from the doubling crystal **(a)** elliptically polarized at 45° to the o- and e-axes, **(b)** plane polarized parallel to the o-axis, and **(c)** plane polarized parallel to the e-axis [10.128]

Efficiency tripling depends on the fundamental and second-harmonic photons emerging from the first crystal in a ratio of 1 : 1 over a broad intensity range. *Craxton* [10.128, 130] has shown that this requirement can be achieved by means of an appropriate choice of polarization angle in the doubler. Figure 10.18 shows three tripling schemes which can lead to high efficiency of the third-harmonic generation. In each case, two out of three photons of the ordinary beam propagating in the doubling crystal at the fundamental wavelength ω, are converted to an extraordinary beam at 2ω. Depending on crystal orientation and thickness, the unconverted photons of the fundamental beam can emerge from the doubling crystal either elliptically polarized at 45° to the o- and e-axes; plane polarized parallel to the o-axis, or plane polarized parallel to the e-axis.

A detailed discussion of the relative merits of the different configurations illustrated in Fig. 10.18 can be found in the above-cited references.

10.1.6 Examples of Harmonic Generation

Examples of several different types of laser systems which incorporate harmonic generators will illustrate the current state-of-the-art of this technology. We will consider high-average-power harmonic generation in cw, medium and high repetition rate pulsed lasers featuring KTP and CD*A crystals, as well as high peak power and energy production in large-aperture, single-shot laser systems employing KDP.

Employing a KTP crystal in a Type-II orientation, *Liu* et al. [10.131] demonstrated harmonic generation with an output power of 5.6 W at 532 nm by intracavity frequency doubling in a Nd : YAG laser oscillator acousto-optically Q-switched at 5 KHz. The crystal was placed inside a Nd : YAG laser oscillator with a YAG rod of 3 mm × 79 mm optically pumped with a cw krypton lamp contained in an elliptical cavity. Both oscillator mirrors separated by 120 cm, were flat and were dielectrically coated for maximum reflection at 1.06 μm. The ouptut mirror was dichroic-coated to allow maximum transmission at 532 nm. The second-harmonic output was optimized by rotating a half-wave plate which was inserted into the resonator.

A schematic of the system is shown in Fig. 10.19, and the one-way second-harmonic output as a function of the lamp current is plotted in Fig. 10.20. More recently, *Perkins* et al. [10.117, 118] achieved over 20 W of average output at 0.532 μm from a similar laser. The laser was also a krypton-arc-lamp cw

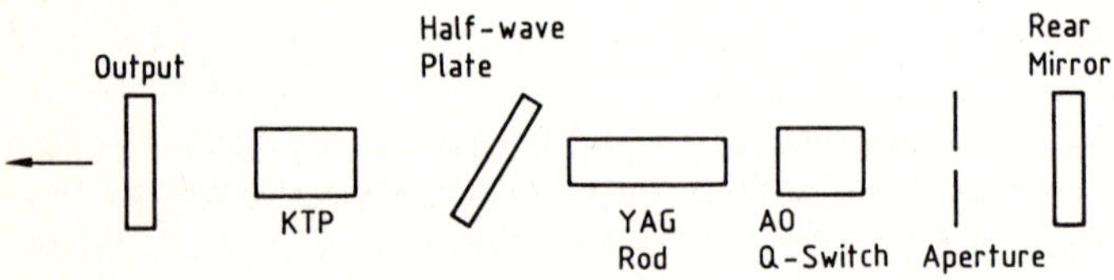

Fig. 10.19. Intracavity second-harmonic generation with KTP

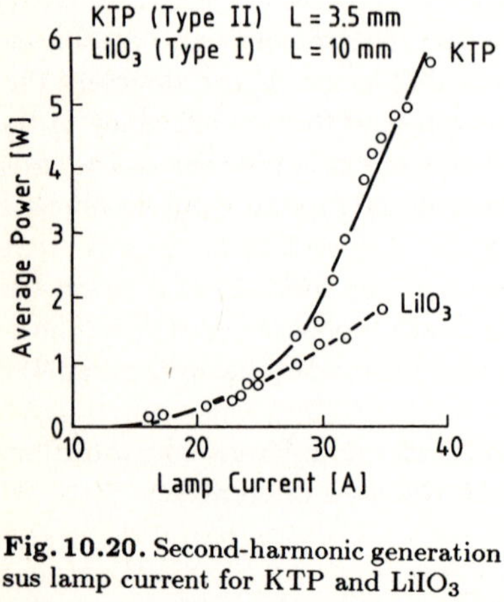

Fig. 10.20. Second-harmonic generation versus lamp current for KTP and $LiIO_3$

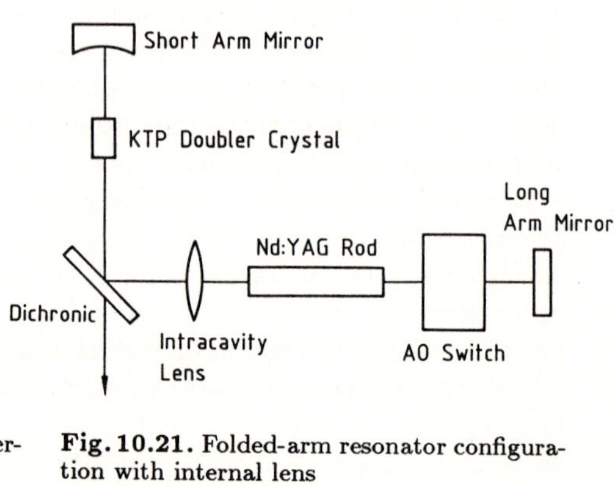

Fig. 10.21. Folded-arm resonator configuration with internal lens

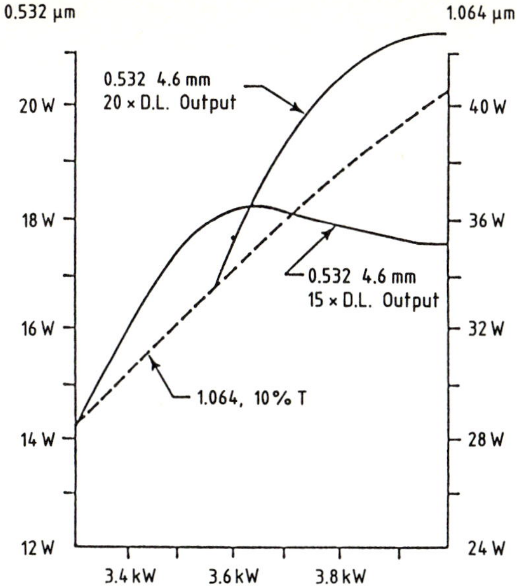

Fig. 10.22. Q-switched output power versus lamp power (25-kHz switching rate)

Nd : YAG system, with an acousto optic Q-switch and internal second-harmonic generation using Type-II KTP. The major difference compared to the system mentioned above is in the resonator design. The doubling crystal is located in a folded-arm resonator of the type shown in Fig. 10.21. The folded resonator permits collection of the harmonic energy in both directions. Also the inclusion of a focusing lens allowed adjustment of the TEM_{00}-mode radius between 60 and 120 μm. A maximum conversion efficiency of 54 % was achieved with a 4.6 mm long KTP crystal, Q-switched at 25 kHz at a lamp input power of 3.6 kW. The Q-switched output power versus lamp power is plotted in Fig. 10.22. Shown is the system output at 1.06 μm with a 10 % output coupler, and the second-harmonic power obtained with different resonator configurations.

The power output increased by changing the resonator to a multimode configuration (15 and 20 times diffraction limited, respectively). Thermal absorption in the doubling crystal leads to considerable nonspherical aberration in the intracavity beam. These aberrations have a stronger deleterious effect on the fundamental mode as compared to a multimode beam. With the same laser, using the intracavity-lens-folded arm configuration an output of 9 W has been demonstrated in cw operation.

Kogan et al. [10.103] developed a 1 J, frequency-doubled Nd : YAG laser consisting of a TEM_{00} mode oscillator, four amplifiers, an optical isolator, and a KD^*P doubler cut for Type-II phase matching at 1064 nm. The system operates at a pulse repetition rate of 10 pps, a pulse width of 17 ns, and a beam divergence of 0.4 mrad in a 12.5 mm diameter beam. At an average power of 10.5 W at 532 nm the conversion efficiency was about 50 %. A schematic of the laser optical system is shown in Fig. 10.23. It was reported that the system delivered over 20 million shots without any damage to the crystal. It should

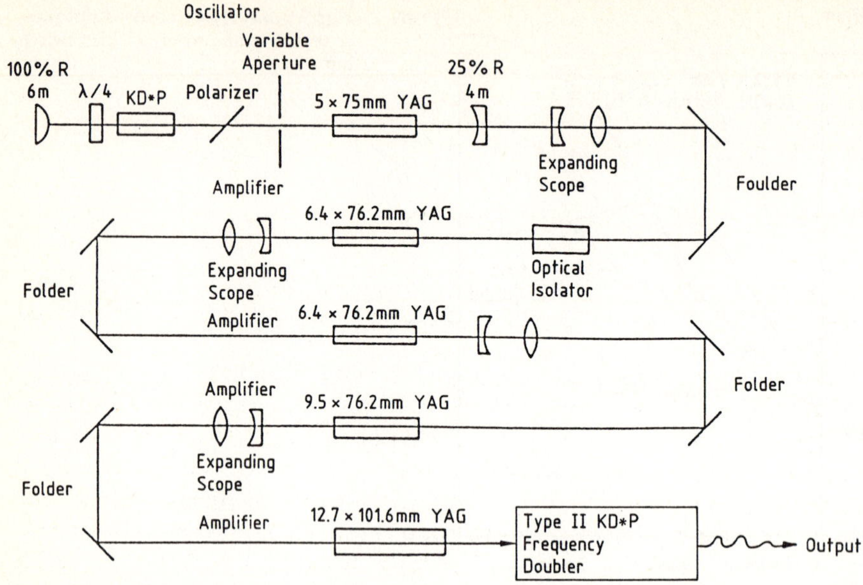

Fig. 10.23. Optical schematic of a high-energy, TEM_{00}-mode Nd : YAG laser

be noted that KD*P is critically phase matched at this wavelength. Also, due to the high sensitivity of the crystal to angular misalignment the 25 mm long crystal had to be operated with a fundamental beam whose divergence is less than 1 mrad. For this reason, a diffraction-limited oscillator was developed.

Hon [10.116, 132, 133] obtained an average frequency-doubled output of 35 W at 532 nm with a cw pumped repetitively Q-switched Nd:YAG laser. The acousto-optic Q-switch produced 150 ns wide pulses at 10 kHz. Harmonic generation was achieved with an intracavity CD*A crystal. The system featured electro-optical compensation for self heating in the crystal and beamshaping to minimize thermal gradients.

CD*A permits noncritical phase matching at about 112° C. A pulsed electric field was used to fine tune the phase-matching condition. Application of an electric field provides a means of shifting the phase-match temperature, shifts equivalent to 2° C temperature changes have been obtained. Absorption of the powerful radiation passing through the crystal leads to heating which detunes the crystal from the phase-match condition. This self heating effect causes a drop of harmonic power and leads to output instability. A closed-loop phase-matching system, using electro-optical tuning provides short- and long-term phase-matching compensation for thermal gradients. The system also incorporated transverse beamshaping with an aspect ratio of about 2.5 : 1. Shaping of the beam's transverse profile into an elliptical format, and utilization of one-dimensional heat flow offer significantly reduced thermal profiles and permit efficient operation at high average power. The temperature profile present across a nonlinear crystal driver at multi-watt power levels is sufficient to cause local

Table 10.5. Frequency conversion efficiencies achieved on the Argus laser

Wavelength [nm]	Conversion efficiency	Maximum energy generated [J]	Output power [GW]
532	83 %	346	495
355	55 %	41	68
266	51 %	50	83

departures from phase match sufficient to degrade performance. By profiling the input beam cross section appropriately, thermal gradients can be reduced to acceptable levels.

The requirement for generation of short wavelengths from inertial-confinement-fusion lasers has prompted the development of large-aperture harmonic converters for doubling, tripling and quadrupling of the Nd : glass laser output. As an example, we will consider conversion experiments originally carried out on the now dismantled Argus laser at the Lawrence Livermore National Laboratory [10.104]. Argus is an image-relayed, spatially filtered laser system capable of delivering up to 1 kJ output energy in a 600 ps pulse. With a 10 cm diameter, and 2.29 cm thick KDP crystal up to 346 J of laser energy at 532 nm was generated. This corresponds to a 83 % conversion efficiency. Tripling and quadrupling yielded efficiencies of 76 % and 70 %, respectively. A summary of the performance of the conversion experiments is shown in Table 10.5. Experimental results of these harmonic-generation experiments are shown in Figs. 10.24–26. The variation of the doubling efficiency vs. input at the fundamental intensity for the Type-I KDP crystal is shown in Fig. 10.24. Since the crystals were uncoated, the internal conversion efficiency refers to the measured overall efficiency corrected for Fresnel reflection losses. An illustration of the tripling experiments is shown in Fig. 10.25. It shows the conversion efficiency as a function of the 1064/532 nm mix ratio. The performance of the quadrupler is illustrated in Fig. 10.26, in which the conversion efficiency is plotted vs. input intensity. As can be seen from this figure, for input intensities above 2 GW/cm^2

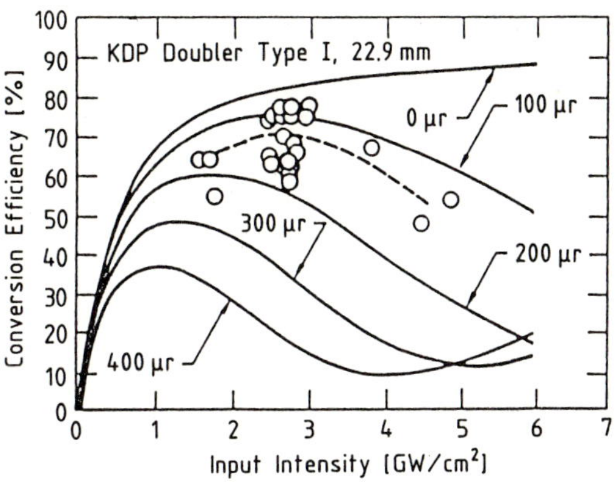

Fig. 10.24. Frequency-doubling efficiency as a function of input power

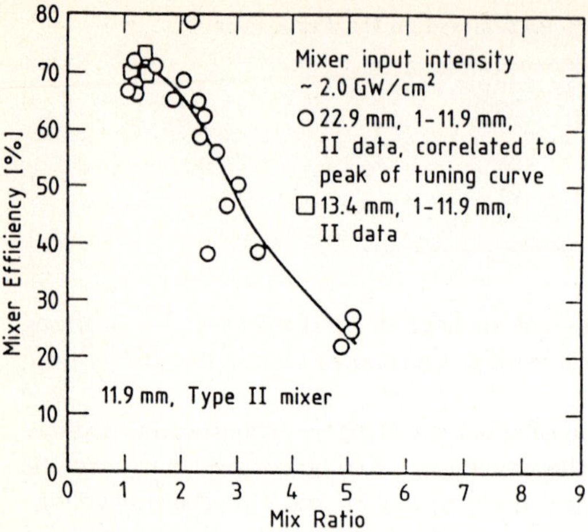

Fig. 10.25. Frequency-tripling efficiency as a function of 1064 nm to 532 nm mix ratio

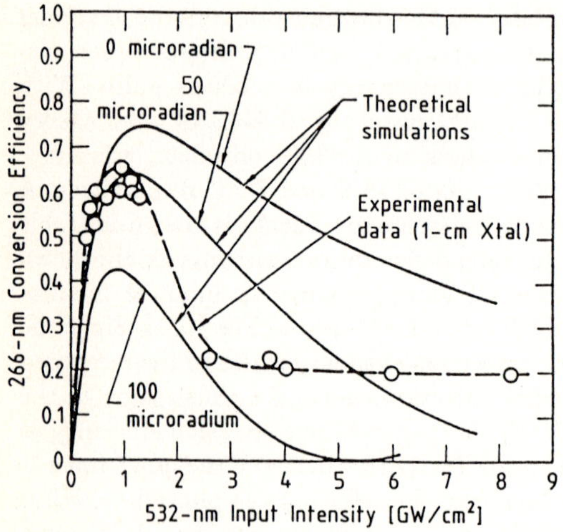

Fig. 10.26. Conversion efficiency from 532 nm to 266 nm

a decline in output was observed due to nonlinear absorption and damage in the quadrupling crystal.

Following the early experiments on the Argus system, KDP crystal arrays with apertures of 74 cm have been built for conversion of the Nova laser output. These large aperture converters are constructed from 15 cm or 27 cm KDP segments, held together by an eggcrate stainless steel support structure [10.134].

10.2 Parametric Oscillators

As was mentioned at the beginning of this chapter, two beams with different frequencies incident on a nonlinear crystal will generate a traveling polarization wave at the difference frequency. Provided the polarization wave travels at the same velocity as a freely propagating electromagnetic wave, culmulative growth will result. For reasons which will soon become clear, the two incident beams are termed "pump" and "signal" waves, having a frequency of ν_p and ν_s, and the resulting third wave is termed "idler" wave with frequency ν_i. Under proper conditions, the idler wave can mix with the pump beam to produce a traveling polarization wave at the signal frequency, phased such that growth of the signal wave also results. The process continues with the signal and idler waves both growing, and the pump wave decaying as a function of distance in the crystal. Energy conservation requires that

$$\nu_p = \nu_s + \nu_i \quad . \tag{10.35}$$

In order to achieve significant parametric amplification, it is required that at each of the three frequencies the generated polarization waves travel at the same velocity as a freely propagating electromagnetic wave. This will be the case if the refractive indices of the material are such that the k vectors satisfy the momentum-matching condition $k_p = k_s + k_i$. For collinearly propagating waves this may be written

$$\frac{n_p}{\lambda_p} = \frac{n_s}{\lambda_s} + \frac{n_i}{\lambda_i} \quad , \tag{10.36}$$

where n_s, n_i, and n_p are the refractive indices at the signal, idler, and pump frequency, respectively. The pump signal is usually provided by a laser and, therefore, λ_p is fixed. However, if the refractive indices are varied, the signal and idler frequencies will tune. Considerable control of the refractive indices, and wide tuning, is possible by making use of the angular dependence of the birefringence of anisotropic crystals, and also by temperature variation. Rapid tuning over a limited range is possible by electro-optic variation of the refractive indices.

A schematic of a typical parametric oscillator is shown in Fig. 10.27. The gain-producing nonlinear material is located inside an optical resonator, which must provide sufficient feedback to overcome the losses in the system. There are several different types of optical parametric oscillators characterized by the form of feedback provided: The doubly resonant oscillator employs feedback at both the signal and idler frequencies; the singly resonant oscillator employs feedback at either the signal or idler, but not both. In both types of oscillators the mirror through which the pump radiation enters the cavity is highly

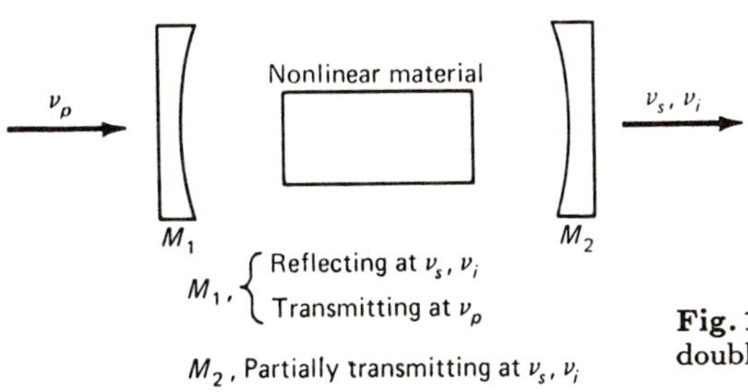

Fig. 10.27. Parametric oscillator with a doubly resonant cavity

transmitting, and the exit mirror may be either transmitting or reflecting at the pump frequency [10.141].

The first successful operation of an optical parametric oscillator was achieved by *Giordmaine* and *Miller* [10.142] in 1965. Their oscillator employed lithium niobate as the nonlinear material and used a 5300-Å pump signal derived by frequency-doubling the 1.06-μm output of a $Nd:CaWO_4$ laser. Since then, work has proceeded rapidly, and it is now possible to tune through most of the visible and near-infrared wavelength region.

The theory of optical parametric oscillators has been described in [10.127, 135–140]. For a given crystal temperature and orientation, maximum parametric gain occurs at that signal and idler frequency where $\Delta k = 0$. For small gain at line center, one obtains [10.140]

$$G \approx \frac{2\omega_s\omega_i\eta_s\eta_i\eta_p|d|^2 l^2 P_p}{A} \quad , \tag{10.37}$$

where η_s, η_i, η_p are the plane-wave impedances of the three waves. It is useful to note that when $\omega_s = \omega_i = \omega_p/2$ (where $\omega_p/2$ is termed the degenerate frequency), that the single-pass gain of an optical parametric amplifier is equal to the conversion efficiency ($P_{2\omega}/P_\omega$) of a second-harmonic generator with a fundamental frequency $\omega_p/2$. If we let $\omega_s = (\omega_p/2)(1+\delta)$ and $\omega_i = (\omega_p/2)(1-\delta)$, that is,

$$\delta = \frac{2\omega_s - \omega_p}{\omega_p} = \frac{2\lambda_p - \lambda_s}{\lambda_s} \quad , \tag{10.38}$$

then the parametric gain off degeneracy is reduced from that at degeneracy by the factor $1-\delta^2$. With (10.38) the single-pass parametric gain for a confocally focused signal, idler, and pump may be written

$$G = \tfrac{1}{2}\omega_p^2\eta^3|d|^2 l^2 \frac{P_p}{\pi w_0^2}(1-\delta^2)^2 \quad , \tag{10.39}$$

where ω_0 is the degenerate frequency, and $w_0^2 = (l\lambda_0)/(2\pi n)$ is the degenerate confocal spot size. For a crystal of $LiNbO_3$ at a degenerate wavelength of 1 μm, this yields a gain under optimum focusing conditions of

$$G = 0.005 l P_p (1-\delta^2)^2 \quad , \tag{10.40}$$

where l is in centimeters and P_p in Watts. Thus 1 W of pumping power in a 4-cm $LiNbO_3$ crystal provides 2 % single-pass gain at a degenerate wavelength of 1 μm.

To construct an oscillator it is necessary to resonate either the signal or the idler, or both. The latter case, where both the signal and idler are resonant, yields the lowest threshold, but poses severe stability problems and mirror requirements. To determine the threshold of the oscillator, we require that the single-pass parametric gain be sufficient to offset the round-trip cavity loss. For doubly-resonant devices with a low loss α_s, α_i at the signal and idler frequency, threshold is reached if

$$G_{\text{double}} \approx \alpha_s \alpha_i \quad . \tag{10.41a}$$

Alternately, if only the signal is resonant and it is assumed that no idler radiation is returned to the crytsal input ($\alpha_i = 1$), one obtains for small α_s

$$G_{\text{single}} \approx 2\alpha_s \quad . \tag{10.41b}$$

The ratio of threshold pump power with the signal only resonant as compared to both signal and idler resonant is $2/\alpha_i$. Thus for a 2% idler cavity loss, 100 times as much power is required for the signal-only-resonant case. For single-pass signal and idler-power losses of 2% each, (10.41b) yields $G = 4 \times 10^{-4}$. Equation (10.40) shows that to attain this gain in a 4-cm crystal of $LiNbO_3$ at a degenerate wavelength of 1 µm requires a pump power of only 20 mW. The possibility of cw parametric oscillators with such low thresholds was first demonstrated by *Smith* et al. [10.143]. Their first oscillator employed a 5-mm crystal of $Ba_2NaNb_5O_{15}$ and was pumped by a frequency doubled 1.06-µm Nd^{3+} : YAG laser. Threshold was observed at 45 mW of multimode power.

The threshold of a pulsed singly-resonant oscillator can be calculated using the results of *Brosnan* et al. [10.144]. The energy density required to achieve threshold is given by

$$J_0 = \frac{2.25\tau}{\kappa g_s \mathcal{L}^2}\left(\frac{L}{2\tau c}\ln\frac{P_n}{P_0} + 2\alpha l + \ln\frac{1}{\sqrt{R}} + \ln 2\right)^2 \quad , \tag{10.42}$$

where $\kappa = 2\omega_s\omega_i d_{\text{eff}}^2 / n_s n_i n_p \varepsilon_0 c^3$. The mode coupling coefficient is $g_s = w_p^2 / (w_p^2 + w_s^2)$ and the effective parametric gain length is $\mathcal{L} = l_w \operatorname{erf}(\sqrt{\pi} l)/2l_w$), where l_w is the walk-off length and l the crystal length. Here L is the optical length of the cavity, τ the pump pulse width, P_n/P_0 the ratio of signal power to noise power at threshold, and R the product of mirror reflectivities.

As noted earlier, with the pump frequency fixed, any process which changes the refractive indices at the signal, idler, or pump wavelength will tune the oscillator. A temperature-tuning curve for an oscillator containing a CDA crystal pumped at 5320 Å is shown in Fig. 10.28 [10.144]. Parametric generation in the 3-cm-long crystal was achieved by focusing the output from a frequency-doubled Q-switched Nd : YAG laser to a power density of 300 MW/cm^2 inside the crystal. An angular tuning curve for $LiNbO_3$ pumped at 1.06 µm by a Q-switched Nd : YAG laser is shown in Fig. 10.29 [10.145]. The oscillator, containing a 5-cm-long crystal, angle-tunes from degeneracy and operates over the entire 1.4- to 4.4-µm range. Output energies of 1 mJ/pulse at 5 pps have been observed with a 15% conversion efficiency.

A commercially available oscillator that is tunable from 0.54 to 3.7 µm using only one 90° cut $LiNbO_3$ crystal and a Nd : YAG laser as pump source has been described in [10.146–148]. The tuning range is obtained by various combinations of laser wavelengths and oscillator mirrors. The internally frequency-doubled Nd : YAG laser, which contains a $LiIO_3$ crystal as nonlinear

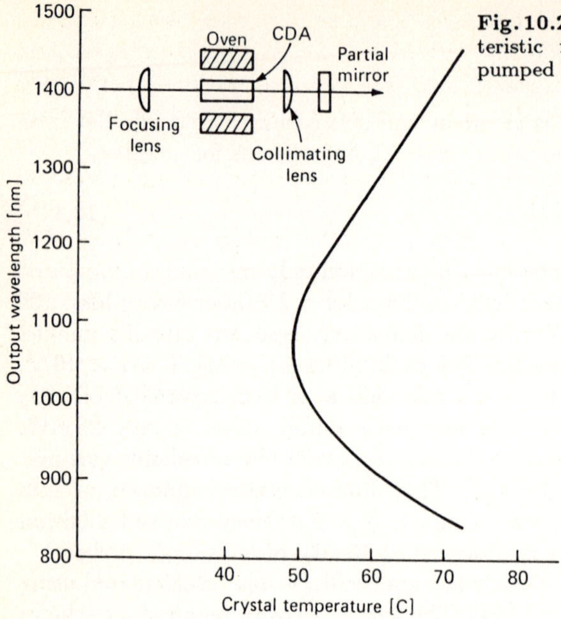

Fig. 10.28. Temperature tuning characteristic for CDA parametric oscillator pumped at 0.53 μm [10.144]

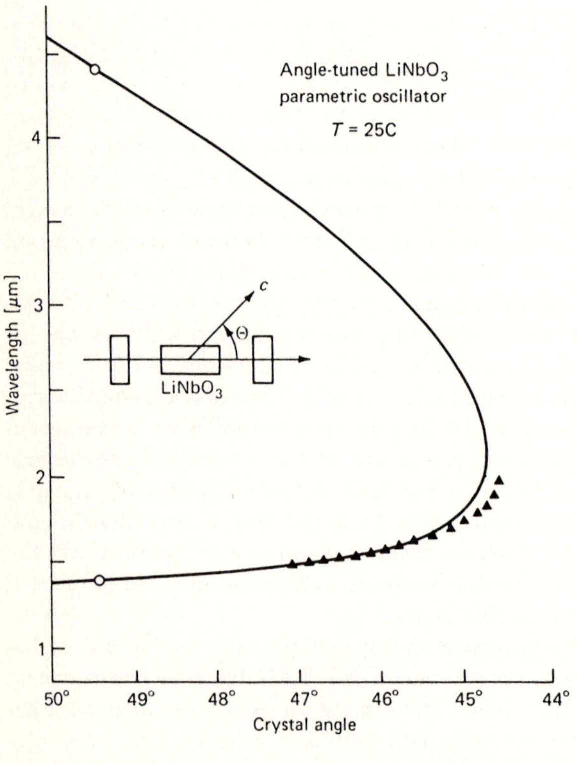

Fig. 10.29. Angle-tuned $LiNbO_3$ parametric oscillator [10.145]

element, emits a total of four different wavelengths. The parametric oscillator consists of a focusing lens, input and output mirrors, and a 4-cm-long lithium niobate crystal. This crystal is mounted inside an oven whose temperature is variable from 50 to 450° C. In this singly resonant oscillator the mirrors are highly reflecting at one of the two output frequencies.

Detailed discussions of parametric oscillators can be found in a number of review articles [10.140, 149–151].

A summary of representative parametric oscillators is given in Table 10.6. The materials used for parametric oscillators must be transparent or low loss at the pump wavelength as well as the signal and idler; they must also have adequate optical quality to allow an oscillator cavity to be formed. Pumping $LiNbO_3$ [10.160] with different visible wavelengths yielded spectral coverage from 0.548 to 3.65 μm. Shorter wavelengths were obtained from ADP pumped by a frequency-quadrupled Nd:YAG laser [10.162]. The tuning range was from 0.42 to 0.73 μm. Proustite (Ag_3AsS_3) pumped in Nd:$CaWO_4$ provided coverage out to 8.5 μm [10.168]; CdSe pumped by $CaF_2:Dy^{2+}$ reached 13.7 μm [10.169]. Hence, most of the infrared and visible spectrum can be covered by parametric oscillators.

The optical parametric oscillator has long held promise as a tunable, high-power, solid-state laser source. This promise has not been fulfilled due to the limitations of available nonlinear materials. Soon after the first operation of the parametric oscillator in the mid 1960s, the theory was quickly developed in the next few years, and numerous nonlinear materials were experimented with by the early 1970s. However, a variety of practical problems with these materials, especially low-optical-damage thresholds, and unavailability of large, high quality crystals, as well as the development of competing tunable-laser sources, particularly the pulsed dye laser, limited interest in parametric oscillators. For example, damage by the high intensity fields, necessary to achieve good conversion, has been a particular problem with $LiNbO_3$, the most useful material for this application.

With the recent introduction of new nonlinear materials such as BBO, KTP, urea and the chalcopyrites, renewed interest has developed in optical parametric oscillators.

Unlike other candidate nonlinear crystals, BBO has a very high damage threshold and this allows the design of the parametric oscillator for pump intensities as high as 1 GW/cm^2. At this intensity level conversion efficiencies approaching 30 % can be expected. Preliminary calculations have shown that tunable output ranging from 400 to 700 nm and from 2 to 3 μm may be achieved if the oscillator is pumped by the third harmonic of a Nd:YAG laser [10.47].

The high nonlinear coefficient and the high damage threshold (although not as high as BBO) make KTP attractive for use in parametric oscillators. Recently, a Nd:YAG pumped parametric oscillator was constructed with a tunable output around 2 μm. Due to the short interaction length, limited by the available crystal size (5 mm cube), only operation slightly above threshold could be achieved [10.172].

Table 10.6. Parametric oscillators

Pump source	λ_p [μm]	Pump power [W]	Tuning technique	Nonlinear material
Nd : $CaWO_4$ (doubled)	0.53	10^5	Temp.	$LiNbO_3$
Nd : glass (doubled)	0.53	–	Angle	KDP
Nd : glass (doubled)	0.53	5×10^4	Angle	$LiNbO_3$
Ruby	0.69	3×10^6	E-O and angle	$LiNbO_3$
Nd : YAG (doubled, cw)	0.53	0.3	Temp.	$Ba_2Na\,Nb_5O_{15}$
Ruby	0.69	2.7×10^5	Fixed	$LiNbO_3$
Ruby	0.69	9×10^5	Fixed	$LiNbO_3$
Ruby	0.69	7.5×10^5	Angle + temp.	$LiNbO_3$
Nd : YAG	1.06	0.2	Temp.	$LiNbO_3$
Nd : YAG (doubled)	0.53 0.66 0.47–0.65	–	Temp.	$LiNbO_3$
Nd : YAG	1.06	–	Temp.	$LiNbO_3$
Nd : YAG (quadrupled)	0.26	2×10^5	Temp.	ADP
Nd : $CaWO_4$	1.06	8×10^4	Angle	Ag_3AsS_3
Nd : YAG	1.06	17×10^6	Angle	$LiIO_3$
Nd : YAG	1.83	2.4×10^3	Angle	CdSe
Nd : $YAlO_3$	1.08	2.6 (average)	Temp.	$LiNbO_3$
Nd : YAG	1.83	5×10^3	Angle	CdSe
Nd : $CaWO_4$	1.065	–	Angle	Ag_3AsS_3
Dy : CaF_2	2.36	10^5	Angle	CdSe
Nd : YAG	1.06	10^6	Angle	$LiNbO_3$
Nd : YAG (doubled)	0.53	10^6	Temp.	CDA
Ruby (doubled)	0.347	5×10^5	Angle	$LiIO_3$
Nd : YAG	355	2×10^6	Angle	Urea
Nd : YAG	1.06	6×10^4	Angle	$AgGaS_2$
Ho : YLF	2.05	1.8×10^5	Angle	$AgGaSe_2$

Competing with these crystals for application in parametric oscillators in the visible and near-infrared regions is the organic crystal urea. A number of devices have been reported based on this material. For example, *Rosker* et al. [10.48, 85] designed a singly-resonant urea optical parametric resonator with a frequency tripled Nd : YAG as pump source which produced output in the 0.5 to 1.23 μm range. The conversion efficiency was about 20 % for pump intensities

S-Singly, D-doubly resonant	Output power [W]	Wavelength covered [μm]	Ref.
D	10^3	0.73–1.9	[10.142] [10.152]
D	5×10^3	0.96–1.18	[10.153]
D	50	0.68–2.35	[10.154]
D	4×10^4	1–1.08	[10.155]
D	1.5×10^3	0.98–1.06	[10.143]
D, S	$6 \times 10.^4$	1.04–2.08	[10.156]
S	0.25×10^6	1.06	[10.157]
S	0.34×10^6	1.6–2.05	[10.158]
		1.2–1.05	
D	17×10^{-3}	2.13	[10.159]
S	(70–105)	0.68–0.77	[10.160]
	$\times 10^{-3}$	0.9–2.0	
		0.55–3.65	
D	0.1	0.92–0.98 (doubled)	[10.161]
–	10^5	0.42–0.73	[10.162]
D	10^3	1.82–2.56	[10.163]
D, S	13×10^3	1.95–2.34	[10.164]
		3.8–4.2	
S	180	2.2–9.6	[10.165]
D, S	1.2	2.1	[10.166]
	0.5	1.8–2.7	
	0.2	2.8–3.7	
S	–	2.2	[10.167]
		9.8–1.04	
S	10^2	1.22–8.5	[10.168]
S	22×10^3	7.9–13.7	[10.169]
		2.8–3.4	
S	5×10^4	1.4–4.4	[10.145]
S	6×10^5	0.85–1.4	[10.144]
D	10^4	0.41–2.1	[10 170]
S	4×10^5	0.5–1.23	[10.85]
S	10^4	1.4–4.0	[10.149]
S	25×10^3	3.3–5.8	[10.171]

three times above threshold. The pump beam had a 1 mm diameter and a peak intensity of 50 MW/cm^2, and the crystal length was 23 mm. The main attribute of urea is the high damage threshold of 1.4 GW/cm^2 at 355 μm, which is at least an order of magnitude greater than $LiNbO_3$.

Recently, a number of parametric oscillators were constructed with crystals of the class of semiconductor chalcopyrites such as $AgGaS_2$ and $AgGaSe_2$. Interest in these crystals stems from the fact that they are transparent from the visible to the far infrared and they possess high nonlinear coefficients. For example, $AgGaS_2$ is transparent from 0.53 to 12 μm and $AgGaSe_2$ has a nonlinear coefficient 90 times higher compared to KDP. Unfortunately, these crystals have very low damage thresholds around 10 MW/cm^2. Parametric oscillators employing $AgGaSe_2$ and $AgGaSe_2$ pumped by Nd: YAG, Ho: YAG and Ho: YLF have been built [10.49, 50, 171]. Tuning from 1.4 to 4.0 μm was demonstrated for Nd: YAG pumping, and continuous tuning between 3.3 and 5.6 μm was achieved with the holmium laser.

Whether any of these new materials will become successful for broader application of optical parametric oscillators remains to be seen. The growth of large, high quality crystals remains difficult. On the other hand, BBO, KTP, urea and $AgGaSe_2$ are still at their early stages of development.

10.3 Raman Laser

Recently, the Raman laser which is based on stimulated Raman scattering (SRS) has shown to represent a practical way to access wavelengths not directly available from solid-state lasers. Stimulated Raman scattering was first observed by *Woodbury* and *Ng* [10.173] in experiments with ruby lasers in 1962. In its basic form, the Raman laser consists of a high pressure gas cell and resonator optics [10.174]. If this completely passive device is pumped by a high-power laser, a fraction of the laser wavelength is shifted to a longer wavelength. The particular wavelength shift depends on the gas in which SRS takes place. Also, if the power level of the laser is increased, additional spectral lines will appear at longer as well as shorter wavelengths with respect to the pump wavelength.

The basic Raman effect is an inelastic light scattering process. The energy levels of interest for Raman scattering are shown in Fig. 10.30. An incident quantum $h\nu_\mathrm{p}$ is scattered into a quantum $h\nu_\mathrm{s}$ while the difference in energy $h(\nu_\mathrm{p} - \nu_\mathrm{s}) = h\nu_\mathrm{R}$ is absorbed by the material scattering center. In Fig. 10.30 u is the upper state of the molecule, and i and f are the initial and final states. In principle, the excitation of the material may be a pure electronic excitation, or

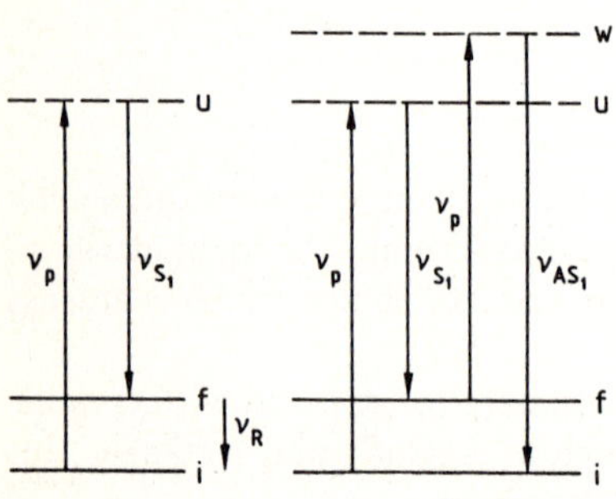

Fig. 10.30a,b. Raman process. Generation of first Stokes light (a) and first anti-Stokes light (b)

a vibrational or rotational excitation of a molecule. Solid-state-laser-pumped Raman lasers typically employ gases such as hydrogen or methane, therefore level i and f are the vibrational levels of the ground state of the molecule. The upper level u can be a real state or a "virtual" upper state. The frequency ν_s is called a Stokes frequency and is smaller than the incident light frequency by an amount equal to the excitation or pump frequency ν_p of the system. The difference between ν_p and ν_s

$$\nu_p - \nu_s = \nu_R \tag{10.43}$$

is the Raman shift which is characteristic of the material in which the Raman process is observed.

If the system is in an excited state to begin with, it may make a transition downward while the light is scattered. In that case the scattered light contains anti-Stokes frequencies which are larger than the incident frequency.

In the stimulated Raman effect, the pump laser at frequency ν_p excites molecules to level u, and if a population inversion exists between levels u and f, it can produce lasing action. In this case the radiation ν_s becomes amplified, while the pump radiation ν_p loses energy. The process has typical laser characteristics, such as pump energy threshold, exponential gain and narrow line width. The emission in Fig. 10.30a is called the first Stokes line, usually written S_1. If a high-power laser is focused into a Raman medium, additional lines will appear at the output. Actually, a single laser frequency interacting with molecules will produce a "comb" of frequencies, each separated from its neighbor by the frequency spacing ν_R. These additional lines will be to the left and right of the wavelength scale with regard to the laser pump wavelength.

The additional lines are produced by parametric four-wave mixing of the various waves propagating in the Raman medium. As an example Fig. 10.30b illustrates the generation of one such line, having a wavelength shorter than the pump wavelength. This so-called anti-Stokes line is the result of the interaction of ν_p and ν_{s1} both propagating in the same direction. The parametric four-wave mixing process does not require a population inversion between w and i, therefore there is no well defined threshold. The simplest way of looking at this interaction is that the two frequencies beat together to produce polarization (induced dipole moments in the molecules) at the difference frequency. This polarization then modulates the laser-molecule interaction and produces light beams at the side frequencies.

Stokes lines have lower frequency (longer wavelength) and anti-Stokes lines have higher frequency (shorter wavelength). In each case the line is labeled first, second, etc., by counting the number of frequency shifts from the pump laser.

The gain equations given below, which are taken from [10.175], provide the basic design parameters of the Raman lasers. A rigorous mathematical treatment of stimulated Raman scattering can be found in [10.176–179]. These references also provide comprehensive introductions to all areas of coherent Raman spectroscopy.

SRS can be described as a nonlinear interaction involving the third-order nonlinear susceptibility χ^3. At a medium's Raman resonance, the third-order susceptibility reduces to the peak Raman susceptibility χ''_R, where the double prime indicates the imaginary part of the total susceptibility.

The growth of the electric field at the Stokes wavelength and depletion of the pump field is governed by the equations:

$$\frac{\partial E_p}{\partial z} = -\frac{\omega_p}{2cn_p}\chi''_R|E_s|^2 E_p \tag{10.44a}$$

$$\frac{\partial E_s}{\partial z} = \frac{\omega_s}{2cn_s}\chi''_R|E_p|^2 E_s \tag{10.44b}$$

where $\omega_p - \omega_s = \omega_R$ are the frequencies, n_s and n_p are the indices of refraction, and c is the velocity of light. For a constant pump field, the Stokes field grows exponentially with a power gain given by

$$P_s(l) = P_s(0)e^{g_s l} \tag{10.45}$$

where g_s is the gain coefficient and l the interaction length in the Raman medium.

$$g_s = \frac{\omega_s \chi''_R |E_p|^2}{n_s c} = \frac{4\pi\chi''_R I_p}{\lambda_s n_s n_p \varepsilon_0 c} \quad . \tag{10.46}$$

If one expresses the third-order Raman susceptibility, χ''_R, in terms of spontaneous Raman scattering cross-section, $d\sigma/d\Omega$, one obtains

$$g_s = \frac{4\lambda_s^2 N(d\sigma/d\Omega)I_p}{n_s^2 \hbar\omega_p \Delta\omega_R} \tag{10.47}$$

where λ_s is the Stokes wavelength and I_p is the pump intensity. N is the population density and $\Delta\omega_R$ is the full-width, half-maximum Raman linewidth.

From these equations follows that the gain for a single-pass Raman medium is proportional to the incident intensity, the active media cross-section (which includes pressure and linewidth dependencies), and the length of the Raman cell.

The threshold of a Raman laser is usually defined as the gain required to achieve an output power at the Stokes wavelength that is of the same order as the pump radiation. For example, to achieve a 1 MW Stokes shifted power output, one requires a gain length product of $g_s l = 36$ in the Raman medium in order for the radiation to build up from the initial spontaneous noise level which is $P_s = h\nu_s \Delta\nu_s \approx 10^{-19}$ W in the visible. Quantitative information such as gain coefficient and Stokes wavelength for many gases, liquids and solids are listed in [10.180]. Table 10.7 summarizes the data for the most important Raman media.

The maximum theoretical conversion efficiency of a Raman laser is

$$\eta = \frac{\nu_p - \nu_R}{\nu_p} \quad .$$

Table 10.7. Stokes shift and Raman scattering cross-section for several gases

Medium	ν_R [cm^{-1}]	$\frac{d\sigma}{d\Omega}$ $\left[\frac{\text{cm}^2}{\text{Ster}}\right]$
H_2	4155	8.1×10^{-31}
CH_4	2914	3.0×10^{-30}
N_2	2330	3.7×10^{-31}
HF	3962	4.8×10^{-31}

If one introduces the numbers given for ν_p in Table 10.7 it is obvious that the conversion efficiency can be very high. For example, a frequency doubled Nd:YAG laser, Raman shifted with CH_4 provides an output at 630 nm. With $\nu_R = 2914\,\text{cm}^{-1}$ and $\nu_p = 18,797\,\text{cm}^{-1}$ one obtains $\eta = 84\,\%$.

Device Implementation

Practical Raman lasers employed to shift the wavelength of solid-state lasers are restricted to gases as the nonlinear medium. In liquids, stimulated Raman scattering is complicated by the onset of Brillouin scattering and by self-focusing which leads to filament formation. Although Raman shifting is simple in principle, there are numerous design parameters to consider in engineering a practical device. SRS in gases compete with optical breakdown and with stimulated Brillouin scattering. In order to avoid these competing nonlinear effects, and to produce significant energy conversion, pump pulselength, input beam diameter, focusing and interaction length – which are all interrelated – must be carefully optimized [10.181]. Raman lasers have been designed in a number of configurations such as the single-pass cell, Raman resonator, oscillator-amplifier and waveguide system. A short description of the salient features of these different designs will be given below.

The simplest gas Raman laser is based on the single-pass emission in a gas filled cell, as shown in Fig. 10.31a. The output beam quality is similar to that of the input pump. While the optics may be simply designed to prevent optical damage to the windows at focal intensities high enough to produce significant energy conversion, many nonlinear processes may occur to limit the conversion efficiency at high energies. Copious second Stokes and anti-Stokes production may occur, as well as stimulated Brillouin scattering and optical breakdown. The single-pass cell does not provide discrimination against these other nonlinear processes at high energy inputs.

By using mirrors at each end, feedback can be selectively enhanced at only the first Stokes wavelength, and the higher-order Stokes processes which would normally be built up by multiple passes would be eliminated. All that is required is that the gain be large enough to start the stimulated scattering process at a low enough threshold to achieve overall high conversion efficiency.

A gas Raman laser utilizing a single resonator, is shown schematically in Fig. 10.31b. The mirror coatings selected allow all the pump light to pass into the cavity, while inducing resonator action at the Raman shifted frequency. The

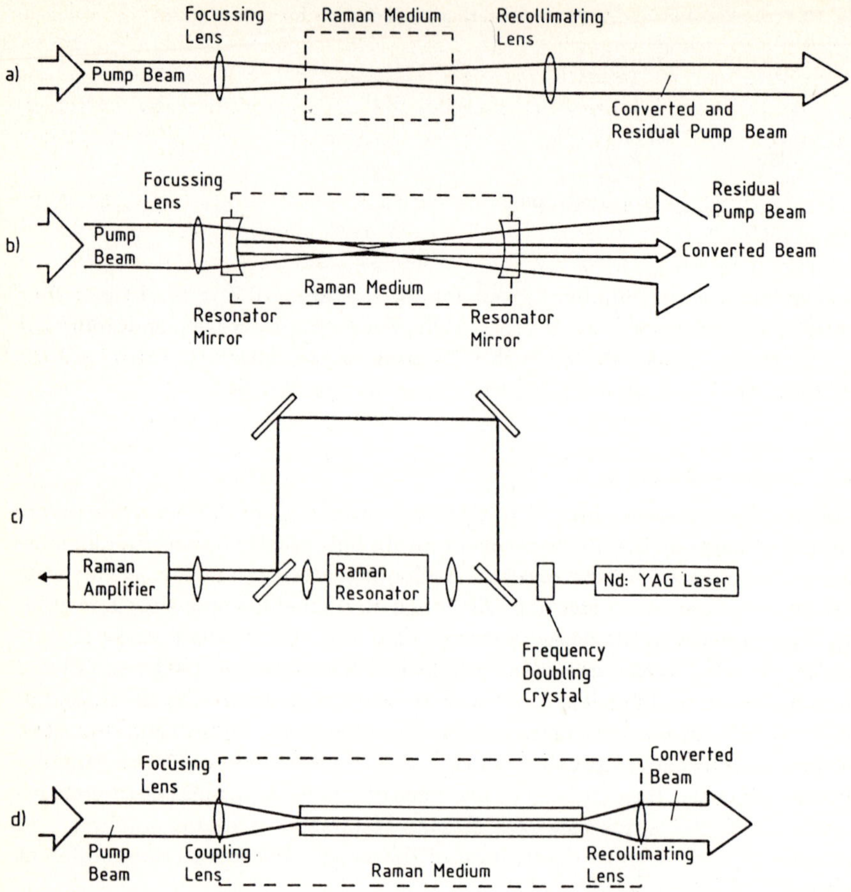

Fig. 10.31. Raman laser configuration **(a)** single pass cell, **(b)** Raman resonator, **(c)** Raman oscillator-amplifier, **(d)** waveguide

resonator transforms a multi-mode pump laser beam into a nearly diffraction-limited output beam with a slightly narrower pulse width. The high-quality beam is a result of keeping the pump intensity below single-pass threshold so that only multiple reflections of the lowest-order mode will achieve sufficient intensity to reach the stimulated scattering threshold.

The resonator length must be balanced between two conflicting requirements. Long cells will allow a large beam size on the cell windows, which is necessary to prevent optical damage. The number of round-trip passes in the cell, however, varies inversely as the length of the cell, so the best resonator action is produced with cells whose length is much shorter than the equivalent pump laser pulse width. Alignment difficulties will also become troublesome at the longest lengths. For these reasons resonators should be limited to lengths less than about 50 cm. The type of resonator is also an important parameter.

In order to have a large spot size on the windows and a small waist at the center of the cell, a nearly concentric arrangement is required. The nearer the resonator is to concentric, however, the smaller the waist becomes, and the larger the laser intensity becomes [10.182]. Higher-order nonlinear effects are then produced which limit the conversion efficiency, particularly at higher energies.

For high-energy applications an oscillator-amplifier arrangement, as shown in Fig. 10.31c may be considered. A portion of the pump beam bypasses the resonator, and, along with the Raman shifted resonator output, enters the amplifier cell. The two input beams entering the amplifier must be propagating precisely parallel or antiparallel. The low-energy oscillator can be designed to produce only the first Stokes wavelength, and discriminate against all others. The amplifier has to be designed for a gain low enough that self-oscillation is prevented, yet it need to saturate at a sufficiently strong Stokes signal.

The use of a glass capillary to confine the pump beam to a long interaction length, as shown schematically in Fig. 10.31d has been exploited in the waveguide Raman laser. Almost complete conversion may be obtained due to the long interaction length, even at relatively low pump intensities. Because of the physical nature of the waveguide, the intensity at the inner walls of the capillary is near zero so there is little danger of damage. Similarly the input and output coupling optics can be in low intensity regions. One has to realize, however, that the long interaction length possible in a waveguide makes it a high gain device not just for the first Stokes component. At input intensities large enough to reach threshold easily in the pump pulse, generated intensities may be large enough to produce second Stokes components, as well as other nonlinear effects.

Examples of Raman-Shifted Lasers

Table 10.8 summarizes the wavelengths and energies available from a Raman laser which consists of a gas cell filled with hydrogen and receives an input of 85 mJ in a 5 ns pulse at 560 nm, generated from a Nd:YAG pumped dye

Table 10.8. Wavelengths, energies and optimum gas pressures of Raman output of one commercial Raman shifter. S_1 denotes ith Stokes wavelength; AS_1 denotes ith anti-Stokes wavelength. The pump laser had 85 mJ of energy in a 5 ns pulse at 560 nm. Gas was hydrogen at room temperature

Wavelength [nm]	Energy [mJ]	Pressure [psi]
195 (AS_8)	0.0031	125
213 (AS_7)	0.0091	125
234 (AS_6)	0.024	110
259 (AS_5)	0.054	115
290 (AS_4)	0.10	145
330 (AS_3)	0.26	160
382 (AS_2)	0.78	190
454 (AS_1)	2.10	200
730 (S_1)	17.00	90
1048 (S_2)	6.20	300
1855 (S_3)	0.60	275

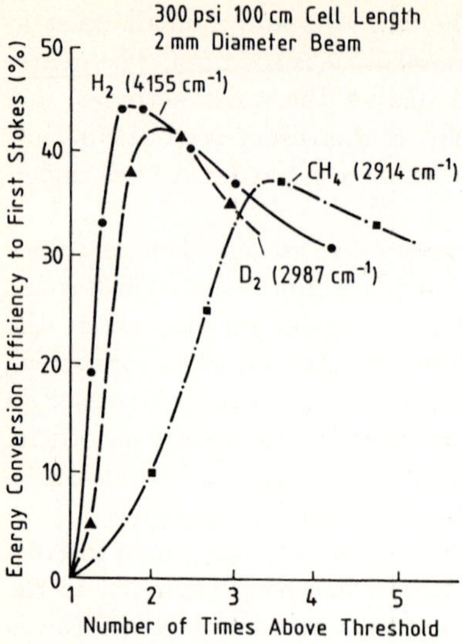

Fig. 10.32. Stimulated Raman conversion efficiency versus number of times above threshold for 0.532 μm pumping

laser [10.183]. The maximum energy is obtained at the first Stokes wavelength which occurs at 730 nm and then falls off rapidly for the additional Stokes and anti-Stokes lines.

Figure 10.32 shows the conversion efficiency of a Raman laser which is pumped by a frequency doubled Nd:YAG laser [10.175]. Plotted is the conversion efficiency to the first Stokes wavelength for three gases versus the normalized pump intensity expressed in times above threshold. The experiments were carried out with a 2 mm diameter beam in a 1 meter long cell. The collimated pump beam at 532 nm was emitted from a frequency doubled Q-switched Nd:YAG laser. Hydrogen has the lowest threshold and also the highest conversion efficiency. Threshold was achieved at an input of 8 mJ with an 8 ns long pulse.

The same set-up was also employed to produce 80 mJ of 1.9 μm output, by passing the 1.064 μm beam directly through the hydrogen filled cell. In this case the beam had a diameter of 6 mm and the input was 240 mJ.

The design and performance of a state-of-the-art hydrogen gas Raman laser which converts the frequency doubled output of a Nd:YAG laser from 532 to 683 nm with an efficiency of up to 40 % will be illustrated below [10.184]. Figure 10.33 shows a block diagram of the pump laser which consists of a Q-switched Nd:YAG oscillator, a single-pass amplifier followed by a frequency doubler. The 532 nm pump beam emerging from the CD*A doubler had an output energy up to 210 mJ, a pulsewidth of 24 ns, a beam diameter of 4.9 mm and a beam divergence of 1 mrad. The repetition rate of the system was 2 Hz. The Raman laser consisted of the gas pressure cell and the resonator optics as shown in Fig. 10.34. The Raman cell is made from a block of aluminum bored with a 16 mm diameter hole, fitted with O-ring grooves and windows held together by end caps. The 25 mm diameter by 12.5 mm thick fused silica windows are antireflection coated at both the pump and the first Stokes wavelengths. The

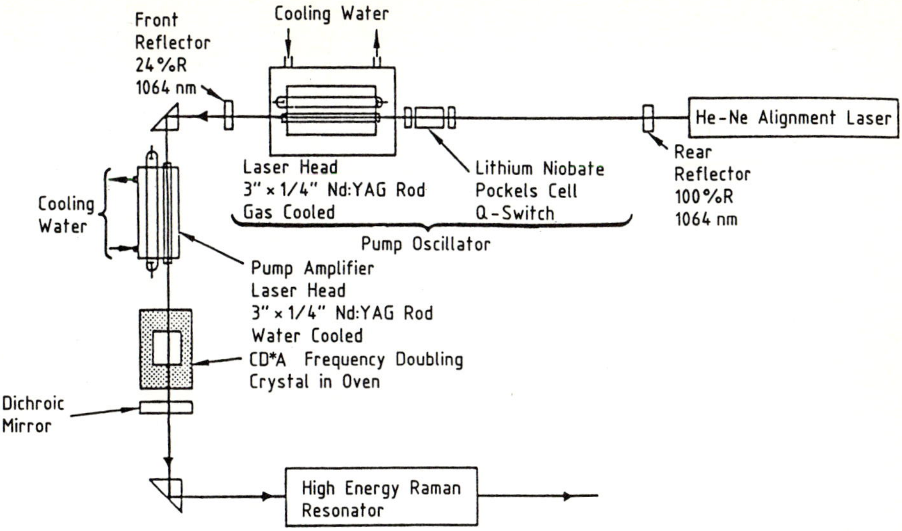

Fig. 10.33. High energy Raman resonator optical schematic

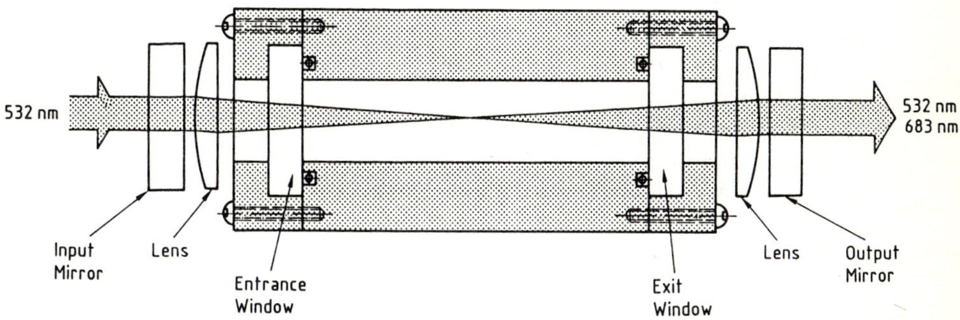

Fig. 10.34. Resonator with intracavity lenses

output of the Raman laser was measured as a function of input power with gas pressure, resonator configuration and length as parameters. Figure 10.34 depicts the concentric resonator configuration comprised of two flat mirrors and a pair of plano-convex lenses. The resonator length was 20 cm. The entrance mirror has high transmission at 532 nm and high reflection at 683 nm. The exit mirror has high transmission for the pump wavelength at 532 nm and 50 % reflectance at the first Stokes wavelength. The reflectance for the second Stokes (954 nm) and first anti-Stokes (436 nm) wavelength was only a few percent at both mirrors.

Figure 10.35 shows the output for different gas pressures and input energies. Optical breakdown leads to a drop in energy at the highest energies and pressures.

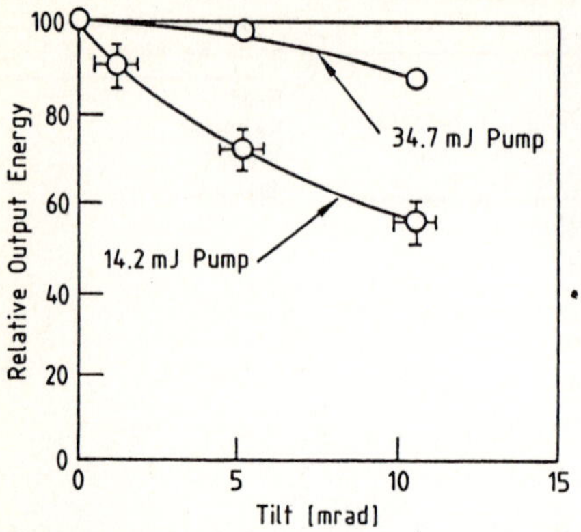

Fig. 10.35. Output energy from tilted concentric resonator. Hydrogen pressure 1000 psi, output reflector 50% at 683 nm, 20 cm resonator

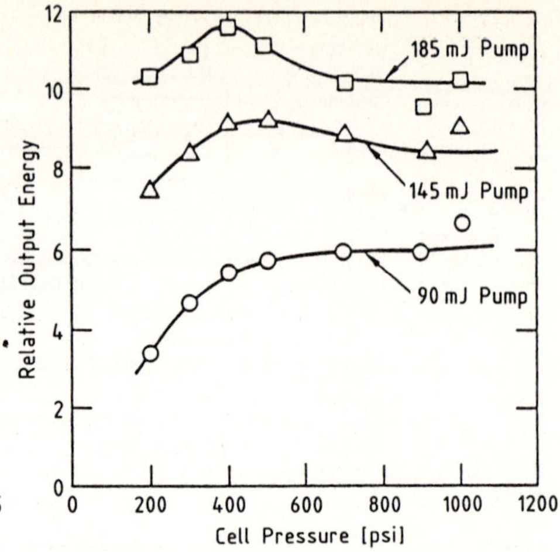

Fig. 10.36. Concentric resonator output energy for different Raman cell pressures and input energies. Output reflector 50% at 683 nm, 20 cm resonator

The misalignment sensitivity of the resonator axis to the pump axis is shown in Fig. 10.36. At low energies, the tilt is important because the gain is low and a good overlap of the beams is necessary. At the higher energies, however, tilt is not as important. It should be noted that the resonator itself was aligned and the misalignment refers only to the tilt with respect to the pump beam axis.

Figure 10.37 shows the output vs. input for the resonator at the exact length for the concentric geometry (solid line) and for slight variations in length. As is apparent from the data, the resonator length becomes more critical for the higher input energies, and a change of (1–2) % in resonator length has a noticeable effect. The output energy of 65 mJ at 683 nm for 185 mJ input at 532 nm was the highest energy achieved with the concentric resonator. Optical breakdown, probably caused by small particles and impurities in the gas set this upper limit.

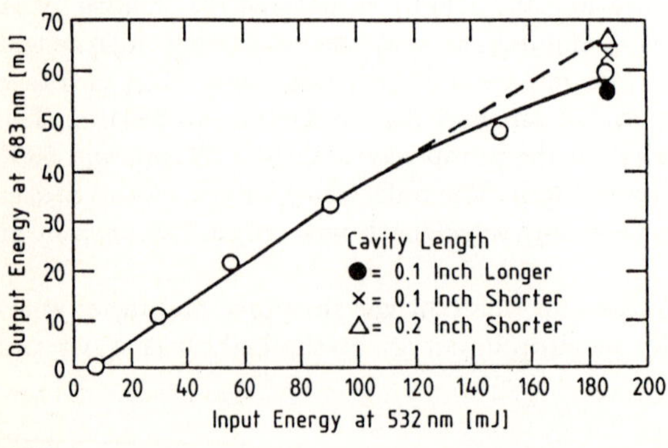

Fig. 10.37. Concentric resonator output energy for different pump input energies and cavity lengths. The solid line corresponds to the concentric position. (Hydrogen pressure: 1000 psi, output reflector 50% at 683 nm, 20 cm resonator)

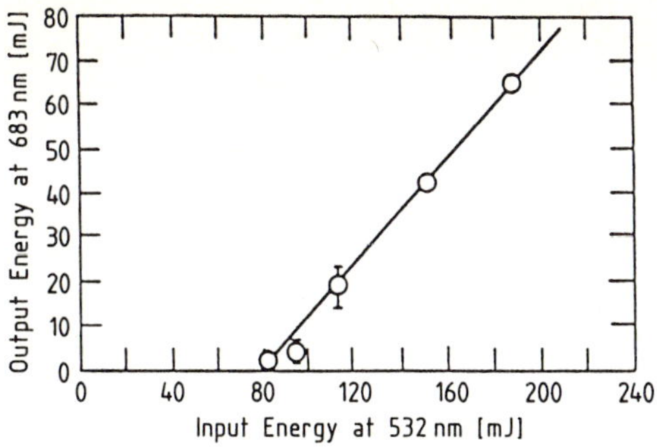

Fig. 10.38. Flat/flat resonator output energy for different input energies. 35 cm flat/flat resonator, output reflector 50 % at 683 nm, hydrogen pressure 1100 psi

By removing the lenses shown in Fig. 10.34 experiments were also performed with a 35 cm long flat/flat resonator. Hydrogen has enough gain to allow high conversion efficiency without an internal focus in the Raman cell. Since there is no focus in the cell, the problem of optical breakdown is reduced.

For the flat/flat resonator configuration the Raman-laser output increased monotomically with gas pressure. The output vs. input energy at 1100 psi, the highest pressure applied, is shown in Fig. 10.38. The average input intensity was 100 MW/cm^2 and the output reflectance at 683 nm was 50 %. The threshold increased from 8 to 80 mJ compared to the concentric resonator, however, the total output at the highest input energy was the same, i.e. 65 mJ. The highest output energy of 78 mJ was achieved at 1700 psi gas pressure and an input energy of 195 mJ. It required also a change in output mirror, which for this case had a reflectivity of 27 % at 683 nm and 86 % at 532 nm.

10.4 Optical Phase Conjugation

Optical phase conjugation, also referred to as wave-front reversal, was first observed by workers in the Soviet Union [10.185, 186] in 1971. For an introduction to the field, and detailed accounts of the many theoretical and experimental studies on the subject, the reader is referred to the extensive literature [10.2, 188–192].

Optical phase conjugation started to generate considerable interest after potential applications have been treated in the open literature (e.g. compensating for the effects of atmospheric turbulence in the propagation of high-energy laser radiation, and for correcting optical train distortions and pointing errors in laser fusion systems). From the standpoint of solid-state laser engineering, phase conjugation via stimulated Brillouin scattering is particularly interesting, because it offers the possibility for compensating optical distortions produced in laser amplifier stages.

Following a brief introduction, we will illustrate the application of optical phase conjugation to high-average-power oscillator-amplifier solid-state laser systems. From a mathematical point of view, phase conjugation can be explained by considering an optical wave of frequency ω moving in the $(+z)$ direction,

$$E(x,y,z,t) = A(x,y)\exp\{\mathrm{j}[kz+\phi(x,y)] - \mathrm{j}\omega t\} \tag{10.48}$$

where E is the electric field of the wave with wavelength $\lambda = 2\pi/k$. The transverse beam profile is given by the function $A(x,y)$ and the phase factor $\phi(x,y)$, indicating how the wave deviates from a uniform, ideal plane wave. In particular, the phase factor carries all the information about how the wave is aberrated.

If the beam given by (10.48) is incident upon an ordinary mirror, it is reflected upon itself and the sign of z changes to $(-z)$, all other terms of the equation remain unchanged. However, if the beam is incident upon a phase conjugate mirror it will be reflected as a conjugated wave E_c given by

$$E_\mathrm{c}(x,y,z,t) = A(x,y)\exp\{\mathrm{j}[-kz-\phi(x,y)] - \mathrm{j}\omega t\} \quad . \tag{10.49}$$

In addition to the sign change of z, the phase term has changed sign, too. The conjugate beam corresponds to a wave moving in the $(-z)$ direction, with the phase $\phi(x,y)$ reversed relative to the incident wave. We can think of the process as a reflection combined with phase or wave-front reversal. The phase reversal expressed by (10.49), for example, means that a diverging beam emitted from a point source, after reflection at a phase-conjugate mirror will be converging and be focused back to the point of origin.

A practical application of phase reversal in a laser system is depicted in Fig. 10.39. Shown is an oscillator which produces an output with a uniform flat wavefront that is distorted in the amplifier medium. An ordinary mirror merely inverts the distortion as it reflects the beam, thereby keeping the distortion

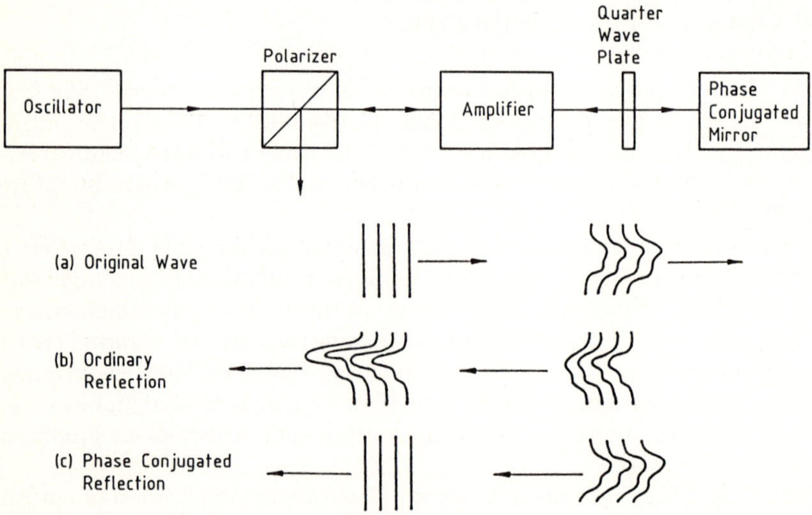

Fig. 10.39. Compensation for optical phase distortions caused by an amplifying medium using nonlinear optical phase conjugation [10.193]

fixed with respect to the propagation direction. With a second pass through the amplifier, the distortion is essentially doubled. A phase-conjugate mirror, on the other hand, reverses the wavefronts relative to the wave propagation direction; hence, the same region of the amplifier that originally created the distortion compensates for it during the second pass through the amplifier.

Phase-conjugate mirrors have been demonstrated in solids, liquids and gaseous media using a number of nonlinear optical interactions, such as three-wave and four-wave mixing, stimulated Brillouin scattering (SBS), stimulated Raman scattering (SRS), and photon echoes [10.2].

While many manifestations of self-pumped phase-conjugate mirrors have been demonstrated, an approach that has proven to be very practical is the use of the phase-conjugating properties of stimulated Brillouin scattering (SBS) in gases [10.2]. The high efficiency which can be obtained with SBS, combined with the absence of induced damage effects in gases, make this approach particularly attractive. Stimulated Brillouin scattering involves the scattering of light by sound waves (i.e., acoustic phonons, or pressure/density waves), which results from optically induced electrostriction. The resultant Stokes scattered light is down-shifted in frequency by a relatively small amount ($\sim 10^9$ to 10^{10} Hz). Stimulated Brillouin scattering, which is a threshold process has a relatively large cross-section provided that the coherence time ($1/\Delta\omega$) of the incident light is long compared to the response time τ of the acoustic phonon.

For pressurized gases, τ is typically on the order of tens of nanoseconds. Therefore, high-efficiency stimulated Brillouin scattering requires linewidth narrowing or preferably single-longitudinal-mode operations.

The gain of the Stokes-shifted wave is generally the greatest in a direction opposite to that of the incident beam. The efficiency of the SBS process (defined as the ratio of the Stokes-shifted, backward-going energy or power, to the incident optical energy or power) can be as high as 80 % in practical laser devices.

In the following section, we will illustrate the correction of optical aberration in amplifier stages by means of phase conjugation. Most of the work has been performed with high-power glass laser systems [10.187] and with Nd : YAG laser [10.194–196], using stimulated Brillouin scattering in methane.

Figure 10.40 shows a block diagram of a diffraction-limited oscillator/amplifier system which employs a phase-conjugate mirror [10.193]. The oscillator produces a 10 mJ TEM_{00} output which is passed through two amplifiers in series. Upon reflection from the phase conjugate mirror, the pulse retraces its

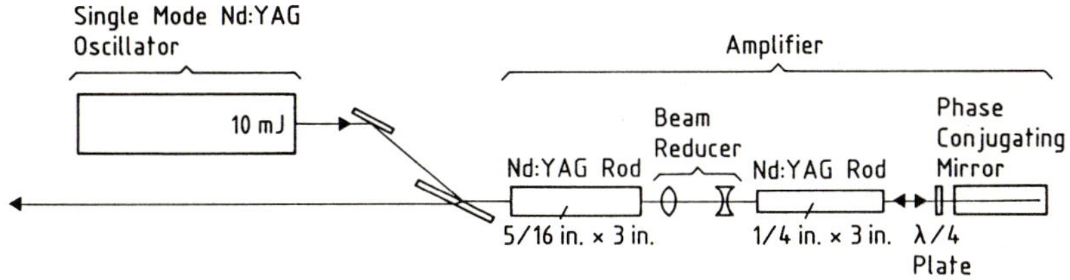

Fig. 10.40. A phase-conjugate Nd : YAG oscillator-amplifier system [10.193]

path through the rods and is amplified to an energy of 850 mJ. The pulse repetition frequency is 10 Hz. Reflectivities of over 80 % have been achieved at 1.06 μm with a gas cell filled with methane at 1300 psi. At this pressure, the threshold for stimulated Brillouin scattering was 1 GW/cm^2.

Besides compensating for thermal lensing, the phase conjugating mirror shown in Fig. 10.40 has some additional interesting features. Misalignment of a conventional rear mirror in a configuration, as shown in Fig. 10.40, caused by mechanical, thermal or vibration effects will lead to beam wander and can be critical in the operation of the system. On the other hand, the beam reflected from the phase conjugated mirror in Fig. 10.40 is self aligning. Also, if an ordinary mirror is substituted for the phase-conjugated mirror, at least one optical isolator would be requird since the arrangement is equivalent to four amplifier stages, and superradiance would prevent any substantial population inversion build-up. This problem is circumvented by employing a phase-conjugated mirror, which has zero reflectivity until threshold is reached.

So far we have addressed compensation of thermal lensing in amplifier rods with a phase-conjugate mirror. However, in most cases, thermally-induced birefringence will degrade performance also and needs to be considered (for example, if the amplifier stages are followed by a harmonic generator).

Even though the spatial wavefront is inverted in the stimulated Brillouin scattering process, the polarization of the backward going beam is not conjugated, i.e. the polarization state of the backward-wave behaves similarly to that of a conventional mirror. Several concepts have been developed, whereby birefringence compensation is introduced prior to the phase conjugation process, and thus present a uniformly polarized wave to the phase conjugate mirror.

If a 90° rotator is placed between two identically pumped rods, as shown in Fig. 10.41 the amplifier birefringence can be largely compensated [10.195]. Figure 10.42 shows the measured percentage of depolarization in a single pass

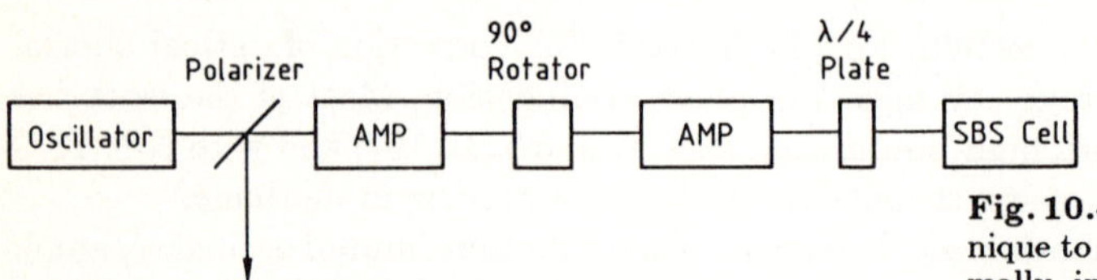

Fig. 10.41. Compensation technique to circumvent the thermally induced birefringence in the two laser amplifier rods, using a 90° polarization rotator [10.195]

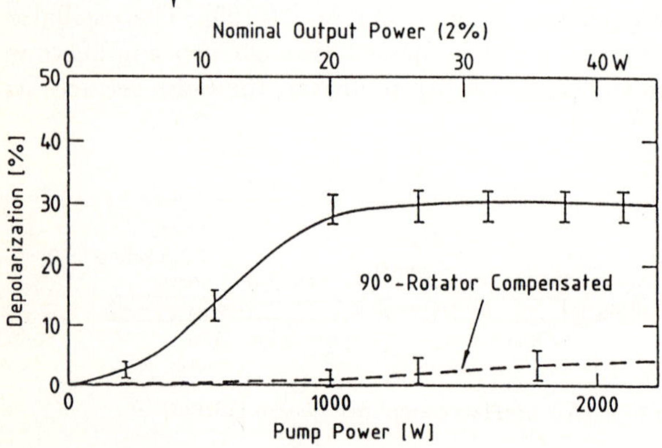

Fig. 10.42. Measured depolarization of two YAG rods as a function of the pump power, with and without a 90° polarization rotator

through the two YAG rods as a function of total pump power with and without the birefringence compensation scheme. The small residual depolarization is a reflection of the practical difficulty of "balancing" the two rods exactly.

In case of a single amplifier, the optical equivalent to the situation depicted in Fig. 10.41 is a two-pass or four-pass arrangement with a Faraday rotator between passes. Implementation of birefringence compensation with single amplifier stages can be found in [10.196].

Another approach investigated by *Basov* et al. [10.197] is to split the beam into two orthogonal polarizations after passing through the amplifier. One of the beams is then rotated 90° so it has the same polarization as the other. The two beams will cooperatively reflect from the SBS phase-conjugate mirror. The back-reflected phase-conjugated beams then pass back through their respective optical paths and are recombined at the beam splitter before propagating through the amplifier for a second time. The beam emerging from the amplifier has the same beam quality and polarization as the original incident beam.

It has been convincingly demonstrated by a number of researchers, that phase conjugation via stimulated Brillouin scattering in CH_4 gas can greatly reduce amplifier aberrations in Nd:YAG oscillator/amplifier systems. Compared to a single-pass oscillator-amplifier system, phase conjugation requires longitudinal-mode control of the oscillator, and a number of optical components such as a polarizer, rotator, waveplate, focusing lens, and the gas cell in the amplifier chain.

It greatly depends on the particular application and average-power requirements of the system, when it is justified to remove the troublesome aberrations associated with the amplifier stages by means of a phase-conjugate mirror.

11. Damage of Optical Elements

Optically induced damage to the laser material and to components of the system generally determine the limit of useful performance of high-power solid-state lasers. Therefore, an understanding of the mechanisms which cause radiation damage to optical components, and a knowledge of the damage threshold of the materials employed in a laser, are of great importance to the engineer who is designing a laser system.

Damage may occur either internally or at the surface of an optical component due to a number of intrinsic and extrinsic factors. Intrinsic processes that limit the optical strength of materials include linear absorption, color-center formation and a variety of nonlinear processes such as self-focusing, multiphoton absorption and electron avalanche breakdown. Extrinsic factors such as impurities, material defects (voids, dislocations, etc.) surface scratches and digs and surface contamination determine the threshold for surface damage. Surface damage is generally the more serious problem because it occurs at irradiance levels well below those that cause bulk damage. The problem of surface damage is compounded when the laser system contains many surfaces.

There are different ways of defining damage threshold. One might define damage as the physical appearance of a defect in the material or by a degradation in the output performance of the laser system. Such a parameter might be lower output energy or power, greater beam divergence, or increased feedback from scattering centers. From the standpoint of the user of the laser system, the performance deterioration is of more importance than the physical appearance of a defect in the material. Very often the occurrence of a small blemish or defect does not alter system performance. Also, one can frequently observe that a defect in the material which appeared at a certain energy level does increase in size after repeated firings at the same or even higher levels.

Since it is very difficult to relate an observed damage center to the degradation in laser performance, it is generally agreed upon to define the damage threshold as the energy or power density at which a change in physical appearance of the test component has been observed. It should be pointed out, however, that damage is highly probalistic in nature [11.1]. At any level, there is a nonzero probability of damage, and this probability increases with increasing optical intensity. The number usually quoted as damage threshold is the single-shot damage level. As the power level is decreased, the probability of damage decreases correspondingly, until a power level is reached at which the material will withstand a very large number of exposures. This power level represents the highest level at which a system can be consistently operated without component damage.

11.1 Surface Damage

Undoubtedly the weakest link in a well-designed laser is the surfaces of the optical components. It has been found that surfaces of bulk materials, even those that are scrupulously cleaned, have a lower damage threshold than the bulk. For example, for laser glass and sapphire the best obtainable surface damage threshold is a factor 4 to 5 lower than in the bulk.

One reason for the lower surface-damage threshold compared to bulk damage stems from the fact that even after the polishing process used in the finishing of optical elements, there are residual scratches, defects, and imperfections on the surface. At these points, the electric field in the light wave is greatly intensified so that the effective field value just inside the surface is much greater than the average field. When breakdown or failure of the surface occurs, it occurs near one of these imperfections [11.2]. Furthermore, as a consequence of the polishing process, minute inclusions of polishing material or other impurities occur at the surface. These impurities may be strongly absorbing (like jeweler's rouge, a commonly used polishing material) and, if so, can provide a nucleus around which damage may occur. Other factors tending to lower the damage threshold at the surface are the tendencies for contamination of the surface due to airborne particles, fingerprints, or absorption of impurities.

From the foregoing it follows that surface damage is caused either by absorption of submicrometer inclusions or by the formation of a plasma at the surface because of electron avalanche breakdown in the dielectric. In the former case, microexplosions of isolated inclusions occur just under the surface [11.3]. In the latter case, the electric field produced by an intense laser beam is sufficiently high that the dielectric material will break down in the optical field [11.4].

This avalanche ionization represents the ultimate failure mode of dielectric materials. In this process, a few electrons, either in shallow traps in the material, or due to easily ionized impurities, gain sufficient energy from the oscillating electric field of the light wave to cause secondary ionization. The ensuing avalanche is usually associated with a luminous plasma formed in the material. The avalanche process is the high frequency analogue of dc breakdown.

Avalanche ionization is the primary cause of surface damage, once absorbing impurities are removed. Due to field intensification effects, avalanche ionization is more likely to occur in the vicinity of pits and scratches at the surface. Super-polishing techniques, such as bowl-feed polishing, ion polishing, and chemical etching, lead to higher damage thresholds than those for conventionally polished optical surfaces.

Surface damage typically takes the form of a pit. The formation of pits in glass have been explained as the result of a standing wave which is formed inside the glass near the surface [11.5]. The standing wave is caused by reflections from the plasma-glass boundary. After plasma initiations there is a layer of plasma several wavelengths thick contiguous to the surface of interest. This plasma is densest at its source, that is, next to the surface. The electric field at the antinodes of this standing wave can become twice as large as the electric field in the incident beam. The first antinode is within one-quarter wavelength of the surface. It is this large electric field at the antinode that initiates pitting at or just inside the surface.

A sample of transparent dielectric material exposed to a collimated beam of light will damage at the exit surface at a lower power level than at the entrance surface due to the effect of Fresnel reflection [11.6]. When a light pulse enters an isotropic optical material at near-normal incidence, there is a reflection at the air-to-sample boundary. Since the index of refraction n of the material is greater than that of air, the reflected light wave suffers a phase shift of 180° with respect to the incident light wave. This phase shift results in a partially destructive interference of the two light waves within a distance of $\lambda/4$ of the entrance surface. As a result the light intensity I_{en} at the entrance surface is related to the incident intensity I_0 by

$$I_{en} = \frac{4}{(n+1)^2} I_0 \quad . \tag{11.1}$$

At the exit surface, the reflected wave suffers no phase shift, so that the intensity at the exit surface I_{ex} is related to the intensity incident on the sample by

$$I_{ex} = \left(\frac{4n}{(n+1)^2} \right)^2 I_0 \quad . \tag{11.2}$$

Thus the ratio of the intensity inside the medium at the exit surface to the intensity inside the medium at the entrance surface is

$$\frac{I_{ex}}{I_{en}} = \left(\frac{2n}{n+1} \right)^2 \quad . \tag{11.3}$$

For $n = 1.55$, this ratio is 1.48.

11.2 Inclusion Damage

Solid-state laser materials tend to exhibit minute inhomogeneities, which generally form in the process of fabricating the material. The inhomogeneities can consist of small bubbles, dielectric inclusions, or platinum particles in the case of the laser glass. In crystalline materials, although some microinhomogeneities do occur during the growth process, there is less evidence that particulate damage is the limiting factor. In glass, however, during the course of melting, crucible material, or material from other portions of the apparatus, finds its way into the melt and upon cooling condenses out in the form of particulate inclusions. This has been a particularly difficult problem in the fabrication of laser glass, which is manufactured in platinum containers. The problem of platinum inclusions has been solved by glass manufacturers either by using crucibles of ceramic and clay materials or by carefully controlling the oxygen content in the melt. If the material contains minute inhomogeneities such as particulate inclusions, especially metallic inclusions, they can absorb the laser radiation, giving rise to local heating and melting of the surrounding laser host material. This in turn causes stress concentrations sufficient to rupture the material. Elaborate theoretical treatments of the problem of stress formation in the vicinity of an absorbing particle have been published [11.7].

To summarize the considerations that apply, let us consider an absorbing particle embedded in laser material, illuminated by a brief and intense pulse of light. If the particle is a sphere of diameter d, it presents an absorbing area to the incident light of $\pi d^2/4$. The total energy absorbed in the particle is approximately

$$\Delta E = It\left(\frac{\pi d^2}{4}\right)[I - \exp(-\alpha d)] \quad , \tag{11.4}$$

where I is the power density, t is the pulse duration, and α is the material absorption coefficient. The maximum temperature rise experienced by the particle will be given by

$$\Delta T = \frac{\Delta E}{C_p(\pi d^3/6)} \quad , \tag{11.5}$$

neglecting heat conduction during the pulse. In the above equation C_p is the specific heat of the inclusion.

Equation (11.5) indicates that the temperature of metal particles subject to a 20-J/cm^2, 30-ns laser pulse can exceed 10,000 K for a range of particle sizes. These high temperatures produce stress in the glass adjacent to the particles which can exceed the theoretical strength of glass and result in failure.

11.3 Self-focusing

Self-focusing is a consequence of the dependence of the refractive index of the material on the light intensity.

As we discussed in Sect. 4.3.1, an intense laser beam propagating in a transparent medium induces an increase in the index of refraction by an amount proportional to the beam intensity. If we express the nonlinear part of the refractive index given in (4.38) by the beam intensity I we can write

$$\Delta n = n_2|E^2| = \gamma I \quad . \tag{11.6}$$

The relationship between γ expressed in [cm^2/W] and n_2 expressed in electrostatic units [esu] is as follows

$$\gamma = 4\pi n_2 / c n_0 \quad \text{or} \tag{11.7}$$

$$\gamma = 4.19 \times 10^{-3} n_2 / n_0 [\mathrm{cm}^2/\mathrm{W}] \quad .$$

For optical glass typical values are

$$n_2 = 1.5 \times 10^{-13}\,\mathrm{esu},$$
$$n_0 = 1.55 \qquad \text{or} \qquad \gamma \approx 4 \times 10^{-7}\,\mathrm{cm}^2/\mathrm{GW} \quad .$$

Thus at a power density of $2.5\,\mathrm{GW/cm^2}$ the fractional index change is 1 ppm. For numerical calculations the following conversion factors between esu and mks units are helpful:

$$n_2[\mathrm{m}^2/V^2] = 0.9 \times 10^9 n_2[\mathrm{esu}] \; ;$$
$$E[\mathrm{esu}] = 3 \times 10^4 E[\mathrm{V/m}] \quad ;$$
$$P[\mathrm{esu}] = 10^{-7} P[\mathrm{W}] \quad .$$

Since laser beams tend to be more intense at the center than at the edge, the beam is slowed at the center with respect to the edge and consequently converges. If the path through the medium is sufficiently long, the beam will be focused to a small filament attempting to transport a high power density, and the medium will usually break down via avalanche ionization.

This convergence is limited only by the dielectric break-down of the medium. Due to this intensity-induced index change, small beams (5 mm or so) tend to collapse as a whole (whole-beam self-focusing), forming a damage track in the material when the intensity exceeds the damage level. Larger beams tend to break up into a large number of filaments, each of which causes damage (small-scale self-focusing) [11.8–10].

A variety of physical mechanisms result in intensity-dependent contributions to the refractive index, such as electrostriction, the Kerr effect, and electronic polarizability [11.11–13]. For short-pulse operation in solids, the electronic contribution, which is caused by the nonlinear distortion of the electron orbits around the average position of the nuclei, usually dominates. In most glasses this polarization comprises 75 to 90 % of the total nonlinear index [11.14].

Whole-Beam Self-focusing

Self-focusing of laser radiation occurs when the focusing effect due to the intensity-dependent refractive index exceeds the spreading of the beam by diffraction. For a given value of n_2, two parameters characterize the tendency of a medium to exhibit self-focusing. The first, the so-called critical power of self-focusing, is the power level that will lead to self-focusing that just compensates for diffraction spreading. The second is the focusing length which represents the distance at which an initially collimated beam will be brought to a catastrophic self-focus within the medium for power levels in excess of the critical power.

A crude description of the phenomenon can be obtained by considering a circular beam of constant intensity entering a medium having an index nonlinearity Δn proportional to power density. We define the critical power P_c to be that for which the angle for total internal reflection at the boundary equals the far-field diffraction angle. The critical angle for total internal reflection is

$$\cos \Theta_c = \frac{1}{1 + (\Delta n / n_0)} \quad . \tag{11.8}$$

If we expand the cosine for small angles and the right-hand side of (11.8) for small $\Delta n/n$, we obtain

$$\Theta_c^2 = 2\Delta n / n_0 \quad . \tag{11.9}$$

The beam expands due to diffraction with an angular divergence according to

$$\Theta_D \approx 1.22 \frac{\lambda_0}{n_0 D} \quad , \tag{11.10}$$

where D is the beam diameter and λ_0 is the wavelength in vacuum. For $\Theta_c = \Theta_D$, and noting that the power density I of a beam propagating in a dielectric medium is related to the rms value $|E|$ of the electric field by

$$I = \left(\frac{c}{4\pi}\right) n_0 |E|^2 \quad , \tag{11.11}$$

the critical power can be expressed as [11.15]

$$P'_{cr} = \frac{(1.22\lambda_0)^2 c}{64 n_2} \quad , \tag{11.12}$$

where P'_{cr} and n_2 are in esu units.

Theoretical analysis of self-focusing based either upon numerical solutions of the nonlinear wave equation, ray tracing, or quasi-optic approximation [11.8–10, 16] lead to formulas describing the behavior of the self-focusing Gaussian beam. The results differ somewhat depending upon which aspect of the phenomenon is under study.

For Gaussian beams the equation for P_{cr} has the same functional form as (11.12) and differs by just a numerical factor. It has been shown [11.17–20] that for a Gaussian beam self-focusing in the center of the beam occurs for power levels above

$$P_{cr} = \frac{\lambda_0^2 c}{32\pi^2 n_2} \quad . \tag{11.13}$$

With the assumption that the shape of the amplitude profile is unchanged under self-focusing, one finds the following relation between power and self-focusing length: For a Gaussian profile whose spot size $w(z)$ changes by a scale factor $f(z) = w(z)/w_0$, the width of the beam, assuming no initial divergence $f'(0) = 0$, varies with z according to [11.8]

$$f^2(z) = 1 - \left(\frac{P}{P_{\text{cr}}} - 1\right)\left(\frac{z^2}{k^2 w^4}\right) \quad , \tag{11.14}$$

where k is the wave number and w is the beam radius at the entrance of the nonlinear material. For $P \ll P_{\text{cr}}$ (11.14) describes the ordinary diffractional divergence of the beam. In a beam with $P = P_{\text{cr}}$ there is a balance of forces and $f(z) = 1$ for all z; this is referred to as self-trapping of the beam. For larger powers $P > P_{\text{cr}}$ the beam is focused at a distance

$$z_{\text{f}} = kw^2 \left(\frac{P}{P_{\text{cr}}} - 1\right)^{-1/2} \quad . \tag{11.15}$$

As can be seen, the self-focusing length is inversely dependent on the square root of the total laser power. Equation (11.15) may be used even when the constant phase surface passing through $z = 0$ is curved, corresponding to a converging or diverging incident beam. One need only replace z_{f} by

$$z_{\text{f}}(R) = (z_{\text{f}}^{-1} - R^{-1})^{-1} \quad , \tag{11.16}$$

where R is the radius of curvature of the incident phase surface ($R < 0$ for converging beams) and $z_{\text{f}}(R)$ is the new position of the self-focus.

Since the power in a laser pulse changes in time, a related change occurs with the instantaneous self-focusing length z_{f} in (11.15). The temporal amplitude change of the incident pulse, causing moving foci, will lead to a string of heavily damaged spots along the track. For example, when a laser beam is weakly focused into a sample of dielectric material, the self-focus first appears at, or near, the geometrical focal point, then splits into two foci, one of which travels down stream and one up stream. The up-stream focus reaches its minimum z_{f} value at the peak power of the pulse. It will dwell at this position for a while and this presumably allows massive damage to occur at the head of the track. As the beam intensity decreases, the focus will tend to move back down stream. Therefore, the damage track of microscopic bubbles observed in a broad range of materials is the fossil record of the moving focus generated by dynamic self-focusing of the incident light. This picture of bulk damage has been given by several authors [11.8–10, 12]; it is in agreement with the observation of very small diameter damage tracks in transparent dielectrics when irradiated with short duration, intense laser pulses [11.12, 21, 22].

Small-Scale Self-focusing

Actual laser beams do not have perfectly smooth envelopes as was assumed in treating whole-beam self-focusing. Experimentally it is found that beams containing amplitude irregularities on propagating in a nonlinear medium eventually break up into separate filaments [11.21, 23–27]. The physical description of the process is that interference between a strong beam and weak beams produces index variations due to the intensity-dependent refractive index and that the diffraction of the strong beam by the resulting phase grating spills energy into the weak beams.

A theory first suggested by *Bespalov* [11.28] treats self-focusing as an instability phenomenon wherein an initial perturbation (e.g., dust, refractive index inhomogeneities), no matter how small, grows to what are ultimately catastrophic proportions. Linear stability theory predicts that a small, one-dimensional intensity modulation of spatial wave number K, superimposed upon a uniform intensity background, will undergo approximately exponential growth with gain coefficient g given by [11.29]

$$g = K\left(\frac{\gamma I}{n_0} - \frac{K^2\lambda^2}{16\pi^2}\right)^{1/2} \quad , \tag{11.17}$$

where λ is the wavelength in the medium, and n_0 is the linear refractive index. For a range of K from 0 to $(4\pi/\lambda)(\gamma I/n_0)^{1/2}$, g is real and the ripples grow exponentially. The maximum growth rate occurs at

$$K_{\rm m} = \left(\frac{2\pi}{\lambda}\right)\left(\frac{2\gamma I}{n_0}\right)^{1/2} \tag{11.18}$$

and has the value

$$g_{\rm m} = \left(\frac{2\pi}{\lambda}\right)\left(\frac{\gamma I}{n_0}\right) \quad . \tag{11.19}$$

An initial perturbation $\delta = |\delta E_0|/E_0$ of the fastest growing instability mode leads to self-focusing at a distance given by [11.23]

$$Z_{\rm f} = (g_{\rm m})^{-1} \ln\left(\frac{3}{\delta}\right) \quad . \tag{11.20}$$

Since the self-focal length of small-scale fluctuations is generally much shorter than that of large-scale intensity variations, the self-focusing of fluctuations determines the effective damage threshold of beams propagating in nonlinear dielectrics.

The validity of the linearized theory of small-scale self-focusing was verified by *Bliss* et al. [11.24] who used a shear plate interferometer to generate a perturbation of known amplitude and spatial frequency on an intense beam. It was found that the theory provided an accurate description of the modulation growth (Fig. 11.1).

The laser designer is primarily interested in the threshold for self-focusing, since it is an effect he seeks to avoid rather than to study. Therefore, particular interest is attached to the values of n_2 for subnanosecond pulses in laser media. Several workers have measured the effective value of n_2 in various materials for pulses from 20 ps to 250 ps duration. It was reported that the results obtained

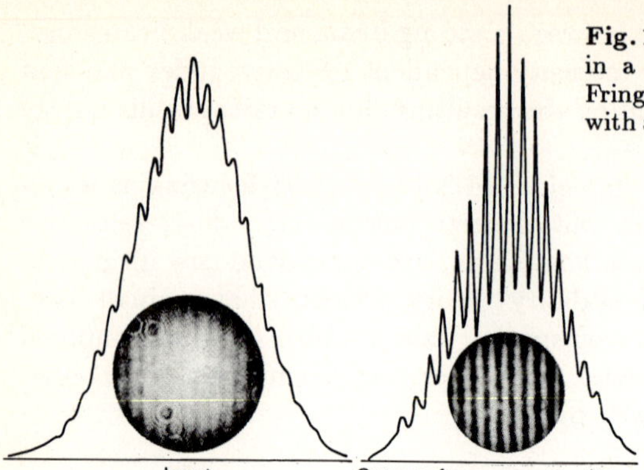

Fig. 11.1. Growth of interference fringes in a 24-cm-long unpumped ED-2 glass. Fringes are spaced 2 mm apart on a pulse with a peak intensity of 5 GW/cm^2 [11.24]

Table 11.1. Linear and nonlinear refractive index for various optical materials

Material	Refractive index n	Nonlinear refractive index n_2 (10^{-13} esu)
FK-6	1.446	1.13
Fused silica	1.456	0.99
BK-7	1.517	1.46
LSO	1.519	1.43
ED-2	1.555	1.52
ED-4	1.567	1.73
EY-1	1.635	2.03
SF-7	1.640	5.83
LAK-3	1.694	2.57
YAG	1.829	4.09
LASF-7	1.914	6.20
CCI-4	1.460	2.50
Ruby	1.75–1.76	1.40
LHG-5	1.531	1.13
LHG-6	1.522	1.04

were not very sensitive to pulse duration. Table 11.1 contains a summary of published values of n_2 for several important optical materials [11.30–32].

Self-focusing is of primary concern only in Nd : glass lasers. Compared to other solid-state lasers, in glass systems beams of extremely high peak powers pass through long sections of dielectric materal. For a typical laser glass with $n_2 = 1.5 \times 10^{-13}$ esu and $\lambda_0 = 1.06\,\mu$m one obtains from (11.12) a value of $P_{cr} = 711$ kW. Typical Q-switched Nd : glass oscillator-amplifier systems are capable of generating peak output powers around 1 GW at beam diameters of 20 mm. Assuming a Gaussian beam radius of $w_0 = 5$ mm and $\lambda = 1.06\,\mu$m, we obtain $z_f = 420$ cm from (11.15) for the self-focusing length in the final amplifier rod. For a mode-locked system producing 1 J in a 20 ps pulse and $w_0 = 5$ mm, one obtains $z_f = 60$ cm. Large-aperture mode-locked Nd : glass laser systems employed for fusion experiments operate at up to 10^5–10^6 times the critical power in the beam. Clearly, in these lasers break-up of the laser beam into filamentary regions with consequent damage to laser components represents the major limitation in achieving higher powers.

For a 1.06 μm laser beam at an intensity of 10 GW/cm^2 and with $n_0 = 1.5$ and $n_2 = 1.5 \times 10^{-13}$ esu, we obtain from (11.18, 19) a gain coefficient of $g_m = 0.16\,\text{cm}^{-1}$ for a disturbance which has a spatial wavelength of $\lambda_{\text{rms}} =$ 0.43 mm. Assuming that the disturbance has a depth of modulation of 1 % ($\delta = 0.01$), then according to (11.20) the perturbation will focus at a distance of $Z_f = 35$ cm.

From these considerations, it follows that spatial beam quality is of paramount importance in high-power systems. Small perturbations are always present in the beam profile, arising from diffraction around dirt and defects on surfaces, or from bubbles or inclusions in the media, surface irregularities, or the random fluctuations of refractive index within the optical media.

In large Nd : glass lasers built for fusion research several approaches are taken to reduce the growth of small-scale instabilities. First, by maintaining a high degree of cleanliness and uniformity on optical surfaces and within the bulk media, the basic level of the perturbation is kept to a minimum. Secondly, by inserting spatial filters at suitable intervals in the amplifier chain, an exponential increase of the perturbation amplitude is prevented. Finally, by choice of suitable materials for both active and passive components in the laser system, the coefficient n_2 is kept to a minimum value. Even with these precautions, system design must be carried out carefully to reduce the effect of the index nonlinearity and the ensuing phase distortion.

11.4 Damage Threshold of Optical Materials

The data presented here on bulk, surface and dielectric thin-film damage levels should only be used as a guide, since damage thresholds depend on a great number of laser and materials parameters, as well as sample cleanliness and prior history. Laser parameters, such as wavelength, energy, pulse duration, transverse and longitudinal mode structure, beam size, location of beam waist all play an important part in the definition of damage threshold. For example, multimode beams usually contain "hot spots" at which the peak power is many times the average value estimated from the pulse energy, pulse duration, and beam cross section.

Experience also shows that the damage threshold can vary greatly even within one particular sample due to compositional variations, localized imperfections, absorption centers, etc. It has also been shown that the prior history of a sample can have a large influence on the damage level. For example, it has been observed that in some cases baking a sample at elevated temperature, or irradiating a sample with laser pulses of increasing fluence will raise the damage threshold.

The results of experimental and theoretical investigations on laser-induced damage in laser materials, optical components and devices are discussed annually at the Boulder Damage Conference held every year since 1969. The proceedings of these symposia [11.33], as well as a summary published annually in Applied Optics [11.34] comprise the main body of literature on laser-damage phenomena.

11.4.1 Scaling Laws

Although the specific threshold at which laser damage is observed depends on a great many factors, empirical scaling laws have been determined which offer a guide to the parametric variation of the thresholds for a given material. For most laser materials, damage seems to vary according to

$$E = E_1(t/t_1)^{0.5} \tag{11.21}$$

where E [J/cm^2] is the damage threshold, t is the pulse duration, and E_1 and t_1 are the reference damage threshold and pulse duration, respectively. For example, Fig. 11.2 shows bulk and surface damage thresholds of optical materials as a function of pulse duration [11.35]. The solid lines correspond to threshold fluences which increase as the square root of pulse duration.

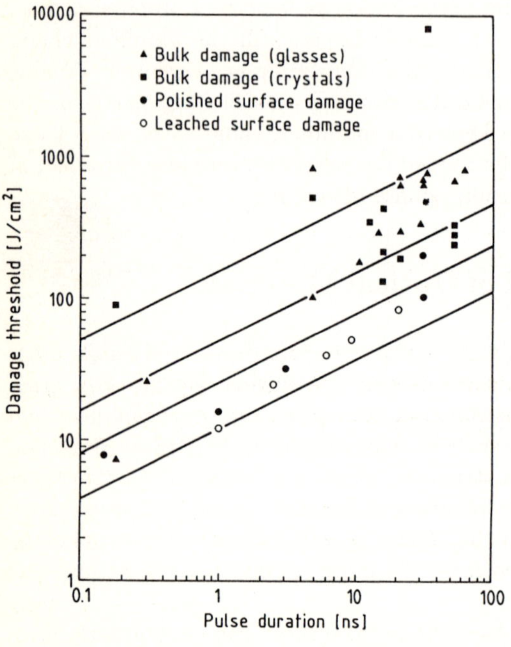

Fig. 11.2. Bulk and surface damage thresholds of optical materials as a function of pulse duration. Measurements made at 1.06 or 0.69 μm [11.35]

Frequently, the reference damage threshold is given for a pulse duration of one nanosecond. In this case (11.21) can be written

$$E_{\text{th}}[\text{J/cm}^2] = E_1\sqrt{t_1[\text{ns}]} \quad \text{or} \tag{11.22}$$

$$I_{\text{th}}[\text{GW/cm}^2] = E_1/\sqrt{t[\text{ns}]} \quad . \tag{11.23}$$

Typical values for E_1 are listed in Table 11.2 for carefully prepared and high quality materials. Equations (11.22, 23) together with the numerical values from Table 11.2 provide order-of-magnitude estimates for a large number of materi-

Table 11.2. Typical damage thresholds for optical materials such as glasses, crystals and coatings for 1 ns pulses in the visible or near-infrared regime

Damage Threshold Fluence E_1 [J/cm^2]	Material
5	Antireflection Coatings
10	Multilayer Dielectric Coatings
10	Surface Damage
50	Bulk Damage

als, wavelengths and pulse width ranging from the picosecond to the millisecond regime.

It should be emphasized that this scaling law is only expected to apply if other parameters are held constant.

It has been further shown that the surface damage on both glasses and half-wave films shows a simple dependence on surface roughness. A scaling law has been proposed, in the form:

$$\delta^m I_{\text{th}} = \text{const} \tag{11.24}$$

where I_{th} is the power density at which damage occurs, δ is the rms surface roughness, and m varies from 1.0 to 1.5. This scaling law is seen to hold for $10\,\text{Å} < \delta < 300\,\text{Å}$, a remarkable range for such a simple relation. Again, its application implies that other factors (laser pulse duration, spot size, etc.) are held constant.

A scaling rule has also been developed for the dependence of laser damage threshold on spot size. If Å is the size of the focal spot of the incident laser beam, the intensity at which damage occurs scales according to the rule:

$$I_{\text{th}}(tA)^{1/2} = J_{\text{th}}(A/t)^{1/2} = \text{const} \tag{11.25}$$

where I_{th} is the intensity threshold, J_{th} the energy density threshold, and t the pulse duration.

By probing with laser spots smaller than the separation of isolated surface defects, several investigators were able to separate uniform damage, which is characteristic of the coating itself, from defect-driven localized failures. Defect damage is measured at much lower fluences than required for uniform failure. A larger spot size will increase the probability that surface defects such as pits, inclusions, or scratches will be within the beam area.

11.4.2 Laser Materials

The scaling law of fluence vs. pulsewidth for surface damage given in the previous subsection holds very well for laser glasses. At a pulse length of 1 ns disk amplifiers routinely operate at fluence levels of 10 J/cm^2 without damage [11.36], and damage-free operation at 5 J/cm^2 for a few hundred picosecond long pulses, and $2\,J/cm^2$ for pulses tens of picoseconds long have been achieved with spatially clean beams [11.36]. For Q-switched Nd: glass lasers, damage threshold of 30 to $40\,J/cm^2$ with 30 ns long pulses have been reported [11.37]. The failure mode of laser crystals is usually associated with surface damage at the ends. In laser rods such as Nd: YAG, Nd: YLF, Nd: Cr: GSGG, ruby, alexandrite, etc. internal damage occurs relatively seldom unless an internal focus is created by a back-reflection from an optical element. Mode-locked Nd: YAG master oscillator-amplifier systems have generated 100 mJ in a 100–ps pulse at power densities of $3\,GW/cm^2$ without damage.

In cw operation, Nd: YAG oscillators can produce up to 25 W TEM_{00} mode. With a typical beam diameter of 1 mm and 90 % reflective output mirror, the circulating power density in the resonator becomes $30\,kW/cm^2$. In none of these lasers is surface damage or internal damage of the Nd: YAG crystal a problem. Recently, surface and bulk damage for crystals of Nd: Cr: GSGG have been measured [11.38]. The surface damage threshold was $14\,J/cm^2$ for a 17 ns long pulse and bulk damage was higher than that of BK-7.

11.4.3 Damage in Optical Glass

Damage studies on a larger number of optical glasses and colored filter glasses are reported in [11.37, 39]. It was found that surface damage, which occurs at lower energy levels than internal damage, is a strong function of glass type and surface finish. For barium crown glass, a damage threshold of $28\,J/cm^2$ for a 25-ns pulse was measured. For SF-4 glass, damage threshold is approximately $20\,J/cm^2$ for Q-switched pulses and $100\,J/cm^2$ for a 100-μs long pulse.

In general, glass materials with lower refractive indices are more resistant to surface damage than high index materials. This effect may be due, in part, to the difficulty in polishing high index glasses, which are generally softer, to the same degree of perfection. The material is also less likely to fail due to multiphoton absorption, electron avalanche breakdown and self-focusing when n is low. The importance of defects at surfaces and within the material is greatly influenced by the material index because the local field enhancement is proportional to n or higher powers of n depending on the defect geometry.

Figure 11.3 shows results of surface damage thresholds for fused silica and a fluorophosphate glass. Damage thresholds similar to that of BK-7, i.e. 20–$25\,J/cm^2$ at 1 ns [11.40].

Bulk damage in highly transparent materials is a very nonlinear process. Figure 11.4 is an example of the results of damage threshold measurement in SiO_2 using tightly focused, 100 ps, 532 nm pulses [11.41]. In such measurements, a small increase in input irradiance causes the sample to change from highly transmitting to highly absorbing with subsequent catastrophic failure. The current consensus is that such failure is due to avalanche breakdown initiated by multiphoton absorption associated with defects within the bandgap.

In measuring damage threshold of filter glasses, an empirical formula was found which relates the absorption of the filter to the damage threshold. It was found that the higher the absorption the lower the damage threshold. For a

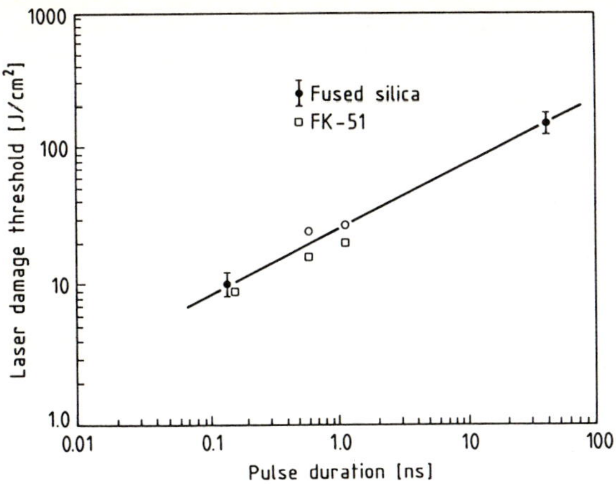

Fig. 11.3. Damage thresholds of conventionally polished bare surfaces as a function of 1064-μm laser pulse duration [11.40]

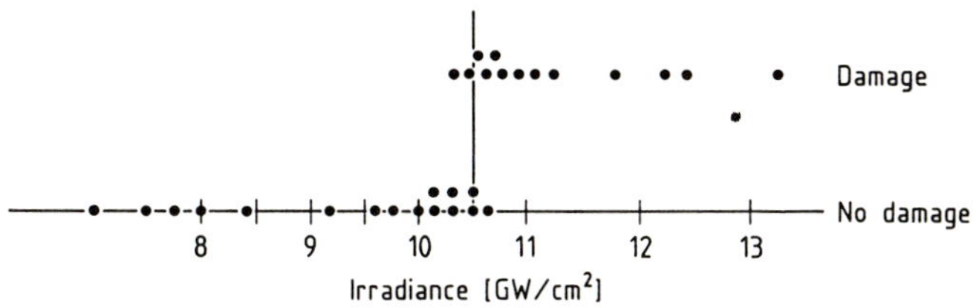

Fig. 11.4. Bulk damage in SiO_2 at 532 nm [11.4]

laser wavelength of 1.06 μm and a 700-μs-long pulse, the best fit was found to be given by [11.37]

$$E_{\text{th}} = \frac{185}{k^{0.74}} \quad , \tag{11.26}$$

where E_{th} is the damage threshold energy and k is the absorption constant [cm^{-1}]. The parameter k is related to the sample thickness d and the internal transmission T by

$$k = \frac{1}{d} \ln \frac{1}{T} \quad . \tag{11.27}$$

The empirical formula is valid for k values between 10^{-2} to 50 [cm^{-1}].

11.4.4 Damage Levels for Nonlinear Materials

From all the optical components and materials employed in solid state laser systems the largest spread in damage thresholds occurs with nonlinear crystals. Materials such as KDP, BBO and KTP have damage thresholds which are relatively high and are on the order of GW/cm^2, whereas $Ba_2Na_5NbO_3$, $AgGaSe_2$, and $AgGaS_2$ have very low thresholds on the order of KW/cm^2.

In addition to exhibiting damage mechanisms operative in other dielectric materials, nonlinear materials are subjected to damage by virtue of their nonlinear properties. It is frequently observed that the damage levels observed in harmonic generation are far lower in the presence of the harmonic than those observed with just the fundamental present. This effect is often caused by multiphoton absorption. Damage thresholds for important nonlinear crystals are summarized [11.42–45].

For KDP crystals, the average threshold for bulk damage is $7\,\mathrm{J/cm^2}$ for 1 ns pulses at $1.06\,\mu\mathrm{m}$ with current state of the art technology [11.46]. The bulk damage consists of isolated microfractures, suggesting the presence of absorbing inclusions.

It has also been shown that the prior history of a sample can have a significant influence on the damage level. For example, the bulk damage threshold of KDP was increased fivefold by baking the crystals at elevated temperature and then irradiating them with laser pulses of increasing fluence [11.46]. Figure 11.5 shows the bulk laser damage threshold of KDP as a function of pulse duration.

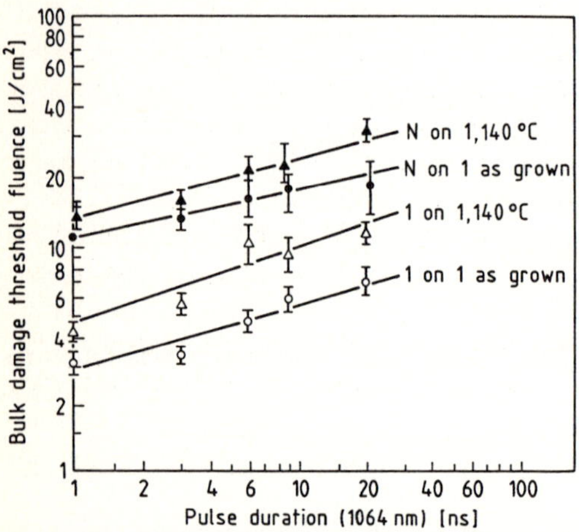

Fig. 11.5. Dependence of the bulk laser damage threshold of KDP on laser pulse duration. Single shot as grown (1 on 1) and after baking (1 on 1, 140° C); and multiple exposures of increasing irradiance as grown (N on 1) and after baking (N on 1, 140° C). [11.46]

For lithium niobate a surface damage threshold of $180\,\mathrm{MW/cm^2}$ for a 10-ns pulse with a Gaussian intensity profile has been reported. For multimode operation, surface damage was observed at $30\,\mathrm{MW/cm^2}$ for a 10-ns-long pulse and $10\,\mathrm{MW/cm^2}$ for a 200-ns-long pulse [11.42, 43].

In lithium iodate, surface damage levels of $400\,\mathrm{MW/cm^2}$ for a 10-ns pulse were reported. During frequency doubling, bulk fractures occurred at $30\,\mathrm{MW/cm^2}$ at $1.06\,\mu\mathrm{m}$ and $15\,\mathrm{MW/cm^2}$ at $0.56\,\mu\mathrm{m}$ for a Gaussian beam. For multimode operation, damage threshold was $2\,\mathrm{MW/cm^2}$ at $0.53\,\mu\mathrm{m}$ [11.42, 43].

11.4.5 Laser Induced Damage in Dielectric Thin Films

Multilayer dielectric films are generally the weakest elements of any laser system. Because of the relative complexity and variety of deposition factors, thin films vary greatly in purity, morphology, uniformity and composition, depending, to a great extent, on the specifics of the fabrication process. In addition, residual stress, substrate cleanliness, and other extrinsic factors tend to reduce the thin-film damage level well below that expected for the same material bulk form.

Damage in dielectric coatings is very sensitive to coating design. Generally, multilayer coatings fail in the high-index component. High-reflectance coatings exhibit damage thresholds which are typically a factor two higher compared to antireflection coatings. The reason for the difference is generally attributed to the fact that in *AR* coatings, the field intensity is near maximum at the substrate surface, where the coating is most vulnerable, due to surface imperfection, imbedded polishing materials and poor film structure.

A hierarchy of dielectric coatings arranged in order of decreasing damage resistance includes the following thin films [11.47]: ThF_4, SiO_2, MgF_2, Al_2O_3, CaF_2, ZrO_2, TiO_2, SiO, LiF, MgO, CeO_2, ZnS. More recent measurements indicated that antireflection coatings of either ThF_4, SiO_2 or Al_2O_3 applied to $LiNbO_3$ as substrate had damage thresholds ranging from 2 to 12 GW/cm^2 for a 9 ns Q-switched pulse from a Nd:YAG laser [11.48]. Tests also revealed that pre-irradiation of thin films with subthreshold laser intensities increases the damage threshold [11.49]. Since the probability of the laser beam striking a defect site will be greater for larger spot sizes, it has been observed that laser-induced damage thresholds of thin film coatings decreased with increasing laser beam spot size and were invariant for spot sizes greater than 150 μm [11.50].

In Fig. 11.6 the damage threshold is plotted for a ThF_4 film as a function of film thickness. The measurements were performed with 5 and 15 ns long pulses at a wavelength of 532 nm [11.51].

Figure 11.7 shows on the left the distribution of damage thresholds for commercial multilayer titania/silica AR coatings produced by electron-gun evap-

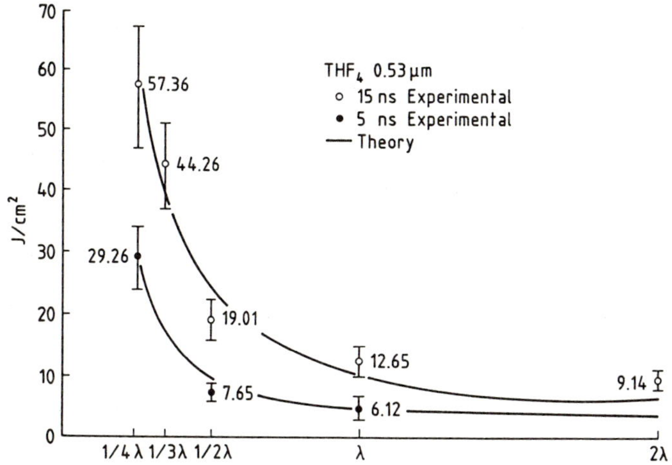

Fig. 11.6. Damage threshold vs film thickness at 0.53-μm laser wavelength [11.51]

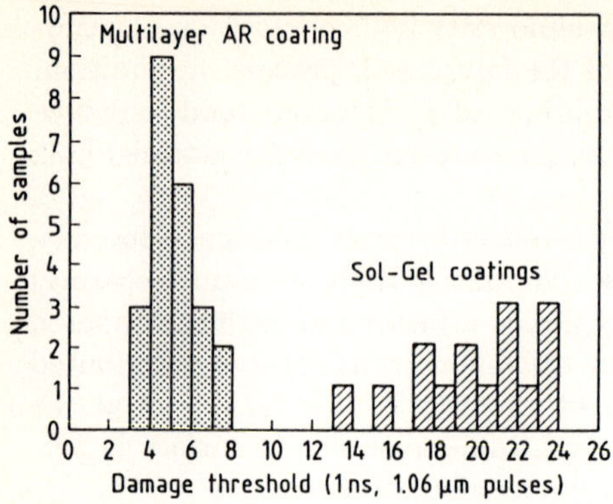

Fig. 11.7. Distribution of damage thresholds for sol-gel coatings and commercial multilayer titania/silica antireflection coatings produced by electron-gun evaporation [11.52]

oration. The thresholds were measured at 1.06 μm with a 1 ns pulse from a Nd : glass laser [11.52]. The histogram shows that most coatings failed at around 5 J/cm^2. A single-layer gradient index AR coating deposited by the Sol-Gel process had a damage threshold three to four times greater than the dielectric AR coatings produced by vapor deposition. Drawbacks of Sol-Gel coatings include their fragile nature and difficulty in cleaning the optical components.

An excellent summary of the issues related to coatings damage has been provided by *Lowdermilk* et al. [11.53].

11.5 System Design Considerations

The designer of a laser system, although he cannot change the damage thresholds of the materials employed in a laser, can reduce the damage susceptibility and improve reliability of his system by the right choice of materials and by proper design and operation of the laser. We will enumerate a few design and operating procedures that can be used to reduce the limitations imposed on system performance by damage.

Choice of Materials

In powerful lasers it is important that only high-damage-resistance AR and multilayer coatings are employed. These coatings usually include ThF_4, SiO_2, TiO_2, and MgF_2 compositions which are applied with an electron beam gun. Beam splitters, lenses, and prisms located in the beam might have to be fabricated from quartz, sapphire, or highly damage-resistant glass, such as ED4 or BK-7 instead of standard optical glass. The surfaces of these components should be pitch-polished to a 5/10 or 0/0 surface quality. Surface conditioning or "superpolishing", such as chemical etching, ion-beam polishing, laser conditioning, or bowl-feed polishing, are very effective in raising the damage threshold.

All components of the laser system should be inspected for surface defects, such as scratches, fingerprints, "orange peel", and bulk defects, such as inclusions, bubbles, striation, schlieren, and internal stress. Standard specifications of the internal quality of optical glass are reviewed in [11.54].

Inclusions in a laser rod can be best checked by illuminating the rod from the side and observing the presence of scatter centers. The fringe pattern obtained from the Twyman-Green interferometer will reveal any density gradients and optical inhomogeneities in the laser rod. Inspection of the laser rod Q-switch crystal, frequency doubling crystal, etc., between cross-polarizers will reveal internal stresses, striation, and schlieren. Optical surfaces are best inspected under a binocular microscope (magnification of 50×) with a strong collimated light beam reflected off the surface.

Design of System

One of the most important considerations in the design of a laser system is the choice of beam size. The basic criterion is to make the beam size and, therefore, the aperture of all optical elements large enough so that the power and energy densities are well below the damage threshold levels for the particular mode of operation. Even if the average power density is below the damage level, if the beam contains hot spots the damage threshold can be easily exceeded in localized areas. Therefore beam uniformity is of great concern in high power systems. The gross difference between single-shot damage thresholds obtained using multimode lasers and those operating in the TEM_{00} mode has been demonstrated by many investigators.

The damage levels listed in this chapter have usually been measured in a carefully controlled laboratory environment, with smooth Gaussian beams, meticulously clean optics and materials of excellent optical quality. The numbers clearly present upper limits. For a typical industrial or military laser employing production type components and coatings the operating fluence and fluxes should be at least an order of magnitude below the numbers given for damage threshold.

A very important design consideration is the avoidance of damage caused by reflections from optical surfaces in the laser system. For example, very often a telescope is employed between amplifier stages to increase the beam diameter. The design and location of the negative lens employed in such a telescope is very critical, because if its reflections from the concave surface are focused into the preceding laser rod, damage will most certainly occur. Therefore, a plano/concave rather than a concave/concave lens should be used and the curved surface should be faced away from the preceding laser rod. In this case the reflected light from this surface is diverging rather than converging.

Reflections from the rear surface of an amplifier rod are very dangerous because the reflected radiation is amplified in passing through the amplifier stage, and optical elements at the input side of the amplifier, which are normally operated at much lower energy densities, can be damaged. This is especially the case if a telescope is employed between the oscillator and amplifier stage. If the beam travels backwards through the amplifier chain, beam diameter decreases for increasing intensities and damage thresholds are easily exceeded. One way to minimize this kind of damage is to tilt the various amplifier stages and separate them as far as possible from each other. A beam traveling backwards through an amplifier will then miss the aperture of the preceding stage. In order to suppress prelasing, which can cause the same type of damage, the amplifier rod should have wedged ends. Damage to optical surfaces can be minimized by employing highly damage-resistant coatings or by arranging the surfaces at Brewster's angle.

Operation of System

From an operational point of view, the most important factors to consider, if damage is to be avoided, are proper alignment of components and absolute cleanliness of all surfaces. Misalignment of the oscillator components can cause a change of the transverse mode structure, hot spots and beam uniformities can occur. Misaligned components outside the resonator cavity can cause undesirable secondary reflection leading to damage.

Surfaces which are contaminated by dust, dirt, or fingerprints will rapidly degrade and cause nonuniform beam intensities, which in turn can cause damage in more expensive optical components. Surfaces should be inspected frequently and cleaned with an airblast or with alcohol. A dental mirror and a flashlight are very convenient for the inspection of optical surfaces in a laser system. Very often components are mounted close together and surfaces are hard to see. For the cleaning of surfaces which are relatively inaccessible, a stream of pressurized clean air or gas should be used. A useful device for cleaning optical surfaces is an ionizing airgun sold by 3M Company. The unit, which contains an α-particle source of polonium isotope 210 in the valve and nozzle sections, can be connected to a gas cylinder. The gas stream containing α-radiation can remove very efficiently statically charged particles from a surface.

The degradation and subsequent damage of optical surfaces by atmospheric dust particles and by cleaning residue has been studied by *Barber* [11.55]. It was found that dust particles consisting of minerals are melted onto the surface, leaving a circular indentation after the particle is dislodged by subsequent pulses.

In another type of degradation, craters are caused by liquids being absorbed a few microns into the glass surface during the cleaning operation, then producing miniature explosions when a strong laser pulse strikes the absorbed liquid. Experiments showed that laser rod ends were badly pitted after liquid cleaning without drying, and no damage occurred on rods which were vacuum-dried.

From the foregoing follows that the working area around a laser should be kept as clean as possible, free from dust and contaminants, such as cigarette smoke, approaching clean room conditions as closely as possible.

Appendix A

Laser Safety

Hazards associated with solid-state lasers can be separated into two broad categories – those hazards related to the laser beam itself and those hazards related to the electronic equipment.

Radiation Hazards

Although high-power solid-state lasers have the capability of inflicting burns on the skin, the possibility of eye injury is of vastly greater importance. The property of lasers that is of primary concern with regard to eye hazards is their high radiance, i.e. the combination of high power density and directionality. The latter property causes the eye lens to focus the parallel beam emitted from a laser to a tiny spot on the retina. It also makes the laser hazardous over long distances.

Figure A.1 shows a cross section of the human eye. The figure shows a collimated beam of visible light impinging onto a cornea, being focused by the lens, passing through the vitreous humor, and impinging onto the retina. At the retina the light is absorbed. Under normal conditions the light energy is converted by the retina into chemical energy, stimulating optical sensations.

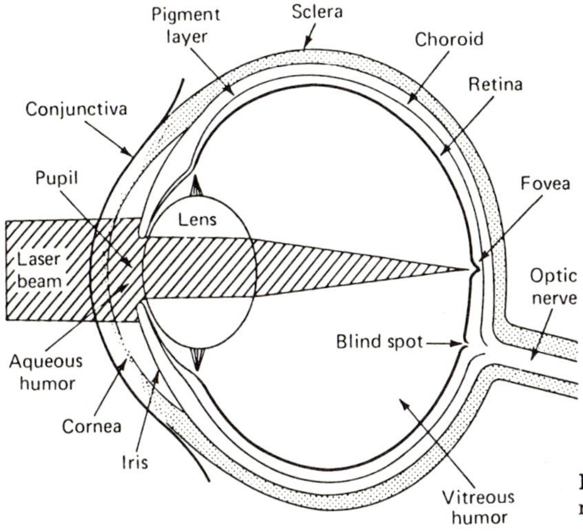

Fig. A.1. Cross-section of the human eye

Figure A.1 illustrates the principal reason for possibe eye injury from a laser. The parallel rays of a laser can be focused through the eye lens to a point image of about 10 to 20 μm in diameter. If the pupil is dilated sufficiently to admit the entire beam of a laser, even lasers having power outputs of only a few milliwatts will produce power densities of kilowatts per square centimeter on the retina. Obviously, power densities of this magnitude may cause severe damage to the retina. For worst-case conditions – the eye focused at infinity and the pupil fully dilated (about 0.7 cm diameter) – the ratio of power density at the eye lens to that at the retina will be about 5×10^5.

By comparison, conventional sources of illumination are extended, they are considerably less bright, and they emit light in all directions. These light sources produce a sizable image on the retina with a corresponding lower power density. For example, the image of the sun on the retina has a power density of about 30 W/cm^2.

The eye hazards for lasers operating outside the visible spectrum are dependent on the spectral transmission of the eye. Ultraviolet light with a wavelength shorter than 0.31 μm and infrared radiation beyond 1.9 μm are completely absorbed by the cornea, while near-infrared radiation with a wavelength longer than 1 μm is absorbed mainly by the crystalline lens. Figure A.2 illustrates the spectral transmission characteristics of the human eye [A.1].

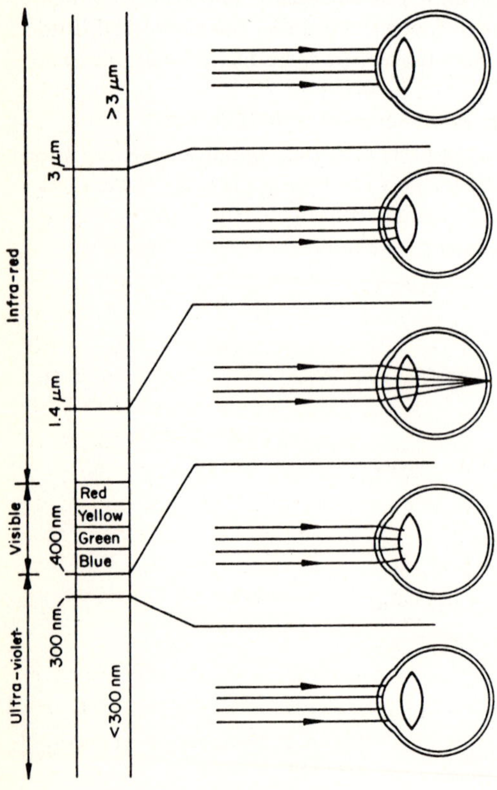

Fig. A.2. Spectral transmission characteristics of the human eye [A.1]

Electrical Hazards

Although the hazards of laser radiation are receiving deserved attention from government agencies, users, and manufacturers, the chief hazard around solid-state lasers is electrical rather than optical. Most solid-state lasers require high-voltage power supplies and the use of energy-storage capacitors charged to lethal voltages. Furthermore, associated equipment such as Q-switches, optical gates, modulators, etc., are operated at high voltages. The power supply and associated electrical equipment of a laser can produce serious shock and burns and, in extreme cases, can lead to electrocution.

Safety Precautions Applicable to Solid-State Lasers

Enclosure of the beam and target in an opaque housing is the safest way of operating a laser. This level of safety precaution is almost mandatory for laser materials-processing systems operated in an industrial environment. In these systems interlocked doors, warning signs and lights, key-locked power switches, and emergency circuit breaker, and like precautions are taken to protect operators and passers-by from electrical and radiation hazards of the laser equipment. Also, viewing microscopes and parts are filtered or blocked or prevent the issuance of laser radiation, and laser impact points are surrounded by shields. At points of access for routine maintenance and set-up, warning signs are displayed prominently, and interlocks prevent firing of the laser while doors or ports are open.

In the laboratory it is often not possible to enclose fully a high-power laser. In these situations the following safety precautions should be observed:

- Do not look into the beam or at specular reflections of the beam.
- Wear adequate eye protection when laser is operating. Laser safety goggles should be shatter-resistant and designed to filter out the specific wavelength generated by the laser.
- Employ a countdown or other audible warning before laser is fired.
- Control access to the laser area and have a flashing red light on the door when laser is in operation. Operation of very large laser systems, such as those used in thermonuclear fusion experiments, usually require the installation of safety interlocks at the entrance of the laser facility to prevent entry of unauthorized personnel into the facility while the laser power supply is charged and capable of firing the laser.
- Provide protection against accidentally contacting charged-up capacitors in energy-storage banks, high-voltage power supplies, etc. These components should be installed in cabinets having interlocked doors. Furthermore, capacitor banks should be equipped with gravity-operated dump solenoids.

The key to a successful safety program is the training and familiarization of the personnel involved with laser hazards and subsequent control measures.

Laser Safety Standards

The potential of lasers for inflicting injury, praticularly to the eyes, was recognised early, and as a consequence extensive studies have been undertaken into the biological mechanisms of laser damage, in an attempt to define safe working levels of optical radiation.

A number of organizations and government agencies have developed laser safety standards for users and manufacturers of laser equipment. In most countries, requirements are being developed based on the standard provided by the International Electrotechnical Commission. This IEC standard specifies requirements for both manufacturers and users [A.2].

In the United States, regulations for manufacturers are published by the Center for Devices and Radiological Health (formerly Bureau of Radiological Health), under the Food and Drug Administration. User requirements in the US are the responsibility of the individual states, but only a few states have such laws.

Under the federal safety standard, four classes of lasers have been created to allow manufacturers to categorize a product based on its potential hazard to a user. These classes are labeled I through IV, from the most safe to the most hazardous. Class I and II are used for lasers of cw outputs of less than 0.39 μW and 1 mW, respectively. Class III and IV include lasers which represent a definite hazard to the eye from direct or scattered radiation. Solid-state lasers generally fall into class III or IV of that standard, which requires that warning signs, labels, and protective housings be installed to prevent human irradiation by levels in excess of a "safe" limit. Safety interlocks must be provided for each portion of the protective housing that can be removed by the user during normal operation or maintenance. In the event that the design of the safety interlock allows the user to defeat the system manually, the manufacturer must incorporate visible or audible warnings of this interlock override during laser operation. In addition, key-lock switches and warning lights are required along with other safety devices.

A document which is very useful as a guide for the safe use of lasers and systems is the standard published by *ANSI* [A.4]. A comprehensive reference work covering every aspect of safety with lasers and other optical sources is the handbook written by D. *Sliney* et al. [A.5]. For manufacturers of laser equipment, a laser safety newsletter [A.6], and a listing of the laser safety requirements in different countries might be of interest [A.7].

The Author found the information presented in Figs. A.3 and 4 quite useful in calculating the safe distance at which accidental direct viewing of the laser beam does not present a hazard [A.5]. These situations occur quite frequently in field tests involving rangefinders and similar equipment. Figure A.3 provides the occupational health and safety limits for laser radiation in the visible and near infrared spectrum. The graphs show the safe flux and fluence limits vs. pulsewidth for the visible and NIR. Compared to the visible regime, the threshold for eye damage is higher in the NIR because the focussed spot on the retina is larger, furthermore higher transmission and reflection losses are

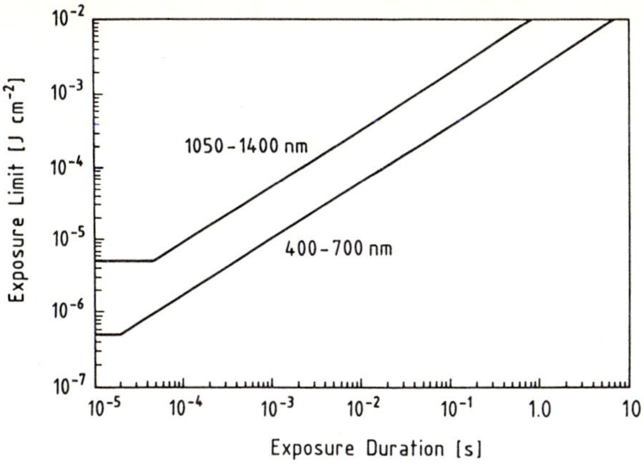

Fig. A.3. Exposure limits for direct viewing of laser beam as a function of pulse width. Ranges are for visible and near-infrared spectral regions [A.5]

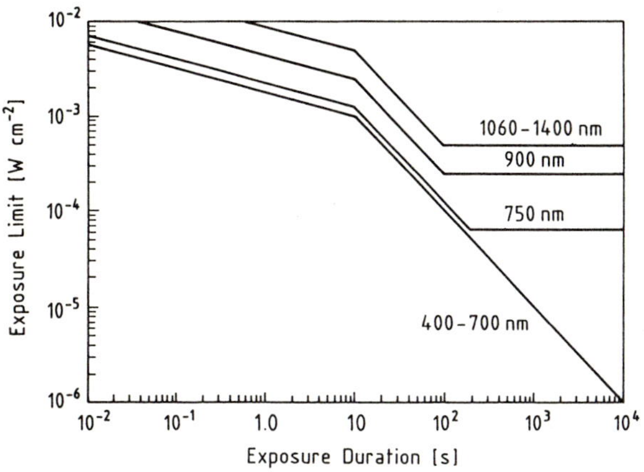

Fig. A.4. Exposure limits for direct viewing of a cw laser beam [A.5]

encountered for the longer wavelength. The increase in fluence level with longer pulses reflects the fact that heat produced at the retina can be more effectively dissipated to the surrounding tissue as the exposure time increases. For exposure times shorter than about 50 μs, heat dissipation is insignificant and the damage threshold does not depend on pulsewidth. The data presented in these figures should only be considered as guidelines. For exposure levels 10 times as high as indicated in Fig. A.3, there is a 50 % probability of the occurrence of ophthalmologically visible retinal lesions [A.5].

Appendix B

Conversion Factors and Constants

In this section we have listed some of the most frequently used conversion factors, constants and definitions.

Physical constants

h	$= 6.626 \times 10^{-34}\,\mathrm{Js}$	Planck constant
e	$= 1.602 \times 10^{-19}\,\mathrm{As}$	Charge of an electron
k	$= 1.381 \times 10^{-23}\,\mathrm{JK^{-1}}$	Boltzmann constant
c	$= 2.998 \times 10^{8}\,\mathrm{ms^{-1}}$	Speed of light in vacuum
ε_0	$= 8.854 \times 10^{-12}\,\mathrm{As\,V^{-1}m^{-1}}$	Permittivity of free space
μ_0	$= 1.257 \times 10^{-6}\,\mathrm{VsA^{-1}\,m^{-1}}$	Permeability of free space
Z_0	$= \sqrt{\mu_0/\varepsilon_0} = 376.7\,\Omega$	Impedance of free space
g	$= 9.81\,\mathrm{ms^{-2}}$	Acceleration due to gravity

Conversion factors

1 J	$= 10^7\,\mathrm{erg} = 0.239\,\mathrm{cal}$
1 eV	$= 1.602 \times 10^{-12}\,\mathrm{erg} = 1.602 \times 10^{-19}\,\mathrm{J}$
1 dyne	$= 1\,\mathrm{cm\,g\,s^{-2}}$
1 erg	$= 1\,\mathrm{dyne\,cm}$
1 W	$= 10^7\,\mathrm{erg\,s^{-1}}$
1 N	$= 1\,\mathrm{mkg\,s^{-2}} = 10^5\,\mathrm{dyne} = 1\ \mathrm{Wsm^{-1}}$
1 bar	$= 10^6\,\mathrm{dyne\,cm^{-2}}$
1 Kp	$= 9.81\,\mathrm{kgms^{-2}}$

Conversion of angles

$1°$	$= 17.45\,\mathrm{mrad};\ 1' = 0.29\,\mathrm{mrad};\ 1'' = 4.85\,\mu\mathrm{rad}$
1 rad	$= 57°\,17'\,45''$, $1\,\mathrm{mrad} = 3'\,26''$
1 rad	$= \sqrt{4/\pi}\,\sqrt{\mathrm{strad}}$

Conversion of transmission, T, to optical density, D;

$$T = 10^{-D}$$

Conversion of linewidth given in wavelength ($\Delta\lambda$) or wavenumber (Δn) to bandwidth ($\Delta\nu$)

$$\Delta\nu/\nu = \Delta\lambda/\lambda = \Delta n/n$$

$$\Delta\nu = c\,\Delta n$$

$$\Delta\lambda = \Delta n\lambda^2 = \Delta\nu\lambda^2/c$$

Amplifier gain

$$g(\mathrm{dB}) = 10\log(E_2/E_1)$$

Optical units

Wave number	$n(1/\mathrm{cm}) \approx 10^4/\lambda\ (\mu\mathrm{m})$
Frequency	$\nu(\mathrm{Hz}) \approx 3 \times 10^{14}/\lambda\ (\mu\mathrm{m})$
Photon energy	$E(\mathrm{J}) \approx 1.987 \times 10^{-19}/\lambda\ (\mu\mathrm{m})$
Photon energy	$E(\mathrm{eV}) \approx 1.24/\lambda\ (\mu\mathrm{m})$

Conversion of English units into the MKS system

1 inch	= 2.540 cm
1 gal	= 3.785 ltr
1 atm	= 14.7 psi
1 Btu	= 1054.3 Ws
$T(\mathrm{C})$	$= \frac{5}{9}[T(\mathrm{F}) - 32]$
$\Delta T(\mathrm{C})$	$= \frac{5}{9}\Delta T(\mathrm{F})$
1 cal/cm Cs	= 242 Btu/hr ft F
1 cal/g	= 1.8 Btu/lb

References

Chapter 1

1.1 From the extensive literature on laser physics, we shall mention only a few recent textbooks:
K. Shimoda: *Introduction to Laser Physics*, 2nd. ed., Springer Ser. Opt. Sci., Vol.44 (Springer, Berlin, Heidelberg 1986)
A.E. Siegman: *Lasers* (University Science Books, Mill Valley, Calif. 1986)
1.2 M. Garbuny: *Optical Physics* (Academic, New York 1965)
1.3 H. Haken: *Laser Theory* (Springer, Berlin, Heidelberg 1984)
1.4 I.I. Sobelman: *Atomic Spectra and Radiative Transistions*, Springer Ser. Chem. Phys., Vol.1 (Springer, Berlin, Heidelberg 1979)
I.I. Sobelman, L.A. Vainshtein, E.A. Yukov: *Excitation of Atoms and Broadening of Spectral Lines*, Springer Ser. Chem. Phys., Vol.7 (Springer, Berlin, Heidelberg 1981)
1.5 H. Statz, G.A. de Mars: In *Quantum Electronics*, ed. by C.H. Townes (Columbia Univ. Press, New York 1960) pp.530-537
1.6 E.L. Steele: *Optical Lasers in Electronics* (Wiley, New York 1968)

Chapter 2

2.1 A.L.Schawlow, C.H. Townes: Phys. Rev. **112**, 1940 (1958)
2.2 T.H. Maiman: Nature **187**, 493 (1960)
2.3 P.P. Sorokin, M.J. Stevenson: Phys. Rev. Lett. **5**, 557 (1960); and *Advances in Quantum Electronics* (Columbia Univ. Press, New York 1961) p.65
2.4 E. Snitzer: Phys. Rev. Lett. **7**, 444 (1961)
2.5 L.F. Johnson, L. Nassau: Proc. IRE **49**, 1704 (1961)
2.6 J.E. Geusic, H.M. Marcos, L.G. Van Uitert: Appl. Phys. Lett. **4**, 182 (1964)
2.7 E. Snitzer, R.F. Woodcock, J. Segre: IEEE J. QE-**4**, 360 (1968)
2.8 E. Snitzer, C.G.Young: In *Lasers*, Vol.2, ed. by A.K. Levine (Dekker, New York 1968) pp.191-256
C.G. Young: Proc. IEEE **57**, 1267 (1969)
Handbook of Lasers, ed. by R.J. Pressley (CRC Press, Cleveland, OH 1971)
K. Patek: *Glass Lasers* (Butterworth, London 1970)
2.9 P.F. Moulton: Paramagnetic ion lasers, in *Handbook of Laser Science and Technology*, ed. by M.J. Weber (CRC Press, Boca Raton, FL 1986) Vol.1, pp.21-295
L.G. DeShazer, S.C. Rund, B.A. Wechsler: Laser Crystals, in *Handbook of Laser Science and Technology*, ed. by M.J. Weber (CRC Press, Boca Raton, FL 1987) Vol.5, pp.281-338
J.C. Walling: Tunable Paramagnetic-Ion Solid-State Lasers, in *Tunable Lasers*, ed. by L.F. Mollenauer and J.C. White, Topics Appl. Phys., Vol.59 (Springer, Berlin, Heidelberg 1987) Chap.9

2.10 L. DeShazer, M. Bass, U. Ranon, T.K. Guka, E.D. Reed, T.W. Strozyk, L.Rothrock: Laser operation of neodymium in YVO_4 and gadolinium gallium garnet (GGG) and of holmium in YVO_4. 8th Int'l Quant. Electr. Conf., San Francisco, CA (1974)
C.D. Brandle, T.C. Vanderleeden: IEEE J. QE-**10**, 67 (1974)
2.11 E.V. Zharikov, N.N. Il'ichev, V.V. Lapter, A.A. Malyutin, V.G. Ostroumov, P.P. Pashinus, I.A. Shcherbakov: JEEE J. QE-**12**, 338-341 (1982);
D. Pruss, G. Huber, A. Beimowski: Appl. Phys. B **28**, 355-358 (1982)
2.12 L.F. Johnson, J.E. Geusic, L.G. Van Uitert: Appl. Phys. Lett. **7**, 127 (1965)
2.13 D.P. Devor, B.H. Soffer: IEEE J. QE-**8**, 231 (1972)
2.14 L. Schearer, M. Leduc: IEEE J. QE-**22**, 756-758 (1986)
2.15 M.J. Weber, M. Bass, K. Andringa: Appl. Phys. Lett. **15**, 342 (1969)
M. Stalder, W. Lüthy: Polarisation of 3 μm laser emission in $YAlO_3$:Er. Preprint (1987)
2.16 R.F. Belt, J.R. Latore, R. Uhrin, J. Paxton: Appl. Phys. Lett. **25**, 218 (1974)
2.17 M.J. Bina, C.R. Jones: J. Opt. Soc. Am. **63**, 463 (1973)
2.18 M.J. Weber: IEEE J. QE-**9**, 1079 (1973)
2.19 R.V. Alves, R.A. Buchanan, K.A. Wickersheim, E.A.C. Yates: J. Appl. Phys. **42**, 3043 (1971)
2.20 R.C. Ohlmann, K.B. Steinbruegge, R. Mazelsky: Appl. Opt. **7**, 905 (1968)
2.21 K.B. Steinbruegge, R.H. Hopkins, G.W. Roland: Increased energy storage neodymium laser material: Silicate oxyapatite. Tech. Report AFAL-TR-72-37, Air Force Avionics Lab., WPAFB (1972)
2.22 K.B. Steinbruegge, G.D. Baldwin: Appl. Phys. Lett. **25**, 220 (1974)
2.23 G.D. Baldwin: Q-switched evaluation of CaLa SOAP:Nd. Tech. Report AFAL-TR-72-334, Air Force Avionics Lab., WPAFB (1972)
2.24 W.W. Krühler, J.P. Jeser, H.G. Danielmeyer: Appl. Phys. **2**, 329 (1973)
2.25 H.P. Weber, T.C. Damen, H.G. Danielmeyer, B.C. Tofield: Appl. Phys. Lett. **22**, 534 (1973)
2.26 H.P. Weber, P.F. Liao, B.C. Tofield: IEEE J. QE-**10**, 563 (1974)
H.P. Weber, P.F. Liao, B.C. Tofield, P.M. Bridenbaugh: Appl. Phys. Lett. **26**, 692 (1975)
2.27 J.G. Gualtieri, T.R. Aucoin: Appl. Phys. Lett. **28**, 189 (1976)
2.28 H.G. Danielmeyer, W.W. Krühler, J.P. Jeser: Appl. Phys. **2**, 335 (1973)
2.29 K.Nassau, A.M. Broyer: J. Appl. Phys. **33**, 3064 (1962)
2.30 L.L. Harper, J.R. Thornton: Increased energy storage Nd laser material: Sodium lanthanum molybdate. Tech. Report AFAL-TR-72-38, Air Force Avionics Lab., WPAFB (1972)
2.31 J.R. O'Connor: Appl. Phys. Lett. **9**, 407 (1966)
2.32 R.A. Fields, M. Birnbaum, C.L. Fincher: Highly efficient diode pumped Nd: crystal lasers. CLEO'87 (Baltimore, MA)
A.W. Tucker, M. Birnbaum, C.L. Fincher: J. Appl. Phys. **52**, 3067 (1981)
T. Chin, R.C. Morris, O. Kafri, M. Long, D.F. Heller: Athermal Nd:BEL. CLEO'86 (San Francisco, CA)
2.33 C.F. Cline, R.C. Morris: Doped Beryllium Lanthanate Crystals, U.S. Patent 3,866,142 (February 1975). See also Data Sheet $La_2Be_2O_5$, Allied Chemical Corp. Morristown, NJ (1975)
2.34 L.F. Johnson: J. Appl. Phys. **33**, 756 (1962); and **34**, 897 (1963)
2.35 E.P. Chickles: Appl. Phys. Lett. **19**, 119 (1971); and Stimulated emission at 0.85 μm in Er^3:YLF. 7th Int'l Quant. Electr. Conf., Montreal, Canada (1972)
D.P. Devor: 2.06 μm laser performance and design options for rangefinder and illuminator applications. IEEE/OSA Conf. Laser Eng. and Appl. (May 1975)
2.36 E.J. Sharp, D.T. Horowitz, T.E. Miller: J. Appl. Phys. **44**, 5399 (1973)
2.37 C.D. Greskovich: Oxide ceramic laser. Report 73 CRD 218, General Electric Comp., Schenectady, NY (July 1973)

2.38 G. Müller, N.Neuroth: J. Appl. Phys. **44**, 2315 (1973)
2.39 M.J. Weber, M. Bass, G.A. deMars: J. Appl. Phys. **42**, 301 (1971)
2.40 W.F. Krupke, J.B. Gruber: J. Chem. Phys. **41**, 1225 (1964)
2.41 Z.J. Kiss, R.C. Duncan: Proc. IRE **50**, 1531 (1962)
2.42 S.A. Pollack: Proc. IEEE **51**, 1793 (1963)
2.43 E. Snitzer: Appl. Phys. Lett. **6**, 45 (1965)
2.44 K.O. White, E.H. Holt: The erbium doped glass laser. Report ECOM-5294, U.S. Army, Fort Monmouth, NJ (1970)
2.45 D.P. Devor: High power, high efficient 2.1 μm laser device, Tech. Report AFAL-TR-71-181, Air Force Avionics Lab., WPAFB (1971)
2.46 T.A. Laird, L.G. DeShazer: IEEE J. QE-**11**, 97 (1975)
2.47 M.J. Weber, M. Bass, E. Comperchio, L.A. Riseberg: IEEE J. QE-**7**, 497 (1971);
N.P. Barnes, D.J. Gettemy: IEEE J. QE-**17**, 1303 (1981)
E.V Zharikov, S.P. Kalitin, V.V. Laptev, V.G. Ostroumov, Z.S. Sadov, V.A. Smirnov, I.A. Shcherbakov: Sov. J. Quantum Electron. **16**, 16 (1986)
E.W. Duczynski, G. Huber, V.G. Ostroumov, I.A. Shcherbakov: Appl. Phys. Lett. **9**, 1562 (1986)
M. Storm, L. Esterowitz, M. Kokta: CLEO'87 (Baltimore, MD)
T.Y. Fan, G. Huber, R.L. Byer, Mitzscherlich: CLEO'87 (Baltimore, MD) paper FL1
2.48 L.M Hobrock, L.G. DeShazer, W.F. Krupke, G.A. Keig, D.E. Watter: Four-level operation of Tm: Cr:$YAlO_3$ laser at 2.35 μm. 7th Int'l Quant. Electr. Conf., Montreal, Canada (1972)
2.49 Z.J. Kiss, R.J. Pressley: Proc. IEEE **54**, 1236 (1966)
2.50 J.R. Thornton, W.D. Fountain, G.W. Flint, T.G. Crow: Appl. Opt. **8**, 1087 (1969)
2.51 L.F. Johnson: *Lasers*, Vol.1, ed. by A.K. Levine (Dekker, New York 1966) p.137
2.52 P.F. Moulton: IEEE J. QE-**21**, 1582 (1985)
M.G. Knights, P.F. Moulton: CLEO'87 (Baltimore, MD) paper Th M2
2.53 R.W.G. Wyckoff: *Crystal Structures* (Wiley, New York 1963)
2.54 K. Nassau: *Applied Solid State Science*, Vol.2, ed. by R. Wolfe (Academic, New York 1970)
2.55 A.A. Kaminskii: *Laser Crystals*, Springer Ser. Opt. Sci., Vol.14 (Springer, Berlin, Heidelberg 1981)
O.C. Cronemeyer: J. Opt. Soc. Am. **56**, 1703 (1964)
2.56 D.M. Dodd, D.L. Wood, R.L. Barns: J. Appl. Phys. **35**, 1183 (1964)
2.57 T.H. Maiman, R.H. Hoskins, I.T. D'Haenens, C.K. Asawa, V. Evtuhov: Phys. Rev. **123**, 1151 (1961)
2.58 P.G. Kriukov, V.S. Letokhov: *Laser Handbook*, Vol.1, ed. by E.T. Arrechi, E.O. DuBois (North-Holland, Amsterdam 1972) pp.561-595
2.59 F.J. McClung, S.E. Schwarz: J. Appl. Phys. **33**, 3193 (1962)
2.60 C.J. Hubbard, E.W. Fisher: Appl. Opt. **3**, 1499 (1964)
2.61 A.I. Mahan, C. Bitterli, S.M. Connor, D.G. Grant: J. Opt. Soc. Am. **59**, 49 (1969)
2.62 D.F. Nelson, M.D Sturge: Phys. Rev. **137** A, 1117 (1965)
2.63 I.J. D'Haenens, C.K. Asawa: J. Appl. Phys. **33**, 3201 (1962)
2.64 A.L. Schawlow: In *Advances in Quantum Electronics* (Columbia Univ. Press, New York 1961) p.50
2.65 M.G. Holland: J. Appl. Phys. **33**, 2910 (1962)
2.66 V. Evtuhov, T.K. Neeland: *Lasers*, Vol.1, ed. by A.K. Levine (Dekker, New York 1966) pp.1-36
2.67 Airtron Division of Litton Industries, Data Sheet: Ruby Laser Crystals
Crystal Products Division of Union Carbide, Data Sheet: Ruby Laser Rods

2.68 W. Koechner: Rev. Sci. Instr. **41**, 1699 (1970)
2.69 C.A. Burrus, J. Stone: Appl. Phys. Lett. **26**, 318 (1975)
2.70 T. Kushida, J.E. Geusic: Phys. Rev. Lett. **21**, 1172 (1968)
2.71 N.P. Barnes, D.J. Gettemy, L. Esterowitz, R.A. Allen: IEEE J. QE-**23**, 1434 (1987)
2.72 T. Kushida, H.M. Marcos, J.E. Geusic: Phys. Rev. **167**, 289 (1968)
2.73 U. Ranon, L.G. DeShazer, J.K. Guha, E.D. Reed: Spectroscopy of Nd^{3+} in ED2 laser glasses and the laser cross section in Nd:YAG. 8th Int'l Quant. Electr. Conf., San Francisco, CA (1974), Digest, paper C.2
2.74 J.K. Neeland, V. Evtuhov: Phys. Rev. **156**, 244 (1967)
M. Birnbaum, J.A. Gelbwachs: J. Appl. Phys. **43**, 2335 (1972)
M.J. Weber, T.E. Varitimos: J. Appl. Phys. **42**, 4996 (1971)
S. Singh, R.G. Smith, L.G. Van Uitert: Phys. Rev. B **10**, 2566 (1974)
2.75 J.A. Koningstein, J.E. Geusic: Phys. Rev. **136**, 711 (1964)
2.76 A.A. Zlenko, V.A. Sychugor, G.P. Shipulo: Sov. J. QE-**2**, 474 (1973)
2.77 P.H. Klein, W.J. Croft: J. Appl. Phys. **38**, 1603 (1967)
2.78 H.F. Mahlein: IEEE J. QE-**6**, 529 (1970)
2.79 C.G. Bethea: IEEE J. QE-**9**, 254 (1973)
2.80 R.G. Smith: IEEE J. QE-**4**, 505 (1968);
S. Singh, R.G. Smith, L.G. VanUitert: Phys. Rev. B **10**, 2566-2572 (1974);
A.A. Kaminskii: Sov. Phys. JETP **37**, 388-399 (1968)
J. Marling: IEEE J. QE-**14**, 56 (1978)
2.81 M. Birnbaum: J. Appl. Phys. **44**, 2928 (1973)
2.82 R.W. Wallace, S.E. Harris: Appl. Phys. Lett. **15**, 111 (1969)
2.83 R.W. Wallace: IEEE J. QE-**7**, 203 (1971)
2.84 M.D. Thomas, G.A. Rines, E.P. Chicklis, W. Koechner: High power 1.3 micorn Nd: YaG laser. CLEO'86 (San Francisco, CA) paper WM4
2.85 S.M. Yarema, D. Milam: IEEE J. QE-**18**, 1941 (1982)
2.86 M.J. Weber: J. Non-Cryst. Solids **42**, 189 (1980)
2.87 S.E. Stokowski, R.A. Saroyan, M.J. Weber: Nd:doped laser glass spectroscopic and physical properties. Lawrence Livermore Nat'l Lab. Report M-095 (Rev.2) (1981)
2.88 M.J. Weber, D. Milam, W.L. Smith: Opt. Eng. **17**, 463 (1978)
2.99 W.W. Simmons, J.T. Hunt, W.E. Warren: IEEE J. QE-**17**, 1727 (1981)
2.90 Hoya Optics, Menlo Park, CA; Schott Optical Glass, Dunyea, PA; Kigre, Toledo, OH, USA
2.91 D. Duston: IEEE J. QE-**6**, 3 (1970)
2.92 M. Michon: Phys. Lett. **19**, 219 (1965)
2.93 R. Dumanchin: IEEE J. QE-**7**, 53 (1971)
2.94 P.C. Magnante: IEEE J. QE-**8**, 440 (1972)
2.95 A.A. Mak, D.S. Prilezhaev, V.A. Serebryakov, A.D. Starikov: Optics and Spectroscopy **33**, 381 (1972)
2.96 *Anonymous*: Fundamentals of Damage in Laser Glass. Report NMAB-271 (National Academy of Sciences, Washington, DC 1970)
2.97 S.E. Stokowski: Glass lasers, in *Handbook of Laser Science and Technology*, ed. by M.J. Weber (CRC Press, Boca Raton, FL 1982) pp.215-264
2.98 C.F.. Rapp: Laser glasses, in *Handbook of Laser Science and Technology*, ed by M.J. Weber (CRC Press, Baco Raton, FL 1987) pp.339-372
2.99 E. Snitzer: Appl. Opt. **5**, 1487 (1966)
2.100 E. Snitzer: IEEE **54**, 1249 (1966)
2.101 W.F. Krupke: IEEE J. QE-**10**, 450 (1974)
2.102 E.V. Zharikov, N.N. Il'ichev, V.V. Laptev, A.A. Malyutin, V.G. Ostroumov, P.P. Pashinus, I.A. Shcherbakov: Sov. J. Quantum Electr. **12**, 338-341 (1982)
2.103 D. Pruss, G. Huber, A. Beimowski: Appl. Phys. B **28**, 355-358 (1982)

2.104 E.V. Zharikov, N.N. Il'ichev, V.V. Laptev, A.A. Mayutin, V.G. Ostroumov, P.P. Pashinin, A.S. Pimenov, V.A. Smirnov, I.A. Shcherbakov: Sov. J. Quantum Electron. **13**, 82-85 (1983)

2.105 A. Beimowski, G. Huber, D. Pruss, V.V. Laptev, I.A.. Shcherbakov, E.V. Zharikov: Appl. Phys. B **28**, 234 (1982)

2.106 E.V. Zharikov, V.A. Zhitnyuk, G.M. Zverev, S.P. Kalitin, I.I. Kuratev, V.V. Laptev, A.M. Onishchenko, V.V. Osiko, V.A. Pashkov, A.S. Pimenov, A.M. Prokhorov, V.A. Smirnov, M.F. Stel'makh, A.V. Shestakov, I.S. Shcherbakov: Sov. J. Quantum Electron., **12**, 1652-1653 (1982)

2.107 D.S. Sumida, D.A.Rockwell: Dependence of Cr:Nd:GSGG pumping on Cr concentration, CLEO'86 (San Francisco, CA) paper WQ3

2.108 P. Fuhrberg, W. Luhs, B. Struve, G. Litfin: Single-mode operation of Cr:doped GSGG and $KZnF_3$, in *Tunable Solid-State Lasers II*, ed. by A.B. Budgor, L. Esterowitz, L.G. DeShazer, Springer Ser. Opt. Sci, Vol.52 (Springer, Berlin, Heidelberg 1987)

2.109 Lawrence Livermore National Laboratory, medium average power solid-state laser. Technical Information Seminar (October, 1985)

2.110 J.Y. Lice, C. Lice, M.G. Cohen: High average power normal mode Cr:Nd:GSGG lasers. CLEO'86 (San Francisco, CA) paper TuK33

2.111 E. Reed: IEEE J. QE-**21**, 1625 (1985)

2.112 J.A. Caird, M.D. Shinn, T.A. Kirchoff, L.K. Smith, R.E. Wilder: Appl. Opt. **25**, 4294-4305 (1986)

2.113 S.E. Stokowski, M.A. Randles, R.C. Morris: CLEO'87 (Baltimore, MD) paper WB4

2.114 E.P. Chicklis: private communication

2.115 M. Knight, G. Rines, J. McCarthy, T. Pollak, K.A. Smith, E.P. Chicklis: High power Nd:YLF laser performance, CLEO'84 (Anaheim, CA) paper WM1

2.116 G.Rines, M. Thomas, M. Knight. E.P. Chicklis: Gain and parasitic suppresion in Nd:YLF, CLEO'84 (Anaheim, CA) paper WM2
M.G. Knights, M.D. Thomas, E.P. Chicklis, G.A. Rines, W. Seka: IEEE J. QE-**24**, 712 (1988)

2.117 T.M. Pollak, W.F. Wing, R.J. Grasso, E.P. Chicklis, H.P. Jenssen: IEEE J. QE-**18**, 159-163 (1982)

2.118 J.E. Murray: IEEE J. QE-**19**, 488-490 (1983)

2.119 E.V. Zharikov, V.I. Zhekov, L.R. Kulevskii, T.M. Mirina, V.V. Osiko, A.M. Prokhorov, A.D. Savel'ev, V.V. Smirnov, B.P. Starikov, M.I. Timoshenko: Sov. J. Quantum Electron. **4**, 1039 (1975)

2.120 K.S. Bagdesarov, V.I. Zhekov, L.A. Kulevskii, V.A. Lobachev, T.M. Mirina: Sov. J. Quantum Electron. **10**, 1127 (1980)

2.121 K.L. Vodop'vanov, L.A. Kulevskii, A.A. Malyutin, P.P. Pashinin, A.M. Prokhorov: Sov. J. Quantum Electron. **12**, 541 (1982)

2.122 M. Bass, Wei-Qiang Shi, R. Kurtz, M. Kokta, H. Diegl: Room temperature operations of the 50 doped Er:YAG laser at 2940 nm. CLEO'86 (San Francisco, CA) Postdeadline paper ThT1-1

2.123 J.Y. Liu, K.L. Liu, M.Y. Cohen: Pulsed Er:YAG laser performance at 2.94 μm. CLEO'87 (Baltimore, MD)

2.124 H.P. Weber, W. Luthy: Erbium laser for medical applications, CLEO (Baltimore, MD 1987) paper ThJ1

2.125 J.G. Manni, G.A. Rines, P.T. Moulton: Characterization of 2.9 μm laser operation in Er-doped crystals. CLEO'87 (Baltimore, MD) paper ThJ3

2.126 G.J. Kintz, R. Allen, L. Esterowitz: CW and pulsed 2.8 μm laser emission from diode-pumped Er^{3+}:YLF at room temperature. CLEO'87 (Baltimore, MD)

2.127 E. Snitzer, R. Woodcock: Appl. Phys. Lett. **6**, 45 (1965)

2.128 E. Snitzer: Proc. IEEE **54**, 1259 (1966)

2.129 Y. Morishige, S. Kishida, K. Washio: Opt. Lett. **9**, 147-149 (1984)
2.130 I.M. Jauncey, L. Reekie, R.J. Mears: Opt. Lett. **12**, 164-165 (1987)
2.131 L.F. Johnson, R.E. Dietz, H.J. Guggenheim: Phys. Rev. Lett. **11**, 318-320, 1963
2.132 L.F. Johnson, R.E. Dietz, H.J. Guggenheim: Appl. Phys. Lett. **5**, 21-22, 1964
2.133 L.F. Johnson, H.J. Guggenheim, R.A. Thomas: Phys. Rev. **149**, 179-185 (1966)
2.134 L.F. Johnson, H.J. Guggenheim: J. Appl. Phys. **38**, 4837-4839, 1967
2.135 L.F. Mollenauer, J.C. White (eds.): *Tunable Lasers*, Topics Appl. Phys., Vol.59 (Springer, Berlin, Heidelberg 1987)
2.136 L.F. Johnson, H.J. Guggenheim: IEEE J. QE-**10**, 442 (1974)
2.137 J.C. Walling, H.P. Jenssen, R.C. Morris, E.W. O'Dell, O.G. Peterson: Annual Meeting of the Optical Society of America (1978)
2.138 J.C. Walling, H.P. Jenssen, R.C. Morris, E.W. O'Dell, O.G. Peterson: Opt. Lett. **4**, 182-183, 1979
2.139 J.C. Walling, O.G. Peterson, H.P. Jenssen, R.C. Morris, E.W. O'Dell: IEEE J. QE-**16**, 1302-1315 (1980)
2.140 J. Buchert, A. Katz, R.R. Alfano: CLEO'83 (Baltimore, MD) paper ThR2
2.141 M.L. Shand, J.C. Walling: IEEE J. QE-**18**, 1829 (1982); CLEO'83 (Baltimore, MD) paper ThR1
2.142 J. Buchert, R.R. Alfano: Laser Focus, 117-123 (Sept. 1983)
2.143 J. Buchert, A. Katz, R.R. Alfano: IEEE J. QE-**19**, 1477 (1983)
2.144 M.L. Shand, S.T. Lai: IEEE J. QE-**20**, 105-108 (1984)
2.145 P.F. Moulton, A. Mooradian, T.B. Reed: Opt. Lett. **3**, 164-166 (1978)
2.146 P.F. Moulton, A. Mooradian: Appl. Phys. Lett. **35**, 838-840 (1979)
2.147 P.F. Moulton, A. Mooradian: Tunable transition-metal-doped solid state lasers, in *Laser Spectroscopy IV*, ed. by H. Walther, K.W. Rothe, Springer Ser. Opt. Sci, Vol.21 (Springer, Berlin, Heidelberg 1979) pp.584-589
2.148 P.F. Moulton: IEEE J. QE-**18**, 1185-1188 (1982);
2.149 P.F. Moulton: Opt. News **8**, 9 (1982)
2.150 P.T. Kenyon, L. Andrews, B. McCollum, A. Lenspicki: IEEE J. QE-**18**, 1189 (1982)
B. Struve, G.Huber: Appl. Phys. B **30**, 117-120 (1983)
2.151 B. Struve, G. Huber: Laser action and broad band fluorescence in Cr^{3}+:GdScGa garnet. 12th Int'l Quantum Electr. Conf. (1982) paper ThR-5
2.152 E.V. Zharikov, V.V. Laptev, E.I. Sidorova, Yu P. Timofeev, I.A. Shchevbakov: Sov. J. Quantum Electron. **12**, 1124 (1982)
2.153 H.P. Christensen, H.P. Jenssen: IEEE J. QE-**18**, 1197-1201 (1982)
2.154 P. Hammerling, A.B. Budgor, A. Pinto (eds.): *Tunable Solid State Lasers*, Springer Ser. Opt. Sci., Vol.47 (Springer, Berlin, Heidelberg 1985);
A.B. Budgor, L. Esterowitz, L.G. DeShazer (eds.): *Tunable Solid State Lasers II*, Springer Ser. Opt. Sci., Vol.52 (Springer, Berlin, Heidelberg 1986);
Proc. Topical Meeting on Tunable Solid State Lasers, Oct. 26-28, 1987, Williamsburg, VA, to be published
W. Künzel, U. Dürr: Appl. Phys. B **28**, 233 (1982)
2.155 U. Brauch, U. Dürr: Optics Commun. **49**, 61 (1984)
P.F. Moulton, A. Mooradian, T.B. Reed: Opt. Lett. **3**, 164 (1978)
P.F. Moulton, A. Mooradian: Appl. Phys. Lett. **35**, 838 (1979)
2.156 D.J. Ehrlich, P.F. Moulton, R.M. Osgood: Opt. Lett. **4**, 184 (1978)
2.157 D.J. Ehrlich, P.F. Moulton, R.M. Osgood: Opt. Lett. **5**, 339 (1980)
2.158 H. Samelson, J.C. Walling, D.F. Heller: SPIE **335**, 85 (1982)
2.159 M. Shand, H.P. Jenssen: IEEE J. QE-**19**, 480 (1983)
S. Guch: CLEO'83 (Baltimore, MA) paper ThR3
2.160 J.C. Walling: Laser Focus, 48 (February 1982)
2.161 R.C. Sam, J.J. Yeh: High repetition rate, narrow band alexandrite laser, 1983 OSA Annual Mtg., New Orleans, LA, paper WS4

2.162 W.R. Rapoport, J.J. Yeh, R.C. Sam: CLEO'83 (Baltimore, MD) paper ThR4
2.163 H. Samelson, D.J. Harter: Output properties of continuous wave pumped alexandrite laser. 1983 OSA Annual Mtg., New Orleans, LA, paper WS5
2.164 H. Samelson: private communication, 1984
2.165 P.F. Moulton: Laser Focus **14**, 83 (May 1983)
2.166 P. Lacovara, L. Esterowitz, R. Allen: Opt. Lett. **10**, 273-275 (1985)
2.167 L.G. DeShazer, K.W. Kangas: CLEO'87 (Baltimore, MD) paper ThQ4

Chapter 3

3.1 W.W. Rigrod: J. Appl. Phys **36**, 2487 (1963)
3.2 A.Y. Cabezas, R.P. Treat: J. Appl. Phys **37**, 3556 (1966)
3.3 R. Dunsmuir: J. Electron. Control **10**, 453 (1961)
3.4 H. Statz, G.A. Mars, D.T. Wilson: J. Appl. Phys. **30**, 1510 (1965)
3.5 D. Röss: *Lasers, Light Amplifiers and Oscillators* (Academic, New York 1969)
3.6 J.W. Strozyk: IEEE J. QE-**3**, 343 (1967)
3.7 R.F. Marshall, D.L. Roberts: Proc. IRE **50**, 2108 (1962)
3.8 D.V. Keller, B.I. Davis: IEEE J. QE-**2**, 178 (1966)
3.9 C.H. Thomas, E.V. Price: IEEE J. QE-**2**, 617 (1966)
3.10 V.I. Mishin: Pri. Tekh. Eksp. **4**, 181 (1971)
3.11 R.V. Lovberg, E.R. Wooding, M.L. Yeoman: IEEE J. QE-**11**, 17 (1975)
3.12 A.M. Bonch-Bruevich, S.E. Potanov, Ya.A. Imas: Sov. Phys. **13**, 44 (1968)
3.13 R.H. Pantell, J. Warszawski: Appl. Phys. Lett. **11**, 213 (1967)
3.14 C.G. Young, J.W. Kantorski, E.O. Dixon: J. Appl. Phys. **37**, 4319 (1966)
3.15 C.F. Padula, C.G. Young: IEEE J. QE-**3**, 493 (1967)
3.16 W. Koechner: IEEE J. QE-**8**, 656 (1972)
3.17 H.G. Danielmeyer: J. Appl. Phys. **41**, 4014 (1970)
3.18 T. Kimura, K. Otsuka: IEEE J. QE-**6**, 764 (1970)
3.19 R. Polloni, O. Svelto: IEEE J. QE-**4**, 481 (1968)
3.20 J.F. Nester: IEEE J. QE-**6**, 97 (1970)
3.21 I.C. Chang, E.H. Lean, C.G. Powell: IEEE J. QE-**6**, 436 (1970)
3.22 G.D. Baldwin, I.T. Basil: IEEE J. QE-**7**, 179 (1971)
3.23 R.B. Chesler: Appl. Opt. **9**, 2190 (1970)
3.24 R.P. Johnson, N.K. Moncur, L.D. Siebert: CLEO'87 (Baltimore, MD) paper FP2
3.25 V. Evtuhov: J. Appl. Phys. **38**, 4051 (1967)
3.26 D. Röss: IEEE J. QE-**2**, 208 (1966)
3.27 Owens-Illinois, Toledo, OH: Data Sheet: Laser Glass
3.28 C.G. Young: Laser Focus **3**, 36 (February 1967)
3.29 R.B. Chesler, D.A. Draegart: Appl. Phys. Lett. **23**, 235 (1973)
3.30 M. Saruwateri, T. Kimura, K. Otuka: Appl. Phys. Lett., **29**, 291 (1976)
3.31 D.L. Sipes: Appl. Phys. Lett. **47**, 74-76 (1985)
3.32 J. Katz: TDA Progress Reports 42-66, Jet Propulsion Laboratory, Pasadena, CA (1981) p.94
3.33 R.A. Fields, M. Birnbaum, C.L. Fincher: CLEO'87 (Baltimore, MD)
3.34 W. Kozlovsky, T.Y. Fan, F.L. Byer: CLEO'86 (San Francisco, CA) Techn. Digest, paper WG4, p.169
3.35 T.Y. Fan, G.J. Dixon, R.L. Byer: Opt. Lett. **11**, 204 (1986)
3.36 R. Allen, L. Esterowitz, L. Goldberg, J.F. Weller: CLEO'86 (San Francisco, CA) paper FC1, Tech. Digest p.390
3.37 T.Y. Fan, R.L. Byer: CLEO'87 (Baltimore, MD) paper FL3
3.38 H. Hemmati: CLEO'87 (Baltimore, MD)
3.39 M.D. Thomas, H.H. Zenzie, T.C. McCarthy, E.P. Chicklis: CLEO'87 (Baltimore, MD) paper ThJ5

3.40 J.J. Smith, C. Ma: CLEO'87 (Baltimore, MD) paper FL6
3.41 W.M. Grossman, R.W. Wallace: Preprint (1987)
3.42 S. Basu, R.L. Byer: CLEO'87 (Baltimore, MD) paper WN3
3.43 R.J. Mears, L. Reekie, S.B. Poole, D.N. Payne: Electron. Lett. **21**, 738 (1985)
3.44 S.B. Poole, D.N. Payne, R.J. Mears, M.E. Fermann, R.I. Laming: IEEE/OSA J. LT-**4**, 870 (1986)
3.45 L. Reekie, R.J. Mears, S.B. Poole, D.N. Payne: IEEE/OSA J. LT-**4**, 956 (1986)
3.46 R.J. Mears, L. Reekie, I.M. Jauncey, D.N. Payne: High-gain rare-earth-doped fiber amplifier at 1.54 μm. 6th Int'l Conf. on Integrated Optics and Optical Fiber Communication (Optical Society of America, Washington, DC 1987) paper W12
3.47 I.M. Jauncey, J.T. Lin, L. Reekie, R.J. Mears: Electron. Lett. 22, 198 (1986)
3.48 R.J. Mears, L. Reekie, S.B. Poole, D.N. Payne: Electron. Lett. 22, 159 (1986)
3.49 S.B. Poole, D.N. Payne, M.E. Fermann: Electron. Lett. **21**, 737 (1985)
3.50 I. Bennion: Electron. Lett. 22, 341 (1986)
3.51 I.D. Miller: 6th Int'l Conf. on Integrated Optics and Optical Fiber Communication (Optical Society of America, Washington, DC 1987) paper W12
3.52 R.J. Mears, L. Reekie, I.M. Jauncey: CLEO'87 (Baltimore, MD) paper WD3
3.53 C.A. Miller, I.D. Miller: CLEO'87 (Baltimore, MD) paper WD2
5.54 D.N. Payne, L. Reekie, R.J. Mears, S.B. Poole, I.M. Jauncey, J.T. Line: CLEO'86 (San Francisco, CA) paper FN1
3.55 C.L. Tang, H. Statz, G.A. DeMars, D.T. Wilson: Phys. Rev. **136**, A1 (1964)
3.56 V.Yu. Petrun'kin, L.N. Pakhomov, S.V. Kruzhalov, N.M. Kozhevnikov: Sov. Phys.-Tech. Phys. **17**, 1222 (1973)
3.57 A.R. Clobes, M.J. Brienza: Appl. Phys. Lett. **21**, 265 (1972)
3.58 A.M. Bonch-Bruevich, V.Yu. Petrun'kin, N.A. Esepkina, S.V. Kruzhalov, L.N. Pakhomov, V.A. Chernov, S.L. Galkin: Sov. Phys.-Tech. Phys. **12**, 1495 (1968)
3.59 T.J. Kane, R.L. Byer: Opt. Lett. **10**, 65 (1985)
3.60 W.R. Trutna, D.K. Donald, M. Nazarathy: CLEO'87 (Baltimore, MD) paper WN2
3.61 T.J. Kane, R.W. Wallace: (private communication)
3.62 G.A. Rines, P.F. Moulton, M.G. Knights: (private communication)
3.63 R.C. Sam: High average power alexandrite lasers. Allied Corp. (Morristown, NJ) private communication

Chapter 4

4.1 L.M. Frantz, J.S. Nodvik: J. Appl. Phys. **34**, 2346 (1963)
4.2 R. Bellman, G. Birnbaum, W.G. Wagner: J. Appl. Phys. **34**, 780 (1963)
4.3 E.L. Steele: J. Appl. Phys. **36**, 348 (1965)
4.4 E.L. Steele: *Optical Lasers in Electronics* (Wiley, New York 1968)
4.5 P.G. Kriukov, V.S. Letokhov: *Laser Handbook I*, ed. by E.T. Arecchi, E.O. Schulz-DuBois (North-Holland, Amsterdam 1972) pp.561-595
4.6 P.V. Avizonis, R.L. Grotbeck: J. Appl. Phys. **37**, 687 (1966)
4.7 C.R. Jones, P.V. Avizonis, P. Sivgals: Experimental investigation of the behavior of neodymium-glass laser amplifiers. NBS Spec. Pub. **341**, 28 (1970)
4.8 J.M. McMahon: Glass laser material testing at naval research laboratory. ASTM Report on Damage in Laser Glass STP-**469**, 117 (1969)
4.9 I.F. Balashov, V.A. Berenberg, V.V. Blagoveshchenskii: Sov. Phys. **14**, 692 (1965)
4.10 M. Michon, R. Auffret, R.Dumanchin: J. Appl. Phys. **41**, 2739 (1970)
4.11 J.E. Geusic, H.E.D. Scovil: In *Quantum Electronics III* (Columbia Univ. Press, New York 1964) pp.1211-1220
4.12 W.R. Sooy, R.S. Congleton, B.E. Dobratz, W.K. Ng: In *Quantum Electronics III* (Columbia Univ. Press, New York 1964) pp.1103-1112

4.13 J.I. Davis, W.R. Sooy: Appl. Opt. **3**, 715 (1964);
P. Labudde, W. Seka, H.P. Weber: Appl. Phys. Lett. **29**, 732 (1976);
J.M. Rinefierd, S.D. Jacobs, D.C. Brown, J.A. Abate, O. Lewis, H. Appelbaum: Laser Induced Damage in Optical Materials, Boulder, CO (1978)
4.14 R. Carman: Laser fusion program. Semiannual Report UCRL-50021-73-1, Lawrence Livermore Lab., Livermore, CA (January-June 1973) p.154
4.15 D.C. Brown: *High-Peak Power Nd: Glass Laser Systems*, Springer Ser. Opt. Sci., Vol.25 (Springer, Berlin, Heidelberg 1981)
4.16 Special issue on lasers for fusion, IEEE J. QE-**17** (1981)
4.17 Lawrence Livermore Lab., Laser Program Annual Reports UCRL-50021-74 (1974) to UCRL-50021-86 (1986)
4.18 M.J. Weber: J. Non-Cryst. Solids **42**, 189 (1980)
4.19 G.J. Linford, R.A. Saroyan, J.B. Trenholme, J.T. Weber: IEEE J. QE-**15**, 510 (1979)
4.20 W.E. Martin, D. Milam: IEEE J. QE-**18**, 1155 (1982)
D.W. Hall, M.J. Weber: IEEE J. QE-**20**, 831 (1984)
D.W. Hall, R.A. Haas, W.F. Krupke, M.J. Weber: IEEE J. QE-**19**, 1704 (1983)
4.21 S.M. Yarema, D. Milam: IEEE J. QE-**18**, 1941 (1982)
4.22 C. Yamanaka, Y. Kato, Y. Izawa, K. Yoshida, T. Yamanaka, T. Sasaki, M. Nakatsuka, T. Mochizuki, J. Kuroda, S. Nakai: IEEE J. QE-**17**, 1639 (1981)
4.23 C.C. Young, J.W. Kantorski: Optical gain and inversion in Nd:glass lasers. Proc. 1st DOD Conf. on Laser Technology (1964) p.75
4.24 R.W. Beck: Damage Threshold studies of glass laser materials. Ownes-Illinois, Tech. Report ARPA, Contract DAHC15-69-C-0303 (January 1970)
4.25 T.G. Crow, T.J. Snyder: Techniques for achieving high power Q-switched operation in YAG: Nd. Final Tech. Report AFAL-TR-70-69, Air Force, WPAFB (1970); see also Laser J. **18** (November/December 1970)
4.26 N.P. Barnes, V.J. Corcoran, I.A. Crabbe, L.L. Harper, R.W. Williams, J.W. Wragg: IEEE J. QE-**10**, 195 (1974)
4.27 E.A. Teppo: Nd: YAG laser technology. NWC Tech. Memo 2534, Naval Weapons Center, China Lake, CA (August 1975) p.193
4.28 W.W. Rigrod: J. Appl. Phys. **34**, 2602 (1963)
4.29 E.O. Schulz-DuBois: Bell Systems Techn. J. **43**, 625 (1964)
4.30 A.Y. Cabezas, R.P. Treat: J. Appl. Phys. **37**, 3556 (1966)
4.31 A.Y. Cabezas, G.L. McAllister, W.K. Ng: J. Appl. Phys. **38**, 3487 (1967)
4.32 J. Bunkenberg, J. Boles, D.C. Brown, J. Eastman, J. Hoose, R. Hopkins, L. Iwan, S.D. Jacobs, J.H. Kelly, S. Kumpan, S. Letzring, D. Lonobile, L.D. Lund, G. Mourou, S. Refermat, W. Seka, J.M. Soures, K. Walsh: IEEE J. QE-**17**, 1620 (1981)
4.33 W.B. Bridges: IEEE J. QE-**4**, 820 (1968)
4.34 J.P. Campbell, L.G. DeShazer: J. Opt. Soc. Am. **59**, 1427 (1969)
4.35 T. Trenholme: A user oriented axially symmetric diffraction code. Semiannual Report UCRL-50021-73-1, Lawrence Livermore Lab., Livermore, CA (January-June 1973) p.46
4.36 D.R. Speck, E.S Bliss, J.A. Glaze, J.W. Herris F.W. Holloway, J.T. Hunt, B.C. Johnson, D.J. Kuizenga, R.G. Ozarski, H.G. Patton, P.R. Rupert, G.J. Suski, C.D. Swift, C.E. Thompson: IEEE J. QE-**17**, 1599 (1981)
4.37 J.T. Hunt, J.A. Glaze, W.W. Simmons, P.A. Renard: Appl. Opt. **17**, 2053 (1978)
4.38 S.A. Akhmanov, R.V. Khokhlov, A.P. Sukhorukov: *Laser Handbook II*, ed. by E.T. Arecchi, E.O. Schulz-DuBois (North-Holland, Amsterdam 1972) p.1151
4.39 R.Y. Chiao, E. Garmire, C.H. Townes: Phys. Rev. Lett. **13**, 479 (1964)
4.40 V.I. Bespalov, V.I. Talanov: JETP Lett. **3**, 307 (1966)
4.41 J. Trenholme: Review of small signal theory. Laser Program Annual Report UCRL-50021-74, Lawrence Livermore Lab., Livermore, CA (1974) p.178

4.42 E.S. Bliss, D.R. Speck, J.F. Holzrichter, J.H. Erkkila, A.J. Glass: Appl. Phys. Lett. **25**, 448 (1974)
4.43 J.F. Holzrichter, D.R. Speck: J. Appl. Phys. **47**, 2459 (1976)
4.44 J. Trenholme: Review of small signal theory. Laser Program Annual Report, UCRL-50021-74, Lawrence Livermore Lab., Livermore, CA (1974) p.178
4.45 J.B. Trenholme: Proc. Soc. Photo-Opt. Instr. Eng. **69**, 158 (1975)
4.46 E.S. Bliss, D.R. Speck, J.F. Holzrichter, J.H. Erkkila, A.J. Glass: Appl. Phys. Lett. **25**, 448 (1974)
4.47 E.S. Bliss, J.T. Hunt, P.A. Renard, G.E. Sommargren, H.J. Weaver: Effects of nonlinear propagation on laser focusing properties. Lawrence Livermore Lab.. Preprint UCRL-77557 (1975)
4.48 E.S. Bliss, G.E. Sommargren, H.J. Weaver: Loss of focusable energy due to small scale nonlinear effects. Conf. on Laser Eng. and Appl., Washington, DC (May 1975) paper 8.2
4.49 J.A. Glaze: Proc. Soc. Photo-Opt. Instr. Eng. **69**, 45 (1975)
4.50 W.W. Simmons, J.T. Hunt, W.E. Warren: IEEE J. QE-**17**, 1727 (1981)
4.51 W.W. Simmons, S. Guch, F. Rainer, J.E. Murray: A high energy spatial filter for removal of small scale beam instabilities in high power solid state lasers. Conf. on Laser Eng. and Appl., Washington, DC (1975) paper 8.4
4.52 M.A. Duguay, L.E. Hargrove, K.B. Jefferts: Appl. Phys. Lett. **9**, 287 (1966)
4.53 F. Shimuzu: Phys. Rev. Lett. **19**, 1097 (1967)
4.54 R.J. Joenk: Phys. Lett. **24**A, 228 (1967)
4.55 F. DeMartini, C.H. Townes, T.K. Gustafson, P.L. Kelley: Phys. Rev. **164**, 312 (1967)
4.56 M.A. Duguay, J.W. Hansen, S.L. Shapiro: IEEE J. QE-**6**, 725 (1970)
4.57 R.A. Fisher: Picosecond optical pulse nonlinear progapation effects, Ph.D. thesis, Univ. California, Berkeley, CA (1971)
4.58 J.A. Fleck: J. Appl. Phys. **36**, 1301 (1965)
4.59 P.V. Avizonis, W.R. Willoughby: J. Appl. Phys. **37**, 682 (1966)
4.60 G.J. Linford, L.W. Hill: Appl. Opt. **13**, 1387 (1974)
4.61 A.A. Mak, B.G. Malinin, V.A. Novikov, D.S. Prilezhaev, A.I. Stepanov, V.I. Ustyugov: Sov. Phys. **14**, 1418 (1970)
4.62 Y.A. Anan'ev, I.F. Balashov, A.A. Mak: Sov. Phys. **11**, 124 (1966)
4.63 L. Tonks: J. Appl. Phys. **35**, 1134 (1964)
4.64 C.G. Young: Proc. IEEE **53**, 1267 (1965)
4.65 J.A. Glaze, S. Guch, J.B. Trenholme: Appl. Opt. **13**, 2808 (1974)
4.66 J.B. Trenholme: Fluorescence amplification and parasitic oscillation limitations in disk lasers. Memo Rep. 2480, Naval Res. Lab., Washington, DC (July 1972)
4.67 G.P. Kostometov, N.N. Rozanov: Sov. J. Quantum Electron. **6**, 696 (1976)
4.68 D.C. Brown, S.D. Jacobs, N. Nee: Appl. Opt. **17**, 211 (1978)
4.69 G.J. Linford, R.A. Sarogan, J.B. Trenholme, M.J. Weber: IEEE J. QE-**15**, 510 (1979)

Chapter 5

5.1 A.G. Fox, T. Li: Bell Syst. Tech. J. **40**, 453 (1961)
5.2 G.D. Boyd, J.P. Gordon: Bell Syst. Tech. J. **40**, 489 (1961)
5.3 G.D. Boyd, H. Kogelnik: Bell Syst. Tech. J. **41**, 1347 (1962)
5.4 H. Kogelnik, T. Li: Appl. Opt. **5**, 1550 (1966); also H. Kogelnik: In *Lasers 1*, ed. by A.K. Levine (Dekker, New York 1966) pp.295-347
5.5 H.K.V. Lotsch: Optik **28**, 65, 328, 555 (1968/1969); Optik **29**, 130, 622 (1969); and Optik **30**, 1, 181, 217, 563 (1969)
5.6 J.S. Kruger: Electro-Opt. Syst. Designs 12 (September 1972)
5.7 R.J. Freiberg, A.S. Halsted: Appl. Opt. **8**, 335 (1969)

5.8 G. Goubau, F. Schwering: IRE Trans. Ant. and Prop. **AP-9**, 248 (1961)
5.9 J.R. Pierce: Proc. Nat. Acad. Sci. **47**, 1808 (1961)
5.10 T. Li: Bell Syst. Tech. J. **44**, 917 (1965)
5.11 L.A. Vainshtein: Sov. Phys. JETP **44**, 1050 (1963); Sov. Phys. JETP **17**, 709 (1963)
5.12 D.E. McCumber: Bell Syst. Tech. J. **44**, 333 (1965)
5.13 B.A. See: Laser resonator, properties of laser beams and design of optical systems. N68-16910, Weapons Research Establishment, Salisbury, South Australia (1967)
5.14 J.S. Kruger: Beam divergence for various transverse laser modes. Report AD-729-299, Harry Diamond Lab., Washington, DC (1971)
5.15 G.L. McAllister, M.M. Mann, L.G. DeShazer: Transverse mode distortion in giant-pulse laser oscillators. IEEE Conf. Laser Eng. and Appl., Washington, DC (1969)
5.16 A.G. Fox, T.Li: IEEE J. QE-**2**, 774 (1966)
5.17 H. Kogelnik: Bell Syst. Tech. J. **44**, 455 (1965)
5.18 J. Steffen, J.P. Lörtscher, G. Herziger: IEEE J. QE-**8**, 239 (1972)
5.19 D.R. Whitehouse, C.F. Luck, C. Van Mertens, F.A. Horrigan, M. Bass: Mode control technology for high performance solid state laser. Rept. TR ECOM-0269-F, US Army Electr. Command, Fort Monmouth, NJ (1973)
5.20 T.J. Gleason: Analysis of complex laser cavities. Rept. HDL-TM-71-5, Harry Diamond Lab., Washington, DC (1971)
5.21 D.C. Hanna: IEEE J. QE-**5**, 483 (1969)
5.22 H.W. Kogelnik: IEEE J. QE-**8**, 373 (1972)
5.23 R.B. Chesler, D. Maydan: J. Appl. Phys. **43**, 2254 (1972)
5.24 A.L. Bloom: Properties of laser resonators giving uniphase wave fronts. Tech. Bulletin No.7, Spectra Physics, Mountain View, CA (1963)
5.25 R.J. Freiberg, A.S. Halsted: Appl. Opt. **8**, 355 (1969)
5.26 I.M. Belousova, O.B. Danilov: Soc. Phys. **12**, 1104 (1968)
5.27 E.A. Teppo: Nd: YAG laser laboratory experiments, Tech. Note 4051-2 (Feb. 1972); Tech. Note 4051-7, Naval Weapons Center, China Lake, CA (August 1973)
5.28 L.W. Davis: J. Appl. Phys. **39**, 5331 (1968)
5.29 J.E. Geusic, H.J. Levingstein, S. Singh, R.C. Smith, L.G. Van Uitert: Appl. Phys. Lett. **12**, 306 (1968)
5.30 L.M. Osterink, J.D. Forster: Appl. Phys. Lett. **12**, 128 (1968)
5.31 W.C. Fricke: Appl. Opt. **9**, 2045 (1970)
5.32 F.A. Levine: IEEE J. QE-7, 170 (1971)
5.33 W.C. Scott, M.DeWit: Appl. Phys. Lett. **18**, 3 (1971)
5.34 L.G. DeShazer, E.A. Maunders: Appl. Opt. **6**, 431 (1967)
5.35 H.K.V. Lotsch: Japan J. Appl. Phys. **4**, 435 (1965)
5.36 W.F. Hagen: J. Appl. Phys. **40**, 511 (1969)
5.37 J.P. Lörtscher, J. Steffen, G. Herziger: IEEE J. QE-**7**, 505 (1975)
5.38 H.P. Kortz, R. Ifflander, H. Weber: Appl. Opt. **20**, 4124 (1981)
5.39 V. Magni: Appl. Opt. **25**, 107 (1986); also Appl. Opt. **25**, 2039 (1986)
5.40 S. De Silverstri, P. Laporta, V. Magni: Opt. Commun. **57**, 339 (1986)
5.41 N.K. Berger, N.A. Deryugin, Y.N. Lukyanov, Y.E. Studenikin: Opt. Spectrosc. USSR **43**, 176 (1977)
5.42 R. Hauck, H.P. Kortz, H. Weber: Appl. Opt. **19**, 598 (1980)
5.43 D.C. Hanna, C.G. Sawyers, M.A. Yuratich: Opt. Quantum Electron. **13**, 493 (1981)
5.44 D.C. Hanna, C.G. Sawyers, M.A. Yuratich: Opt. Commun. **37**, 359 (1981)
5.45 P.H. Sarkies: Opt. Commun. **31**, 189 (1979)
5.46 M. Born, E. Wolf: *Principles of Optics* (Pergamon, New York 1964)
5.47 F.A. Jenkins, H.E. White: *Fundamentals of Optics* (McGraw-Hill, New York 1957)

5.48 K. Iizuka: *Engineering Optics*, 2nd. ed., Springer Ser. Opt. Sci., Vol.35 (Springer, Berlin, Heidelberg 1986)
5.49 J.K. Watts: Appl. Opt. 7, 1621 (1968)
5.50 M. Hercher: Appl. Opt. **8**, 1103 (1969)
5.51 B.L. Booth, S.M. Jarrett, G.C. Barker: Appl. Opt. **9**, 107 (1970)
5.52 R.J. Collier, C.B. Burckhard, L.H. Lin: *Optical Holography* (Academic, New York 1971)
5.53 L.D. Siebert: Appl. Opt. **10**, 632 (1971)
5.54 D.A. Kleinman, P.P. Kisliuk: Bell Syst. Tech. J. **41**, 453 (1962)
5.55 H.K.V. Lotsch: A modified Fabry-Perot interferometer as a discrimination filter and a modulator for longitudinal modes. Sci. Rept. No.2, US Air Force Contract AFI9(604)-8052(1962)
5.56 S.A. Collins, G.R. White: Appl. Opt. 2, 448 (1963);
5.57 G. Magyar: Opt. Techn. (November 1969) p.231
5.58 M. Hercher: Appl. Phys. Lett. **7**, 39 (1965)
5.59 D.G. Peterson, A. Yariv: Appl. Opt. **5**, 985 (1966)
5.60 R.M. Schotland: Appl. Opt. **9**, 1211 (1970)
5.61 M. Daehler, G.A. Sawyer, E.L. Zimmermann: J. Appl. Phys. **38**, 1980 (1967)
5.62 G. Magyar: Rev. Sci. Instr. **38**, 517 (1967)
5.63 M.M. Johnson, A.H. LaGrone: Appl. Opt. **12**, 510 (1973)
5.64 W.B. Tiffany: Appl. Opt. **7**, 67 (1968)
5.65 F.J. McClung, D. Weiner: IEEE J. QE-**1**, 94 (1965)
5.66 P.W. Smith: Proc. IEEE **60**, 422 (1972)
5.67 W. Wiesemann: Appl. Opt. **12**, 2909 (1973)
5.68 E. Snitzer: Appl. Opt. **5**, 121 (1966)
5.69 N.M. Galaktionova, G.A. Garkavi, V.F. Egorova, A.A. Mak, V.A. Fromzel: Opt. Spectrosc. **28**, 404 (1970)
5.70 W. Culshaw, J. Kannelaud: IEEE J. QE-**8**, 381 (1971)
5.71 W. Culshaw, J. Kannelaud, J.E. Peterson: IEEE J. QE-**10**, 253 (1974)
5.72 M.G. Danielmeyer, W.N. Leibolt: Appl. Phys. **3**, 193 (1974)
5.73 D. Röss: Appl. Phys. Lett. **8**, 109 (1966)
5.74 M.P. Vanyukov, V.I. Isaenko, L.A. Luizova, A. Shorokhov: Opt. Spectrosc. **20**, 535 (1966)
5.75 H.G. Danielmeyer: IEEE J. QE-**6**, 101 (1970)
5.76 R. Carman: Semi-annual Rept. **27**, Lawrence Livermore Lab., Livermore, CA (1972)
5.77 B.H. Soffer: J. Appl. Phys. **35**, 2551 (1964)
5.78 B.B. McFarland, R.H. Hoskins, B.H. Soffer: Nature **207**, 1180 (1965)
5.79 J.E. Bjorkholm, R.H. Stolen: J. Appl. Phys. **39**, 4043 (1968)
5.80 K. Yoshino, K. Kawabe, Y. Inuiski: Jpn. J. Appl. Phys. **8**, 1168 (1969)
5.81 R.F. Wuerker: SPIE J. **9**, 122 (1971)
K.A. Arunkumar, J.D. Trolinger: Opt. Engineeing **27**, 657 (1988)
5.82 W.R: Sooy: Appl. Phys. Lett. **7**, 36 (1965)
5.83 V. Daneu, C.A. Sacchi, O. Svelto: IEEE J. QE-**2**, 290 (1966)
5.84 D.C. Hanna: Electron. Lett. **8**, 369 (1972)
5.85 D.C. Hanna, B. Luther-Davis, H.N. Rutt, R.C Smith: Opto-Electronics **3**, 163 (1971)
5.86 D.C. Hanna, B. Luther-Davis, R.C. Smith: Opto-Electronics **4**, 249 (1972)
5.87 C.L. Tang, H. Statz, G.A. DeMars, D.T. Wilson: Phys. Rev. **136**, A1 (1964)
5.88 A.M. Bonch-Bruevich, A.Yu. Petrun'kin, N.A. Esepkina, S.V. Kruzhalov, L.N. Pakhomov, V.A. Chernov, S.L. Galkin: Sov. Phys. **12**, 1495 (1968)
5.89 A.R. Clobes, M.J. Brienza: Appl. Phys. Lett. **21**, 265 (1972)
5.90 V.Yu. Petrun'kin, L.N. Pakhomov, S.V. Kruzhalov, N.M. Kozhevnikov: Sov. Phys. **17**, 1222 (1973)
5.91 A. Biraben: Opt. Commun. **29**, 353 (1979)

5.92 O.E. Nanii, A.N. Shelaev: Sov. J. Quant. Electron. **14**, 638 (1984)
5.93 G.A. Rines, P.F. Moulton, M.G. Knights: CLEO '87 (Baltimore, MD) paper ThQ1
5.94 T.J. Kane, R.L. Byer: Opt. Lett. **10**, 65 (1985)
5.95 T.J. Kane, R.W. Wallace: preprint
5.96 D.A. Draegert: IEEE J. QE-**8**, 235 (1972)
5.97 H.G. Danielmeyer: Appl. Phys. Lett. **16**, 124 (1970)
5.98 H.G. Danielmeyer: Appl. Phys. Lett. **17**, 519 (1970)
5.99 W.R. Trutna, D.K. Donald, M. Nazarathy: CLEO '87 (Baltimore, MD) paper WN2
5.100 B. Zhou, T.J. Kane, G.T. Dixon, R.L. Byer: Opt. Lett. **10**, 62 (1985)
5.101 Y.K. Park, G. Giuliani, R.L. Byer: Opt. Lett. **5**, 96 (1980)
5.102 T. Kedni, D. Treves: Appl. Opt. **20**, 2108 (1981)
5.103 Y.L. Sun, R.L. Byer: Opt. Lett. **7**, 408 (1982)
5.104 F.J. McClung: Appl. Opt. **9**, 103 (1970)
5.105 J.F. Nester: IEEE J. QE-**6**, 97 (1970)
5.106 W. Koechner: IEEE J. QE-**8**, 656 (1972)
5.107 R.B. Chesler: Appl. Opt. **9**, 2190 (1970)
5.108 G.D. Baldwin: IEEE J. QE-**7**, 179 (1971)
5.109 T. Kimura: IEEE QE-**6**, 764 (1970)
5.110 H.G. Danielmeyer: J. Appl. Phys. **41**, 4014 (1970)
5.111 H. Nagai: IEEE J. QE-**8**, 857 (1972)
5.112 H.G. Danielmeyer: IEEE J. QE-**6**, 101 (1970)
5.113 G.R. Hanes, B.P.Stoicheff: Nature **195**, 587 (1962)
5.114 A.P.Veduta, A.M. Leontovich, V.N. Smorchkow: Sov. Phys.-JETP **21**, 59 (1965)
5.115 T.P. Hughes, K.M. Young: Nature **196**, 332 (1962)
5.116 D.J. Bradley, G. Magyar, M.C. Richardson: Nature **212**, 63 (1966)
5.117 D.A. Berkley, G.J. Wolga: J. Appl. Phys. **38**, 3231 (1967)
5.118 A. Flamholz, G.J. Wolga: J. Appl. Phys. **39**, 2723 (1968)
5.119 D.J. Bradley, M.S. Engwell, A.W. McCullough, G. Magyar, M.C. Richardson: Appl. Phys. Lett. **9**, 150 (1966)
5.120 D.J. Bradley, M.S. Engwell, A.W. McCullough: Phil. Trans. R. Soc. (London) Ser. A **263**, 225 (1968)
5.121 D. Pohl: Appl. Phys. Lett. **26** A, 357 (1968)
5.122 E. Ritter: Optical coatings and thin-film techniques, in *Laser Handbook*, ed. by F.T. Arrechi, E.O. Schulz-DuBois (North-Holland, Amsterdam 1972), pp. 897-921
5.123 P. Baumeister: *Handbook of Optical Design* (US Government Printing Office, Washington, DC 1963)
5.124 D.S. Heavens: *Optical Properties of Thin Solid Films* (Butterworth, London 1955)
5.125 H.A. Macleod: *Thin-Film Optical Filters*, 2nd ed. (Macmillan, New York 1986)
5.126 A. Vasicek: *Optics of Thin Films* (North-Holland, Amsterdam 1960)
5.127 A.E. Siegman, Proc. IEEE **53**, 227 (1965)
5.128 A.E. Siegman, R. Arrathon: IEEE J. QE-**3**, 156 (1967)
5.129 A.E. Siegman: Appl. Opt. **13**, 353 (1974)
5.130 W.H. Steier: Unstable resonators, in *Laser Handbook*, ed. by M. Stitch (North-Holland, Amsterdam 1979) Vol.3
5.131 R.L. Herbst, M. Komine, R.L. Byer: Opt. Commun. **21**, 5 (1977)
5.132 Yu.A. Ananev, G.N. Vinokurov, L.V. Kovalchuk, N.A. Sventsitskaya, V.E. Sherstobitov: Sov. Phys.-JETP **31**, 420 (1970)
5.133 R.L. Byer, R.L. Herbst: Laser Focus 48-57 (July 1978)
5.134 W.F. Krupke, W.R. Sooy: IEEE J. QE-**5**, 575 (1969)

5.135 C. Giuliani, Y.K. Park, R.L. Byer: The radial birefringent element and its application to laser resonator design. 11th Int'l. Quant. Electr. Conf. (Boston, MA 1980) postdeadline paper
5.136 G. Giuliani, Y.K. Park, R.L. Byer: Opt. Lett. **5**, 491 (1980)
5.137 J.M. Eggleston, G. Giuliani, R.L. Byer: J. Opt. Soc. Am. **71**, 1264 (1982)
5.138 D.T. Harter, T.C. Walling: Opt. Lett. **2**, 706 (1986)
5.139 Courtesy of McDonnell Douglas Astronautics Comp., St. Louis Div.
5.140 S. De Silvestri, P. Laporta, V. Magni, O. Svelto: Opt. Lett. **12**, 894 (1987)
5.141 D. Andreou: Rev. Sci. Instrum. **49**, 586 (1978)
5.142 T.F. Ewanizky, J.M. Craig, Appl. Opt. **15**, 1465 (1976)
5.143 P.G. Gobbi, S. Morosi, G.C. Reali, A.S. Zarkasi: Appl. Opt. **24**, 26 (1985)
5.144 J.M. Yarborough, J. Hobart: Conf. on Laser Eng. Appl., Washington, DC (1973) postdeadline paper
5.145 A.L. Bloom: J. Opt. Soc. Am. **64**, 447 (1974)
5.146 I.J. Hodgkinson, J.I. Vukusic: Opt. Commun. **24**, 133 (1978)

Chapter 6

6.1 J.H. Goncz, P.B. Newell: J. Opt. Soc. Am. **56**, 87 (1966)
6.2 J.H. Goncs: Instr. Soc. Am. Trans. **5**, 1 (1966)
6.3 L. Noble, C.B. Kretschmer: Optical pumps for lasers. Rept. AD-742, ILC (Sunnyvale, CA 1972)
6.4 J.L. Emmet, A.L. Schawlow: J. Appl. Phys. **35**, 2601 (1964)
6.5 J.R. Oliver, F.S. Barnes: Proc. IEEE **59**, 638 (1971)
6.6 J.H. Goncz, W.J. Mitchell: IEEE J. QE-**3**, 330 (1967)
6.7 J. Oliver, F.S. Barnes: IEEE J. QE-**5**, 225 (1969)
6.8 W.D. Fountain, L.M. Osterink, J.D. Foster: IEEE J. QE-**6**, 232 and 684 (1970)
6.9 M.B. Davies, P. Scharman, J.K. Wright: IEEE J. QE-**4**, 424 (1968)
6.10 C. Noble, J. Moffat, L. Reed, J. Richter: Optical pumps for lasers. Final Rept. TR-ECOM-0035-F, U.S. Army Electr. Comm., Fort Monmouth, NJ (1971)
6.11 K.R. Lang, F.S. Barnes: J. Appl. Phys. **35**, 107 (1964)
6.12 J.H. Goncz: J. Appl. Phys. **36**, 742 (1965)
6.13 J.P. Markiewicz, J.L. Emmett: IEEE J. QE-**2**, 707 (1966)
J.F. Holzrichter, J.L. Emmett: Appl. Opt. **8**, 1459 (1969)
6.14 D.E. Perlman: Rev. Sci. Instr. **37**, 340 (1966)
6.15 J.H. Rosolowski, R.J. Charles: J. Appl. Phys. **36**, 1792 (1965)
6.16 R.A. Dugdale, R.C. McVickers, S.D. Ford: J. Nuclear Mat. **12**, 1 (1964)
6.17 H.E. Edgerton, J.H. Goncz, J. Jameson: Xenon flashlamps, limits of operation. Proc. 6th Int'l Congr. on High Speed Photography, Haarlem, Netherlands (1963) p.143
6.18 L. Waszak: Microwaves **130** (May 1969)
6.19 EG&G Linear Xeonon Flash Tubes, Data Sheet 1002-B
6.20 An introduction to flash tubes. ILC Tech. Bull. No.1 (1984)
6.21 B. Newell, J.D. O'Brian: IEEE J. QE-**4**, 291 (1968)
6.22 B. Newell: Private communication
6.23 N.L. Yeamans, J.E. Creedon: Long-life high repetition-rate flash tubes. Techn. Rept. ECOM-3043, U.S. Army Electr. Comm., Ft. Monmouth, NJ (1968)
6.24 K.B. Ward: Lamp-pumped Nd:YAG laser. Final Rept. NASA-CR-132821, Holobeam, Inc., Paramus, NJ (1973)
6.25 J.F. Holzrichter, N. Dobeck, A. Pemberton: Flashlamp development. Laser Progr. Ann. Rept. UCRL-50021-74, Lawrence Livermore Lab. (1974) pp.107–115
6.26 An verview of flashlamps and CW arc lamps. Techn. Bull. No.3, ICL Technology, Sunnyvale, CA (1986)

6.27 B. Smith: Laser Focus 22, 58 (September 1986)
6.28 J. Richards, D. Rees, K. Fueloep, B.A. See: Appl. Opt. 22, 1325 (1983)
6.29 J.H. Kelly, D.C. Brown, K. Teegarden: Appl. Opt. **19**, 3817 (1980)
6.30 K. Yoshida, Y. Kato, H. Yoshida, C. Yamanaka: Rev. Sci. Instr. **55**, 1415 (1984)
6.31 T.B. Read: Appl. Phys. Lett **9**, 342 (1966)
6.32 W. Koechner: Laser Focus **5**, 29 (September 1969)
6.33 M. Grasis, L. Reed: Long-life krypton arc lamp for pumping Nd:YAG lasers. Techn. Rept. AFAL-TR-73-156, ILC Technologies, Sunnyvale, CA (1973)
6.34 W.E. Thouret: Illum. Eng. **55**, 295 (1960)
6.35 W. Koechner, L. DeBenedictis, E. Matovich, G.E. Mevers: IEEE J. QE-**8**, 310 (1972)
6.36 S. Yoshikawa, K. Iwamoto, K. Washio: Appl. Opt. **10**, 1620 (1971)
6.37 L. Noble, R. Maynard, I. Reed: Useful light output comparisons for small bore krypton lamps and alkali metal vapor lamps. Tech. Rept. R-ILC-72-5, ILC Technologies, Sunnyvale, CA (1971)
6.38 D. Smart, Korad Div., Hadron, Inc., Santa Monica, CA; Private communication
6.39 L.M. Osterink: Efficient high power Nd:YAG laser characteristics. 1969 IEEE Conf. on Laser Eng. Appl., Washington, DC (1969)
6.40 J.E. Anderson, R.C. Eschenbach, H.H. Troue: Appl. Opt. **4**, 1435 (1965)
6.41 M.E. Mack: Appl. Opt. **13**, 46 (1974)
6.42 J.E. Jackson, D.M. Yenni: CW plasma arc pumped laser. Tech. Rept. AFAL-TR-66-363, Air Force Avionics Lab., WPAFB, OH (1966)
6.43 E.F. Vincent: Coaxial gas discharge lamp with a hollow center for pumping lasers. US Patent 3,337,762 (1967)
6.44 Yu A. Kalinin: Sov. J. Opt. Techn. **37**, 1 (1970)
6.45 D.A. Huchital: RF-excited krypton arc lamps for pumping Nd:YAG lasers. Rept. AFAL-TR-74-135, Air Force Avionics Lab., WPAFB, OH (1974); also IEEE J. QE-**12**, 1 (1976)
6.46 W.Elenbaas: *The High Pressure Mercury Vapor Discharge* (Interscience, New York 1961)
6.47 V. Evtuhov, J.K. Neeland: Appl. Phys. Lett. **6**, 75 (1965)
6.48 D. Roess: Electronics **39**, 115 (1966)
6.49 G. Reiling: J. Opt. Soc. Am. **54**, 532 (1964)
6.50 J.F. Waymouth, W.C. Gungle, J.M. Harris, F. Koury: Illum. Eng. **85** (February 1965)
6.51 I. Liberman, D.A. Larson, R.G. Young, C.H. Church, L. Armstrong, R. Liebermann: Optical pumps for lasers. Final Rept. ECOM-02097-F, U.S. Army Electr. Comm., Ft. Monmouth, NJ (1968)
6.52 I. Liberman, D.A. Larson, C.H. Church, IEEE J. QE-**5**, 238 (1969)
6.53 L. Noble: Optical pumps for lasers. Rept. AFAL-TR-72-50 (1972), Rept. AFAL-TR-74-107 (1974), USAF Avionics Lab., WPAFB, Dayton, OH; Alkali vapor lamps for lower power Nd:YAG lasers. Eng. Note 10, ILC Technologies, Sunnyvale, CA (1972)
6.54 J.D. Foster, R.F. Kirk: Space qualified Nd:YAG laser. Rept. NASA-CR-1771, NASA, Washington, DC (1971)
6.55 R.C. Ohlmann, W. Culshaw, K.K. Chow, H.V. Hance, W.B. Leonard, J. Kanneland: High efficiency, single frequency laser and modulator study. Rept. AD-740325, Lockheed Missiles and Space Co., Palo Alto, CA (1971)
6.56 W.E. Forsythe, E.Q. Adams: J. Opt. Soc. Am. **35**, 108 (1945)
6.57 F.J. Studer, R.F. VanBeers: J. Opt. Soc. Am. **54**, 945 (1964)
5.58 C.H. Church, I. Liberman: J. Appl. Opt. **5**, 1 (1967)
6.59 W. Koechner: *Solid-State Laser Engineering*, 1st ed., Springer Ser. Opt. Sci., Vol.1 (Springer, Berlin, Heidelberg 1976)

6.60 N.P. Barnes: J. Appl. Phys. **44**, 230 (1973)
6.61 J.P. Budin, M. Neubauer, M. Rondot: IEEE J. QE-**14**, 831 (1978)
6.62 W. Koechner, H. Verdun: Laser-diode array pumped solid state laser systems. Final Rept., Naval Ocean Systems Command, Contract No N00014-85-C-0174 (1986)
6.63 CLEO'86 (San Francisco, CA) papers WG1-4; and CLEO'87 (Baltimore, MD) papers WN1-5, WS1-4, FL1-6
6.64 D. Botez: IEEE Spectrum **22**, 43 (June 1985)
6.65 M. Ettenberg: Laser Focus **21**, 86 (May 1985)
6.66 D. Botez: Laser Focus **23**, 68 (March 1987)
6.67 Y. Suematsu: Phys. Today **38**, 32 (May 1985)
6.68 P.S. Cross, R.R. Jacobs, D.R. Scifers: Photonics Spectra **18**, 79 (September 1984)
6.69 J. Hecht: Lasers and Applications **3**, 61 (January 1984)
6.70 D.S. Chemla: Phys. Today **38**, 57 (May 1985)
6.71 Y. Arakawa, A. Yariv: IEEE J. QE-22, 1887 (1986)
6.72 D.R. Scifers, R.D. Burnham, W. Streifer: Appl. Phys. Lett. **41**, 118 (1982)
6.73 W. Streifer, D.R. Scifres, G.L. Harnagel, D.F. Welch, J. Berger, M. Sakamoto; Advances in diode laser pumps. Preprint (1988)
6.74 J. Ungar, N. Bar-Chaim, I. Ury: Electron. Lett. 22, 279 (1985)
6.75 McDonnell Douglas Astronautics Comp., Final Report N66001-83-C-0072. Diode Pumped Slab Laser Study, 1986, prepared for U.S. Navy Naval Ocean System Center, San Diego, CA
6.76 P.S. Cross, D.Scifers: The development and production of ultra-high power, two dimensional diode laser arrays. Spectra Diode Laboratories (1986)
6.77 G.L. Harnagel, D. Welch, P. Cross, D. Scifers: Lasers and Applications **5**, 135 (June 1986)
6.78 G. Harnagel, P. Cross, D. Scifers, D. Woland: Electr. Lett. 22, 231 (1986)
6.79 M. Ettenberg: IEEE J. QE-**16**, 186 (1980)
6.80 P. Cross, R. Jacobs, D. Scifers: Lasers and Applications **4**, 89 (April 1985)
6.81 E.A. Teppo: Nd:YAG laser laboratory experiments. Tech. Note 4051-2, Naval Weapons Center, China Lake, CA (1972)
6.82 M. Toda, T.J. Zamerowski, I. Ladony, R.U. Martinelli: Laser materials for the 0.67 to 2.5 μm range. NASA Contractor Rept. 4050, HDL-CR-86-351-1 (March 1987)
6.83 C.G. Young: Appl. Opt. **5**, 993 (1966)
6.84 N.A. Kozlov, A.A. Mak, B.M. Sedov: Sov. J. Opt. Techn. **33**, 549 (1966)
6.85 J. Bordogng, W. Hannan, C. Reno, R. Tarzaiski: Solar-pumped laser. Final Rept. N66-19516, Contract NAS9-3671 (1966)
6.86 J. Falk, L. Huff, J.P. Taynai: Solar-pumped, mode-locked, frequency doubled Nd:YAG laser. Conf. on Laser Eng. and Appl., Washington, DC (1975)
6.87 C.J. Reed, E.P. Chicklis, C. Carney, M.G. Knights: Combustion Pumped Solid-State Lasers (1985) paper ThX1
6.88 E.T. Makhrov: Sov. J. Quantum Electron. **10**, 1552 (1980)
6.89 L.C. Yang, J.R. Cook: Optically pumped Nd:glass lasers using chemical flash-bulbs. Rept. IHTR851, Naval Ordinance Station, Indian Head, MD (1984)
6.90 J.F. Waymouth, J.W. Shafer: Investigation of the feasibility of pumping high energy Hd:glass lasers by chemical flashlamps. Rept. UCRL-15627, Lawrence Livermore Lab. (May 1984)
6.91 C.L. Smith, E. Homentowski, C. Stokes: Appl. Optics **6**, 1130 (1967)
6.92 G.N. Glascoe, J.V. Lebacqz: *Pulse Generators* (McGraw-Hill, New York 1948)
6.93 W.E. Brown: Microwaves 7, 83 (May 1968)
6.94 L.C. Yang: Laser Focus **9**, 37 (July 1973)
6.95 N.P. DePratti: J. Phys. E. Sci. Instr. Great Britain **4**, 1 (1971)
6.96 V.J. Corcoran, R.W. McMillan, S.K. Barnoske: IEEE J. QE-**10**, 618 (1974)

6.97 F. Benjamin: Electro-Optical System Designs, 32 (March 1975); W.L. Gagnon, G. Allen: Power conditioning. Laser Progr. Ann. Rept. UCRL-50021-74, Lawrence Livermore Lab. (1974)
6.98 R.H. Dishington, W.R. Hook, R.P. Hilberg: Appl. Opt. **13**, 2300 (1974)
6.99 B.R. Hayworth: Laser Focus **7**, 33 (August 1971)
6.100 W.R. Hook, R.H. Dishington, R.P. Hilberg: IEEE Trans. ED-**19**, 308 (1972)
6.101 I. Liberman, D.A. Larson, R.G. Young, C.H. Church, L. Armstrong, R. Liebermann: Optical pumps for lasers. Phase II, Final Rept. ECOM-02097-F, U.S. Army Electr. Comm., Ft. Monmouth, NJ (1968)
6.102 O. Jensen: Microwaves **13**, 54 (January 1974)
6.103 J.L. Wilkerson, R.S. Witte: Microwaves **9**, 75 (May 1970)
6.104 R.L. Stephens: Laser Focus **8**, 38 (July 1972)
6.105 C.G. Young: Proc. IEEE **57**, 1267 (1969)
6.106 D. Roess: Appl. Opt. **3**, 259 (1964)
6.107 I.P. Lesnick, C.H. Church: IEEE J. QE-**2**, 16 (1966)
6.108 Yu A. Kalinin: Sov. J. Opt. Techn. **37**, 1 (1970)
6.109 D. Roess: Microwaves **4**, 29 (1965); also IEEE J. QE-**2**, 208 (1966)
6.110 C.H. Church, I. Liberman: Appl. Opt. **6**, 1966 (1967)
6.111 G.J. Fan, C.B. Smoyer, J. Nunez: Appl. Opt. **3**, 1277 (1964)
6.112 D.L. Sipes: Appl. Phys. Lett. **47**, 74 (1985)
6.113 J. Berger; D.F. Welch, D.R. Scifers, W. Streifer, P.S. Cross: Appl. Phys. Lett. **51**, 1212 (1987); also Electron. Lett. **23**, 669 (1987)
6.114 T.Y. Fan, R.L. Byer: CLEO'87 (Baltimore, MD) paper FL3
6.115 W.J. Kozlovsky, T.Y. Fan, R.L. Byer: CLEO'86 (San Francisco, CA) paper WG4
6.116 R.A. Fields, M. Birnbaum, D.L. Fincher: CLEO'87 (Baltimore, MD) paper FL4
6.117 G.J. Kintz, R. Allen, L. Esterowitz: CLEO'87 (Baltimore, MD) paper FL2
6.118 R. Allen, L. Esterowitz, L. Goldberg, J.F. Weller, M. Storm: Electron. Lett. **22**, 947 (1986)
6.119 M. Ettenberg: Laser Focus **22**, 86 (May 1985)
6.120 T.Y. Fan, G. Huber, R.L. Byer, P. Mitzscherlich: CLEO'87 (Baltimore, MD) paper FL1
6.121 J. Hecht: Lasers and Applications **3**, 61 (January 1984)
6.122 F. Hanson, D. Haddock: Appl. Opt. **27**, 80 (1988)
6.123 S.B. Schuldt, R.L. Aagard: Appl. Opt. **2**, 509 (1963)
6.124 V. Evtuhov, J.K. Neeland: Appl. Opt. **6**, 437 (1967)
6.125 J.G. Edwards: Appl. Opt. **6**, 837 (1967)
6.126 K. Kamiryo, T. Lano, K. Matsuzawa: Jap. J. Appl. Phys. **5**, 1217 (1966)
6.127 D. Fekete: Appl. Opt. **5**, 643 (1966)
6.128 C. Bowness: Appl. Opt. **4**, 103 (1965)
6.129 K. Kamiryo: Proc. IEEE **53**, 1750 (1965)
6.130 D.M. Camm: Appl. Opt. **23**, 601 (1984); F. Docchio, L. Pallaro, O. Svelto: Appl. Opt. **24**, 3752 (1985); F. Docchio: Appl. Opt. **24**, 3746 (1985)
6.131 Yu. A. Kalinin, A.A. Mak: Opt. Tech. **37**, 129 (1970)
6.132 E.J. Seppi: New Monte Carlo method and review of present methods for calculating the characteristics of exciation geometries for solid state lasers. Paper P-655, Contract DAGC15-67-C0011, Inst. for Defense Analysis, Sci. and Tech. Div. (1970)
6.133 D.R. Skinner, J. Tregellas-Williams: Australian J. Phys. **19**, 1 (1966)
6.134 J. Whittle, D.R. Skinner: Appl. Opt. **5**, 1179 (1966)
6.135 D.R. Skinner: Appl. Opt. **8**, 1467 (1969)
6.136 I.H. Wenzel: *Lasers and Applications* (Ohio State Univ. Press, Columbus, 1963) p.115

6.137 K. Kamiryo, T. Kano, H. Matsuzawa: Proc. IEEE **55**, 1630 (1967)
6.138 W.R. Sooy, M.L. Stitch: J. Appl. Phys. **34**, 1719 (1963)
6.139 C.H. Cooke, J. McKenna, J.G. Skinner: Appl. Opt. **3**, 957 (1964)
6.140 D. Röss: *Lasers, Light Amplifiers and Oscillators* (Academic, New York 1969) p.426
6.141 V. Daneu, C.C. Sacchi, O. Svelto: Alta Frequenza, 758 (Nov. 1964)
6.142 H.U. Leuenberger, G. Herziger: Appl. Opt. **14**, 1190 (1975)
6.143 W. Koechner: Appl. Opt. **9**, 1429 (1970)
6.144 J. Trenholme: Optimizing the design of a kilojoule laser amplifier chain. Laser Fusion Program, Semiannual Rept., Lawrence Livermore Lab. (1973) p.60;
P. Laporta, V. Magni, O. Svelto: IEEE J. QE-**21**, 1211 (1985)
M.S. Mangir, D.A. Rockwell: IEEE J. QE-22, 574 (1986);
6.145 R.L. Greene, J.L. Emmett, A.L. Shawlow: Appl. Opt. **5**, 350 (1966)
6.146 K. Mori, Jap. J. Appl. Phys. **12**, 325 (1973)
6.147 Yu.H. Hahn, Yu.B. Hahn: Opt. Spectra 11 (Febr. 1975)
H.T. Powell, K.S. Jancaitis, J.B. Trenholme, C.E. Urich, G.R. Wirtenson: CLEO'85 (Baltimore. MA) Paper ThX3
6.148 D.D. Bhawalkar, L. Pandit: IEEE J. QE-**9**, 43 (1973)
6.149 W.W. Morey: IEEE J. QE-**8**, 818 (1972)
6.150 A.N. Fletcher, R.A. Hollins, R.F. Kubin, R.A. Henry, T.M. Atienza-Moore, M.E. Pietrak: Appl. Phys. B **30**, 195 (1983)
A.N. Fletcher: Appl. Phys. B **37**, 31 (1985)
6.151 G.F. Albrecht, J.M. Eggleston, J.J. Ewing: IEEE J. QE-22, 2099 (1986)
6.152 J.M. Eggleston, G.F. Albrecht, R.A. Petr, J.F. Zumdieck: IEEE J. QE-22, 2092 (1986)
6.153 Medium average power solid state lasers. Techn. Inform. Seminar, Lawrence Livermore Lab. (October 1985)
6.154 M.R. Siegrist: Appl. Opt. **15**, 2167 (1976)
6.155 J.L. Emmett, W.F. Krupke, J.B. Trenholme: The future development of high-power solid-state laser systems. Rept. UCRL-533444, Lawrence Livermore Lab. (1982)

Chapter 7

7.1 H.S. Carslaw, J.C Jaeger: *Conduction of Heat in Solids* (Oxford Univ. Press, London 1948) p.191
7.2 S.T. Hsu: *Engineering Heat Transfer* (Van Nostrand, Princeton, NJ 1963) p.274
7.3 W. Koechner: Appl. Opt. **9**,1429 (1970)
7.4 W. Koechner: J. Appl. Phys. **44**, 3162 (1973)
7.5 S. Timoshenko, J.N. Goodier: *Theory of Elasticity* (McGraw-Hill, New York 1951)
7.6 W. Koechner: J. Appl. Phys. **2**, 279 (1973)
7.7 M. Born, E. Wolf: *Principles of Optics* (Pergamon, London 1965)
7.8 J.F. Nye: *Physical Properties of Crystals* (Oxford Univ. Press, London 1964)
7.9 R.W. Dixon: J. Appl. Phys. **38**, 5149 (1967)
7.10 J.D. Foster, L.M. Osterink: J. Appl. Phys. **41**, 3656 (1970)
7.11 W. Koechner, D.K. Rice: IEEE J. QE-**6**, 557 (1970)
7.12 H. Kogelnik: Bell Syst. Tech. J. **44**, 455 (1965)
7.13 D.A. LaMarre: High performance laser research. Rept. AD 840913, American Opt. Corp. (June 1968)
7.14 W. Koechner: Appl. Opt. **9**, 2548 (1970)
7.15 K.B. Steinbruegge, T. Henningsen, R.H. Hopkins, R. Mazelsky, N.T. Melamed, E.P. Riedel, G.W. Roland: Appl. Opt. **11**, 999 (1972)
7.16 K.B. Steinbruegge, G.D. Baldwin: Appl. Phys. Lett. **25**, 220 (1972)

7.17 A. Stein: CW YAG laser techniques. Rept. AD 743979, US Army Electronics Command, Fort Monmouth, NJ (May 1972)
7.18 M.A. Karr: Appl. Opt. **10**, 893 (1971)
7.19 L.M. Osterink, J.D. Foster: Appl. Phys. Lett. **12**, 128 (1968); N. Barnes, S.J. Scalise: Appl. Opt. **17**, 1537 (1978)
7.20 W.C. Scott, M. de Wit: Appl. Phys. Lett. **18**, 3 (1971)
7.21 W. Koechner, D.K. Rice: J. Opt. Soc. Am. **61**, 758 (1971)
7.22 S.D. Sims, A. Stein, C. Roth: Appl. Opt. **5**, 621 (1966)
7.23 R.L. Townsend, C.M. Stickley, A.D. Maio: Appl. Phys. Lett. **7**, 94 (1965)
7.24 H. Welling, C.J. Bickart, H.G. Andresen: IEEE J. QE-**1**, 223 (1965)
7.25 G.D. Baldwin, E.P. Riedel: J. Appl. Phys. **38**, 2726 (1967)
7.26 A.P. Veduta, A.M. Leontovich, V.N. Smorchkov: Sov. Phys. JETP **21**, 59 (1965)
7.27 S. Epstein: J. Appl. Phys. **38**, 2715 (1967)
7.28 A.Y. Cabezas, L.G. Komai, R.P. Treat: Appl. Opt. **5**, 647 (1966)
7.29 J.W. Carson: Dynamic optical properties of laser materials. Final Techn. Rept. No.P66-135, Office of Naval Research, Washington, DC (1966)
7.30 D.C. Burnham: Appl. Opt. **9**, 1727 (1970)
7.31 D. White, D. Gregg: Appl. Opt. **4**, 1034 (1965)
7.32 F.W. Quelle: Appl. Opt. **5**,633 (1966)
7.33 E.P. Riedel, G.D. Baldwin: J. Appl. Phys. **38**, 2720 (1967)
7.34 E. Snitzer, C.G. Young: *Lasers*, ed. by A.K. Levine (Dekker, New York 1968) Vol.2, p.191
7.35 E. Matovich: The axial gradient laser. Proc. DOD Conf. on Laser Technology, San Diego, CA (1970) p.311
7.36 M.K. Chun, J.T. Bischoff: IEEE J. QE-**7**, 200 (1971)
7.37 K. Richter, W. Koechner: Appl. Phys. **3**, 205 (1974)
7.38 V.I. Danilovskaya, V.N. Zubchaninova: Temperature stresses forming in cylinders under the effect of a luminous flux. US Government Res. & Dev. Rept. 70, No.AD-704-020 (1970)
7.39 B.A. Ermakov, A.V. Lukin: Sov. Phys. **15**, 1097 (1971)
7.40 S.D. Sims, A. Stein, C. Roth: Appl. Opt. **6**, 579 (1967)
7.41 R.F. Hotz: Appl. Opt. **12**, 1834 (1973)
7.42 T.J. Gleason, J.S. Kruger, R.M. Curnutt: Appl. Opt. **12**, 2942 (1973)
7.43 G. Benedetti-Michelangeli, S. Martelluci: Appl. Opt. **8**, 1447 (1969)
7.44 W. Rundle: Korad Division, Hadron, Inc., unpublished results
7.45 E.A. Teppo: Nd:YAG laser technology. NWC Techn. Memo 2534, Appendix C (1975); Techn. Note 4051-2 (1972), Naval Weapons Center, China Lake, CA
7.46 J.D. Foster, R.F. Kirk: Rept. NASA-CR-1771, Washington, DC (1971)
7.47 R.A. Kaplan: Conductive cooling of a ruby rod. Technical Note No.109, TRG, Melville, NY (1964)
7.48 W.F. Hagen: Techn. Rept. AFAL-TR-73-111, Air Force Avionics Lab., Wright Patterns AFB, Ohio (1973)
7.49 W.F. Hagen, C.G. Young, J. Keefe, D.W. Cuff: Segmented Nd:glass lasers. Proc. DOD Conf., San Diego, CA (1970) p.363
7.50 E. Matovich, G.E. Mevers: 1KW axial gradient Nd:YAG laser. Final Rept. N00014-70-C-0406, ONR, Boston (1971)
7.51 M.M. Heil, D.L. Flannery: A review of axial gradient laser technology, Proc. DOD Conf., San Diego, Calif. (1970) p.287
7.52 E. Matovich: Segmented ruby oscillator-amplifier. Rept. AFAL-TR-69-317, Air Force Avionics Lab, Wright Patterson AFB, Ohio (1970)
7.53 J.M. Eggleston, T.J. Kane, K. Kuhn, J. Unternahrer, R.L. Byer: IEEE J. QE-**20**, 289 (1984); T.J. Kane, J.M. Eggleston, R.L. Byer: IEEE J. QE-**21**, 1195 (1985)
7.54 W.S. Martin, J.P. Chernoch: US Patent no.3, 633, 126 (January 1972)

7.55 J.P. Chernoch, W.S. Martin, J.C. Almasi: Performance characteristics of a face-pumped, face-cooled laser, the mini-FPL. Tech. Rept. AFAL-TR-71-3, Air Force Avionics Lab., Wright Patterson AFB, Ohio (1971)

7.56 W.B. Jones, L.M. Goldman, J.P. Chernoch, W.S. Martin: IEEE J. QE-**8**, 534 (1972);
G.J. Hulme, W.B. Jones: Total internal reflection face pumped laser, Proc. Soc, Photo-Optical Instr. Eng. **69**, 38 (1975)

7.57 J.M. Eggleston, T.J. Kane, T. Unternahrer, R.L. Byer: Opt. Lett. **9**, 405 (1982)

7.58 J.M. Eggleston, G.F. Albrecht, R.A. Petr, J.F. Zumdieck: IEEE J. QE-**22**, 2092 (1986)

7.59 M. Reed, K. Kuhn, J. Unternahrer, R.L. Byer: IEEE J. QE-**21**, 412 (1985)

7.60 T. Henningsen: Evaluation of CaLaSOAP:Nd for use in a one joule, 30 Hz laser designator. Res. Rep. 72-ICI-LAMAT-RI, Westinghouse Research Lab., Pittsburgh, PA (1972)

7.61 J.P. Chernoch: High power Nd-YAG Mini-FPL, Final report AFAL-TR-75-93, Air Force Avionics Lab., WPAFB, Ohio (1975)

7.62 T.J. Kane, R.C. Eckardt, R.L. Byer: IEEE J. QE-**19**, (1983)

7.63 Y.S. Liu, W.B. Jones, J.P. Chernoch: Recent developmetn of high power visible laser sources employing solid state slab lasers and nonlinear harmonic conversion techniques. Tech. Rep., 81-CRD104, General Electric, Schenactady, NY (1981)

7.64 G.F. Albrecht, J.M. Eggleston, J.J. Ewing: IEEE J. QE-22, 2099 (1986)

7.65 D.L. Begley, D.J. Krebs, M. Ross: Proc. SPIE **742**, 42 (1986)

7.66 Diode pumped slab laser study - risk reduction phase. Final Rept. N 66001-83-C-0072, McDonnell Douglas (1986)

7.67 W.B. Jones: Laser Focus/Electro Optics 107 (1983)

7.68 S. Basu, T.J. Kane, R.L. Byer: IEEE J. QE-22, 2052 (1986)

7.69 D.C. Brown, K.L. Kotik: CLEO'84 (Anaheim, CA) paper WE4
D.C. Brown: CLEO'84 (Anaheim, CA) paper WE5

7.70 T.C. Almasi, W.S. Martin: US Patent .3, 631, 362 (December 1971)

7.71 D.C. Brown, J.M. Kelly, J.A. Abate: IEEE J. QE-**17**, 1755 (1981)

7.72 J. A. Abate, L. Lund, D. Brown, S. Jacobs, S. Refermat, J. Kelly, M. Gavin, J. Waldbillig, O. Lewis: Appl. Opt. **20**, P351 (1981);
D.C. Brown, J.A. Abate, L. Lund, J. Waldbillig: Appl. Opt. **20**, 1588 (1981)

7.73 D.C. Brown: *High Peak Power Nd:glass Laser Systems*, Springer Ser. Opt. Sci., Vol.25 (Springer, Berlin, Heidelberg 1981)

7.74 S.R. Bowman, L.M. Ding, C.O. Alley: CLEO'87 (Baltimore, MA)

7.75 J.H. Kelly, D.L. Smith, J.C. Lee, S.D. Jacobs, D.J. Smith, J.C. Lambropoulos: CLEO Conf. 1987

7.76 D.C. Brown, K.L. Kotik, J. Kuper: Tech. Digest, CLEO'84 (Anaheim, CA) Paper WE3

7.77 D.C. Brown, K. Lee, G. Linford: J. Opt. Soc. Am. **73**, 1970A (1983)

7.78 W.W. Simmons, D.R. Speck, L.J. Hunt: Appl. Opt. **17**, 999 (1978)

7.79 W.E. Martin, J.B. Trenholme, G.T. Linford, S.M. Yarema, C.A. Hurley: IEEE J. QE-**17**, 1744 (1981)

7.80 Lawrence Livermore National Lab., Medium Average Power Solid State Laser Technical Information Seminar, October 29, 1985, Livermore, Calif.

Chapter 8

8.1 R.W. Hellwarth: In *Advances in Quantum Electronics* (Columbia University Press, New York 1961) p.334

8.2 F.J. McClung, R.W. Hellwarth: Proc. IRE **51**, 46 (1963)

8.3 W.G. Wagner, B.A. Lengyel: J. Appl. Phys. **34**, 3040 (1963)

8.4 R.B. Kay, G.S. Waldman: J. Appl. Phys. **36**, 1319 (1965)
8.5 R.H. Dishington: Energy extraction optimization in lasers, Proc. Soc. Photo-Optical Instr. Eng. **69**, 135 (1975)
8.6 J.E. Midwinter: Br. J. Appl. Phys. **16**, 1125 (1965)
8.7 A.R. Newberry: Br. J. Appl. Phys. **1**, 1849 (1968)
8.8 G.D. Baldwin: IEEE J. QE-**7**, 220 (1971)
8.9 R.B. Chesler, M.A. Karr, J.E. Geusic: Proc. IEEE **58**, 1899 (1970)
8.10 R.J. Collins, P. Kisliuk: J. Appl. Phys. **33**, 2009 (1962)
8.11 R.C. Benson, M.R. Mirarchi: IEEE Trans. Milit. Electr. MIL-**8**, 13 (1964)
8.12 W. Buchman, W. Koechner, D. Rice: IEEE J. QE-**6**, 747 (1970)
8.13 I.W. Mackintosh: Appl. Opt. **8**, 1991 (1969)
8.14 E.L. Steele, W.C. Davis, R.L. Treuthart: Appl. Opt. **5**, 5 (1966)
8.15 R. Daly, S.D. Sims: Appl. Opt. **3**, 1063 (1964)
8.16 J.E. Geusic, M.L. Hensel, R.G. Smith: Appl. Phys. Lett. **6**, 175 (1965)
8.17 E.J. Woodbury: IEEE J. QE-**3**, 509 (1967)
8.18 R.G. Smith, M.F. Galvin: IEEE J. QE-**3**, 406 (1967)
8.19 D. Findlay, A.f. Fray: Opto-Electronics **2**, 51 (1970)
8.20 M. Born, E. Wolf: *Principles of Optics*, 2nd ed. (Macmillan, New York 1964)
8.21 I.P. Kaminow, E.H. Turner: Appl. Opt. **54**, 1374 (1966)
8.22 C.L. Hu: J. Appl. Phys. **38**, 3275 (1967)
8.23 B.H. Billings: J. Opt. Soc. Am. **39**, 797 (1949)
8.24 B.H. Billings: J. Opt. Soc. Am. **39**, 802 (1949)
8.25 R.O'B. Carpenter: J. Opt. Soc. Am. **40**, 225 (1950)
8.26 B.H. Billings: J. Opt. Soc. Am. **42**, 12 (1952)
8.27 R. Goldstein: Laser Focus, 21 (February 1968)
8.28 J.T. Milek, S.J. Welles: Linear electro-optic modulator materials, Report AD 704-556, Hughes Aircraft Co., Culver City, Calif. (January 1970)
8.29 F. Zernike, Jr.: J. Opt. Soc. Am. **54**, 1215 (1964)
8.30 J.H. Ott, T.R. Sliker: J. Opt. Soc. Am. **54**, 1442 (1964)
8.31 T.R. Sliker, S.R. Burlage: J. Appl. Phys. **34**, 1837 (1963)
8.32 M. Yamazaki, T. Ogawa: J. Opt. Soc. Am. **56**, 1407 (1966)
8.33 R.A. Phillips: J. Opt. Soc. **56**, 629 (1966)
8.34 D. Milam: Appl. Opt. **12**, 602 (1973)
8.35 W.R. Hook, R.P. Hilberg: Appl. Opt. **10**, 1179 (1971)
8.36 W.W. Buchman: Pockels cell potential distribution and its effects on design, Report KER 00001, Korad Div. Hadron, Inc. (1969)
8.37 L.L. Steinmetz, T.W. Pouliot, B.C. Johnson: Appl. Opt. **12**, 1468 (1973)
8.38 M. Dore: IEEE J. QE-**3**, 555 (1967)
8.39 C.H. Clayson: Electr. Lett. **2**, 138 (1966)
8.40 B. Trevelyan: J. Sci. Instr. **2**, 425 (1969)
8.41 J.M. Lev: Electr. Lett. **2**, 425 (1969)
8.42 M. Okada: IEEE J. QE-**6**, 526 (1970)
8.43 B. Stadnik: ACTA Technica CSAV **1**, 65 (1970)
8.44 M.B. Davies, P.H. Sarkies, J.K. Wright: IEEE J. QE-**4**, 533 (1968)
8.45 A.S. Bebchuk, L.A. Kulevskiy, V.V. Smirnov, Yu. N. Solovyeva: Rad. Eng. and Electr. Phys. **14**, 919 (1969)
8.46 R.P. Hilberg, W.R. Hook: Appl. Opt. **9**, 1939 (1970)
8.47 M.K. Chun, J.T. Bischoff: IEEE J. QE-**8**, 715 (1972)
8.48 W.R. Hook, R.P. Hilberg, R.H. Dishington: Proc. IEEE **59**, 1126 (1971)
8.49 J.F. Ney: *Physical Properties of Crystals* (Oxford Univ. Press, London 1964)
8.50 A.W. Warner, M. Onoe, G.A. Coquin: J. Acoust. Soc. Am. **42**, 1223 (1967)
8.51 E.A. Teppo: Nd:YAG Laser Lab. Experiments, January 1972 to June 1973, Technical Note 4051-7, Naval Weapons Center, China Lake, Calif. (August 1973)
8.52 R.V. Lovberg, E.R. Wooding, M.L. Yeoman: IEEE J. QE-**11**, 17 (1975)

8.53 M.J.P. Payne, H.W. Evans: CLEO'84, paper TUB16 Technical Digest
8.54 W.E. Schmid: IEEE J. QE-**16**, 790 (1980)
8.55 M.K. Chun, E.A. Teppo: Appl. Opt. **15**, 1942 (1976)
8.56 M.B. Rankin, G.D. Ferguson: Proc. Soc. Photo-Opt. Instr. Eng. **160**, 67 (1978)
8.57 J. Richards: Appl. Opt. **22**, 1306 (1983)
8.58 H.W. Mocker: Appl. Opt. **5**, 1829 (1966)
8.59 I.F. Balashov, V.A. Berenberg, B.A. Ermakov: Sov. Phys.-Tech. Phys. **13**, 699 (1968)
8.60 M.A. Duguay, J.W. Hansen: Appl. Phys. Lett. **15**, 192 (1971)
8.61 A.R. Newberry: Br. J. Appl. Phys. **1**, 1849 (1968)
8.62 E.I. Gordon: Proc. IEEE **54**, 1391 (1966)
8.63 R. Adler IEEE Spectrum **4**, 42 (1967)
8.64 C.F. Quate, C.D. Wilkinson, D.K. Winslow: Proc. IEEE **53**, 1604 (1965)
8.65 R.W. Dixon, IEEE J. QE-**3**, 85 (1967)
8.66 N. Uchida, N. Niizeki: Proc. IEEE **61**, 1073 (1973)
8.67 D. Maydan: IEEE J. QE-**6**, 15 (1970)
8.68 T. Nowicki: Electro-Opt. Syst. Design **24** (January 1974)
8.69 M. Cohen: Electro-Opt. Syst. Design **23** (April 1972)
8.70 D.E. Flinchbaugh: Electro-Opt. Syst. Design **24** (January 1974)
8.71 D.A. Pinnow: IEEE J. QE-**6**, 223 (1970)
8.72 R.W. Dixon: J. Appl. Phys. **38**, 5149 (1967)
8.73 Isomet Data Sheet, Acousto-Optic Q-switch, Model 453 (May 1973)
8.74 M.G. Cohen: Optical Spectra **32** (November 1973)
8.75 M.C. Cohen, R.T. Daly, R.A. Kaplan: IEEE J. QE-**7**, 58 (1971)
8.76 R.B. Chesler, D.A. Pinnow, W.W. Benson: Appl. Opt. **10**, 2562 (1971)
8.77 P.P. Solokin, J.J. Luzzi, J.R. Lankard, G.D. Pettit: IBM J. Res. Div. **8**, 182 (1964)
7.78 P. Kafalas, J.I. Masters, E.M.E. Murrary: J. Appl. Phys. **35**, 2349 (1964)
8.79 B.H. Soffer. J. Appl.Phys. **35**, 2551 (1964)
8.80 R. Wuerker, L.O. Heflinger: J. Soc. Photo-Optical Instr. Eng. **9**, 122 (1971)
8.81 J.A. Armstrong: Saturable optical absorption in phthalocyanine dyes, Research Paper RC-1253, IBM Watson Res. Center (August 1964); W.R. Sooy: Appl. Phys. Lett. **7**, 36 (1965)
8.82 B.B. McFarland, R.H. Hoskins, B.H. Soffer: Nature **207**, 1180 (1965)
8.83 Mitsubishi Electr. Corp., Plastic Ruby Laser Q-switch, Models 100 to 170 RA, Data Sheet
8.84 G.B. Michelangeli, M. De Maria, G. Giuliani, E. Palange: Selenia S.p.A. Via del Castelli Romani 2, 00040 Rome, Italy, Preprint
8.85 T.T. Basiev, I.Ya. Itskhoki, B.G. Lysol, S.B. Mirov, O.B. Cherednichenko: Sov. J. QE-**13**, 370 (1983)
8.86 Z. Guifen, S. Meidong, C. Xijie: Chin. Phys. **3**, 460 (1983)
8.87 D.S. Sumida, S.C. Rand, D.A. Rockwell: Technical Digest, CLEO Conference, 1986, paper WM5, San Francisco, Calif.
8.88 A.A. Vuylsteke: J. App. Phys. **34**, 1615 (1963)
8.89 W.R. Hook, R.H. Dishington, R.P. Hilberg: Appl. Phys. Lett. **9**, 125 (1966)
8.90 W.J. Rundle: J. Appl. Phys. **39**, 5338 (1968)
8.91 W.J. Rundle: IEEE J. QE-**5**, 342 (1969)
8.92 W.R. Hook: Proc. IEEE **54**, 1954 (1966)
8.93 W. Rundle, W.K. Pendleton: IEEE J. QE-**12**, (1976)
8.94 A.E. Siegman: IEEE J. QE-**9**, 247 (1973)
8.95 D. Milam, R.A. Bradbury, A. Hordvik, H. Schlossberg, A. Szoke: IEEE J. QE-**10**, 20 (1974)
8.96 D. Maydan, R.B. Chesler: J. Appl. Phys. **42**, 1031 (1971)
8.97 R.B. Chesler, D. Maydan: J. Appl. Phys. **42**, 1028 (1971)
8.98 D. Cheng: IEEE J. QE-**9**, 585 (1973)

8.99 D. Maydan: J. Appl. Phys. **41**, 1522 (1970)
8.100 H.A. Kruegle, L. Klein: Appl. Opt. **15**, 466 (1976)
8.101 R.H. Johnson: IEEE J. QE-**9**, 255 (1973)
8.102 C.W. Reno: Appl. Opt. **12**, 883 (1973)
8.103 J.R. Bettis, A.H. Guenther: IEEE J. QE-**6**, 483 (1970)
8.104 A.H. Guenther, J.R.Bettis, R.E.Anderson, R.V.Wick: IEEE J.QE-**6**, 492 (1970)
8.105 D. Milam, C.C. Callagher, R.A. Bradbury, E.S. Bliss: Rev. Sci. Instr. **43**, 1482 (1972)
8.106 D. von der Linde, O. Bernecker, A. Laubereau: Opt. Commun. 2, 215 (1970)
8.107 J.P. Letellier: Parallel plate transmission line Pockels cell. NRL Report 8463, Naval Res. Lab., Washington, DC (1972)
8.108 A.J. Alcock, M.C. Richardson, K. Leopold: Rev. Sci. Instr. **41**, 1028 (1970)
8.109 P.D. Morgan, N.J. Peacock: J. Phys. E. Sci. Instr. **4**, 677 (1971)
8.110 F. Deutsch: Br. J. Appl. Phys. **1**, 1711 (1968)
8.111 V.J. Corcoran, R.W. McMillan, P.M. Rushworth: Appl. Opt. **14**, 643 (1975)
8.112 R.C. Hyer, H.D. Sutphin, K.R. Winn: Rev. Sci. Instr. **46**, 1333 (1975)
8.113 G.A. Kimber, P.J. Bateman: A Faraday isolator system for use at wavelengths between 0.4 and 0.9 μm. Tech. Report 66153, Royal Aircraft Estab. (1966)
8.114 K. Eidmann, P. Sachsenmaier, H. Salzmann, R. Sigel: J. Phys. E. Sci. Instr. **5**, 56 (1972)
8.115 Hadron/Cilas Data Sheet, Model RF 70, Faraday Effect Optical Isolator
8.116 C.F. Padula, C.G. Young: IEEE J. QE-**3**, 493 (1967)
8.117 G. Leppelmeier, W. Simmons: Faraday Isolators, Semiannual Report UCRL-50021-73-1, p.78, and UCRL-50021-73-2, p.50 (Lawrence Livermore Lab., Livermore, CA 1973)

Chapter 9

9.1 R. Bracewell: *The Fourier Transform and Its Applications* (McGraw-Hill, New York 1965)
9.2 A.J. DeMaria: Proc. IEEE **57**, 2 (1969)
9.3 P.S. Smith: Proc. IEEE **58**, 1342 (1970)
9.4 A. Laubereau, W. Kaiser: Opto-Electr. **6**, 1 (1974)
9.5 D. von der Linde: Appl. Phys. **2**, 281 (1973)
9.6 S.L. Shapiro (ed.): *Ultrashort Light Pulses*, Topics Appl. Phys., Vol.18 (Springer, Berlin, Heidelberg 1977)
9.7 W. Kaiser (ed.): *Ultrashort Laser Pulses*, Topics Appl. Phys., Vol.60 (Springer, Berlin. Heidelberg 1988)
9.8 H.W. Mocker, R.J. Collins: Appl. Phys. Lett. 7, 270 (1965)
9.9 A.J. DeMaria, D.A. Stetser, H. Heyman: Appl. Phys. Lett. **8**, 174 (1966)
9.10 J.A. Leck: Phys. Rev. B **1**, 84 (1970)
9.11 P.G. Kriukov, V.S. Letokhov: IEEE J. QE-**8**, 766 (1972)
9.12 N.G. Basov, Yu.A. Drozhbin, P.G. Kriukov, V.B. Lebedev, V.S. Letokhov, Yu.A. Matveetz: JETP Lett. **9**, 256 (1969)
9.13 S.V. Chakalin, P.G. Kriukov, Yu.A. Matveetz, O.B. Shatberashvili: Opto-Electr. **6**, 249 (1974)
9.14 V.S. Letokhov: Sov.Phys.JETP **28**, 562 (1969); Sov.Phys.JETP **28**, 1026 (1969)
9.15 T.I. Kuznetsova: Sov. Phys. JETP **30**, 904 (1970)
9.16 E.B. Treacy: Phys. Lett. **28** A, 34 (1968)
9.17 R.R. Cubeddu, O. Svelto: IEEE J. QE-**5**, 495 (1969)
9.18 D.J. Bradley, G.H.C. New, S.J. Caughey: Phys. Lett. **30** A, 78 (1969)
9.19 S.L. Shapiro, M.A. Duguay: Phys. Lett. **28** A, 698 (1969)
9.20 M.A. Duguay, J.W. Hansen, S.L. Shapiro: IEEE J. QE-**6**, 725 (1970)
9.21 R.C. Eckardt: Opto-Electr. **6**, 67 (1974)

9.22 R.C. Eckardt: IEEE J. QE-**10**, 48 (1974)
9.23 V.V. Korobkin, A.A. Malyutin, A.M. Prokhorov: JETP Lett. **12**, 150 (1970)
9.24 R.C. Eckardt, C.H. Lee, J.N. Bradford: Appl. Phys. Lett. **19**, 420 (1971)
9.25 D. von der Linde: IEEE J. QE-**8**, 328 (1972)
9.26 D. von der Linde, O. Bernecker, W. Kaiser: Opt. Commun. **2**, 149 (1970)
9.27 D.J. Bradley, G.H.C. New: Proc. IEEE **62**, 313 (1974)
9.28 H.E. Rowe, T. Li: IEEE J. QE-**6**, 49 (1970)
9.29 M.A. Duguay, J.W. Hansen, S.L. Shapiro: IEEE J. QE-**6**, 725 (1970)
9.30 S.D. Fanchenko, B.A. Frolov: JETP Lett. **16**, 101 (1972)
9.31 M.A. Duguay, J.W. Hansen: Appl. Phys. Lett. **15**, 192 (1969)
M.A. Duguay, A.T. Mattick: Appl. Opt. **10**, 2162 (1971)
9.32 M. Maier, W. Kaiser, J.A. Giordmaine: Phys. Rev. Lett. **17**, 1275 (1966)
9.33 R.C. Eckardt: Appl. Phys. Lett. **15**, 425 (1969)
9.34 J.A. Giordmaine, P.M. Rentzepis, S.L. Shapiro, K.W. Wecht: Appl. Phys. Lett. **11**, 216 (1967)
9.35 R. Harrach, G. Kachen: J. Appl. Phys. **39**, 2482 (1968)
9.36 H. Weichel: J. Appl. Phys. **44**, 3635 (1973)
9.37 E.S. Bliss, D. Milam: Laser damage study with subnanosecond pulses. Report AFCRL-72-0233, AF Cambridge Research Lab. (1972)
9.38 P.N. Everett: Rev. Sci. Instr. **41**, 1495 (1970)
9.39 A.J. DeMaria: Science **156**, 1557 (1967)
9.40 D.J. Bradley, G.H.C. New, S.J. Caughey: Opt. Commun. **2**, 41 (1970)
9.41 D.J. Bradley, T. Morrow, M.S. Petty: Opt. Commun. 2, 1 (1970)
9.42 M.E. Mack: IEEE J. QE-**4**, 1015 (1968)
9.43 G. Girard, M. Michon: IEEE J. QE-**9**, 979 (1973)
9.44 D. von der Linde, K.F. Rodgers: IEEE J. QE-**9**, 960 (1973)
9.45 K.H. Drexhage, G.A. Reynolds: IEEE J. QE-**10**, 720 (1974)
9.46 H. Jelinkova, K. Hamal, N. Novotny, M. Vrboba: IEEE J. QE-**10**, 724 (1974)
9.47 R. Cubeddu, R. Polloni, C.A. Sacchi, O. Svelto: IEEE J. QE-**5**, 470 (1969)
9.48 E.G. Arthurs, D.J. Bradley, I.S. Ruddock, T.J. Glynn: IEEE J. QE-**10**, 720 (1974)
9.49 A.R. Clobes, M.J. Brienza: IEEE J. QE-**6**, 651 (1970)
9.50 B. Fan, B. Leskovar, C.C. Lo, G.A. Morton, T.K. Gustafson: IEEE J. QE-**10**, 654 (1974)
9.51 K.H. Drexhage: In *Dye Lasers*, ed. by F.P. Schäfer (Springer, Berlin, Heidelberg 1973)
9.52 M.W. McGeoch: Opt. Commun. **7**, 116 (1973)
9.53 R.L. Carman, M.E. Mack: Phys. Rev. A **5**, 341 (1972)
9.54 R.L. Carman, J. Fleck, L. James: IEEE J. QE-**8**, 586 (1972)
9.55 I.K. Krasyuk, P.P. Pashkin, A.M. Prokhorov: JETP Lett. **7**, 89 (1968)
9.56 A.R. Clobes, M.J. Brienza: Appl. Phys. Lett. **14**, 288 (1969)
9.57 J.P. Letellier: Passively mode-locked Nd:YAG oscillator stability study. NRL Mem. Report 2684, NRL. Washington, DC (1973)
9.58 Laser Fusion Program. Semi-Annual Report Jan.-June (1972), UCRL-50021-72-1 and July-Dec. (1972), UCRL-50021-72-2, University of California, Lawrence Livermore Lab., Livermore, Calif.
9.59 Data Sheet: Model 1501C, Korad Div. Hadron, Inc. Santa Monica, Calif.
9.60 R.C. Eckardt, J.L. DeRosa, J.P. Lettellier: IEEE J. QE-**10**, 620 (1974)
9.61 M.W. McGeoch: Opt. Commun. **7**, 116 (1973)
9.62 D. von der Linde, K.F. Rodgers: Opt. Commun. **8**, 91 (1973)
9.63 D.J. Bradley, W. Sibbett: Opt. Commun. **9**, 17 (1973)
9.64 N.G. Basov, P.G. Kryukov, V.S. Letokhov: Direct picosecond structure exploration of mode-locked Nd-glass laser radiation. Lebedev Phys. Inst., Moscow, USSR, Preprint No.82 (1972)
9.65 P.C. Magnante: J. App. Phys. **40**, 4437 (1969)

9.66 A. Penzkofer: Opto-Electr. **6**, 87 (1974)
9.67 J.E. Murray, W.H. Lowdermilk: J. App. Phys. **51**, 3548 (1980)
9.68 A.N. Zherikhin: Sov. Phys. JETP **39**, 52 (1974)
9.69 J.E. Murray: IEEE J. QE-**17**, 1713 (1981)
9.70 W. Rundle: Private communication, Hadron, Inc., Korad Div., Santa Monica, Calif.
9.71 J.M. McMahon, J.L. Emmett: Development of high power Nd:glass laser systems. Proc. 11th Symp. Electr. Ion and Laser Beam Tech., San Francisco, Calif. (1971) pp.269-278
9.72 R.J. Harrach, T.D. MacVicar, G.I. Kachen, L.L. Steinmetz: Laser beam filtering by saturable dyes. Report UCRL-51008, Lawrence Livermore Lab., University of California, Livermore, Calif. (1971)
9.73 A.E. Siegman, D.J. Kuizenga. Appl. Phys. Lett. **14**, 181 (1969)
9.74 D.J. Kuizenga, A.E. Siegman: IEEE J. QE-**6**, 694 and 709 (1970)
9.75 A.E. Siegman, D.J. Kuizenga: Opto-Electr. **6**, 43 (1974)
9.76 D.J. Kuizenga, D.W. Phillion, T. Lund, A.E. Siegman: Opt. Commun. **9**, 221 (1973)
9.77 D.J. Kuizenga, D.W. Phillion, A.E. Siegman: IEEE J. QE-**10**, 721 (1974)
9.78 M. DiDomenico, J.E. Geusic, H.M. Marcos, R.G. Smith: Appl. Phys. Lett. **8**, 180 (1966)
9.79 C.J. Kennedy: IEEE J. QE-**10**, 528 (1974)
9.80 L.M. Osterink, J.D. Foster: J. Appl. Phys. **39**, 4163 (1968)
9.81 H.H. Boyden: Microwaves 58 (1971)
9.82 D.J. Kuizenga, A.E. Siegman: IEEE J. QE-**6**, 694 (1970)
9.83 D.J. Kuizenga: Opt. Commun. 22, 156 (1977)
9.84 D.J. Kuizenga: IEEE J. QE-**17**, 1694 (1981)
9.85 L.M. Osterink, J.D. Foster: Laser spectral control techniques, Final Tech. Report AD-941118, Air Force Syst. Comm., Griffis Air Force Base, Rome, NY (1968)
9.86 P.W. Smith: Proc. IEEE **58**, 1342 (1970)
9.87 G.R. Huggett: Appl. Phys. Lett. **13**, 186 (1968)
9.88 A.E. Siegman: *Lasers* (University Science Books, Mill Valley, CA 1986)
9.89 T.S. Kinsel: IEEE J. QE-**9**, 3 (1973)
9.90 K. Gürs: IEEE J. QE-**3**, 175 (1967)
9.91 T. Deutsch: Appl. Phys. Lett. **7**, 80 (1965)
9.92 A.J. DeMaria, C.M. Ferrar, G.E. Danielson: Appl. Phys. Lett. **8**, 22 (1966)
9.93 M. Michon, J. Ernest, R. Auffret: Phys. Lett. **23**, 457 (1966)
9.94 K. Washio, T. Yamane, S. Yoshikawa, K. Koizumi, Y. Jkeda: IEEE J. QE-**9**, 807 (1973)
9.95 P. Bado, J.S. Coe: 1987 Conf. on Lasers and Electro-optics (Baltimore, MA) Tech. Digest, paper FR1
9.96 P.H. Chiu, J. Weston, P. Pax: Tech. Digest, Conf. on Lasers and Electro-optics, Baltimore, Maryland, May 1987, paper THQ5
9.97 M.F. Becker: IEEE J. QE-**8**, 687 (1972)
9.98 K. Otsuka, T. Kimura: IEEE J. QE-**8**, 23 (1972)
9.99 C.W. Reno: Appl. Opt. **12**, 883 (1973)
9.100 C.B. Hitz, L.M. Osterink: Appl. Phys. Lett. **18**, 378 (1971)
9.101 T.R. Gurski: Appl. Phys. Lett. **15**, 5 (1969)
9.102 M.Z. Lewis, J.T. Knudtson: Appl. Opt. **21**, 2897 (1982)
9.103 G.F. Albrecht, L. Lund, D. Smith: Appl. Opt. **22**, 1276 (1983)
9.104 H. Vanherzeele, J.L. VanEck, A.E. Siegman: Appl. Opt. **20**, 3484 (1981)
9.105 W. Seka, J. Bunkenburg: J. Appl. Phys. **49**, 2277 (1978)
9.106 G.F. Albrecht, J. Bunkenburg: Opt. Commun. **38**, 377 (1981)
9.107 J.E. Murray, D.J. Kuizenga: Appl. Phys. Lett. **37**, 27 (1980)
9.108 D.J. Kuizenga, D.W. Phillion, T. Lund, A.E. Siegman: Opt. Commun. **21**, 327 (1977)

Chapter 10

10.1 P.A. Franken, A.E. Hill, C.W. Peters, G. Weinreich: Phys. Rev. Lett. 7, 118 (1961)
10.2 C.R. Giuliano: Phys. Today **34**, 27 (1981); Laser Focus **19**, 55 (1983)
10.3 N. Bloembergen: *Nonlinear Optics* (Benjamin, New York 1965)
10.4 P.D. Maker: Phys. Rev. Lett. **8**, 21 (1962)
10.5 J.A. Giordmaine: Phys. Rev. Lett. **8**, 19 (1962)
10.6 M. Born, E. Wolf: *Principles of Optics* (Pergamon, London 1959)
10.7 J.E. Midwinter, J. Warner: Brit. J. Appl. Phys. **16**, 1135 (1965)
10.8 H.P. Weber, E. Mathieu, K.P. Meyer: J. Appl. Phys. **37**, 3584 (1966)
10.9 G.D. Boyd, A. Ashkin, J.M. Dziedzic, D.A. Kleinman: Phys. Rev. **137**, 1305 (1965)
10.10 M.V. Hobden: J. Appl. Phys. **38**, 4365 (1967)
10.11 F. Zernike, J.E. Midwinter: *Applied Nonlinear Optics* (Wiley, New York 1973)
10.12 D.A. Kleinman, A. Ashkin, G.D. Boyd: IEEE J. QE-2, 425 (1966)
10.13 G.D. Boyd, D.A. Kleinman: J. Appl. Phys. **39**, 3597 (1968)
10.14 D. Hon: In *Laser Handbook*, ed. by M. Stitch, (North-Holland, New York 1979) Vol.3, pp.421-456
10.15 G.C. Baldwin: *An introduction to Nonlinear Optics* (Plenum, New York 1969)
10.16 D.A. Kleinman: In *Laser Handbook*, ed. by F.T. Arecchi, E.O. Schulz-DuBois (North-Holland, Amsterdam 1972) Vol.2, pp.1229-1258
10.17 R.W. Minck, R.W. Terhunme, C.C. Wang: Proc. IEEE **54**, 1357 (1966)
10.18 N. Bloembergen: Proc. IEEE **15**, 124 (1963)
10.19 F.A. Jenkins, H.E. White: *Fundamentals of Optics* (McGraw-Hill, New York 1957)
10.20 N.F. Nye: *Physical Properties of Crystals* (Clarendon, Oxford 1960)
10.21 W.F. Hagen, P.C. Magnante: J. Appl. Phys. **40**, 219 (1969)
10.22 K. Kato: IEEE J. QE-**10**, 622 (1974)
10.23 K. Kato: IEEE J. QE-**10**, 616 (1974)
10.24 J.H. Boyden, E.G. Erickson: Second harmonic generation. Semi-annual Report AD729.682 Naval Research Dept. Washington DC (1971)
10.25 R.C. Miller: Phys. Lett. **26A**, 177 (1968)
10.26 A. Ashkin, G.D. Boyd, J.M. Dziedzic: Phys. Rev. Lett. **11**, 14 (1963)
10.27 G.E. Francois: Phys. Rev. **143**, 597 (1966)
10.28 M.W. Dowley: Appl. Phys. Lett. **13**, 395 (1968)
10.29 D. Eimerl: IEEE J. QE-**23**, 1361 (1987)
10.30 F.R. Nash, G.D. Boyd, M. Sargent, P.M. Bridenbaugh: J. Appl. Phys. **41**, 2564 (1970)
10.31 M. Okada, S. Ieiri: IEEE J. QE-7, 468 (1971)
J.P. van der Ziel, N. Bloembergen: Phys. Rev. **135**, A1662 (1964)
10.32 M. Okada, S. Ieiri: J. QE-7, 560 (1971)
R.A. Phillips: J. Opt. Soc. Am. **56**, 629 (1966)
10.33 W.K. Ng, E.J. Woodbury: Appl. Phys. Lett **18**, 550 (1971)
R.S. Adhav, R.W. Wallace: IEEE J. QE-**9**, 855 (1973)
10.34 R.G. Smith: J. Appl.Phys. **41**, 3014 (December 1970)
R.S. Adhav: Electro-Opt. Syst. Design **20** (1974)
10.35 D. Eimerl: The potential for efficient frequency conversion at high average power using solid state nonlinear optical materials. Lawrence Livermore National Laboratory Report UCID.20565, (October 1985)
D. Eimerl: Laser Program Annual Report, Lawrence Livermore National Laboratory, Ca., Report UCRL-50021-83, (1983) pp.6-69
10.36 V.S. Suvorov, A.S. Sonin, I.S. Rez: Sov. Phys. JETP **26**, 33 (1968)
10.38 V.S. Suvorov, I.S. Rez: Opt. Spectrosc. (USSR) 27, 94 (1969)

Quantum Technology Ltd., Agincourt, Canada, Second Harmonic Generation Data Sheet No. 707 (1974)
10.37 K. Kato: Appl. Phys. Lett. **25**, 342 (1974)
10.39 Y.D. Golyaev, V.G. Dmitriere, I.A. Itskhoki, V.N. Krasnyanskaya, I.S. Rez, E.A. Shalaev: Sov. J. Quant. Electr. **3**, 72 (1973)
G. Nath, S. Haussuhl: Appl. Phys. Lett. **14**, 154 (1969)
10.40 J.E. Bjorkholm: IEEE J. QE-**4**, 970 (1968)
10.41 J.E. Geusic, H.J. Levinstein, J.J. Rubin, S. Singh, L.G. van Uitert: Appl. Phys. Lett. **11**, 269 (1967); Appl. Phys. Lett. **12**, 224 (1968); Appl. Phys. Lett. **12**, 306 (1968)
10.42 Airtron, Division of Litton, Data Sheet KTP, 1984, Morris Plains, New Jersey
10.43 J.C. Jacco, G.M. Loiacono: Final Report under contract DAAK20-83-C-0139, 1986, Night Vision and Electro-Optics Lab. Fort Belvoir, VA 22060
10.44 T.E. Gier, F.C. Zumsteg: KTP Crystals for second harmonic generation, Final Report 1978, Air Force Avionics Lab., Report No: AFAL-Tr-78-208, Wright Patterson Air Force Base, Ohio 45433
10.45 Y.S. Liu, D. Dentz, R. Belt: Opt. Lett. **9**, 76 (1984)
10.46 J.T. Lin, C. Chen: Lasers and Optronics **6**, 59 (November 1987)
10.47 J.T. Lin: Analyses of frequency conversion and application of nonlinear crystals, Proc. Int. Conf. Lasers (STS, Arlington, VA 1986) p.262
10.48 M.J. Rosker, K. Cheng, C.L. Tang: IEEE J. QE-**21**, 1600 (1985)
10.49 Y.X. Fan, R.C. Eckardt, R.L. Byer, R.K. Route, R.S. Feigelson: Appl. Phys. Lett. **45**, 313 (1984)
10.50 N.P. Barnes, D.J. Gettemy, J.R. Hietanen, R.A. Iannini: *Conf. Lasers and Electro-Optics (CLEO)*, Baltimore, Maryland 1987, paper ThA3
10.51 J.T. Milek, S.J. Welles: Linear electro-optic modulator materials, Report AD704556, Hughes Aircraft Corp., Culver City, Calif. (1970)
10.52 F. Zernike: J. Opt. Soc. Am. **54**, 1215 (1964)
10.53 R.S. Adhav, A.D. Vlassopoulos: Laser Focus, **47**, (May 1974)
10.54 K. Kato: Opt. Commun. **9**, 249 (1973)
10.55 K. Kato, S. Nakao: Jap. J. Appl. Phys. **13**, 1681 (1974)
10.56 R.S. Craxton, S.D. Jacobs, J.E. Rizzo, R. Boni: IEEE J. QE-**17**, 1782 (1981)
10.57 Lasermetrics, Electro-Optics Div., Englewood, New Jersey, Data Sheet 8701, Optical Harmonic Generating Crystals (February 1987)
10.58 Y.S. Luh, M.M. Fejers, R.S. Feigelson, R.L. Byer: CLEO'87 (Baltimore, Md) paperTuH2
10.59 K. Nassau, H.J. Levinstein, G.M. Loiacono: J. Phys. Chem. Solids **27**, 983 (1966)
10.60 A. Ashkin, G.D. Boyd, J.M. Dziedzic, R.G. Smith, A.A. Ballman, H.J. Levinstein, K. Nassau: Appl. Phys. Lett. **9**, 72 (1966)
10.61 G.D. Boyd, R.C. Miller, K. Nassau, W.L. Bond, A. Savage: Appl. Phys. Lett. **5**, 234 (1964)
10.62 R.C. Miller, A. Savage: Appl. Phys. Lett. **9**, 169 (1966)
J.W. Yao, T.S. Fahlen: J. Appl. Phys. **55**, 65 (1984)
T.Y. Fan, C.E. Huang, B.Q. Hu, R.C. Eckardt, Y.X. Fan, R.L. Byer, R.S. Feigelson: Appl. Opt. **26**, 2390 (1987)
K. Kato: IEEE J. QE-**24**, (1988)
10.63 M. Bass: IEEE J. QE-**7**, 350 (1971)
10.64 R.B. Andreev, V.D. Volosov, A.G. Lakintsev: Sov. J. Quant. Electr. **2**, 529 (1973)
10.65 A.M. Glass, D. von der Linde, T.J. Negran: Appl. Phys. Lett. **25**, 233 (1974)
10.65 B.T. Levine, C.G. Bethea: Appl. Phys. Lett **20**, 272 (1972)
10.66 J.L. Jackel, C.E. Rice: Appl. Phys. Lett. **41**, 508 (1982)
10.66 J.E. Pearson, G.A. Evans, A. Yariv: Opt. Commun. **4**, 366 (1972)

10.67 J.L. Jackel, C.E. Rice, J.J. Veselka: Appl. Phys. Lett. **41**, 607 (1982)
10.68 S. Singh, D.A. Draegert, J.E. Geusic: Phys. Rev. B **2**, 2709 (1970)
10.69 R.R. Rice: J. Electrochem. Soc. **16**, 839 (1969)
10.70 G. Nath, S. Haussuhl: Phys. Lett. A**29**, 91 (1969)
10.71 G. Nath, H. Mahmanesch, M. Gsander: Appl. Phys. Lett. **17**, 286 (1970)
10.72 U. Deserno, G. Nath: Phys. Lett. A **30**, 483 (1969)
C. Chen, B. Wu, G. You, A. Jiang, Y. Huang: 13th Int'l. Quantum Electr. Conf. 1984, paper MCC5
B.G. Huth, Y.C. Kiang: J. Appl. Phys. **40**, 4976 (1969)
10.73 H. Ito, H. Naito, H. Inaba: IEEE J. QE-**10**, 247 (1974)
10.74 S. Singh, W.A. Bonner, J.R. Potopowicz, C.G. van Uitert: Appl. Phys. Lett. **17**, 292 (1970)
10.75 B.K. Vainshtein: *Modern Crystallography I*, Springer Ser. Solid-State Sci., Vol.15 (Springer, Berlin, Heidelberg 1981)
10.76 J.C. Jacco, G.M. Loiacono, M. Jaso, G. Mizell, B. Greenberg: J. Cryst. Growth **70**, 484 (1984)
10.78 C. Chen, Y.X. Fan, R.C. Eckardt, R.L. Byer: CLEO'86 (San Francisco, CA) paper ThQ4
10.79 K.C. Liu, M. Rhoades: CLEO'87 (Baltimore, MD)
10.80 K. Kato: IEEE J. QE-22, 1013 (1986)
10.81 K. Jain, G.H. Hewig, Y.Y. Cheng, J.I. Crowley: IEEE J. QE-**17**, 1593 (1981)
10.82 J.P. Hermann, J. Ducuing: J. Appl. Phys. **45**, 5100 (1974)
10.83 J.L. Oudar, R. Hierle: J. Appl. Phys. **48**, 2699 (1977)
10.84 J.M. Halbout, S. Blit, W. Donaldson, C.L. Tang: IEEE J. QE-**15**, 1176 (1979)
10.85 M.J. Rosker, C.L. Tang: J. Opt. Soc. Am. B **2**, 691 (1985)
10.86 C.C. Tang, A.F. Garito: Phys. Rev. B **28**, 6766 (1983)
10.87 B.F. Levine: J. Appl. Phys. **50**, 2523 (1979)
10.88 R.J. Twieg, K. Jain: in *Nonlinear Optical Properties of Organic and Polymeric Materials*, ed. by D.J. Williams (ACS Symposium Series, Washington, DC 1983) Chap. 3
10.89 J.C. Baumert, R.T. Twieg, G.C. Bjorklund, J.A. Logan, C.W. Dirk: CLEO'87 (Baltimore, MD) paper ThA5.
10.90 P. Gunter, C. Bosshard, K. Sutter, C. Medrano, H. Arend, G. Chapuis, R.T. Twieg, D. Dobrowolski: CLEO'87 (Baltimore, MD) paper TuH1
10.91 M.W. Dowley, E.B. Hodges: IEEE J. QE-**4**, 552 (1968)
10.92 J.E. Murray, R.J. Pressley, J.H. Boyden, R.B. Webb: IEEE J. QE-**10**, 263 (1974)
10.92 G.A. Massey: Nd:$YAlO_3$ laser device development, Final Report AFAL-TR-72-285, Air Force Avionics Lab., Wright-Patterson AFB, Dayton, Ohio (1972)
10.93 M. Bass, H.H. Barrett: IEEE J. QE-**8** 338 (1972)
10.94 M. Bass: Development of an optimized second harmonic generator cell, Final Report NASA-CR-86335 (1970)
10.95 J.T. Lin: *Conf. on Lasers and Electro-Optics, 1987*, Baltimore, Maryland, paper TuH4, Technical Digest
10.96 J.T. Lin, C. Chen: Lasers and Optronics **6**, 59 (November 1987)
10.98 International Laser Systems Inc., Orlando, Fla., Frequency doubled YAG laser, Data Sheet Model NT-612 (1974)
10.99 N.G. Basov, A.R. Zaritskii, S.D. Zakharov, P.G. Kryukov, Yu.A. Matveets, Yu.V. Senatskii, A.I. Fedosimov, S.V. Chekalin: Sov. J. Quant. Electr. 2, 533 (1973)
10.100 W.H. Hagen: Industrial Res. **48** (1972)
10.101 W. Rundle: Korad Div., Hadron, Inc., private communication
10.102 D.T. Hon: IEEE J. QE-12, 148 (1976)
10.103 R.M. Kogan, T.G. Crow: Appl. Opt. **17**, 927 (1978)

10.104 G.J. Linford, B.C. Johnson, J.S. Hildum, W.E. Martin, K. Snyder, R.D. Boyd, W.L. Smith, C.L. Vercimak, D. Eimerl, J.T. Hunt: Appl. Opt. **21**, 3633 (1982)
10.105 J.E. Geusic, H.J. Levinstein, S. Singh, R.G. Smith, L.G. Van Uitert: Appl. Phys. Lett. **12**, 306 (1968)
10.106 C.B. Hitz, J. Falk: Frequency doubled neodymium laser. Rept. AFAL-TR-12, Air Force Avionics Lab., Wright-Patterson AFB, Dayton, Ohio (1971)
10.107 J.M. Yarborough, J. Falk, C.B. Hitz: Appl. Phys. Lett. **18**, 70 (1971)
10.108 C.B. Hitz: Final Report, Contract NASA-20967, G.C. Marshall Space Flight Center, Huntsville, Ala. (July 1970)
10.109 W. Culshaw, J. Kannelaud, J.E. Peterson: IEEE J. QE-**10**, 253 (1974)
10.110 R.G. Smith: IEEE J. QE-**6**, 215 (1970)
10.111 J.E. Murray, S.E. Harris: J. Appl. Phys. **41**, 609 (1970)
10.112 A. Stein, R.A. Kaplan: Appl. Phys. Lett. **16**, 338 (1970)
10.113 R.B. Chesler, M.A. Karr, J.E. Geusic: J. Appl. Phys. **41**, 4125 (1970)
10.114 M.A. Karr: J. Appl. Phys. **42**, 4517 (1971)
10.115 G.A. Massey, J.M. Yarborough: Appl. Phys. Lett. **18**, 576 (1971)
10.116 D.T. Hon, IEEE J. QE-**13**, 99D (1977)
10.117 P.E. Perkins, T.S. Fahlen: J. Opt. Soc. Am. B **4**, 1066 (1987)
10.118 P.E. Perkins, T.A. Driscoll: J. Opt. Soc. Am. B **4**, 1281 (1987)
10.119 C.B. Hitz, L.M. Osterink: Appl. Phys. Lett. **18**, 378 (1971)
10.120 T.R. Gurski: Appl. Phys. Lett. **15**, 5 (1969)
10.121 J.H. Boyden, E.G. Erickson, R. Webb: Mode-locked frequency doubled neodymium laser. Tech. Report AFAL-TR-70-214, Air Force Avionics Lab., Wright-Patterson, Dayton, Ohio (1970)
10.122 R.R. Rice, G.H. Burkhart: Appl. Phys. Lett. **19**, 225 (1971)
10.123 C.B. Hitz, J. Falk: Frequency doubled neodymium laser. Rept. AFAL-TR-72-12, Air Force Avionics Lab., Wright-Patterson AFB, Dayton, Ohio (1972)
10.124 C.J. Kennedy: IEEE J. QE-**10**, 528 (1974)
10.125 J. Falk: IEEE J. QE-**11**, 21 (1975)
10.126 O. Bernecker: IEEE J. QE-**9**, 897 (1973)
10.127 J.A. Armstrong, N. Bloembergen, J. Ducuing, P.S. Pershan: Phys. Rev. **127**, 1918 (1962)
E.O. Ammann, P.C. Montgomery: J. Appl. Phys. **41**, 5270 (1970)
10.128 R.S. Craxton: Opt. Commun. **34**, 474 (1980)
10.129 W. Seka, S.D. Jacobs, J.E. Rizzo, R. Boni, R.S. Craxton: Opt. Commun. **34**, 469-473 (1980)
10.130 R.S. Craxton: IEEE J. QE-**17**, 1771 (1981)
10.131 Y.S. Liu, D. Dentz, R. Belt: Opt. Lett. **9**, 76 (1984)
10.132 D.T. Hon: IEEE J. QE-**12**, 148 (1976)
10.133 D.T. Hon, S. Guch, F.Y. Wu, H.W. Bruesselbach: Hughes Aircraft Company, Report AFAL-TR-78-131, (June 1978), Air Force Avionics Lab., Wright Patterson Air Force Base, Ohio
10.134 B.C. Johnson, T. Marchl, J. Mihoevich, W.L. Smith, J.E. Swain, R. Wilder, J.D. Williams: Int. Quant. Electr. Conf. (May 1983), Paper TUE3
10.135 R.G. Smith: IEER J. QE-**9**, 530 (1973)
10.136 G.D. Boyd, A. Ashkin: Phys. Rev. **146**, 187 (1966)
10.137 G.D. Boyd, D.A. Kleinman: J. Appl. Phys. **39**, 3597 (1968)
10.138 G.D. Boyd, F.R. Nash: J. Appl. Phys. **42**, 2815 (1971)
10.139 R. Basu, W.H. Steier: IEEE J. QE-**8**, 693 (1972)
10.140 S.E. Harris: Proc. IEEE **57**, 2096 (1969)
10.141 J.E. Bjorkholm, A. Ashkin, R.G. Smith: IEEE J. QE-**6** 797 (1970)
10.142 J.A. Giordmaine, R.C. Miller: Phys. Rev. Lett. **14**, 973 (1965)
10.143 R.G. Smith, J.R. Geusic, H.J. Levinstein, J.J. Rubin, S. Singh, L.G. van Uitert: Appl. Phys. Lett. **12**, 308 (1968)

10.144 G.A. Massey, R.A. Elliott: IEEE J. QE-**10**, 899 (1974)
S.J. Brosnan, R.L. Byer: IEEE J. QE-**15**, 415 (1979)
10.145 R.L. Herbst, R.N. Fleming, R.L. Byer: Appl. Phys. Lett. **25**, 520 (1974)
10.146 R.W. Wallace, S.E. Harris: Laser Focus **6**, 42 (1970)
10.147 Chromatix Inc., Mountain View, Calif., Optical parametric oscillator, Model 1020, Data Sheet
10.148 R.W. Wallace, S.E. Harris: Appl. Phys. Lett. **15**, 111 (1969)
10.149 R.G. Smith: *Laser Handbook I* ed. by F.T. Arecchi, E.O. Schultz-DuBois (North-Holland, Amsterdam 1972) pp.837-895
10.150 E.R. Nichols, J.C. Corbin, V.L. Donlan: A review of parametric oscillators and mixers and an evaluation of materials for 2 to 6μm applications. Rept. AFAL-TR-74-161, Air Force Avionics Lab., Wright-Patterson AFB, Dayton, Ohio (1974)
10.151 R.L. Byer: Optical parametric oscillators, Report 2140, Stanford Microwave Lab. (1973), in *Treatise in Quantum Electronics*, ed. by C.L. Tang (Academic, New York 1973)
10.152 J.A. Giordmaine, R.C. Miller: Appl. Phys. Lett. **9**, 298 (1966)
10.153 S.A. Akhmanov, A.I. Kovrygin, A.S. Piskarskas, V.V. Fadeev, R.V. Khoklov: Zhur. Eksp. i. Theoret. Fiz. Pisima **3**, 372 (1966)
10.154 R.C. Miller, W.A. Nordland: Appl. Phys. Lett. **10**, 53 (1967)
10.155 L.B. Kreuzer: Appl. Phys. Lett. **10**, 336 (1967)
10.156 J.E. Bjorkholm: Appl. Phys. Lett. **13**, 53 (1968)
10.157 J.E. Bjorkholm: Appl. Phys. Lett. **13**, 399 (1968)
10.158 J. Falk, J.E. Murray: Appl. Phys. Lett. **14**, 245 (1969)
10.159 E.O. Ammann, M.K. Oshman, J.D. Foster, J.M. Yarborough: Appl. Phys. Lett. **15**, 131 (1969)
10.160 R.W. Wallace: Appl.Phys. Lett. **17**, 497 (1970)
10.161 E.O. Ammann, J.M. Yarborough, J. Falk: J. Appl. Phys. **42**, 5618 (1971)
10.162 J.M. Yarborough, G.A. Massey: Appl. Phys. Lett. **18**, 438 (1971)
10.163 D.C. Hanna, B. Luther-Davies, H.N. Rutt, R.C. Smith: Appl. Phys. Lett. **20**, 34 (1972)
10.164 L.S. Goldberg: A repetitively pulsed $LiIO_3$ internal optical parametric oscillator. VIII Int'l Quant. Electr. Conf., Montreal, Canada (1972)
10.165 R.L. Herbst, R.L. Byer: CdSe infrared parametric oscillator. VIII Int'l Quant. Electr. Conf., Montreal, Canada (1972)
10.166 E.O. Ammann: Three to five micron parametric oscillator. Rept. AFAL-TR-72-210, Air Force Avionics Lab., Wright-Patterson AFB, Dayton. Ohio (1972)
10.167 R.L. Herbst, R.L. Byer: Appl. Phys. Lett. **21**, 189 (1972)
10.168 D.C. Hanna, B. Luther-Davies, R.C. Smith: Appl. Phys. Lett. **22**, 440 (1972)
10.169 A.A. Davydov, L.A. Kulevskii, A.M. Prokhorov, A.D. Savel'ev, V.V. Smirnov, A.V. Shirkov: Opt. Commun. **9**, 234 (1973)
10.170 G. Nath, G. Pauli: Appl. Phys. Lett. **22**, 75 (1973)
10.171 R.C. Eckardt, Y.X. Fan, R.L. Byer, C.L. Marquardt, M.E. Storm, L. Esterowitz: Appl. Phys. Lett. **49**, 608 (1986); CLEO'86 (San Francisco, CA) paper MH2
10.172 R. Burnham: Private communication, Fibertek. Inc. (1988)
10.173 E.J. Woodbury, W.K. Ng: Proc. IRE **50**, 2367 (1962)
10.174 E.J. Woodbury, G.M. Eckhardt: US Patent no. 3,371,265 (27 February 1968); G. Eckhardt, R.W. Hellwarth, F.J. McCluny, S.E. Schwartz, D. Weiner, E.J. Woodbury: Phys. Rev. Lett. **9**, 455 (1962)
10.175 R.L. Byer: Frequency conversion via stimulated Raman Scattering, Electro-Optical Systems Design 12:2:24. (1980)
10.176 Y.R. Shen: Stimulated Raman scattering, in *Light scattering in Solids I*, 2nd ed., ed. by M. Cardona, Topics Appl., Phys.8 (Springer, Berlin, Heidelberg, 1983) Chap.7

10.177 G.L. Eesley: *Coherent Raman Spectroscopy* (Pergamon, New York 1981)
A. Owyoung: CW stimulated Raman spectroscopy, in *Chemical Applications of Nonlinear Raman Spectroscopy*, ed. by B. Harvey (Academic, New York 1981) pp.281-320
10.178 W. Kaiser, M. Maier: Stimulated Rayleigh, Brillouin and Raman spectroscopy, in *Laser Handbook Vol.II*, ed. by F.T. Arecchi, E.O. Schulz-Dubois (North-Holland, Amsterdam 1972)
10.180 F.P. Milanovich: *Handbook of Laser Science and Technology, Vol.III* ed. by M.J. Weber (CRC, Boca Raton, Florida 1986) p.283
10.181 R.L. Byer, R.L. Herbst: Parametric oscillation and mixing, in *Nonlinear Infrared Generation*, ed. by Y.R. Shen, Topics Appl. Phys., Vol.16 (Springer, Berlin, Heidelberg 1977) pp.81-137
10.183 B.E. Perry: Photonics Spectra (1984) p.45; Data sheet for Quanta-Ray Model RS-1
10.184 D.G. Bruns, D.A. Rockwell: High energy Raman resonator. Hughes Aircraft Comp. Culver City, California, Final Report 1981, Report FR-81-72-1035
10.185 B.I. Stepanov, E.V. Ivakin, A.S. Rubanov: Sov. Phys. Doklady **16**, 46 (1971)
10.186 B.Ya. Zel'dovich, V.I. Popovichev, V.V. Ragul'skii, F.S. Faizullov: Sov. Phys. JETP **15**, 109 (1972); O.Yu. Nosach, V.I. Popovichev, V.V. Ragul'skii, F.S. Faizullov: Sov. Phys. JETP **16**, 435 (1972)
10.187 D.M. Pepper: Optical Engineering, **21**, 156-286 (1982) Special issue on Nonlinear optical phase conjugation
10.188 R.A. Fisher (ed.): *Optical Phase Conjugation* (Academic, New York 1983)
10.189 B.Ya. Zel'dovich, N.F. Pilipetsky, V.V. Shkunov: *Principles of Phase Conjugation*, Springer Ser. Opt. Sci., Vol.42 (Springer, Berlin, Heidelberg 1985)
10.190 D.M. Pepper: Nonlinear optical phase conjugation, in *The Laser Handbook, Vol.4*, ed. by M.L. Stitch, M. Bass (North-Holland, Amsterdam 1985)
10.191 V.V. Shkunov, B.Ya. Zel'dovich: Optical phase conjugation. Scientific American **253**, 54-59 (December 1985)
10.192 D.M. Pepper: Applications of optical phase conjugation. Scientific American **254**, 74-83 (January 1986)
10.193 D.M. Pepper, D.A. Rockwell, H.W. Bruesselbach: Photonics Spectra????, 95 (August 1986)
10.194 I.G. Zubarev, A.B.Mironov, S.I. Mikhailov: Sov. J. QE-**10**, 1179 (1980)
10.195 D.T. Hon: Opt. Eng. **21**, 252-256 (1982)
10.196 I.D. Carr, D.C. Hanna: Appl. Phys. B **36**, 83 (1985)
10.197 N.G. Basov, V.F. Efimkov, I.G. Zubarev, A.V. Kotov, S.I. Mikhailov, M.G. Smirnov: JETP Lett **28**, 197 (1978)

Chapter 11

11.1 M. Bass, H.H. Barrett: Appl. Opt. **12**, 690 (1973)
11.2 N. Bloembergen: Appl. Opt. **12**, 661 (1973)
11.3 N.L. Boling, G. Dubé: Appl. Phys. Lett: **23**, 658 (1973)
11.4 N. Bloembergen: IEEE J: QE-**10**, 375 (1974)
11.5 N.L. Boling, M.D. Crisp, G. Dubé: Appl. Opt. **12**, 650 (1973)
11.6 M.D. Crisp, N.L. Boling, G. Dubé: Appl. Phys. Lett. **21**, 364 (1972)
11.7 R.W. Hopper, D.R. Uhlmann: J. Appl. Phys. **41**, 4023 (1970)
11.8 S.A. Akhmanov, R.V. Khokhlov, A.P. Sukhorukov: Self-focusing, self-defocusing and self-modulation of laser beams, in *Laser Handbook*, ed. by F.T. Arecchi, E.O. Schulz-DuBois (North-Holland, Amsterdam 1972) p.1151
11.9 E.S. Bliss: Nonlinear propagation studies, in *NBS Spec. Publ.* No. 414, p.7 (US Gov. Print. Office, Washington, D.C. 1974)

11.10 J.H. Marburger: Theory of self-focusing in fast nonlinear response, in *NBS Spec. Publ.* No. 356, pp.51-59 (US Gov. Print. Office, Washington, D.C. 1971)
11.11 A. Feldman, D. Horowitz, R.M. Waxler: IEEE J. QE-**9**, 1054 (1973)
11.12 E.L. Kerr: Laser beam self-focusing and glass damage caused by electrostrictively driven acoustic waves, in *ASTM Spec. Techn. Publ.* No. 469, pp.23-33 (ASTM, Philadelphia Pa. 1969); Electro-strictive laser beam focusing in glass and small-scale track formation, in *NBS Spec. Publ.* No. 356, pp.61-75 (US Gov. Print. Office, Washington, D.C., 1971)
11.13 M.A. Duguay, J.W. Hansen: Measurement of the nonlinear index n_2 of glass using picosecond pulses, in *NBS Spec. Publ.* No. 341, pp. 45-50 (US Gov. Print. Office, Washington, D.C. 1970)
11.14 R. Hellwarth, J. Cherlow, T.T. Yang: Phys. Rev. **B11**, 964 (1975)
11.15 R.Y. Chiao, E. Garmire, C.H. Townes: Phys. Rev. Lett. **13**, 479 (1964)
11.16 P.L. Kelley: Phys. Rev. Lett. **15**, 1005 (1965)
11.17 E.L. Dawes, J.H. Marburger: Phys. Rev. **179**, 862 (1969)
11.18 W.G. Wagner, H.A. Haus, J.H. Marburger: Phys. Rev. **175**, 256 (1968)
11.19 G.M. Zverev, V.A. Pashkov: Sov. Phys. JETP **30**, 616 (1970)
11.20 E.L. Kerr: Phys. Rev. **A4**, 1195 (1971)
11.21 J. Davit: Filamentary damage in glasses, in *ASTM Spec. Techn. Publ.* No. 469, p.100-109 (ASTM Philadelphia, Pa. 1969); NBS Spec. Publ. No. 341, pp.37-44 (US Gov. Print. Office, Washington, D.C. 1970)
11.22 C.R. Guiliano: Time evolution of damage tracks in sapphire and ruby, in *NBS Spec. Publ.* No. 356, pp.44-50 (US Gov. Print. Office, Washington, D.C. 1971)
11.23 A.J. Campillo, S.L. Shapiro, B.R. Suydam: Appl. Phys. Lett. **24**, 178 (1974); Appl. Phys. Lett. **23**, 628 (1973)
11.24 E.S. Bliss, D.R. Speck, J.F. Holzrichter, J.H. Erkkila, A.J. Glass: Appl. Phys. Lett. **25**, 448 (1974)
11.25 J.A. Fleck, C. Layne: Appl. Phys. Lett. **22**, 467 (1973)
11.26 J.R. Jokipii, J. Marburger: Appl. Phys. Lett. **23**, 696 (1973)
11.27 M.M.T. Loy, Y.R. Shen: IEEE J. QE-**9**, 409 (1973)
11.28 V.I. Bespalov, V.I. Talanov: JETP Lett. **3**, 307 (1966)
11.29 J.A. Glaze: High energy glass lasers: Proc. Soc. Photo-Opt. Instr. Eng. **69**, 45 (1975)
11.30 A. Owyoung: IEEE J. QE-**9**, 1064 (1973)
11.31 E.S. Bliss, D.R. Speck, W.W. Simmons: Appl. Phys. Lett. **25**, 728 (1974)
11.32 A. Glass: Refractive index nonlinearity, in *Laser Program Annual Report* UCRL-50021-74, Lawrence Livermore Laboratory, University of California, Livermore, Calif., p.255 (1974)
11.33 Laser Induced Damage in Optical Materials., in. *Natl. Bur. Stand. US*, ed. by H.E. Bennett, A.H. Guenther, D. Milam, B.E. Newnam. Some of the more recent publications in this series are: Spec. Publ. 688 (1985); 669 (1984); 638 (1983); 620 (1981); 568 (1979)
11.34 H.E. Bennett, A. H. Guenther, D. Milam, B.E.Newnam: Appl. Opt. **26**, 826 (1987); **25**, 258 (1986); **23**, 3782 (1984); **22**, 3276 (1983); **20**, 3003 (1981); **19**, 2375 (1980)
11.35 In, *Lawrence Livermore National Laboratory Laser Program, Annual Report* UCRL-50021-82 (1982) p.15
11.36 In, *Laser Program, Annual Reports*, Lawrence Livermore Lab., Livermore, Ca, Reports UCRL 50021-74 (1974) to 50021-86 (1986)
11.37 N. Neuroth, R. Hasse, A. Knecht: Damage by laser radiation of improved neodymium-activated laser glass, colored glasses and optical glasses, in *NBS Spec. Publ.* No. 356, pp.3-14 (US Gov. Print. Office, Washington, D.C. 1971)
11.38 H.E. Bennett, A.H. Guenther, D. Milam, B.E. Newman: Appl. Opt. **26**, 813 (1987)

11.39 C. Yamanaka, T. Sasaki, M. Hongyo: Investigation of damage in laser glass, in *NBS Spec. Publ.* No. 356, pp.104-112 (US Gov. Print. Office, Washington, D.C. 1971)
11.40 A.J. Glass, A.H. Guenther: Appl. Opt. **18**, 2112 (1979)
11.41 M.J. Soileau: In, *Photonics Spectra* (1987) p.109
11.42 M. Bass, D.W. Fradin, L.H. Holway: Experimentation and investigation of optical-irradiation-induced surface damage in optically nonlinear materials. Rept. AFCRL-72-0714 (Raytheon Corp., Waltham, Mass. 1973)
11.43 W.D. Fountain, L.M. Osterink, G.A. Massey: *Optically induced physical damage to $LiNbO_3$, Proustite, and $LiIO_3$* (US Gov. Print. Office, Washington, D.C. 1971) pp.91-97
11.44 R. Webb: Catastrophic surface damage produced in $Ba_2NaNab_5O_{15}$ crystals during intra-cavity frequency doubling, in *NBS Spec. Publ.* No. 356 pp.98-103 (US Gov. Print. Office, Washington DC 1971)
11.45 R.S. Adhav, R.W. Wallace: IEEE J: QE-**9**, 855 (1973)
11.46 J.E. Swain, S.E. Stokowski, D. Milam: Appl. Phys. Lett. **41**, 12 (1982)
11.47 A.F. Turner: Ruby laser damage thresholds in evaporated thin-films and multilayer coatings, in *NBS Spec. Publ.* No. 356, pp.119-123 (US Gov. Print. Office, Washington DC 1971)
11.48 S.C. Seitel, J.B. Franck, C.D. Marrs, G.D. Williams: IEEE J. QE-**19**, 475 (1983)
11.49 H.E. Bennett, A.H. Guenther, D. Milam, B.E. Newnam: Appl. Opt. **23**, 3782 (1984)
11.50 L.D. De Shazer, B.E. Newnam, K.M. Leung: Appl. Phys. Lett. **23**, 607 (1973)
11.51 H.E. Bennett, A.H. Guenther, D. Milam, B.E. Newnam: Appl. Opt. **22**, 3276 (1983)
11.52 H.E. Bennett, A.J. Glass, A.H. Guenther, B.E. Newnam: Appl. Opt. **19**, 2375 (1980)
11.53 W.H. Lowdermilk, D. Milam: IEEE J: QE-**17**, 1888 (1981)
11.54 W.B. Alexander: Electro-Opt. Syst. Design **7**, 12 (1975)
11.55 T.L. Barber: Rev. Sci. Instr. **40**, 1630 (1969)

Appendix

A1 In *A.R. Henderson, Optics and Laser Technology* (1984) p.75
A2 International Electrotrechnical Committee, Geneva, Switzerland. IEC Publication 825, Radiation Safety of Laser Products, Equipment Classifications, Requirements and User's Guide (1984)
A3 Center for Devices and Radiological Health (HFZ-300) 8757 Georgia Avenue, Silver Spring MD 20910. Performance Standards for Laser Products, Regulation 21CFR1040 (latest revision August 1985)
A4 American National Standards Institute 1430 Broadway, New York NY 10018 ANSI Standard for the Safe Use of Lasers ANSIZ-136.1-1986 (1986)
A5 D. Sliney, M. Wolbarsht: *Safety with Lasers and Other Optical Sources* (Plenum, New York 1980)
A6 R. Weiner: *Laser Safety Update* (Weiner, 544-23rd Street, Manhattan Bean CA 90266)
A7 R. Weiner: Status of Laser Safety Requirements Laser & Applications, in *1987 Buyers Guide* p.277

Subject Index

Absorption 5, 8, 13, 51, 57, 59, 72, 84, 89
— coefficient 14, 15, 43, 44
— cross section 15, 42–44
Acoustooptic Q-switch 431–437, 444–446
Actinide 29, 38
Active mirror amplifier 398, 399
ADP 418, 490, 493, 494, 500–503
Airy disc 174
Alexandrite laser 68, 71–76, 124, 125
Alignment sensitivity of resonator 185–188
Alkali vapor lamp 271, 272
— *See also* Metal vapor discharge lamp
Aluminate 31
Amplifier, pulse operation 131–150
— Nd:glass 141–147
— Nd:YAG 147–150
— ruby 135–141
Amplifier, steady-state operation 150–155
— Nd:glass 154, 155
— ruby 152–154
Amplitude fluctuation *see* Output fluctuation
Angular divergence *see* Beam divergence
Arc lamp *see* Flashlamp
Asymmetric resonator 127
Atomic energy level *see* Energy level
Atomic lineshapes 8–12
Axial gradient laser 388, 389
Axial mode control 215–226
Axial modes 168, 203, 207, 208, 451–454
— *See also* Longitudinal modes

Ba_2NaNbO_3 481, 493, 496, 500–502, 506, 507, 524
Bandwidth *see* Linewidth
Beam divergence 173, 174, 182
BEL, laser host 33, 34
Beryllate, laser host 33, 34
Beta-barium borate 498, 500, 502, 504, 505, 523
Biaxial crystal 482, 486, 496
Birefringence
— electrically induced 414–430
— thermally induced 350, 361–365, 379
Birefringent filter 244–246
Blackbody radiation 2, 3, 249, 255
Bleachable dye *see* Saturable absorber
Bohr's frequency relation 1
Boltzmann distribution 3, 4
Bragg angle 434
Bragg scattering 434
Brewster angle 45, 54, 392, 399
Brightness, definition 129, 189
Brillouin scattering 477
Broadening of atomic transitions *see* Line broadening effects

Cavity dumping 444–446
— *See also* Pulse-transmission mode
CDA 418, 481, 487, 490, 491, 493, 494, 522
Ceramic, host material 34
Cerium 38
Charging unit, power supply 292–298
Chemically pumped laser 289, 290
Chromium lasers 68–70
— *See also* Ruby laser; Alexandrite laser
Circulating power 87, 88
Cladding of laser rods 165
Closed coupled pump cavity 311, 312, 330, 331
Coatings *see* Dielectric thin films
Coherence length
— frequency doubling 480–482, 486, 487
— laser radiation 213–215
Collision broadening 10
Concave-convex resonator 177, 191, 192, 197–200
Conductive cooling 386–388
Confocal parameter 173
Confocal resonator 176, 177
Confocal unstable resonator 237, 238, 240–242
Continuous arc lamp 248, 250, 265–270
Conversion factors 480, 544, 564, 565
Cooling equipment 384, 385
Cooling fluids 164, 341, 383
Cooling techniques 382–401
— conductive cooling 386–388
— cylindrical laser rod 382–388
— gases, air 385, 386
— liquids 382–384
— noncylindrical laser material 388–401
Cr:GSGG laser 68
Cr:$KZnF_3$ laser 69, 70

Critical phase matching 485, 486
Cross section, stimulated emission 15, 44, 49, 54, 68, 69, 71, 72, 77
Cryptocyanine 439, 458
Czochralski process for crystal growth 39, 51, 52

Damage
— definition 540
— mechanisms 541–549
— threshold 550–556
Damage susceptibility, reduction of 556–558
Decay time, resonator 24, 25, 81–83
Degeneracy, energy levels 4
Degenerate frequency 520, 521
Depolarization loss 363–366
Depopulation *see* Depumping processes
Depumping processes 162–166
Detonation pumped laser 290
Dichroic mirrors 340, 506, 507, 508, 519
Dielectric constant, definition 478
Dielectric thin films 233–235, 554–556
Diffraction effects 156, 157
Diffraction limited system performance 193–195
Diffraction losses, resonator 180, 181
Diffuse reflecting pump cavity 313, 328, 329, 347
Dipolar broadening 10
Disk amplifier 310, 399–401
— *See also* Slab laser
Dispersion of air 507
Doppler broadening 11
Doubly resonant oscillator 519, 525
Dye Q-switch 220, 221, 437–442
Dysprosium 38

Eastman Kodak dyes 439, 458, 462
— *See also* Saturable absorber
Efficiency factors 84, 89–94, 96, 97, 107, 111, 268, 321, 337
Einstein coefficients 4–7
Electric dipole transition 7, 8, 21
Electronic feedback loop 230, 471–476
Electronic polarizability 544
Electrooptic effect 414–422
Electrostriction 479, 544
Ellipsoidal pump cavity 313, 328
Elliptical cylinder, pump cavity 311, 322–328, 331–334, 341, 344–346
Emerald laser 68, 69
Emissivity 255
End-pumped lasers 314–317
Energy level
— diagram 16, 18, 40, 49, 50, 56, 61, 64, 73
— nomenclature 34, 35
Energy storage 133, 139
Erbium 37, 62–66
Er:glass laser 64–66, 123
Er:YAG laser 63, 64
Etalon 210, 219, 220, 222, 224–226

Europium 38
Excited state absorption 69, 70, 73, 76
Explosion energy, flashlamp 259–262
External triggering 306–308

Fabry-Perot interferometer 210, 211
Fabry-Perot resonator 79, 203–211
— *See also* Resonator, optical
Face-pumped active material 310, 321, 389–401
Faraday effect 449, 450, 479
Faraday rotator 127, 128, 449, 450
Feedback loop *see* Electronic feedback loop
Fiber laser 123, 124
Filament lamp *see* Tungsten filament lamp
Finesse 205
Flashlamp 251–264
— construction 251–253
— cooling 250–253
— electrical characteristic 257, 258
— failure modes 259–264
— life expectancy 261, 262
— operating condition 264
— optical characteristic 253–257
Flooded cavity 345–347
Fluorescence 26, 84
Fluorescent lifetime 5, 45, 49, 54, 61, 65, 72, 77, 86
Fluorides, laser host 34
Forbidden transition 21
Forced gas cooling 348, 385, 386
Fourier transform 453
Four-level laser 19, 20
Fraunhofer diffraction 174
Frequency changes, laser output 229–231
Frequency doubling *see* Second-harmonic generation
Frequency modulation 160, 161
Frequency stability 229–231
Fresnel diffraction rings 238, 240, 241
Fresnel number 180, 181, 189, 236, 241
Fringe visibility 213, 214
Frustrated total internal reflection 412
Fundamental mode 170, 172–174
— *See also* Gaussian beam

G parameters, resonator 179, 183–185
GaAlAs laser diodes 94–97, 119–124, 274–289
Gadolinium 38
Gain coefficient 80, 85, 87, 89, 92, 132–134, 151
Gain saturation 84–87, 151, 156
Garnet, laser host 31, 68
— *See also* Nd:YAG, Er:YAG, Cr:GSGG, Nd:Cr:GSGG lasers
Gaussian beam 172–174
Gaussian lineshape 12, 213
Gaussian temporal profile 453
Giant pulse *see* Q-switch

Glass, laser host 30, 54–57, 64–66
Ground level 17–20
GSGG, laser host 47, 57–60, 67, 68

Harmonic generation *see* Second-harmonic generation
Heat removal *see* Cooling techniques
Heat-transfer coefficient 352
Helical flashlamp *see* Flashlamp
Hemispherical resonator 176, 177, 506
Hermite polynomial 169, 170
Hole burning *see* Spatial hole burning
Holmium 37, 38
Homogeneous broadening 9
Host materials 29–34

Idler wave 519–521
Inclusion damage 542, 543
Index matching 482–486
Index surface 416, 483
Indicatrix 482
Inhomogeneous broadening 11
Injection locking 223, 224
Injection trigger 306, 307
Interferometer *see under* Fabry-Perot, Michelson, Twyman-Green
Intracavity frequency doubling 504–511
Isolator *see* Optical isolator

KDP 416–420, 448, 482, 487, 491, 493–496, 500–503, 512, 516, 517, 524
Kerr cell Q-switch 429, 430
Kerr effect 402, 429, 430, 460, 479, 544
Krypton arc lamp 227, 250, 265–270
— electrical characteristic 267
— mechanical design 265
— performance 267–269
— spectral data 268, 269
— thermal consideration 266
KTP 493, 497, 498, 500, 510, 514, 523

Laguerre polynomial 169, 170
Large-radius mirror resonator 176, 177
Laser amplifier *see* Amplifier
Laser diode efficiency 95
Laser diode pumping 94–97, 119–124, 127
Laser diodes 274–289
— *See also* GaAlAs laser diodes
Laser safety 559–561
— standards 562–563
Laser threshold *see* Oscillator, threshold condition
Lifetime broadening 9
$LiIO_3$ 497, 499, 500
$LiNbO_3$ 470, 471, 496, 499, 500–504, 523, 524
Linear flashlamp *see* Flashlamp
Line broadening effects 8–12
Linewidth 8, 12, 46, 49, 54, 77, 83, 206
Longitudinal mode selection 215–226
Longitudinal modes 168, 203–208, 451–453
Long term stability 226
Loop gain 80
Lorentzian lineshape 10, 11
Loss modulation 466–468

Mercury arc lamp 270, 271
Metal vapor discharge lamp 248, 270–272
Metastable level 20–22
Michelson interferometer 213
Mirror mounts 232, 233
Mode beating 98, 228, 508
Mode locking 451–455
Mode locking, active 466–476
— system design 469–476
— theory 466–469
Mode locking, passive 455–466
— saturable dye 461–462
— system design 455, 460, 461
— system performance 462–466
— theory 451–454
Mode patterns 170, 171, 194
Mode radius, resonator modes 171, 172, 175, 182
Mode selection
— longitudinal 215–226
— transverse 189–196
Molybdates, laser host 33
Moving focus, laser damage 546
Multiple-beam lasers 141, 142

Nd:BEL laser 122
Nd:Cr:GSGG laser 47, 57–60
Nd:glass laser
— amplifier 141–147, 155
— damage threshold 549–551
— material properties 47, 54–57
— oscillator-amplifier system 141, 145
— oscillator performance 111–114
— thermal effects 371, 379
Nd:SOAP laser 32, 356, 463
Nd:YAG laser
— amplifier 147–150
— birefringence 355, 356, 361–366
— cw pumped oscillator 114–118, 193–195, 227–229, 410, 411, 444–446
— frequency doubled 504–510
— material properties 47, 48–53
— mode locked oscillator 469, 472–475
— oscillator-amplifier system 516
— oscillator performance 114–122
— pulse-pumped oscillator 264, 348, 424, 516
— thermal effects 357–360
Nd:YAlO laser 31, 32, 367
Nd:YLF laser 47, 60–62
Nd:YVO_4 laser 122
Negative branch unstable resonator 242–244
Neodymium 36
— *See also* Nd:glass, Nd:YAG, Nd:YAlO, Nd:YLF, Nd:YVO_4, Nd:Cr:GSGG lasers
Nitrobenzene 429

Noble gas discharge lamp 247, 248, 251–264
Noncritical phase matching 486
Noncylindrical laser material 388–401
Nonlinear coefficient 480, 491, 493, 499
Nonlinear crystal 492–499
Nonlinear optical effect 477–479
Nonlinear refractive index 159, 543–549
Nonspherical aberration 368
Nonuniform pumping 155

Optical benches 232
Optical coatings *see* Dielectric thin films
Optical gate 446–449
Optical glass, damage threshold 552
Optical isolator 126, 449, 450
Optical mixing 478
Optical phase conjugation 535–539
Optical resonator *see* Resonator
Optical transmission, quartz 255
Optically induced damage 540–555
Organic dye *see* Saturable absorber
Organic nonlinear materials 498, 499, 523, 524
Oscillator, threshold condition 80–84
Oscillator loop, feedback 471–476
Output coupling, optimum 91, 117
Output coupling efficiency 91
Output efficiency 97
Output fluctuation 97–98, 226–229
Output vs input calculation 88–97
Oxysulfide, laser host 32

Parametric oscillator 518–526
Parasitic modes 165, 166
Parasitic noise 98
Passive mode locking
see Mode locking, passive
Passive Q-switch *see* Dye Q-switch
Pentaphosphates 33
PFN *see* Pulse-forming network
Phase coherence, stimulated emission 7, 8
Phase matching *see* Index matching
Phase mismatch 480, 481
Phase modulation 468, 472, 473
Phosphate, laser host 32
Piezooptic effect 423
Planck's constant 1
Planck's law 2
Plane-parallel resonator 177, 178
Plane wave impedance 480
Plano-concave resonator 177
Plastic Q-switch 441
Pockels cell 414–428, 442–444, 478–480
Pockels cell Q-switch 414–428, 442–444
Polarization, induced 477–479
Polarizers 363–366, 422, 427–429, 441, 442, 448, 518, 519
Population inversion 1, 15–22, 83, 132, 138
Postlasing 422–424
Potassium titanyl phosphate 497, 498, 500, 501, 504, 505, 510, 523
Power supplies
— cw pump sources 290–292
— flashlamps 292–298
Praseodymium 38
Prelasing 165, 422–424
Pulse-extraction switch *see* Optical gate
Pulse-forming network 295, 296, 298–305
Pulse-reflection mode Q-switch 402, 403
Pulse-shape distortion 160, 161
Pulse stacker
Pulse-transmission mode Q-switch 442–446
Pump band 17–20
Pump cavity 309–349
— configuration 309–321
— construction 335–349
— cooling technique 342–349
— coupling efficiency 321–331
— energy balance 336, 337
— pump light distribution 331–335
Pumping rate 24, 83, 84, 95
Pseudosimmer triggering 308

Q-spoiling *see* Q-switch
Q-switch
— acoustooptical 431–437
— color center 441
— dye 437–442
— electrooptical 414–430
— mechanical 412–414
Q-switch electronics 430, 431
Q-switch methods
— pulse-reflection mode 402–442
— pulse-transmission mode 442–446
Q-switch theory 403–412
— fast switch 403–406
— repetitive switch 408–412
— slow switch 407, 408
Q-switching *see* Q-switch
Quality factor Q 80, 81
Quantum efficiency 45, 46, 77
Quantum noise 98, 102
Quartz, optical transmission 255

Radial birefringent element 238, 239
Radiation lifetime 5, 6
Raman laser 526–535
Raman-Nath scattering 432–434
Rare earth ions 35–38
Rate equations 22–27, 131, 403, 404
Rayleigh criterion 174
RDA 418, 493, 494, 500
RDP 418, 494, 500, 502
Rectangular slab laser 389–401
Reflectivity of metals 338, 339
Relaxation oscillation 97–100, 108, 117–119, 226–229
Resonant charging 293, 296
Resonant reflector 208–210, 216–218

Resonator, optical
— configuration 174–178, 196–203
— diffraction loss 180–183
— hardware design 232–235
— modes 169–172
— sensitivity 185–188
— unstable 236–244
— with internal lens 183, 184, 200–203
Resonator modes, definition 169–172, 203–208
Ring laser 126, 127, 223
Rotating prism Q-switch 412–414
Ruby laser
— amplifier 135–141, 152–154, 372, 381
— material property 31, 38–47, 94
— oscillator performance 103–111, 216, 224–226, 329, 406, 424, 425, 442–444, 463
— thermal effect 371, 372, 378

Safety *see* Laser safety
Samarium 38
Sapphire 31, 39
Saturable absorber 437–442
— *See also* Dye Q-switch
Saturation density 77, 85, 86, 92, 132, 136, 140, 149, 153, 154, 156, 438
Scandium borate laser 70
Scraper mirror 238
Second-harmonic generation 479–511
— index matching 482–486
— internal 504–511
— material properties 492–504
— system parameters 487–492
— theory 479–482
Self-focusing 543–549
Semiconductor, pump source 247, 248, 274–289, 314–321
Sensitivity factor, thermal lensing 360
Sensitizer 30, 37, 38, 57–60, 64
Series injection triggering 294–296, 304–307
Servo loop *see* Electronic feedback loop
Short-term amplitude stability 226–228
Shutter, electrooptic 447–449
Side-pumped active material 317–321, 388–401
Signal distortion in amplifiers 155–161
Silicate, laser host 32
Simmer triggering 307, 308
Singly resonant oscillator 519–526
Slab laser 240, 319, 320, 389–401
Slope efficiency 91, 110
Small-scale ripple growth 159, 160, 547, 548
Small-scale self-focusing 546–549
SOAP, laser host 32
Spark gap, laser triggered 447–449
— *See also* Shutter, electrooptic
Spatial coherence 189
Spatial hole burning 127, 223
Spectral characteristic, laser output 211–215
Spherical pump cavity 313
Spherical resonator 177
Spike suppression 100–102
Spiking *see* Relaxation oscillation
Spinning prism Q-switch 412, 413
Spontaneous decay 162, 163
Spontaneous emission 5
Spotsize, definition 170, 172, 173, 176, 177, 178, 183, 184
Stability diagram for resonator 178, 179, 184, 185, 186, 198
Steady-state amplification 150–152
Stefan-Boltzmann equation 3
Stimulated emission 6, 7, 13, 86
Stimulated emission cross section 15, 42
Stimulated Raman scattering 477
Stokes factor 96
Stokes shift 477
Streak camera 460
Sun, pump source 248, 289
Superradiance 162–164
Surface damage 541, 542

Telescopic resonator 200–203
TEM_{mnq}, TEM_{plq} modes, definition 169–174, 206, 207
Temporal distortion, amplifier 160, 161
Thermal beam distortion 157, 192, 193, 355–381
Thermal broadening 10
Thermal effect, laser rod
— birefringence 361–368, 379
— lensing 157, 183, 184, 350, 357–361, 366, 380
— photoelastic effect 355–357
— stress 353–355
— temperature distribution 351–353, 369–378
Thermally compensated resonator 192, 193
Third-harmonic generation 511–518
Three-level laser 17–19
Threshold condition *see* Oscillator, threshold condition
Thulium 38
Titanium laser 70, 76–78
Torsional rod Q-switch 412
Transition metal 29, 38
Transverse mode selection *see* Mode selection, transverse
Transverse modes 168–172, 194
Traveling wave oscillator 126, 127
Trigger circuit, flashlamp 294, 295, 306–308
Tunable lasers 66–78
Tungstate, laser host 33
Tungsten filament lamp 248, 250, 272–274
Two-photon absorption 460
Twyman-Green interferometer 46, 52
Type I, II phase matching 484, 485, 513

Uniaxial crystal 483
Unstable resonator 236–244
Upper state efficiency 95–97

Vanadate, laser host 33
Vanadium laser 70
Verdet constant 450
Vortex stabilized arc lamp 248, 270
Vibronic lasers 66–78, 244–246

Waist, Gaussian beam 173
Wavefront distortion 159, 543–548
Wavelength selection 244–246
Wien's displacement law 3

Xenon arc, spectral data 253–257

YAG, laser host 29, 31, 48–53, 63, 64
— *See also* Nd:YAG, Er:YAG lasers
YAP, YAlO, laser host 31, 32
YLF, laser host 34, 60–62
Ytterbium 30, 31, 38

Zig-zag slab laser 237, 391, 392